2015 IFIP/IEEE International Conference on Very Large Scale Integration (VLSI-SoC 2015)

AA001054

Daejeon, South Korea
5 – 7 October 2015

IEEE Catalog Number: CFP15LSI-POD
ISBN: 978-1-4673-9141-2

**Copyright © 2015 by the Institute of Electrical and Electronic Engineers, Inc
All Rights Reserved**

Copyright and Reprint Permissions: Abstracting is permitted with credit to the source. Libraries are permitted to photocopy beyond the limit of U.S. copyright law for private use of patrons those articles in this volume that carry a code at the bottom of the first page, provided the per-copy fee indicated in the code is paid through Copyright Clearance Center, 222 Rosewood Drive, Danvers, MA 01923.

For other copying, reprint or republication permission, write to IEEE Copyrights Manager, IEEE Service Center, 445 Hoes Lane, Piscataway, NJ 08854. All rights reserved.

***This publication is a representation of what appears in the IEEE Digital Libraries. Some format issues inherent in the e-media version may also appear in this print version.**

IEEE Catalog Number: CFP15LSI-POD
ISBN 13:   978-1-4673-9141-2
ISSN: 2324-8432

Additional Copies of This Publication Are Available From:

Curran Associates, Inc
57 Morehouse Lane
Red Hook, NY 12571 USA
Phone: (845) 758-0400
Fax: (845) 758-2633
E-mail: curran@proceedings.com
Web: www.proceedings.com

2015 IFIP/IEEE International Conference on Very Large Scale Integration (VLSI-SoC)

October 5 - 7, 2015
Daejeon, Korea

Sponsored by:

International Federation for
Information Processing

The Institute of Electrical and
Electronics Engineers

IEEE Council on Electronic
Design Automation

IEEE Circuits and
Systems Society

Special Interest Group on
Design Automation

Korean Institute of Information and
Scientists and Engineers

Embedded Systems and
Research Center

Table of Contents

Message from the General Chairs VIII

Message from the Technical Program Chairs. IX

Organizing Committee. X

Program Committee XII

External Reviewers XIV

Keynote Presentations XV

CAD: Physical Design and Test
Session Chair: Jun-Sung Yang (Sungkunkwan Univ)

An Incremental Timing-Driven Flow Using Quadratic Formulation for Detailed Placement 1
Guilherme Flach, Jucemar Monteiro, Mateus Fogaça, Julia Puget, Paulo F. Butzen, Marcelo Johann and Ricardo Reis

Flare Reduction in EUV Lithography by Perturbation of Wire Segments 7
Sudipta Paul, Pritha Banerjee and Susmita Sur-Kolay

Analysis and testing on delays with two time frames 13
Masahiro Fujita

Contactless Transmission of Intellectual Property Data to Protect FPGA Designs 19
Lilian Bossuet, Pierre Bayon and Viktor Fischer

Embedded: Analysis and Optimization for Embedded Systems
Session Chair: Soontae Kim (KAIST) and JeongGil Ko (ETRI)

Tailoring Instruction-Set Extensions for an Ultra-Low Power Tightly-Coupled Cluster of OpenRISC Cores. 25
Michael Gautschi, Michele Scandale, Andreas Traber, Antonio Pullini, Alessandro Di Federico, Michele Beretta, Giovanni Agosta and Luca Benini

Design of Asynchronous RISC CPU Register-File Write-Back Queue. 31
Matthew Kim, Paul Beckett and Karl Fant

Modular Performance Analysis of Multicore SoC-based Small Cell LTE Base Station 37
Manikantan Srinivasan, C Siva Ram Murthy and Anusuya Balasubramanian

Embedded Low Power Analog CMOS Fuzzy Logic Controller Chip for Industrial Applications. 43
Manikandan Pandiyan and Geetha Mani

VLSI-SoC 2015 Table of Contents

Special Session A: The Brain-Inspired Computing: from Circuits to Software
Session Chair: Gi-Joon Nam (IBM)

Digital CMOS Neuromorphic Processor Design Featuring Unsupervised Online Learning 49
  Jae-Sun Seo and Mingoo Seok

An Overview on Memristor Crossabr Based Neuromorphic Circuit and Architecture 52
  Zheng Li, Chenchen Liu, Yandan Wang, Bonan Yan, Chaofei Yang, Jianlei Yang and Hai Li

Software Synthesis and Applications on Neuromorphic Systems. . . . . N/A
  Pallab Datta

Low power: Power and Error Management
Session Chair: Helen Li (University of Pittsburgh)

An Equation-Based Battery Cycle Life Model for Various Battery Chemistries 57
  *Alberto Bocca, Alessandro Sassone, Donghwa Shin, Alberto Macii, Enrico Macii and
  Massimo Poncino*

Power-Management High-Level Synthesis 63
  Dominik Macko, Katarína Jelemenská and Pavel Čičák

An Optimal Operating Point By using Error Monitoring Circuits with An Error-Resilient
Technique. 69
  Jaemin Lee, Seungwon Kim, Youngmin Kim and Seokhyeong Kang

Dynamic Error Tracking and Supply Voltage Adjustment for Low Power 74
  Pierre Nicolas Nicolaz and Kiyoung Choi

Special Session B: Directed SelfAssembly Lithography
Session Chair: Seongbo Shim (Samsung)

Directed self-assembly for contact patterning: alphabet concept and design strategy . . . . N/A
  Maryann Tung

Physical Design and Mask Optimization for Directed Self-Assembly Lithography (DSAL) 80
  Seongbo Shim and Youngsoo Shin

Calibration and application of a DSA compact model at Mentor Graphics . . . . N/A
  Jiwan Han

Memory: Non Volatile Memory Architecture
Session Chair: Kwanyeob Chae (Samsung) and Rahul Rao (IBM)

Qualifying Non-Volatile Register Files for Embedded Systems through
Compiler-directed Write Minimization and Balancing. 86
  Chengmo Yang and Maria Ruiz Varela

VLSI-SoC 2015   Table of Contents

Non-Volatile Memories in FPGAs: Exploiting Logic Similarity to Accelerate Reconfiguration and
Increase Programming Cycles..92
   Yuan Xue, Patrick Cronin, Chengmo Yang and Jingtong Hu

Prefetch-based Dynamic Row Buffer Management for LPDDR2-NVM Devices.................98
   Jaehyun Park, Donghwa Shin and Hyung Gyu Lee

Design Space Exploration of Row Buffer Architecture for Phase Change Memory with
LPDDR2-NVM interface..104
   Jaehyun Park, Donghwa Shin and Hyung Gyu Lee

Architecture I: 3D Circuits and Architecture
Session Chair: Kiyoung Choi (SNU)

Architecture Exploration of 3D FPGA to minimize internal layer connection...................110
   *Motoki Amagasaki, Yuto Takeuchi, Qian Zhao, Masahiro Iida, Morihiro Kuga and
Toshinori Sueyoshi*

Compact Interconnect Approach for Networks of Neural Cliques Using 3D Technology.........116
   *Bartosz Boguslawski, Hossam Sarhan, Frédéric Heitzmann, Fabrice Seguin, Sebastien
Thuries, Olivier Billoint and Fabien Clermidy*

A Thermal Estimation Model for 3D IC Using Liquid Cooled Microchannels and Thermal TSVs...122
   Surajit Roy, Supriyo Mandal, Chandan Giri and Hafizur Rahman

An Equilibrium Partitioning Method for Multicast Traffic in 3D NoC Architecture..............128
   Lin Wei and Lei Zhou

Circuits & Systems I: System-on-Chips for Signal Processing and Communications
Session Chair: Per Larsson-Edefors (Chalmers Univ)

An Integrated SoC for Science Data Processing in
Next-Generation Space Flight Instruments Avionics......................................134
   *Xabier Iturbe, Didier Keymeulen, Emre Ozer, Patrick Yiu, Daniel Berisford, Kevin Hand and
Robert Carlson*

Hardware Architecture and Optimization of Sliding Window Based Pedestrian
Detection on FPGA for High Resolution Images by Varying Local Features...................142
   Asim Khan, Muhammad Umar Karim Khan, Muhammad Bilal and Chong-Min Kyung

10Mbps Human Body Communication SoC for BAN 149
   Hyung-Il Park, Ingi Lim, Sungweon Kang and Whan-Woo Kim

Integrating Wearable Low Power CMOS ECG Acquisition SoC with Decision Making
System for WSBN Applications..154
   Manikandan Pandiyan, Dr. Geetha Mani, Dr. Jovitha Jerome and Natarajan S

IV

Variability: Design and Test for Mitigating Variability and Reliability Issues
Session Chair: Jing-Jia Liou (NTHU) and Daijoon Hyun (Samsung)

Slack-aware Timing Margin Redistribution Technique Utilizing Error Avoidance
Flip-Flops and Time Borrowing 159
   Mini Jayakrishnan, Jose Pineda de Gyvez and Kim Tae-Hyoung

Circuit Performance Optimization for Local Intra-die Process Variations using a Gate
Selection Metric. 165
   Victor Champac, Alejandra Nicte-Ha Reyes and Andres Gomez

Scalable Algorithm for Structural Fault Collapsing in Digital Circuits 171
   Raimund Ubar, Lembit Jürimägi, Elmet Orasson and Jaan Raik

Cost Reduction of System-level Tests with Stressed Structural Tests and SVM. 177
   Jing-Jia Liou, Meng-Ta Hsieh, Jun-Fei Cherng and Harry H. Chen

Architecture II: CPU/GPU, Cache, and Interconnect
Session Chair: Soontae Kim (KAIST)

Energy-Efficient Exclusive Last-Level Hybrid Caches Consisting of SRAM and STT-RAM 183
   Namhyung Kim, Junwhan Ahn, Woong Seo and Kiyoung Choi

JAIP-MP: A Four-Core Java Application Processor. 189
   Chun-Jen Tsai, Tsung-Han Wu and Hung-Cheng Su

Locality-Aware Vertex Scheduling for GPU-based Graph Computation. 195
   Hyunsun Park, Junwhan Ahn, Eunhyeok Park and Sungjoo Yoo

Traffic-Aware Buffer Reconfiguration in on-Chip Networks 201
   Ramin Bashizade and Hamid Sarbazi-Azad

Circuits & Systems II: Circuits and Systems for Signal Processing and Communications
Session Chair: Jongsun Park (Korea Univ)

Design Optimization of Polyphase Digital Down Converters for Extremely High Frequency
Wireless Communications 207
   Gain Kim, Raffaele Capoccia and Yusuf Leblebici

Dual-Mode Double Precision / Two-Parallel Single Precision Floating Point Multiplier
Architecture. 213
   Manish Kumar Jaiswal and Hayden K.-H So

Hardware Implementation of Real-Time Multiple Frame Super-Resolution 219
   Kerem Seyid, Sebastien Blanc and Yusuf Leblebici

Timing and Robustness Analysis of Pulsed-Index Protocols for Single-Channel IoT
Communications 225
   Shahzad Muzaffar and Ibrahim M. Elfadel

Analog: Ultra-Low-Power Analog Circuits for IoT Applications
Session Chair: Jaeha Kim (SNU)

A Time Interleaved DAC Sharing SAR Pipeline ADC for Ultra-Low Power Camera
Front Ends. 231
   Anvesha Amaravati, Manan Chugh and Arijit Raychowdhury

High-Efficiency Voltage Regulation Stage in Energy Harvesting Systems. .237
   Sung-Eun Kim, Taewook Kang, Sungweon Kang, Kyunghwan Park and Myung-Ae Chung

A Fully On-Chip 25MHz PVT-Compensation CMOS Relaxation Oscillator 241
   Hamed Abbasizadeh, Behnam Samadpoor Rikan and Kang-Yoon Lee

Prototyping: Modeling, Validation, and Debug
Session Chair: Dominique Borrione (TIMA Laboratory) and
Anupam Chattopadhyay (Nanyang Technological Univ)

A time-window based approach for dynamic assertions mining on control signals 246
   Alessandro Danese, Francesca Filini and Graziano Pravadelli

Physical-based Modeling and Fast Simulation of Wireline Links 252
   Jun Guo, Peng Liu and Weidong Wang

Trace Signal Selection Methods for Post Silicon Debugging 258
   Shridhar Choudhary, Amir Masoud Gharehbaghi, Takeshi Matsumoto and Masahiro Fujita

Timing Attack on NEMS Relay Based Design of AES 264
   Samah Saeed, Bodhisatwa Mazumdar, Sk Subidh Ali and Ozgur Sinanoglu

Posters
Session Chair: Jae-Joon Kim (POSTECH)

A High Efficiency Rectifier for Inductively Power Transfer Application 270
   Qiong Wei Low, Liter Siek and Mi Zhou

Design and Analysis of Search Algorithms for Lower Power Consumption and Faster
Convergence of DAC Input of SAR-ADC in 65nm CMOS. 274
   Ananthanarayanan Parthasarathy

Efficient Signature-Based Sub-Circuit Matching 280
   Amir Masoud Gharehbaghi and Masahiro Fujita

Reversible Circuit Rewriting with Simulated Annealing 286
   Nabila Abdessaied, Mathias Soeken, Gerhard W. Dueck and Rolf Drechsler

A Hybrid Embedded Compression Codec Engine for Ultra HD Video Application 292
   Seongmo Park, Kyungjin Byun and Nak-Woong Eum

VLSI-SoC 2015   Table of Contents

A New Sizing Approach for Lifetime Improvement of Nanoscale Digital Circuits due to
BTI Aging 297
   Andres Gomez Chacon and Victor Champac

Virtual Prototype Based on Aldebarn CPU Core 303
   Jae-Jin Lee, Chan Kim, Kyungjin Byun and Nakwoong Eum

A Generic Clock Controller for Low Power Systems: Experimentation on an AXI Bus 307
   *Chadi Al Khatib, Claire Aupetit, Cyril Chevalier, Chouki Aktouf, Gilles Sicard and
Laurent Fesquet*

Fast Global Interconnnect Driven 3D Floorplanning 313
   Artur Quiring, Markus Olbrich and Erich Barke

Filtering Dirty Data in DRAM to Reduce PRAM Writes. 319
   Hyunsun Park, Chanha Kim, Sungjoo Yoo and Chanik Park

On the estimation of assertion interestingness 325
   Tara Ghasempouri and Graziano Pravadelli

Hardware/Software Partitioning of Embedded System-on-Chip Applications. 331
   Jia Wei Tang, Yuan Wen Hau and Muhammad Nadzir Marsono

Exploiting Scalable CGRA Mapping of LU for Energy Efficiency using the LAYERS
Architecture 337
   *Zoltán Endre Rákossy, Dominik Stengele, Gerd Ascheid, Rainer Leupers and Anupam
Chattopadhyay*

Dynamic Migratory Selection Strategy for Adaptive Routing In Mesh NoCs. 343
   John Jose, Joe Augustine and Sijin Sebastian

A Cluster-Based Reliability- and Thermal- Aware 3D Floorplanning using redundant
STSVs 349
   Ying-Jung Chen and Shanq-Jang Ruan

Trace Buffer Attack: Security versus Observability Study in Post-Silicon Debug 355
   Yuanwen Huang, Anupam Chattopadhyay and Prabhat Mishra

Author Index 361

Message from the General Chairs

On behalf of the Organizing Committee, we welcome you to the 23rd IFIP/IEEE International Conference on Very Large Scale Integration (VLSI-SoC) in Daejeon, the City of Science and Technology of Korea.

For many years, VLSI-SoC has been sponsored by the IFIP TC10 WG 10.5, IEEE CEDA, and IEEE CASS, and held in cooperation with the ACM SIGDA. This year, VLSI-SoC is also sponsored by the Korean Institute of Information Scientists and Engineers (KIISE). We also pay special thanks to the academic sponsors, the Korea Advanced Institute of Science and Technology (KAIST) and Embedded Systems Research Center (ESRC) at Seoul National University. The 23rd VLSI-SoC is specially collocated with the 23rd IFIP World Computer Congress (WCC) at Daejeon Convention Center.

VLSI-SoC covers a broad range of topics in the areas that surround VLSI and SoC (System-on-Chip). It provides a forum to exchange ideas and showcase research and development results in Electronics Design Automation (EDA), design methodology, test, design, verification, devices, process, systems issues, and application domains of VLSI and SoC.

The 23rd VLSI-SoC provides three timely and outstanding Keynote talks. Dr. Kookyeon Kwak, Head of Advanced Standard R&D Lab at LG Electronics, addresses the SoC requirements for post smart devices era. Prof. Enrico Macii, Politecnico di Torino, introduces synthesis of graphene logic circuits including models, implementation styles, algorithms, and tools. Prof. Hidetoshi Onodera, Kyoto University, talks about design challenges and solutions in the era of Internet of Things (IoT).

We would like to thank all the members of the Organizing Committee for their contributions. VLSI-SoC 2015 would not happen without their efforts and commitment. Special thanks are given to the Program Co-Chairs, the Vice Program Chair, and the Technical Program Committee Members for their time and efforts to make the comprehensive technical program of the conference. Finally, we would like to thank all the authors, speakers, poster presenters, and the attendees for their contributions to the conference.

We hope that you all enjoy the conference and your stay at the City of Science and Technology.

Naehyuck Chang and Kiyoung Choi
General Co-Chairs

Message from the Technical Program Chairs

On behalf of the Technical Program Committee of the 23rd IFIP/IEEE International Conference on Very Large Scale Integration (VLSI-SoC 2015), we welcome you to Daejeon, Korea and thank you for joining us at this important event.

This year the technical program consists of 3 keynote presentations, 11 lecture sessions, 2 special sessions, a poster session, and a Ph.D. forum. Through a peer-review process, 44 papers out of 117 submissions were accepted as regular papers and will be presented; additional 17 submissions were accepted as posters. Authors from 28 different countries and PC members from 24 different countries have made it possible to prepare a high quality technical program. We would like to thank all the people who contributed, especially those who submitted papers and participated in the reviewing process.

The wide international participation makes the symposium a remarkable occasion for interaction among the experts in the areas related to Very Large Scale Integration (VLSI) and System-on-Chip (SoC). Three high profile speakers will deliver keynote presentations in their respective areas: Dr. Kookyeon Kwak (LG Electronics, Korea), Prof. Enrico Macii (Politecnico di Torino, Italy) and Prof. Hidotoshi Onodera (Kyoto University, Japan).

We wish you an exciting and enjoyable experience in VLSI-SoC 2015 and, beyond the technical contents of the symposium, we hope you will enjoy your stay in Daejoen.

Youngsoo Shin and Chi-Ying Tsui
Technical Program Co-Chairs

Organizing Committee

General Chairs
Naehyuck Chang, KAIST, Korea
Kiyoung Choi, Seoul National Univ, Korea

Program Chairs
Youngsoo Shin, KAIST, Korea
Chi-Ying Tsui, HKUST, Hong Kong

Vice Program Chair
Jae-Joon Kim, POSTECH, Korea

Special Session Chair
Gi-Joon Nam, IBM, USA

Finance Chair
Youngmin Yi, Univ of Seoul, Korea

Local Arrangement Chairs
Ji-Hoon Kim, Chungnam National Univ, Korea
Seokhueong Kang, UNIST, Korea

Publication Chairs
Yoonjin Kim, Sookmyung Women's Univ, Korea
Jongeun Lee, UNIST, Korea

PhD forum Chairs
Srinivas Katkoori, University of South Florida, USA
Jason Xue, City University of Hong Kong, Hong Kong

Publicity Chairs
Tsung-Yi Ho, National Chiao Tung Univ, Taiwan
Nak Woong Eum, ETRI, Korea
Hiroshi Nakamura, Univ of Tokyo, Japan
Jose L. Ayala, Complutense Univ of Madrid, Spain

Registration Chairs
Jaeyoung Chung, Incheon National Univ, Korea
Hyung Gyu Lee, Daegu Univ, Korea

Web Chair	Donghwa Shin, Yeungnam Univ, Korea
Steering Committee	Manfred Glesner, TU Darmstadt, Germany
	Matthew Guthaus, UC Santa Cruz, USA
	Salvador Mir, TIMA, France
	Ricardo Reis, UFRGS, Brazil
	Michel Robert, U. Montpellier, France
	Luis Miguel Silveira, INESC ID, Portugal
	Chi-Ying Tsui, HKUST, Hong Kong

Program Committee

Analog and mixed-signal IC design Track

Co-Chairs:
Jaeha Kim, Seoul National University, Korea
Tai-Cheng Lee, National Taiwan University, Taiwan

Ke-Horng Chen, National Chiao-Tung University, Taiwan
Kenichi Okada, Tokyo institute of Technology, Japan
Sai-Weng Sin, University of Macau, China
Michiel Steyaert Steyaert, KU Leuven
Jose M. de La Rosa, Instituto de Microelectrónica de Sevilla, IMSE-CNM (CSIC)
Jaehyouk Choi, Ulsan National University of Science and Technology, Kore

System architectures NoC, 3D, multi-core, and reconfigurable Track

Co-Chairs:
Yuan Xie, UC Santa Barbara, USA
Nam Sung Kim, University of Wisconsin, USA

Jishen Zhao, University of California, Santa Cruz
Jiang Xu, Hong Kong University of Science and Technology, Hong Kong
Myoung Jung, UT dallas, USA
Ulya Karpuzcu, University of Minnesota, USA
Radu Teodorescu, Ohio State University, USA
Leandro Indrusiak, University of York, USA
Ian O'Connor, Lyon Institute of Nanotechnology, France
Michael Huebner, Ruhr-University Bochum, Germany

CAD synthesis and analysis Track

Co-Chairs:
Minsik Cho, IBM
Masahiro Fujita, University of Tokyo, Japan

Bei Yu, UT Austin, USA
Duo Ding, Oracle Microelectronics
Myung-Chul Kim, IBM Corporation
Takashi Kambe, Kinki University, Japan
Tiziano Villa, Universita' di Verona, Italy
Ricardo Reis, Universidade Federal do Rio Grande do Sul, Brazil
Zhiru Zhang, Cornell University, USA

Circuits and systems for signal processing and communications Track

Co-Chairs:
Oscar Gustafsson, Linköping University, Sweden
Per Larsson-Edefors, Chalmers University, Sweden

Hyeon-Min Bae, KAIST, Korea
Liam Marnane, University College Cork, Ireland
Tobias Noll, RWTH Aachen University, Germany
Jongsun Park, Korea University, Korea
Christoph Studer, Cornell University, USA
Dajiang Zhou, Waseda University, Japan
Fatih Ugurdag, Ozyegin University, Turkey
Luc Claesen, Universiteit Hasselt, Belgium

Embedded system architecture, design, and software Track

Co-Chairs:
Vijaykrishnan Narayanan, Penn State University, USA
Jason Xue, City University of Hong Kong, Hong Kong

Ingchao Lin, National Cheng Kung University, Taiwan
Wang Yu, Tsinghua University, China
Zili Shao, Hong Kong Polytechnic University, Hong Kong
Lar Bauer, Karlsruhe Institute of Technology, Germany
Koji Inoue, Kyushu University, Japan
Sri Parameswaran, University of New South Wales, Australia
Akash Kumar, National University of Singapore, Singapore

Low-power and thermal-aware design Track

Co-Chairs:
Massimo Poncino, Politecnico di Torino, Italy
Tadahiro Kuroda, Keio University, Japan

Jose L. Ayala, Complutense Univ of Madrid, Spain
Aida Todri-Sanial, French National Center for Scientific Research, France
Mirko Loghi, Università di Udine, Italy
Donghwa Shin, Yeungnam University, Korea
Chia-Lin Yang, National Taiwan University, Taiwan
Masaaki Kondo, The University of Electro-Communications, Japan

VLSI-SoC 2015 — Program Committee

Memory technology, circuit, and system Track

Co-Charis:
Yiran Chen, University of Pittsburg, USA
Rahul Rao, IBM

Minki Cho, Intel,
Swaroop Ghosh, Intel,
Jingtong Hu, Oklahoma State University, USA
Kwanyeob Chae,
Nitin Chandrachoodan, IIT Madras, India
Chengmo Yang, University of Delaware, USA
Lionel Torres, LIRMM, France

Prototyping, verification, modeling, and simulation Track

Co-Chairs:
Graziano Pravadelli, University of Verona, Italy
Swarup Bhunia, Case Western Reserve University, USA

Daniel Grosse, University of Bremen, Germany
Pierre-Emmanuel Gaillardon, Ecole polytechnique federale del Lausanne (EFPL), Switzerland
Anupam Chattopadhyay, Nanyang technological university, Singapore
Prabhat Mishra, University of Florida, USA
Sandip Ray, Intel
Laurence Pierre, TIMA, France
Florian Letombe, Synopsys,
Adam Pawlak, Silesian University of Technology, Poland

Design for variability, reliability, and test Track

Co-Chairs:
Chris Kim, Univ. of Minnesota, USA
Jing-Jia Liu, National Tsing-Hua University, Taiwan

Matteo Sonza Reorda, Politecnico di Torino, Italy
Swaroop Ghosh, Intel
Victor Champac, INAOE, Mexico
Tony Kim, Nanyang Technological University, Singapore
Xiaofei Wang, University of Minnesota, USA
Satoshi Ohtake, Oita University, Japan

Security Track

Co-Chairs:
Ozgur Sinanoglu, New York University Abu Dhabi
Srinivas Katkoori, University of South Florida, USA

Debdeep Mukhopadhyay, IIT Kharagpur, India
Mohammad Tehranipoor, University of Connecticut, USA
Paolo Maistri, TIMA Laboratory, France
Joseph Zambreno, Iowa State University, USA
Siddharth Garg, New York University, USA
Yier Jin, The University of Central Florida, USA

External Reviewers

Aghaeikhouzani, Hoda
Al Kadi, Muhammed
Ates, Ozgur

Bayram, Ismail
Bell, Chris
Bhattacharya, Sarani

Chakraborty, Abhishek
Chen, Yi-Jung
Choi, Woong
Conceição, Calebe

Damschen, Marvin
Danese, Alessandro
Dziurzanski, Piotr

Fu, Chenchen

Ghasempouri, Tara
Grimm, Tomas
Güzel, Aydın Emre

Huang, Jiun-Lang

Janßen, Benedikt
Jayakrishnan, Mini
Jung, Gihoon

Kerekare, Srinivas Rao
Knechtel, Johann

Levent, Vecdi Emre
Lewandowski, Matt
Li, Ang
Li, Zheng
Lora, Michele

Mazumdar, Bodhisatwa
Miyase, Kohei
Mori, Jones Yudi

Palaz, Okan
Pan, Xiang
Patranabis, Sikhar
Pham, Nam Khanh
Pontié, Simon
Puranik, Aditya

Sahin, Onur
Schwiegelshohn, Fynn
Shihab, Mustafa
Singh, Amit
Studer, Christoph
Suresh, Chandra

Tang, Liang
Temizkan, Fatih
Thomas, Renji
Tosun, Mustafa
Tunc, Cihan

Villacorta, Hector

Wen, Wujie
Wu, I-Peng

Xue, Yuan

Yan, Bonan
Yasin, Muhammad
Yildiz, Abdullah

Zhang, Hongyan
Zhang, Jie
Zhang, Yaojun
Zhou, Li

Keynote Presentations

Monday, October 5, 11:30-12:30

SoC Requirements for Post Smart Devices Era

Kookyeon Kwak
EVP, Head of Advanced Standard R&D Lab, LG Electronics

Abstract:

The cycle of new technology trend becomes even shorter due to the development of IT technology. Smart devices era has begun with the advent of iPhone, had a big impact on mobile industry as well as home appliance, automotive and various industries. This smart device era that seemed to start just now has already prepared to evolve into next era. First, TV will be developed focusing on not smart feature but picture quality, original value of TV for customer to be able to feel reality on TV. Secondly Mobile will enhance differentiated UX like virtual reality and augmented reality. Especially, the accumulated technologies in these fields will be extended to the IoT and smart car. The evolution direction of conventional smart devices (TV, mobile) and development of new fields (IoT, smart car), including the key SoC technology for this upcoming post smart devices era will be addressed in the keynote speech.

Bio:

Dr. Kookkyeon Kwak received the B.S. from Seoul National University, Korea, the M.S. from KAIST, Korea, and Ph.D. from Polytechnic University, USA. He has been with LG Electronics since 1979 holding various management positions; he is currently senior research fellow (EVP) and head of Advanced Standard R&D Lab in the company. His major achievements include standardization of mobile DTV, development of 5th generation VSB chip, development of satellite and terrestrial DMB receiver chip, and development of CD-ROM drive. He has received 2009 Ungbi Medal in Order of Science and Technology Merit, 2004 Presidential Citation in 49th Information and Communication Day, 2002 Presidential Award in 7th New Technology Electronic Components Contest, and so on.

VLSI-SoC 2015   Keynote Presentations

Tuesday, October 6, 9:30-10:30

**Synthesis of Graphene Logic Circuits: Models,
Implementation Styles, Algorithms and Tools**

**Enrico Macii
Politecnico di Torino**

Abstract:

Graphene has attracted, in recent years, the attention of the research community as a potential vehicle for the implementation of logic functions and several attempts have been made to establish design flows that allow the adoption of graphene devices as the basis of new integrated circuits. This talk introduces an efficient design style, the Pass-XNOR Logic (PXL), which allows the implementation of adiabatic logic circuits with ultra low-power features. In particular, issues such as device modeling, device architectures, synthesis algorithms and tool flow integration are discussed in some details in order to demonstrate the viability of the PXL design style.

Bio:

Enrico Macii is a Full Professor of Computer Engineering at Politecnico di Torino. From 1991 to 1997 he was also an Adjunct Faculty at the University of Colorado at Boulder. He holds a PhD degree in Computer Engineering from Politecnico di Torino (1995). Since 2007, he is the Vice Rector for Research and Technology Transfer at Politecnico di Torino, and since 2012 also the Rector's Delegate for International Affairs. He was the Editor-in-Chief of the IEEE TCAD for the term 2006-2009. He is a Member of the Board and the Chief Technology Officer of ST-POLITO s.c.a.r.l. He was the National FP7 ICT Delegate from 2011 until 2013, and one of the Italian Members of the Public Authorities Board of the ENIAC and ARTEMIS Joint Undertakings from 2009 until 2013. His research interests are in the design of electronic digital circuits and systems, with particular emphasis on low-power consumption aspects. He has extended his research activities to the broad area of bioinformatics with various types of data analyses for different applications in the biomedical domain.

XVI

VLSI-SoC 2015   Keynote Presentations

Wednesday, October 7, 09:30-10:30

Design Challenges and Solutions in the era of IoT

Hidetoshi Onodera
Kyoto University

Abstract:

The talk stars by the exploration of integrated circuits in the era of IoT predicted by empirical laws that have been correctly predicting the trend. These laws include the Moore's law for exponential growth of integration scale, the Makimoto's wave for 10-year cycles of standardization and customization, and the Bell's law for the birth and death of computer classes. Then, based on a few examples of IoT applications, together with the predictions of ITRS, design challenges in the era of IoT will be reviewed. Among them, this talk will focus on energy and reliability challenges and address possible solutions. The key for energy reduction is supply and threshold voltage scaling. A practical scaling strategy will be explained. For a reliable system with soft error resilience, we have been developing a dependable-VLSI platform. A key concept of the platform will be explained followed by a demonstration of the robustness against soft-errors.

Bio:

Dr. Onodera received the B.E., and M.E., and Dr. Eng. degrees in Electronic Engineering, all from Kyoto University, Kyoto, Japan. He joined the Department of Electronics, Kyoto University, in 1983, and currently a Professor in the Department of Communications and Computer Engineering, Graduate School of Informatics, Kyoto University. His research interests include design technologies for Digital, Analog, and RF LSIs, with particular emphasis on low-power design, design for manufacturability, and design for dependability. He served as a Program Chair and a General Chair of ICCAD and ASP-DAC. He was a Chairman of the IPSJ SIGDA, the IEICE Technical Group on VLSI Design Technologies, the IEEE SSCS Kansai Chapter, the IEEE CASS Kansai Chapter, and IEEE Kansai Section. He served as an Editor-in-Chief of IEICE Trans. on Electronics and IPSJ Trans. on System LSI Design methodology.

An Incremental Timing-Driven Flow Using Quadratic Formulation for Detailed Placement

Guilherme Flach, Jucemar Monteiro, Mateus Fogaça, Julia Puget, Paulo Butzen*, Marcelo Johann, Ricardo Reis

Universidade Federal do Rio Grande do Sul (UFRGS) - Instituto de Informática - PGMicro/PPGC

Universidade Federal do Rio Grande (FURG)*

{gaflach, jucemar.monteiro, mateus.fogaca, julia.puget, paulobutzen*, johann, reis}@{inf.ufrgs.br, furg.br*}

Abstract—In this work, we present a flow for the Incremental Timing-Driven Placement problem. Given a legal placement, the aim is to reduce the circuit's timing violations without changing significantly the cell density, subject to a maximum displacement constraint. Our flow consists of two core steps: useful clock skew optimization and critical path fine tuning. During useful clock skew optimization, sequential cells are replaced, seeking to minimize clock skew. After that, a quadratic formulation is used to further reduce critical path delays. An incremental legalization tool is also presented, which supports the methods developed in this work. Our Incremental Timing-Driven Placement flow can achieve, on average, 0.3%, 26.2%, 8.7% and 23.7% of the normalized quality score improvement compared to state-of-the-art algorithms.

Keywords—Timing Optimization, Detailed Placement, Quadratic Placement, Legalization

I. INTRODUCTION

Placement has been an important research subject for a long time in EDA. Proved to be NP-Hard [1], it is one of the most challenging problems in physical synthesis, due to technology restrictions and the constant increasing in the number of cells integrated into ICs. Despite the many works that promoted tremendous improvements in placement quality in the last decade, this is still a challenging problem for next generation circuits [2], specially to address timing closure [3], [4], [5], [6].

During placement, legal positions for the circuit cells are defined with the primary goal of minimizing the total wirelength connecting the cells. Other common placement objectives are congestion reduction and timing closure, which are only indirectly taken into account by wirelength minimization. Typically, the placement step is divided into three phases: global placement, legalization and detailed placement [7].

In the global phase the non-overlapping constraint is relaxed and cells are evenly spread over the placement area and an initial rough position to the cells is set. The remaining overlap is removed during the legalization step, where cells are moved to legal positions trying to minimize the total cell movement. After legalization, the detailed placement makes fine adjustments to cell positions, performing optimizations that are hard to be seen during the global phase.

Besides the traditional wirelength minimization, a timing-driven placement seeks to keep the delay of the circuit paths under a user specified value. This can be accomplished indirectly, by increasing the weight of connections on critical paths, or, more directly, by interleaving cell position changes with timing analysis.

In this work, a flow for incremental timing-driven detailed placement is presented. Our flow, UFRGS/FURG Brazil, obtained the first place in the Incremental Timing-Driven Placement problem at ICCAD 2014 Contest[8]. An overview of it is shown in Figure 1. It consists of an incremental legalization engine, called Jezz, a timing analysis tool and two core optimization steps: clock skew exploration and critical path fine tuning.

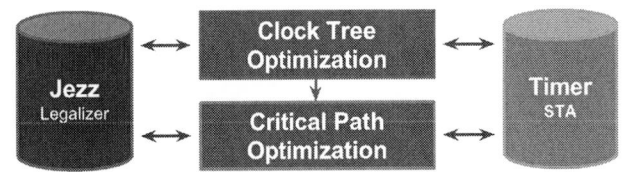

Figure 1. Flow Overview

The optimization methods rely on the legalization engine to make changes in the placement solution, keeping it legalized. The engine allows a cell to be inserted in any valid position trying to reduce the impact on other cells. The timer engine performs full and incremental static timing analysis (STA) of the circuit. The flow starts by performing an useful clock skew optimization where new positions for sequential cells are defined.After that, a quadratic formulation is used to further reduce critical path delays.

The main contributions of this work are: an efficient flow for Incremental Timing-Driven Placement considering the infrastructure provide by ICCAD 2014 Contest [8], an incremental legalization tool called Jezz, used during detailed placement, and a quadratic placement formulation for critical path reduction.

II. PROBLEM DEFINITION

The goal is to evaluate the impact of incremental timing-driven placement after initial wirelength/routability-driven placement [8]. More specifically, given an already legalized solution, the objective is to improve timing violations without changing significantly the cell density, subject to a maximum displacement constraint, keeping the placement legalized. The circuit timing is computed by means of static timing analysis, and the cell density is measured using the ABU metric [9].

Timing arc delays are modeled by parametric functions, and interconnection delays are modeled using Elmore delay. The interconnections are estimated using Steiner trees generated by the FLUTE algorithm [10].

The quality of a placement solution is measured by the Equation (1)

$$Q = 100 \times w_{abu}\left(abu' - abu\right)\left[\sum_{i \in \{\text{tns, wns}\}} w_i\, Q_i\right] \quad (1)$$

where

$$Q_{wns} = \sum_{j \in \{\text{early, late}\}} w_j\left(1 - \frac{wns'_j}{wns_j}\right) \quad (2)$$

indicates how much the worst negative slack is improved compared to the initial solution and

$$Q_{tns} = \sum_{k \in \{\text{early, late}\}} w_k\left(1 - \frac{tns'_k}{tns_k}\right) \quad (3)$$

indicates how much the total negative slack is reduced. For the infrastructure, the weighting factors are set as follows: $w_{tns} = 2.0$, $w_{wns} = 1.0$, $w_{abu} = 1.0$, $w_{early} = 1.0$, and $w_{late} = 5.0$. They were defined by ICCAD 2014 Contest organization. The maximum quality score is 1800. The final score is given by Equation (4)

$$Q \times (1 + \alpha) \quad (4)$$

where α is a runtime factor that, for every 2x speed up, increases by 5% the original quality score bounded at $\pm 20\%$.

III. JEZZ LEGALIZATION ENGINE

Jezz is an incremental legalization engine used to keep the placement solution legalized when some perturbation (e.g. cell swap, a single cell repositioning) is performed enabling detailed placement optimization. While inserting a cell in a row, Jezz seeks to reduce the total number of legalized cells that need to be shifted in order to open room to the new one. Note, however, that Jezz does not take into account any density metric as ABU to select how the cells should be shifted.

Although Jezz can perform both full and incremental legalization, only the incremental legalization is used in this work, as the initial placement is already legalized. It intrinsically handles cell-to-site alignment and has blockage support. A cache system is used for fast look-up during incremental legalization, allowing Jezz to be used as incremental legalization in detailed placement algorithms. Jezz's name is inspired from the 1992 game JezzBall [11].

A. Jezz Data Structures

In standard-cell based designs, the placement area is divided into same-height rows, and a row is divided into same-width sites. A legalized placement solution must have no overlapping among cells, and they must be aligned to the rows and sites.

Jezz defines three types of nodes: (1) whitespace, (2) blockage and (3) cell. Jezz Cell node's represents movable standard-cells, fixed standard-cells or obstacles (e.g. macro-blocks or blockage region). Jezz explicitly represents whitespace. Blockage and cell nodes have constant widths, while whitespace ones can be shrunk, enlarged, split, merged or removed. Whitespace and cell nodes are free to be moved, while blockage nodes are fixed.

A row is a linked list of nodes, and, by definition, the sum of node widths within a row equals the row width. If a row has no standard cells and no blockages, it has a single whitespace node filling all of it.

Jezz places cells only in integer positions, which intrinsically handle the cell-to-site alignment. Currently, only single-row height cells is supported.

B. Incremental Legalization

When a node is inserted in a row, it overlaps one or more nodes. If the node is entirely enclosed by a whitespace, the withespace node is simply split in two, and the node, inserted in the middle, as shows Figure 2.

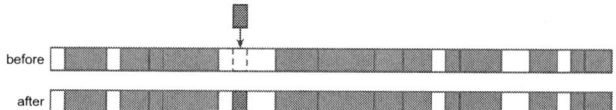

Figure 2. Cell insertion in a whitespace.

In any other case, the new node is supposed to be inserted in between the node that overlaps its left edge and the immediate right neighbor as shown in Figure 3. Starting at the left and right adjacent nodes, two sequential searches are then performed to check the number of cells that need to be shifted in order to make room for the new cell.

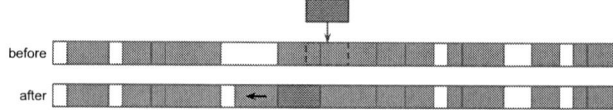

Figure 3. Cell insertion between two nodes.

The search algorithm sequentially sweeps the nodes seeking for whitespaces. The search stops when the accumulated width of whitespaces is greater or equal to the width of the new node, or when there is no more space left. This procedure is outlined in Algorithm 1. If the combined number of spaces that can be opened to left and right is less than the node width, the node cannot be inserted in the current row and surrounding rows maybe tested.

Jezz then opens the maximum available space to the direction with the lowest number of impacted cells. If space still needs to be opened, the other direction is used.

C. Cache System

During incremental legalization, Jezz needs to find out which node (cell, whitespace, blockage) within the row overlaps or is overlapped by the cell being inserted. Since the nodes are stored using a linked list, only a sequential search can be

Algorithm 1: Computation of the Impact of a Cell Insertion

```
1  node = reference node;
2  disrupt = 0;
3  overflow = w;
4  while (node is not null and overflow > 0) do
5      if node is blockage then
6          break;
7      end
8      if node is whitespace then
9          offset = width(node) > overflow? overflow :
                 width(node);
10         overflow = overflow - offset;
11     else
12         disrupt = disrupt + 1;
13     end
14     node = previous(node);
15 end
```

used to look-up the overlapping node. To reduce the number of nodes visited during the sequential search, Jezz implements a cache system with pointers to nodes in the middle of the linked list.

Each row is divided into same-width regions, and, for each region a pointer to a node inside that region is stored, as shown in Figure 4. Pointers are not updated all the time, to avoid unnecessary overhead. If the node is displaced horizontally, the cache pointer may point to a node outside of its respective region. For small displacement, that should be fine, as we still get a node close to the aimed region anyway. If the node was moved to another row, the pointer gets invalid, and Jezz looks at the pointer in neighboring regions. Cache pointers are updated when a look-up at that region is performed.

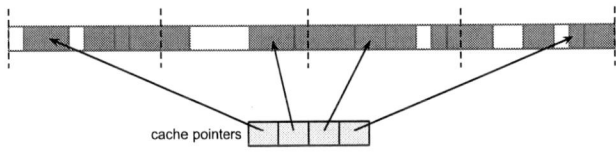

Figure 4.   Jezz Cache

IV.   CLOCK SKEW REDUCTION

Clock tree plays a crucial role for the correct operation of a digital circuit. It is one of the largest networks in the design [12]. The ICCAD 2014 contest adopts the FLUTE Steiner tree generator to estimate the clock network in the same way as any other signal network. Since FLUTE does not necessarily generate balanced trees, this may lead to large clock skew. Therefore, given the problem formulation, it is important to use a strategy to reduce the clock skew.

FLUTE is very sensitive to changes in pin positions, so that small changes may generate completely different tree structures that can be more suitable in terms of clock skew. Along with this, the maximum displacement constraint encourages the development of an approach with focus on local movement of sequential gates.

Algorithm 2 presents the strategy for clock skew optimization applied in this work. The input is a list of all sequential cells and their pins' timing information, the minimum gain in Quality Score to accept any change, the maximum number of consecutive failures when improving the solution and minimum quality score. The latter two are used in the stop criteria.

As shown in line 1, all elements on the list are sorted in descending order by their late slack. The algorithm's main loop is presented at lines 2-14. First, the algorithm tries to swap the gate with maximum late slack with the right and then with the left nodes (Lines 4-5). Note, that a node can be both a cell or a whitespace. The gain for each swap is stored. Then, it compares the two swaps and applies only the best one, if the gain is positive. At the end of each iteration, the current register is removed from the list. The loop stops when the Quality Score achieves the target threshold or the current solution stops improving (Line 2).

Every time a swap is performed, the nets connecting the swapped cells are re-routed using FLUTE and then an incremental static timing analysis is performed. With the updated timing information, the quality score is computed. When the swap is undone, a cached version of the previous routing is retrieved saving runtime.

Algorithm 2: Clock Skew Reduction Algorithm

Data: clockedGates, minGainSwap, maxFails, minQualityScore

```
1  sort( clockedGates );
2  while computeQualityScore() < minQualityScore
    or countLastConsecutiveFailures() > maxFails do
3      gate = clockedGates.head();
4      gainSwapLeft = trySwapLeft( gate );
5      gainSwapRight = trySwapRight( gate );
6      if gainSwapLeft > 0 and
        gainSwapLeft > gainSwapRight then
7          swapLeft( gate );
8      else
9          if gainSwapRight > 0 and
            gainSwapRight > gainSwapLeft then
10             swapRight( gate );
11         end
12     end
13     clockedGates.remove( gate );
14 end
```

V.   CRITICAL PATH OPTIMIZATION USING QUADRATIC FORMULATION

In this work, a quadratic placement algorithm was adapted from global placement to operate only on critical paths for timing-driven detailed placement. Therefore, the quadratic algorithm moves only a small fraction of the total number of cells in the circuit.

The main idea behind the critical path optimization algorithm is to use a quadratic global placement algorithm with quadratic formulation to find a good balance for the timing propagation of each net on the critical path.

978-1-4673-9141-2/15 $31.00 © 2015 IEEE

If a timing-driven algorithm optimizes only locally, one net at a time, in critical path, the impact of the connections on remaining nets is not properly considered. On the other hand, if a critical path algorithm optimizes the connections of all nets in the critical path, the impact of connections among near-critical cells is not easily considered.

In the proposed technique, timing propagation for net-to-net and for all nets of the critical path is optimized at the same time with a quadratic formulation. Therefore, net delay is balanced among neighbor nets and for entire critical paths.

By construction, a quadratic placement algorithm places all cells at the barycenter of their neighbors. In a single-fanout chain of cells, all cells are always placed half way from their drivers and sinks. This naturally minimizes the interconnect delay, which is proportional to the square of the wire length. However, many cells drive more than one sink, so the path minimization feature is only indirectly taken into account.

In our approach, we focus on a few paths to take advantage of the wire delay minimization property of the quadratic formulation. Therefore only cells and their direct neighbors on top most critical paths are considered movable.

A. Quadratic Technique

The formulation of the quadratic placement requires a differentiable and continuous objective function to be efficiently optimized. However, the most common metric to estimate the wirelength is the half-perimeter (HPWL) of the minimum bounding box enclosing all cells of a net, which is neither differentiable nor continuous.

To overcome that, quadratic formulation estimates the wirelength using the squared Euclidean distance of each pair of connected cells as shown in Equation 5, where $\omega(i,j)$ is the connection weight and (x_i, yi) and (x_j, y_j) are the cell positions.

$$\sum \omega(i,j) \times \left|(x_i - x_j)^2 + (y_i - y_j)^2\right| \qquad (5)$$

The quadratic formulation also requires multiple-fanout nets to be decomposed into pin-to-pin connections. Nets are decomposed with a specific net model, such as star, clique, hybrid [13] or Bound2Bound (B2B) [14].

The function $\phi(x,y)$ (Equation 6) represents the total squared Euclidean wirelength of the circuit. The entry i,j in the matrices Q_x and Q_y represents the weight of the connection between the cell i and cell j. A zero weight indicates no connection at all. The matrices are sparse as, usually, each cell connects only to a few other cells. They are also symmetric and positive definite [14].

$$\phi(x,y) = \frac{1}{2}x^T Q_x x + c_x^T x + \frac{1}{2}y^T Q_y y + c_y^T y + const \qquad (6)$$

The minimum cost of the $\phi(x,y)$ is obtained by setting its derivative to zero as presented in Equation 7. The minimum point of $\phi(x,y)$ function is equivalent to obtaining the equilibrium state or state of minimum energy of a spring system.

Consequently, the circuit connections (decomposed nets) can be seen as springs.

$$\nabla\phi(x,y) = Q_x x + c_x + Q_y y + c_y = 0 \qquad (7)$$

B. Timing-Driven Quadratic Algorithm for Detailed Placement

The nets in the timing-driven quadratic algorithm are decomposed using the clique net model [13]. This net model has no significant impact on runtime, since only a small fraction of the total number of nets is being considered during quadratic timing-driven detailed placement.

Some cells may violate the maximum displacement constraint after the linear system is solved. This violation is fixed by moving them back inside the maximum displacement region. The last stage in the detailed quadratic algorithm is to legalize all cells moved by our algorithm using Jezz legalizer in incremental mode.

In Algorithm 3, our quadratic timing-driven detailed placement method is presented. Our algorithm receives the circuit boundary area, the netlist and the number of critical paths as inputs.

Algorithm 3: Critical Path Quadratic Optimization Algorithm

Data: placement boundary, circuit netlist, number of paths x
1 Initialize Jezz in incremental mode;
2 Backup cells position;
3 Set all register as fixed elements in Jezz legalizer;
4 Get all combinational cells for x highest critical path;
5 Get all combinational cells in the first deep logic level from ones in critical path;
6 Set remaining cells as fixed elements;
7 Create the linear system;
8 Solve the linear system;
9 Fix displacement violation;
10 Legalize moved cells;
11 Update evaluation metrics;
12 Update quality score;
Result: legal placement

In Line 1, Jezz legalizer is initialized in incremental mode. The initial position of each cell of the circuit is stored (Line 2). They are used to detect maximum displacement violation. All sequential cells are set as fixed elements (Line 3). That prevents Jezz from moving registers while it legalizes combinational cells. Combinational cells for the top x more critical paths are obtained (Line 4), and, for each combinational cell in the critical path, the neighbour combinational cells are obtained (Line 5). All circuit cells are changed to fixed in our algorithm (Line 6), except ones that are part of critical paths (Lines 4 and 5). A linear system is built and solved (Lines 7 and 8) in the same way as it is built and solved by the quadratic algorithm in the global placement stage. The displacement distance is measured and, if it violates maximum displacement, the cell is moved back inside the maximum displacement region (Line 10). Finally, the evaluation metrics,

978-1-4673-9141-2/15 $31.00 © 2015 IEEE

such as slack timing, density utilization, etc., are updated (Line 11), and the quality score is computed (Line 12). The quadratic solution is accepted if the quality score is improved, otherwise, it is rejected.

VI. EXPERIMENTAL RESULTS

A. Experimental Configuration

The Detailed Timing-driven placement flow was coded in C++11 and compiled with G++ 4.8.3. The experimental results were obtained running the static binary in ICCAD 2014 Contest [8] infrastructure. It is composed by a machine with CPU 32 x 64-bit Intel(R) Xeon(R) E5-2650 v2 2.60GHz, 20480 KB of cache, 64 GB of RAM and CentOS release 6.2 (Final) operating system with kernel 2.6.32-220.el6.x86_64. The benchmarks are composed by seven circuits already legalized and with timing violations. Five of then were released during the contest.

The ICCAD 2014 benchmark circuits have from 130k up to 958k gates and some of them have blockage areas, which are regions in the circuit where it is forbidden to place cells. The maximum density allowed is around 70% of the area limit of utilization, and two maximum displacement values are defined for each of them, short and long. The characteristics of the benchmarks are shown in Table I. The evaluation criteria proposed by ICCAD 2014 Contest [8] use a compromise between placement quality score improvement and runtime factor.

In ICCAD 2014 Contest [8], the total runtime and quality score of the 5 best ranked teams were used to normalize the final placement quality score. The teams were ranked, from the first to the fifth place, respectively, UFRGS/FURG-Brazil (1374.38), VDA-TP (1370.43), CUHK-ITP (1264.45), NTU-ITP (1111.34), and UFSC-Brazil (1089.02) by average of the normalized quality score improvement from short and long displacement.

B. Discussion of the Experimental Results

In Table II, the normalized results of the 5 best ranked algorithms are presented, for both short and long displacement limits. For the short displacement limit, our algorithm obtained the best normalized quality score for b19, vga_lcd, and leon2, compared to the remaining teams. For the long displacement limit, our algorithm obtained the best normalized quality score improvement for b19, vga_lcd, leon3mp, and leon2, and it also obtained the best average normalized quality score improvement overall.

For short displacement limits, our algorithm obtained the best normalized quality score improvement, 1683.88, in leon2 circuit, and the worst normalized quality score improvement, 502.39, in mgc_matrix_mult. On average, our algorithm obtained 1294.24 of normalized quality score improvement, that was the second best improvement.

For long displacement limits, our algorithm obtained the best normalized quality score improvement, 1708.05, in b19 circuit, and the worst normalized quality score improvement, 757.11, in mgc_edit_dist. On average, our algorithm obtained 1454.51 of normalized quality score improvement, that was the best normalized quality score improvement. For mgc's

benchmarks, the useful clock skew algorithm was unable to explore the skew to reduce significantly timing violation.

In Table III, the ratio among our algorithm and the remaining four best ranked algorithms is shown. The ratio is computed by dividing our normalized quality score by the normalized quality score of the remaining teams. Our algorithm obtained, on average, 19%, and 18% higher normalized quality score than, respectively, UFSC-Brazil, and NTU-ITP teams for all circuits for short displacement limit. On the other hand, our algorithm is, on average, 2% and 6% worst than, respectively, VDA-TP, and CUHK-ITP teams in normalized quality score for all circuits in the short displacement limit.

For long displacement limits, our algorithm obtained, on average, higher normalized quality score than the remaining teams. On average, our algorithm has 3%, 33%, 26% and 30% higher normalized quality score than, respectively, VDA-TP, UFSC-Brazil, CUHK-ITP and NTU-ITP teams.

On average, considering the final quality score, our algorithm has 0.3% better quality score than VDA-TP Team. Comparing our algorithm with the remaining three algorithms, we have 26.2%, 8.7% and 23.7% better quality score than, respectively, UFSC-Brazil, CUHK-ITP, and NTU-ITP teams.

In the Table IV is shown the impact only of the quadratic incremental timing-driven algorithm. The results were measured after the useful skew clock optimization for circuits that still have timing violation. Our quadratic algorithm moves less than 1% of the total number of cells in benchmarks. The ABU increases significantly for netcard, mgc_edit_dist and mgc_atrix_mult after critical path optimization.

Table IV. IMPACT OF THE QUADRATIC ALGORITHM

Circuits	Moved Cells	Execution AVG. (s)	ABU - Short $(\times 10^{-2})$	ABU - Long $(\times 10^{-2})$
vga_lcd	50	2.28	2.17	2.99
b19	77	2.40	2.59	2.66
leon3mp	-	-	-	-
leon2	-	-	-	-
netcard	38	23.58	28.66	28.98
mgc_edit_dist	66	1.35	60.29	60.36
mgc_matrix_mult	147	1.25	66.46	66.62

VII. CONCLUSION AND FUTURE WORK

In this paper, we presented our Incremental Timing-Driven Placement flow. Our flow aims at timing violation reduction given an already placed and legalized circuit. We implemented a fast static timing analysis tool and also optimized FLUTE [10] algorithm to improve its runtime, which can take up to 70% of the total runtime of our algorithm.

The Timing-Driven Detailed Placement flow first optimizes clock skew, then, if timing violation still exits, it optimizes combinational gates in the critical paths using a quadratic placement formulation. Experimental results showed that our algorithm obtained the best final average normalized quality score improvement for the ICCAD 2014 Contest benchmarks.

Considering the contest formulation, the clock tree network plays an important role in the timing violation reduction, as it is estimated using a regular, possibly non-balanced, Steiner tree. As a future work, we plan to use a more accurate estimation of the clock tree and a weighting scheme based on the global critically and density utilization of cells.

978-1-4673-9141-2/15 $31.00 © 2015 IEEE

Table I.   BENCHMARKS INFRASTRUCTURE FEATURES

Circuits	# Cells	# Pins	Blockage	Dimensions (μm)	Max. Density	Max. Disp.	Clock (ns)
vga_lcd	164891	184	No	898.6 x 898	0.70	10 / 200	4
b19	219268	47	No	1187.2 x 1188	0.76	20 / 200	5
leon3mp	649191	333	No	1989.2 x 1990	0.70	30 / 300	35
leon2	794286	700	No	2086.4 x 2086	0.70	40 / 400	64
netcard	958792	1846	Yes	2520.6 x 2520	0.72	50 / 400	42
mgc_edit_dist	130674	2574	Yes	843.4 x 844	0.75	30 / 200	5
mgc_matrix_mult	155341	4802	Yes	895.2 x 896	0.70	30 / 200	4.4

Table II.   NORMALIZED QUALITY SCORE IMPROVEMENT

Circuits	Quality Improvement (%) for Short Displacement					Quality Improvement (%) for Long Displacement				
	VDA-TP	UFSC-Brazil	CUHK-ITP	NTU-iTP	UFRGS/FURG-Brazil	VDA-TP	UFSC-Brazil	CUHK-ITP	NTU-iTP	UFRGS/FURG-Brazil
vga_lcd	1367.92	1289.96	1492.01	1423.37	1494.24	1347.54	1296.55	1413.61	1497.48	1523.01
b19	1620.61	1271.34	1347.80	1410.61	1648.62	1695.26	1269.70	1566.32	1505.85	1708.05
leon3mp	1513.95	1180.94	1861.95	1201.30	1582.38	1512.60	1204.62	1859.56	1102.38	1581.03
leon2	852.15	1263.74	1253.90	1126.11	1683.88	857.14	1289.20	266.94	1055.04	1678.35
netcard	1344.15	1322.19	1561.13	1562.61	1386.53	1347.35	1303.20	1464.87	1566.27	1387.01
mgc_edit_dist	1367.28	745.19	1156.70	550.90	761.67	1347.45	745.19	1048.00	674.36	757.11
mgc_matrix_mult	1218.41	532.21	925.14	434.04	502.39	1794.26	532.21	484.38	448.40	1547.05
AVG.	1326.35	1086.51	1371.23	1101.28	1294.24	1414.51	1091.52	1157.67	1121.40	1454.51

Table III.   RATIO OF THE NORMALIZED QUALITY SCORE IMPROVEMENT

Circuits	Ratio for Short Displacement				Ratio for Long Displacement			
	VDA-TP	UFSC-Brazil	CUHK-ITP	NTU-iTP	VDA-TP	UFSC-Brazil	CUHK-ITP	CUHK-iTP
vga_lcd	1.09	1.16	1.00	1.05	1.13	1.17	1.08	1.02
b19	1.02	1.30	1.22	1.17	1.01	1.35	1.09	1.13
leon3mp	1.05	1.34	0.85	1.32	1.05	1.31	0.85	1.43
leon2	1.98	1.33	1.34	1.50	1.96	1.30	6.29	1.59
netcard	1.03	1.05	0.89	0.89	1.03	1.06	0.95	0.89
mgc_edit_dist	0.56	1.02	0.66	1.38	0.56	1.02	0.72	1.12
mgc_matrix_mult	0.41	0.94	0.54	1.16	0.86	2.91	3.19	3.45
AVG.	0.98	1.19	0.94	1.18	1.03	1.33	1.26	1.30

REFERENCES

[1] N. Sherwani, *Algorithms for VLSI physical design automation*, 3rd ed. Kluwer Academic Publishers, 1999.

[2] C. Alpert, Z. Li, G.-J. Nam, C. Sze, N. Viswanathan, and S. Ward, "Placement: Hot or not?" in *Computer-Aided Design (ICCAD), 2012 IEEE/ACM International Conference on*, Nov 2012, pp. 283–290.

[3] D. Papa, T. Luo, M. Moffitt, C. Sze, Z. Li, G.-J. Nam, C. Alpert, and I. Markov, "Rumble: An incremental timing-driven physical-synthesis optimization algorithm," *Computer-Aided Design of Integrated Circuits and Systems, IEEE Transactions on*, vol. 27, no. 12, pp. 2156–2168, Dec 2008.

[4] T. Luo, D. Newmark, and D. Pan, "A new lp based incremental timing driven placement for high performance designs," in *Design Automation Conference, 2006 43rd ACM/IEEE*, 2006, pp. 1115–1120.

[5] W. Choi and K. Bazargan, "Incremental placement for timing optimization," in *Computer Aided Design, 2003. ICCAD-2003. International Conference on*, Nov 2003, pp. 463–466.

[6] S. Dutt and H. Ren, "Discretized network flow techniques for timing and wire-length driven incremental placement with white-space satisfaction," *Very Large Scale Integration (VLSI) Systems, IEEE Transactions on*, vol. 19, no. 7, pp. 1277–1290, July 2011.

[7] A. Kahng, *VLSI Physical Design: From Graph Partitioning to Timing Closure*, 1st ed.   Springer London, Limited, 2011.

[8] M.-C. Kim, J. Huj, and N. Viswanathan, "Iccad-2014 cad contest in incremental timing-driven placement and benchmark suite: Special session paper: Cad contest," in *Computer-Aided Design (ICCAD), 2014 IEEE/ACM International Conference on*, Nov 2014, pp. 361–366.

[9] M.-C. Kim, N. Viswanathan, C. J. Alpert, I. L. Markov, and S. Ramji, "Maple: Multilevel adaptive placement for mixed-size designs," in *Proceedings of the 2012 ACM International Symposium on International Symposium on Physical Design*, ser. ISPD '12.   New York, NY, USA: ACM, 2012, pp. 193–200.

[10] C. Chu and Y.-C. Wong, "Flute: Fast lookup table based rectilinear steiner minimal tree algorithm for vlsi design," *Computer-Aided Design of Integrated Circuits and Systems, IEEE Transactions on*, vol. 27, no. 1, pp. 70–83, Jan 2008.

[11] Wikipedia, "Jezzball," 2014, [Online; accessed 08-Nov-2014]. [Online]. Available: http://en.wikipedia.org/wiki/JezzBall

[12] D. J. Lee, "High-performance and low-power clock network synthesis in the presence of variation," Ph.D. dissertation, The University of Michigan, 2011.

[13] N. Viswanathan and C. C.-N. Chu, "Fastplace: Efficient analytical placement using cell shifting, iterative local refinement and a hybrid net model," *Proceedings of the 2004 International Symposium on Physical Design*, pp. 26–33, 2004.

[14] P. Spindler, U. Schlichtmann, and F. Johannes, "Kraftwerk2 - a fast force-directed quadratic placement approach using an accurate net model," *Computer-Aided Design of Integrated Circuits and Systems, IEEE Transactions on*, vol. 27, no. 8, pp. 1398–1411, Aug 2008.

[15] L.-T. Wang, Y.-W. Chang, and K.-T. T. Cheng, *Electronic design automation: synthesis, verification, and test*.   Morgan Kaufmann, 2009.

978-1-4673-9141-2/15 $31.00 © 2015 IEEE

Flare Reduction in EUV Lithography by Perturbation of Wire Segments

Sudipta Paul*[†], Pritha Banerjee[†] and Susmita Sur-Kolay*

*Advance Computing and Microelectronics Unit, Indian Statistical Institute, Kolkata, India
Email: sudiptapaul2@gmail.com, ssk@isical.ac.in
[†]Dept. of Computer Science and Engineering, University of Calcutta, India
Email: banerjee.pritha74@gmail.com

Abstract—**With growing demand for complex and high density integrated chips (IC), optical lithography with 193 nm immersion technology has become a bottleneck in the chip manufacturing industry. IC fabrication industry is looking forward to next generation lithography methods, for example, Extreme Ultraviolet Lithography (EUVL). While EUVL is capable of printing with a wavelength of 13.5 nm, it suffers from a major drawback called flare, due to the scattering of light on blank surfaces. Large flare and/or its large variation cause critical dimension (CD) violations. In this paper, we propose an Integer Linear Programming based method to mitigate the effects of flare in the post routing step through perturbation of wire segments. Experimental results on a set of synthetic circuits show significant reduction of flare and its standard deviation across the chip surface.**

I. INTRODUCTION

Advancement of technology has scaled down the process geometry to the nanometer range. Nowadays IC foundries have to use an optical lithography system with a larger wavelength of light to print the layout features (patterns) having much smaller dimensions on the wafer. This is referred as the sub-wavelength lithography gap [1] but it leads to huge CD distortion for printed metal features. One of the popular ways of printing at sub 20nm process technology nodes using 193i lithography system is referred as multiple patterning [9] which requires decomposition of the layout into multiple layout with larger values of the minimum spacing in order to make it compatible with the larger wavelength. It can be observed from the various works [9] [10] on multiple patterning that it needs multiple masks for printing a single layout thereby leading to increased mask cost. Multiple patterning also has the problem of CD distortion. EUVL, a next generation technique using light beam having wavelength of 13.5nm, is thus convenient to print modern technology nodes with smaller feature sizes. However 13.5nm wavelength of light is not transmitted but absorbed by most of the materials which is the main disadvantage of EUVL system. To solve this problem, reflective optical components and clear-field masks are used to print patterns. The layout patterns are formed by the absorbers on the clear-field mask, while reflective materials remain in the vacant regions. Incident light gets absorbed by the absorbers and the patterns get printed, whereas light reflects from the vacant regions.

Although reflective materials in the clear-field mask can

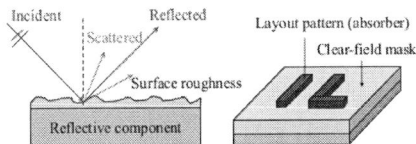

Fig. 1. Light scatters from the photo reflective coating on the mask and patterns are printed using the light absorbers (Courtesy [3])

prevent the light from getting absorbed, undesirable scattered light reflected due to surface roughness, reduces the contrast between bright (vacant regions) and the dark (layout pattern) regions on exposure to EUV light. Such undesired scattered light reflected from the vacant regions on the surface is called *flare* in EUVL. Figure 1 shows the clear-field mask and the scattering of light on the uneven surface of the photo reflective substance. The reduction of contrast between patterns and vacant regions causes high flare and CD distortion. Since flare is proportional to the surface roughness of the optical system and inversely proportional to square of the wavelength, EUVL suffers from rather high levels of flare compared to traditional lithography technologies [3]. Flare is the most prominent challenge for EUVL.

There are few strategies to compensate the effect of flare in the literature, such as optical proximity correction (OPC) and dummy fills for CMP. It was observed that one percent of change in flare may cause a change of 10 nm in CD at the 22 nm technology node and may be considerably larger for more advanced technology nodes [4]. The non-uniformity of CD caused by large flare may not be fully compensated by global CD resizing techniques [3]. Instead, previous works [3] [5] on flare reduction have used dummification for reducing flare and its variation. Layout regions with less *pattern density* contribute more to the generation of flare than the regions with higher pattern density. Thus, vacant regions of the layout are filled with dummy (non-functional) metal features to increase the pattern density there, and reduce the flare. However dummification requires a large number of dummy features to be added to the layout which increases the mask cost. Moreover, large amount of dummy features may increase the coupling capacitance [6] between metals which can affect the circuit speed. Additional constraints are required to bound the coupling capacitance of metals (wire segments) within a threshold. In this paper, we propose the first work on minimization of flare at the post-routing stage without dummification as well as any performance degradation.

This work has been partially funded by India-Taiwan Programme in Science and Technology (Project GITA/DST/TWN/P-43/2013).

It was shown in [4] that uniform pattern density does not induce uniform flare distribution. Typically, the flare in the central region of the chip is much higher than that in the peripheral regions. This phenomenon is referred as the *flare periphery effect* [3]. It is reported that conforming the *pattern density map* to the *flare map* minimizes the maximum as well as the average value of flare. In this work we propose a wire perturbation based strategy to minimize the peak flare, the mean flare and its standard deviation implying better control on CD uniformity without sacrificing the delay significantly. Previous works [13] [14] have used wire perturbation for conflict and stitch-awareness in Double Patterning Lithography. Wire perturbation provides only local changes in layout. Even though flare is a global phenomenon but can be reduced by judicial modification in the post layout stage. Wire perturbation can be an acceptable approach for layout modification.

In the rest of this paper, Section II details out how to compute flare, Section III describes our flare minimization problem and Section IV describes our proposed method. Experimental results appear in Section V and finally Section VI presents the concluding remarks.

II. PRELIMINARIES

This section describes the existing model[11] [12] of flare computation in EUVL.

Computation of Flare: Flare in EUVL system is traditionally modeled as a *Point Spread Function* (PSF). The flare map is obtained by convolving the PSF with the original image intensity I_o of the layout. However, due to a large computation overhead, I_o is approximated to be the pattern density at every cell on a gridded layout as in [11] [12] to compute flare. As vacant regions on clear-field masks contribute to flare, vacancy density map is used in computation of flare map instead of pattern density. After dividing the entire layout into suitably sized rectangular grid cells, the flare $F(x, y)$ corresponding to each cell (x, y) is computed as

$$F(x, y) = D_v(x, y) \otimes PSF(x, y) \qquad (1)$$

Here, $D_v(x, y)$ and $PSF(x, y)$ are respectively the vacancy density and PSF of the cell (x, y).

III. PROBLEM FORMULATION.

In this work, the layout for a particular interconnect layer is considered where the interconnect patterns are essentially arrangements of vertical and horizontal wire segments. This interconnect layer is referred as the layout henceforth, which is divided into rectangular fixed sized grid cells. The pattern density of each cell $D_p(x, y)$ is defined to be the ratio of the total area of its existing patterns and the maximum area of patterns that can be printed in this cell without violating the design rule. The vacancy density of a cell is thus defined as $D_v(x, y) = 1 - D_p(x, y)$. The initial flare map F for the layout can then be computed using Equation 1.

The horizontal and vertical segments of the patterns in a layout can be perturbed to change the pattern density map, and thereby the flare map. Here, perturbation of a wire segment refers to the movement of the entire segment onto a predefined

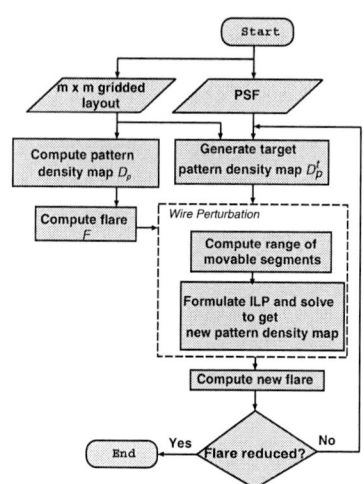

Fig. 2. Overview of flare optimization by wire perturbation

routing track in a direction orthogonal to its orientation. However, moving a segment to a new location may disconnect the net, or even if it remains connected, may increase the wirelength. Hence, we define $S = \{s_i : 1 \le i \le n\}$ as the set of movable segments where each segment s_j can be perturbed within a *perturbation range* r_j depending on the geometry of pattern containing the segment. As flare depends on the pattern density map D_p of the layout, a target pattern density map D_p^t is computed so that it conforms with the flare map F of the gridded layout. The problem of flare minimization is defined as follows:

Given a grid based layout, determine S, the set of movable segments and perturb the segments to new locations such that the maximum flare and the variance of the flare map is minimized with minimal increase in wirelength.

In other words, the problem can be formulated as:

Given a flare conforming target density map D_p^t for a gridded layout, determine $S = \{s_j\}$, the set of movable segments with the perturbation range of r_j of each segment $s_j \in S$, and reassign these segments to the grid cells such that $\sum |D_p^t(x, y) - D_p(x, y)|$ over all grid cells is minimized.

IV. PROPOSED METHOD

In this section we discuss our proposed method of minimization of flare by wire segment perturbation. Section IV-A describes the overall flow of the proposed method. Section IV-B illustrates the computation of flare conforming target pattern density map. Section IV-C elaborates the range computation for movable segments and Section IV-D presents the Integer linear programming (ILP) based formulation for the layout perturbation.

A. Overview

Figure 2 shows the overall flow of our flare minimization method. First, the pattern density map and the corresponding flare map are computed. Flare conforming target pattern density map D_p^t is estimated following the method described in Section IV-B. The perturbation range for each movable segment is computed using the method described in Section

978-1-4673-9141-2/15 $31.00 © 2015 IEEE

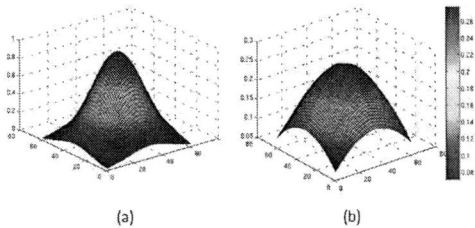

Fig. 3. (a) A 2D Gaussian function and (b) the flare map of a circuit layout

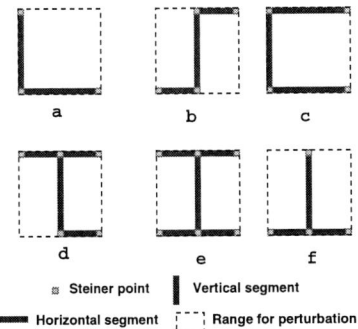

Fig. 4. Six types of patterns which can be perturbed

IV-C below. An ILP is formulated and solved to relocate the wire segments within their respective ranges. Next, the new pattern density and flare maps are computed, to check flare reduction with respect to the initial flare map. If the maximum flare has not improved, then the entire process can be repeated with a different target pattern density map.

B. Estimation of target pattern density

Flare periphery effect [3] causes higher flare at the center of the layout than the periphery in a EUVL system as discussed earlier. This phenomenon is modeled as a 2D Gaussian distribution [11]. Figures 3(a) and (b) illustrate this with a 2D Gaussian distribution and a flare map of a layout of size 61×61 grid cells respectively. As the flare of a layout always follows the Gaussian characteristics, a pattern density conforming to the flare map needs to be dense in the central region and gradually sparse towards the periphery. Thus, the target pattern density D_p^t is modeled as a Gaussian distribution with a standard deviation of σ, not necessarily same as that of the PSF.

C. Perturbation Range

In a layout, the possible patterns consisting of connected horizontal and vertical wire segments which can be perturbed with bounded increase in wirelength, are classified into six types of primitive patterns [2] as shown in Figure 4. In Figure 4 each segment is defined with two steiner points (vias) on either side. These vias are either movable, or fixed points that are connected to the pins of the blocks (terminals). This implies that a via connected to a terminal should remain fixed to maintain the connectivity with the block pin, but a via connecting two wire segments only can be moved during perturbation.

The perturbation range r_j of a wire segment s_j of a pattern is defined as the smallest enclosing rectangle containing the

entire pattern. Note that the segment s_j can be moved only to a predefined routing track within the rectangular region $r_j = \{(X_{bl}, Y_{bl}), (X_{tr}, Y_{tr})\}$, where (X_{bl}, Y_{bl}) and (X_{tr}, Y_{tr}) are the co-ordinates of the bottom left and top right corners respectively, within which the perturbation of any wire segment should be confined in order to obtain a bounded increase of wirelength.

Figure 5 shows the perturbation range, and the effect of perturbation on wirelength for the pattern types shown in Figure 4(b) and (c). For the perturbation range corresponding to Figure 5(a), the Figures 5(a1), (a2) and (a3) show three possible cases of perturbation and their effects on wirelength. Let the rectangle $MNOP$ be the perturbation range r for a segment s, where N and P correspond to (X_{bl}, Y_{bl}) and (X_{tr}, Y_{tr}) respectively. The dotted arrows show the possible direction of the movement of segment s. Let, A and B be the two vias connecting h_1 and h_2 with s. The length of h_1 and h_2 may change depending on the type of the via, and the direction of the movement of s during perturbation. A perturbation with no change in wirelength is shown in the Figure 5(a1) for the pattern type in Figure 5(a), where A and B are both representing movable vias, and s is moved to the right by a distance of L_1. The length of h_2 decreases by L_1, and the length of h_1 increases by an equal amount. As a result, the total length remains same assuring no change in wirelength. The vias A and B are both moved and placed in the new positions. Similarly, s may be moved leftwards when it is connected to movable vias on both sides. In this case, the length of h_1 instead of h_2 decreases, and the length of h_2 increases by an equal amount.

Figures 5(a2) and (a3) describe the situations when the vertical wire segment has vias at the end which are connected to terminals. The pattern shown in Figure 5(a2) has via A attached to a terminal and cannot be moved. However, B is a movable via. The segment s is moved leftwards by L_2. The length of h_2 increases by L_2, and A remains intact to maintain the connection with the terminal, while a new movable via A_1 is used to keep the connection between s and h_2. The length of h_1 decreases by L_2 compensating the increase in h_2. A situation where a change in wirelength during perturbation is possible is illustrated in the Figure 5(a3). Here the type of vias A and B are same as in Figure 5(a2) but the segment s is moved rightwards by L_3 along with the movable via B connected to it. The length of h_1 increases by L_3. The length of h_2 remains same to maintain the connectivity with the terminal through fixed via A, and in order to retain the connection between s and h_2 a new movable via A_1 has to be introduced. This perturbation scenario shows increase in the total wirelength by L_3. However, this increment is restricted within the defined perturbation range.

In Figure 5(b1) and (b2) depicts the different possible situations during perturbation of the pattern shown in the Figure 5(b) in the presence of fixed and movable vias. The leftward movement of s is prohibited because any position to the left of s is outside the rectangle $MNOP$ defined for its range. Figure 5(b1) shows the rightward perturbation of s by L_4 where both A and B are movable vias. Both h_1 and h_2 decrease by L_4 which in turn decreases the wirelength. In Figure 5(b2), both A and B are fixed and s is moved rightwards by L_5 resulting in the need for two new vias A_1 and

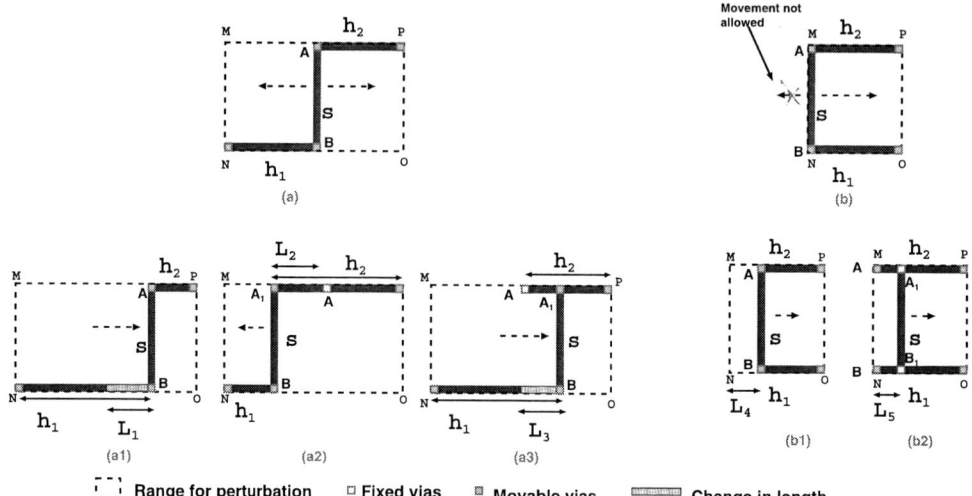

Range for perturbation ☐ **Fixed vias** ▨ **Movable vias** ▥ **Change in length**

Fig. 5. Pattern (a) Type b, and (b) Type c of Fig. 4, and their respective feasible cases of perturbation within the defined perturbation range in the presence of movable and fixed vias

B_1 to connect s with h_1 and h_2 respectively. The lengths of both h_1 and h_2 remain unchanged. The concept of perturbation within a defined range described here can be applied to all the six types of patterns shown in Figure 4 with restricted change in the wirelength.

D. Layout Perturbation

Given the current pattern density map D_p, flare conforming target pattern density map D_p^t and the set of movable segments $S = \{s_j\}$ with each s_j having a perturbation range r_j, we formulate an ILP to minimize the flare of a given layout. The objective is to minimize the cumulative sum of the difference between D_p^t and the new pattern density D_p obtained by perturbation of wire segments, i.e, minimization of $\sum |D_p^t(i,j) - D_p(i,j)|$.

The movable wire segments have specific height and width in a cell. Therefore, moving a segment from one cell to another causes pattern area of the segment to be transferred from one cell to the other. This leads to approximation of all the density maps to area maps. Multiplying D_p^t with the total pattern area allowed by design rules in the corresponding cell, gives the area map A_p^t. Then, the target pattern area of a cell $A_i^{t'}$ is computed as its available area A_i^v pro-rated by the ratio of its A_i^t to the maximum of A_p^t over all grid cells.

The minimization of flare is next formulated as an ILP for each row of cells in the gridded layout. We have performed segment perturbation in two different ways. While the first does not have any constraint over the ordering of the wire segments within a row of grid cells, the second one retains the initial order of the wire segments during perturbation. These are illustrated in the Section IV-D1 and IV-D2. Let

- n_c : Number of cells in any row.

- G_R : Sequence of cells in any row from left to right, $(g_1, g_2, \ldots, g_{n_c})$

- n_s : Number of segments in any row.

- seg_i : Sequence of segments in i^{th} cell.

- seg_R : Sequence of segments $(s_1, s_2, \ldots, s_{n_s})$ in any row.

- A_j^s : Area of the jth segment.

- A_i^v : Area available for allocation of the segments excluding the area of fixed segments in i^{th} cell.

- A_i^t : Pattern area for the ith cell generated from D_p^t.

- A_{max}^t : Maximum of A_i^t over all grid cells.

- c_j^r : Subsequence of G_R such that c_j^r lies within the range r_j of the jth segment.

- x_{ij} : 0-1 variable; 1 if segment j is assigned to i^{th} cell and 0 otherwise.

- A_i^a : Total area of the segments allocated to the ith cell computed by Equation 2.

- $A_i^{t'}$: Target pattern area for the ith cell computed by Equation 3.

- α : User defined parameter.

$$A_i^a = \sum_{j=1}^{n_s} x_{ij} A_j^s \mid \forall i, g_i \in G_R \qquad (2)$$

$$A_i^{t'} = A_i^v \frac{A_i^t}{A_{max}^t} \mid \forall i, g_i \in G_R \qquad (3)$$

The objective of minimizing the sum of differences of the target pattern area $A_i^{t'}$ and the new area A_i^a of ith cell after the perturbation is formulated as below.

1) Unordered Perturbation of Wire Segments: Here we consider the case when the segments need not preserve their ordering while perturbation.

$$maximize : \sum_i A_i^a - \alpha(\sum_i |A_i^a - A_i^{t'}|) \qquad (4)$$

978-1-4673-9141-2/15 $31.00 © 2015 IEEE

subject to:

$$A_i^a \le A_i^v \mid \forall i, g_i \in G_R \qquad (5)$$

$$\sum_{i=1}^{n_c} x_{ij} = 1 \mid \forall j, s_j \in seg_i \qquad (6)$$

The objective function in Equation 4 is essentially minimizing the difference between the total assigned area A_i^a and the target area $A_i^{t'}$ in each cell i. Equation 5 ensures that the area of segment assigned to a cell i never exceeds the available area A_i^v in it. The constraint 6 guarantees that each segment is assigned to exactly one cell.

This objective function with the modulus being non-linear, a standard technique is used to transform it into a linear one [7].

2) Constraints for Segment Ordering: Traditionally horizontal and vertical segments are on different layers. A conflict between two horizontal wire segments of same metal layer may occur if the initial ordering of the vertical wire segments connected to them are reversed by perturbation. This leads to design rule violation by two overlapping horizontal segments of different nets.

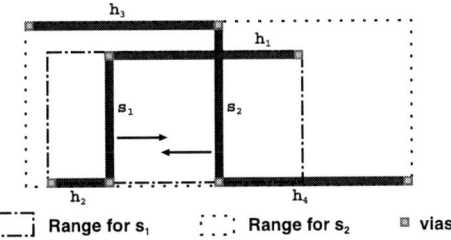

Fig. 6.   Conflict between h_2 and h_4 occurs due to the movement of either in the direction shown by the corresponding arrow

Figure 6 depicts the possibility of conflict between two horizontal segment due to the change in the order of the vertical segments connected to them. In the Figure 6 any one or both of the vertical wire segments s_1 and s_2 can be perturbed towards the direction shown by the arrows. However, such perturbation can generate a conflict between horizontal segments h_2 and h_4. The lower ends of s_1 and s_2 connecting h_2 and h_4 respectively on the same routing track cause a conflict. In order to avoid such a conflict, s_2 cannot be placed anywhere in front of s_1 and vice-versa. So, we add another constraint to our ILP formulation to maintain the segment ordering, as given below.

$$x_{kp} + x_{lq} = 1, \qquad (7)$$

- $\forall p, q$ s.t. $s_p \in G_R, s_q \in G_R$ and $p \le q$
- $g_k \in c_p^r$
- $g_l \in (g_1 ... g_{k-1})$ and $g_l \in c_q^r$

Constraint 7 preserves the order between each pair of vertical wire segments by restricting the perturbation range for each segment.

TABLE I.   Specifications of Circuits

Circuit	$W \times H$	$tr_h \times tr_v$	#Seg	$g_h \times g_v$	Range	
					Min	Max
ckt_A15	47 X 47	15 X 15	30	5 X 5	1	4
ckt_B15			33		1	4
ckt_C15			33		1	4
ckt_D15			34		1	4
ckt_E15			37		1	4
ckt_F15			36		1	4
ckt_A30	92 X 92	30 X 30	172	10 X 10	1	9
ckt_B30			146		1	9
ckt_C30			166		1	9
ckt_D30			141		1	9
ckt_E30			155		1	9
ckt_F30			146		1	9
ckt_A45	137 X 137	45 X 45	344	15 X 15	1	14
ckt_B45			338		1	14
ckt_C45			372		1	14
ckt_D45			337		1	14
ckt_E45			328		1	14
ckt_F45			339		1	14
ckt_G45			330		1	14
ckt_H45			318		1	14
ckt_A60	182 X 182	60 X 60	617	20 X 20	1	19
ckt_B60			600		1	19
ckt_C60			608		1	19
ckt_D60			579		1	19
ckt_E60			611		1	19

V.   Experimental Results

We have implemented our method using C on a system with Intel(R) Core(TM) i3-3217U CPU @ 1.80 GHz processor with 4GB memory. We used lp_solve package [7] for solving the ILP and MATLAB [8] for performing convolution during flare computation. We have performed our experiments on a set of randomly generated synthetic layouts described in Table I, focusing on perturbation in a layer with the reserved direction being vertical only. The input layouts are divided into equal sized cells such that each cell consist of three routing tracks. In Table I, the columns named Circuit, $W \times H$, $tr_h \times tr_v$, #Seg and $g_h \times g_v$ are the circuit name, width and height of the circuit, routing track resolution maintaining the design rule, number of movable vertical segments and cell resolution respectively. *Min* and *Max* define the minimum and maximum range possible in terms of the number of cells. The segments which increases the wirelength during perturbation are considered to be non-movable. Hence Wirelength remained same even after perturbation.

We have modeled our PSF as a Gaussian function with the peak at the center of the layout, and the standard deviation σ captures the effect of flare within 2σ distance around the center. An ILP is formulated for each row. User defined parameter α is set to be 0.5. The central area of the layout is more significant in terms of flare, thus perturbation starts from the middle row and proceeds towards the periphery. The pattern density of the input layout is modified based on the solution of the ILP satisfying the objective for reduction of flare. The values of flare computed from the initial and the new vacancy density maps are compared to report the reduction in the peak, mean and standard deviation of the flare map. Figure 7(a) and (b) show the initial and the final (post-ILP based perturbation) vacancy density maps and the corresponding flare maps for the circuit ckt_B60 respectively. The reduction in maximum flare in the final flare map is 18% for unordered formulation.

In Table II, the columns labeled Max, $StdDev.$ and $Mean$ show the maximum flare, standard deviation of flare and mean

978-1-4673-9141-2/15 $31.00 © 2015 IEEE

TABLE II.   COMPARISON OF FLARE LEVEL : INITIAL VS. OUR FINAL OUTPUT

Circuit	Initial			unordered					ordered				
	Max	Mean	Std Dev	Max	Mean	Std Dev	%Perturbed	Time(s)	Max	Mean	Std Dev	%Perturb	Time(s)
ckt_A15	0.608	0.378	0.27205	0.563	0.353	0.25592	70.00	0.02	0.565	0.354	0.25947	56.67	0.01
ckt_B15	0.556	0.345	0.24708	0.503	0.317	0.22542	57.58	0.01	0.513	0.323	0.23150	48.48	0.01
ckt_C15	0.552	0.343	0.24659	0.525	0.329	0.23594	51.52	0.02	0.528	0.330	0.23726	39.39	0.02
ckt_D15	0.532	0.335	0.24178	0.501	0.316	0.22825	44.12	0.02	0.511	0.323	0.23176	26.47	0.02
ckt_E15	0.496	0.309	0.22054	0.459	0.291	0.20958	70.27	0.01	0.460	0.290	0.20624	43.24	0.03
ckt_F15	0.508	0.317	0.22993	0.475	0.297	0.22349	44.44	0.02	0.482	0.049	0.54946	27.78	0.02
ckt_A30	0.143	0.096	0.05269	0.121	0.091	0.04741	63.37	0.05	0.133	0.093	0.04865	43.60	1.78
ckt_B30	0.177	0.118	0.06535	0.163	0.112	0.05921	69.86	0.04	0.167	0.115	0.06107	44.52	1.52
ckt_C30	0.159	0.107	0.05691	0.146	0.101	0.05264	58.43	0.04	0.148	0.102	0.05330	30.12	1.28
ckt_D30	0.197	0.130	0.06956	0.163	0.122	0.06160	65.96	0.04	0.189	0.126	0.06727	43.26	0.83
ckt_E30	0.171	0.117	0.06138	0.152	0.111	0.05656	59.35	0.05	0.159	0.112	0.05730	48.39	1.15
ckt_F30	0.188	0.124	0.06570	0.165	0.119	0.06144	65.07	0.04	0.167	0.119	0.06120	49.32	0.72
ckt_A45	0.199	0.134	0.06034	0.173	0.129	0.05442	71.51	0.18	0.187	0.131	0.05790	38.66	35.83
ckt_B45	0.214	0.140	0.06468	0.179	0.133	0.05685	67.75	0.14	0.192	0.134	0.05994	54.44	117.75
ckt_C45	0.193	0.126	0.05836	0.154	0.119	0.05034	68.28	0.16	0.172	0.119	0.05301	45.70	132.74
ckt_D45	0.215	0.141	0.06559	0.185	0.133	0.05758	73.29	0.19	0.183	0.134	0.05819	49.85	87.99
ckt_E45	0.209	0.141	0.06341	0.184	0.136	0.05809	68.29	0.16	0.187	0.136	0.05892	47.26	52.46
ckt_F45	0.210	0.140	0.06414	0.186	0.133	0.05749	69.91	0.16	0.193	0.135	0.06060	44.25	98.07
ckt_G45	0.216	0.144	0.06597	0.180	0.135	0.05747	70.91	0.13	0.191	0.137	0.05977	46.36	48.28
ckt_H45	0.216	0.145	0.06589	0.197	0.139	0.05987	68.24	0.16	0.188	0.138	0.05936	55.35	218.00
ckt_A60	0.167	0.115	0.04657	0.142	0.110	0.04123	73.58	0.31	0.149	0.110	0.04258	52.84	1087.00
ckt_B60	0.176	0.120	0.04916	0.145	0.113	0.04258	73.33	0.36	0.152	0.114	0.04369	56.17	3042.86
ckt_C60	0.176	0.118	0.04907	0.144	0.111	0.04186	73.85	0.34	0.158	0.113	0.04462	50.16	3136.62
ckt_D60	0.185	0.124	0.05122	0.155	0.118	0.04491	74.78	0.35	0.166	0.120	0.04709	56.13	2617.30
ckt_E60	0.182	0.119	0.05036	0.147	0.112	0.04287	73.32	0.35	0.160	0.114	0.04534	52.70	1232.00
Geo Mean	0.242	0.159	0.08266	0.212	0.151	0.07466	-	-	0.221	0.152	0.07679	-	-
Normalized	1	1	1	0.874	0.90323	0.947	-	-	0.912	0.957	0.92904	-	-

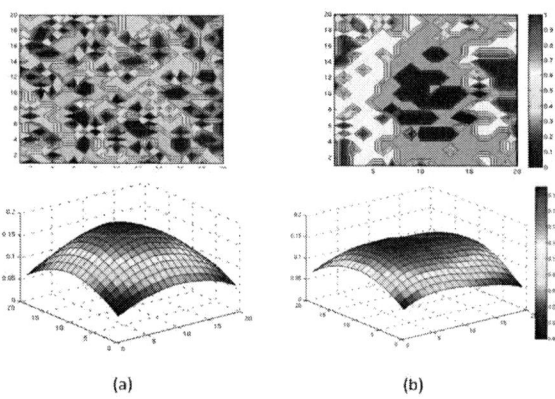

Fig. 7.   (a) Initial vacancy density map (top) and flare map (bottom) of ckt_B60, and (b) final maps after ILP-based perturbation

flare. The percentage of the total number of segments which are perturbed and the CPU time in seconds appear under %$Perturb$ and $Time$. Experimental results show a reduction of 12.6% and 8.8% in maximum flare, 5.3% and 4.3% in flare mean flare, and 9.7% and 7.1% in the standard deviation of flare for the cases of unordered and ordered formulations respectively. Flare reduction is lower in the ordered formulation than that for the unordered because in the ordered case the perturbation range for each wire segment is more restricted.

VI.   CONCLUSION

This paper presents the first method on reduction of flare using wire segment perturbation in EUVL system without dummification, thus reducing the mask cost. The dummification after wire perturbation will lead to lesser dummies while reducing more flare. The impact of dummification, timing and cross-talk after perturbation will be studied in future.

ACKNOWLEDGMENT

We thank Prof. Yao-Wen Chang of National Taiwan University and Prof. Shao-Yun Fang of National Taiwan University of Science and Technology for their valuable advice.

REFERENCES

[1]   H.-Y. Chen and Y.-W. Chang, "Routing for Manufacturability and Reliability", IEEE Circuits and Systems Magazine, Vol-9 , pp. 20 - 31, 2009

[2]   S.-R Pan and Y.-W. Chang, "Crosstalk-Constrained Performance Optimization by Using Wire Sizing and Perturbation", Proc. of Computer Design, pp. 581 - 584, 2000

[3]   S.-Y. Fang and Y.-W. Chang, "Simultaneous Flare Level and Flare Variation Minimization with Dummification in EUVL", Proc. of 49th Annual Design Automation Conference (DAC), pp. 1179-1184, 2012

[4]   J. Lee, K. Song et al., "A study of flare variation in extreme ultraviolet lithography for sub-22nm line and space pattern", Jpn. J. Appl. Phys., pp. 06GD09, Jan. 2010.

[5]   C.-Y. Liu , H.-J. K. Chiang , Y.-W. Chang and J.-H. R. Jiang, "Simultaneous EUV Flare Variation Minimization and CMP Control with Coupling-Aware Dummification", Proc. of 51st Annual Design Automation Conference (DAC), pp. 1-7, 2014

[6]   H. Xiang , L. Deng , R. Puri , K.-Y. Chao and M. D. F. Wong, "Dummy Fill Density Analysis with Coupling Constraints", Proc. of ISPD, pp, 1-7, 2007

[7]   http://lpsolve.sourceforge.net/5.5

[8]   http://in.mathworks.com/products/matlab

[9]   R. S. Ghaida, T. Sahu, P. Kulkarni and P. Gupta, "A Methodology for the Early Exploration of Design Rules for Multiple-Patterning Technologies", Proc. of International Conference on Computer-Aided Design (ICCAD), pp. 50-56, 2012

[10]   J. Kuang and E. F. Y. Young, "An Efficient Layout Decomposition Approach for Triple Patterning Lithography", Proc. of Design Automation Conference (DAC), pp. 1-6, 2013

[11]   F. M. Schellenberg, J. Word and O. Toublan, "Layout compensation for EUV flare", Proc. of SPIE 5751, Emerging Lithographic Technologies IX, pp. 320-329, 2005.

[12]   C. Zuniga, M. Habib et al., "EUV Flare and Proximity Modeling and Model-based Correction", Proc. of SPIE 7969, Extreme Ultraviolet (EUV) Lithography II, pp. 79690T, 2011

[13]   S. Y. Chen and Y. W. Chang, "Native-conflict-aware wire perturbation for double patterning technology", Proc. of International Conference on Computer-Aided Design (ICCAD), pp. 556-561, 2010

[14]   S. Y. Chen and Y. W. Chang, S. Y. Fang, S. Y. Chen and Y. W. Chang, "Native-Conflict and Stitch-Aware Wire Perturbation for Double Patterning Technology" IEEE Transactions on Computer-Aided Design of Integrated Circuits and Systems (TCAD), pp. 703-716, 2012

978-1-4673-9141-2/15 $31.00 © 2015 IEEE

Analysis and testing on delays with two time frames

Masahiro Fujita

The University of Tokyo, Tokyo, JAPAN - Email: fujita@ee.t.u-tokyo.ac.jp

Abstract—We discuss additional delays caused by various variations that may change overall "observed" behavior of circuits. First we analyze functional changes caused by such additional delays on the inputs of each gate in the circuit. We show that unlike stuck-at faults, such additional delays can introduce many more faulty functions on a gate, and we propose a functional delay fault model with two time frames for such changed behaviors caused by additional delays. The fault model can be examined by tester equipments utilizing the scan paths in the chips. As additional delays by variation and other reasons naturally happen in multiple locations simultaneously, there can be exponentially many multiple fault combinations to be considered. It is not at all easy to analyze it with traditional automatic test pattern generation (ATPG) methods which rely on fault dropping with explicit representation of fault lists. So in the second part of the paper, we present an ATPG method based on implicit representation of fault lists which is formulated as part of the SAT problem for ATPG. As faults are represented implicitly, even if numbers of simultaneous faults are exponentially large, we may still be able to successfully perform ATPG processes. Experimental results have shown that even for ISCAS89 large circuits, complete sets of test vectors for all multiple combinations of the proposed functional delay fault are successfully generated in a couple of hours or so.

I. INTRODUCTION

As the semiconductor technology continues to shrink, we have to expect more and more varieties of variations in the process of manufacturing especially for large chips. Such variations, especially ones on delays in circuits, can change circuits' "observed" behaviors, which is generally called delay fault. In this paper, we discuss such changed functionality caused by additional delays due to variation and other reasons. There have been works on testing whether such delays cause any changes in the behaviors of the circuit, which is called delay testing. Most of them try to measure delays of the circuit being tested by checking signal propagation path delays, such as testing longest paths [1], analyzing accumulation of small delays in gates [2], and many others. As there may be so many signal propagation paths in large circuits and delays could vary a lot depending on variations, delay estimation may have to have large ranges in values. As a result, those delay testing methods may not work well, because appropriate threshold delays for delay testing can not be easily defined. In this paper instead of trying to measure or estimate delays through delay testing, we concentrate on what are possible functional changes due to such distributed delays with wide ranges of values. Our proposed delay fault model is to define the possible functional situations where inputs of some gates in a circuit could get the values of previous cycles instead of current cycles. For a signal in a circuit, if the value in

the previous cycle is the same as the one in the current cycle, such additional delay will not introduce any changes in terms of functionality. On the other hand, if they are different, "observed" functionality may change, or may not change depending on internal don't cares derived from the fanout regions from the faulty locations. If we assume that a gate in a circuit may use the values in the previous cycles as their inputs, the observed and resulting functions realized by the gate can vary in many ways as discussed in the following section. For example, there are possibly "16" different functions which can be realized by a two-input AND/OR gate with such additional delays. That is, all possible functions with two-inputs may potentially be observed with a two-input AND/OR gate with additional delays. This may suggest that it may make sense to model faulty behaviors caused by distributed and additional delays as general functional faults rather than structurally defined faults, such as stuck-at faults. Please note that in the above discussion, for example, AND gate is assumed to be doing the correct operations all the time but its input values can become partially or totally wrong due to delays, which is observed as functional changes.

The functional delay fault model proposed in this paper is defined over two time frames of sequential circuits, as for an input of each gate in the circuit we need to refer to its previous value as well as its current value. Please note that the test vectors generated for the fault model are for two consecutive cycles and can be given to the circuits under test through scan paths, which is exactly the same as the combinational testing with scan paths.

In the proposed fault model, in order to analyze the delay related faulty behaviors, it is essential to deal with multiple faults rather than single faults. As there is no specific assumption on the variation which causes delay changes, here we assume each input of a gate may have independent accumulated delays from primary inputs and inputs from flipflops. Such additional delays can happen in multiple locations simultaneously. It may be the case where most of the gates in a circuit may get the values for the previous cycle rather than current cycle. In such cases, from the viewpoint of faults caused by the delays, there can be many, such as hundred, thousands or more simultaneous faults in the circuit. As a result, when we are generating test vectors for such combinations of faults, we need to manage ultra large lists of fault combinations. In general, ATPG (Automatic Test Pattern Generation) processes use fault simulators to eliminate all of the faults combinations which can be detected with the current set of test vectors (called fault dropping process). Traditionally in almost all cases, fault combinations are explicitly represented in fault

978-1-4673-9141-2/15 $31.00 © 2015 IEEE

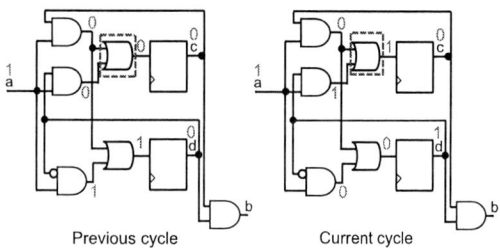

Previous cycle   Current cycle

Fig. 1. Values on inputs and output of a gate in a circuit

lists, as that is an easy and simple way for their manipulations. For functional and multiple faults, however, explicit representation is no longer feasible. For example, if there are 16 faults possible with a gate and we need to consider up to 10 simultaneous and multiple faults, the size of the fault list in explicit representations is in the order of 16^{10} or more. This is the case when we consider only one particular set of 10 faulty locations. In general there are many such sets of locations in a circuit. No explicit representation can keep such large numbers of instances. Instead we need to represent them with some implicit methods. This is in some sense a similar problem to so called "state explosion" problem [10] in model checking and formal analysis in general. In such fields, implicit representations are commonly used in order to deal with larger problems. In this paper, we show an ATPG method based on such implicit representation of fault lists. By formulating the ATPG process as an incremental Satisfiability (SAT) solving, fault lists are naturally represented and processed with implicit representation in terms of formulae [12]. We define circuits based on multiplexers in order to represent the delay fault model discussed above. Those circuits have parameter variables, and the values of parameter variables determine which faults exist or do not exist in the target circuit. Such a circuit for fault modeling is introduced to each possibly faulty location, i.e., all inputs of all gates in the target circuit. Therefore, the parameter variables altogether show how multiple faults exist in the circuit. This is an implicit way to represent multiple faults. As faults are represented implicitly, even if numbers of simultaneous faults are large, such as $2^{(ten\ thousands)}$, we can still successfully perform ATPG processes as shown in the experiments below.

The rest of the paper is organized as follows. In the next section we discuss possible functional faults or wrong operations caused by widely distributed and additional delays. The proposed fault model based on two time frames is introduced in the following section. Then we present an ATPG method based on incremental SAT formulations which represents fault lists implicitly. The experimental results are shown next, and the final section gives concluding remarks.

II. Functional faults caused by distributed additional delay

As we discussed in the introduction, additional delays due to variation and others can let a gate in a circuit receive possibly incorrect values in the previous cycles rather than the correct ones in the current cycles, which may result in wrong computation by the gate compared with the original functionality of the gate with the correct input values. Please

note that the functionality of the gate still remains correct, but the values it uses for computation may be wrong due to additional delays. Let us discuss these issues using an example shown in Figure 1.

There is an OR gate surrounded by dash lines in the sequential circuit. Let us assume that in the previous cycle, the values of the flipflops are both 0 and the value of the primay input, a, is 1. The values of internal signals are shown in the left of Figure 1. So the input values of the OR gate are both 0, and so its output is 0. Now in the current cycle, the values of the flipflops are 0 and 1, as these are the values at the inputs of the flipflops in the previous cycle. Let us assume the value of the input, a, in the current cycle is 1. Then the values of internal signals becomes the ones shown in the right of Figure 1. So the inputs values of the OR gate are 0 and 1, and so its output is 1. In our two time frame fault model, the input values of the OR gate in the current cycle are the ones in the previous cycles instead of the current cycle. Therefore, the input values of the OR gate are both 0 which are different from the ones in the current cycle, and the output of the OR gate becomes 0 under the fault. Please note that the OR gate is still functioning correct with respect to the input-output relation. The input values of the OR gate becomes incorrect due to the fault.

In this section we discuss how functionality of a gate may look like changed due to such delay increase. In general, the value of a signal can be the one for the current or the one in the previous cycle due to delays, and so there are possibly four combinations for the current and previous values, i.e., (previous, current) = (0, 0), (0, 1), (1, 0), and (1, 1) for an input of a gate. Obviously if the current and previous values are the same, there will not be any changes in the observed function. So the cases to be analyzed are the ones where (previous, current) = (0, 1) and (1, 0). Also, depending on the values of the other inputs of the gate, the observed functionality of the gate may or may not change. In order to change the functionality, the other inputs need to be so called non-controlling values, i.e., 0 for OR gate and 1 for AND gate, or those other inputs must also change values due to additional delays simultaneously.

For simplicity, in this paper we assume that inputs to the combinational part of a sequential circuit can have any possible value combinations. This is, in general, not true, as values provided by flipflops are only the ones for reachable states from initial states, which may not be all states. Accurately speaking, as we are dealing with sequential circuits, we need to manage which are "reachable" states and which are not in order to precisely compute effects of the additional delays. As reachability computation is very expensive for practical sizes of designs, here we simply assume all states are reachable. There are ways to compute supersets of reachable states, such as the ones using techniques for property directed reduction [8], [9], but utilization of such techniques within our proposed method is a future topic and out of the scope of this paper.

Now let us discuss how additional delays can affect the observed functionality of AND gate. Figure 2 shows partial

x	y	NoFault	zx1	zx2	zx3	zy1	zy2	zy3	zxy1	zxy2	zxy3
0	0	0	0	0	0	0	0	0	1	0	1
0	1	0	1	0	1	0	0	0	0	0	0
1	0	0	0	0	0	1	0	1	0	0	0
1	1	1	1	0	0	1	0	0	1	0	0
Resulting function		AND	y	$x\bar{y}$	$\bar{x}y$	x	0	$x\bar{y}$	ENOR	0	$\bar{x}\bar{y}$

Fig. 2. Functionality changes of AND gate due to input delays

possible behaviors of an AND gates with additional delays. The column, "NoFault" shows truth table values of the correct AND operation. For each value combination of x and y, that is, for each row of the truth table, the gate may get the previous values of x and/or y instead of the current values. The column, "zx1" shows the case where only in the second row of truth table, the AND gate gets the value of x in the previous cycle and that value is 1 which is different from the value in the current cycle. Due to this incorrect value, the output of the AND gate becomes 1 which is wrong. Assuming that this is the only error in the output of the AND gate, as shown in the truth table, the resulting observed logic function at the output of the AND gate is y instead of $x \land y$. The column, "zx2", shows the case where only the fourth row of the truth table changes its value due to the late arrival of x-input of the AND gate. Here we assume that such late arrival value, which is 0, is different from the correct current value, which is 1. So the resulting observed function becomes $x \land \neg y$ instead of simple AND. The column, "zx3" shows the case where these two errors happen simultaneously. As we mentioned above, we assume additional delays in the circuits can happen in distributed and independent ways. Columns, "zy1", "zy2", and "zy3", show the corresponding cases where values of y-input of the AND gate arrive late and their previous and incorrect values are used by the AND gate. As seen from the figure, the resulting observed functions are, x, 0 (constantly 0 function), and $x \land \neg y$. Moreover, values of both x-input and y-input may arrive late, which are the cases shown in columns, "zxy1", "zxy2", and "zxy3. As seen from the figure, the resulting observed functions are, $exclusive - nor$, 0 (constantly 0 function), and $\neg x \land \neg y$. Figure 2 shows that there are seven incorrect functions possibly observable at the output of the AND gate, if inputs of the gate arrive late and the previous wrong values are used.

Please note that here we have analyzed only a subset of possible behaviors. It is possible that the values of x and y can be independently chosen to be the previous values. By observing the truth tables shown in Figure 2, we can realize that each row of the truth table for AND function could change its value with appropriate late arrival of inputs and values in the previous cycle independently . This means that in our models, essentially all possible logic functions with two-inputs can potentially be realized by the delays due to variation. Therefore, for the analysis of faulty behaviors caused by late arrival of signal values, all functional faults, which are 15 in total in the case of two-input gates, should be taken into account. Please note that this discussion is true only if we can freely chose the values in the previous cycles. In real sequential circuits, however, the values in the previous cycles are determined by the sequential circuits themselves and can not be freely chosen. So all of 15 functions may not be realized

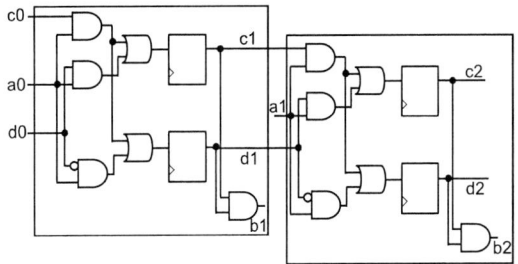

Fig. 3. Two time frame expanded circuit from Figure1

in sequential circuits due to this.

III. THE PROPOSED FUNCTIONAL DELAY FAULT MODEL

Based on the discussions in the previous section, in this section we present a functional delay fault model with two time frame, FDF2. Please note that in our functional delay fault model, it is essential to deal with "multiple" faults, as additional delays are most likely wide-distributed in a circuit and the values in many internal signals may change values simultaneously.

Because we need information on the values of signals in the current cycle as well as the ones in the previous cycle, given sequential circuits must be time frame expanded by two times. For example, an example sequential circuit and its two time frame expanded one are shown in Figure 1 and Figure 3 respectively. Please note that although there are flipflops in the expanded circuit shown in Figure 3, those should be analyzed as pure buffers with no delays for the following mathematical analysis. That is, in our analysis, circuits are considered as pure combinational circuits with no delays in flipflops of the circuits.

Our first fault model caused by additional delays is called functional delay fault model with two time frames, FDF2, and it assumes that under faults, the values of inputs of a number of gates are the ones in the previous cycle instead of the current cycle. This can be represented with a multiplexer for each input of a gate in the second time frame of the expanded circuit. The 0-input of the multiplexer is connected to the original source whereas the 1-input is connected to the corresponding signal of the gate in the first time frame of the expanded circuit. Example insertions are shown in Figure 4. Please note that for easiness of drawing figures, only one gate in the second time frame is converted to have multiplexers in its inputs. In actual modeling inputs of all gates in the second time frame of the expanded circuit should have multiplexers. These multiplexers allow the inputs of the gates in the second time frame to get either values in the current cycle or the ones in the previous cycle depending on the control signals, $v1, v2$, of the multiplexer. Those control signals are called parameter variables and represent which faults are active in the circuit. Please note that if both of them are 0, there is no fault on that gate. Therefore, if the summation of the numbers of inputs of all gates is m in the combinational part of the given sequential circuit, there are totally $2^m - 1$ multiple fault combinations in the circuit. As we said, it is essential to deal with all of these fault combinations, or as many as possible, when we perform

978-1-4673-9141-2/15 $31.00 © 2015 IEEE

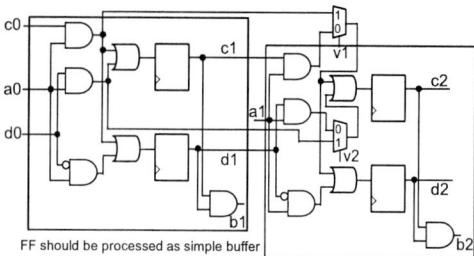

Fig. 4. Multiplexers are added to a gate in the circuit shown in Figure 3

ATPG for functional delay fault testing.

IV. ATPG METHODS BASED ON INCREMENTAL SAT FORMULATION

As we assume all states of flipflops are reachable in this paper, values of pseudo inputs coming from flipflops, that is, inputs, $c0$ and $d0$, in Figure 3, are assumed to be able to have all combinations of values. All possible fault combinations under our model can be represented by all value combinations of the control inputs of multiplexers, except for all 0 which represent the non-faulty case. Such control inputs are called parameter variables in this paper. This is a implicit way to represent multiple faults just like the state encoding with state variables in model checking [10]. The number of possible faults is exponential with respect to the number of multiplexers, which is the same as the number of inputs of all gates. With implicit representation, very large numbers of possible simultaneous faults are represented with exponentially small numbers of variables.

Let x be the set of inputs to the one time frame or two time frame circuit, and v be the set of control signals of multiplexers, that is, parameter variables. Also, let $NoFault(x)$ and $Faulty(v, x)$ be the logic functions realized at the outputs by the circuit without and with multiplexers, respectively. An example of formula, $NoFault(x)$, can be generated from the circuit shown in Figure 3, and an example of formula, $Faulty(v, x)$, can be generated from the circuits shown in Figure 4 assuming that all inputs of gates have multiplexers.

Then an ATPG process for one fault can be formulated as the following SAT problem:

$$\exists v, x.Faulty(v, x) \neq NoFault(x) \qquad ...(1)$$

Please note that this is a normal SAT problem and says some fault can be detected by some input vector, as under that input vector the two circuits behave differently. Let the solution values of variables, (v, x), be (v_1, x_1) respectively. Now we have found that the fault corresponding to v_1 can be detected by the input, x_1.

In traditional ATPG processes, fault simulators are used for input vector, x_1, to eliminate the detectable faults from the target remaining faults (fault dropping process). In our case, this approach does not work as there are so many possible fault combinations which can never be manipulated explicitly. Please remind that multiple faults are essential in order to deal with faults caused by distributed and additional delays. So the question is how to eliminate faults detectable by a test vector "implicitly" ?

We formulate it as a SAT problem in the following way:

$$\exists v.Faulty(v, x_1) \neq NoFault(x_1)$$

where x_1 is one of the solutions for (1). All the faults corresponding to the values of v, which are the solution of the SAT problem, can be detected by the test vector, x_1. Therefore, in order to eliminate the detected faults by the test vector, x_1, when generating next test vectors, we should add the following constraint to (1):

$$Fauty(v, x_1) = NoFault(x_1)$$

This constrains that values of v should be the ones which behave correctly with test vector, x_1, that is, undetectable faults.

So the next step of our ATPG process is to solve the following SAT problem:

$$\exists v, x.(Faulty(v, x) \neq NoFault(x))$$
$$\wedge (Faulty(v, x_1) = NoFault(x_1)) \qquad ...(2)$$

where x_1 is the solution of (1) above.

Let the solution values of variables, (v, x), for (2) be (v_2, x_2) respectively. Then x_2 becomes the second input test vector. It detects some fault which cannot be detected by the test vector, x_1.

We keep doing this until there is no more solution. Here we assume that the following SAT problem has a solution

$$\exists v, x.(Faulty(v, x) \neq NoFault(x))$$
$$\wedge (Faulty(v, x_1) = NoFualt(x_1))$$
$$\wedge (Faulty(v, x_2) = NoFault(x_2)) \wedge ...$$
$$\wedge (Faulty(v, x_{n-1}) = NoFault(x_{n-1})) \quad ...(3)$$

but the following SAT problem has no solution, that is, unsatisfiable,

$$\exists v, x.(faulty(v, x) \neq NoFault(x))$$
$$\wedge (Faulty(v, x_1) = NoFault(x_1))$$
$$\wedge (Fauty(v, x_2) = NoFault(x_2)) \wedge ...$$
$$\wedge (Faulty(v, x_n) = NoFault(x_n)). \quad ...(4)$$

As (3) has a solution and (4) does not have a solution, the input test vectors, $x_1, x_2, ..., x_n$ can detect all of the detectable faults, as the unsatisfiability of the formula (4) guarantees that there is no more detectable fault. So they become a set of complete test vectors for our multiple fault model exclusive of redundant faults. Please note that redundant faults are automatically excluded from the target faults, as redundant faults have no valid test vectors, which means there is no solution for the SAT problem.

The numbers of test vectors required to detect all faults, or in other words, the performance of the ATPG algorithm depends on how many times the formula (3) becomes satisfiable. Please note that each test vector is generated explicitly whereas detectable faults by the current set of test vectors are implicitly and automatically excluded from the target fault combinations.

As seen from the above, the SAT problems to be solved are pure "incremental SAT" problem. The formulae are updated to have more constraints, that is, the following formula is a super set of the previous formulae. Therefore, all learning and

978-1-4673-9141-2/15 $31.00 © 2015 IEEE

backtracks made so far in case-split based SAT solvers, which are common nowadays, are guaranteed to be all valid in the following formulae, and reasoning in the previous formula can simply be continued, not restarted, in the following formula. In reasoning about the formula (1) above, after some numbers of backtracking, a SAT solver finds a solution (v_1, x_1). The next formula to be checked is (2) where (v_1, x_1) is not a solution, and so the SAT solver simply backtracks without any reasoning required. After some numbers of backtracks, the SAT solver finds another solution, (v_2, x_2). This reasoning continues until the expanded formula becomes unsatisfiable, which means case-splitting has covered all cases implicitly.

Now in order to illustrate the ATPG process more clearly, we show an example run for FDF2 testing of a small ISCAS89 circuit, s27. First the formula of (1) above for s27 is solved by a SAT solver. This formula has 103 clauses. After 14 conflicts/backtracks the SAT solver generate a first test vector. In the next step, the formula of (2) above for s27 becomes the target. That has 136 clauses which are 33 more than the previous (first) formula. This additional clauses come from $(Faulty(v, x_1) = NoFault(x_1))$ part of (2) above as well as the learned clauses in the first SAT solving. In the second run of the SAT solver, it generates the second test vector without additional conflict/backtrack. The formula for the third run of the SAT solver has 160 clauses, which are 24 clauses more than the second run, in order to exclude the faults detectable by the second test vector efficiently also with newly learned clauses. The third run finds the third test vector with one additional conflict/backtrack. This process continues and after six iterations, the resulting SAT formula becomes unsatisfiable. In total six test vectors are generated and the final formula is unsatisfiable. This unsatisfiability can be made sure with 21 conflicts/backtracks in total. That is, the total number of conflicts/backtracks required for all seven (the number of test vectors plus 1 for the final UNSAT problem) SAT solving for s27 is 21. Please note that the final problem is unsatisfiable and needs 21 conflicts/backtracks in total to prove its unsatisfiability for s27.

As can be seen from the above execution trace, the problem is an incremental SAT problem as a whole. Or we can say that we are solving an unsatisfiable problem totally, but start with satisfiable ones and add more constraints incrementally. That is, the set of the SAT problems (or formulae) can be considered as a single SAT problem, which should be unsatisfiable eventually. So the overall process of the proposed ATPG method is just to solve a single SAT problem to make sure it is unsatisfiable, allowing dynamic addition of more constraints during the SAT reasoning process. That is, each time we find a new test vector, we add new constraints which exclude that test vector from the remaining solution space. They also include learned clauses in the previous run. By slightly modifying existing (case-split based) SAT solvers, we can realize the proposed ATPG method inside SAT solvers.

One remark in our formulation is that ATPG for single, or double, or triple faults, and so on, can easily be formulated within our SAT based ATPG with implicit representation of fault lists. We can add constraints to restrict how many parameter variables can be simultaneously one. If only one parameter variable can be one at a time, it is ATPG for single faults. In the experiments below, we compare the numbers of test vectors for complete multiple faults (there are $2^m - 1$ fault combinations where m is the number of potential faulty locations) and single faults.

The above discussions can also be casted to non-SAT based ATPG techniques with learning, such as [3], [4], if we introduce additional circuits with parameter variables to represent faults. As ATPG tools are well developed utilizing various circuit-related techniques and reasoning, such ATPG tools with the above method for the representation of detectable faults as circuits can potentially realize very efficient ATPG tools for our fault model as well. This will be one of our future directions.

V. EXPERIMENTAL RESULTS

We have implemented the proposed ATPG methods for the proposed functional delay fault, FDF2, on top of ABC tool [5]. For easiness of experiments, all ISCAS89 circuits are first converted into AIG format where there are only two-input AND gates and inverters. So all faults of FDF2 are defined on inputs of those AND gates. The characteristic of the converted ISCAS89 circuits are shown in Table I.

The ATPG results with our fault model are shown in Table II. The column of #Test shows the numbers of generated test vectors. One test vector for FDF2 consists of two time frames. Time is the processing time in seconds on a server computer having Linux kernel 2.6.32 64-bit, Dual Xeon E5-2690 2.9GHz, 128GB memory. The column of #Untest shows the numbers of redundant (untestable) multiple faults combinations. We performed two sets of experiments. The first one is to generate test vectors assuming faults can happen at any inputs of the gates and flipflops. The results are marked "Multiple" in the table.

As can be seen from the results, the numbers of test vectors are not so large, one thousands to two thousands for circuits having around 10,000 gates. Please note that these test vectors can detect all non-redundant multiple fault combinations, which are exponentially many with respect to the numbers of possible fault locations. So this is a little bit surprising. Also, the results show the there are many redundant (untestable) faults for large circuits. This is interesting as this means that there are many cases where additional distributed delays may not cause logical error (as some of them are redundant). Using these informations, circuits may be able to be optimized for delays.

The second sets of experiments is to generate test vectors assuming faults can happen only at the inputs of flipflops and there is no fault at the inputs of the AND gates. These are marked as "Multiple FF only" in Table II. Practically speaking if the values of flipflops do not change, the behaviors of the sequential circuits in terms of finite state machines do not change, although some primary output may generate wrong values. As the numbers of multiple fault combinations

Circuit	PI	PO	FF	AND	Circuit	PI	PO	FF	AND
s27	4	1	3	8	s953	16	23	29	347
s298	3	6	14	102	s1196	14	14	18	477
s344	9	11	15	105	s1238	14	14	18	532
s349	9	11	15	109	s1423	17	5	74	462
s382	3	6	21	140	s1488	8	19	6	663
s386	7	7	6	166	s1494	8	19	6	673
s400	3	6	21	148	s5378	35	49	179	1,389
s444	3	6	21	155	s9234	19	22	228	1,958
s510	19	7	6	213	s13207	31	121	669	2,719
s526	3	6	21	203	s15850	14	87	597	3,560
s641	35	24	19	146	s35932	35	320	1,728	11,948
s713	35	23	19	160	s38417	28	106	1,636	9,219
s820	18	19	5	345	s38584	12	278	1,452	12,400
s832	18	19	5	356					

TABLE I

CHARACTERISTICS OF ISCAS89 AFTER CONVERTED INTO AIG

Circuit	Multiple			Multiple FF only		
	#Test	Time	#Untest	#Test	Time	#Untest
s27	6	0.01	0	3	0.01	0
s208.1	42	0.02	0	2	0.01	0
s298	32	0.02	3	9	0.01	0
s344	37	0.02	0	8	0.01	0
s349	31	0.02	1	8	0.01	0
s382	38	0.03	7	10	0.01	0
s386	66	0.04	3	3	0.01	0
s400	36	0.03	63	13	0.01	0
s444	30	0.03	2,047	9	0.01	0
s510	63	0.06	0	4	0.01	0
s526	50	0.05	>10,000	11	0.01	0
s641	70	0.05	0	15	0.01	0
s713	84	0.07	127	14	0.01	0
s820	105	0.14	31	2	0.01	0
s832	139	0.03	127	2	0.01	0
s953	115	0.20	0	15	0.02	0
s1196	155	0.40	0	12	0.01	0
s1238	159	0.46	0	12	0.01	0
s1423	120	0.42	15	35	0.02	0
s1488	141	0.41	0	4	0.01	0
s1494	140	0.43	0	4	0.01	0
s5278	304	9.42	>10,000	52	0.06	0
s9234	662	60.61	>10,000	71	0.14	>10,000
s13207	731	142.81	>10,000	204	4.18	>10,000
s15850	619	174.32	>10,000	258	7.01	>10,000
s35932	426	580.99	>10,000	424	115.98	0
s38417	1,329	1,744.78	4,095	509	84.82	0
s38584	1,705	4,630.01	>10,000	635	78.52	>10,000
Total	7,435	7,345.88	–	2,348	290.94	–

TABLE II

COMPLETE TEST GENERATION RESULTS FOR FDF2 ON ISCAS89 CIRCUITS

Circuit	FDF2 + Stuck-at Multiple		Circuit	FDF2 + Stuck-at Multiple	
	#Add	Time		#Add	Time
s27	2	0.01	s820	71	7.11
s298	14	0.19	s832	103	4.94
s344	15	0.27	s953	43	6.83
s349	17	0.31	s1196	174	42.10
s382	26	0.39	s1238	143	55.37
s386	66	1.86	s1423	52	5.83
s400	31	0.45	s1488	52	37.16
s444	22	0.64	s1494	69	23.89
s510	21	1.60	s5278	286	588.59
s526	22	0.54	s9234	415	9,794.26
s641	50	1.11	s13207	286	28,064.38
s713	61	1.28	s15850	363	24,321.51

TABLE III

TEST VECTOR GENERATION FOR MULTIPLE STUCK-AT FAULTS STARTING WITH THE ONES FOR FDF2

as the techniques introduced in [6], [7], although the goals are different. We are working on new functional delay fault models and targeting complete test vectors for all multiple faults which is not achieved so far for ISCAS89 circuits as long as we know. The largest ISCAS89 circuits have more than ten thousands gates, which means that there are more than $2^{(ten\ thousands)}$ multiple fault combinations. For such large numbers of fault combinations, according to our experiments, a couple of thousands of test vectors are sufficient to detect all of them exclusive of redundant faults. This is a very important result, as our functional delay fault models make good sense if we can deal with wide varieties of multiple faults.

REFERENCES

[1] W. Qiu, D. M. H. Walker: An Efficient Algorithm for Finding the K Longest Testable Paths Through Each Gate in a Combinational Circuit, *IEEE International Test Coference (ITC)*, pp. 592-601, 2003.

[2] Matthias Sauer, Stefan Kupferschmid, Alexander Czutro, Ilia Polian, Sudhakar M. Reddy, Bernd Becker: Functional test of small-delay faults using SAT and Craig interpolation, *IEEE International Test Coference (ITC)*, 2012.

[3] Michael H. Schulz, Erwin Trischler, Thomas M. Sarfert: SOCRATES: A Highly Efficient Automatic Test Generation System, *IEEE Transaction on Computer Aided Design*, pp. 126- 137, Jan. 1988.

[4] John Giraldi, Michael L. Bushnell: Search State Equivalence for Redundancy Identification and Test Generation, *International Test Conference (ITC)*, pp. 184-193, 1991.

[5] Robert K. Brayton, Alan Mishchenko: ABC: An Academic Industrial-Strength Verification Tool, *22nd International Conference on Computer Aided Verification (CAV 2010)*, pp.24–40, 2010.

[6] Satoshi Jo, Takeshi Matsumoto, Masahiro Fujita: SAT-Based Automatic Rectification and Debugging of Combinational Circuits with LUT Insertions, *Asian Test Symposium (ATS)*, pp.19-24, Nov. 2012.

[7] Masahiro Fujita, Satoshi Jo, Shohei Ono, Takeshi Matsumoto: Partial synthesis through sampling with and without specification, *International Conference on Computer Aided Design (ICCAD)*, pp787-794, Nov. 2013.

[8] Aaron R. Bradley: SAT-Based Model Checking without Unrolling, *Verification, Model Checking, and Abstract Interpretation (VMCAI)*, 2011.

[9] Niklas En, Alan Mishchenko, Robert K. Brayton: Efficient implementation of property directed reachability, *Formal Methods in Computer-Aided Design (FMCAD)*, 2011.

[10] Edmund M. Clarke, William Klieber, Milos Novcek, Paolo Zuliani: Model Checking and the State Explosion Problem, *LASER Summer School*, pp. 1-30, 2011.

[11] Stephan Eggersglus, Robert Wille, Rolf Drechsler: Improved SAT-based ATPG: more constraints, better compaction, *International Conference on Computer Aided Design (ICCAD)*, pp.85-90, Nov. 2013.

[12] Masahiro Fujita, Alan Mishchenko: Efficient SAT-based ATPG techniques for all multiple stuck-at faults, *International Test Conference (ITC)* Oct. 2014.

are significantly smaller, the ATPG takes much shorter time. Therefore, it is expected that much larger circuits than IS-CAS89 can be processed.

As the final experiments, in order to see the relationships between FDF2 and stuck-at faults, starting with the test vectors generated for FDF2 (Multiple on Table II), additional test vectors are generated for multiple stuck-at faults. Each test vector for FDF2 has two time frames, and so it becomes two test vectors when generating test vectors for stuck-at faults. The results are shown in Table III. Complete sets of test vectors for multiple stuck-at faults have been generated. As some numbers of additional test vectors are required, FDF2 has somehow different characteristics from stuck-at faults in terms of test vectors. Please note that the test vectors for stuck-at faults cannot be directly used for FDF2, as FDF2 needs test vectors for two time frames.

VI. CONCLUDING REMARKS

We have shown fault models caused by delay variations and its associated ATPG methods with implicit representation of fault lists. We are recognizing that the ATPG method shown here is essentially doing the same or very similar

Contactless Transmission of Intellectual Property Data to Protect FPGA Designs

L. Bossuet, V. Fischer

Laboratoire Hubert Curien, CNRS UMR 5516
Université Jean Monnet
42000 Saint-Etienne, France
Lilian.bossuet@univ-st-etienne.fr

P. Bayon

Brightsight
Delft, 2628, The Netherlands
bayon@brightsight.com

Abstract— **Over the past 10 years, the designers of intellectual properties (IP) have faced increasing threats including illegal copy or cloning, counterfeiting, reverse-engineering. This is now a critical issue for the microelectronics industry, mainly for fabless designers and FPGA designers. The design of a secure, efficient, lightweight protection scheme for design data is a serious challenge for the hardware security community. In this context, this paper presents the first ultra-lightweight transmitter using side channel leakage based on electromagnetic emanation to send embedded IP identity discreetly and quickly. In addition, we present our electromagnetic test bench and a coherent demodulation method using slippery window spectral analysis to recover data outside the device. The hardware resources occupied by the transmitter represent less than 0.022% of a 130 nm Microsemi Fusion FPGA. Experimental results show that the demodulation method success to provide IP data with a bit rate equal to 500 Kbps.**

Keywords—IP protection; side channel; electromagnetic emanation analysis;

I. INTRODUCTION

For digital circuit design the re-use of embedded intellectual properties (IP) is more and more important due to prohibitive cost of ASIC design. Nevertheless the IP business suffers from a lack a security due to the intrinsic form of IPs sales and exchanges. Many dedicated threats target the IP life cycle and result to revenues losses for the IP designers [1]. The IP threat model includes illegal re-use, illegal sales, cloning (illegal copy) of the IP. The extent of threats targeting IPs is linked to the type of IP: software IPs (typically hardware description language files), firmware IPs (*synthesized* netlist), and hardware IPs (FPGA bitstream or physical layout).

One of the solutions for the IP designers to protect their intellectual property is to be able to detect the presence of a copy of an IP embedded in a digital device by using IP identification. Works on IP watermarking and IP fingerprinting try to provide the IP identification service. But, most of the time the published solutions are not practical mainly because of the complexity of the watermarking/fingerprinting verification scheme [2]. Efficient IP identification scheme needs to be contactless, rapid and ultra-lightweight. Up to now, these three characteristics are not available in the state-of-the-art. To meet these requirements, in this paper we propose an ultra-lightweight

binary frequency shift keying (BFSK) transmitter to forward IP identity (that could generated for example by a feedback shift-register or a physical unclonable function [1]) discreetly using an electromagnetic channel. Such circuit is usually called "spy circuitry". Using the electromagnetic channel, it is possible to contactless check the presence of an IP inside a digital device.

The rest of this paper is organized as follows. In Section 2, we present related works. In Section 3, we detail the proposed electromagnetic communication of data (i.e. IC/IP information). In Section 4, we present a method to analyze the electromagnetic spectrum to BFSK signal demodulation. In Section 5, we compare published spy circuitries that use a side-channel for IP protection and hardware Trojans. In Section 6, we present two industrial scenarios for the use of the proposed IP protection scheme and in Section 7 we present our conclusions.

II. RELATED WORK

Well-known threat in cryptographic engineering is side channel attacks [3], [4]. Most of the dynamic characteristics of both hardware and software implementations of cryptographic primitives can be used for side channel analysis: computation time, power consumption, electromagnetic radiation, optical radiation, even the sound produced during computation. These side channels can be used as transmission channels to send intellectual property data from a device or an embedded IP. For example in [5], the thermal channel representing a contactless communication was used to transfer information from an embedded tag to a remote receiver. However the embedded thermal tag used in this commercial solution requires a relatively large area (255 Spartan-3 slices). In [6], the authors propose using two shift registers to generate a recognizable signature-dependent power consumption pattern to reveal the IP signature. Power consumption was also used in [7] to communicate the IP identity. To reinforce such work, the authors of [8] propose using the power supply signal of an IP as a physical hash function for fingerprinting.

Related works can be found also in the malicious hardware design such as hardware Trojans. Such systems use side channels to forward secret information such as a symmetric cipher key from cryptographic hardware implementation [9],

[10], but also to cause or amplify side-channel leakage of cryptographic hardware [11].

Expect [5], all the related works use power consumption as a communication channel which is not contactless. Unlike the proposed solution, all the related works are not lightweight and rapid as the section 5 of this paper will show.

III. PROPOSED EM COMMUNICATION OF IP DATA

Electromagnetic radiation is an efficient side channel since, unlike measurement of power consumption, electromagnetic radiation can be measured locally. One of the main advantages of this side channel is that it is impossible to hide the leak concerning electromagnetic radiation by using a global countermeasure. Moreover the electromagnetic test bench is not expensive (less than US$ 10K without an oscilloscope, which is the most expensive component). Last but not least, a spectral analysis of the electromagnetic radiation provides information on the oscillating structure such as a ring-oscillator [12]. For all these reasons, we use the electromagnetic channel for our IP identification scheme. To this end, we designed an ultra-lightweight BFSK transmitter which could be activated outside the device by an ID checker or internally by a specific event (e.g. specific input sequence, internal data value, system state). Note that an enable signal is required to reduce the power consumed by the ring oscillator. Moreover, a permanently activated transmitter could be detected more easily by a spectral analysis of electromagnetic emanations of the device and could also cause local heating and premature aging of the chip.

BFSK is one of the common modulation schemes used in digital communication. The binary data are sent using a sinusoidal carrier at two frequency tones f_0 and f_1, representing high ('1') and low ('0') logic levels. The binary data arriving at the transmitter input at certain bitrates determine the commutation of the tones at the transmitter output. The proposed BFSK transmitter uses a dedicated configurable ring-oscillator, as shown in Fig. 1. The configurable ring-oscillator is designed using one multiplexor, $N+K$ delay elements, and a feedback chain controlled by a NAND gate for activation of transmission to reduce power consumption. Actually, the transmitter is used only during a short time when the enable signal is high, and it consumes power only during this small piece of time. The power consumption of this transmitter is thus completely negligible.

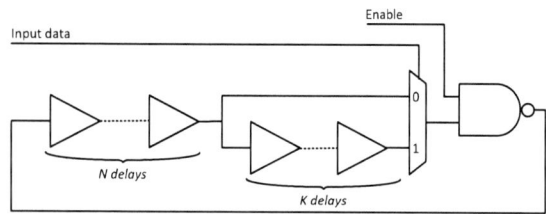

Fig. 1. Architecture of the ultra-lightweight digital BFSK transmitter based on a configurable ring oscillator.

Input data controls the multiplexor, as shown in Fig. 1. When input data is low, the ring oscillator uses N delays and

its oscillation frequency is f_0. When input data is high, the ring oscillator uses $N+K$ delays and its oscillation frequency is f_1. Since the ring oscillator's oscillation frequency decreases with an increase in the number of delay elements, frequency f_0 is higher than frequency f_1. These two frequencies have to be selected according to the bandwidth of the electromagnetic analysis platform, which is used to acquire and measure the transmitted signal. The bandwidth of our test bench, which is described in Section 4, was limited to 100 MHz and 1 GHz by the low-noise amplifier.

The proposed BFSK transmitter was implemented in Microsemi FUSION flash based FPGA (130 nm CMOS technology) containing 600K logic gates (M7AFS600). The device contains 13 824 tiles, each tile can be used to implement one D-flip-flop or one configurable multiplexor-based logic block implementing any 3-input logic function.

The configurable number of delays in the ring oscillator of the proposed BFSK transmitter makes it possible to select precisely the two frequencies f_0 and f_1 using parameters N and K. Table I lists the ring oscillator frequencies and the number of Fusion tiles used by the BFSK transmitter for five values of N and K, with N ranging from 0 to 4, and K ranging from 1 to 5. According to Table I, f_0 can be chosen between 119 MHz ($N=4$) and 385 MHz ($N=0$) and f_1 can be chosen between 70 MHz ($N=4$, $K=5$) and 280 MHz ($N=0$, $K=1$). The exact value of f_0 depends on the number of delay elements, but also on the placement and routing of the transmitter. For the values N and K listed in Table I, the frequency variation was less than 1.7% (the maximum frequency deviation in Table I is 2 MHz when $N = 4$).

TABLE I. HARDWARE IMPLEMENTATION RESULTS OF THE BFSK TRANSMITTER

N	K	f_0 (MHz)	f_1 (MHz)	Fusion Tiles	LUT4	EG
0	1	385	280	3	2	4.67
	2	383	210	4	3	5.34
	3	384	151	5	4	6.01
	4	385	130	6	5	6.68
	5	381	111	7	6	7.35
1	1	272	189	4	3	5.34
	2	272	156	5	4	6.01
	3	270	120	6	5	6.68
	4	271	106	7	6	7.35
	5	269	93	8	7	8.02
2	1	168	144	5	4	6.01
	2	169	124	6	5	6.68
	3	169	100	7	6	7.35
	4	168	91	8	7	8.02
	5	168	79	9	8	8.69
3	1	146	128	6	5	6.68
	2	147	112	7	4	7.35
	3	146	92	8	5	8.02
	4	145	84	9	6	8.69
	5	144	74	10	7	9.36
4	1	123	110	7	6	7.35
	2	121	98	8	7	8.02
	3	122	83	9	8	8.69
	4	121	77	10	9	9.36
	5	119	70	11	10	10.03

The number of tiles used by the BFSK transmitter is very low, i.e. from 3 tiles ($N=0$, $K=1$) to 11 tiles ($N=4$, $K=5$). These values are equivalent to less than 0.022% and less than 0.080% of the total number of tiles included in the targeted 600K-gate FUSION FPGA, respectively. This very small number of tiles is very promising for good dissimulation of the BFSK transmitter inside the sea of gates/tiles. In order to estimate the number of resources needed for implementation with Xilinx SRAM FPGA or Altera SRAM FPGA, Table I gives the number of 4-input look-ups (LUT4) used by the BFSK transmitter with such FPGAs.

To evaluate the logical resources needed by the BFSK transmitter in ASIC implementations, the right hand column in Table I gives the number of equivalent gates (EG) in the transmitter. The gate count was estimated using the Virtual Silicon standard cell library based on the UMC L180 0.18 μm 1P6M Logic process (UMCL18G212T3 [13]). The delay gates are replaced by more efficient standard NOT gates. The gate count of a standard NOT gate is 0.67 EG, and that of the standard multiplexor, 2.33 EG. The standard NAND gate uses 1 EG. So the number of gates of the whole BFSK transmitter ranges from 4.67 EG ($N = 0$, $K = 1$) to 10.03 EG ($N = 4$, $K = 5$). Note that one flip-flop requires between 5.33 EG and 12.33 EG to store a single bit [13].

Such a transmitter is clearly ultra-lightweight in both FPGA and ASIC implementations. The small logical resources requirement of the proposed spy circuitry makes reverse engineering it harder, although not impossible [15]. Even with recent CMOS technologies, the attacker can reverse engineer ICs using a scanning electron microscope and an automatic tool for circuitry extraction [15], [16]. Nevertheless, the smaller the piece of hardware used for BFSK transmitter the harder it is to detect during reverse engineering. Detection of the transmitter using standard Trojan detection methods [17], [18] is not feasible because the transmitter does not change the data path of the circuit and because of the ultra-low signal-to-noise ratio on the electromagnetic channel, as shown in our experimental results below (Section 4).

IV. EXPERIMENTAL RESULTS

The electromagnetic radiation of the device was evaluated using the near-field electromagnetic analysis test bench described in [12]. The border between the far field and the near field can be considered to be about 23 mm from the device, depending on the hardware concerned. The most important part of the test bench is the acquisition chain. It determines the signal to noise ratio and measurement precision.

The chain, as presented Fig. 2, is composed of:

- A Langer magnetic probe with a frequency range of from 30 MHz to 3 GHz and a spatial resolution of approximately 500 μm.
- A Miteq low-noise amplifier with a frequency range of from 100 MHz to 1 GHz.
- A LeCroy oscilloscope with a frequency range of up to 3.5 GHz and a sampling rate of up to 40 GS/s.

As presented in Fig. 2, the device to be tested (the board) is fixed to a XYZ table with repeatability of movement of 1 μm. The test bench, including the acquisition chain, XYZ table, FPGA configuration and power supply variations, is controlled by a computer.

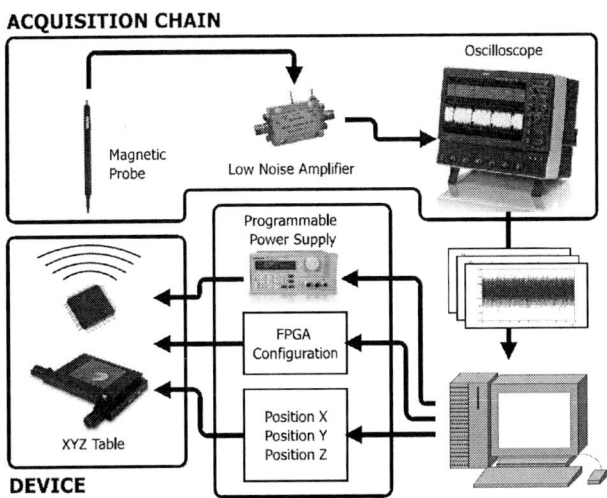

Fig. 2. Near-field Electromagnetic analysis test bench.

Electromagnetic analysis of IC is contactless, local, and can be spatial or/and temporal. This last point makes it possible to perform frequency analysis of the electromagnetic emanation. In the your bench the spectral range is limited to 100 MHz and 1 GHz. Standard devices aimed at direct BFSK demodulation cannot be used for these relatively high frequencies. Available integrated BFSK demodulators are limited to a few dozen megahertz. For this reason, we developed a dedicated BFSK demodulation scheme for our needs, in which a spectral analysis of the low noise amplifier output (a component of the test bench) is performed to measure the f_0 and f_1 spectral contribution. The transmitted high (low) level is detected when f_1 spectral contribution is higher (lower) than that of f_0.

Fig. 3 illustrates the spectral analysis of the BFSK transmitter's electromagnetic emission when $N = 1$ and $K = 1$, which corresponds to a small transmitter with high frequencies in Table I. For the spectral analysis, a 16 384 points FFT was computed. Fig. 3 presents a zoom (X and Y axis) on the global spectrum of the local EM emanation of the circuit when the BFSK transmitter sends a '0' (*in blue*) and when the BFSK transmitter sends a '1' (in red). Notice also that we placed a small antenna in the close vicinity of the ring. The amplitude of the spectral rays at 180, 200, 220, 240, 260, 280 MHz are very high compared to the two spectral rays at 189.2 MHz and 272.2 MHz. However, we have cut the upper part of the spectrum in order to see the two interesting rays (at 189.2 MHz and 272.2 MHz). These frequencies correspond to the spectral contribution of the BFSK frequencies f_0 and f_1. Actually, according to Table I, for Microsemi FUSION FPGA, the frequency f_0 is around 272 MHz and f_1 is around 189 MHz. Slight variations in the frequency values (around the values presented in Table I) are due to environmental variations (mainly temperature). These variations are too small to jeopardize the success of demodulation because the BFSK designer knows the chosen parameters N and K and targeted frequencies f_0 and f_1 for transmission of low and high levels.

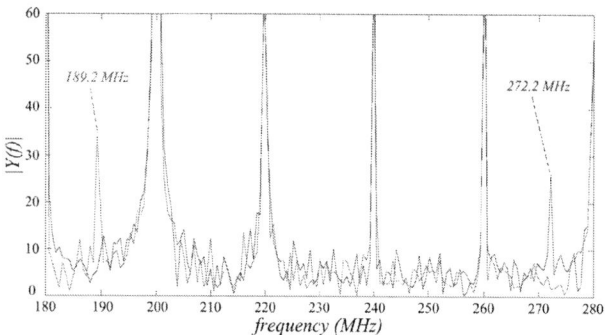

Fig. 3. Parts (zoom in x-axis and y-axis) of the electromagnetic emanation spectral analysis of the proposed BFSK transmitter when it sends a high level at f_1=189.2MHz (in red) and when it sends a low level at f_0=272.2 MHz (in blue).

Without knowledge of the BFSK parameters, the electromagnetic transmission cannot be easily detected because it cannot be distinguished from spectral noise. The signal-to-noise ratio of the BFSK transmission in Fig. 3 is -135 dB for a 1 GHz bandwidth. Such an ultra-low SNR represents efficient protection against unwanted BFSK transmitter detection via a side channel. However, knowing the N and K parameters, the BFSK designer can calibrate the demodulation (determine the two frequencies) by electromagnetic analysis of the ring oscillators based on the differential spectral analysis as described in [12].

The spectral contribution of the two BFSK frequencies during transmission (which is limited by the transmitter enable signal) at the low-noise amplifier output is measured to determine the transmitted bit sequence. In order to apply the demodulation technique described here, called coherent

demodulation, theoretically the bitstream rate (bitrate) is limited to 0.5 times the frequency difference between f_0 and f_1. In the case shown in Fig. 3, theoretically the maximum bitrate is 41.5 Mbps (theoretically limited to 53 Mbps when $N = 0$ and $K = 1$). The coherent demodulation scheme cannot be used for higher bitrates, instead more sophisticated and more expensive demodulation techniques should be used. To reduce the size of the BFSK transmitter, a time-interleaving transmitter should be used to increase the bitrate. However, the high transmission rate is not so important, because data (e.g. IP identifier or stolen data) can be stored at a higher data rate before being transmitted at a slower rate. The small area and unobtrusiveness of the BSFK transmitter are much more important properties than transmission rate.

For the coherent demodulation of the electromagnetic radiation, we propose a slippery window spectral analysis. Indeed, overall spectral analysis masks the effects of the nonstationarity of the signal and therefore provides no information about its temporal evolution. Slippery window spectral analysis is a three-dimensional representation of the signal: amplitude, frequency, and time. It requires two quantities Fw, the width of the FFT window frame and the difference $\Delta\tau$ between two frames. For our experiment, we chose Fw equal to 16 384 points (2^{14}-point FFT) and $\Delta\tau$ equal to 100 points. For each frame, the FFT provides the software demodulator with the amplitude of signals f_0 and f_1 which enables the demodulator to distinguish between a transmitted '1' or '0'.

To illustrate data transmission from the circuit via the EM channel, we used a shift register that stored the following 16-bit sequence: "1011110111011100". The clock frequency of the shift register is 500 kHz. When the enable signal of the transmitter is given, the sequence is sent cyclically to the BFSK transmitter, which transmits it via the electromagnetic channel. Fig. 4 gives the result of the coherent demodulation obtained at a 500 Kbps bit rate, which served as a proof of concept.

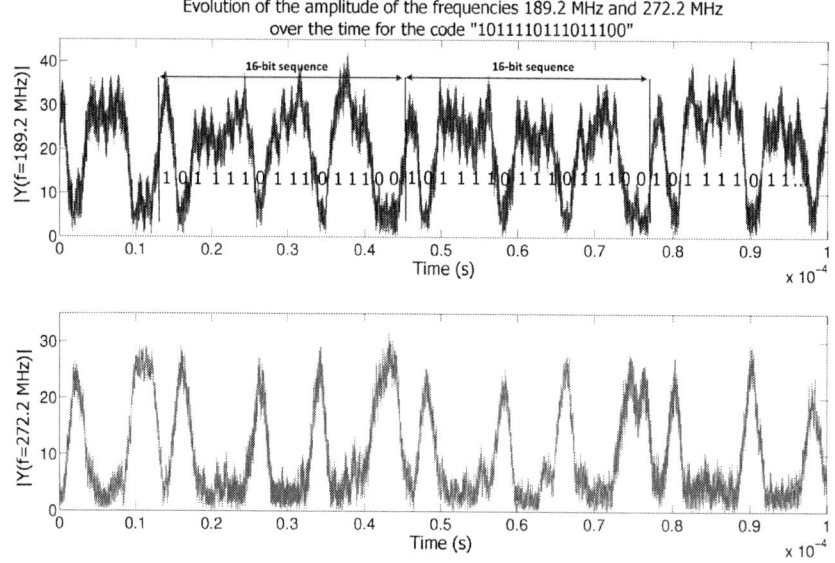

Fig. 4. Coherent demodulation of the 16-bit code word cyclically sent by the BFSK transmitter at 500 Kbps, the top (bottom) graph corresponds to the spectral contribution of signal at f_1 (f_0).

V. COMPARISON WITH STATE OF THE ART SPY CIRCUITRIES USING A SIDE-CHANNEL

Table II compares the implementation of the proposed ultra-lightweight BFSK transmitter with other recently published state of the art methods. Table II gives the spy circuitry application (*App.*) for each reference; this may be IP protection (*IPP*) or hardware Trojan (*HT*) or both (for the presented work, *PW*). In addition, Table II gives the year of publication (*YoP*), the side channel used, the hardware resources required only for the leakage generator (for example we do not take the hardware used for IP watermark generation or the Trojan's payload into account). Unfortunately, the principles compared do not use the same hardware. For the sake of correctness, we give the implementation results as they are presented in the referenced papers. Nevertheless, the implementation bitrate of these previously published works can be roughly compared with our proposed solution. Based on published data, we computed the bitrate of all the proposals by using the number of clock cycles needed to send information via the side channel. For all the references presented in this table, the bitrate was computed using a 500 KHz frequency for data synchronization (same frequency is used for data transmission in Fig. 4).

TABLE II
SUMMARY OF CHARACTERISTICS OF SPY CIRCUITRIES EXPLOITING SIDE-CHANNELS

App.	Ref.	YoP	Side channel	Hardware resources	Target	Bit rate (@500 KHz)
IPP	[5]	2008	Thermal emanation	255 Spartan-3 slices	Xilinx Spartan-3	7.10^{-3} bps
IPP	[6]	2008	Power consumption	16 * 16-bit circular shift-registers	Xilinx Spartan-3 and Virtex-II	200 bps
IPP	[7]	2010	Power consumption	16-bit circular shift-register	Xilinx Virtex-II Pro	500 bps
HT	[9]	2009	Power consumption	8 parallel D-flip-flops or 16-bit circular shit register	Xilinx Spartan-3E and Virtex-II Pro	485 bps
HT	[10]	2013	Power consumption	16-bit circular shift-registers per bit	Xilinx Virtex-5	976 bps
Both	PW	2015	Electro-magnetic emanation	1 configurable ring-oscillator (like a D-flip-flop in ASIC)	Microsemi Fusion	500 Kbps

As can be seen in Table II, the proposed work reaches the highest bitrate. The reason for such a good performance is first that we use a spectral analysis of the local electromagnetic leak based on a simple frequency modulation. Except for [4], all the other solutions use a global measurement of power consumption, which reduces the signal-to-noise ratio of the information leaked via the side channel. Our proposal is clearly the smallest spy circuitry ever published. Although solutions based on circular shift-registers are well adapted to last generation FPGA families, since the 16-bit shift registers can be designed using only one look-up table, they are not suitable for ASIC technologies. Currently, an ASIC implementation of a 16-bit shift register requires 16 flip-flops whereas the solution we propose occupies an area equivalent

to only one D-flip-flop.

In this paper, we present the proposed spy circuitry for IP protection, but it can also be used for hardware Trojan. Most of the other proposals could also be used for both applications. Note that in 2012, Kasper et al. proposed to use the work initiated in [9] for hardware Trojan or IP watermarking implementation [19]. However, by using electromagnetic emanation and a configurable ring oscillator, the proposed solution is the most convincing for industrial applications (e.g., those aimed at IP protection) because of its very small area and high bitrate.

VI. INDUSTRIAL SCENARIOS USING THE PROPOSED IP PROTECTION

According to the previous section, in comparison with other works, our propose goes clearly towards using a spy circuit in an industrial context for IP protection. Two industrial scenarios are presented in the following.

The first scenario is the identification of embedded IP in the supply chain. This identification is used in order to be sure to don't use counterfeiting (fake) devices. In recent years, the issue of counterfeiting of integrated circuits has increased considerably. For example, the number of counterfeit electronic circuits collected by U.S. Customs from 2001 to 2011 has increased by around 700 [20]. Between 2007 and 2010, U.S. Customs collected 5.6 million counterfeit electronic products [21]. Overall, the estimation of counterfeiting is about 7% of the semiconductor market [22] which represented a loss of about $ 22 billion in 2014 for the legal industry.

It is therefore crucial and strategic to be able to detect counterfeit IC as soon as possible in the supply chain (this is particularly crucial for military and space grad devices). Fig. 5 shows a possible framework to manage the device under test (control the enable signal) and check the IP identification by using an EM probe, an amplifier and a dedicated acquisition system including a spectral analysis and the proposed demodulation method. Due to the high bit rate of the proposition solution the identification of the ID requires less than 1ms (with 500Kps bit rate as in Fig 4.). This counterfeiting detection could be completed by other physical (invasive or not) and optical inspection [23].

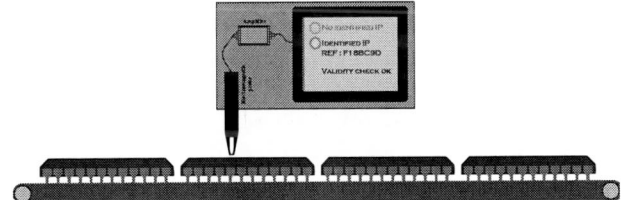

Fig. 4. Rapid and contactless IP identification in the supply chain by using EM transmission of IP ID.

The second scenario occurs when a IP designers would like to verify the illegal presence of its IP inside a device (ASIC or FPGA). In this case the proposed transmitter provides to the identification scheme a data like a PUF [24] or a watermarking. Watermarking is a technique of steganography which provides the ownership of an IC (or an IP) by checking for the presence of hidden information called the watermark [3]. Most of the watermarking methods proposed in the literature need a complex verification scheme. Nevertheless it is possible to use power consumption as proposed in [8] and [3] but it is easy and cheap to use global countermeasure in order to mask the power consumption due to the watermark [25]. Using electromagnetic emanation in this scenario is better because as electromagnetic emanation is local it is really hard to mask it by using a global countermeasure. Moreover, in this paper we have shown that due to the SNR of BFSK signal, it is unrealistic to try to detect it without the precise knowledge of the used frequencies for data transmission.

VII. CONCLUSION

IP protection has become crucial topics for hardware security due to the lack of trust in IP market. In this paper, we have presented an ultra-lightweight transmitter of IP identity using the electromagnetic side channel. Based on a configurable ring oscillator, our solution exploits a BFSK signal to transmit information by way of the electromagnetic channel. By performing a slippery window spectral analysis of the near field electromagnetic emanations captured locally over the BFSK transmitter circuitry, the proposed transmission achieves a high bitrate (experimentally 500 Kbps and theoretically limited to 53 Mbps with a Microsemi Fusion FPGA), which has not been achieved before. Moreover, the transmitter occupies very small area. For the highest frequency and data rate, our solution requires 4.67 equivalent gates, which it is less than the requirement of a small D-flip-flop. Such a small requirement of logical resources makes reverse engineering of the chip very difficult and detection of the transmitter using standard Trojan detection methods is not feasible.

ACKNOWLEDGMENT

The work presented in this paper was realized in the frame of the SALWARE project number ANR-13-JS03-0003 supported by the French "Agence Nationale de la Recherche"and by the French "Fondation de Recherche pour l'Aéronautique et l'Espace".

REFERENCES

[1] B. Colombier, L. Bossuet, "Survey of hardware protection of design data for integrated circuits and intellectual properties," Computers & Digital Techniques, IET, vol.8, no.6, pp.274,287, 2014.

[2] C. Marchand, L. Bossuet, E. Jung, "IP watermark verification based on power consumption analysis," In Proceedings of the 27th IEEE International System-on-Chip Conference, SOCC 2014, pp. 330-335, 2014.

[3] P. Kocher, J. Jaffe, B. Jun, "Differential Power Analysis", in Wiener M. (Ed.), Proceedings of the 19th Annual International Cryptology Conference, CRYPTO 1999, Springer, Lecture Note on Computer Science, vol. 1666, pp. 388-397, 1999.

[4] N. Kamoun, L. Bossuet, A. Gazel, "Experimental Implementation of 2ODPA attacks on AES design with flash-based FPGA Technology," in proceedings of the 22nd IEEE International Conference on Microelectronics, IMC 2010, pp. 407-410, 2010.

[5] C. Marsh, T. Kean, D. Mclaren, "Protecting designs with a passive thermal tag," In Proceedings of the 15th IEEE International Conference on Electronics, Circuits and Systems, ICECS 2008, pp.218-221, 2008.

[6] D. Ziener, J. Teich, "Power Signature Watermarking of IP Cores for FPGAs," Journal of Signal Processing Systems, Springer, vol. 51, pp. 123-136, 2008.

[7] G. T. Becker, M. Kasper, A. Moradi and C. Paar, "Side-channel based watermarks for integrated circuits," In Proceedings of the IEEE International Symposium on Hardware-Oriented Security and Trust, HOST 2010, pp. 30-35, 2010.

[8] S. Kerckhof, F. Durvaux, F.X. Standaert, and B. Gérard, "Intellectual Property Protection for FPGA Designs with Soft Physical Hash Functions: First Experimental Results," In Proceedings of the IEEE International Symposium on Hardware-Oriented Security and Trust, HOST 2013, pp. 7-12, 2013.

[9] L. Lin, M. Kasper, T. Güneysu, C. Paar, W. Burleson, "Trojan Side-Channels: Lightweight hardware Trojans through Side-Channel Engineering," In Proceedings of Workshop on Cryptographic Hardware and Embedded Systems, CHES 2009, Springer, Lecture Notes in Computer Science, vol. 5747, pp. 382-395, 2009.

[10] S. Kutzner, A. Poschmann, and M. Stöttinger, "TROJANUS: An Ultra-Lightweight Side-Channel Leakage Generator for FPGAs", In Proceedings of International Conference on Field-Programmable Technology, ICFPT 2013, pp. 160-167, 2013.

[11] J.F. Gallais, J. Großschädl, N. Hanley, M. Kasper, M. Medwed, F. Regazzoni, J.M. Schmidt, S. Tillich, and M. Wójcik, "Hardware Trojans for Inducing or Amplifying Side-Channel Leakage of Cryptographic Software," In Proceedings of the Second International Conference on Trusted Systems, INTRUST 2010, pp. 253-270, 2010.

[12] P. Bayon, L. Bossuet, A. Aubert, V. Fischer. "EM leakage analysis on True Random Number Generator: Frequency and localization retrieval method", in Proceedings of the Asia Pacific International Symposium and Exhibition on Electromagnetic Compatibility, APEMC 2013, 2013.

[13] Virtual Silicon Inc. 0.18 μm VIP Standard Cell Library Tape Out Ready, Part Number: UMCL18G212T3, Process: UMC Logic 0.18 μm Generic II Technology: 0.18μm, 2004.

[14] P. Kitos, Y. Zhang (Eds.), "RFID Security – Techniques, Protocols and System-on-Chip (1st ed.)" Spinger Publishing compagny, 2008.

[15] R. Torrance, and D. James, "The state-of-the-art in semiconductor reverse engineering," In Proceedings of the 48th Design Automation Conference, DAC 2011, ACM/EDAC/IEEE , pp. 333-338, 2011

[16] P. Subramanyan, N. Tsiskaridze, W. Li, A. Gascon, W. Tan, A. Tiwari, N. Shankar, S. Seshia, and S. Malik, "Reverse Engineering Digital Circuits Using Structural and Functional Analyses," in IEEE Transactions on Emerging Topics in Computing, 2013.

[17] D. Agrawal, S. Baktir, D. Karakoyunlu, P. Rohatgi, and B Sunar, "Trojan Detection using IC Fingerprinting" in Proceedings of the IEEE Symposium on Security and Privacy, pp. 296-310, 2007.

[18] Y. Jin, and Y. Makris, "Hardware Trojan Detection using Path Delay Fingerprint" in IEEE International Workshop on Hardware-Oriented Security and Trust, HOST 2008, pp. 51-57, 2008.

[19] M. Kasper, A. Moradi, G.T. Becker, O. Mischke, T. Güneysu, C. Paar and W. Burleson, "Side Channels as Building Blocks", In Journal of Cryptography Engineering, Springer, vol. 2, no. 3, pp. 143-159, 2012.

[20] C. Gorman, "Counterfeit Chips on the Rise," IEEE Spectrum, June 2012.

[21] AGMA, Alliance for Gray Markets and Counterfeit Adatement, http://www.agmaglobal.org

[22] M. Pecht, and S. Tiku, "Bogus! Electronic manufacturing and consumers confront a rising tide of counterfeit electronics," IEEE Spectrum, May 2006.

[23] M. Tehranipoor, U. Guin, D. Forte. Counterfeit Integrated Circuits - Detection and Avoidance. Springer, 2015.

[24] L. Bossuet, X. T. Ngo, Z. Cherif, V. Fischer. A PUF based on a transient effect ring oscillator and insensitive to locking phenomenon. IEEE Transactions on Emerging Topics in Computing, Vol. 2, Issue 1, pp. 30-36, 2014.

[25] N. Kamoun, L. Bossuet, and A. Ghazel, "Correlated Power Noise Generator as a Low Cost DPA Countermeasure to Secure Hardware AES Cipher," In Proceedings of the International Conference on Signals, Circuits and Systems, SCS 2009, pp. 1-6, 2009.

Tailoring Instruction-Set Extensions for an Ultra-Low Power Tightly-Coupled Cluster of OpenRISC Cores

Michael Gautschi*, Andreas Traber*,
Antonio Pullini*, Luca Benini*
*Integrated Systems Laboratory, ETH Zurich,
Gloriastrasse 35, 8092 Zurich, Switzerland.
{gautschi, atraber, pullinia, lbenini}@iis.ee.ethz.ch

Michele Scandale§, Alessandro Di Federico§,
Michele Beretta§, Giovanni Agosta§
§Dipartimento di Elettronica, Politecnico di Milano,
Piazza Leonardo da Vinci 32, 20133 Milano, Italy.
{michele.scandale,alessandro.difederico}@polimi.it,
{michele.beretta,giovanni.agosta}@polimi.it

Abstract—Baseline RISC instruction sets for ultra-low power processors are constantly being tuned to reduce cycle count when executing computation-intensive applications. Performance improvements often come at a non-negligible price in terms of area and critical path length and imply deeper pipelines and complex memory interfaces. This penalizes control-intensive code execution and significantly increases cost and complexity of building multi-core clusters. In addition, some extensions are not easily exploited by compilers and may increase code development effort, especially when considering parallel applications. In this paper we describe our efforts in enhancing a baseline open ISA (OpenRISC) and its LLVM compiler back-end to significantly reduce execution cycles while minimizing the impact on core micro-architecture complexity, number of pipeline stages, area and power. In addition, we improved the core micro-architecture to streamline its integration in a tightly-coupled cluster, sharing instruction cache and data memory, thereby further enhancing parallel execution efficiency. The combined effect of ISA, compiler and micro-architecture evolution gives an average energy efficiency boost of 59% on vector intensive code and 41% otherwise, at an area and power increase of 2.3% and 18% on a four-core processor cluster.

I. INTRODUCTION

The increasing performance requirements in computing intensive and mobile devices raise the need for powerful, but also ultra-low power (ULP) processors and computing architectures. Since power scales quadratically with the supply voltage, it is more energy efficient to operate a digital circuit near the threshold voltage of transistors [1]. The reduced speed can be compensated with multiple processors working in parallel at lower voltages [2]. To maximize energy efficiency, single processors can be turned off depending on the current workload, while resources like memories and caches can be shared among the cores. To form an energy-efficient multi-core cluster capable of processing computation intensive applications, the processing element (PE) needs to fulfill certain requirements like having a small area footprint, being energy efficient and having a shallow pipeline allowing fast inter-core communication and data exchange to allow fine-grained data and task-level parallelism. Moreover, great attention must be paid to integration of the cores to avoid the creation of a bottleneck in the system due to the instruction and data interface. We claim, that it is not enough to just use the most energy efficient 32 bit processor on the market, the ARM Cortex M0+, to construct an energy efficient multi-core cluster, but point out the requirement of careful optimization for an operation in a tightly-coupled cluster. First, the ARM RTL

code is not publicly available which makes an efficient cluster integration unfeasible. And second, while its energy efficiency of only $9.39\,\mu$W/MHz [3] would suite very well for ULP-operation, it is not meant to be used for computation intensive or parallel applications. Other architectures, like the ARM Cortex M4 are more powerful in this domain but also show $3-4\times$ higher requirements in area and power consumption [3]. This increase in area and power consumption comes from moving to a deeper pipeline and from enriching the instruction set with DSP like features and more advanced ALU operations. In order to keep the area and power figures under control a careful evaluation of additional features and instructions must be done, in particular in ultra-low power contexts.

In our work we have chosen the OpenRISC architecture which is a) open source and suitable to be optimized to work in a multi-core environment, and b) shows a low area footprint with only 35.5 kGE which allows the core to achieve an energy efficiency of $25.8\,\mu$W/MHz in 65 nm technology, which is competitive with an ARM Cortex M4 with an energy efficiency of $32.8\,\mu$W/MHz at similar area costs in a 90 nm technology [3]. While the computational efficiency of the OpenRISC architecture has been improved in terms of instructions per cycle (IPC) [4], it is missing important DSP-like features which allow a Cortex M4 to execute a program in less cycles due to its enriched instruction set. Enhancing an instruction set with very specific instructions is the key to increase performance in application-specific computing [5].

In order to push the OpenRISC architecture to higher performance and energy efficiency, we have looked into possible ISA-extensions which a) allow the core to significantly reduce the number of executed instructions to run computation-intensive applications, b) do not significantly increase the core's area and power consumption, c) can be efficiently integrated in a multi-core cluster, and d) are sufficiently general such that the compiler can automatically map their intermediate representation to the proposed instructions. Throughout this paper we discuss the benefits and costs of ISA-extensions, like hardware loops, extended addressing modes, vector ALU support and others in a multi-core cluster. Further, we present solutions how to efficiently implement these ISA-extensions in an ULP-cluster by taking into account the performance of each PE and its joint functionality with the cluster components, such as the shared instruction cache and the shared data memory. On our benchmarks, the execution of computation intensive benchmarks shows $1.1-5\times$ faster execution than with the initial ISA, and $1.1-4.5\times$ faster execution than on

978-1-4673-9141-2/15 $31.00 © 2015 IEEE

a Cortex M4. The faster execution and slightly higher power consumption combine to an overall energy efficiency boost of 47.8% compared to the initial ISA. The area of the extended core increases by 25% and the power efficiency per core of 33.8 μW/MHz on average is comparable with an ARM Cortex M4 (32.8 μW/MHz in 90 nm [3]). Integrated in the multi-core cluster the area overhead diminishes to only 2.3%.

The rest of this paper is organized as follows. In Section II we present the PULP cluster and OpenRISC architecture. In Section III we introduce the proposed ISA-extensions and its costs. Section IV describes several aspects to efficiently include the extensions in a multi-core cluster. In Section V we report area, power, and timing results of the ISA-extended OpenRISC cores on a four core cluster environment and in Section VI we draw our conclusions.

II. OVERVIEW OF THE PULP-PLATFORM

PULP (Parallel processing Ultra-Low Power platform) represents an effort to design a many-core platform responding to the demands of heavily-constrained embedded applications. Such applications would greatly benefit from a low-power, flexible computing fabric that is able to provide significant performance when needed and remain in a very low-consumption state otherwise. To achieve these goals, PULP features *clusters* of simple PEs that can be used to exploit both coarse- and fine-grain data- or task-level parallelism. At the same time, voltage and frequency scaling can be controlled at a fine granularity to achieve high energy efficiency when the performance constraints are more relaxed or when the power budget is tighter. Fig. 1 shows the PULP cluster architecture. In its current configuration it consists of four PEs, a shared 4-way associative I$ with four cache banks of 1 KB each and a L0 buffer of 128 bits per core holding the most recent cacheline to reduce access contentions at the cache banks [6]. The refill port of the shared I$ connects to the system bus together with an L2 memory, and several peripherals (not shown). The PEs have a multi-banked tightly coupled data memory (TCDM) acting as a shared scratchpad memory, instead of private data caches to avoid memory coherency overhead. TCDM is further divided into 16x4 KB SRAM banks, and 16x0.5 KB standard cell based memory (SCM) banks allowing the cluster to operate at ultra low voltages to achieve maximum energy efficiency [7]. Intra-cluster communication is based on a high bandwidth, low-latency interconnect, implementing a word-level interleaving scheme to reduce the access contention to TCDM banks. A lightweight, low-programming-latency, multi-channel DMA enables fast and flexible communication with other clusters, the L2 memory and external peripherals. In the following we focus on a single core and its integration in the cluster.

A. The OpenRISC Processing Element

The OpenCores community [8] developed the OpenRISC architecture, an open-source processor using a GCC-based toolchain. A RISC architecture is well-suited to be integrated in a tightly-coupled multi-core cluster because of its low area footprint and the low pipeline depth allowing to interact with other processors in a single cycle. In a previous work we developed OR10N, a complete redesign of the micro-architecture in order to balance pipeline stages, and increase IPC [4]. The redesigned core is divided into four pipeline stages, *instruction fetch (IF)*, *instruction decode (ID)*, *execute (EX)*, and *write back (WB)* and achieves near-optimal IPC values of 1. All operations can be completed in a single cycle except for

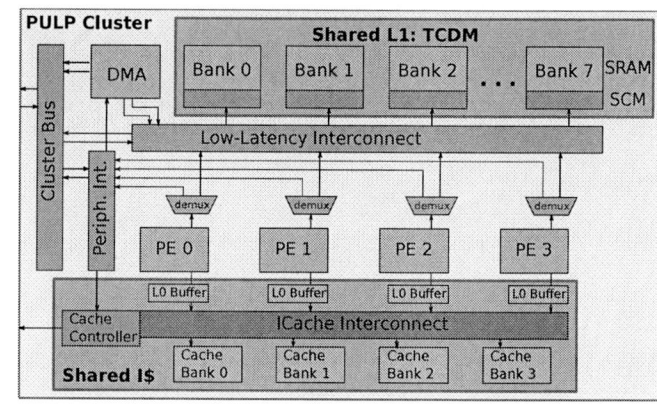

Fig. 1. PULP cluster featuring four OpenRISC processor cores, a shared instruction cache, eight TCDM banks utilized as L1 memory and a DMA for fast, concurrent data movements.

multiplications which are pipelined once, and can lead to stalls if the result is used in the subsequent cycle. While the core was being attached to an instruction and data memory [4], it has now been integrated in a PULP-cluster by connecting it to an I$ and a low-latency interconnect. Implemented in the cluster, the OR10N core utilizes 35.5 kGE. In this work we focus on increasing the efficiency of the generated machine code by introducing zero-overhead hardware loops, auto-incrementing memory operations, a more efficient multiplier architecture, and vectorial ALU operations. As we will show, getting rid of control code in small loops can lead to speedups of up to 2×. Auto-incrementing memory operations allow to get rid of instructions to maintain counters and addresses, by storing the updated memory address back to the latch-based register file. Ultimately, a vector unit that allows to concurrently process four 8-bit or two 16-bit values has been implemented in order to increase the throughput of the core. However, such instruction extensions are only useful if the compiler is able to produce such instructions. Therefore, we have modified the backend of the LLVM compiler to automatically generate code for the proposed ISA extensions and used it to evaluate the costs and benefits of the introduced ISA-extensions.

III. ISA-EXTENSIONS FOR OPENRISC

A. Hardware Loops

Zero-overhead hardware loops are a common feature in many processors, especially DSPs. Basically, a hardware loop is an implementation of a countable loop that avoids the need to explicitly test the loop counter and perform the branch. This is achieved by providing the hardware with information about the trip count and the beginning and end address of a loop, which are then used by specialized hardware in the computation of the next program counter (PC). The impact of hardware loops can be amplified by the presence of a loop buffer, i.e. a specialized cache holding the loop instructions, which removes any fetch delay [9]. In OR10N, we evaluated up to four nested hardware loops through the instructions shown in Tab. I. Each hardware loop has associated 3 special purpose register: HWLP_START and HWLP_END for the start and end address of the loop, and HWLP_COUNT for the loop counter.

Hardware loop setup can be performed either explicitly, by initializing each SPRs using the *lp.start*, *lp.end* and *lp.count* (or *lp.counti*) instructions, or initializing them in a single instruction using *lp.setup* (or *lp.setupi*).

Hardware loops are implemented in the micro-architectural

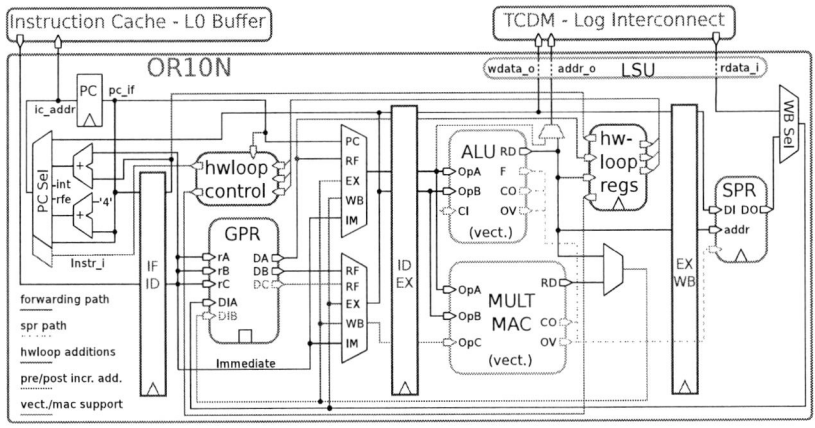
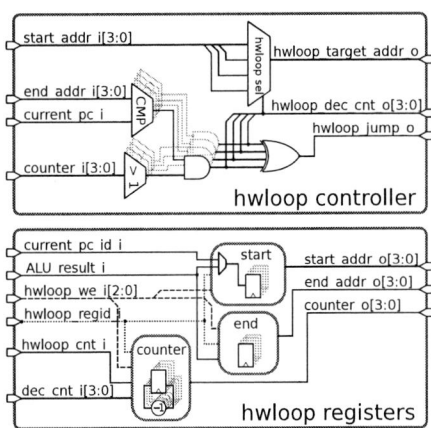

Fig. 2. Simplified block diagram of the OpenRISC micro-architecture, highlighting the additional hardware for hardware loop support.

Fig. 3. Detailed implementation of the hardware loop control and register block.

TABLE I. ZERO-OVERHEAD HARDWARE LOOP OPERATIONS

Each hardware loop is identified with an ID: L0-L3. The notation `HWLP_START[J]` is used to refer to the `HWLP_START` register of the Jth hardware loop.

Instruction format and Opcode	Semantics
`lp.start J, S` (eg. lp.start L0, 10) `000010 000 JJ SSSSSSSSSSSSSSSSSSSSS`	`HWLP_START[J]=sext(S*4)+PC`
`lp.end J, S` (eg. lp.end L0, -8) `000010 001 JJ EEEEEEEEEEEEEEEEEEEEE`	`HWLP_END[J]  =sext(E*4)+PC`
`lp.counti J, C` (eg. lp.counti L0, 8) `000010 010 JJ CCCCCCCCCCCCCCCCCCCCC`	`HWLP_COUNT[J]=zext(C)`
`lp.count J, rA` (eg. lp.count L0, r5) `000010 011 JJ AAAAA----------------`	`HWLP_COUNT[J]=[rA]`

`lp.setupi J, E, C (eg. lp.setupi L0, 4, 8)`
`000010 100 JJ CCCCCCCCCCCCEEEEEEEEE`
	`HWLP_START[J]=PC+4`
	`HWLP_END[J]  =zext(E*4)+PC`
	`HWLP_COUNT[J]=zext(C)`

`lp.setup J, E, rA (eg. lp.setup L0, 8, r5)`
`000010 101 JJ AAAAAEEEEEEEEEEEEEEEEE`
	`HWLP_START[J]=PC+4`
	`HWLP_END[J]  =zext(E*4)+PC`
	`HWLP_COUNT[J]=[rA]`

level with only two additional blocks. A controller, shown in the upper part of Fig. 3 and a set of registers to store the hardware loop variables. The controller is a purely combinational block which checks if the current PC matches one of the end addresses, and if the counter is greater than 1. In case both checks are true, the hwlp-controller informs the main controller to set the next PC to the corresponding start address of the loop. If two or more end points are equal, the controller gives priority to the lowest hwlp-ID (i.e., L0 has the highest priority). Since the performance gain is maximized when the loop body is small, it is not beneficial to support a lot of register sets. Therefore, we introduced two register sets in order to allow two nested loops. The support of additional hardware loops would bring marginal performance improvements at a non-negligible cost in terms of area (≈ 1.5 kGE per register set).

B. Extended Addressing Modes

Along with *Zero-overhead hardware loops*, we evaluated the performance and area impact of extended addressing modes. In the basic OR10N implementation only one type of load and store (*ld/st*) was available, in which the effective address was computed by adding a base address stored in a register and an offset encoded as an immediate value. The OR10N core was extended by implementing *ld/st* with both base and offset in register and *ld/st* with pre- and post-increment with both immediate and register offset.

Supporting all the variants of those instructions requires the addition of 65 new opcodes. To avoid saturation of available opcodes in the OpenRISC ISA, the new instructions are encoded using only 4 main opcodes and a sub-opcode field, which imposes a limitation on the immediate size from 16 to 11 bits. Out of those 65 new instructions, 3 instructions (1 for word, 1 for half word, 1 for byte transfer) are dedicated to each type of store, 6 instructions, coding different data sizes and sign extension, for each type of load with overall 5 new types of *ld/st*. These instructions are encoded reusing as much as possible the encoding scheme of the OpenRISC, keeping the source and destination registers at the same positions as shown in Tab. II.

TABLE II. LOAD/STORE ADDRESSING MODES.

Register-register addressing mode is expressed with the notation `rX(rY)`. Auto-incrementing addressing modes are identified with a ! before (preincrement) or after (postincrement) the base address. MMMMM in the sub-opcode is used to indicate bits that are dedicated to differentiate each type of load/store.

Old Instruction format	Opcode
`l.s[bhw] I(rA), rB`	`MMMMMM IIIII AAAAA BBBBB IIIIIIIIIII`
`l.l[bhw][zs] rD, I(rA)`	`MMMMMM DDDDD AAAAA IIIII IIIIIIIIIII`

New Instruction format	Opcode
`l.s[bhw] rD(rA), rB`	`010101 DDDDD AAAAA BBBBB -----0 1MMMM`
`l.l[bhw][zs] rD, rB(rA)`	`010111 DDDDD AAAAA BBBBB -----0 1MMMM`
`l.s[bhw] I(rA!), rB`	`010100 IIIII AAAAA BBBBB IIIIII 0MMMM`
`l.s[bhw] rD(rA!), rB`	`010100 DDDDD AAAAA BBBBB -----1 1MMMM`
`l.l[bhw][zs] rD, I(rA!)`	`010110 DDDDD AAAAA IIIII IIIIII 0MMMM`
`l.l[bhw][zs] rD, rB(rA!)`	`010110 DDDDD AAAAA BBBBB -----1 1MMMM`
`l.s[bhw] I(!rA), rB`	`010101 IIIII AAAAA BBBBB IIIIII 0MMMM`
`l.s[bhw] rD(!rA), rB`	`010101 DDDDD AAAAA BBBBB -----1 1MMMM`
`l.l[bhw][zs] rD, I(!rA)`	`010111 DDDDD AAAAA IIIII IIIIII 0MMMM`
`l.l[bhw][zs] rD, rB(!rA)`	`010111 DDDDD AAAAA BBBBB -----1 1MMMM`

In OR10N the effective address is calculated in the ALU and then used to access the memory during regular *ld/st* and pre-increment operations. In case of a post-increment the address generation is bypassed and the memory is addressed with the base address. Loads with pre- or post-increment need to write two registers at the same time (the data read from memory and the incremented address pointer) and this required the addition of an extra write port to the register file. Due to the non criticality of this path, storing the incremented address in the write back stage can be avoided and the register file can be written directly in the current cycle. To support stores with the offset or increment in a register an extra read port was required, in fact, 3 different values have to be fetched from the register file at the same time (base address, offset and data to write).

978-1-4673-9141-2/15 $31.00 © 2015 IEEE   27

The additional write and read port costs less than 1 kGE due to its non critical timing and its latch based implementation.

C. ALU Vector Support

Aiming for a more efficient processing of 8, and 16 bit data leads to the introduction of a vectorized ALU where the datapath is segmented into two, or four parts and allows to compute up to four bytes in parallel leading to a speedup of up to a factor of four. Such operations are also known as subword parallelism [10], packed-SIMD or micro-SIMD [11] instructions. However, the 32-bit operations remain the norm, meaning that extending a processor with such vector capabilities is only promising if the area and power overhead can be kept at a minimum. In order to extend our 32-bit OpenRISC ISA with subword parallelism, we have added two new opcodes for vector operations, one for ALU operations and one for vectorial comparisons. Each operation is available in three different vector modes for halfword (h) and byte (b) operations:

- lv.inst.{h,b}   rD, rA, rB,
- lv.inst.{h,b}.sc rD, rA, rB,
- lv.inst.{h,b}.sci rD, rA, I,

where the first is a register to register operation, and the other two are vector operations with a register rA and a scalar replication of the register rB, or an immediate. The operations based on an adder and shifter have been realized by splitting the data path in four segments. The full 32 bit result is computed by chaining the four adder results with the carry. Multiplications and MACs on the other hand, are complicated because of the additional muxing to support four concurrent multiplications.

The OpenRISC ISA supports full 64 bit result multiplications and MAC operations, which cannot be implemented within a single cycle without impacting the maximum frequency. In a previous implementation [4], the multiplication was realized with a two cycle multiplier and MAC operations were based on a special purpose accumulator which is accessible through special instructions. Given the fact that the full 64 bit result is often not required, and that the compiler is not always able to group instructions in a way such that no stalls occur, we have decided to simplify the multiplication by only generating the 32-bit results. This allows to support vectorial multiplications with subword selection [10] in a single cycle. Notice, that the full 64-bit product can still be generated by the addition of four partial products which can be generated in sequence. While in the original implementation three cycles were required to generate a 64-bit result, with subword selection it is possible to create it with 10 instructions. Reducing the multiplications to 32-bit makes the 64-bit accumulation register useless. Instead of the large accumulator, a normal register can be used as accumulation register which leads to two major advantages: a) it is possible to concurrently maintain multiple accumulation registers and b) the additional delay of moving data back and forth from the accumulation special register to a general purpose register vanishes since the accumulator is placed in the GPR in the first place.

Implemented in the micro-architecture of OR10N as depicted in Fig. 2, the vectorial ALU and fused MULT/MAC unit increase the core area by 6 kGE which accounts for 13% of the core but less than 1% of the complete cluster. The additional read port of the register file is shared with the advanced addressing modes and costs less than 1 kGE.

In Section V we show the benefit of the new multiplier with and without vectorial support.

D. Other Extensions

Along with the presented instruction extensions, several small ALU extensions such as *min, max, avg, abs* have been implemented. The area penalty of these extensions is less than 1%. These instructions are seldom used, however they can speed up significantly operations, such as normalization.

IV. IMPLEMENTATION IN THE ULP-CLUSTER

To keep the hardware complexity at a minimum, the introduced ISA-extensions have been implemented to a large extent with existing resources. The combined costs of all extensions, including all vectorial operations, are 9 kGE which is 25% of the core area. As pointed out, a good core architecture is not enough for an efficient integration in a multi-core cluster which is why we optimized the data and instruction interface without increasing the critical path. In the following we highlight three cluster-specific integration details which allow the processor to work more efficiently in a multi-core cluster.

A. Reducing Cache Accesses with a L0 Buffer

The amount of energy consumed by an instruction cache is not to be underestimated. In fact, a core with a very high IPC is accessing the instruction cache every cycle. Keeping this in mind and the fact that hardware loops are most effective if the loop body is small and contains no branches, adding a small L0 buffer between the instruction cache and the fetch interface (shown in Fig. 1) is a promising approach to lower power consumption by reducing the cache accesses. We have chosen a 128 bit wide L0 buffer, which is capable of holding the most recent cacheline. Hence, this architecture benefits if the compiler is capable of placing hardware loops aligned with 128 bits. In particular, when paired with a shared instruction cache, this approach is very effective since it significantly reduces access contention at the cache banks [6] and allows a loop to be fetched only once if the size does not exceed four instructions. Even though the buffer can only hold 4 instructions, this is not a restriction on the loop size. Larger loops would still benefit from the L0 buffer because only every fourth request has to be forwarded to the cache controller.

B. Unaligned Memory Access

Data structures are not always word aligned, and even if the compiler is aware of a vector unit, it is not always possible to align data. For example, in a simple 2D-filter on 8 bit data types, as depicted in Fig. 4, it is not possible to read aligned data in every cycle, thus leading to unnecessary masking and shifting or resorting to byte-wise loads. To support unaligned memory accesses in the tightly-coupled cluster, we have two options: 1) extend the data interface to 64 bit — 2) implement the unaligned memory access in 2 cycles.

Since the critical path in the cluster is already on the return path of the data memory, and that reading 64 bit of data means doubling the size of the interconnect and also the number of banks, the former option is not the most promising. In fact, by doubling the size of the interconnect, the depth increases due to its logarithmic tree. Moreover, since the delay of the interconnect depends on the depth, it would impact the length of the cluster's critical path, leading to a lower energy efficiency. For this reason we did not further evaluate this solution. The second approach enables unaligned memory

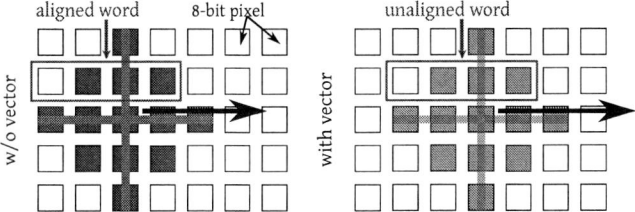

Fig. 4. Example showing the benefit of unaligned access when computing a stencil over an image with 8 bit pixels.

access with two subsequent memory requests, which are then merged into a single word by the load store unit. This hardware implementation does not add to the critical path and brings no additional hardware complexity to the core. Since the TCDM serves requests within a single cycle, it is possible to read unaligned data in only two cycles. On the other hand, to perform the same load operations, an architecture without support for unaligned data requests would require at least five instructions. Two cycles to read the two data items, two cycles to shift data and one cycle to combine them in a single register.

Our implementation of unaligned memory requests does allow to use the vector unit more often and is much more efficient in code size, and latency than a software only support. Compared to a single cycle load operation, our implementation does not require significant additional hardware resources and is therefore very well suited for the multi-core cluster.

C. Branch Prediction to Balance the Paths to I$

Branching was initially performed by using the forwarded flag from the ALU, thus allowing to branch without stalling the pipeline or having to predict and probably revert the pipeline. This was particularly important when processing loops. With hardware loop support this argument falls apart and stall free branching is no longer required for efficient loop handling. Moreover, since the L0 buffer delays the instruction request path, branch prediction becomes a requirement to prevent slowing down the system. Specially at lower voltages, where the cluster becomes extremely energy efficient, this path becomes critical. Synthesizing the cluster without branch prediction and an L0 buffer of 128 bits in a 28 nm process leads to a frequency degradation of 13% at 0.6 V. Branch prediction would split this path, but also slightly increase the runtime due to mispredictions. Given the fact that we want to operate at ultra-low power, and that the misprediction penalty is only one cycle in our flat pipeline, we have chosen the most simple branch predictor, which is to always take backward branches, never take forward branches.

The branch prediction removes the critical path to the I$, while increasing the number of cycles in our benchmarks by only 1.32 % on average, with a maximum of 3.4 %. In combination with hardware loops, this number decreases to only 0.7 % additional cycles. Hence, we conclude that even a very simple branch predictor, if paired with hardware loops, does only increase the number of cycles by 0.7 % while allowing the core to be clocked 13% higher at low voltages.

V. PERFORMANCE, AREA, AND POWER RESULTS

In the following we are discussing the performance, area, and power impact of the introduced ISA-extensions. The execution time in number of cycles has been measured in RTL-simulations, while the area increase was determined with Synopsys Design Compiler in topographical mode, and

the power estimations were performed on a back-annotated placed&routed netlist at nominal voltage of 1.2 V and a frequency of 50 MHz. The complete design, including a final tapeout of the PULP chip, was done in UMC 65 nm technology. Unless otherwise specified, all power and area numbers are related to this technology.

A. Execution Time

Each proposed instruction set extension has been analyzed individually by enabling it in the compiler through dedicated flags. Specifically, we compared the basic ISA with 4 different configurations, which, respectively, enable 2 sets of hardware loops (H), pre- and post-increment addressing modes (I), the new single cycle multiplication and three operand MAC unit (M) and vector operations with unaligned access (V). Finally, we also performed a comparison with the ARM Cortex M4 (Cortex M4). Fig. 5 incrementally shows the benefit of each ISA-extension on our benchmark applications which range from basic matrix multiplications, based on 8, 16 and 32 bits, through convolutions, filters and cryptographic algorithms. In particular matrix multiplications are presented in two forms: the classical implementation with row-by-column products, and an optimized version with row-by-row products preceded by the transposition of the second matrix, being a good candidate for auto-vectorization. All benchmarks are executed on a single core. Hardware loops and automatic addressing updates bring speedups up to $1.75\times$. The MAC unit has a good impact on computational intensive benchmarks like convolution and matrix multiplications and allows to process those benchmarks up to $2.25\times$ faster. The use of the vector unit can bring an additional boost when it is applicable. E.g., on the matrix multiplication on 8 bit data, our vector optimization achieves a $5\times$ speedup with respect to the original ISA. Besides the impact of the ISA extensions, this figure results from the ability to vectorize row-by-row products. Similarly, on the 16 bit optimized matrix multiplication we achieve an overall $3.5\times$ speedup. On the classical row-by-column implementation on 8 bit data, data access on the column cannot be efficiently vectorized as there is no hardware support for either strided or gather/scatter load/store operations. On 16 bit data matrix multiplication the compiler does not apply vectorization as it is detected as not effective by the heuristics.

B. Area and Power Efficiency

Fig. 7 shows the increase in area and power with each additional feature. We can conclude that even though the area per core increases by 25% to 44.5 kGE, the overhead at cluster level remains small with only 2.3% of overhead due to the presence of interconnects, peripherals, and large amount of memory in the cluster. The total power consumption of one core running at 50 MHz with all features enabled is 1.688 mW, which translates to an average energy efficiency of 33.8 μW/MHz. The core shows a higher power consumption during the execution of vector heavy code where an energy efficiency of 47.8 μW/MHz is achieved.

At the maximum frequency of 362 MHz, which is not affected by the ISA-extensions, the four-core cluster is capable of processing 1.4 GOPS at a power budget of 93 mW of which 55% is consumed by the cores. With respect to a single core implementation, the four-core cluster can process four times more operations, at a power increase of only 67%. Thus, moving from a single core to a four-core cluster increases the power-efficiency by a factor of 2.4.

978-1-4673-9141-2/15 $31.00 © 2015 IEEE

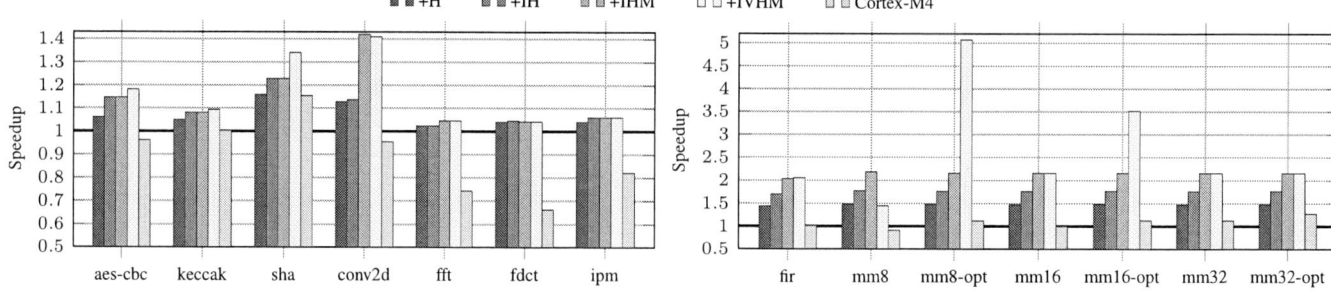

Fig. 5. Comparison of the speedup with the introduced ISA-extensions versus the initial ISA. Number of cycles are measured in RTL simulations for the OR10N core and in case of the Cortex M4 on the real hardware. All the results are normalized w.r.t. the number of cycles on the plain OpenRISC ISA.

C. Core and Cluster Energy Efficiency

In Fig. 6 the total amount of energy required to run each benchmark is shown for the original ISA, the extended ISA with and without vector unit, and a Cortex M4. To calculate the numbers for the Cortex M4, an energy efficiency of 16.7 μW/MHz is assumed, which is scaled down to 65 nm from a 90 nm technology [3]. In normal execution, where no vector code is used, the energy savings are 41.1% on average. Due to the additional speedup of the vector unit, the savings increase to 59.6% in the vector intensive matrix multiplications. When comparing the efficiency to an ARM Cortex M4, we observe that the extended ISA is up to 48% more energy efficient when processing vector code due to the vector instructions. If no vector code can be used, the performance of the proposed ISA shows no clear trend towards Cortex M4 or OpenRISC.

Taking into account all cluster resources, a cluster consisting of the improved cores consumes 18% more power with respect to a cluster with the initial micro-architecture. Taking the complete cluster power into account, the energy savings of the cluster based on the extended ISA with all features enabled ranges from 39% to 66%. On average the cluster is 47.8% more energy efficient than the initial architecture.

VI. Conclusions

We have evaluated several instruction extensions for the OpenRISC-based OR10N core targeting an efficient integration in a multi-core cluster with a shared memory and instruction cache. The proposed ISA-extensions include hardware loops, pre- and post-increment addressing modes, new MAC instructions, and a set of vector instructions. The introduced ISA-extensions allow the core to compete with an ARM Cortex M4 in terms of runtime in cycles and in energy efficiency. Compared to the initial ISA, the support of the extended ISA increases the area of the OR10N core by 25% but at the same

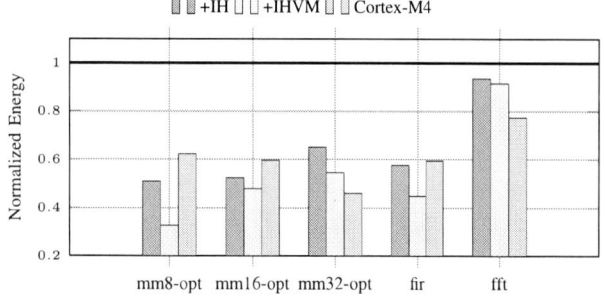

Fig. 6. Energy efficiency of the core with the optimized ISA integrated in the multi-core cluster in UMC 65 nm. All the results are normalized w.r.t. the absorbed energy on the plain OpenRISC ISA.

time allows to execute computation intensive applications up to 5$\times$ faster. The combined effort of ISA-extensions and smart cluster integration results in a total energy improvement of 47.8% on average while increasing the area and power of the four-core cluster by only 2.3% and 18%.

Acknowledgment

This work was partially supported by the IcySoC RTD project evaluated by the Swiss NSF and funded by Nano-Tera.ch with Swiss Confederation financing.

References

[1] R.G. Dreslinski et al., "Near-threshold computing: Reclaiming moore's law through energy efficient integrated circuits," *Proceedings of the IEEE*, vol. 98, no. 2, pp. 253–266, Feb 2010.

[2] A.Y. Dogan et al., "Low-power processor architecture exploration for online biomedical signal analysis," *Circuits, Devices Systems, IET*, vol. 6, no. 5, pp. 279–286, Sept 2012.

[3] ARM, "ARM Cortex M0+/M4," http://www.arm.com/products/processors/cortex-m/, [Online; accessed 7-May-2015].

[4] M. Gautschi, et al., "SIR10US: A tightly coupled elliptic-curve cryptography co-processor for the OpenRISC," in *IEEE Int. Conf. Appl.-specific Syst. Archit. Processors (ASAP)*, June 2014, pp. 25–29.

[5] M. Arnold and H. Corporaal, "Designing domain-specific processors," in *Int Symp. Hardw./Softw. Codesign*, 2001, pp. 61–66.

[6] I. Loi, D. Rossi, G. Haugou, M. Gautschi, and L. Benini, "Exploring Multi-banked Shared-L1 Program Cache on Ultra-Low Power, Tightly-Coupled Processor Clusters," *CF '15*.

[7] P. Meinerzhagen, C. Roth, and A. Burg, "Towards generic low-power area-efficient standard cell based memory architectures," in *MWSCAS*, Aug 2010, pp. 129–132.

[8] OpenCores Community, "OpenRISC Community Portal," http://opencores.org/or1k/Main_Page, 2014, [accessed 18-September-2014].

[9] G.-R. Uh et al., "Techniques for effectively exploiting a zero overhead loop buffer," in *Compiler Construction*. Springer, 2000, pp. 157–172.

[10] R. Lee, "Subword parallelism with max-2," *Micro, IEEE*, vol. 16, no. 4, pp. 51–59, Aug 1996.

[11] A. Shahbahrami, B. Juurlink, and S. Vassiliadis, "A comparison between processor architectures for multimedia applications," in *ProRISC*, 2004, pp. 138–152.

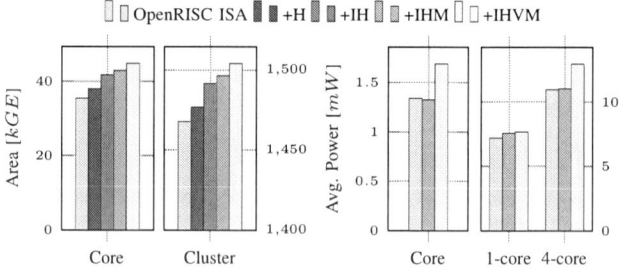

Fig. 7. Area and power comparison with different ISA-extensions on core and cluster level. Power is shown on a cluster with 1 and 4 cores.

Design of Asynchronous RISC CPU Register-File Write-Back Queue

Matthew M. Kim
Electrical & Computer Engineering
RMIT University,
Melbourne, Australia
myungha.kim@rmit.edu.au

Karl M. Fant
Theseus Research
fant@theseusresearch.com

Paul Beckett
Electrical & Computer Engineering
RMIT University,
Melbourne, Australia
pbeckett@rmit.edu.au

Abstract - **This paper presents the comparison results of Area, Performance and Power of FIFO and Data-Queue on a logically determined Null Convention Logic RISC CPU register file Write-Back circuit. A shift register block implemented using Delay-Insensitive techniques operates in a way that is identical to a FIFO. In this paper, we illustrate the architectures of the Delay-Insensitive Asynchronous Data-Queue and FIFO and analyze the characteristics of these circuits. This comparison results can be also used to the other buffering unit of the CPU such as scoreboard for Dynamic scheduling or Cache Controller memory interface circuits.**

Keywords—Null Convention Logic; Asynchronous; Register File; Write-Back Queue; RISC CPU

I. INTRODUCTION

Over the past several decades, many universities and industry groups [1-6] have developed microprocessor architectures using asynchronous digital techniques. The intention has always been to take advantage of the general benefits of asynchronous design compared to conventional synchronous design, particularly the potential for lower power consumption [7-9]. As a result, asynchronous CPU design may be particularly applicable to mobile and wearable systems and Internet of Things (IoT) technologies which all require high performance networked functions with Multi-tasking Operation Systems and Network Protocol Stack Software, all operating with extremely low power.

Asynchronous technology especially Delay-Insensitive circuits naturally exhibits event-driven behavior [10] in which circuits consume power only when they are required to switch. There are currently three approaches to asynchronous design: Bounded-delay (Micro-pipeline) [11] and Quasi Delay-insensitive (QDI) [12] and logically determined design Null Convention Logic (NCL) [10]. The fundamental operational unit (gate) in NCL is a general threshold gate with state holding behavior. As a result, NCL circuits can be somewhat simpler to design than other asynchronous approaches. Application specific systems-on-chip may be formed from libraries of standard cells and prebuilt IP blocks that can be connected in a "correct by construction" manner with minimal timing constraints.

The concept of pipelining is central to any high performance microprocessor design whether synchronous or otherwise. Previous asynchronous designs [1-3] have all employed pipelining to increase their processing performance. Other issues that asynchronous CPU designs share with conventional microarchitecture include hazards (i.e., structural, control and data hazards). These have to be treated carefully in an NCL based microprocessors although, in a similar way to conventional RISC processors, the compiler can be set up to avoid them.

NCL naturally exhibits strong data flow behavior. Thus, those parts of the architecture that exhibit similarly strong data streaming characteristics from their inputs to outputs will be easier to design. These include the Instruction-Decoder and Arithmetic Logic Unit. Conversely, it will be more complex to implement a Register File because the block has separate outputs (to obtain the source operands) and input (Result Forwarding, Write-Back) interfaces in the one module. In this case, the inputs and outputs operate with different protocols, especially in the case of pipeline processing architectures. This will be a challenge for NCL microprocessor architectures.

In an asynchronous technology such as NCL the execution time of each processor block, ALU, multiplier etc., will always be different and even the input to output delay in the same execution block may differ substantially as the delay is input data dependent. Overall, these exhibit average-case delays rather than worst case, which can further complicate their design. This mandates the use of a suitable Result-forwarding circuit on the Write-Back side of the register File to guarantee its correct operation. The layout of this result-forwarding circuit must include careful buffer sizing to maximize its performance.

In this paper, we present the design of an NCL based Write-Back Unit for an asynchronous RISC CPU. The design is based on the observation that a shift-register (SR) chain in a DI asynchronous circuit behaves identically to an asynchronous FIFO in that its input data will propagate (combinationally) through the entire circuit until it arrives at the end of the chain. Data therefore fills from the output end backwards towards the input, determined by the separate rates with which data values enter and exit the FIFO. By careful use of this SR buffer structure, the performance, power and area of the Write-Back Unit can be optimized.

The remainder of this paper is organized as follows: Section II addresses the basic principles of asynchronous logic and outlines the NCL design style. Section III describes the

978-1-4673-9141-2/15 $31.00 © 2015 IEEE

architecture of the NCL based Register File and Write-Back Unit. Section IV shows the NCL based Write-Back Buffer Design. Section V explains the comparison and analysis results. Finally, in Section VI we conclude the paper and identify future work.

II. ASYNCHRONOUS LOGIC DESIGN AND NCL

In this section, we briefly describe main differences between conventional synchronous and asynchronous design and introduce the characteristics of Null Convention Logic that are important to this application.

Conventional synchronous design uses a centralized clock signal to control the data flow through combinational logic between edge-triggered register stages. In contrast, asynchronous design employs localized hand-shaking signals (e.g., Request and Acknowledge) to control its data flow. Of the three types of asynchronous models in current use (micro-pipelines, QDI and NCL), NCL has the least sensitive delay constraint [10]. In the Bounded Delay (BD) model, delays in both gates and wires are bounded and delays are evaluated based on worst-case scenarios to avoid signal hazard conditions [11]. QDI is constrained by the isochronic fork. NCL is constrained by the orphan path which is considerably less constraining than the isochronic fork.

Null Convention Logic [10] is based on threshold logic gates exhibiting internal state holding in each gate. It is a symbolically complete logic system that exhibits delay insensitive behavior within some minor constraints. Unlike other DI techniques, NCL gates rely on one simple timing assumption: that the feedback path for its state holding elements must be faster than the forward delay through the gate. This limitation is usually straightforward to overcome.

Input completeness requires that the output signal of the gate may not transition to DATA until all inputs have transitioned to DATA. Similarly, the output will not transition to NULL until all inputs have reached NULL. Input Completeness is achieved by adding the value NULL to the basic logic values (TRUE and FALSE), to represent a status of "no data". Output Data will only be valid when all input signals have transitioned from NULL to DATA. An NCL circuit consists of an interconnection of primitive modules known as M-of-N threshold gates with hysteresis. All functional blocks, including both combinational logic and storage elements, are constructed out of these same primitives. To represent these three values in this work we use a dual rail coding style which expresses the value '00' (NULL), '01' FALSE and '10' TRUE. A NULL wave front separates two DATA wave fronts.

III. ASYNCHRONOUS CPU AND REGISTER FILE WRITE-BACK UNIT

A number of asynchronous CPU designs including a NCL based CPU have been previously proposed and analyzed. Earlier asynchronous techniques for resolving register dependencies include: the register locking mechanism on the AMULET 1 processor [13], which the ARM core implements using Bundled Data techniques; register locking plus a "last result" register used in the AMULET2 processor [14]; the last

result bypass mechanism of the asynchronous MIPS R3000 from Caltech [3]; the scoreboard-like Data Hazard Detection Table (DHDT) of the Asynchronous MIPS Processor from University of Birmingham [15]; the counter flow pipeline architecture and the asynchronous "queue" FIFO for the AMULET3 processor [1]. The "queue" in the AMULET3 is an efficient solution to the problems of results forwarding and exception handling within an asynchronous pipeline. However, it has the disadvantage of being a full-custom design implemented using matched-delay-based, bundled-data encoding, which limits the possibility of design-space exploration and technology portability [7]. The AMULET3 uses an asynchronous "queue" FIFO (AQF) to implement the result forwarding unit [1]. The nanoSpa processor [2], which used Dual-Rail Delay-Insensitive technology, from the same organization as AMULET, also includes a nanoSpa forwarding unit (nFU) as their result forwarding logic. The AQF is a circular buffer that acts both as a forwarding unit and a reorder buffer whereas nFU is not used as a reorder buffer because nanoSpa does not execute instructions out of order. In a similar way to the AMULET3 (AQF, queue on their Writeout unit), nanoSpa (nFU, nanoForward Unit) and Caltech MIPS R3000 (which uses a unit-number queue on their Write-Back unit), our NCL based RISC processor uses a queue on the Write-Back circuit.

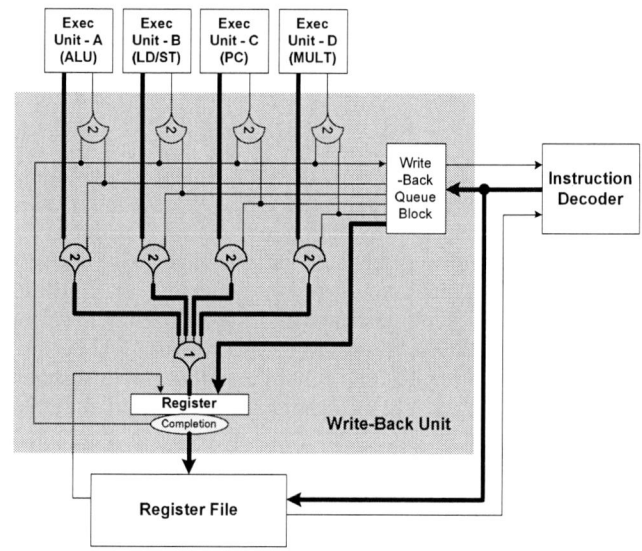

Figure 1 Write-Back Unit Circuit Diagram

Figure 1 shows the simplified circuit diagram of a NCL based RISC CPU Write-Back Unit. The controls are derived from the Instruction Decoder and separate 32bit Dual-rail results emerge from each of the four separate Execution Units. The Acknowledge hand-shaking signals are connected from the completion logic back to the Execution Units. A NCL fan-in steering circuit executes the multiplex function here. The block shown in yellow in Figure 1 is the Write-Back Queue, the subject of this paper. The Write-Back Queue can be partitioned for each Execution Units for out-of-order execution but in that case, significantly more complex instruction order control logic needs to be added. This paper does not consider out-of-order execution.

IV. NCL BASED PROCESSOR WRTE-BACK QUEUE DESIGN

As we discussed previously, the Execution Unit delay time on the NCL based processor varies widely and its execution time depends on the input data. As a result, it is easy to lose the correct instruction order. A First-In/First-Out (FIFO) structure (such as shown in Figure 1 is a good solution to maintain the correct instruction order. As also noted above, a register chain built using NCL naturally exhibits First-In/First-Out behaviour. In this section we will illustrate the main differences between a conventional Register-based Queue (the "Data-Queue") and a NCL based FIFO.

A. NCL Register-Chain and Data-Queue

Figure 2 illustrates the basic Data Flow of the NCL Register-Chain. Bold lines represent "DATA (Value 1)" and the Light lines are "NULL (Value 0)". (T1) is the Reset state and all the completion inverters are being reset so that the data flow and the acknowledge flow status are all NULL.

In (T2) the reset is negated and all of the Acknowledge signals become DATA simultaneously. Note that the forward data path is still entirely NULL.

At State (T3) an input DATA value is presented. As a result, this input value flows all the way through the registers until the end of the Buffer. In the meantime, the DATA value is held active by the Ack Input line.

At States (T4), (T5) and (T6) the input becomes respectively NULL, DATA and finally NULL again. By (T6)

the Buffer Full status is asserted. The buffer undergoes a NULL-DATA-NULL-DATA status and the first input DATA will become present at the output when the Ack Input signal is asserted.

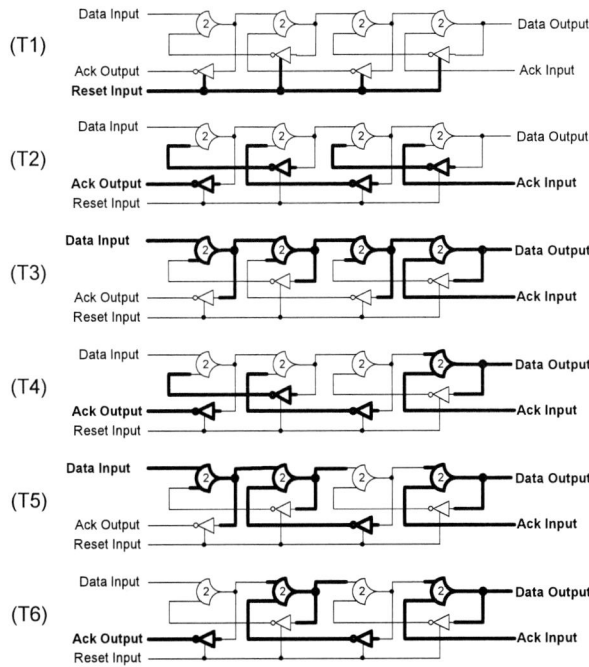

Figure 2 NCL Data-Queue Register Data Flow

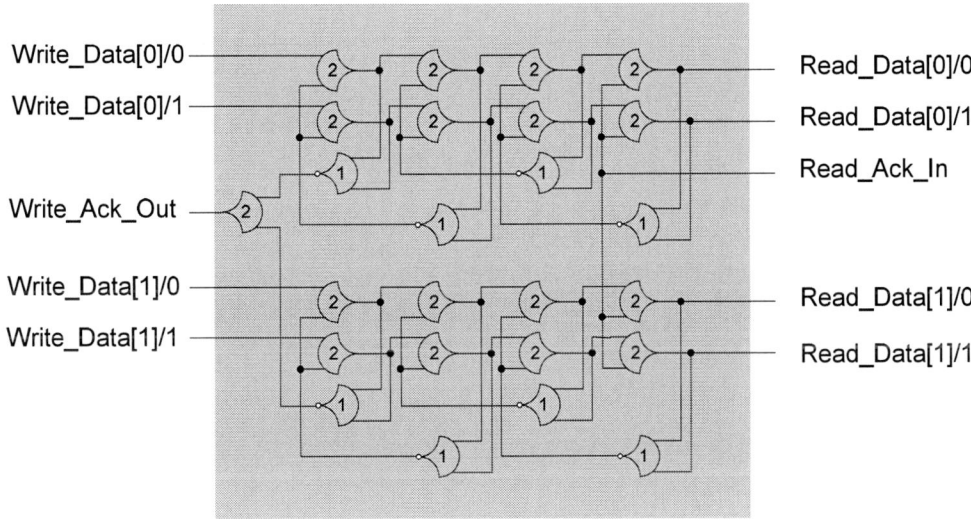

Figure 3 2x2 NCL Data-Queue

B. NCL-based Data-Queue

Figure 3 shows the circuit diagram of NCL Data-Queue. This has 2 bit dual-rail width and 2 depth buffering function. As in Figure 2, the circuit is Register-Chain and working as a Data-Queue, After Reset, if all the acknowledge signals are

DATA then input data flow-through to the end of the buffer therefore the input DATA and NULL stacked in the Buffer until the Read_Ack_In signal is switching.

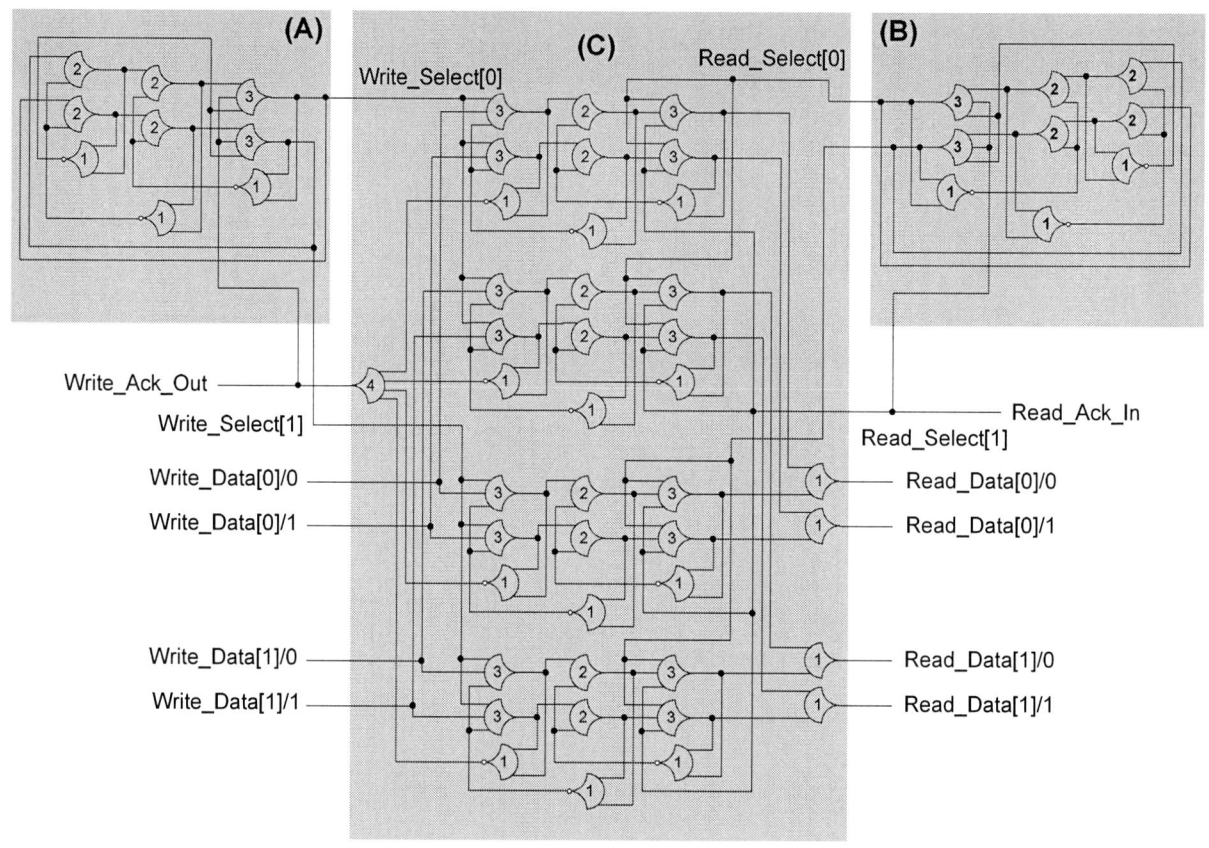

Figure 4 2x2 NCL FIFO

C. NCL based FIFO

Figure 4 shows the circuit diagram of NCL FIFO. In this example the FIFO is two bits wide and two deep, dual-rail encoded. Block (A) shows the Write-Pointer which is comprised of a two stage State-Sequencer [10]. The output of this sequencer enables a specific register for writing. Block (C) is a memory (Register) block which contains the data to be written. In this 2x2 dual-rail FIFO, we can maintain 8 bits of information. Block (B) is a Read-Pointer. In the same way as the Write-Pointer the output signal of this block points to the register to read. The Write and Read Pointers are controlled using the input and output hand-shaking signals (Write_Ack_Out and Read_Ack_In).

The State-Sequencers can be replaced with the Dual-rail Binary-Counters [16] with Binary to One-hot decoders but the size of this combination is not very area efficient for this small size buffers. In this paper, we presented Dual-rail examples but these can be easily converted to Single-rail control signal Buffers. Generally for the Write-Back Queue Block, we need to use both Dual-rail and Single-rail encoding.

V. COMPARISON ANALYSIS AND RESULTS.

In this section, we illustrate the comparative results for area, performance and power for each buffer type. These simulations were performed using 28nm UTBB-FDSOI [17] device models with a supply set to 1V. The circuits were designed using Verilog-HDL and imported to the Cadence Virtuoso® system and simulated using Cadence Ultrasim® and NC-Verilog®. The results were obtained for pre-layout, transistor-level simulation.

A. Area comparison result and analysis

Table 1 and Figure 5 present the comparative results of the Area between Data-Queue and FIFO of the 8-bit Buffer. Cell (Gates) represents the NCL Threshold gate count and Area is based on an estimate of the transistor sizes of each gate normalized to a simple inverter of area 1. The Data-Queue uses much small area compare to FIFO even though it contains both DATA and NULL values. As expected in this pre-layout case, area is almost proportional to gate count.

B. Performance comparison result and analysis

Table 2 and Figure 6 show the comparison results for performance. When the depth of the buffer is small, the input to output delay of the Data-Queue is smaller than that of the FIFO. Delay increase with depth and exceeds the FIFO delay for depths greater than 5. We note that the input to output delay of deep FIFOs (e.g., depth = 100) makes it unacceptable as a Write-Back buffer. Propagation delay is particularly important in the case where the Register File Write-Back comes quickly after an Operand Register value is read. If the delay of the Buffer is greater than the execution time, it will cause a pipeline stall. The Cycle Time of the Data-Queue is faster than

978-1-4673-9141-2/15 $31.00 © 2015 IEEE

that of the FIFO but if the delay is smaller than the shortest cycle time of the CPU then this will not be a problem. The shortest instruction cycle usually occurs for the NOP (No Operation) instruction but the CPU cycle time of asynchronous processors vary widely with input data, and fast cycles can result from data patterns that cause short carry propagation paths within the ALU.

Table 1 Area Comparison Results

Compare Type	Depth x Width	Data Queue	FIFO
Cells(Gates)	3x8	130	248
	5x8	234	392
	10x8	494	798
	100x8	5174	8290
Area	3x8	477.50	1107.50
	5x8	859.50	1822.50
	10x8	1814.50	3679.00
	100x8	19004.50	37372.00

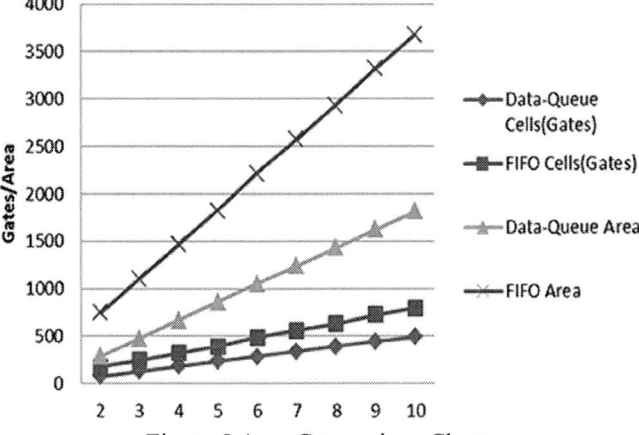

Figure 5 Area Comparison Chart

Table 2 Performance Comparison Results

Compare Type	Depth x Width	Data Queue	FIFO
Input to Output Delay	3x8	240ps	469ps
	5x8	437ps	512ps
	10x8	912ps	591ps
	100x8	9,588ps	869ps
Cycle Time	3x8	505ps	648ps
	5x8	506ps	712ps
	10x8	514ps	889ps
	100x8	514ps	1,435ps

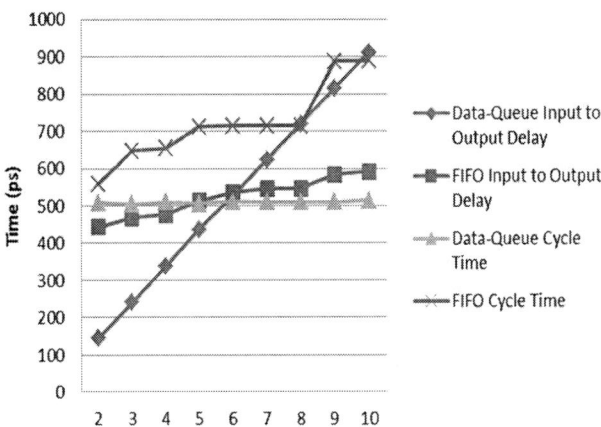

Figure 6 Performance Comparison Chart

C. *Power comparison result and analysis*

Table 3 and Figure 7 show the comparison results for power consumption. The average power was measured with a 100MHz (10nS) input transition rate applied under the same conditions to each architecture.

Table 3 Power Comparison Results

Compare Type	Depth x Width	Data Queue	FIFO
Average Power Consumption	3x8	17.45uW	15.80uW
	5x8	28.38uW	18.39uW
	10x8	61.19uW	23.79uW
	100x8	653,35uW	103.02uW

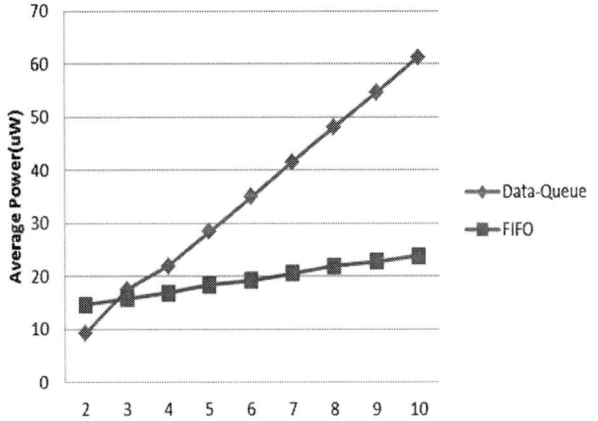

Figure 7 Power Comparison Chart

The Write-Back Unit is one of the most active module in this NCL based RISC CPU therefore the issue of Power Consumption is important for this buffer design and exhibits a standard area-power trade-off. The NCL FIFO uses more area but it is more power efficient compared to the Data-Queue when the Buffer depth is greater than 3. The State-sequencers control the enable signals of the buffer inputs and outputs.

Thus in the memory block (C) of Figure 4, only the selected register is switching during each instruction cycle, resulting in increased power efficiency.

Both of the buffers (Data-Queue and FIFO) are controlled using the input and output Acknowledge signals. The depth of the Buffer depends on the depth of the pipelining in the Execution Units such as ALU, Multiplier and Floating Point Processing Unit. For example, if we use Floating Point multiplier, it will require deeper queues – in the order of 15. In this case, the Data-queue circuit will be unsuitable for that purpose if the system is to maintain high performance (e.g., overall cycle times below 1nS).

VI. CONCLUSION

In this paper we have described the Write-Back circuit of the NCL based RISC CPU especially the Write-Back Queue. In this paper, we limited our analysis to a comparison between the Data-Queue and FIFO design styles. The lack of detailed test data for previous Write-Back units makes it difficult to contrast the different asynchronous design styles (BD, QDI). Further, there are no prior examples of NCL based Register File Write-Back unit design. Using these comparison results, designers can select the appropriate buffer types depending on their pipelining depth requirements and target performance of the circuit. We can apply this approach in a similar manner way when we design other buffer types for the other processor blocks in the NCL based Asynchronous microprocessors and System on Chip (SoC) design.

Future work will include performance, area and power optimization of this circuit and, in addition, Out-of-Order execution function will be implemented to explore its effect on the overall performance of the processor.

REFERENCES

[1] S. B. Furber, D. A. Edwards, and J. D. Garside, "AMULET3: a 100 MIPS asynchronous embedded processor," in *Computer Design, 2000. Proceedings. 2000 International Conference on*, 2000, pp. 329-334.

[2] L. Tarazona, L. Plana, and D. Edwards, "Architectural enhancements for a synthesised self-timed processor," in *Proceedings of the 19th UK Asynchronous Forum, http://cas.ee.ic.ac.uk/AsyncForum19/*, 2007.

[3] A. J. Martin, A. Lines, R. Manohar, M. Nystroem, P. Penzes, R. Southworth, *et al.*, "The design of an asynchronous MIPS R3000 microprocessor," in *Advanced Research in VLSI, Conference on*, 1997, pp. 164-164.

[4] A. Bink and R. York, "ARM996HS: the first licensable, clockless 32-bit processor core," *Micro, IEEE*, vol. 27, pp. 58-68, 2007.

[5] A. Mokhov, M. Rykunov, D. Sokolov, and A. Yakovlev, "Design of processors with reconfigurable microarchitecture," *Journal of Low Power Electronics and Applications*, vol. 4, pp. 26-43, 2014.

[6] M.-H. Oh, Y. W. Kim, S. Kwak, C.-H. Shin, and S.-N. Kim, "Architectural Design Issues in a Clockless 32 Bit Processor Using an Asynchronous HDL," *ETRI Journal*, vol. 35, pp. 480-490, 2013.

[7] L. A. Tarazona, D. A. Edwards, and L. A. Plana, "A Synthesisable Quasi-Delay Insensitive Result Forwarding Unit for an Asynchronous Processor," in *Digital System Design, Architectures, Methods and Tools, 2009. DSD '09. 12th Euromicro Conference on*, 2009, pp. 627-634.

[8] S. Keller, A. J. Martin, and C. Moore, "DD1: A QDI, Radiation-Hard-by-Design, Near-Threshold 18uW/MIPS Microcontroller in 40nm Bulk CMOS," 2015.

[9] H. van Gageldonk, K. van Berkel, A. Peeters, D. Baumann, D. Gloor, and G. Stegmann, "An asynchronous low-power 80C51 microcontroller," in *Advanced Research in Asynchronous Circuits and Systems, 1998. Proceedings. 1998 Fourth International Symposium on*, 1998, pp. 96-107.

[10] K. M. Fant, *Logically determined design: clockless system design with NULL convention logic*: John Wiley & Sons, 2005.

[11] I. E. Sutherland, "Micropipelines," *Commun. ACM*, vol. 32, pp. 720-738, 1989.

[12] A. Martin, "Compiling communicating processes into delay-insensitive VLSI circuits," *Distributed Computing*, vol. 1, pp. 226-234, 1986/12/01 1986.

[13] J. V. Woods, P. Day, S. B. Furber, J. D. Garside, N. C. Paver, and S. Temple, "AMULET1: an asynchronous ARM microprocessor," *Computers, IEEE Transactions on*, vol. 46, pp. 385-398, 1997.

[14] S. B. Furber, J. D. Garside, P. Riocreux, S. Temple, P. Day, L. Jianwei, *et al.*, "AMULET2e: an asynchronous embedded controller," *Proceedings of the IEEE*, vol. 87, pp. 243-256, 1999.

[15] Q. Zhang and G. Theodoropoulos, "Towards an asynchronous MIPS processor," in *Advances in Computer Systems Architecture*, ed: Springer, 2003, pp. 137-150.

[16] S. C. Smith and J. Di, "Designing asynchronous circuits using NULL convention logic (NCL)," *Synthesis Lectures on Digital Circuits and Systems*, vol. 4, pp. 1-96, 2009.

[17] *Circuits Multi-Projects*. Available: http://cmp.imag.fr/

978-1-4673-9141-2/15 $31.00 © 2015 IEEE

Modular Performance Analysis of Multicore SoC-based Small Cell LTE Base Station

Manikantan Srinivasan[†] and C Siva Ram Murthy[†]
[†]Department of Computer Science & Engineering
Indian Institute of Technology Madras (IITM), Chennai.
mani@cse.iitm.ac.in, murthy@iitm.ac.in

Anusuya Balasubramanian[*]
[*]CEWiT India
IITM Research Park, Chennai.
anu@cewit.org.in

Abstract—Densely deployed small cell base stations are a key factor to enable 5G networks. Multicore System-on-a-Chip (MCSoC) solutions supporting small cell LTE base station functionality are commercially available. An MCSoC has to be integrated with necessary software components leveraging specific accelerators for creating a complete LTE base station. System architects have a tough challenge in determining whether a small cell MCSoC would meet their intended small cell base station's feature, functionality and performance demands. This paper discusses the application of modular performance analysis (MPA) using Real-Time Calculus (RTC) on an MCSoC-based LTE base station implementation. Our study validates its applicability as well as its ability to identify new designs that will improve the system performance. This study is first in its kind to propose an analytical framework for analyzing an LTE base station's system design and performance.

Keywords—Small Cell, System-on-a-Chip, Modular Performance Analysis, Real-Time Calculus, LTE eNodeB.

I. Introduction

5G (Fifth Generation) networks, the next generation to the 4G (Long Term Evolution (LTE)) cellular networks are aimed for deployment by 2020. 5G networks aim to provide 1000 times more capacity and connection speed to the users in comparison to 4G. Challenges exist to enable the 5G networks with 1000 times capacity by 2020, and densely deployed small cell base stations are perceived as a key solution to achieve this goal [1]. Small Cell LTE Base Station equipment vendors can leverage the various Multicore System-on-a-Chip (MCSoC) designs available in the market. The MCSoC designs vary in features and capabilities and by correct SoC selection, a Micro, Pico or a Femto cell Base Station (BS) can be designed. The intended BS architecture must be realizable using the SoC features, which requires an evaluation. One option is to perform a complete evaluation by porting an existing software solution on to a SoC enabled hardware evaluation system and verifying the behavior. The pros of such an evaluation is a complete system validation, while the cons are demands on time and the knowledge of the system. Current days where time is of essence to bring out newer solutions, analytical evaluation models that can capture the system behaviour are helpful. The analytical models must be able to verify whether the system can (i) meet the time limits in terms of delay performance, and (ii) support necessary buffers for efficient operation. In this paper, an analytical framework to evaluate an LTE Base Station (BS), using Real-Time Calculus (RTC) [2], that meets the analytical modeling requirements is proposed.

The ability to support the computationally intensive LTE physical layer processing using general purpose hardware platforms has been explored with successful prototypes [3], [4], [5]. Kai et al. in [3] prove that with Single Instruction Multiple Data (SIMD) architecture and multicore processing, an LTE BS can be developed on a commodity hardware platform. Feng Tao et al.'s prototype uses an Intel i7 processor based platform to meet the high demands of LTE processing [4]. Zheng et al. show that a GPU is an efficient choice for handling highly parallel LTE BS workloads [5]. These are systems specifically designed to meet the real-time high throughput signal processing demands of an LTE system. Shan et al. present a two-stage approach to implement an SoC supporting the baseband processing for an LTE small cell base station in [6]. Their suggested approach could be applicable in case of new LTE supported SoC implementations.

LTE system design studies have been very specific, focused toward LTE functionality implementation on certain platforms, and can help solution developers. System architects may need to evaluate different platforms or may have platform migration requirements when developing their next generation systems. The LTE BS design must meet the requirements specified in the 3GPP Technical Report 25.913 [7]. 3GPP TR 25.913 section 6.2.1.1 specifies that, at least 200 users per cell should be supported in the active state for spectrum allocations up to 5 MHz, and at least 400 users for higher spectrum allocation. Understanding an LTE BS design's peak capacity is critical. An analytical framework to evaluate alternate system designs and to determine tighter performance bounds is required. This study intends to bridge the gap by proposing an analytical framework for LTE BS design validation.

Real-Time Calculus (RTC) [2] based on Network Calculus [8] helps in effective evaluation of real-time embedded system design and performance [9]. This study proposes the application of modular performance analysis (MPA) of a Small Cell LTE BS using RTC. The LTE test bed developed at CEWiT using Texas Instruments SoC TMS320TCI6614, is used for the study [10]. With RTC based analysis the limitation in the existing design was identified and improvements explored with new architecture designs. This study is first of its kind to the best of the authors' knowledge, in proposing a rigorous analytical framework for an LTE BS's performance analysis. This study will help both MCSoC developers who wish to offer SoC solutions for LTE BSs and equipment vendors who develop complete LTE BS solutions using MCSoCs.

The remainder of the paper is organized as follows: Section II outlines the necessary background on MPA using RTC. Section III presents the current CEWiT's LTE test bed design.

978-1-4673-9141-2/15 $31.00 © 2015 IEEE

Section IV details the derivation of the performance parameters in terms of input event curves and resource demands, followed by RTC based analysis for the existing design. Section V proposes newer architecture designs and analyses the performance improvement. Section VI concludes and highlights future work.

II. MPA USING RTC

A system architect designs a system such that, the intended functionality in a given architecture will meet all the necessary requirements. The architect's primary question is, whether the requirements are fully satisfied. As proposed by Wandeler et al., a system can be characterized as a set of incoming and outgoing event arrival sequences, data sizes, and execution demands [9]. The computing and communication resources' capacities can also be specified. RTC can then be used to compute the system's hard upper bound and lower bound performance values [2]. Since the model is abstracted at a very high level, the performance calculation is done very efficiently. RTC is based on Network Calculus [8]. Network Calculus can be viewed as deterministic queuing theory, and helps to determine the timing properties of the data flows in queuing networks. RTC extends the concepts of Network Calculus to the real-time embedded system domain. RTC uses a generic event and resource model, allows hierarchical scheduling and also takes computing and communication resources for calculations. A detailed explanation on MPA using RTC is given in [9], and the necessary aspects for completeness are given below.

Event Arrival Curves and Resource-based Arrival Curves: Let *R(t)* denote the number of events that arrive on an event stream in the time interval [0,t). Let $\overline{\alpha}^u$ and $\overline{\alpha}^l$ denote the upper most (maximum) and lower most (minimum) arrival events. R, $\overline{\alpha}^u$ and $\overline{\alpha}^l$ are related by,

$$\overline{\alpha}^l(t-s) \leq R(t) - R(s) \leq \overline{\alpha}^u(t-s), \forall s < t \quad (1)$$

with $\overline{\alpha}^l(0) = \overline{\alpha}^u(0) = 0$. For performance analysis, the focus is on the demands made by the events on the processing elements. The resource-based arrival curves for *any* time interval $\triangle$, are denoted as $\alpha(\triangle) = [\alpha^u(\triangle), \alpha^l(\triangle)]$. Arrival curves represent arrival patterns that are sporadic, periodic or periodic with jitter. Arrival curves corresponding to any finite length stream traces, are determined using a sliding window approach.

Resource Curves: Let *C(t)* denote the number of processing or communication cycles available from a resource over the time interval [0,t). Let β^u and β^l denote the upper most (maximum) and lower most (minimum) resource availability. C, β^u and β^l are related by,

$$\beta^l(t-s) \leq C(t) - C(s) \leq \beta^u(t-s), \forall s < t \quad (2)$$

with $\beta^l(0) = \beta^u(0) = 0$. The abstract resource curves for *any* time interval $\triangle$, are denoted as $\beta(\triangle) = [\beta^u(\triangle), \beta^l(\triangle)]$. The resource curves are determined using data sheets or from analytically derived properties or by measurements.

Processor and Communication Models: RTC helps in modeling processors and interprocessor communication mechanisms. RTC enables the analysis of tasks assigned to the processor in one of three different modes: tasks processed with strict priority are treated as Greedy Processing Component

(GPC), tasks processed as per Earliest Deadline First (EDF) and tasks processed in the First In First Out (FIFO) order. Interprocessor communication or bus based communication can be modeled using the Time Division Multiple Access (TDMA) method or by means of strict priority.

Abstract Component Model and MPA with RTC: In a real-time system the events are processed by a sequence of hardware/software components. In MPA, a hardware/software component is mapped as an abstract component. The inputs provided for the RTC are $\alpha(\triangle)$ - Abstract input event stream and $\beta(\triangle)$ - Abstract input resource. The outputs generated upon applying the RTC are, (i) Abstract output event stream: $\alpha'(\triangle) = [\alpha'^u(\triangle), \alpha'^l(\triangle)]$ and (ii) Abstract remaining resource: $\beta'(\triangle) = [\beta'^u(\triangle), \beta'^l(\triangle)]$. The outputs for fixed priority processing model are obtained by using the following relations (3 - 6) and operators:

$\otimes$ - min-plus convolution, $\oslash$ - min-plus deconvolution
$\overline{\otimes}$ - max-plus convolution, $\overline{\oslash}$ - max-plus deconvolution

$$\alpha'^u = min\{(\alpha^u \otimes \beta^u) \oslash \beta^l, \beta^u\} \quad (3)$$
$$\alpha'^l = min\{(\alpha^l \oslash \beta^u) \otimes \beta^l, \beta^l\} \quad (4)$$
$$\beta'^u = (\beta^u - \alpha^l) \overline{\oslash} 0 \quad (5)$$
$$\beta'^l = (\beta^l - \alpha^u) \overline{\otimes} 0 \quad (6)$$

Real-Time Calculus Toolbox: The RTC Toolbox is provided by the Computer Engineering and Networks Laboratory (TIK), ETH Zurich, Switzerland [11]. It is a Matlab toolbox for system-level performance analysis of distributed real-time and embedded systems. The RTC Toolbox provides a library of functions for MPA using RTC. The RTC Toolbox is used in the current study.

RTC benefits: RTC is well suited for performing modular level performance analysis of a complex embedded system such as an LTE base station. In such a system, input received at one interface/level is processed and handed over to the next interface/level. The end-to-end processing time (delay) for such a sequence must be within system defined bounds. In addition to analyzing the performance behavior based on buffer and load variations, system bottlenecks can be identified. The current study analyzes the LTE physical downlink shared channel (PDSCH) physical layer transmit (PHY Tx) chain processing function for varying load conditions.

III. CEWiT'S TEST BED

MCSoC solutions enabling faster LTE BS design and deployment are available now [12], [13]. SoC architectures support many accelerators to enable efficient design of LTE BSs. While SoC provides efficient processing capabilities for the Physical layer, integration with suitable Layer2 (L2), Layer3 (L3) and Management components is required for a complete LTE BS implementation. With a suitable processing sequence using the accelerators, the system architecture and design is critical for a successful LTE BS implementation.

CEWiT's LTE test bed (CEWIT-TB) is designed for implementing and performing studies on LTE/LTE-A functions. CEWIT-TB uses TI SoC TMS320TCI6614 (SoC6614) based platforms [14]. Currently, the CEWIT-TB operates in TDD

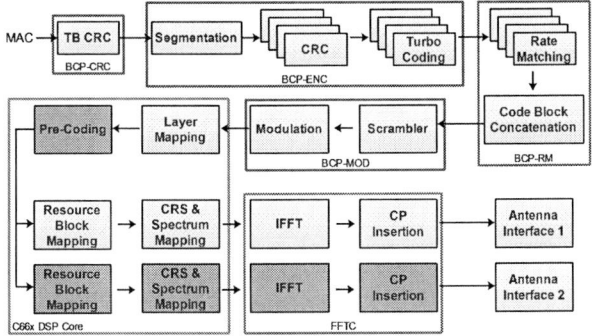

Figure 1. Functional placement of PDSCH PHY Tx Chain in SoC6614

Table I. PHY Tx CHAIN MIMO PROCESSING DEMANDS

Modulation	64QAM		16QAM		QPSK	
Tasks	Proc. Cycles	Time (μs)	Proc. Cycles	Time (μs)	Proc. Cycles	Time (μs)
Pre BCP	9052	7.54	8998	7.5	8558	7.13
BCP	19636	49.09	18861	47.15	17279	43.2
Post BCP	8912	7.43	8883	7.4	8641	7.2
CRS	111742	93.13	111856	93.2	110903	92.42
LM	24181	20.15	24084	20.07	24117	20.1
PC	66663	55.55	66672	55.56	67100	55.92
RBM	22035	18.36	21834	18.19	21717	18.1
SM	89101	74.25	89437	74.53	93575	77.98
Pre IFFT	19295	16.08	19020	15.85	19015	15.85
IFFT	13825	34.56	13766	34.42	13715	34.29

mode and uses 10 MHz bandwidth. It uses a TDD switching time of 10ms and supports the uplink(UL)-downlink(DL) configuration type 3, thereby supporting 6 DL and 3 UL subframes [15]. The CEWIT-TB supports proportionally fair scheduler for DL, and complete L1 and L2 functionality. Support for L3 and other features will be added in the future. Current implementation uses one Digital Signal Processing (DSP) core for DL Tx processing and one DSP core for UL Rx processing. A simplified DL and UL scheduling along with L2 functions are performed in one DSP and one DSP core is reserved for future extensions.

Figure 1 shows the functional placement of the CEWiT's Physical Downlink Shared Channel (PDSCH) Transmit (Tx) chain in the SoC6614. CRC calculation, Code block segmentation, Turbo coding, Rate matching, Code block concatenation, Scrambling and Modulation are performed using Bit Rate Coprocessor (BCP). These functions are executed as a sequence for every Transport Block (TB) handed over by the MAC layer. Layer Mapping (LM), Precoding (PC), Resource Block Mapping (RBM), Cell specific Reference Signals (CRS) generation and Spectrum Mapping are parts of CEWiT's proprietary implementation. They are currently enabled in one of the TMS320C66x DSP Core Subsystems. The functions LM, PC and RBM are performed for every TB. CRS, SM and Inverse Fast Fourier Transform (IFFT) are done once for the entire LTE subframe. IFFT is performed in one of the Fast Fourier Transform Co-processors (FFTC). The data exchange between these processing elements is supported by the SoC's high speed switching fabric - TeraNet.

Figure 2 shows the task placements in the processors and the intertask communication mechanism. In the current design, both BCP and FFTC support one task each while DSP core supports six tasks. BCP task (T2) performs all the functions in a sequence as shown in Figure 1. The DSP tasks (T1) and

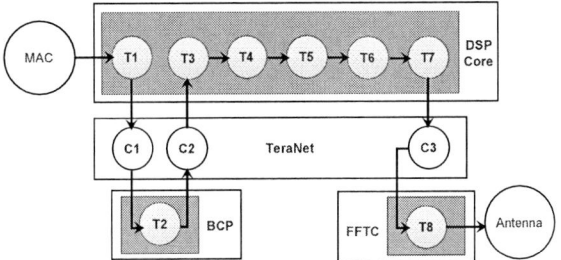

Figure 2. PDSCH PHY Tx chain tasks modeled for MPA

Table II. PHY Tx CHAIN SISO PROCESSING DEMANDS

Modulation	64QAM		16QAM		QPSK	
Tasks	Proc. Cycles	Time (μs)	Proc. Cycles	Time (μs)	Proc. Cycles	Time (μs)
Pre BCP	11895	9.91	10207	8.51	9797	8.16
BCP	15984	39.96	14907	37.27	13132	32.83
Post BCP	5250	4.37	5205	4.34	5286	4.41
LM	8147	6.79	8149	6.79	8141	6.78
RBM	13267	11.06	13277	11.06	13569	11.31
CRS	98494	82.08	98310	81.92	98300	81.92
SM	51950	43.29	51880	43.23	51886	43.24
Pre IFFT	19280	16.07	19143	15.95	19329	16.11
IFFT	14022	35.05	14109	35.27	13503	33.76

(T3) do the required preprocessing and post-processing, before and after the BCP operation. The DSP task (T4) performs the functions of LM, PC, and RBM. The DSP tasks (T5) and (T6) perform CRS generation and Spectrum mapping, respectively. The DSP task (T7) does the preprocessing before the IFFT operation. The tasks T3 to T7 are executed in sequence, with the completion of one task triggering the next task processing. The output from DSP core is acted upon by the FFTC task (T8) which does IFFT computation and Cyclic Prefix insertion. Communication tasks C1 - C3 model the data transfers between the processors.

The number of clock cycles expended for the different modules using Multiple Input Multiple Output (MIMO) and Single Input Single Output (SISO) transmission modes are given in Table I and Table II, respectively. This includes the processing effort for the three modulations - 64QAM, 16QAM, and QPSK. The demands for the three modulations are more or less equal for both MIMO and SISO. Also, the execution demands for the SISO are less compared to MIMO and the processing effort 'Precoding' does not exist in case of SISO. The processing time in microseconds is arrived at by multiplying the clock cycles with the corresponding processor speeds. The time to transfer the data across the TeraNet switch fabric is negligible, hence it is not explicitly considered in the execution demands.

IV. CURRENT PDSCH PHY Tx CHAIN DESIGN ANALYSIS

The MAC layer periodically provides TBs to the PDSCH PHY Tx chain at every Transmit Time Interval (TTI) of one millisecond. The number of TBs provided depends on the number of UEs scheduled for transmission every one millisecond. Since the scheduling is done every millisecond, the TBs from the MAC layer must be processed by the PHY Tx chain into suitable format, and transfered to the radio interface within one millisecond. The PHY Tx chain's end-to-end processing delay, which is the sum of processing delay incurred at the eight tasks (Fig. 2) must be within one millisecond, as given in

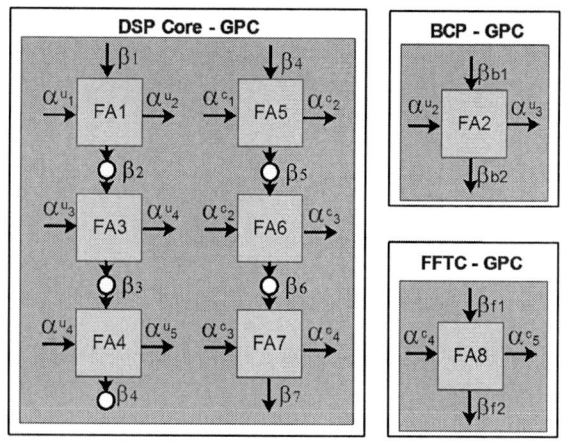

Figure 3. PDSCH PHY Tx Chain abstract performance analysis model

Table III.   PDSCH Phy Tx Chain: Input Curves in MPA RTC Model

α_1^u : Event Arrival Curves (EAC) reflecting number of users scheduled in 1ms.

α_2^u : EAC after BCP preprocessing task (T1) completion.

α_3^u : EAC after BCP processing (T2) completion at BCP.

α_4^u : EAC after BCP post-processing task (T3)completion.

α_5^u : EAC after Layer mapping, Precoding, and Resource Block Mapping functions (T4) in DSP core.

α_1^c : EAC reflecting a LTE subframe scheduled every 1ms.

α_2^c : EAC after CRS generation (T5) completion.

α_3^c : EAC after symbol mapping (T6) completion.

α_4^c : EAC after IFFT preprocessing task (T7) completion.

α_5^c : EAC after IFFT processing (T8) completion at FFTC.

β_1 : Resource curves (RC) reflecting initial resources available at DSP core.

β_2 : RC reflecting residual resources after T1 processing.

β_3 : RC reflecting residual resources after T3 processing.

β_4 : RC reflecting residual resources after T4 processing.

β_5 : RC reflecting residual resources after T5 processing.

β_6 : RC reflecting residual resources after T6 processing.

β_7 : RC reflecting residual resources after T7 processing.

β_{b1} : Resource curves reflecting initial resources available at BCP.

β_{b2} : RC reflecting residual resources after T2 processing.

β_{f1} : Resource curves reflecting initial resources available at FFTC.

β_{f2} : RC reflecting residual resources after T8 processing.

Equations 7 and 8. Figure 3 shows the abstract model applied for evaluating the PDSCH PHY Tx chain implementation. The functionally abstracted elements FA1 to FA8 correspond to the tasks T1 to T8 shown in Figure 2. The input and output, event and resource curves for each of the FA elements shown in Figure 3 is described in Table III. Matching arrow labels indicate sequential task execution. Given the nature of the tasks being executed in sequence, the FA elements are modeled as Greedy Processing Components (GPC). The GPC model is used in system designs involved with preemptive tasks running under fixed priority scheduling policy.

$$Del_{TxChain} = \sum_{i=1}^{8}(Del_{FAi}) \qquad (7)$$

$$Del_{TxChain} <= 1ms \qquad (8)$$

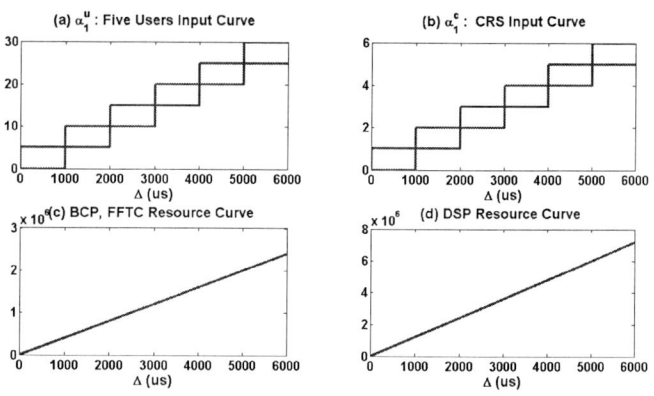

Figure 4.   PDSCH PHY Tx chain MPA: Event and Resource Input Curves

Table IV.   RTC vs Implementation Measured Values (μS)

Number of users	64QAM		16QAM		QPSK	
	RTC	Impl.	RTC	Impl.	RTC	Impl.
1	376	381	374	378	372	373
2	527	539	522	530	517	523
3	677	697	671	690	661	675
4	846	855	838	838	824	825
5	997	1009	986	997	969	979

The end-to-end delay (E2E) analysis is performed for different scenarios, i.e. varying the number of active users supported by the system and the number of users scheduled in every TTI. For both MIMO and SISO transmissions, the minimum number of active users in the system is considered as 6 with one user scheduled every LTE TTI. The maximum number of active users considered is 96 and 150 (for the available 6 DL subframes), with 16 users and 25 users scheduled every LTE TTI for MIMO and SISO, respectively. Variation in the number of users reflect in the number of periodic arrival of events for tasks T1 to T4. The active user count variation is done in steps of 6, so that the number of users scheduled in every TTI (in each of the 6 DL sub frames) is incremented by 1. Figure 4 (a) shows the event input curve α_1^u (for 30 active users in the system) applicable for task T1. The CRS, SM and IFFT are performed for the entire TTI subframe and is done once every one millisecond. The input curve for the tasks T5 to T8 that compute CRS, SM and IFFT would have a unit event arriving every millisecond. Figure 4 (b) shows the event input curve α_1^c (for common function) applicable for task T5. One TMS320C66x Corepac DSP core, BCP and one FFTC comprise the three different processing units involved in the PHY transmission. The DSP core operates at 1.2GHz, while the co-processors operate at 0.4GHz. Operating at 1.2GHz, the DSP core generates resource at the rate of 1200 cycles every microsecond. The co-processors operating at 0.4GHz generate resources at the rate of 400 cycles every microsecond. Figure 4 (c) shows the resource curve for BCP and FFTC and Figure 4 (d) shows the resource curve for DSP core used in the analysis.

As a first step in the analysis, we observed the delays calculated using Equation 7 satisfying the timing constraint Equation 8. Table IV shows the RTC based calculated delay values and the values obtained from the implementation, when MIMO based transmission is done for 1 to 5 users. This validates the functionally abstracted task model and RTC-based analysis as a suitable approach for the LTE BS performance

Table V.   EXISTING ARCH: MAX USERS PER TTI & END-TO-END DLY

Delay	Mode	MIMO			SISO		
(ms)	Modu-lation	64 QAM	16 QAM	QPSK	64 QAM	16 QAM	QPSK
<1	MU	5	5	5	12	13	14
	T (us)	997	986	969	946	977	975
<1.5	MU	6	6	6	20	21	22
	T (us)	1148	1134	1113	1449	1453	1418

Table VI.   SPLIT ARCH: MAX USERS PER TTI & END-TO-END DLY

Delay	Mode	MIMO			SISO		
(ms)	Modu-lation	64 QAM	16 QAM	QPSK	64 QAM	16 QAM	QPSK
<1	MU	7	7	8	15	17	18
	T (us)	867	852	978	931	987	976
<1.5	MU	13	13	13	25	25	25
	T (us)	1465	1438	1390	1442	1374	1265

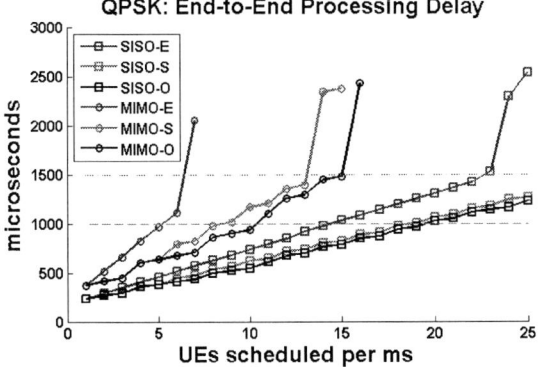

Figure 5.   PHY Tx processing delay applying QPSK for UE load variation.

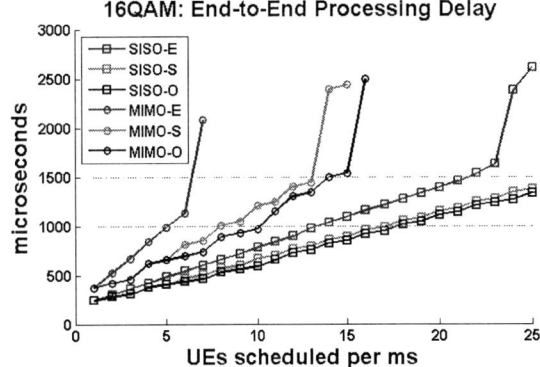

Figure 6.   PHY Tx processing delay applying 16QAM for UE load variation.

evaluation. The benefit of RTC-based analysis is its ability to understand the system behavior for varying inputs. In the current analysis, the end-to-end processing delay within the PHY transmit chain is done by varying the number of active users scheduled in a given TTI. The number of users scheduled per subframe is varied from one to a maximum number such that the end-to-end system delay incurred is finite. The maximum number is 16 and 25 for MIMO and SISO scenarios, respectively. The user data is processed for the three different modulations (64QAM, 16QAM, and QPSK) in both MIMO and SISO scenarios. The calculated end-to-end delay values using RTC for the existing architectural design implementation identified with the legends "MIMO-E" and "SISO-E", are shown in Figures 5, 6 and 7 for the three different modulations. Table V shows the existing design's system capacity in terms of the maximum number of users who can be scheduled for every LTE TTI, and the end-to-end processing delay is less than one millisecond. The current system can support 30 active users in MIMO (any modulation), while it can support 72, 78, and 84 users in SISO using 64QAM, 16QAM and QPSK modulations, respectively. Since the current system uses TDD mode and cycle switching every 10ms, with suitable buffer and data structure design, one can prepare the user data such that the processed data is available at every TTI for the 6 DL frames. In such cases, the processing delay for an LTE subframe can be tolerated until 1.5 milliseconds (satisfying Equation 9). The system with a suitable buffer architecture can support 36 active users in MIMO (any modulation), while it can support 120, 126, and 132 active users in SISO using 64QAM, 16QAM and QPSK modulations, respectively. This assumes a user is scheduled only once in one of the LTE subframes. When the number of users scheduled per subframe increases beyond 7 in case of MIMO and 25 in case of SISO, the combined processing delay is infinite, i.e. the data for transmission cannot be prepared within the required TTI.

$$Del_{TxChain} <= 1.5ms \qquad (9)$$

V.   MODIFIED ARCHITECTURE AND ANALYSIS

Split Architecture: From the RTC-based calculated values, we observe, the delay increases linearly and quickly. Also, the system design does not scale to meet an increase in the number of users. The use of single DSP core is observed to be the bottleneck in the current design as processing is required for every user (i.e. every TB), and total requirement increases as the number of users increases. In the current design, one of the DSP cores is kept reserved for future extensions. We propose a design modification known as "Split Architecture", by which the DSP core processing for the users can be done in two DSP cores (Say DSP-C1 and DSP-C2) instead of a single DSP core. The total number of users *UeTotal* to be scheduled in a subframe will be split into two equal groups *UeGroup1* and *UeGroup2*. In this Split Architecture, DSP-C1 will support tasks T1, T3 and T4 and DSP-C2 will support tasks T1, T3 to T7. The CRS, SM, and preprocessing for IFFT will be done in DSP-C2. The calculated end-to-end delay values using RTC for the "Split Architecture" design, identified with the legends "MIMO-S" and "SISO-S" are shown in Figures 5, 6 and 7. Table VI shows the results observed for the system designed for the "Split Architecture" and this system can support 42, 42, and 48 active users in MIMO using 64QAM, 16QAM and QPSK modulations, respectively, while it can support 90, 102, and 108 users in SISO using 64QAM, 16QAM and QPSK modulations, respectively. We observe 40% to 60%, and 25% to 28% increase in the system capacity for MIMO and SISO, respectively, between the existing design and the proposed "Split Architecture" design. The "Split Architecture" system with a suitable buffer architecture can support 78 active users in MIMO (any modulation) and 150 active users in SISO (any modulation).

Optimized Architecture: The "Split Architecture" design shows that it is possible to increase the active user support when the processing is spread across two DSP cores. However, when comparing the processing loads across the two DSP cores, it is observed that DSP-C1 is less loaded compared to

Table VII.   Opt. Arch: Max users per TTI & End-to-End dly

Delay	Mode	MIMO			SISO		
(ms)	Modu-lation	64 QAM	16 QAM	QPSK	64 QAM	16 QAM	QPSK
<1	MU	10	10	10	17	17	19
	T (us)	990	970	935	991	948	962
<1.5	MU	13	14	15	25	25	25
	T (us)	1367	1496	1113	1400	1335	1226

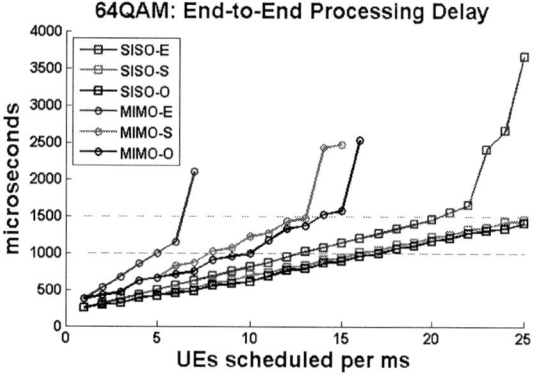

Figure 7.   PHY Tx processing delay applying 64QAM for UE load variation.

DSP-C2, since the common CRS, SM, and preprocessing for IFFT requires a significant effort. We propose an "Optimized Architecture" design, where the DSP cores are assigned groups of users that will optimize the DSP core utilization. Algorithm 1 shows the logic applied in the optimized UE group formation. This logic is adopted since the number of users to be scheduled in a subframe is less than or equal to 25 and the user grouping is to effect equal processing loads in the DSP cores. The calculated end-to-end delay values using RTC for the "Optimized Architecture" design, identified with the legends "MIMO-O" and "SISO-O" are shown in Figures 5, 6 and 7. Table VII shows the results observed for the system designed for the "Optimized Architecture" and this system can support 60 active users in MIMO (using any modulation), while it can support 102, 102, and 114 users in SISO using 64QAM, 16QAM, and QPSK modulations, respectively. We observe 100% and 35% to 41% increase in the system capacity for MIMO and SISO, respectively, between the existing design and the proposed "Optimized Architecture" design. The "Optimized Architecture" system with a suitable buffer architecture can support 78, 84, and 90 active users in MIMO using 64QAM, 16QAM, and QPSK modulations, respectively and 150 active users in SISO (any modulation).

VI.   Conclusion

Modular performance analysis using Real-Time Calculus is identified as a valid analytical method to determine whether a system can meet the necessary performance demands. With suitable abstract task models, input event models, processor resource models and execution demands, it is shown that MPA using RTC can be easily applied on a complex real-time embedded system such as a Small Cell LTE base station designed using MCSoCs. While not restricting to verifying the performance requirements with strict bounds, the ability to identify and validate alternative architectures with minimal effort and less time is observed as a major benefit. MPA using RTC for an LTE PDSCH PHY Tx chain implementation is dealt in this paper and in future this will be applied to other functions supported in an LTE base station.

Algorithm 1 User processing: Optimized load balance

1: **procedure** OptimalLoadBalance($UeTotal$)
2:     **if** $UeTotal >= 16$ **then**
3:         $UeGroup1 = ROUND((UeTotal * 9)/16)$
4:     **else**
5:         **if** $UeTotal >= 12$ **then**
6:             $UeGroup1 = ROUND((UeTotal * 3)/5)$
7:         **else**
8:             $UeGroup1 = ROUND((UeTotal * 2)/3)$
9:         **end if**
10:    **end if**
11:    $UeGroup2 = UeTotal - UeGroup1$
12:    Return $UeGroup1, UeGroup2$
13: **end procedure**

Acknowledgment

This research work was supported by the Department of Science and Technology (DST), New Delhi, India.

References

[1] N. Bhushan, J. Li, D. Malladi, R. Gilmore, D. Brenner, A. Damnjanovic, R. Sukhavasi, C. Patel, and S. Geirhofer, "Network densification: The dominant theme for wireless evolution into 5G." *IEEE Communications Magazine*, vol. 52, no. 2, pp. 82–89, 2014.

[2] L. Thiele, S. Chakraborty, and M. Naedele, "Real-Time Calculus for scheduling hard real-time systems," in *Proceedings of the IEEE International Symposium on Circuits and Systems*, vol. 4, 2000, pp. 101–104.

[3] N. Kai, S. Jianxing, H. Zhiqiang, and K. K. Chai, "LTE eNodeB prototype based on GPP platform," in *Proceedings of the IEEE Globecom Workshops*, 2012, pp. 279–284.

[4] X. Tao, Y. Hou, K. Wang, H. He, and Y. J. Guo, "GPP-Based Soft Base Station Designing and Optimization," *Journal of Computer Science and Technology*, vol. 28, no. 3, pp. 420–428, 2013.

[5] Q. Zheng, Y. Chen, R. Dreslinski, C. Chakrabarti, A. Anastasopoulos, S. Mahlke, and T. Mudge, "Architecting an LTE base station with graphics processing units," in *Proceedings of the IEEE Workshop on Signal Processing Systems (SiPS)*, 2013, pp. 219–224.

[6] T. Shan, Z. Ziyuan, and S. Yongtao, "System-level Design Methodology Enabling Fast Development of Baseband MP-SoC for 4G Small Cell Base Station," in *Proceedings of the Conference on Design, Automation & Test in Europe, DATE '14*, 2014, pp. 198:1–198:6.

[7] 3GPP Technical Report, "25.913 Requirements for Evolved UTRA (E-UTRA) and Evolved UTRAN (E-UTRAN)."

[8] J.-Y. Le Boudec and P. Thiran, *Network Calculus: A Theory of Deterministic Queuing Systems for the Internet*. Berlin, Heidelberg: Springer-Verlag, 2001.

[9] E. Wandeler, L. Thiele, M. Verhoef, and P. Lieverse, "System architecture evaluation using modular performance analysis: A case study," *International Journal on Software Tools for Technology Transfer*, vol. 8, no. 6, pp. 649–667, 2006.

[10] Center of Excellence in Wireless Technology, CEWiT India. [Online]. Available: http://www.cewit.org.in

[11] E. Wandeler and L. Thiele, "Real-Time Calculus (RTC) Toolbox," http://www.mpa.ethz.ch/Rtctoolbox.

[12] Enabling multistandard wireless base stations with TI's KeyStone SoCs. [Online]. Available: http://www.ti.com/lit/wp/spry160/spry160.pdf

[13] Jump Start Your Small Cell Equipment Design. Freescale's solutions from the air to the core. [Online]. Available: http://cache.freescale.com/files/32bit/doc/brochure/BRSMALLCELLS.pdf

[14] TMS320TCI6614 Communications Infrastructure KeyStone SoC, Data Manual. [Online]. Available: http://www.ti.com/lit/ds/symlink/tms320tci6614.pdf

[15] 3GPP Technical Specifications, "36.211 Evolved Universal Terrestrial Radio Access (E-UTRA); Physical channels and modulation."

Embedded Low Power Analog CMOS Fuzzy Logic Controller Chip for Industrial Applications

Manikandan Pandiyan, *IEEE Student Member*, Geetha Mani,

Abstract— The paper presents an analog low power CMOS fuzzy logic controller (FLC) chip implementation and its application to an industrial chemical reactor. This approach employs a simple architecture that contains three stages namely Fuzzification, Inference Engine and Defuzzification. The Fuzzifier circuit generates high precision membership functions (MFs) that are easily tunable by changing the voltages in IC pins. The FLC chip has three input MFs and five output MFs. The two input minimum circuits were used to design the Min-Product inferencing for inference engine. Defuzzification circuit is well designed to reduce the size of the chip. The proposed chip has power consumption of about 10mW. The whole chip area is less than $0.5mm^2$. The resulted chip shows simple structure, good performance in processing speed, also enough flexible to compete digital fuzzy approaches, area consumption and accomplished with less reconfigurable rules, a speed of up to about 8 MFLIPS (Fuzzy Logic Inference Per Second) has been achieved. The results show a good functionality of controller in response to confirm the success of the design. The application of the system to an industrial chemical reactor in a feedback loop is considered.

Keywords— FLC, Inference engine, MF, Chemical Reactor

I. INTRODUCTION

For a wide array of applications, designers of industrial process control systems have worked together develop, and deploy a complete and optimized fuzzy logic controller [1]. There are two general approaches for using fuzzy logic in control systems that is called Fuzzy Control Software, and Hardware. Software-based fuzzy controllers are limited to a slow-speed operation. Therefore they are not competent and efficient for real-time applications in industries, whereas fuzzy hardware control is quite faster than the software approach. Hardware based Fuzzy Control can be made as discrete/integrated circuits. Fuzzy chips are the most ideal form of hardware solutions for fuzzy controllers. In this way, depending on the design techniques employed in three realizations such as digital [8], analog [1,3] and mixed-signal realization (analog/digital combination) [2].

From 1985, there have been many successful developments of analog implementations of fuzzy logic-based systems. Analog circuits are preferable for fuzzy chips since they excel in speed, chip size and power consumption [3]. The main drawback of analog implementations is their relatively low accuracy that doesn't severe limitation in view of the typical demands of many fuzzy control applications. The third realization of fuzzy chips is mixed-signal realization [2], where both analog and digital circuits are used together.

The design is made to simulate an analog realization of three main blocks used in fuzzy controllers. Fuzzifier, in which non-fuzzy or crisp input variables are converted to fuzzy variables, compromises a membership function generator (MFG) and a complementary circuit.

The synthesis of these two circuits that fuzzify input variables is a novel structure [3]. This structure has a suitable range and is calibrated depending on each control application by setting some control voltages.

Generally in setting input membership functions a desired attribute of a general-purpose fuzzy chip that is useable in different industrial control systems. We use Min-Product method in inference engine. Therefore, a min circuit is needed for which we use a simple structure used in [5]. For the defuzzification process, we propose a novel circuit with a small area and good precision.

The synthesis of designed circuits as 2-input, 1-output, 9-rules controller is designed. The operation speed of inferencing in this controller is about 8 MFLIPS. Also it is designed in a single-poly, double-metal CMOS process and occupies an active area of $0.5mm^2$. Finally, realization of FLC is tested in a feedback control loop to control a chemical reactor system with undesired transient properties, such as, high overshoot, large settling time, and its success is demonstrated.

This paper is organized as follows. In Section II, the prerequisites are presented, followed by the description of each sub-block of Fuzzy Logic Controller chip in Section III. Section IV shows FLC chip implementation details and Section V & VI discusses realization of FLC, experimental setup respectively. Conclusion is drawn in Section VII.

II. PREREQUISITES

The kernel of any fuzzy logic controller is fuzzy inference engine. Its dynamic behaviour is generally characterized by a set of fuzzy rules of the form:

If (a set of conditions are satisfied)

Then (a set of consequences can be inferred)

The *if*-clause, an antecedent, is a condition in the application domain; the *then*-clause, a consequent, is a control action given to the process under control. With a set of fuzzy rules the fuzzy inference engine is able to derive a control action for a given set of input values. In other words, a control action is determined by the observed inputs which represent the state of the process to be controlled with the control rules. The approach used in fuzzy control is based on the approximate reasoning method of the generalized modus ponens (GMP). For example, for a two-input single-output n-rule fuzzy system, the GMP states:

Premise: x is A and y is B

Implication R_1: *if x is A_1 and y is B_1 then z is C_1*

Also R_i: *if x is A_i and y is B_i then z is C_i*

Also R_n: *if x is A_n and y is B_n then z is C_n*

Conclusion: z is C

In which x, y and z are linguistic variables, represent two inputs (process states or sensor measurements) and one output (control action). A_i, B_i and C_i are fuzzy sets defined on the appropriate universe U, V and W, respectively, with $i = 1,2,...n$. The fuzzy conditions in the antecedents are combined by the connective "*and*", while

---

Manikandan P was with Dept. of Instrumentation and Control systems Engineering, PSG College of Technology, India (2014 B.E. Batch).

Geetha M is with Dept. of Instrumentation and Control systems Engineering, PSG College of Technology, India. (Email ID: vanajapandi@gmail.com)

the sentence connective "*also*" links the rules into a rule set, or equivalently a fuzzy rule base. It should be noted that A^i, B^i, C^i, as well as A', B' are *apriori* known but C' will be deduced. Basically a fuzzy inference process involves two concepts, the fuzzy implication and the compositional rule of inference. For the fuzzy rule "*if x is A_i and y is B_i then z is C_i*", the fuzzy relation can be expressed as:

$$R^i = (A_i \times B_i) \rightarrow C_i$$

Where x represents the Cartesian product; $\rightarrow$ denotes an operator for fuzzy implication. For an *n*-rule system, the fuzzy relation R is therefore an *n*-ary matrix:

$$R = \{R^1, R^2, \dots\dots R^i, \dots\dots R^n\}$$

If we have the observations A' and B', then the fuzzy conclusion C' can be inferred by

$$C' = (A', B') \blacksquare R$$

Where "$\blacksquare$" represents the compositional operator. This fuzzy reasoning is called the compositional rule of inference.

It is proved [7] that if the sentence connective "*also*" in the rule base is interpreted as union operation

$$R = \bigcup_{i=1}^{n} R^i$$

Then the fuzzy control action inferred from the complete set of fuzzy control rules is equivalent to the aggregated results derived from individual control rule:

$$C' = (A', B') \blacksquare \bigcup_{i=1}^{n} R^i = \bigcup_{i=1}^{n}(A', B') \blacksquare R^i$$

Within the numerous inference strategies, Mamdani's technique is the most commonly used in the existing fuzzy control systems owing to its simplicity. In this method, the minimum operation is adopted for computing a fuzzy implication relation, and the *max-min* as the compositional rule of inference. If we interpret the sentence connective "*and*" in the antecedent of the fuzzy rule as the intersection operation, the Cartesian product is realized by the minimum operation:

$$\mu_{A_i \times B_i}(x, y) = \mu_{A_i}(x) \wedge \mu_{B_i}(y)$$

The membership function for the ith fuzzy relation can then be calculated as:

$$\mu_{R^i}(x, y, z) = \mu_{A_i}(x) \wedge \mu_{B_i}(y) \wedge \mu_{C_i}(z)$$

The conclusion deduced from the ith rule C_i' for the inputs A' and B' can be computed with the max-min compositional rule of inference:

$$\mu_{C_i'}(z) = \max_{x \in U} \max_{y \in V} \{\mu_{A'}(x) \wedge \mu_{B'}(y) \wedge \mu_{R^i}(x, y, z)\}$$
$$= \alpha_i \wedge \mu_{C_i}(z)$$

Where

$$\alpha_i = \max_{x \in U} \{\mu_{A'}(x) \wedge \mu_{A_i}(x)\} \wedge \max_{y \in V} \{\mu_{B'}(y) \wedge \mu_{B_i}(y)\}$$

α_i Is called the firing strength (or weight) of the ith rule for the inputs A' and B'. In practical applications, the inputs are usually singletons, e.g., the measurements from sensors, $A' = x_0$ and $B' = y_0$. The related membership functions $\mu_{A'}(x)$ and $\mu_{B'}(y)$ are equal to zero except at the point $x = x_0$, and $y = y_0$, where $\mu_{A'}(x_0) = 1$, and $\mu_{B'}(y_0) = 1$. In this case, the firing strength can be reduced to:

$$\alpha_i = \mu_{A_i}(x_0) \wedge \mu_{B_i}(y_0)$$

The *n* fuzzy rules in the rule base are aggregated with the union operation. The overall output is then computed by combining the individual result from each rule in the rule base:

$$\mu_{C_i'}(z) = \bigcup_{i=1}^{n} \mu_{C_i}(z)$$

For simplification in hardware implementations, a set of singletons is usually adopted in the consequent part of the fuzzy rule. Assume this set consists of *k* terms, i.e., $\mu_{C_i}(z) = \mu_C(z_j) = 1$ (i = 1, 2, ..., n, while j = 1, 2, ..., or k), then the inferred control action as a response to the actual input (x_0, y_0) can be written as

$$\mu_{C'}(z_j) = \bigcup_{i=1}^{n} \alpha_{ij} \ (j = 1, 2, \dots, k)$$

$\alpha_{ij} = \alpha_i \wedge \mu_{C_i}(z)|_{z=z_j} = \alpha_i$ Is the contribution of the ith rule to the jth term in the consequent part.

Experiments and theoretical investigations proved that Mamdani's technique yields better control results than that of other methods in fuzzy control applications. Moreover, the operators used in this approach are very easy to implement in hardware.

III. ANALOG FUZZY FUNCTIONAL BLOCKS

The main functional blocks of fuzzy classifier chip is described in detail. It consists of three parts: fuzzier, inference engine and defuzzier. In the fuzzier, input variables (non-fuzzy) are mapped to input membership functions. Inference engine is responsible for fuzzy inference depending on the type of inference method. In this work we used Min-Product inference method. Finally, defuzzier is used for converting fuzzy output resulting from inference engine to non-fuzzy values.

A. Fuzzy Interface

In fuzzy logic controller, Fuzzier interface has three stages. There are many solutions proposed to build fuzzier circuit using analog circuits. The structure we implemented is tunable by control voltages larger than those reported in the literatures. An efficient circuit is used for ramp function generation. In this interface, three basic circuits are used: a ramp generator (RG) circuit [6], minimum circuit, and fuzzy complementary circuit.

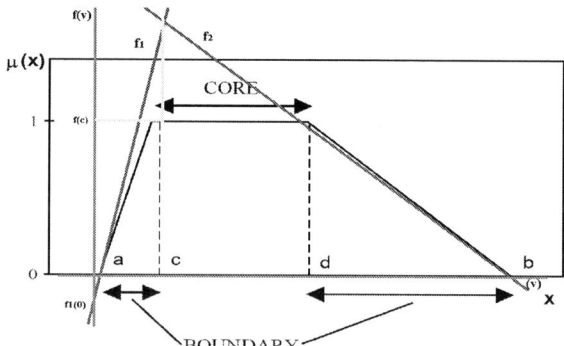

Fig.1. Representation of a trapezoidal membership functions with its parameters.

1) Ramp Generator

The membership functions are generated for input and output variables of a controller. In Fig. 1, a trapezoidal membership function with its parameters is shown. It shows that two ramp functions are necessary to build this membership function. The triangular membership function is a special case of trapezoidal membership function when c and d coincide.

In general, the slope of the ramp functions and position of the membership functions must be tunable. Changing the parameters a, b, c and d, allows construction of different membership functions. Here we assumed that all fuzzy sets are normalized. Thus, Supremum$_x$ (x) = 1. Figure 2 shows the RG circuit designed. In this circuit the output current i_0 is a function of v_1 and v_2. Considering I_{ss} fixed, and M3, M4, M5 and M6 are matched,

We have
$$i_0 = i_3 + i_5 - (i_4 + i_6)$$
$$= g_{m3}\frac{v_r}{2} - g_{m5}\frac{v_r}{2}\left(-g_{m4}\frac{v_r}{2} + g_{m6}\frac{v_r}{2}\right)$$
$$= \frac{v_r}{2}\left(\frac{I_1}{v_{gso3} - v_t} - \frac{I_2}{v_{gso5} - v_t} + \frac{I_1}{v_{gso4} - v_t} - \frac{I_2}{v_{gso5} - v_t}\right)$$
$$= \frac{v_r}{2}\left(\frac{2I_1}{v_{gso3} - v_t} - \frac{2I_2}{v_{gso5} - v_t}\right),$$

Where, all v_{gso}'s are gate to source voltages in quiescent point, and also

$$I_1 = \frac{I_{ss}\frac{v_c}{2}}{v_{gso1} - v_t}, \quad I_2 = \frac{-I_{ss}\frac{v_c}{2}}{v_{gso2} - v_t}$$

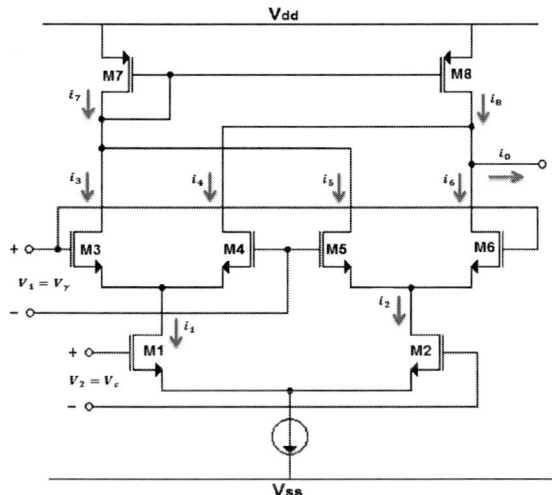

Fig.2.Ramp Generator Circuit

Using pervious equations, and assuming that M1 and M2 are also matched then

$$i_0 = \frac{I_{ss}}{v_{gso1} - v_t}\left(\frac{1}{v_{gso3} - v_t} + \frac{1}{v_{gso5} - v_t}\right)v_r v_c$$

Note that v_c will be set to a constant value hence $v_{gso1} - v_t$ is constant. Also, changes of $v_{gso3} - v_t$ and $v_{gso5} - v_t$ are in opposite directions, and so $\left(\frac{1}{v_{gso3} - v_t} + \frac{1}{v_{gso5} - v_t}\right)$ is approximately constant. Therefore, $i_0 = kv_1 v_2$
Where

$$k = \frac{I_{ss}}{v_{gso1} - v_t}\left(\frac{1}{v_{gso3} - v_t} + \frac{1}{v_{gso5} - v_t}\right)$$

The RG circuit can generate both a positive and negative slope ramp function. Suppose $v_{c-} = 0$, v_{c+} and v_{r-} are constant, then

$$i_0 = k(v_{r+} - v_{r-})(v_+) = kv_{c+}v_{r-} - kv_{c+}v_{r-}$$

$$= mv_{r+} - n$$
Where
$$m = kv_{c+}, \quad n = kv_{c+}v_{r-}$$
Thus a positive slope ramp is generated.
Similarly suppose $v_{c+} = 0$, and let v_{c-}, v_{r-} be constant, then
$$i_0 = k(v_{r+} - v_{r-})(-v_{c-}) = -kv_{c-}v_{r+} + kv_{c-}v_{r-}$$
$$= -mv_{r+} - n$$
Where
$$m = kv_{c-}, \quad n = kv_{c-}v_{r-}$$
Therefore, by changing v_{r+}, v_{r-} and v_c different ramp functions are generated. It is evident that output voltage of RG circuit $v_0 = R_{out}i_{out}$, where R_{out} is output resistance. Figure shows two ramp functions f1 and f2 with positive and negative slopes, respectively. Assuming RG circuit has generated these functions, using Equations, we can write
$$f_1 = m_1 v - n_1$$
$$f_2 = n_2 - m_2 v$$
In order to construct a tunable membership function, from Equations, and we can write
$$f_1(0) = -n_1 = -kv_{r1-}v_{c1+}$$
$$f_2(0) = n_1 = kv_{r2-}v_{c2-}$$
$$a = v|_{f_1=0} = \frac{n_1}{m_1} = \frac{kv_{r1-}v_{c1+}}{kv_{c1+}} = v_{r1-},$$
$$b = v|_{f_2=0} = \frac{n_2}{m_2} = \frac{kv_{r2-}v_{c2-}}{kv_{c2+}} = v_{r2-},$$
Since at value c, f1=f2, we can write, $m_1 c - n_1 = -n_2 + m_2 c$
Therefore
$$c = \frac{n_1 - n_2}{m_1 - m_2} = \frac{v_{r1-}v_{c1+} - v_{r2-}v_{c2-}}{v_{c1+} - v_{c2-}}$$
With this value for c we find
$$f(c) = f_1(c)$$
$$= f_2(c) = \frac{kv_{c1+}v_{c2-}}{v_{c1+} - v_{c2-}}(v_{r1-} - v_{r2-})$$
To construct a fuzzifier with desirable capabilities let $v_{r1-} = v_{r2-}$, a reverse triangular in the positive region of the vertical axis will be formed as shown in figure. By clipping this triangular with a constant value E, a controllable membership function will result. In this case, we have
$$a = v|_{f_2=E} = \frac{n_2 - E}{m_2} = v_{r2-} - \frac{E}{kv_{c2-}},$$
$$b = v|_{f_1=E} = \frac{E - n_1}{m_1} = \frac{E}{kv_{c1+}} - v_{r1-},$$
Thus changing v_{c1} and v_{c2} results in a change in a and b. For $v_{r1} = v_{r2}$, it will be variable also. The position of the membership function can be changed. Note that a trapezoidal membership function results when $v_{c+} > v_{c-}$ and in this case $c = v_{c-}$ and $d = v_{c+}$.

Considering these points it is necessary that c and d equal to v_{r1} and v_{r2}, should be selected respectively, and then by changing v_{c1} and v_{c2}, a and b are determined according to the above equations. In order to cancel the extra parts we used diodes. Since two ramp functions are needed, two RG circuits are used for each membership function. Fig. 3 shows the typical diagram complementary membership function generator (CMFG) circuit [3]. Considering Fig. 3, we can write:

978-1-4673-9141-2/15 $31.00 © 2015 IEEE

$$i_0 = \begin{cases} i_1 = k(v_{1+} - v_{in})v_{c1+}; \ v_{r1+}fv_{in} \\ i_2 = k(v_{in} - v_{r2-})v_{c2-}; \ v_{r2-}pv_{in} \end{cases}$$

The output voltage will be $v_{out} = Ri_o$. A minimum circuit is used in order to clip the curve, and then reverse it by a fuzzy complementary circuit, which will be described in the following sections.

2) Minimum Circuit

Fig. 4 shows the basic structure of a two-input minimum circuit [3]. The two inputs are connected to the gates of the transistors. The output voltage v_{omin} always takes the smaller value of the two inputs v_{in1} and v_{in2} with a positive offset voltage v_{offset}

$$v_{min} = \min(v_{in1}, v_{in2}) + v_{offset}$$

Using more than two transistors in this configuration, multi-input minimum circuits result. Multi input minimum circuit is needed for applications such as three-input fuzzy controllers. However, in our work, there is a two-input fuzzy controller, hence a two-input minimum circuit is adequate.

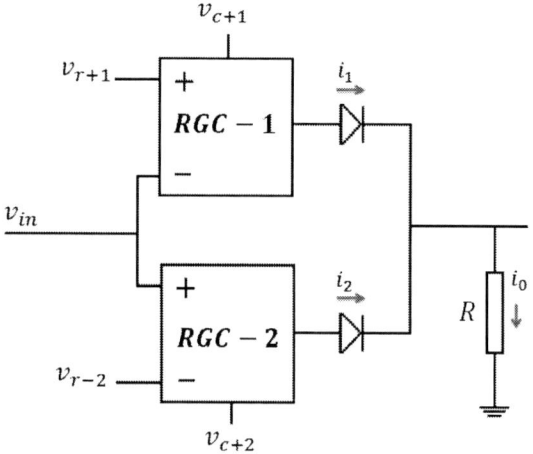

Fig.3. Complementary membership function generator circuit

In order to cancel the offset voltage, v_{off} a negative level shifter circuit is necessary. This shift will be provided by the next stage that is called fuzzy complement circuit with a reference voltage of E.

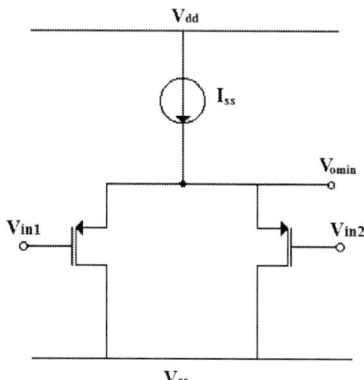

Fig.4. Two input minimum circuit

3) Fuzzy Complementary Circuit

In the earlier subsections, generation of the complement of membership functions and clipping circuit are described. Similarly, the complementary membership function must be converted to an ordinary membership function. A fuzzy

complementary circuit achieves this. The fuzzy complementary circuit consists of two g_m circuits.

In this circuit, assuming E = 0, we have

$$v_{comp} = R_i = R(i_1 + i_2) = R(-g_m v_{in} - g_m v_{comp}),$$

$$v_{comp} = \frac{-Rg_m}{1 + Rg_m} v_{in},$$

$$v_{comp} = -v_{in} \ if \ Rg_m \gg 1$$

Therefore, this is opposite to the input voltage. The complement of the membership function is $E \neq 0$ in Figure, we can write

$$v_{comp} = R(g_m E - g_m v_{in} - g_m v_{comp}),$$

$$v_{comp} = E - v_{in} \ if \ Rg_m \gg 1$$

Fig. 5 shows CMOS realization of the complementary fuzzy block. Note that in the final fuzzifier structure E must be set to cancel offset voltage associated with minimum circuit. In this circuit, due to limited gain of Rg_m, the attenuation is equal to $-Rg_m=1 + Rg_m$. This attenuation is the same in each one of the fuzzifiers used in the controller and the error due to attenuation does not affect fuzzy processing considerably. It is evident that if the input signal has a negative dc value, then E must be chosen to be greater than the Supremum value by |Vdc|.

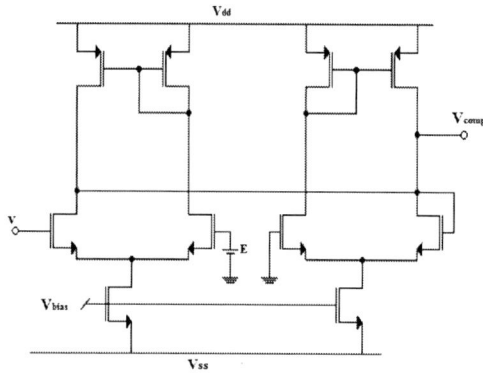

Fig.5. CMOS realization of the complementary fuzzy block

IV. DESIGN OF FLC CHIP

In this section, fuzzy functional blocks, which are described in the previous sections, are combined and their performances are discussed as a fuzzy controller chip. This architecture is a two-input one-output fuzzy controller. Fig. 6 presents the block diagram of the controller. Each input has three membership functions or linguistic terms abbreviated as N (Negative), Z (Zero) and P (Positive), while the output variable is characterized by three singletons. All membership functions have triangular or trapezoidal shape. Parameters used for determining membership functions (V_{r+}, V_{r-}, V_{c+}, V_{c-}) are calibrated by voltages that are set on IC pins. Since input membership functions have three parts, architecture is a 3×3 fuzzy controller and nine rules are accessible. Control voltages set on IC pins, depending on various applications, can change these rules.

The controller in Fig. 6 is constructed with MFG, MIN circuits and a defuzzifier (D blocks) circuit. In this diagram, MFGs are the same Membership Function Generators described in the fuzzifier section. Also, MIN and D units are minimum and defuzzifier circuits. W11 to W33 are weighted values or singleton values that specify fuzzy rules.

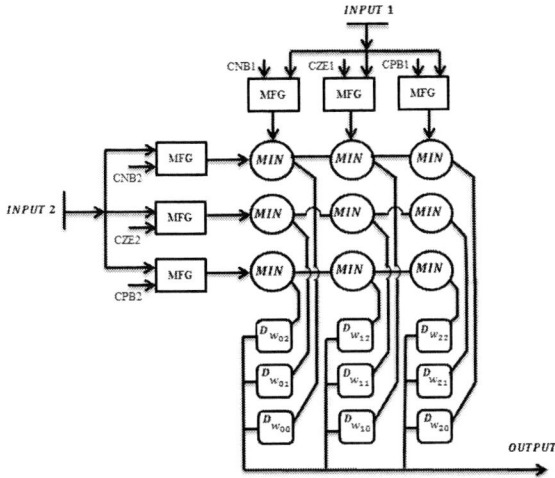

Fig.6. Block diagram of CMOS FLC

Numerical subscribes determine rule number,
For example,

IF Input-1 is N **AND** Input-2 is N **THEN** Output is W11
IF Input-1 is P **AND** Input-2 N **THEN** Output is W31

Note that for simplicity in hardware design usually output membership functions are considered in singleton form. Therefore, we use singleton fuzzy computation method. In this method each output membership function is a constant value on output discourse. Thus one can consider many membership functions for output variable as singleton. Here we used three singleton membership functions in output and each one of them is referenced as N (Negative), Z (Zero) and P (Positive). In this controller, the setpoint and output voltage ranges are set to [1-5] Volts, which can be scaled to range of process variable.

The input range is called inference size of fuzzy controller. Since only the relative values of MFGs output voltages are significant in defuzzification, the MFG output voltage range corresponding to the fuzzy grade interval [0, 1] need not be equal to the output voltage range of the defuzzifier circuit.

V. RESULTS AND DISCUSSION

A level loop of Continuous Stirred Tank Reactor (CSTR) setup is considered for designed FLC chip. Fig.7 shows the closed loop response of realization of FLC controller for the CSTR liquid level process. A set-point of 400 mm was given at the zeroth instant and after the system settled. The process output was found to stay much closer to the set-point and settles much faster. It is found that once the disturbance is given to the FLC by varying the outflow of the CSTR liquid level process, the Process variable settles down with a very slight deviation from the set-point. Thus, it is observed that a FLC chip controller serves both servo and regulatory response with a minimized error simultaneously.

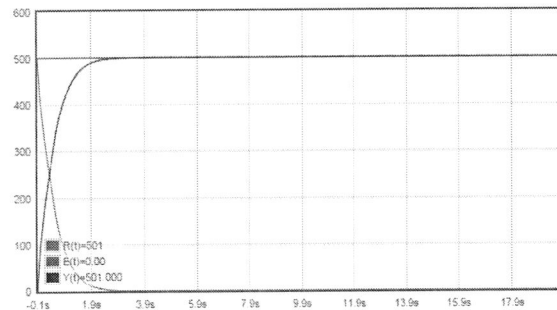

Fig.7. Closed loop response
(Y-axis – Level in mm; X-axis – Time in 100ms)

VI. FUTURE WORK AND FEASIBILITY ANALYSIS

The feasibility of the work has been analyzed and a dedicated industrial control system is proposed. A real time experimental setup for highly nonlinear tank is constructed. The process control system is interfacing DAQ module to the Personal Computer (PC). The laboratory set up for this system is shown in Fig.8. The CSTR system consists of a tank, a water reservoir, pump, rotameter, a differential pressure transmitter, an electro pneumatic converter (I/P converter), a pneumatic control valve, an interfacing DAQ module and a Personal Computer (PC).

The differential pressure transmitter output is interfaced with computer using DAQ module in the RS-232 port of the PC. The pneumatic control valve is air to close, adjusts the flow of the water pumped to the CSTR tank from the water reservoir. The level of the water in the tank is measured by means of the differential pressure transmitter and is transmitted in the form of (4-20) mA to the interfacing DAQ module to the Personal Computer (PC).

Fig.8. CSTR Setup

After computing the control algorithm in FLC, signal is transmitted to I/P converter in the form of current signal (4-20) mA, which passes the air signal to the pneumatic control valve. The pneumatic control valve is actuated by this signal to produce the required flow of water in and out of the tank.

The output of the FLC system is given to the control valve which is the final control element for controlling the level inside the tank. Both the input from the level transmitter and the output to the final control element corresponds to (1-5) Volt. Graphical User Interface (GUI) is done in the virtual instrumentation for the proposed design.

VII. CONCLUSION

This work presents a low power tunable and robust fuzzy logic controller using analog circuits with high speed, low power and small size in CMOS technology. In this proposed industrial FLC, inference engine is designed using parallel configuration. Hence inference speed is independent of the number of rules. A fuzzy controller made with such blocks which occupies a very small area and low-power consumption using nonlinearity property of active devices in an analog technique. Simple circuitry and parallel configuration enable to reach a processing speed of about 8 MFLIPS and power consumption of 10mW. The proposed design works irrespective of applications. Because fuzzy rules and needed inputs membership functions in various applications are different that can be changeable through control and reference pins.

ACKNOWLEDGMENTS

We would like to thank all the researchers, academicians and industrialists, who work and share their knowledge to the VLSI research community. Also we would like to express our gratitude to VLSI SoC 2015 for the support and encouragement.

REFERENCES

[1] C. Y.Chen, Y. T. Hsieh and B. D. Liu; "Circuit implementation of linguistic-hedge fuzzy logic controller in current-mode" IEEE Transaction on Fuzzy Systems, Vol. 11, pp.624-646, 2003.

[2] R. Amirkhanzadeh, A. Khoei, Kh. Hadidi, "A mixed-signal current-mode fuzzy logic controller", Int. Journal of Electronics and Communications, pp. 177-184, 2005.

[3] Hamed Peyravi, Abdollah Khoei, Khayrollah Hadidi, "Design of an analog CMOS fuzzy logic controller chip", Fuzzy Sets and Systems 13(2), pp.245 – 260, 2002.

[4] C. Dualibe, P. Jespers, and M. Verleysen; "A 5.26 MFLIPS programmable Analogue Fuzzy Logic Controller in a Standard CMOS 2.4µ Technology", ISCAS, pp. 377-380, May 2000.

[5] S. Guo, L. Peters, H. Surmann, Design and application of an analog fuzzy logic controller, IEEE Trans. Fuzzy Systems 4 (4) (1996) 429–438.

[6] H.J. Zimmerman, Fuzzy Set Theory and Its Applications, 2nd Edition, Kluwer Academic Publishers, Boston, 1991.

[7] A. Naderi, A. Khoei, Kh. Hadidi; "A New High Speed and Low Power 4-Quadrant CMOS Analog Multiplier in Current-Mode", International Journal of Electronics and Communications, Elsevier, Vol. 63, Issue 9, 2009.

[8] S. Aminifar,A. Khoei, Kh. Haidi, Gh.Yosefi,"A digital CMOS fuzzy logic controller chip using new fuzzifier and max circuit", Int. J. Electron. Commun. (AEÜ), 60, pp.557 – 566, 2006.

[9] Manikandan, Pandiyan and Mani Geetha. "Takagi Sugeno fuzzy expert model based soft fault diagnosis for two tank interacting system."*Archives of Control Sciences* 24, no. 3, pp.271–287 (2014).

[10] Manikandan, P., M. Geetha, and Jovitha Jerome. "Weighted fuzzy fault tolerant model predictive control." *IEEE International Conference on Fuzzy Systems (FUZZ-IEEE)*, pp. 83-90. IEEE, 2014.

[11] Geetha, M.; Manikandan, P.; Jerome, J., "Soft computing techniques based optimal tuning of virtual feedback PID controller for chemical tank reactor," *2014 IEEE Congress on Evolutionary Computation (CEC)*, pp.1922,1928, 6-11 July 2014

[12] Mokarram, M.; Khoei, A.; Hadidi, K, "CMOS fuzzy logic controller supporting fractional polynomial membership functions", Fuzzy Sets Syst, 263, pp.112–126., 2015

Digital CMOS Neuromorphic Processor Design Featuring Unsupervised Online Learning

Jae-sun Seo[1] and Mingoo Seok[2]

[1] School of Electrical, Computer, and Energy Engineering, Arizona State University, Tempe, AZ, USA
[2] Department of Electrical Engineering, Columbia University, New York, NY, USA
jaesun.seo@asu.edu

Invited paper

Abstract—**The compute-intensive and power-efficient brain has been a source of inspiration for a broad range of neural networks to solve recognition and classification tasks. Compared to the supervised deep neural networks (DNNs) that have been very successful on well-defined labeled datasets, bio-plausible spiking neural networks (SNNs) with unsupervised learning rules could be well-suited for training and learning representations from the massive amount of unlabeled data. To design dense and low-power hardware for such unsupervised SNNs, we employ digital CMOS circuits for neuromorphic processors, which can exploit transistor scaling and dynamic voltage scaling to the utmost. As exemplary works, we present two neuromorphic processor designs. First, a 45nm neuromorphic chip is designed for a small-scale network of spiking neurons. Through tight integration of memory (64k SRAM synapses) and computation (256 digital neurons), the chip demonstrates on-chip learning on pattern recognition tasks down to 0.53V supply. Secondly, a 65nm neuromorphic processor that performs unsupervised on-line spike-clustering for brain sensing applications is implemented with 1.2k digital neurons and 4.7k latch-based synapses. The processor exhibits a power consumption of 9.3μW/ch at 0.3V supply. Synapse hardware precision, efficient synapse memory array access, overfitting, and voltage scaling will be discussed for dense and power-efficient on-chip learning for CMOS spiking neural networks.**

Keywords—*neuromorphic computing; CMOS; digital circuits; unsupervised learning; on-chip learning; low-voltage; low-power; spiking neural networks;*

I. INTRODUCTION

Conventional von Neumann architecture has benefited from CMOS scaling to the utmost, driving the semiconductor industry towards high-performance and low power computing for the past several decades. However, with CMOS scaling reaching its physical limits below 10nm node, unconventional computing and architecture are being explored that can overcome the power and processor-memory bottleneck. Using the computationally-intensive yet low-power (~20W) brain as a source of inspiration, neuromorphic computing and its hardware implementation [7-13] have recently gained a great amount of attention.

During the past several years, a number of recognition and classification algorithms have steeply advanced the state of the art using deep learning and artificial neural networks [1-3]. Enhanced recognition accuracy close to human-level perception is being achieved, but this requires iterative training using a well-defined large data set, (i.e., ImageNet [4]) operating on a complex deep learning architecture (>15 layers and >600 million parameters [2]).

Particularly, the common training method employed in deep neural networks is back-propagation, which tunes the weights based on the gradient of the error function, but this requires a known output value for each input value (supervised learning). However, it would be difficult to use back-propagation or supervised learning rules to train real-time sensory input data that are unlabeled. By adopting the unsupervised learning algorithms such as spike-timing dependent plasticity (STDP) found in biological nervous systems, we anticipate to improve learning in robotic or defense applications with the massive amount of unlabeled data, similar to how we analyze and associate sensory input data.

In this work, we propose to implement the core learning principles found in biological nervous systems with advanced CMOS technology in a new architecture. We focus on employing digital logic and memory circuits for the neuromorphic processor design, since they can exploit transistor scaling and dynamic voltage scaling, to realize dense and low-power hardware, respectively. We present two neuromorphic processors that integrate plastic synapses and integrate-and-fire neurons for pattern recognition and data clustering applications. Particular challenges we discuss include the learning capability with synaptic precision, low-voltage synapse array operation, and integration of learning and classification in the same hardware.

II. SPIKE-TIMING DEPENDENT PLASTICITY

In spiking neural networks, neurons are computing elements whose membrane potentials change with incoming spikes and decays over time due to a leak. Neurons communicate with each other via binary messages called spikes, and the synapses store the connection strength between neurons. The strength of the synapses can adapt based on the spiking activity of neurons, and this is referred to as synaptic plasticity. The major difference between how the synapses (weights) update in spiking neural networks (compared to artificial neural networks) is the fact that they depend on the relative spike timing of the pre-synaptic and post-synaptic neurons [5], such that it involves temporal coding (compared to rate coding).

Due to co-dependence on spiking rate or depolarization, spike timing does not solely govern the plasticity of synapses, but it is certainly one important thing that affects synaptic

This research was partially sponsored by the DARPA under contract No. HR0011-09-C-0002.

978-1-4673-9141-2/15 $31.00 © 2015 IEEE

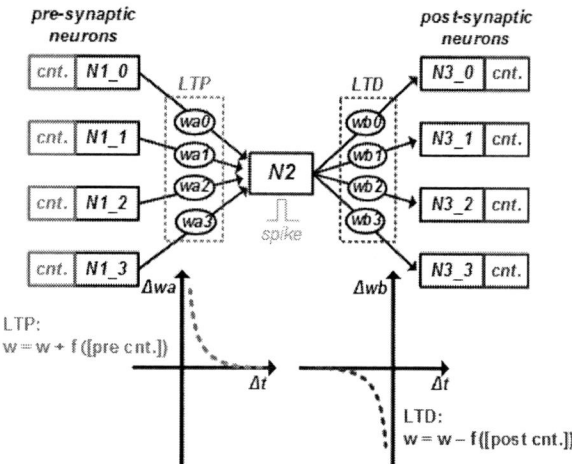

Figure 1. Simplified hardware implementation of STDP learning is shown. When neuron N2 spikes, based on the pre-synaptic (post-synaptic) neuron counter values, LTP on wa* synapses (LTD on wb* synapses) is determined.

plasticity [6]. To that end, without losing generality, we focus on spike-timing dependent plasticity (STDP) as the core learning mechanism for spiking neural networks, and employ on-chip STDP learning for hardware implementation.

Figure 1 illustrates a simplified implementation of STDP learning in hardware. Each neuron holds a counter value (denoted as "cnt.") that represents how much time elapsed since it last spiked. Whenever a neuron spikes, its counter resets to its maximum value, and decreases linearly or exponentially every time step (~ms). The maximum value of the counter represents the number of time steps during which STDP learning is valid for. When neuron N2 spikes, it uses the counter values of pre-synaptic neurons (N1*) for long-term potentiation (LTP) of wa* synapses, while it refers to the counter values of post-synaptic neurons (N3*) for long term depreciation (LTD) of wb* synapses. Note that N3* neuron spikes will cause LTP of wb* synapses. The shape of the learning curve, or how much synapse weights should change based on the counter values, could be individually configured for each neuron [7].

Considering a multi-bit synapse implementation, to compute synapse updates based on STDP, the current synapse value as well as the pre-synaptic or post-synaptic neuron counter value need to be either locally stored or communicated through the network, to perform a read-modify-write process.

III. DIGITAL NEUROMORPHIC PROCESSOR WITH ON-CHIP LEARNING

A. Fully Recurrent Network with Efficient On-Chip Learning in 45nm SOI CMOS

We integrated highly configurable 256 neurons with 64K SRAM synapses with fully recurrent connectivity [7]. When a neuron spikes, LTD requires a row update in the SRAM synapse array, while LTP requires a column update. To perform efficient learning, we employed a transposable SRAM cell that enables both row-by-row and column-by-column update, as shown in Figure 2.

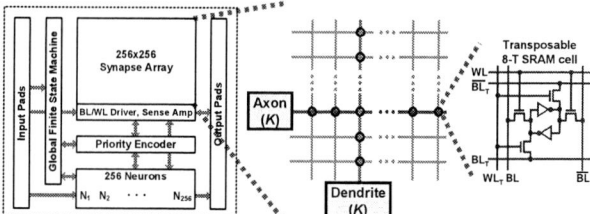

Figure 2. A transposable SRAM cell design enables efficient row-by-row and column-by-column update for LTD and LTP update in synapse SRAM arrays.

We also designed chip variants with 1-bit synapses using probabilistic update and with 4-bit synapses that can represent a richer connectivity between neurons. When 4-bit synapses were employed, the evident increase in synapse area was traded off by the improved learning capability, as shown in Figure 3. It can be seen that 4-b synapses could learn more number of image patterns with smaller number of training iterations, compared to the binary synapse scheme. Furthermore, deterministic read-modify-write is performed on 4-b synapses, thus eliminating the need for a LFSR design in the 1-bit synapse scheme, which was required for the stochastic STDP update.

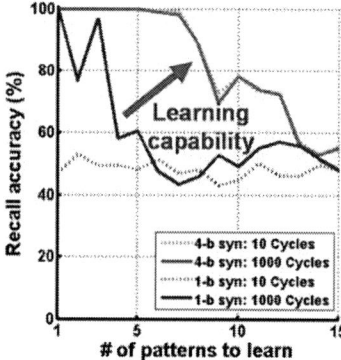

Figure 3. Comparison on learning capability between 1-bit synapse design and 4-bit synapse design is shown. 4-b synapses can learn more number of patterns with smaller number of training iterations.

B. Neuromorphic Spike-Clustering Processor with STDP learning in 65nm CMOS

For invasive deep brain sensing and stimulation applications, we designed a two-layer spiking neural network (Figure 4) hardware with feedback inhibition [8]. The synapse weights between the neurons in different layers were updated using STDP in an unsupervised manner. To mitigate the over-fitting issue during training and also reduce training time, neurons are randomly selected for training with 50% probability, similar to the Dropout technique proposed in deep neural networks [14]. Furthermore, the probability of increasing the synapse weights were reduced with higher spike timing differences, as suggested by STDP learning rules. After training is complete, the receptive fields from the output neurons closely reflect the corresponding cluster template, as shown in Figure 5.

In this small-scale SNN design with 4,864 synapses, latch-based synapses were used. Although the latched-based synapses

978-1-4673-9141-2/15 $31.00 © 2015 IEEE

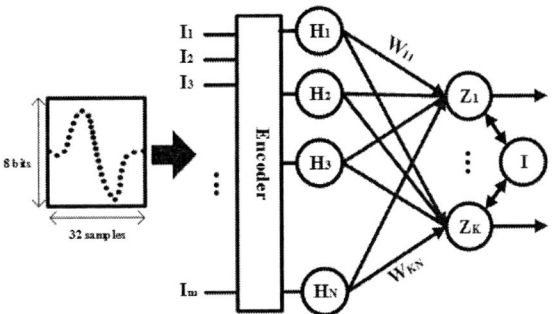

Figure 4. Neurmorphic spike-sorting processor architecture based on Sohonen self-organizing map (H: input neuron, Z: output neuron, I: inhibition neuron). Our design uses m=256, N=1216, and K=4.

consume more area than SRAM synapses, one important merit is that the supply voltage scales much more favorably for latch-based synapses. As a result, the proposed spike-sporting processor could scale down to 0.3V supply voltage and 9.3μW/ch.

In fact, in large-scale spiking neural networks where synapse memory would likely consume a large portion of the overall area, the minimum operating supply voltage of SRAM arrays would become a bottleneck in terms of supply voltage scaling and therefore overall system power consumption. Off-the-shelf large SRAM arrays typically do not scale well below 0.6-0.7V, preventing near-threshold or sub-threshold operation of the digital neuromorphic processor. For large-scale networks that employ a large number of synapses, SRAM arrays that operate down to low voltages with small area overhead will be crucial to reduce overall power consumption.

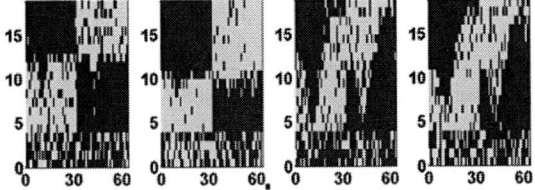

Figure 5. Receptive fields of network after training, which show four distinct spikes could be classified.

IV. CONCLUSION

In this work, we presented digital CMOS neuromorphic processor designs that feature on-chip STDP learning. Synaptic precision, synapse SRAM array access, and low voltage operation were discussed as design techniques. Proceeding towards very large scale systems that can learn and classify on-chip, we envision neuromorphic architecture and circuits that are dynamically configurable for a variety of biological spike-based learning rules. Efficient on-chip communication through intra/inter-layer crossbars as well as on-chip voltage scaling will be crucial for power-efficiency of the overall system

As new learning mechanisms are being found in neuroscience literature [15-17], and different spiking neural

network algorithms are actively being developed [18-19], it is important for hardware researchers to follow such learning algorithms and understand the computation, memory, and communication requirements to efficiently implement online learning hardware.

REFERENCES

[1] A. Krizhevsky, *et al.*, "ImageNet classification with deep convolutional neural networks," *Advances in Neural Processing Information Systems (NIPS)*, 2012.

[2] K. Simonyan and A. Zisserman "Very Deep Convolutional Networks for Large-Scale Image Recognition," *International Conference on Learning Representations (ICLR)*, 2015.

[3] K. He, *et al.*, "Delving Deep into Rectifiers: Surpassing Human-Level Performance on ImageNet Classification," *CoRR*, abs/1502.01852, 2015.

[4] J. Deng, W. Dong, R. Socher, L.-J. Li, K. Li, and L. Fei-Fei, "Imagenet: A large-scale hierarchical image database," *IEEE Computer Vision and Pattern Recognition (CVPR)*, 2009.

[5] S. Song, *et al.*, "Competitive Hebbian learning through spike-timing-dependent synaptic plasticity," *Nature Neuroscience*, pp. 919-926, 2000.

[6] D. E. Feldman, "The spike-timing dependence of plasticity," *Neuron*, vol. 75, no. 4, pp. 556-571, August 2012.

[7] J. Seo, *et al.*, "A 45nm CMOS neuromorphic chip with a scalable architecture for learning in networks of spiking neurons," *IEEE Custom Integrated Circuits Conference (CICC)*, 2011.

[8] B. Zhang, *et al.*, "A Neuromorphic Neural Spike Clustering Processor for Deep-Brain Sensing and Stimulation Systems," *IEEE International Symposium on Low Power Electronics and Design (ISLPED)*, 2015.

[9] P. Knag, *et al.*, "A sparse coding neural network ASIC with on-chip learning for feature extraction and encoding," *IEEE Journal of Solid-State Circuits (JSSC)*, vol. 50, no. 4, pp. 1070-1079, Apr. 2015.

[10] J. K. Kim, *et al.*, "A 640M pixel/s 3.65mW sparse event-driven neuromorphic object recognition processor with on-chip learning," *Symp. of VLSI Circuits*, June 2015.

[11] J. Park, *et al.*, "A 65k-neuron 73-Mevents/s 22-pJ/event asynchronous micro-pipelined integrate-and-fire array transceiver," *IEEE Biomedical Circuits and Systems Conference (BioCAS)*, pp. 675-678, October 2014.

[12] P. Merolla, *et al.*, "A million spiking-neuron integrated circuit with a scalable communication network and interface," *Science*, vol. 345, no. 6197, pp. 668-673, August 2014.

[13] S. Venkataramani, *et al.*, "AxNN: energy-efficient neuromorphic systems using approximate computing," *IEEE International Symposium on Low Power Electronics and Design (ISLPED)*, pp. 27-32, August 2014.

[14] N. Srivastava, et al., "Dropout: a simple way to prevent neural networks from overfitting," Journal of Machine Learning Research, pp. 1929-1958, January 2014.

[15] F. Zenke, *et al.*, "Diverse synaptic plasticity mechanisms orchestrated to form and retrieve memories in spiking neural networks," *Nature Communications*, vol. 6, April 2015.

[16] G. Rachmuth, *et al.*, "A biophysically-based neuromorphic model of spike rate- and timing-dependent plasticity," Proceedings of the National Academy and Sciences of the Unites States of America, vol. 108, no. 49, E1266–E1274, December 2011.

[17] Y. Zuo, *et al.*, "Complementary Contributions of Spike Timing and Spike Rate to Perceptual Decisions in Rat S1 and S2 Cortex," Current Biology, vol. 25, no. 3, pp. 357-363, February 2015.

[18] S. R. Kheradpisheh, *et al.*, "Bio-inspired Unsupervised Learning of Visual Features Leads to Robust Invariant Object Recognition," *CoRR*, abs/1504.03871, 2015.

[19] P. U. Diehl and M. Cook, "Unsupervised learning of digit recognition using spike-timing-dependent plasticity," *Frontiers in Computational Neuroscience*, vol. 9, August 2015.

978-1-4673-9141-2/15 $31.00 © 2015 IEEE

An Overview on Memristor Crossabr Based Neuromorphic Circuit and Architecture

(Invited Paper)

Zheng Li, Chenchen Liu, Yandan Wang, Bonan Yan, Chaofei Yang, Jianlei Yang, and Hai (Helen) Li
Department of Electrical and Computer Engineering
University of Pittsburgh, Pittsburgh, PA 15261
Email: http://www.ei-lab.org

Abstract—As technology advances, artificial intelligence becomes pervasive in society and ubiquitous in our lives, which stimulates the desire for embedded-everywhere and human-centric intelligent computation paradigm. However, conventional instruction-based computer architecture was designed for algorithmic and exact calculations. It is not suitable for handling the applications of machine learning and neural networks that usually involve a large sets of noisy and incomplete natural data. Instead, neuromorphic systems inspired by the working mechanism of human brains create promising potential. Neuromorphic systems possess a massively parallel architecture with closely coupled memory and computing. Moreover, through the sparse utilizations of hardware resources in time and space, extremely high power efficiency can be achieved. In recent years, the use of memristor technology in neuromorphic systems has attracted growing attention for its distinctive properties, such as nonvolatility, reconfigurability, and analog processing capability. In this paper, we summarize the research efforts in the development of memristor crossbar based neuromorphic design from the perspectives of device modeling, circuit, architecture, and design automation.

Keywords—*Neuromorphic computing, neuromorphic circuit and architecture, memristor, crossbar array, resistive memory.*

I. INTRODUCTION

The demand of high performance computing continuously increases as artificial intelligence becomes pervasive in society and ubiquitous in our lives. However, traditional von Neumann computer architecture designed for algorithmic and exact calculations becomes less efficient and scalable. Neuromorphic hardware systems inspired by the working mechanism of human brains [1] potentially can provide the capabilities of biological perception and cognitive information processing within a compact and energy-efficient platform. Therefore, the development of neuromorphic systems has gained a great deal of attention in recent years. Besides conventional CPUs, GPUs, or FPGAs [2][3], the use of emerging technologies such as resistive memory devices (a.k.a. memristor) in neuromorphic design has also been studies [4][5].

As early as in 1971, Professor Chua predicted the existence of memristor based on circuit theory [6]. Forty years later in 2008, the physical realization of memristor was demonstrated through a TiO_2 thin-film [7]. Afterwards, many memristive materials and devices have been rediscovered [8]. A memristor can record its total electrical flux as memristance (M). The feature is highly similar to weighting function of a biological synapse. Moreover, the two-terminal think-film device

structure can be easily integrated into crossbar arrays. It can provide a large number of signal connections within a small footprint and conduct the weighted combination of input signals, making it very promising for massively-parallel, large-scale neuromorphic systems [9].

In this paper, we give an overview on the research efforts in developing neuromorphic circuit and architecture design that leverage memristor crossbar structure. A comprehensive view including device modeling, circuit, architecture, and design automation will be covered in the following sections.

II. MEMRISTOR DEVICE MODELING

Fig. 1(a) illustrates the memristor device realized on a $Pt/TiO_2/Pt$ thin-film structure [7]. The memristive function is achieved through the doping front movement, which can be controlled by external voltage excitation. And its overall memristance is determined by the ratio of the stoichiometric TiO_2 with low conductivity and the semiconductor-alike oxygen-deficient titanium dioxide (TiO_{2-x}). Thus, it can be modeled as a coupled variable resistor model shown in Fig. 1(b), which is equivalent to two series-connected resistors such as

$$M(\alpha) = R_L \cdot \alpha + R_H \cdot (1 - \alpha). \tag{1}$$

Here α ($0 \leq \alpha \leq 1$) is the ratio of doping front position over the total thickness of TiO_2 thin-film, represented by the relative doping front position. The velocity of doping front movement $v(t)$, which is driven by the voltage applied across the memristor $V(t)$ can be expressed as

$$v(t) = \frac{d\alpha}{dt} = \mu_v \cdot \frac{R_L}{h^2} \cdot \frac{V(t)}{M(\alpha)}, \tag{2}$$

where μ_v is the equivalent mobility of dopants, h is the total thickness of the TiO_2 thin-film; and $M(\alpha)$ is the total memristance which is a function of α.

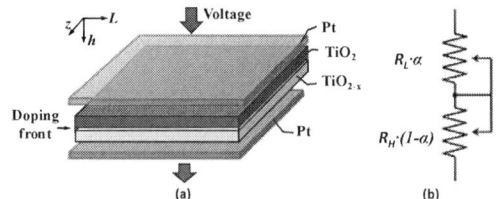

Fig. 1. TiO_2 thin-film memristor. (a) structure, and (b) equivalent circuit.

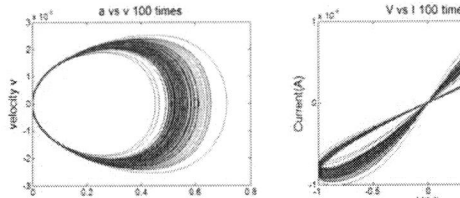

Fig. 2. The impact of process variations on TiO$_2$ thin-film memristors. left: v vs. α ,right: $I-V$ characteristics. (The blue curves are from 100 Monte-Carlo simulations, and red lines are the ideal condition.)

Note that the above bulk model is derived from the mathematical definition of memristor, which assumes a flat doping front moving up or down. In reality, however, filamentary conduction has been observed in nano-scale semiconductors: the current travels through some high conducting filaments rather than evenly passes the entire device [10]. Moreover, as process technology shrinks down, device parameter fluctuations incurred by process variations severely affecting the electrical characteristics. The situation in a memristor could be even worse: the parameter variations can result in the shift of electrical responses, which in turn affect the memristance since the total charge through a memristor indeed is the historic behavior of its current profile.

An approach to converge the difference between the bulk and filament models was to divide a TiO$_2$ thin-film into many tiny filaments and adopt the bulk model to the small flat doping front in each filament [11]. The implications of memristor parameters to the circuit design was explored by taking into account the impact of memristor geometry variations. Fig. 2 shows the dynamic responses of 100 Monte Carlo simulations which can visually demonstrate the overall impact of process variations on the memristive behavior.

Moreover, metal oxide based memristor behaves stochastically and hence even a single memristive device demonstrates large variations in performance. More specific, the static states of a single memristor are not fixed, but have large variations with skewed distributions and heavy tails [12]. The switching mechanism of a memristor, that is, its dynamic behavior, performs as a stochastic process [13]. A stochastic behavior model which bypasses material-related parameters while directly linking the device analog behavior to stochastic

functions was presented to better facilitate the exploration of memristors in hardware implementation. Fig. 3 shows the time dependencies of ON and OFF switching probability at different applied voltages. The results have high approximation to the experimental results [13].

III. NEUROMORPHIC CIRCUIT DESIGN

The applications of memristor crossbar in acceleration of scientific and neuromorphic computing have been studied. For example, the matrix-vector computation can be conducted through crossbar arrays by using voltage/current magnitudes to represent the data [5] or through a spiking neural network [14].

Fig. 4 depicts an overview of the spiking computing design that leverages the compact memristor crossbar structure [14]. It adopts the rate coding model and represents data using the frequency of spikes [15]. Through different bitlines (BLs) in the crossbar, the synaptic weighting functions of different entries are executed in parallel. The *integrate and fire circuits* (IFC) as post-neurons generate output spikes based on the strength of the weighted pre-neuron signals from the crossbar.

A single-layer neural network with N pre-neurons and M post-neurons can be implemented using a N × M crossbar. First, the activity pattern of pre-neurons $\mathbf{x_{N \times 1}}$ is transferred into a set of pulses to wordlines (WLs). The number of spikes on WL$_i$ within the computation period ($n_{x,i}$) corresponds to $x_i \in \mathbf{x}$. The weight from the jth pre-neuron and the ith post-neuron maps to the conductance g_{ij} at the crosspoint of WL$_i$ and BL$_j$. The total weighted signal to post-neuron j is transferred to the current flowing through BL$_j$ and accumulated on a capacitor C_m in IFC. Once the voltage on C_m reaches to a predefined threshold V_{th}, the IFC fires an output spike and resets C_m. The activity function of post-neurons $\mathbf{y}_{M \times 1}$ is represented by a set of spike numbers such as $[n_{y,0}, n_{y,1}, \cdots, n_{y,M-1}]^T$.

Under ideal condition without taking into account the realistic factors in circuit implementation, the spike number produced at the jth post-neuron $n_{y,j} \propto \sum_{i=0}^{N-1} g_{ij}\delta_i$, where δ_i corresponds the spike occurrence at WL$_i$. The assumption $\sum_{i=0}^{N-1} g_{ij} \to 0$, however, is satisfied only when all the resistive devices are at (or close to) the high resistance state. This cannot be generalized as a common condition in applications. Moreover, the delay overhead of IFC to generate pulses and reset C_m cannot be ignored.

The delay of IFC is a critical parameter determining the performance of the spiking neuromorphic system. Fig. 5(a)

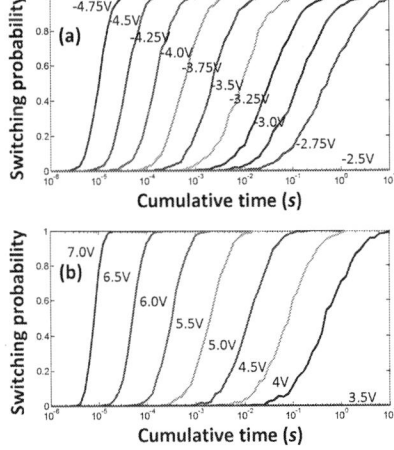

Fig. 3. The time dependency of ON (a) and OFF (b) switching at different external voltage V.

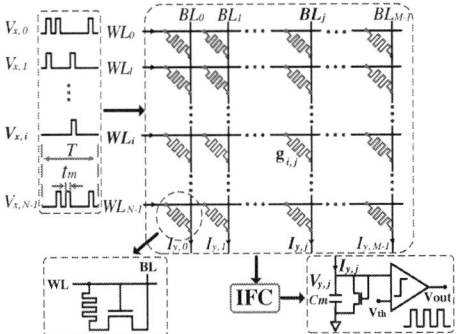

Fig. 4. The spiking neuromorphic design with a memristor crossbar array.

978-1-4673-9141-2/15 $31.00 © 2015 IEEE

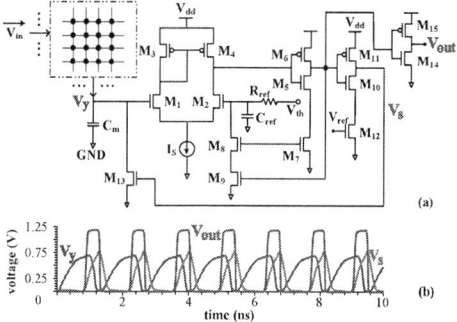

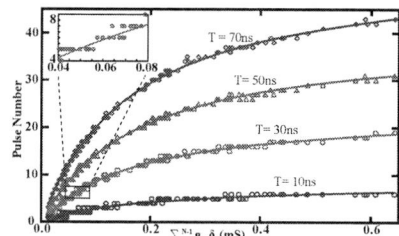

Fig. 5. The IFC circuit (a) the schematic and (b) the simulation waveforms.

Fig. 6. $n_{y,j}$ vs. $\sum_{i=0}^{N-1} g_{ij}\delta_i$ under various T.

depicts the schematic of a new IFC design featuring high speed and low power consumption [14]. During the operation, the BL voltage V_y continues increasing until it reaches V_{th}. Then the differential pair (M_1–M_4) together with the following two cascaded inverters (M_5–M_7 & M_{10}–M_{12}) generates a high voltage at V_s, which in turn enables the discharging transistor M_{13}. Consequently, V_y decreases quickly and eventually turns off M_{13}. As such, the firing of one output spike at V_{out} is completed and a new iteration of integrate-and-fire starts. To shorten the intrinsic operation delay and therefore improve the IFC throughput, a positive feedback loop (M_7–M_9) was deployed based on the traditional comparator. Another approach was to minimize the discharge time of C_m once a spike is fired out, i.e., using a large M_{13} to provide sufficient discharging current.

Fig. 5(b) shows the waveforms of V_y, V_s, and V_{out} of the IFC design using IBM 130nm technology, under the fastest firing frequency (568.2M spikes/sec). The design area is $175.3\mu m^2$, which is compatible to that of traditional designs, e.g., $120\mu m^2$ at 65nm technology in [16]. Its energy consumption is $0.48pJ$-per-spike, which is about a quarter of the one in [16] ($2pJ$-per-spike).

The computational accuracy of the spiking neuromorphic design was evaluated based on a 32×32 crossbar array. Here, the system computational accuracy is defined as the linearity between the obtained output spike number $n_{y,j}$ and actual computation on the crossbar $\sum_{i=0}^{N-1} g_{ij}\delta_i$. Assume that an input spike has a $2ns$ period with 50% utilization rate (that is, $t_m = 1ns$). The resistance values and the input pulse numbers are randomly assigned to cover the entire input range. T varies from $10ns$ to $80ns$ at a step of $10ns$ to examine the temporal scalability of the design.

It can be seen from Fig. 6 that as $\sum_{i=0}^{N-1} g_{ij}\delta_i$ increases, the rising rate of $n_{y,j}$ becomes smaller. This is because a larger $\sum_{i=0}^{N-1} g_{ij}\delta_i$ and therefore a bigger $I_{y,j}$. It results in a faster switching of $V_{y,j}$ from $0V$ to V_{th}, making the impact of the IFC delay overhead t_0 more prominent. Nonetheless, a good computational accuracy (i.e., output linearity) is obtained when $\sum_{i=0}^{N-1} g_{ij}\delta_i$ is small (i.e., $< 0.15mS$). In fact, our investigation at application level also show that most of the operations of neural network implementations fall into this small range [14]. Furthermore, for different combinations of inputs and resistive array patterns with the same $\sum_{i=0}^{N-1} g_{ij}\delta_i$, the generated pulse number may be slightly different (no more than ± 1). Such a fluctuation comes from the difference in $I_{y,j}$'s waveform and amplitude generated by these combina-

tions.

Such a spiking neuromorphic system is designed mainly for learning and classification applications whose algorithms naturally tolerate the low resolution and variability in the computations. Moreover, the imperfect output linearity shown in Fig. 6 can also be compensated during circuit implementation as long as the output spike and the weighted input spikes have a monolithic mapping relation.

IV. ARCHITECTURE

Recently, a highly-efficient reconfigurable neuromorphic computing accelerator with on-chip memristor-based crossbar (MBC) arrays as perceptron network, named as *RENO*, was proposed aiming at the acceleration of ANNs computations [17]. Unlike the spike-based computations where the data is represented by the pulse signals with different frequencies and amplitudes [18], the design adopts a hybrid method in data representation: the computation within MBCs and the signal communications among MBCs are conducted in analog form, while the control information remains as digital signals.

Fig. 7 illustrates the RENO architecture. It works as a complimentary functional unit to CPU and particularly accelerates ANN-relevant executions. In the design, *memristor-based crossbar* (MBC) arrays are used to perform high efficient analog neuromorphic computing. And a *mixed-signal interconnection network* (M-net) is developed to connect the MBCs and conduct the topological reconfiguration of RENO. To receive command/data and send back result in digital form to processor, *input, output* and *configuration FIFOs* are located at the interface of RENO.

MBC arrays are arranged in a *centralized mesh* (CMesh) manner to minimize the cost of the interconnection network [19]. The example in the figure includes four array groups, each of which is formed with four MBC arrays connected through a group router. A MBC array is partitioned into four sub-crossbars to implement the multiplication of the combination of the signed signals and the signed synaptic weights.

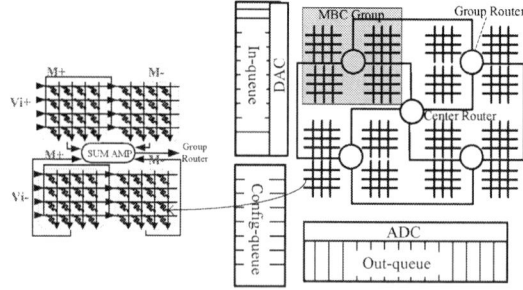

Fig. 7. The RENO architecture.

978-1-4673-9141-2/15 $31.00 © 2015 IEEE

The optimal MBC design contains 64 rows and 64 columns which offers a good compromise between performance and reliability. Moreover, this array scale covers the majority of learning applications, 80% of which have less than 60 neurons in the input layer [20]. Applications requiring larger connection matrices can be partitioned into smaller tasks and executed on multiple MBC arrays simultaneously or sequentially.

In this centralized hierarchical architecture, the data communication is performed at both inter-group and intra-group levels. The *central router* shown in Fig. 7 connects to CPU and all *group routers*. Each group router talks to the four local MBC arrays within the group, three other group routers, and the central router. Such a centralized scheme maximizes the number of parties that each router communicates with, minimizes the effective communication distance and the hop count, mitigates the bottleneck effect of the central router, and simplifies the control complexity.

The signal transmission within RENO can be realized in either digital or analog form. Digital signal transfer has good controllability and supports high-frequency operations. However, as the computation of MBC arrays is in analog form, *digital-to-analog/analog-to-digital* (DA/AD) conversions are required at the interface of MBC arrays and routers, which inevitably degrades the signal precision and results in significant area and power overheads. The small footprint of the MBC arrays limits the data communication distance, e.g., within $0.53mm$, making it possible to transfer signals in analog form. Moreover, the impact of signal distortion generated during the analog signal transmission on computation reliability can be tolerated by the intrinsic high fault resistance of ANN algorithms. Instead, a mixed-signal interconnection network called *M-Net* is used to assist the task mapping and data migration in the MBC arrays. M-Net maintains the data in analog form while transfers the control and routing information in digital form so as to simplify the synchronization and communication between CPU and RENO.

A frontend scheme can be used to assist preparing RENO for ANN computation [21]. As illustrated in Fig. 8, the system frontend is composed of all the preparation steps. The codes that are already or can be implemented with ANN are identified first. Note that here the Boolean function XOR is used just for illustration purpose and the realistic target codes can be much more sophisticated. Based on the characteristics and complexity of the target codes, the topology of ANN, including the number of layers and the number of neurons at each layer etc., is decided and the ANN is trained offline. After that, the trained ANN is mapped to the NCA structure through the reconfiguration logic. During the NCA-aware compilation, the target codes are modified with the annotations of NCA IO instruction generation; then the compiler takes the topology of the trained ANN as the input parameters to generate NCA configuration instructions.

V. DESIGN AUTOMATION FRAME

Although memristor crossbar is believed to be a game changing technology for neuromorphic system realization, how to efficiently design such a system with minimized (or even practical) hardware cost is still a research topic barely touched.

In application layer, for example, large neural networks are usually very sparse. In LDPC coding based on message passing algorithm, for example, the network sparsity is higher than 99% [22]. Here the sparsity of a network is defined as one minus the ratio between the number of actual connections and all possible connections in the network. In fact, such a high sparsity is also close to the biological facts that in neocortex, neurons are typically connected to only 10^{-9} to 10^{-7} of all the neurons and these connections are limited in the neighborhood of $1cm^2$ of the tissue [9].

However, when the sparsity of a network is high, using memristor crossbars to implement such a network becomes inefficient, as the utilization rate of the connections in the crossbar will be low. It may be more efficient to realize these sparse connections using smaller-size crossbars or even discrete synapses. The tradeoffs between the selection of the crossbars with different sizes, the crossbar utilization rates and the impacts on physical design cost inspire this work.

An EDA flow called as AutoNCS was proposed to design a custom memristor-based neuromorphic computing systems (NCS) [23]. It is an iterative process based on spectral clustering algorithm to consolidate synapse connections into clusters and map them to memristor crossbars for high utilization rate of the connections in the crossbars. Note that implemented design can still perform various tasks because the function of a NCS can be trained by tuning the weights of the connections.

Fig. 9 depicts the overview of the design automation frame for large-scale neuromorphic computing system. It consists of the following four components:

1) *Modified spectral clustering* (MSC) that groups the connections in a network into dense clusters that can be efficiently mapped to memristor crossbars;
2) *Greedy cluster size prediction* (GCP) that constrains the largest cluster size within the maximum available crossbar scale;
3) *Iterative spectral clustering* (ISC) that repeatedly performs clustering on the networks to group the connections into clusters, and minimize the outliers that need to be mapped to discrete synapses; and
4) a customized physical design method to realize the neuromorphic systems based on the clustering result.

A testbench of sparse Hopfield network was used to evaluated AutoNCS. The network with a size of 500 was trained for

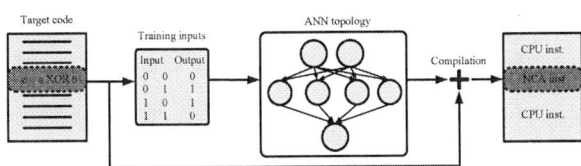

Fig. 8.   The frontend for data preparation.

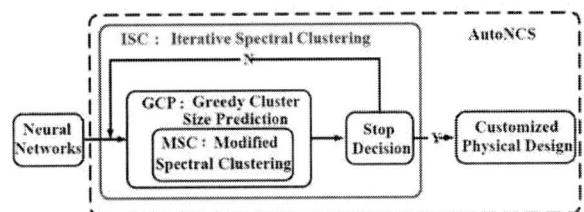

Fig. 9.   The overview of the EDA frame.

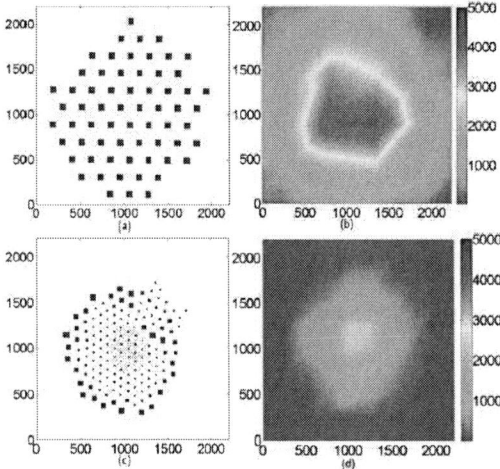

Fig. 10. The placement and routing results of the Hopfield testbench without clustering are shown in (a) and (b). The results with AutoNCS are shown in (c) and (d).

recognition of 30 patterns. While offering a recognition rate above 90%, the sparsity of the network is 94.39%. Fig. 10 compares the optimal placement and routing results in full crossbar (FullCro) and AutoNCS. In optimal FullCro, crossbars with the maximum size are uniformly placed, resulting in heavy wire congestion in the center. However, in AutoNCS, large crossbars on the periphery realized the majority of connections, leaving only sparse connections implemented by small crossbars and discrete synapses in the inner place. This topology reduces wirelength, area and aver-age delay substantially.

VI. CONCLUSION

The emerging memristor technology has demonstrated great potential in neuromorphic system design for its similar behavior to biological synapse, nonvolatile data storage, reconfigurability, analogy process capability, as well as the extreme high connectivity. This paper gives a brief summary on the research activities in utilizing the memristor crossbar arrays for neuromorphic design. Holistic effects across different areas shall be integrated and there are still many problems to be solved to obtain a practical neuromorphic hardware for large-scale applications.

ACKNOWLEDGMENT AND DISCLAIMER

This work is supported in part by NSF 1337198 and DARPA D13AP00042. Any opinions, findings and conclusions or recommendations expressed in this material are those of the authors and do not necessarily reflect the views of NSF, DARPA, or their contractors.

REFERENCES

[1] J. Partzsch and R. Schuffny, "Analyzing the scaling of connectivity in neuromorphic hardware and in models of neural networks," *IEEE Transactions on Neural Networks (TNNLS)*, vol. 22, no. 6, pp. 919–935, June 2011.

[2] S. Li, C. Wu, H. Li, B. Li, Y. Wang, and Q. Qiu, "Fpga acceleration of recurrent neural network based language model," in *2015 IEEE 23rd Annual International Symposium on Field-Programmable Custom Computing Machines (FCCM)*,, May 2015, pp. 111–118.

[3] B. Li, E. Zhou, B. Huang, J. Duan, Y. Wang, N. Xu, J. Zhang, and H. Yang, "Large scale recurrent neural network on gpu," in *2014 International Joint Conference on Neural Networks (IJCNN)*, July 2014, pp. 4062–4069.

[4] M. Hu, H. Li, Q. Wu, and G. Rose, "Hardware realization of bsb recall function using memristor crossbar arrays," in *49th ACM/EDAC/IEEE Design Automation Conference (DAC)*, June 2012, pp. 498–503.

[5] M. Hu, H. Li, Y. Chen, Q. Wu, G. Rose, and R. Linderman, "Memristor crossbar-based neuromorphic computing system: A case study," *IEEE Transactions on Neural Networks and Learning Systems*, vol. 25, no. 10, pp. 1864–1878, Oct 2014.

[6] L. Chua, "Memristor-the missing circuit element," *IEEE Transactions on Circuit Theory*, vol. 18, no. 5, pp. 507–519, Sep 1971.

[7] R. Williams, "How we found the missing memristor," *IEEE Spectrum*, vol. 45, no. 12, pp. 28–35, Dec 2008.

[8] X. Wang, Y. Chen, H. Xi, H. Li, and D. Dimitrov, "Spintronic Memristor Through Spin-torque-induced Magnetization Motion," *IEEE Electron Device Letters*, vol. 30, pp. 294–297, 2009.

[9] S. H. Jo, T. Chang, I. Ebong, B. B. Bhadviya, P. Mazumder, and W. Lu, "Nanoscale memristor device as synapse in neuromorphic systems," *Nano Letters*, vol. 10, no. 4, pp. 1297–1301, 2010.

[10] D. Kim, S. Seo, S. Ahn, D. Suh, M. Lee, B. Park, I. Yoo, I. Baek, H. Kim, E. Yim *et al.*, "Electrical observations of filamentary conductions for the resistive memory switching in NiO films," *Applied physics letters*, vol. 88, no. 20, pp. 202 102–202 102, 2006.

[11] M. Hu, H. Li, Y. Chen, X. Wang, and R. Pino, "Geometry variations analysis of TiO_2 thin-film and spintronic memristors," in *Proceedings of the 16th Asia and South Pacific Design Automation Conference*, 2011, pp. 25–30.

[12] W. Yi, F. Perner *et al.*, "Feedback Write Scheme for Memristive Switching Devices," *Applied Physics A*, vol. 102, no. 4, pp. 973–982, 2011.

[13] G. Medeiros-Ribeiro, F. Perner, R. Carter, H. Abdalla, M. D. Pickett, and R. S. Williams, "Lognormal Switching Times for Titanium Dioxide Bipolar Memristors: Origin and Resolution," *Nanotechnology*, vol. 22, no. 9, p. 095702, 2011.

[14] C. Liu, B. Yan, C. Yang, L. Song, Z. Li, B. Liu, Y. Chen, H. Li, Q. Wu, and H. Jiang, "A spiking neuromorphic design with resistive crossbar," in *Proceedings of the 52nd Annual Design Automation Conference*, 2015, pp. 14.1–14.6.

[15] W. Gerstner and W. M. Kistler, *Spiking Neuron Models.* Cambridge University Press, 2002.

[16] A. Joubert *et al.*, "Hardware spiking neurons design: Analog or digital?" in *IJCNN*, June 2012, pp. 1–5.

[17] X. Liu, M. Mao, B. Liu, H. Li, Y. Chen, B. Li, Y. Wang, H. Jiang, M. Barnell, Q. Wu, and J. Yang, "Reno: A high-efficient reconfigurable neuromorphic computing accelerator design," in *Proceedings of the 52Nd Annual Design Automation Conference*, 2015, pp. 66.1–66.6.

[18] A. S. Cassidy, P. Merolla, J. V. Arthur, S. Esser, B. Jackson, R. Alvarez-Icaza, P. Datta, J. Sawada, T. M. Wong, V. Feldman, A. Amir, D. B. dayan Rubin, E. Mcquinn, W. P. Risk, and D. S. Modha, "Cognitive computing building block: A versatile and efficient digital neuron model for neurosynaptic cores," in *IJCNN*, 2013.

[19] J. Balfour and W. J. Dally, "Design tradeoffs for tiled cmp onchip networks," in *ICS*, 2006, pp. 187 – 198.

[20] O. Temam, "A defect-tolerant accelerator for emerging high-performance applications," in *ISCA*, 2012, pp. 356–367.

[21] X. Liu, M. Mao, H. Li, Y. Chen, H. Jiang, J. Yang, Q. Wu, and M. Barnell, "A heterogeneous computing system with memristor-based neuromorphic accelerators," in *2014 IEEE High Performance Extreme Computing Conference (HPEC)*, 2014, pp. 1–6.

[22] "Ieee standard for information technology–telecommunications and information exchange between systems–local and metropolitan area networks–specific requirements part 11: Wireless lan medium access control (mac) and physical layer (phy) specifications amendment 10: Mesh networking," *IEEE Std 802.11s-2011*, pp. 1–372, Sept 2011.

[23] W. Wen, C.-R. Wu, X. Hu, B. Liu, T.-Y. Ho, X. Li, and Y. Chen, "An eda framework for large scale hybrid neuromorphic computing systems," in *52nd ACM/EDAC/IEEE Design Automation Conference (DAC)*, 2015, pp. 1–6.

An Equation-Based Battery Cycle Life Model for Various Battery Chemistries

Alberto Bocca, Alessandro Sassone, Donghwa Shin[†], Alberto Macii, Enrico Macii and Massimo Poncino

Politecnico di Torino, Corso Duca degli Abruzzi 24, 10129 Italy
{alberto.bocca, alessandro.sassone, alberto.macii, enrico.macii, massimo.poncino}@polito.it
[†] Yeungnam University, 280 Daehak-Ro, Gyeongsan, Gyeongbuk 712-749, Republic of Korea
donghwashin@yu.ac.kr

Abstract—The evaluation of the cycle life of batteries is an essential task in the assessment of the reliability and cost of battery-operated devices. Several compact cycle life models have been proposed in the literature, that exhibit a general trade-off between generality and accuracy. Some models are based on a compact equation derived from experimental data and try to extract a general relationship between cycle life and the relevant parameters (mostly the depth of discharge), but suffer from poor accuracy. At the other extreme, more accurate models, based on incorporating the aging effect into an equivalent circuit, tend to be focused on a specific device and are seldom applicable to another battery.

In this work we propose an equation-based model that tries to overcome the accuracy limits of previous similar models. The model parameters are obtained by fitting the curve based on information reported in datasheets, and can be adapted (with different accuracy levels) to the amount of available information.

We applied the model to various commercial batteries for which full information on their cycle life is available. Results show an average estimation error, in terms of the number of cycles, generally smaller than 10%, which is consistent with the typical tolerance provided in the datasheets, and much lower than previous equation-based models.

Index Terms—Battery cycle life, Modeling.

I. INTRODUCTION

Secondary (i.e., rechargeable) batteries have become an essential component in many applications like mobile telecommunication, aerospace, renewable energy applications, and electric vehicles. The possibility of an early verification of these systems, including the exchange of energy between energy storage devices and the other components, requires accurate and efficient battery models.

Several models, exhibiting tradeoffs between generality and accuracy, have been proposed in the literature. In electronic design, the most common approach in battery modeling consists of the definition of a generic model template in terms of an *equivalent electric circuit* (e.g., [1], [2]), which is then populated either using data obtained from direct measurements on actual devices or by extrapolation of battery characteristics available from datasheets (e.g., [3]). These kinds of models are typically generated for a specific battery chemistry and show a high degree of accuracy. This accuracy may significantly degrade if these models are applied to different battery chemistries.

In other contexts (e.g., automotive, aerospace, smart grids), conversely, designers often rely on simpler mathematical macromodels, such as Peukert's law [4], as a quick estimator for the sizing of the battery sub-system or for preliminary what-if analysis. Such macromodels try to infer a general relationship between the battery runtime and the most relevant parameters, like the Depth of Discharge (DOD) of a battery. This relationship can then be applied to different kinds of batteries with different chemistries, but the estimation results have, in general, a very low degree of accuracy. This is mainly due to the fact that these models tend to primarily focus on a single charge/discharge cycle of a battery, and are not meant to provide information about the "lifetime" of a battery, i.e, its decrease in performance due to long-term *inter-cycle* effects, such as the fading of the total capacity due to aging or to repeated cycling.

The literature provides several studies on these effects, proposing mathematical models that are based on the electrochemical properties or the physics of the batteries and are therefore strongly bound to specific battery materials and chemistry (e.g., [5]–[9]). Although some other aging models, such as those proposed in [10]–[15], are empirically characterized onto a pre-defined equation template, they are still derived by measurements and therefore are not general enough to support different battery chemistries.

In this work, the aim is to derive an aging model that shares the basic features of a Peukert-like equation, that is, (i) analytical, but that can be empirically populated, and (ii) general enough to support different battery chemistries. Specifically, we propose a mathematical model for estimating the number of cycles with respect to the related capacity fade of batteries.

The accuracy of the approach proposed is demonstrated by applying this model to various commercial batteries of different chemistries, for which the manufacturers provide information on the long-term effects in their datasheets. Results show an average estimation error, in the number of cycles, generally within 10%, which is consistent with the typical tolerance provided in various datasheets (e.g., [16]).

The paper is organized as follows. Section II reports related works on battery modeling, while section III describes the

978-1-4673-9141-2/15 $31.00 © 2015 IEEE

proposed mathematical model for estimating the number of cycles of batteries. Section IV reports the experimental results, while Section V draws some conclusions.

II. BACKGROUND AND MOTIVATIONS

A. Battery Aging Issues

A battery, during its lifetime, deteriorates due to irreversible physical and chemical changes that mostly occur during its usage. The main visible effect of such deterioration is that the battery capacity appears to decrease with multiple charge/discharge cycles. This *capacity fading* occurs not only during cycling, but also under storage (i.e., long-term battery storage).

Many researchers have studied aging issues in batteries and have devised various types of models [5]–[7]. The key feature of these models is that they mostly relate capacity fading to battery "usage". The most significant factors that determine battery aging are the following:

- **Temperature**. As with other typical reliability mechanisms, aging increases at higher temperatures (typically according to an Arrhenius-type equation).
- **Depth-of-Discharge (DOD)**. The DOD is the % of battery capacity that has been discharged before starting a new charge phase. A DOD of 100% implies that a battery is fully discharged before re-charging it. Aging is increased by deeper discharge cycles (i.e., higher DOD values)
- **Discharge current**. It is usually measured in C-rate, a current normalized to the one necessary to discharge the nominal battery capacity in one hour. Aging is increased by higher discharge currents.
- **Number of cycles (N)**. Aging obviously worsens as the number of charge/discharge cycles increases.

B. Battery Aging Models

Concerning cycle life estimation, numerous researchers have proposed analytical models capturing the main aging mechanisms and capacity fading based on the electrochemical properties of the batteries and even including full-physics based models (e.g., [8] for lithium-ion batteries).

In [9] the authors proposed a model to calculate the usable number of cycles N of a battery based on the following equation:

$$N = N_1 \cdot e^{\alpha \cdot (1-DOD')} \qquad (1)$$

where DOD' is the normalized depth of discharge ($0 \leq DOD' \leq 1$), α is a characteristic constant of the battery and N_1 is the number of cycles at $DOD' = 1$. This model is empirically characterized for lead-acid, nickel-cadmium (Ni-Cd) and nickel-metal hydride (Ni-MH) batteries, whose cycle-life vs. DOD curve has an exponential shape. It is not, however, suitable for many lithium-based cells, whose cycle-life vs. DOD curve sometimes exhibits a more linear behavior (e.g., for $LiFePO_4$ cells).

A slightly different relationship between cycle-life and DOD was introduced in [10]:

$$N = N_{0.8} \cdot DOD' \cdot e^{\alpha \cdot (1-DOD')} \qquad (2)$$

where $N_{0.8}$ is the cycle life at $DOD = 80\%$, while α is a constant whose value is, respectively, 3 and 2.25 for lead-acid and nickel metal hydride (Ni-MH) tested battery packs (nowadays, Ni-MH batteries have mostly been substituted by lithium-ion batteries as they have higher energy density).

Thaller [11] has defined another relationship for battery cycle life after considering excess capacity F, with respect to the rated capacity, and a penalty factor due to the DOD, by including the P parameter, as reported in (3), which gives this mathematical prediction model for a general battery:

$$N = \frac{1 + F - DOD'}{A \cdot (1 + P \cdot DOD') \cdot DOD'} \qquad (3)$$

In our work F is always considered equal to 0, so that each analysis is performed after starting from the rated capacity of any commercial cell or cell string. The product $A \cdot DOD'$ represents the irreversible capacity loss in each cycle. Values of the parameter A were originally declared to be in the range $0.000 \div 0.002$ [11].

These previous models estimate the cycle life of a battery, always after considering a fixed irreversible capacity fading (e.g, 20%, that is, when the total maximum available capacity reaches 80% of the nominal one).

In [12] the authors introduce a complex cycle life model consisting of different equations, one for each stress factor considered, i.e., C-rate, T and DOD. Despite its high accuracy, the model derivation requires extensive empirical measurements and the model itself lacks the compactness and the generality of a Peukert-like equation.

Another analytical method for battery life prediction is based on the *Amp-hour–throughput*, i.e., the total energy supplied by the battery during its life [13], also called "charge life". The charge life Γ_R in Amp-hours (Ah) is defined as:

$$\Gamma_R = L_R \cdot DOD' \cdot C_R \qquad (4)$$

where C_R is the rated capacity in Ah at a rated discharge current I_R, and L_R is the maximum number of cycles referring to a given normalized depth of discharge DOD' and a discharge current I_R. In the model presented in [14], the authors proposed calculating an equivalent Ah weighted-throughput parameter. They claimed the C-rate effect on aging is negligible in Li-ion cells, since hybrid electric vehicles exhibit relatively large C-rates (±4C). This assumption cannot however be generalized for other applications.

The model proposed in [15] adopted this approach to estimate the cycling capacity fade through a modified definition of the Arrhenius equation, characterized by a square root time dependence.

C. Motivations of the Work

In spite of the various differences, all the above models are built by extracting parameter values through measurements on the batteries under test. Although the generated models are typically very accurate, this approach is quite time-consuming (especially when multiple cycles are involved) and requires expensive laboratory instrumentation.

For this reason, methods that rely only on available data (e.g., datasheets) to derive the capacity fade in batteries using analytical models (e.g., [17]) have been reported in the literature in recent years. Clearly, the accuracy of these models depends on the amount of available information reported into battery datasheets.

Before starting to describe the analytical model for battery capacity fade that we propose in this paper, an important consideration must be drawn: if the battery datasheet provides a cycle-life vs. DOD curve, as the one depicted in Figure 1, one can evaluate the battery cycle life without resorting to any estimation model.

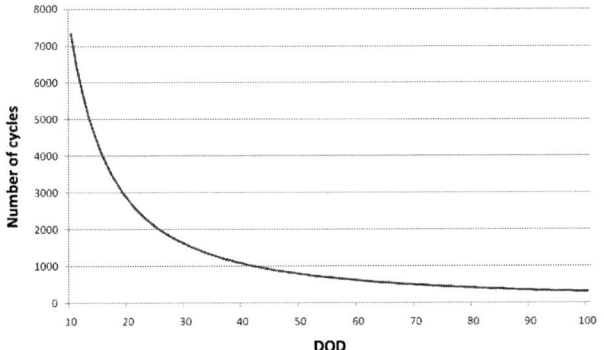

Fig. 1.   A typical plot of *Number of cycles vs. DOD*.

Unfortunately, this kind of information is seldomly present in datasheets. Hence, in these cases, the battery cycle life can be extracted either by means of empirical characterizations, as explained in the previous Section, or by an analytical model able to exploit available information to derive the battery capacity fade.

Many manufacturers provide information about capacity fade in the form of a *Capacity vs. Number of cycles* curve (e.g., [16]) as also depicted in Figure 2. From these plots, it is no simple matter to perform the battery cycle life evaluation, since the data about the number of cycles are available for a given number of DODs only and, furthermore, sometimes they might even show an uncertainty that may range from 8 to 10%, or even higher.

In this work we propose an analytical model for capacity fade that, using only datasheet information, allows the cycle life of a battery of any chemistry or type to be estimated, while having a simple mathematical form similar to Peukert's law.

III. MODELING METHODOLOGY

A. Model Definition

The model proposed in this work somehow mimics the shape of Peukert's law, as expressed by (5), which models the non-linear dependency between capacity and the discharge current:

$$t = \frac{C}{I^k} \tag{5}$$

where C is the capacity of the battery, I is the discharge current, and t is the time for totally discharging the battery; k is the Peukert's coefficient; typical values of k depend on

the battery chemistry and the manufacturing process and they typically range from 1.1 to 1.3.

As a matter of fact, the curves describing the capacity vs. number of cycles exhibit a similar non-linear relationship.

Our objective is therefore to derive a model expressing battery cycle life in a compact mathematical form similar to Peukert's law, and describing the general non-linear relationship between the capacity fade and the DOD.

In the case of capacity fade, the non-linearity concerns both the number of cycles N as well as the DOD, and the actual relationship among these quantities depends also on the value of the target capacity degradation (i.e., the behavior for a 20% capacity fade will be different from that for a 30% capacity fade). In order to model this non-linearity we need to define a new parameter that characterizes the battery performance during the cycling.

The proposed mathematical model is shown in (6); it allows to estimate the number of charging-discharging cycles N for a given battery based on four main parameters.

$$N = L \cdot \frac{C_{fade}}{DOD^h} \tag{6}$$

- L (called the *linear factor*) is the parameter that characterizes the battery performance over cycling,
- C_{fade} is the percentage of capacity loss,
- DOD is the depth of discharge expressed as a percentage $(0, \ldots, 100)$;
- h is the coefficient that models the nonlinear relationship between N and DOD for a certain C_{fade}.

The similarity with Peukert's law is evident. N, considered as an inter-cycle "lifetime" parameter, is obtained as the ratio of capacity and a weighted metric of discharge current (DOD^h). There are however two relevant differences: (1) factor L is used to scale the "lifetime" across multiple cycles, and (2) h is not constant, but depends on C_{fade}. This makes our approach more general with respect to previous models and allows one to adapt it to the available manufacturer's data.

Equation 6 defines a generic model template, which is empirically populated based on the extraction of data from typical capacity fade vs. # of cycles plots, as described in the next section.

B. Extraction of Model Parameters

Besides the "physical" quantities (C_{fade} and DOD) the model includes two other scale parameters, i.e., the linear factor L and the binding coefficient h, which have to be determined by fitting empirical data derived from available information (e.g., datasheet).

The actual parameter identification depends on the amount of available information. As discussed in Section II, our model is meaningful if the battery under analysis only provides information in the form of two or more curves in the (capacity, number of cycles) plane, each corresponding to a different DOD.

Let us assume that there are M such curves available in a datasheet or in a measured set of data. Obviously the larger

M, the more accurate the fitting process will be. Figure 2 exemplifies this scenario.

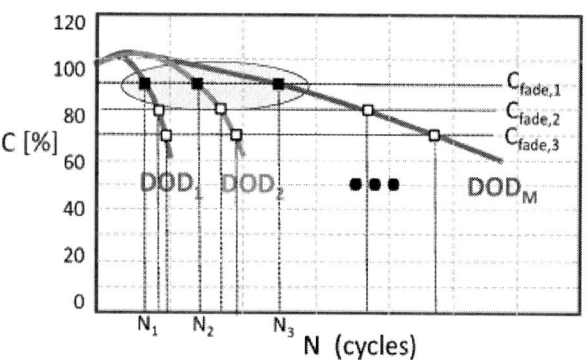

Fig. 2. Model Extraction Scenario.

Since we need to determine two parameters from the curve(s) (h and L), and given the limited number of samples points to be considered, it is feasible to derive them from an exhaustive exploration for all C_{fade} and DOD points, as the values of h and L that minimize the maximum error with respect to the curves. However, an exploration requires a feasible range for these two parameters, which is not easy to determine because they are only weakly linked to "physical" quantities. Of the two, L is the one with some physical interpretation since it can be regarded as a correction factor of the number of cycles N. Therefore, we can assume that L ranges between 1 and a value L_{max}, determined by inspection of the datasheet. As a rule of thumb, it is usually near to the largest value of N reported in the datasheet curves. Conversely, we have no insight of possible values of h. For this reason, we implement the search as a two-phase process, as described by Algorithm 1.

The search is organized into of two main iterations over L. In the first one (Lines 1–7), for all values of C_{fade} (assumed to be discretized into P values) and of the M DOD values it computes the resulting value of h using (7), which is simply a re-arrangement of (6) expressing h instead of N, and determines thus a feasible range $\mathcal{H} = [h_{min}, h_{max}]$ for h.

$$h = \frac{log(L \cdot \frac{C_{fade}}{N})}{log(DOD)} \quad (7)$$

Now that we have a feasible range for h, in the second iteration (Lines 10–26), we determine the optimal values of h and L, as follows. In the outer loop over L (Line 10), the optimal value of h is calculated first; for each value of h (using some discretization step), C_{fade} and DOD, N is computed using the model equation (6) (Line 16), and the error between this value and the one extracted from the datasheet is evaluated. The value of h that yields the least average error is stored as the best for a given value of L into an array $\mathbf{h}$, together with the relative errors (array $\mathbf{Err}$, Lines 22-23).

At the end of the iteration over L, the value of L corresponding to the smallest error is selected as single L_{opt} for the model (Lines 27-28), which is used as an index in $\mathbf{h}$ to determine h_{opt} for each C_{fade}.

Algorithm 1 Search for the best value of L

1: **for all** $L \in [1, L_{max}]$ **do**
2:     **for all** $C_{fade} = 1 \ldots P$ **do**
3:         **for all** $DOD = 1 \ldots M$ **do**
4:             Compute h by (7)
5:         **end for**
6:     **end for**
7: **end for**
8: $\mathcal{H} \leftarrow [h_{min}, h_{max}]$
9: $MinMaxErr \leftarrow \infty$.
10: **for all** $L = 1 \ldots L_{max}$ **do**
11:     $MaxErr \leftarrow 0$.
12:     **for all** $h \in \mathcal{H}$ **do**
13:         $TotErr \leftarrow 0$, $MinAvgErr \leftarrow \infty$.
14:         **for all** $C_{fade} = 1 \ldots P$ **do**
15:             **for all** $DOD = 1 \ldots M$ **do**
16:                 Calculate N using (6) and compute the absolute error E
17:                 $TotErr \leftarrow TotErr + E$
18:             **end for**
19:         **end for**
20:         $AvgErr \leftarrow TotErr/(P * M)$
21:         **if** $AvgErr < MinAvgErr$ **then**
22:             $\mathbf{H}[L] \leftarrow h$
23:             $\mathbf{Err}[L] \leftarrow AvgErr$
24:         **end if**
25:     **end for**
26: **end for**
27: $L_{opt} \leftarrow argmin(\mathbf{Err})$
28: $h_{opt} \leftarrow \mathbf{H}[L_{opt}]$

IV. MODEL VALIDATION

The validation of the proposed model is performed after considering batteries of various chemistries produced by different manufacturers. Although the type of aging data differs from one datasheet to another, we have collected the available information and translated it into the tabular format described in Section III; using these data, we ran the search algorithm to populate the model for each battery under analysis.

A. VRLA Batteries

We start our evaluation from Valve Regulated Lead Acid (VRLA) batteries, which have a more evident nonlinear aging behavior with respect to other chemistries. Moreover, datasheets for most VRLA batteries include more detailed information on aging, typically in the form of the plot of *Capacity vs. Number of cycles* (e.g., Figure 2).

Table I reports the extracted manufacturer data and the resulting model parameters for the AGM-VRLA XTV1272 battery by CSB. The first three columns represent the data extracted from the datasheet [16] for three different C_{fade} points, namely 10, 20, and 40%. The last four columns report the parameters obtained by the search algorithm, the resulting number of cycles N_m from the model, and the estimation error. After comparing N_m against the cycle life extracted from the datasheet (i.e., N_d), the absolute maximum error is 12.33% and the mean value is 9.97%.

TABLE I
EXTRACTED PARAMETERS AND NUMBER OF CYCLES ESTIMATION FOR THE CSB XTV1272 BATTERY.

CSB XTV1272 VRLA battery						
Datasheet			Model			
N_d	DOD	C_{fade}	L	h	N_m	Error(%)
681	30				597	-12.33
305	50	10		1.093621	342	12.13
151	100				160	5.96
861	30				770	-10.57
374	50	20	2464	1.222672	412	10.16
186	100				177	-4.84
1130	30				1021	-9.65
459	50	40		1.343506	514	11.98
231	100				203	-12.12

Although the error is not negligible, it is worth emphasizing that the datasheet for this battery reports a possible range of the number of cycles rather than a single curve, to indicate the intrinsic uncertainty of the estimation. The spread of the values actually increases for increasing DODs. From the datasheet, we found that the possible variation of the cycle life (measured as the difference between the minimum or maximum value with respect to the average) might even be up to 10, 11, and 16% for $C_{fade} = 10, 20$, and 40%, respectively. Hence, the absolute **maximum** estimation error obtained by the proposed model (i.e., around 12, 11, and 12%, respectively) is comparable with the maximum tolerance given by the manufacturer.

B. Other Battery Chemistries

Evaluation of other battery chemistries is complicated by the fact that in general only the manufacturers of VRLA batteries provide plots of *Capacity vs. Number of cycles*, for different DODs. In particular, datasheets usually report only a single *Capacity vs. Number of cycles* curve referring to a single DOD value for lithium-ion (Li-ion) batteries. The availability of just one DOD reference, however, would yield a model with little practical use in this case, since the calibration for discharge patterns would be different from that used for characterization.

Therefore, in order to have a more meaningful assessment of the accuracy of the proposed model, we only selected those batteries whose datasheets report the *Number of cycles vs. DOD* characteristic, even just for a single C_{fade} value. In any case, values of *DOD* below 10% are not used for the derivation of the model because (i) they are not representative of typical battery usage and (ii) they are not statistically representative. It is worth noticing that the number of cycles should approach infinity as $DOD \rightarrow 0\%$; therefore, as *DOD* gets smaller it would be correct to consider a range of values rather than a precise value. Of course, all the characteristics given by the manufacturers always refer to certain operating and working conditions (e.g., charge/discharge current and temperature), which are usually different from one brand to another. In this work, we do not consider the differences among these

conditions, in order to firstly validate the basic proposed model. The parameters and estimation errors for these batteries are reported in Tables II and III, which also report, for a more comprehensive validation, the results of the application of the existing and most meaningful analytical models [9], [11]. As (1) requires the number of cycles at DOD=100% as input parameter, the evaluation of that previous model was not possible for two batteries because this value is not available in their datasheets, as reported in Table II. On the other hand, as the model proposed by [11] is useless for $DOD' = 1$ (in this case, N in (3) would be equal to zero), the analysis was re-performed by considering the maximum DOD=80% as reported in Table III.

In Table II, the largest absolute estimation error of the model occurs for a $LiFeMgPO_4$ battery, almost 20%, while the maximum mean value is 11.35% for the Alpha® one. However, the total average error of the maximum errors for the 10 batteries in the table is 10.66%. The mean errors are obviously smaller, in general less than 10%, and in one case 11.35%. In general, the proposed model shows robustness and accuracy for different types of electric storage devices. For the Li-ion battery by Saft Evolion the linear factor L is very high with respect to any other battery. In fact, the linear factor usually depends on the battery properties of cycling, while the range of the h parameter strictly depends on the linearity of the cycle life with respect to the DOD. The lowest h coefficient found in the model validation is 0.225627 for the Discover 22-24-700 battery, whereas the highest h is 2.000414 for the Saft Evolion.

The chart in Figure 3 reports a comparison of the estimation models after applying each of them to the benchmarks. For a comprehensive report, it also includes the main results obtained for the analysis of the model by [10], whose estimation errors are too great to be reported. Furthermore, for the here proposed model, this chart considers the worst case (i.e., data reported in Table II). Although the previous models have

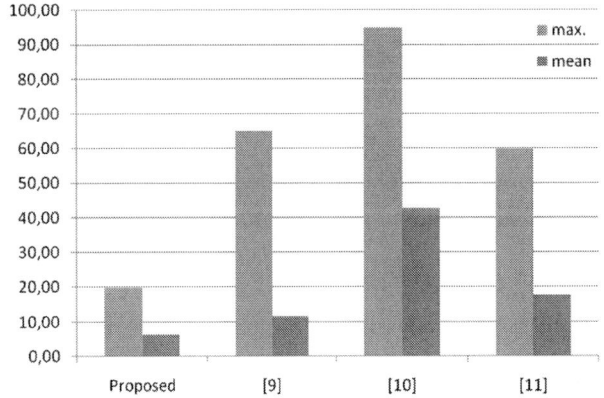

Fig. 3. Maximum and mean estimation errors given by the models for all the selected benchmarks.

two parameters (i.e., coefficients) in their expressions, one of them always strictly depends on the battery properties. In the here proposed model, both parameters L and h can be characterized, resulting in higher accuracy thanks to an additional degree of freedom in the modeling process.

TABLE II

BATTERY DATA, PREDICTION MODEL PARAMETERS, AND ESTIMATION ERROR OF THE CYCLE LIFE FOR VARIOUS BATTERIES WHOSE MANUFACTURERS PROVIDE THE NUMBER OF CYCLES VS. DOD CHARACTERISTIC.

Producer	Code	Type	Model							
			Proposed				[9]			
			L	h	Abs. error [%]		N_1	α	Abs. error [%]	
					max	mean			max	mean
EnerSys	65-PC1750	AGM-VRLA	9083	1.393212	12.34	8.05	330	2.488793	63.03	34.49
Concorde	Sun Xtender	AGM-VRLA	4629	1.176563	15.20	8.79	354	2.644044	28.56	15.73
Sonnenschein	A600	Gel-VRLA	3874	1.020317	2.03	0.92	718	1.747624	21.12	12.84
Alpha Tech.	KL, KM, KH types	$NiCd$	31107	1.587189	18.10	11.35	463	2.412794	54.04	28.38
C&D Tech.	LI TEL 48-170 C	Li-ion	109882	1.420135	6.27	3.60	2987	2.022832	2.53	1.22
Saft	Evolion	Li-ion	1157452	2.000414	13.84	8.15	n.a.	-	-	-
Seiko (SII)	MS621	$Mn\ Si\ Li-ion$	986	0.995693	0.90	0.38	202	1.712398	20.29	12.07
Maxell	ML2016	Li/MnO_2	2393	1.566125	11.28	6.49	39	2.743101	65.11	36.81
Discover	22-24-6700	$LiFePO_4$	671	0.225627	6.99	4.36	n.a.	-	-	-
Valence	U-CHARGE	$LiFeMgPO_4$	153425	1.491094	19.66	9.21	2679	2.764444	19.10	12.31

Note. n.a.: not available

TABLE III

BATTERY DATA, PREDICTION MODEL PARAMETERS, AND ESTIMATION ERROR OF THE CYCLE LIFE FOR VARIOUS BATTERIES WHOSE MANUFACTURERS PROVIDE THE NUMBER OF CYCLES VS. DOD CHARACTERISTIC. THE MAXIMUM DOD IS 80% FOR ALL THE ANALYSES.

Producer	Code	Type	Model							
			Proposed				[11]			
			L	h	Abs. error [%]		A	P	Abs. error [%]	
					max	mean			max	mean
EnerSys	65-PC1750	AGM-VRLA	9083	1.393212	12.34	7.49	0.00140	-0.436228	36.57	22.68
Concorde	Sun Xtender	AGM-VRLA	4629	1.176563	15.19	8.81	0.00180	-0.953029	8.37	5.44
Sonnenschein	A600	Gel-VRLA	3874	1.020317	2.03	0.82	0.00140	-1.010028	3.64	0.99
Alpha Tech.	KL, KM, KH types	$NiCd$	31107	1.587189	18.10	10.70	0.00110	-0.674032	13.67	7.29
C&D Tech.	LI TEL 48-170 C	Li-ion	109882	1.420135	6.26	4.21	0.00020	-0.864030	9.94	4.43
Saft	Evolion	Li-ion	1157452	2.000414	13.84	8.15	0.00010	-0.452045	59.81	34.84
Seiko (SII)	MS621	$Mn\ Si\ Li-ion$	986	0.995693	0.90	0.40	0.00500	-0.999028	0.99	0.42
Maxell	ML2016	Li/MnO_2	2393	1.566125	11.28	6.53	0.00500	1.228006	48.20	32.71
Discover	22-24-6700	$LiFePO_4$	671	0.225627	6.99	4.36	0.00060	-1.200934	52.83	38.52
Valence	U-CHARGE	$LiFeMgPO_4$	153425	1.491094	19.66	8.80	0.00020	-0.967028	27.47	15.69

V. CONCLUSION

A novel mathematical model for estimating the number of cycles of a battery with respect to an expected capacity fade, has been proposed. The related equation describes the cycling behavior of batteries of different chemistries, and it demonstrates the possibility to obtain a very fast and also accurate exploration of battery lifespan. The characterization of the long-term effects for a specific battery only requires two parameters: a linear factor L and the exponential h coefficient. Results show an estimation mean error generally within 10%.

Future works will include temperature and current rates in the model, in order to analyze the estimation error after considering different operating and working conditions with respect to the reference ones.

REFERENCES

[1] L. Benini et al., "Discrete-Time Battery Models for System-Level Low-Power Design," *IEEE Transactions on Very Large Scale Integration (VLSI) Systems*, vol. 9, no. 5, pp. 630-640, Oct. 2001.

[2] M. Chen and G.A. Rincón-Mora, "Accurate Electrical Battery Model Capable of Predicting Runtime and I-V Performance," *IEEE Transactions on Energy Conversion*, vol. 21, no. 2, pp. 504–511, June 2006.

[3] M. Petricca et al., "An Automated Framework for Generating Variable-Accuracy Battery Models from Datasheet Information," *International Symposium on Low Power Design (ISLPED 2013)*, pp. 365–370, Sept. 2013.

[4] W. Peukert, "Über die Abhängigkeit der Kapazität von der Entladestromstärke bei Bleiakkumulatoren," *Elektrotechnische Zeitschrift*, vol. 20, pp. 20–21, 1897.

[5] A. Millner, "Modeling Lithium Ion battery degradation in electric vehicles," *IEEE Conf. Innov. Technol. Efficient Rel. Elect. Supply (CITRES 2010)*, pp. 349–356, Sept. 2010.

[6] P. Ramadass et al., "Mathematical modeling of the capacity fade of li-ion cells," *Journal of Power Sources*, vol. 123, no. 2, pp. 230–240, 2003.

[7] L. Lam and P. Bauer, "Practical Capacity Fading Model for Li-Ion Battery Cells in Electric Vehicles," *IEEE Transactions On Power Electronics*, vol. 28, no. 12, pp. 5910–5918, Dec. 2013.

[8] V. Ramadesigan et al, "Parameter Estimation and Capacity Fade Analysis of Lithium-Ion Batteries Using Reformulated Models," *Journal of the Electrochemical Society*, vol. 158, no. 9, pp. A1048–A1054, 2011.

[9] H.N. Seiger, "Effect of Depth of Discharge on Cycle Life of Near-Term Batteries," *Proc. 16th Intersociety Energy Conversion Engineering Conference (IECEC)*, pp. 102–110, Aug. 1981.

[10] A.F. Burke, "Cycle Life Considerations for Batteries in Electric and Hybrid Vehicles," *SAE Technical Paper* (No. 951951), 1995.

[11] L.H. Thaller, "Expected Cycle Life vs. Depth of Discharge Relationships of Well-Behaved Single Cells and Cell Strings," *Journal of Electrochemical Society*, vol. 130, no. 5, pp. 986–990, 1983.

[12] N. Omar et al., "Lithium iron phosphate based battery – Assessment of the aging parameters and development of cycle life model," *Applied Energy*, vol. 113, pp. 1575–1585, 2014.

[13] P. Symons, "Life Estimation of Lead-Acid Battery Cells for Utility Energy Storage," *Proc. 5th Conf. Batteries for Utility Storage*, July 1995.

[14] V. Marano et al., "Lithium-ion batteries life estimation for plug-in hybrid electric vehicles," *Vehicle Power and Propulsion Conference 2009 (VPPC '09)*, pp. 536–543, 7-10 Sept. 2009.

[15] J. Wang et al., "Cycle-life model for graphite-LiFePO4 cells," *Journal of Power Sources*, vol. 196, no. 8, pp. 3942–3948, Aug. 2011.

[16] CSB Battery Co., Ltd.
www.csb-battery.com/upfiles/dow01404206487.pdf
(accessed December 30, 2014).

[17] R. Spotnitz, "Simulation of capacity fade in lithium-ion batteries," *Journal of Power Sources*, vol. 113, no. 1, pp. 72–80, 2003.

Power-Management High-Level Synthesis

Dominik Macko, Katarína Jelemenská, Pavel Čičák

Faculty of Informatics and Information Technologies
Slovak University of Technology
Bratislava, Slovakia
dominik.macko@stuba.sk, katarina.jelemenska@stuba.sk, pavel.cicak@stuba.sk

Abstract—**Power management is an integral part of almost every new system design. It enables to keep the power under constrains, implementing such power-reduction techniques as power gating, multi-voltage design, or voltage and frequency scaling. Due to the complexity of modern designs, the system level of abstraction is adopted as a design starting point. However, the power management is not yet fully adopted at such abstraction level. In the previous research, we have proposed the abstract power-management specification, simplifying its adoption by an order of magnitude. This paper targets the power-management high-level synthesis, closing thus the gap between the system-level power management and its standard form at lower abstraction levels. Such design automation enables to reduce a number of human errors, potentially introduced by manual design. The presented experimental results validate the proposed approach.**

Keywords—design automation; high-level synthesis; low power; power management; specification; system level

I. INTRODUCTION

Power density in highly integrated CMOS (Complementary Metal-Oxide Semiconductor) circuits presents serious concerns about reliability of the devices. Because of the limited battery capacity in mobile devices, chip packaging and cooling aspects, or simply the energy consumption, the power needs to be managed in every new chip design [1].

Power management has been established as an application method for various power-reduction techniques that have been developed to alleviate the power problem. These techniques include, for example, clock gating, power gating, voltage scaling, or frequency scaling. In complex systems, the power management integration is rather difficult, and therefore systematic low-power design flows have been standardized. Two key industrial standards are CPF (Common Power Format) [2] and UPF (Unified Power Format) [3]. These standards enable to introduce low-level details (e.g. power-supply ports, supply nets, power-management elements) into a functional HDL (Hardware Description Language) model, and thus enable to verify power management at the RTL (Register-Transfer Level). They have soon gained popularity and are supported by the EDA (Electronic Design Automation) industry today. Using these standard formats, a designer can split the system into several power domains. Power domain is a set of system blocks operating in the same power state (the same supply voltage and operation frequency). Power states

are specified using special circuitry – the power-management elements, such as power switches (enabling to power down the domain or to switch its power-supply net), level shifters (adjusting the voltage level of logic signals between domains), isolation, or retention cells. Thus, a power state is defined by a unique combination of control-signals values for power-management elements inside a certain power domain. The designer specifies allowed voltages for supply ports and nets, and for a verification purpose, creates a power-state table, specifying allowed combinations of states among these elements.

Since the complexity of modern designs caused adoption of the system level (ESL) in the design process (as suggested by [4]), the abstraction offered by the current low-power standards is not sufficient. Therefore, the research around extension of low-power design flows offered by these standards to the system level is gaining attention.

A. Related Work

The method [5] augments a transaction-level model with abstract UPF-based power-management concepts. It requires annotating power information to the system-level model in order to proceed with power-architecture exploration. Downsides of this method are that the RTL implementation is manual, the power annotation is time consuming, and architecture exploration does not take into account other important parameters, such as area or performance. Another method [6] automatically fills the power information in the system-level model based on the low-level simulation and technology libraries. Thus, it is highly dependent on a design-reuse concept. This method enables to generate a standard UPF specification, and therefore lower-level verification is easily accomplished using the existing EDA tools. However, the power management offered by this method contains a similar amount of details as the UPF standard; therefore, the abstraction in the system-level model is not sufficient.

Similar methods [7-9] also offer system-level power modelling, supporting the power management. Although they enable power-architecture exploration, they also do not take into account other design parameters (area, performance). Moreover, these methods are not based on standard low-power concepts; therefore, the equivalency between ESL and RTL power management is somewhat difficult to verify. A method [10] puts high-level synthesis to the forefront. A clock-gating ESL specification is passed to the high-level synthesizer, which automatically inserts clock-gating cells into the RTL

This work was partially supported by the Slovak Science Grant Agency (VEGA 1/1008/12 and VEGA 1/0616/14), Slovak University of Technology Bratislava (ASRSS) and COST Action IC 1103 MEDIAN.

978-1-4673-9141-2/15 $31.00 © 2015 IEEE

model. The exact location and activation of clock gating has to be specified in the abstract model, what requires a manual effort. Another disadvantage is that this method does not support other efficient power-management techniques.

B. Paper Overview

Based on the identified benefits and drawbacks of the existing methods for ESL power management, we have proposed a novel low-power design methodology [11, 12]. The methodology provides an abstract power-management specification based on UPF concepts (similarly to [5]), utilizes UPF specification at the lower levels to enable verification using existing EDA tools (similarly to [6]), and offers high-level synthesis for multi-parameters trade-off and RTL design analysis for more accurate results (similarly to [10]). Fig. 1 provides an overview of the proposed design flow (the contributing parts are shown in dark-grey color).

This paper presents the power-management high-level synthesis process augmenting the functional high-level synthesis used in the industry today. Based on the abstract power-management specification at the ESL, the standard UPF specification is generated at the RTL as well as the power-management unit driving the control signals for power-management elements. Section II provides an overview of the existing power-reduction techniques. Section III introduces the basic principles of the proposed abstract power-management specification. Section IV describes the developed power-management high-level synthesis and Section V presents verification of the synthesized aspects. Before the conclusion, the experimental results are provided.

II. POWER-MANAGEMENT/REDUCTION TECHNIQUES

Many power-reduction techniques have been developed [2], but only some of them can be used for power management during the runtime. Power management has the ability to change the power states of the system components and thus enables to save the power or to temporarily increase the system performance. Power management as a way to design a low-power device should implement such an algorithm that gives the components just enough power to perform a certain task and powers them down when not needed. Moreover, all unnecessary switching activities should be prevented.

- *Clock gating* – disables the clock signal to stop loading the same value in some register. From the power-management perspective, it can be used to temporarily stop the operation of a synchronous block of the system. In such a way, the dynamic power consumed by unnecessary switching is saved.
- *Power gating* – shuts down the power to a system block when not needed. It saves both the static and dynamic power, but it takes time to power down and power up the block.
- *Operand isolation* – isolates unneeded datapath elements when not required, reducing unnecessary switching and thus saving dynamic power. Used as a power-management technique, it can be used to isolate inputs of an idle system block.
- *Voltage scaling* – enables to operate a system block at various voltage levels – e.g. a high level for high performance and a low level for power saving. Usually, each power domain is scaling its own supply voltage depending on its current requirements.
- *Frequency scaling* – enables to change the clock frequency of a synchronous system block. Higher frequency consumes more dynamic power, but enables to perform a task faster. This technique is usually combined with the voltage scaling, temporarily lowering the performance (task is not time-critical) to save power.
- *Substrate biasing* – enables to temporarily bias a substrate and thus raise the threshold voltage. This technique reduces the leakage current and thus reduces the static power.
- *Multiple voltages, multiple thresholds, gate sizing, logic restructuring, pin swapping* – these techniques are automatically used by a synthesis tool to select a suitable combination of library cells with various parameters to meet preset constraints.
- *Memory partitioning, bus segmentation, hardware acceleration* – these techniques target architectural decisions. Unfortunately, no component can be considered the best for each system, thus the architecture exploration process is commonly taken to select the optimal components.

Some of these techniques require special elements to be added to the design. When powering a system block down, its outputs should be isolated to prevent floating signals to corrupt data in the powered blocks. For such purpose, various isolation cells can be used – from a simple AND gate to a latch-based isolation elements. The communication between the blocks operating at different voltage levels should be level-shifted. Level shifters are added at the inputs of the receiving block. There exist cells that work as both, the isolators and level shifters. For gating the clock signal, clock-gating logic is used. It is similar to the isolation, but the timing impact should be carefully considered when dealing with the clock signal. To operate a block at different voltage levels, a power switch is needed to switch between power-supply networks or to simply shut the power off. Often, the registers state is required to be retained when powering-off some system block. Thus, some retention cells are used.

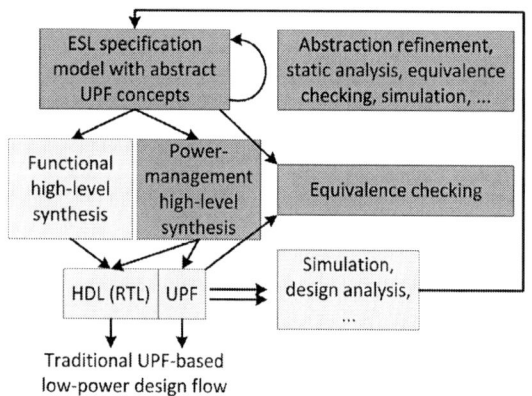

Fig. 1. Overview of the proposed low-power design flow.

978-1-4673-9141-2/15 $31.00 © 2015 IEEE

III. ABSTRACT POWER-MANAGEMENT SPECIFICATION

The abstract power-management specification is also using power states, but in slightly different manner. An abstract power state is not representing control-signals values for power-management elements, since these are not present in the ESL specification. It represents a voltage-frequency pair (i.e. a performance level) of the power domain, to which it is assigned. All power-reduction techniques applicable by means of power management can be introduced by specific power states. The abstract power states include the following:

- *NORMAL* – system blocks in the power domain are operating at the basic voltage and frequency level.
- *HOLD* – all blocks in the domain stop their operation, but stay powered.
- *DIFF_LEVEL#* – blocks of the power domain in this state operate at the voltage and/or frequency level different from the basic one; # is an ordinal number – enables to specify multiple such levels.
- *OFF* – the whole domain is powered off.
- *OFF_RET* – registers values of system blocks in the power domain are retained, while the domain is powered off.

These states specify that certain architectural power-reduction techniques will be integrated into the corresponding domains at lower abstraction levels. The *NORMAL* state means that no explicit architectural power-reduction technique will be applied. The *HOLD* state represents the clock gating and the operand isolation – any switching activity at the power-domain blocks inputs is prevented. The *DIFF_LEVEL* states mean that multiple performance levels are used in the power domain. These enable voltage and frequency scaling and a multi-voltage design. The *OFF* and *OFF_RET* states refer to the application of power gating without and with state retention.

The system is usually set to a specific operating mode in order to execute some task – power domains are in specific power states for such a mode. It is useful to specify possible power modes – i.e. to specify combinations of power states among power domains. Such a practice reduces the verification overhead – only the possible power modes need to be verified. A designer can then easily switch the system power mode to the most feasible one for some specific task, without explicitly specifying which power-domain states need to be changed.

Thus, the abstract power-management specification involves a specification of power domains with a list of possible power states, specification of performance levels for active power states (*NORMAL* and *DIFF_LEVEL*), assignment of system blocks to power domains, specification of possible power modes of the whole system, and switching between the specified power modes. For an illustration, a sample of abstract power-management specification is provided in Fig. 2. We may notice a special *POWER_MODE* variable, representing the current power mode. It needs to be initialized to a certain specified power mode. The power-mode switching is specified by assigning another specified power mode to this variable in the functional design.

```
power_domain1 (off,normal,diff_level1);
diff_level1 (1 V, 50 MHz);
system_block instance1(power_domain1);
power_mode1 (off,hold);
POWER_MODE = power_mode1;

if (control_condition)
  POWER_MODE = power_mode2;
else
  POWER_MODE = power_mode1;
```

Fig. 2. A sample of abstract power-management specificaiton.

This abstract power-management specification is sufficient for high-level synthesis to generate a standard UPF specification. The required power-management elements (isolators, level shifters, power switches, retention elements) can be implicitly deduced from the abstract specification and system-blocks relations. The high-level synthesis has to comply the following rules.

- Inputs of blocks in a power domain in the *HOLD* state coming from outside of the domain have to be isolated.
- Communication between blocks in different power domains that are operating at different voltage levels has to be level-shifted.
- Communication between blocks in different power domains that are operating at different frequencies has to be synchronized.
- The clock signal to system blocks in a power domain that is powered-down has to be stopped.
- Outputs of blocks in a powered-down power domain have to be isolated.
- Power-supply network of a power domain in the *OFF* or *OFF_RET* state has to be switched off.
- Power-supply network of a power domain operating in multiple power states with different voltage levels has to be switchable.

The offered abstraction simplifies the specification, thus reducing the time required for its description. A designer does not have to keep in mind these rules – they should be followed by the high-level synthesis process.

IV. POWER-MANAGEMENT SYNTHESIS

The power-management synthesis consists of two distinguished processes. The first one is the synthesis of power-management specification to the UPF standard form. The other one is the synthesis of a controller, called the power-management unit, generating control signals for power-management support logic – power switches, isolation cells, and retention cells.

A. Power-Intent Specification Synthesis

During this process, power intent is extracted from the abstract power-management specification and subsequently analyzed in order to determine which power-management elements need to be specified in the UPF.

The first step is to create power domains in the UPF specification. Since the power domains are specified already in the ESL abstract specification, this step represents their rewriting in the UPF style. During creation of power domains, system blocks are assigned to them according the ESL specification. Beside the specified power domains, there is a top domain created, representing a main power domain of the whole system. Components that are not explicitly assigned to any power domain belong to the top domain.

The second synthesis step creates power-supply ports according to the required voltages in the design. These are determined based on performance-level assignments in the abstract specification. The next step is to create required supply nets and connect them to supply ports. How and which supply nets are created is determined based on power-domains need for power switch. If a power-domain's supply needs to be switchable, it requires a dedicated net. Otherwise, it can reuse a top-domain supply net. As a part of this step, primary power and ground nets are assigned to each power domain. After the supply network is created, the power-management elements can be specified. Firstly, we create a power switch for each power domain with switchable supply. Whether a power domain needs a switchable supply depends on the abstract power states, in which it can operate. If there are at least two voltage states (a power state with a specific voltage level, including off state) specified for a domain, it needs to have a switchable supply net.

The next step is to create the required isolation in the design. Based on the analysis of abstract power states, the power domains with the *HOLD*, *OFF* and *OFF_RET* states are pinpointed. If a power domain can be in *HOLD*, input isolation is created. If a power domain can be powered down, output isolation is created to prevent floating signals. However, in order to reduce unnecessary switching activity in powered down domain, its inputs are also isolated. The abstract specification of performance levels enables to identify communication between power domains operating at different voltages. This communication is regulated using level shifters. Level shifters are placed at the domain boundary. Input level shifters are located inside the power domain, output level shifters are located inside a parent domain (the top domain). In each power domain that contains *OFF_RET* in abstract specification, the retention is set. It retains the value of all registers inside system blocks, located in such a domain.

As the last step, voltage states of created ports are specified. These ports include the created supply ports and the power-switches output ports. These states are determined based on abstract power states specified for power domains, to which primary supply nets these ports connects. An important part of this step is to create a power-state table (PST) based on the power modes, specified at ESL. One of the differences is that the states are specified for individual ports instead of power domains. Another difference is that these states represent voltage states, not the power states. The last difference is specification of additional voltage modes (analogously to voltage states) that are required for correct power management functionality. For example, if a power domain changes its state from *NORMAL* to *OFF*, it has to be firstly isolated and only after that it can be powered down. Thus, for each transition between power modes in the abstract specification, additional intermediate power modes are required (beside the specified power modes). These additional power modes are translated to voltage modes and if any of them is not present in the PST yet, it is added to the UPF specification.

The generated UPF specification for power switches, isolation, and retention contains the control signals. Isolation and retention requires only one signal for each element, but power switch can require multiple signals – it depends on how many voltages the power switch has to switch between. These control signals are driven by the power-management unit.

B. Power-Management Unit Synthesis

The power-mode changes are modelled in the functional ESL design using the *POWER_MODE* variable. After the extraction of power-intent information from the functional model, the *POWER_MODE* variable is changed to the enumerated type. Enumerators of this variable are the identifiers of the specified power modes. The actual values, which these enumerators represent, are unsigned integer values. It means that the value of the first specified power mode is 0, the next power mode is 1, and so on. An existing functional high-level synthesizer (e.g. Catapult or C-to-Silicon Compiler) generates an RTL implementation of the ESL specification. The *POWER_MODE* variable is preserved or it is synthesized into a register, depending on the support of enumerated type in the target language. Whether it remains an enumerated-type variable or it is a register, *POWER_MODE* drives the input of the power-management unit. This value is passed to a power-mode determination entity (see Fig. 3), in which a control-signal based encoding of the power mode is determined. Such a target power mode is passed to a power-state machine, handling the power-mode transition through a sequence of intermediate power modes.

Since the power-intent specification synthesis generates the required power-management elements in UPF, we are aware of the required control signals to be driven by the power-management unit. Thus, there is no problem to create an encoder that is encoding the power states of some power mode into the combinations of control signals. The problematic part of the power-management unit is the power-state machine. It has to implement the transitions between all possible power modes, including the intermediate modes. The intermediate power modes are transparent to the power-mode determination entity, i.e. these cannot come as a target power mode at the power-state machine inputs. Using these intermediate power modes, the power-state machine creates the correct control sequences, following the rules below.

- A block has to stop its operation before it is powered-down.
- Inputs and outputs of a block have to be isolated before it is powered-down.
- The state of a block is retained before the block is powered-down, if the retention is required.

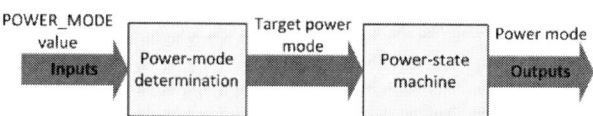

Fig. 3. The synthesized power-management unit architecture.

- A block has to power-up before its state is restored, in case the state was retained.
- A block has to power-up before the isolation is disabled.
- Isolation for a block is activated before the state retention, in case the retention is required.
- The state of a block is restored before the isolation is disabled.
- If the power state of a block is changing from a high-performance state to a low-performance state, the frequency has to be lowered prior to the voltage.
- If the power state of a block is changing from a low-performance state to a high-performance state, the voltage has to be increased prior to the frequency.

The abstract state machine, specified by a designer through power-mode changes in the abstract specification, is thus transformed into a detailed power-state machine containing additional (intermediate) power modes and generating control signals for power-management executive logic. Beside for the power-management elements, the control signals are also driven for a clock-frequency generator, enabling the frequency scaling for system blocks in power domains.

V. SYNTHESIZED POWER-MANAGEMENT VERIFICATION

The abstract power-management specification is already verified at the ESL for presence of syntactical, semantical, and basic structural errors, ensuring its completeness and consistency with the functional specification. Although the automation of the proposed power-management high-level synthesis prevents human-errors introduction, the generated UPF specification should be verified whether the power intent was preserved. Therefore, we have developed an equivalence-checking method between the generated UPF and the abstract power-management specifications. In addition to the power-intent verification, the synthesized power-management unit should be also verified whether it generates correct control sequences and does not violate any of the rules, stated in the previous section. For this purpose, we use the assertion-based verification integrated into the RTL functional verification process.

A. Equivalence Checking

The power intent is extracted from the generated UPF and verified whether corresponds to the ESL specification. Since UPF contains only voltage aspects of the power management a common representation is used for the comparison. This representation includes a list of power domains, a list of power modes, and lists of system blocks in each power domain. Power domains in the common representation are given by their identifiers and voltage states they can reach. Power modes are represented analogously. Thus, the abstract power states are translated to voltage states according to the performance-level assignments. Translation of the UPF to the common representation is achieved using a quasi-reverse process to the high-level synthesis. The first step involves comparison of power domains. The common representation of information extracted from the ESL specification has to contain all UPF-extracted power domains except for the top

domain. The lists of assigned system blocks to each power domain have to be the same. At last, the power states of power domains and power modes are compared. Each ESL-extracted voltage state has to be present in the UPF-extracted common representation. It means that each voltage state given by the ESL power management is possible in the generated RTL power management. Error messages, produced by this verification process, drive a designer to the error source, speeding-up the debugging process.

B. Assertion-Based Verification

We also propose another synthesis process – a synthesis of assertions, checking whether the control sequences are correct during runtime (simulation). The generation of these assertions uses state-space exploration method. They ensure that the control-sequences rules are not violated and if they are, assertions pinpoint what went wrong. The generated assertions also provide a suitable coverage-measurement support, driving a designer to create test stimuli covering unverified power modes or transitions. In order to speed-up the RTL verification preparation, a power-aware test-bench skeleton is created, containing the generated assertions and required UPF constructs. There is also a simple pseudorandom-verification approach used, randomly switching *POWER_MODE* value, and thus testing various power-mode transitions. However, a designer should create a proper test-bench for functional verification of the whole synthesized RTL model, not just the power-management unit.

C. Power-Aware Verification

The proposed high-level synthesis generates the standard UPF power-intent specification and the power-management description in the VHDL standard language. Thus, existing EDA tools should be used for full power-aware verification. A formal power-aware static analysis should be used to validate the generated UPF specification and a power-aware simulation should be used to verify the correct functionality of the power-managed system. This step is also used for power analysis, determining whether the power constraints have been met. If they are not, the previously proposed methodology assumes that the abstract power-management specification is modified and the currently proposed power management high-level synthesis is used to rapidly create another power architecture at the RTL. It enables relatively easy power-architecture exploration.

VI. EXPERIMENTAL RESULTS

To validate the proposed synthesis method, we have performed an experiment with 10 randomly generated samples of abstract power-management specifications with various complexities. The goal was to synthesize these samples through the proposed high-level synthesis processes and validate the generated RTL power-management aspects in the existing professional EDA tool. For this purpose, we have used Modelsim SE 10.2c simulation environment, especially its UPF static analysis and power-aware simulation option.

The parameters of the generated samples are provided in Table I. They involve the number of power modes in the

TABLE I.   PARAMETERS OF THE GENERATED SAMPLES

#	Power Modes	Power Domains	Power States	Blocks	ESL Power Management
1	2	1	2	3	313
2	3	2	2	2.5	500
3	3	4	1.75	3	751
4	10	3	3	2	862
5	5	3	4	2	643
6	10	5	2.4	2	1275
7	7	4	3.5	2	953
8	7	5	3	2	1051
9	5	10	2.1	3	1939
10	3	10	2.2	2	1402

TABLE II.   THE OBSERVED PARAMETERS OF SYNTHESIZED SAMPLES

#	UPF	VM	PMU	PM	Control Signals	SVA	Coverage Statements	Directive Coverage
1	1680	2	3300	4	3	3863	10	100 %
2	2706	3	4452	5	4	4749	17	100 %
3	4658	5	8658	12	7	9488	29	100 %
4	5557	9	63304	49	10	17779	81	100 %
5	6035	15	35205	46	13	25912	103	98 %
6	7443	15	199866	106	13	48111	146	89 %
7	8778	28	142992	114	16	52330	175	81.1 %
8	9399	25	131478	101	17	45150	156	96.1 %
9	12232	27	214122	132	19	91620	188	82.4 %
10	13397	25	67691	65	21	50911	126	99.2 %

abstract specification, the number of power domains, the average number of power states per domains, and the average number of system blocks per domain. The last column in the table represents the number of characters required for the abstract power-management specification for individual samples. Some interesting observed parameters and results after high-level synthesis are provided in Table II. The second column (*UPF*) represents the number of characters required for the synthesized UPF specifications. *VM* refers to the number of voltage modes, generated in UPF. *PMU* represents the number of characters required for description of the generated power-management units. *PM* refers to the number of generated power modes, which include additional intermediate modes. The next column represents the number of control signals, required for individual samples. *SVA* refers to the number of characters required for description of the generated assertions. The next column represents the number of generated explicit coverage statements. The last column provides the achieved coverage using the automatically generated test-benches with pseudorandom power-mode switching in 10 ms simulation runtime. All of the synthesized UPF samples have successfully passed through the proposed equivalence checking and have been statically analyzed by the Modelsim power-aware static checks without any error. The generated power-management units have been exercised during short simulation runtime, achieving high coverage of the power modes. No assertion has been violated, meaning that all control-sequences rules have been followed during simulation. The compilation in Modelsim also proved that the generated power-management information is syntactically and semantically correct. The provided results illustrate that many potential human errors are prevented by this process and a lot of time is saved, considering an amount of automatically generated data. There are up to thirteen thousand characters generated for UPF specification, up to two hundred thousand characters for power-management unit, and up to ninety thousand characters for assertions. The manual effort would require days or even weeks (compared to the seconds) for description of the correct power management at the RTL.

VII. CONCLUSIONS AND FURTHER WORK

In this paper, we have proposed the power-management high-level synthesis. Based on the ESL abstract specification, it synthesizes an RTL power-management model, consisting of the standard UPF specification and the functional description of the power-management unit. The proposed verification steps ensure that the initial power intent is preserved after the synthesis. The proposed synthesis also generates many power-management assertions, usable in RTL functional verification. This automated process significantly speeds-up low-power systems development and avoids many potential human errors. Our further work will involve automated partitioning of system blocks into power domains with power-states assignment. It could enable a complete abstraction from power-reduction techniques, applicable through power management.

REFERENCES

[1] M. Keating, D. Flynn, R. Aitken, A. Gibbons and K. Shi, Low Power Methodology: For System-on-Chip Design, Springer, 2007.

[2] Power Forward Initiative, A practical guide to low power design: User experience with CPF. Power Forward, 2012.

[3] IEEE standard for design and verification of low power integrated circuits. IEEE, 2013. IEEE Std 1801-2013.

[4] The international technology roadmap for semiconductors: Design. ITRS, 2011 edition, 2011.

[5] O. Mbarek, A. Pegatoquet, and M. Auguin, "Using unified power format standard concepts for power-aware design and verification of systems-on-chip at transaction level," IET Circuits, Devices & Systems, vol. 6, no. 5, pp. 287-296, 2012.

[6] J. Karmann and W. Ecker, "The semantic of the power intent format UPF: Consistent power modeling from system level to implementation," in 2013 23rd international workshop on power and timing modeling, optimization and simulation (PATMOS), 2013, pp. 45-50.

[7] Y. Xu, R. Rosales, B. Wang, M. Streubühr, R. Hasholzner, C. Haubelt, and J. Teich, "A very fast and quasi-accurate power-state-based system-level power modeling methodology," in ARCS'12 Proceedings of the 25th international conference on architecture of computing systems, 2012, pp. 37-49.

[8] H. Lebreton and P. Vivet, "Power modeling in SystemC at transaction level, Application to a DVFS architecture," in IEEE Computer society annual symposium on VLSI, 2008, pp. 463-466.

[9] T. Bouhadiba, M. Moy, and F. Maraninchi, "System-level modeling of energy in TLM for early validation of power and thermal management," in DATE '13 Proceedings of the conference on design, automation and test in Europe, 2013, pp. 1609-1614.

[10] S. Ahuja, High level power estimation and reduction techniques for power aware hardware design, Faculty of the Virginia Polytechnic Institute and State University, 2010. Dissertation thesis.

[11] D. Macko and K. Jelemenská, "Managing digital-system power at the system level," in IEEE Africon 2013 Sustainable Engineering for a Better Future, 2013, pp. 179-183.

[12] D. Macko, K. Jelemenská and P. Čičák, "Power-Management Specification in SystemC," in 2015 IEEE 18th International Symposium on Design and Diagnostics of Electronic Circuits & Systems, 2015, pp. 259-262.

An Optimal Operating Point By using Error Monitoring Circuits with An Error-Resilient Technique

Jaemin Lee*, Seungwon Kim*, Youngmin Kim† and Seokhyeong Kang*

*School of Electrical and Computer Engineering
Ulsan National Institute of Science and Technology, Ulsan, Korea 689-798
†Department of Computer Engineering
Kwangwoon University, Seoul, Korea 139-701
jm3430202@unist.ac.kr, kskyh002@unist.ac.kr, youngmin@kw.ac.kr, shkang@unist.ac.kr

Abstract—For applications related to human, such as Internet of Things (IoT) and wearable devices, near threshold voltage (NTV) technology has been proposed for the trade-off between performance and energy consumption. However, error-resilient techniques are required in the circuits to improve reliability of the NTV operation.

In this paper, we propose a low-overhead error-resilient system and a design flow for NTV operations. We use a new monitoring circuit, which can detect timing errors and find an optimal operation point of the system. Also, we propose two different methodologies, which are slack-based methodology and sensitivity-based methodology.

From the proposed monitoring system and the sensitivity-based sorting algorithm, benchmark results show that the optimal designs provide up to 46% monitoring area reduction maintaining similar error detection ability of the conventional error-resilient design.

Index Terms—Error-Resilient Techniques, Optimal Operating Point, Adaptive Voltage Scaling, NTV.

I. INTRODUCTION

Nowadays, low power System-on-Chip (SoC) design has been emphasized in IoT and wearable devices. NTV operation is a promising technique to improve energy efficiency by an order of magnitude. However, significant performance degradation and timing errors caused by process, voltage and temperature (PVT) variations hinder the expansion of the NTV circuits [1].

To lower the operating voltage to the NTV level (400 - 600 mV), error resilient techniques should be adapted for reliable operations [2], [3], [4], [5], [6], [7]. Razor [2] is one of the well-known techniques for timing error detection and correction.

Razor detects timing violations by supplementing error-tolerant flip-flops (Razor flip-flops) as shown in Fig. 1. A Razor flip-flop (FF) consists of a conventional main FF and a shadow latch. It exploits an additional delayed clock to capture the correct data. Timing difference between the regular clock and the delayed clock is set to the safety margin.

The main FF reads the data at the positive edge of the clock, and the shadow latch captures the data during the delayed clock is high. An error occurs when the data from the main FF differ from the data captured by the shadow latch. The error can be corrected by using the correct value in the shadow latch.

By changing the conventional FF to the Razor FF, we detect and correct errors of the circuits. However, Razor FFs increase

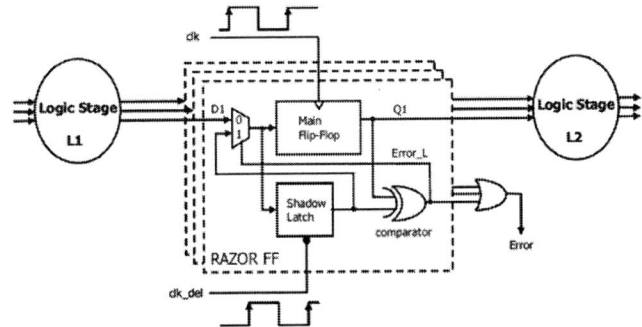

Fig. 1. The architecture of Razor FF

area and power of the design, and all the FFs do not need to be changed into Razor FFs. In this study, we propose a simple error monitoring circuit and a design flow to insert the monitoring circuits, which find an optimal operating point in NTV. The main contributions of our work are summarized as follows.

- We propose a low-overhead error-resilient system for NTV operation.
- We introduce a methodology that inserts the monitoring circuits by sorting FFs based on sensitivity.
- We compare the design overheads in area and the value of underestimation from each error-resilient technique.

The rest of the paper is organized as follows. Section II presents the related works about NTV technology, error resilient system, monitoring circuits, and adaptive voltage scaling technique. Section III shows the methodologies proposed in this study and the experimental setup. The simulation results and the analysis are explained in Section IV. Section V summarizes and concludes the paper.

II. RELATED WORK

To make a low-overhead error-resilient system, NTV technology, error recovery system, monitoring circuits, and adaptive voltage scaling technique are needed. Each related technique and previous works are explained in this section.

A. Near Threshold Voltage (NTV) Technology

NTV is the technology that lowers the operating voltage to the level of the device's threshold voltage. But the major problem of NTV technology includes the severe delay change and timing violations caused by PVT variations. The enough margin is needed to solve this timing issue, but then, the performance will be degraded. As technology scales down, this margin should be increased accordingly and the supply voltage becomes limited. Therefore, error-resilient techniques and systems are proposed to improve the performance and to provide reliable operation at lower voltage.

B. Error Recovery System

[2], [3], [8], [9] have proposed error recovery techniques by using of error-tolerant registers such as Razor and Razor II. The conventional registers (or FFs) are replaced by the specialized FFs (i.e., Razor) which can detect timing errors and correct the value using a shadow latch with a delayed clock. Razor II [3] is the advanced version of Razor FF which reduces the power, energy, and area.

Timing speculation is an example of performance improvement by using of the error-resilient system. [10] shows the way to find a critical path that exercises most frequently during gate-level simulation. It also proposes a method to optimize the path until the error rate is lower than the target error rate. Therefore, not only timing improvement with resilient designs but also power and area reduction can be achieved. [11] proposed a design approach, in which the processor runs at a target error rate instead of correct operation.

C. Monitoring Circuits

Nowadays, many monitoring circuits and methodologies are studied and published. Simple inverter-based ring oscillators (ROs) [12], process-sensitive ROs (PSROs) [13], and alternative monitoring circuits such as phase-locked loop (PLLs) [14] are examples of generic monitors. But delay estimation with generic monitors is not accurate and timing margins become large.

Design dependent monitors have been announced in various ways. A method to synthesize a single representative critical path (RCP) was proposed in [15]. Critical path monitors have been suggested in [16]. They are high-bandwidth sensors, which give timing margin accurately within two FO2 inverter delays to the voltage & frequency control loop. [17], [18] showed a method to design of process-dependent monitors.

Various design dependent monitors that estimate the performance of the circuits were proposed and analyzed in [19], [20], [21], [22], [23]. Path-based ring oscillator (Path-RO), an on-chip path delay measurement methodology is proposed in [20]. This method creates an oscillator from a targeted path for which it is used to measure the path delay under the impact of process variations.

D. Adaptive Voltage Scaling

As the transistor scales down (i.e., sub-32nm technology), the bias temperature instability (BTI) effect becomes a major issue. In very large scale integrated (VLSI) circuits, the BTI effect increases the threshold voltage and it degrades a time-dependent timing [24], [25].

Performance degradation induced by the BTI effect is compensated by adaptive voltage scaling (AVS) by increasing the supply voltage [26], [27], [28]. An advanced negative bias

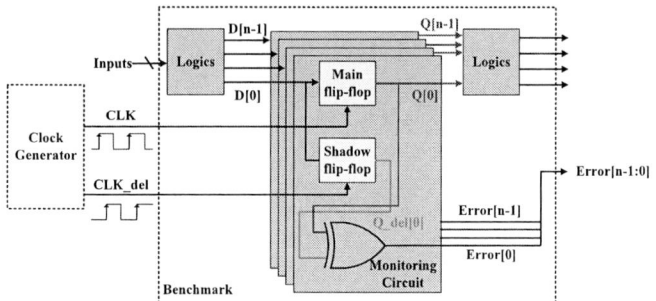

Fig. 2. Proposed error correction system

temperature instability (NBTI) model that allows to predict NBTI degradation on a chip with negligible dependency to measurement degradation is presented in [26]. This method is used to make a low-cost NBTI-aware dynamic voltage frequency scaling (DVFS) framework, which reduces the energy consumption of the processor. Scheduled voltage scaling, a technique that gradually increasing the operating voltage of integrated circuits, is proposed in [28]. This methodology can increase the lifetime of integrated circuits by 46% with no increase in power consumption or temperature. Aging-aware timing sign-off problem when using AVS technique was analyzed in [29].

III. OPTIMAL OPERATING POINT WITH ERROR MONITORING CIRCUITS

A. Proposed Methodology

Our proposed error correction system is presented in Fig. 2. Some or all FFs in the benchmark circuits are changed into the monitoring circuits for error detection.

The FFs in the designs are swapped into the monitoring circuits according to the proposed sorting methods. When the Razor FF is used instead of the normal FF, area and power consumption increase for the sake of performance improvement. Therefore, finding an optimal operating point between the performance and the area (and power) is required.

Fig. 3 shows the simulation flow. First, the benchmark circuits are synthesized, and automatic placement and routing process is conducted for the physical implementation. Then, FFs are extracted and sorted based on certain criteria. The sorting methodologies used in this study are (i) sorting by a timing slack of each FF in ascending order, and (ii) sorting by sensitivity of each FF in descending order. The sensitivity is defined as

$$Sensitivity = Switching\ Activity/Slack,$$

where the FF cell is more sensitive when the timing slack of the FF input port is small and switching activity of the FF is large. Thus, the FFs with higher sensitivity have higher probability of functional errors, and become better candidates for swapping to Razor counterparts.

B. Experimental Setup

To test the error correction system, each design has been implemented in Verilog and synthesized with TSMC 65-nm

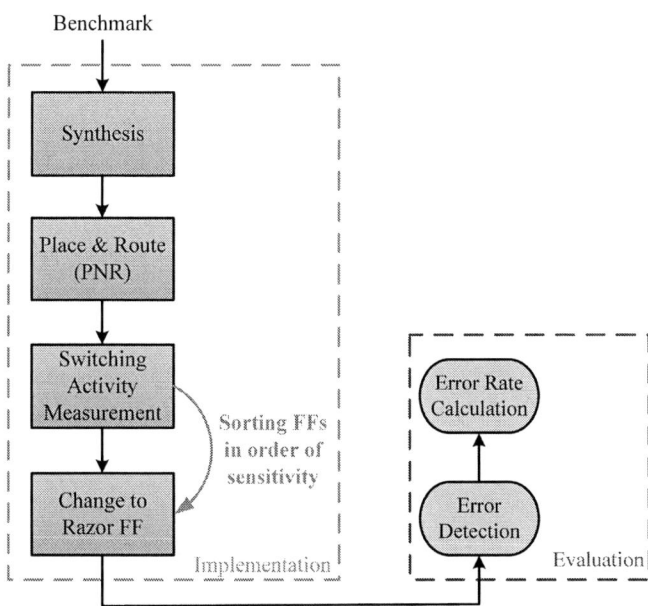

Benchmark

Fig. 3. Simulation flow of the proposed system

cell library using *Cadence RTL Compiler* [30]. After synthesis, the place and route (PNR) stage is conducted to remove hold violations by inserting hold buffers using *Cadence Encounter* [31]. The physical implemented netlist after PNR becomes the reference file of our simulations. We conduct gate-level simulations using *Cadence NC-Sim* [32]. To evaluate the performance of the proposed system, AES Cipher and JPEG Encoder are used as benchmark circuits. There are total 530 FFs in the AES Cipher and 4332 FFs in the JPEG Encoder.

The gate delay is obtained from the standard delay format (SDF) file which is extracted by *Synopsys PrimeTime* [33]. SDF files are characterized and generated at various voltage range from 0.5 V to 1.2 V in 0.01 V increments and these SDF files are used in the gate-level simulations. During the simulation, outputs of the main FF and the shadow FF are compared by using XOR gates. Simulations are conducted with random input patterns for 10,000 cycles.

By changing the clock period constraint in the Synopsys design constraints (SDC) file, different synthesized Verilog files are obtained with *Cadence RTL Compiler* [30]. By comparing the simulation results with various synthesized files, optimal constraints can also be found.

The information about the switching activities of each FF are extracted in the switching activity interchange format (SAIF) by *Synopsys PrimeTime* [33]. We check (i) whether the error bit (i.e., the output of the XOR gate in Fig. 2) of each monitoring circuit has been toggled or not and (ii) how many times the error data has been toggled.

IV. EXPERIMENTAL RESULT

A. 8×8 Multiplier

To check the functionality of the proposed method, we have conducted experiments with a multiplier coupled with

TABLE I
EXPERIMENTAL RESULT WITH 8×8 MULTIPLIER

Clock constraint (*ns*)	1.8			2.0			2.2		
Error rate (%)	5	10	15	5	10	15	5	10	15
Area (*um²*)	1521			1509			1518		
Power (*uW*)	89.4	83.4	80.9	91.1	86.2	83.2	91.3	89.2	83.1
Voltage (V)	0.69	0.67	0.66	0.70	0.68	0.67	0.70	0.69	0.67

Clock constraint (*ns*)	2.4			2.6			2.8		
Target Error rate (%)	5	10	15	5	10	15	5	10	15
Area (*um²*)	1515			1483			1478		
Power (*uW*)	93.8	89.2	86.4	93.9	89.5	86.0	98.1	90.2	85.8
Voltage (V)	0.71	0.69	0.68	0.71	0.69	0.68	0.72	0.70	0.68

input and output FFs. The target error rate, area, power, and the operating voltage are evaluated with different clock constraints.

The implementation and simulation results with a multiplier are summarized in Table I. The design is synthesized tighter with a smaller clock constraint (1.8 *ns*) and looser with a larger clock constraint (2.8 *ns*).

As shown in the table, for the same constraints, the voltage and the power decrease as the target error rate increases.

B. Benchmark Circuits

1) Error Correction System: We have used the AES Cipher and the JPEG Encoder to validate the benefit of our proposed method. The safety margin and the initial clock is 20 ps and 1.2 ns for AES Cipher, and 20 ps and 1.4 ns for JPEG Encoder, respectively.

We have made a gate-level simulation after adding monitoring circuits into FFs (from 100% to 40%). To find an optimal operating point, we have inserted the monitoring circuits based on the timing slack only or sensitivity explained in Section III.

The area and power become smaller as the number of monitoring circuits is reduced in various number of FF swapping cases. However, a further in-depth analysis is required to check the error detecting properties of the designs having smaller number of monitoring circuits since the errors will be ignored when monitoring circuits are not replacing the conventional FFs.

2) The Number of Errors: To find the trend of the number of errors and to check the error detecting ability in a design for different number of monitoring circuits, we have conducted simulations depicted in Fig. 3 and found the number of errors in various cases of different number of monitoring circuits swapped by two different sorting methodologies, slack-based and sensitivity-based.

Fig. 4 (a) shows the simulation results of the benchmark, AES Cipher, with FF sorting by an ascending order of the slack. As the slack is small, the monitoring circuit is required for error detection and correction. So the FF, which has a smaller slack is swapped first with the monitoring circuit counterpart.

As shown in Fig. 4 (a), first, the 40% monitoring case cannot detect errors properly compared to other swapping cases (i.e.,

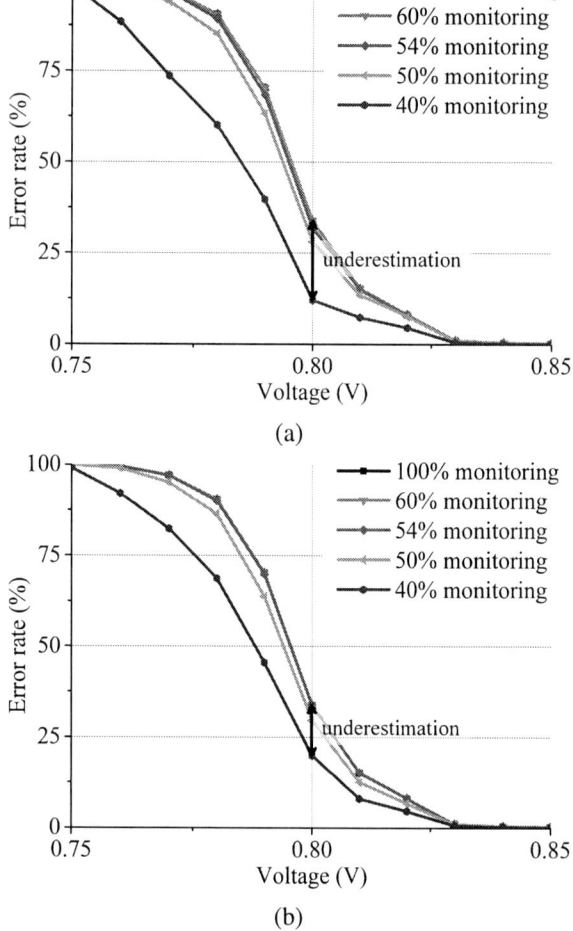

Fig. 4. Experimental results with AES Cipher: The error rate for different number of FF swapping from the sorted FFs list by (a) the slack (b) the sensitivity

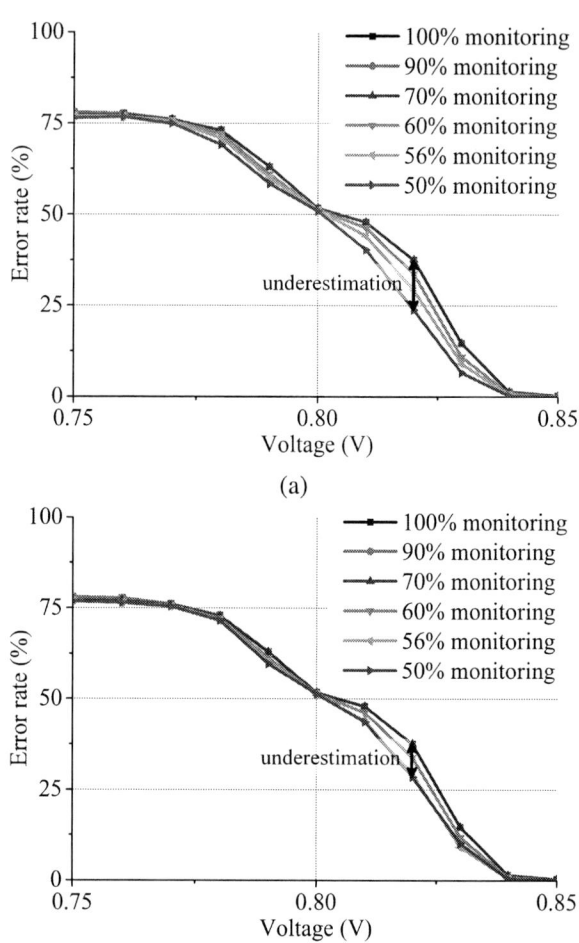

Fig. 5. Experimental results with JPEG Encoder: The error rate for different number of FF swapping from the sorted FFs list by (a) the slack (b) the sensitivity

TABLE II
AES CIPHER: UNDERESTIMATION AT A 0.80 V WHEN SORTING BY SLACK AND SORTING BY SENSITIVITY

	Number of FF swapping				
	100%	60%	54%	50%	40%
Slack-based	0%	0%	5.97%	17.24%	64.28%
Sensitivity-based	0%	0%	0%	12.36%	41.01%

TABLE III
JPEG ENCODER: UNDERESTIMATION AT A 0.82 V WHEN SORTING BY SLACK AND SORTING BY SENSITIVITY

	Number of FF swapping				
	100%	90%	70%	60%	56%	50%
Slack-based	0%	0%	0%	10.46%	22.18%	36.71%
Sensitivity-based	0%	0%	0%	9.82%	22.04%	24.07%

more monitoring cases). When FFs are replaced by monitoring circuits lower than 40% of all FFs in a design, the circuit cannot detect errors correctly and underestimates the error occurrence at the same voltage. Second, even though there are small discrepancies for up to 54% swapping case from the 100% monitoring case (i.e., reference design with 100% monitoring circuits), the overall error detecting ability seems similar among them.

Not only the slack but also the switching activity of the FFs are major parameters in analyzing functional errors of the FFs. To select the appropriate FFs to swap which are more susceptible to the functional error, we propose the sensitivity-based sorting methodology explained in Section III.

The simulation results based on the sensitivity-based sorting are shown in Fig. 4 (b). The FFs are sorted in a descending order of the sensitivity. Higher sensitivity means that the FF switches more often and the slack is small. Thus swapping the FFs with higher sensitivity is more beneficial to find an optimal design.

As shown in Fig. 4 (b), up to 54% monitoring case reveals very similar error rates of the 100% monitoring case. Therefore, we can conclude that the 54% monitoring swapping case provides the best area and power saving without compromising the error detecting ability shown in 100% monitoring case.

We can reduce up to 46% in area of sequential logics by the proposed method with an optimal number of FF swapping.

Table II shows the underestimation at a 0.80V supply voltage when sorting by slack and sorting by sensitivity for the benchmark, AES Cipher. As shown, the sensitivity-based sorting methodology provides much smaller underestimation than the simple slack-based sorting.

The simulated error rates of the benchmark, JPEG Encoder, are plotted in Fig. 5. As shown in Fig. 5 (a), up to 70% monitoring case, the design provides similar error detecting ability with 100% monitoring case. However, the system cannot detect errors properly when less than 50% FFs are swapped.

As shown in the Fig. 5 (a) and (b), both methodologies result in similar behavior in this specific benchmark circuit. When using the sensitivity-based methodology, 70% monitoring case can detect similar errors in almost every voltage. We can reduce up to 30% in area of sequential logics by the proposed method with an optimal number of FF swapping.

Table III shows the underestimation at a 0.82V supply voltage when sorting by slack and sorting by sensitivity for JPEG Encoder. Sorting by sensitivity results in less underestimation than sorting by slack.

V. CONCLUSION

This paper proposes a low-overhead error-resilient system and a design methodology for finding an optimal number of error detecting logics such as Razor in NTV operations. We use a new monitoring circuit, which can detect timing errors, and find an optimal operating point of the system. Our proposed technique can reduce up to 46% of area compared to the conventional error-resilient designs in two benchmark circuits.

Our ongoing work seeks to analyze the proposed method for finding optimal designs in various benchmarks. As shown in simulation results, different benefits are expected in different circuits. Improved sorting algorithms to choose the better candidates to swap are also interesting to investigate.

REFERENCES

[1] R. G. Dreslinski, M. Wieckowski, D. Blaauw, D. Sylvester, and T. Mudge, "Near-threshold Computing: Reclaiming Moore's Law Through Energy Efficient Integrated Circuits," in *Proc. IEEE*, 98(2) (2010), pp.253-266.

[2] D. Ernst, N. S. Kim, S. Das, S. Pant, R. Rao, T. Pham, C. Ziesler, D. B.laauw, T. Austin, K. Flautner, and T. Mudge, "Razor: A Low-Power Pipeline Based on Circuit-Level Timing Speculation," in *Proc. MICRO-36*, 2003, pp.7-18.

[3] S. Das, C. Tokunaga, S. Pant, W.-H. Ma, S. Kalaiselvan, K. Lai, D. M. Bull, and D. T. Blaauw, "Razor2: In Situ Error Detection and Correction for PVT and SER Tolerance," in *Proc. IEEE ISSCC*, 2008, pp. 400-401.

[4] M. Choudhury, V. Chandra, K. Mohanram, and R. Aitken, "TIMBER: Time Borrowing and Error Relaying for Online Timing Error Resilience," in *Proc. IEEE DATE*, 2010, pp. 1554-1559.

[5] S. Ghosh, S. Bhunia, and K. Roy, "CRISTA: A New Paradigm for Low-Power and Robust Circuit Synthesis Under Parameter Variations Using Critical Path Isolation," *IEEE Trans. on CAD* 26(11) (2007), pp. 1947-1956.

[6] B. GReskamp and J. Torrellas, "Paceline: Improving Single-Thread Performance in Nanoscale CMPS through Core Overclocking," in *Proc. PACT*, 2007, pp. 213-224.

[7] V. Subramanian and A. Somani, "Conjoined Pipeline: Enhancing Hardware Reliability and Performance through Organized Pipeline Redundancy," in *Proc. IEEE PRDC*, 2008, pp. 9-16.

[8] N. D. P. Avirneni and A. K. Somani, "Low Overhead Soft Error Mitigation Techniques for High-Performance and Aggressive Designs," in *IEEE Trans. on Computers* 61(4) (2012), pp. 488-501.

[9] S. Kim, I. Kwon, D. Fick, M. Kim, Y. Chen, and D. Sylvester, "Razorlite: A Side-Channel Error-Detection Register for Timing-Margin Recovery in 45nm SOI CMOS," in *Proc. IEEE ISSCC*, 2013, pp. 264-266.

[10] B. Greskamp, L. Wan, W. R. Karpuzcu, J. J. Cook, J. Torrellas, D. Chen, and C. Zilles, "BlueShift: Designing Processors for Timing Speculation from the Ground Up," in *Proc. HPCA*, 2009, pp. 213-224.

[11] A. B. Kahng, S. Kang, R. Kumar, and J. Satori, "Recovery-Driven Design: A Power Minimization Methodology for Error-Tolerant Processor Modules," in *Proc. IEEE/ACM DAC.*, 2010, pp. 825-830.

[12] M. Bhushan, A. Gattiker, M. Ketchen, and K. K. Das, "Ring Oscillators for CMOS Process Tuning and Variability Control," *IEEE Trans. on SM* 19(1) (2006), pp. 10-18.

[13] I. A. K. M. Mahfuzul, A. Tsuchiya, K. Kobayashi, and H. Onodera, "Variation-Sensitive Monitor Circuits for Estimation of Global Process Parameter Variation," *IEEE Trans. on SM* 25(4) (2012), pp. 571-580.

[14] K. Kang, S. P. Park, K. Kim, and K. Roy, "On-Chip Variability Sensor using Phase-Locked Loop for Detecting and Correcting Parametric Timing Failures," *IEEE Trans. on VLSI* 18(2) (2010), pp. 270-280.

[15] Q. Liu and S. S. Sapatnekar, "Capturing Post-Silicon Variations using A Representative Critical Path," *IEEE Trans. on CAD* 29(2) (2010), pp. 211-222.

[16] A. Drake, R. Senger, H. Singh, G. Carpenter, and N. James, "Dynamic Measurement of Critical-Path Timing," in *Proc. IEEE ICICDT*, 2008, pp. 249-252.

[17] L. M. Burns, L. Dauphinee, R. A. Gomez, and J. Y. C. Chang, "Process Monitor for Monitoring and Compensating Circuit Performance," U.S. Patent 7375540, 2008.

[18] D. J. Philling and C. Talledo, "In-Situ Monitor of Process and Device Parameters In Integrated Circuits," U.S. Patent 7583087,. 2009.

[19] T. Black, "A Critical Path Based Parametric Ring Oscillator," in *Master's Thesis*, Texas Tech University, 2000.

[20] X. X. Wang, M. Tehranipoor and R. Datta, "Path-RO: a Novel On-Chip Critical Path Delay Measurement Under Process Variations," in *Proc. ICCAD*, 2008, pp. 640-646.

[21] D. Fick, N. Liu, Z. Foo, M. Fojtik, J. -S. Seo, D. Sylvester and D. Blaauw, "In Situ Delay-Slack Monitor for High-Performance Processors Using An All-Digital Self-Calibrating 5 ps Resolution Time-to-Digital Converter," in *Proc. IEEE ISSCC*, 2010, pp. 188-189.

[22] H. C. Ngo, G. D. Carpenter, A. J. Drake and J. B. Kuang, "Circuit Timing Monitor Having a Selectable-Path Ring Oscillator," *U.S.Patent* No. US7810000B2, 2010.

[23] K. Shaik, "Implementation of a Critical Path Based Parametric Ring Oscillator," *BSEE Thesis*, Texas Tech University, 2011.

[24] V. Huard, N. Ruiz, F. Cacho, and E. Pion, "A bottom-up approach for system-on-chip reliability," *Microel. Reliab.*, 51(9-11) (2011) pp. 1425-1439.

[25] B. Kaczer, S. Mahato, V. V. de A. Camargo, M. T.-Luque, Ph. J. Rousell, T. Grasser, F. Catthoor, P. Dobrovolny, P. Zuber, G. Wirth, and G. Groeseneken, "Atomistic Approach to Variability of Bias-Temperature Instability in Circuit Simulations," in *Proc. IEEE IRPS*, 2011, pp. XT.3.1-XT.3.5.

[26] M. Basoglu, M. Orhansky, and M. Erez, "NBTI-Aware DVFS: A New Approach to Saving Energy and Increasing Processor Lifetime," in *Proc. ISLPED*, 2010, pp. 253-258.

[27] S. V. Kumar, C. H. Kim, and S. S. Sapatnekar, "Adaptive Techniques for Overcoming Performance Degradation due to Aging in Digital Circuits," in *Proc. ASP-DAC*, 2009, pp. 284-289.

[28] I. Zhang and R. P. Dick, "Scheduled Voltage Scaling for Increasing Lifetime in the Presence of NBTI," in *Proc. ASP-DAC*, 2009, pp. 492-497.

[29] T.-B. Chan, W.-T. J. Chan, A. B. Kahng, "On Aging-Aware Signoff for Circuits With Adaptive Voltage Scaling," in *Proc. DATE*, 2013, pp. 1683-1688.

[30] *Cadence RTL Compiler User's Manual*. http://www.cadence.com .

[31] *Cadence SoC Encounter User's Manual*. http://www.cadence.com .

[32] *NC-Sim User's Manual*. http://www.cadence.com .

[33] *PrimeTime User's Manual*. http://www.synopsys.com .

Dynamic Error Tracking and Supply Voltage Adjustment for Low Power

Pierre Nicolas-Nicolaz
Department of Electrical and Computer Engineering
Seoul National University
Seoul, South Korea
pierre.nicolas@dal.snu.ac.kr

Kiyoung Choi
Department of Electrical and Computer Engineering
Seoul National University
Seoul, South Korea
kchoi@snu.ac.kr

Abstract — Amongst many techniques to reduce power consumption of chips, lowering the supply voltage is known to be the most effective one. However, lowering the supply voltage of chips too much down to near the threshold voltage of transistors causes the logic delay to vary exponentially with intrinsic and extrinsic variations such as process variations, temperature variations, and aging, and thus forces the designer to set increased timing margin or use more advanced techniques such as adaptive voltage scaling, where the supply voltage is adjusted by tracking timing errors due to the variations. This paper proposes a technique for adaptive supply voltage adjustment that minimizes power consumption in the near-threshold voltage region while satisfying a given constraint on error rate. It can be used in signal processing applications where intermittent errors are tolerated. The technique employs a current sensing completion detector for tracking errors and increases/decreases the voltage if the error rate is too high/low. We show that, for the average case, our approach tracks errors better than other approaches. We also show that it achieves 42% and 54% power savings for an error rate of 0.1% and 5%, respectively, for the TT corner at 25°C

Keywords— Low-Power; Adaptive Voltage Scaling; Near Threshold Voltage; Current Sensing Completion Detector

I. INTRODUCTION

Constant decrease in transistor size, advancement in CAD tools, and ever-increasing demand for more functionalities in a system have led to an exponential increase of the number of transistors integrated in a single device. From the thousands of transistors in an Intel 8080 chip in the 70's, we succeeded at cramming more than several billions of transistors in modern chips (e.g., the IBM Power 7 with 80MB of L3 cache contains more than two billion transistors). This increase in processing power came at the cost of increased power consumption. While several techniques have been developed to reduce power consumption (e.g., power gating), the main design parameter that chip manufacturers used in the past to improve power performance was reducing the nominal supply voltage. From the 0.25μm technology node down to the 45nm node, the supply voltage was reduced by a factor of 2.7. Since the 45nm node, however, the supply voltage has stagnated around 1V and thus reducing total energy of a chip becomes more and more difficult ([1]).

To overcome this difficulty, it has been proposed to reduce the supply voltage of chips below their nominal values. While it is possible to use a supply voltage which goes even below the threshold voltage of a transistor (e.g., see the original work in [2]), the speed of the transistor decreases exponentially and its sensitivity to intrinsic and extrinsic variations (variation of oxide thickness, doping, transistor gate width, temperature, supply voltage variations, aging, etc.) is severely increased. Furthermore, the cost of leakage energy due to decreased transistor speed offsets the savings in dynamic energy at very low supply voltage, which pushes designers to design in the Near-Threshold Voltage (NTV) region, thus making systems that perform Near-Threshold Computing (NTC). The increased sensitivities to variations still force designers to use excessive timing margin and cause huge performance losses (e.g., see [3]).

We devised a technique to automatically adjust the supply voltage of a logic block to match a certain timing, despite the excessive variation and changing computational load. In traditional approaches, the designer must use an overly conservative supply voltage margin to meet the timing constraint. A better technique, called Adaptive Voltage Scaling (AVS), dynamically adjusts the supply voltage according to variations. However, at a fixed clock frequency, this technique doesn't take into account the intensity of the computation; while some input data can lead to near-to-zero slack time, most input data may lead to big slack time and thus wasting unnecessary energy. Our technique adjusts the supply voltage for the circuit to properly run at a given constant clock frequency under variations and changing computational load. It allows the designer to make the circuit to meet the timing constraint without considering variations too much, and adjusts the supply voltage automatically at runtime. We deliberately chose to allow errors in our design, making it especially suitable for signal processing applications. We also propose a way to bound the error rate and keep the actual error rate close to that bound to further minimize power.

II. BACKGROUND & RELATED WORK

We first quickly describe NTC and its impact on power consumption, then describe several timing error detection techniques, and finally explain some previous AVS techniques.

978-1-4673-9141-2/15 $31.00 © 2015 IEEE

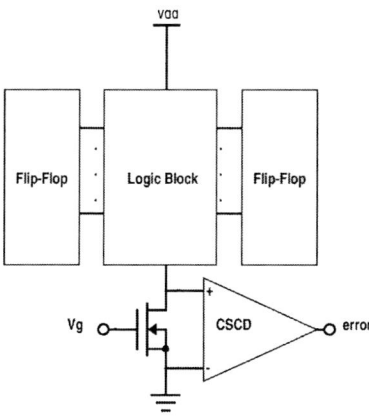

Fig. 1.   Possible architecture for a CSCD device.

A. Near Threshold Computing (NTC)

As explained before, a key technique to reduce power consumption in chips is to reduce the supply voltage since

$$E_T = E_{DYN} + E_L = \alpha C_{load} V_{dd}^2 + I_{leak} T_D V_{dd} \quad (2.1)$$

where α is the switching probability of the transistors, C_{load} is the load capacitance, I_{leak} is the leakage current associated with the technology used, and T_D is the leakage time. By reducing the supply voltage, dynamic energy is reduced according to this square law allowing quadratic power savings since the second term of the equation is very small compared to the first.

However, when the supply voltage is reduced under its nominal value, the transistors become slower; this increases T_D and therefore the leakage energy. Thus the supply voltage cannot be reduced indefinitely. Instead, there exists an energy minimum point which balances the savings in dynamic energy and the increase in leakage energy (see [4] for a more in-depth analysis).

B. Timing error detection

Lowering so much supply voltage causes excessive timing variation. A circuit using NTC has to track its error if it would not lower its frequency too much. We describe here commonly used error tracking methods, and then the method we used in our technique.

1) Canary Path – Critical Path Replica/Monitor technique

The canary path technique or its variant is the most commonly used technique for detecting timing errors. [5] is a good example of such a technique. A replica of the critical path is created, usually with NOT, NOR, or NAND CMOS logic gates. If we assume that the critical path and the replica are subject to the same amount of variation, by counting the number of gates a signal propagates within one cycle time, we can evaluate the worst-case slack time according to variations.

2) Razor

The razor technique was originally proposed in [6], which adds to each flip-flop a *shadow latch* controlled by a delayed clock, usually delayed by 20%~50% of the clock period. The shadow latch will latch the data after the clock delay, and if the data latched by the main flip-flop and the shadow latch are different, then it is considered as an error. The result can be corrected during the next clock cycle using the shadow latch output, but in that case, the cycle should be stalled.

3) Current Sensing Completion Detector (CSCD)

The current sensing completion detection was first introduced as a way to efficiently replace complex handshaking protocols in self-timed (asynchronous) logic. The idea is to measure the current drawn by the CMOS logic (or the current flowing to the ground) to see whether the CMOS logic is computing (current flowing) or not computing (no current flowing). Thus, by monitoring the current, it can detect when the CMOS logic has finished its computation.

Recently, the CSCD concept has been studied to be used in synchronous circuits at near threshold voltage in [7]. We used a modified version of the circuit presented in [7] for our work. The basic idea is represented in Figure 1, where we measure the current flowing to the ground through a transistor. The CSCD block basically evaluates the small voltage difference applied to its input through a dynamic sense amplifier. It will be explained in more details in section III.C.

C. . Adaptive Voltage Scaling (AVS)

The 25 year old idea of AVS is to adjust supply voltage according to intrinsic and/or extrinsic variations. Although the idea was first proposed a long time ago, it is attracting interests recently because advances in technology cause a stark increase in the amount of delay variation. It should be noted that AVS is usually considered as a way to adapt the supply voltage to variations for a given clock frequency of the system such as the one in [8].

Some recent researches focus on the use of Dynamic Voltage Scaling in near- or subthreshold-voltage designs. Examples include the ones in [9] and [10]. The researches in [11] and [12] also consider the possibility of having errors in their designs. However, all these researches consider a variant of the critical path replica technique to detect errors. With such a technique alone, they cannot exploit the fact that the actual delay changes depending on the computational load (or input data pattern) and thus the average delay is typically much shorter than the critical path delay to further reduce the voltage and thus power consumption. In our work, we adapt the supply voltage to variations and the change of computational load for a fixed frequency. This is allowed by the use of the CSCD device which enables us to obtain the actual slack time.

III. PROPOSED TECHNIQUE

As stated before, we use the CSCD module to track errors and change the supply voltage accordingly to minimize slack

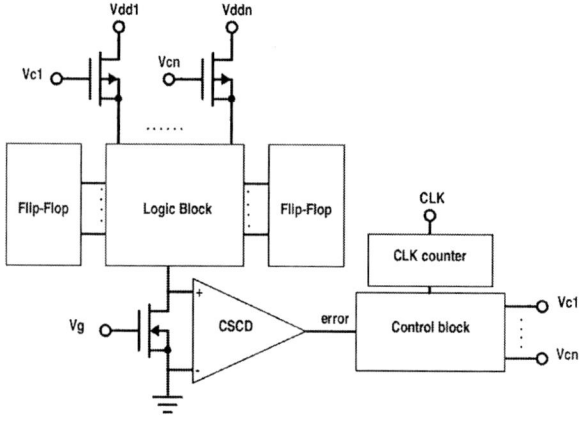

Fig. 2. General scheme for our supply voltage adjustment technique.

time. First we will describe the general architecture of our solution, then describe how we used the CSCD module, and finally explain how we control the supply voltage according to the error rate.

A. General description

We have depicted a scheme of our technique in Figure 2. We included our CSCD design together with switch transistors that select one of possible supply voltage lines. Note that, instead of using multiple supply lines, we could use a DC-DC converter (thus sharing a voltage domain across multiple blocks for area reasons). This would allow to reduce the area overhead due to using multiple supply lines and multiple switch transistors. However, we do not further consider how to generate multiple supply voltages on a chip and just assume we have at our disposal multiple voltages such as in [12].

After the CSCD analyzes the current flow to decide whether or not there is an error in the actual cycle, we use a control block to determine which supply voltage to use. The goal of the control block is to keep the error rate close to a given bound and minimize the power. We will describe it with the clock counter block in more detail in section D.

Finally, note that the switch transistors, the control block, and the CSCD can be shared amongst several logic blocks being placed next to each other on the chip and in the same pipeline. Since they are placed next to each other on the chip, they are going to be subject to the same amount of variation. And since they share the same pipeline, their computational load tend to be quite similar (consider, for example, manipulating input data with small integer values), meaning that they can share the same supply voltage without degrading much.

B. Choosing the right error tracking device

Most of the AVS techniques use a critical path replica technique to track errors. However, due to place and route restrictions, the critical path can be far away from the logic block on the chip, and the actual process and temperature variations can be different between the real path and its replica. Another issue is aging, where the replica is going to age a lot

faster than the real circuit. These issues can lead to discrepancies between the measured slack time and the actual worst case slack time. Also, in contrast to other error tracking techniques presented here, this technique doesn't intrinsically take into account the change in computational load; if the input stays constant for 2 cycles, the actual slack time is going to be close to the cycle time but the measured slack time won't relay that information.

Compared to the canary path technique, Razor takes into account input data and realistically reflects intrinsic and extrinsic variations. However, the main issue with Razor is that the shadow latch must be able to effectively latch the data when we have a rising edge (we assume triggering at the rising edge) on the shifted clock. This implies two problems:

- First, the amount of shift we apply to our clock defines the amount of variation we can recover an error from. This can be problematic in the NTV regime where variations can cause delay of the order of multiple clock cycles.

- Secondly, the output of the logic must not change until the rising edge (for the previous cycle) of the shifted clock. Razor usually solves this problem by adding a minimum-path length constraint to the logic block. For paths that can finish before the rising edge of the shadow latch, buffers are added to increase their timing so that they don't disturb the output for the previous clock cycle. The more buffers are added, the bigger can be our error detection window. However, it will increase the area as well as the overall power consumption. More importantly, increasing the time taken by the quick paths may decrease the slack time for some inputs that do not sensitize any of longer paths and thus blocks further reducing the supply voltage which would have been possible for the original delays of the quick paths.

For these reasons, we do not adopt Razor, but choose to use a CSCD device to implement our technique. To the best of our knowledge, no other previous work uses such a technique to dynamically control the supply voltage.

C. Use of the CSCD with multiple voltage sources

Our CSCD device is strongly inspired from [7]. However, we measure the current flowing to the ground instead of measuring the current flowing from the source. This is because it is easier and more stable since we use multiple voltage sources.

Since we measure the current flowing to the ground, we need to take the dual of the offset-calibrated quantizer circuit used in [7]. We depicted in Figure 3(a) the quantizer used in our experiment. Note that for the sake of simplicity, we replaced the transistors calibrating the threshold by a simple current source, and we calibrated the CSCD module by changing the value of the current source. The current source can be replaced with the circuit presented in [7]. If the current is over some threshold defined through calibration at a rising edge of the clock, we consider that the previous operation has

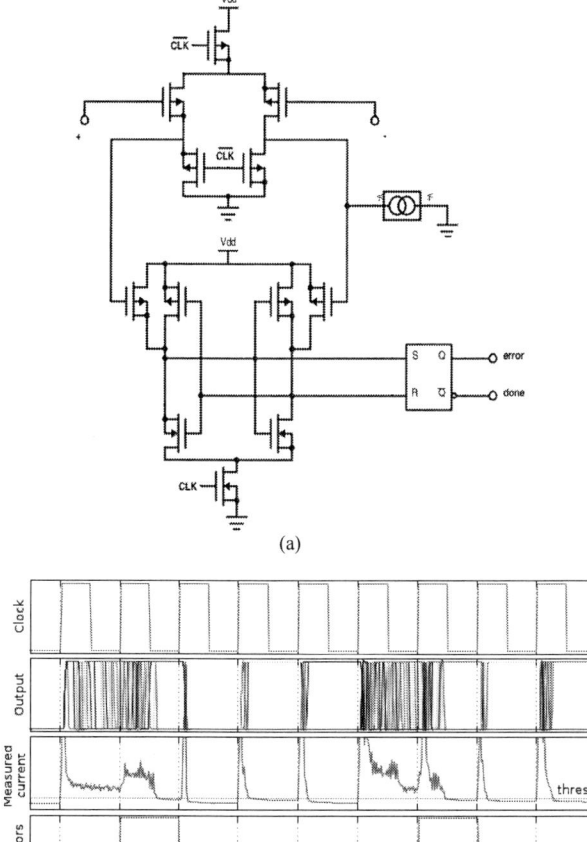

(a)

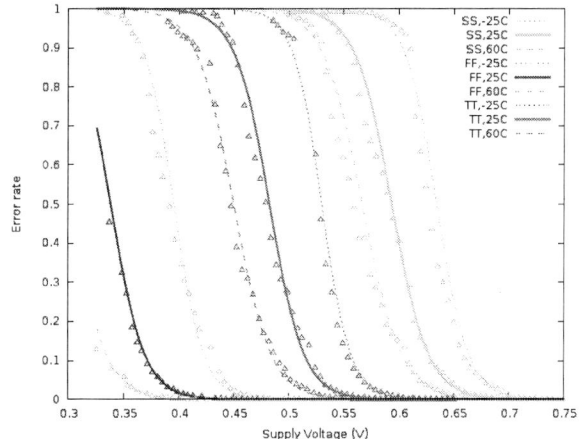

Fig. 4. Supply voltage vs. error rate for a static supply voltage on a 32-bit adder, 200MHz (28nm process).

(b)

Fig. 3 (a) Schematic diagram of the dual of the offset-calibrated quantizer and (b) Timing diagram of the CSCD device.

not been finished yet and we raise the error signal. This is what happens for the second and seventh clock cycle in Figure 3(b).

The gating transistor used for the CSCD causes energy loss and slower the logic block. Its size also defines the amount of the voltage peak, and thus the amount of variations that the CSCD module can be calibrated for. According to our experiments, we could set the parameters of the CSCD for various process corners, for temperature going from -25°C to 60°C while keeping the speed loss under 10% and having a negligible power consumption for the CSCD module (switch transistors included).

D. Error rate boundaries

As we can see in Figure 4, the error rate varies in an exponential fashion according to the supply voltage (we regressed our data on sigmoid functions). In the NTV regime, for the TT corner at 25°C, the minimum energy point is at 0.47V in our experiments, which corresponds roughly to 40% errors according to Figure 4. Thus, an implicit higher bound

for the error rate is 40%: if the error rate is higher, power consumption will increase. However, in practical, we want to keep the error rate below 5% as more errors would give a non-useable result most of the time. Given ourselves a specific error rate between 0.1% and 5%, we propose a method to keep the actual error rate close to the chosen rate.

Our method to bound the error rate by selecting the appropriate supply voltage is to sample the error rate during a sample period T_S and

- decrease the supply voltage if the error rate is lower than a lower threshold L_D

- increase the supply voltage during the sampling period as soon as the error rate being measured is higher than an upper threshold L_U

This allows to keep the error rate constrained between two bounds, to have hysteresis, and to forbid the error rate to be greater than the upper bound L_U. To choose T_S and the distance D between L_U and L_D, one should consider the amount of power required to switch from one voltage to another. If this cost is high, we don't want to change our supply voltage often and we will have to increase T_S and D. On the other hand, the more we increase T_S and D the less the supply voltage will depend on the change of computational load.

The implementation of the control block is pretty straightforward. First, the clock counter in Figure 2 sends a signal to the control block to know when a new sampling starts. It can be shared amongst all logic blocks on the chips which allows a significant power reduction. Then, one register (of at least $log_2(M*T_s)$ where M is the maximum error rate) is used to store the number of errors in the current sampling cycle, and one up-down counter is used to store the supply voltage level. When we have a rising clock on the sampling initialization signal, we initialize the error register to 0 and increase it at each rising edge of the error signal. If that register becomes greater than L_U*T_S, we increment the up-

978-1-4673-9141-2/15 $31.00 © 2015 IEEE

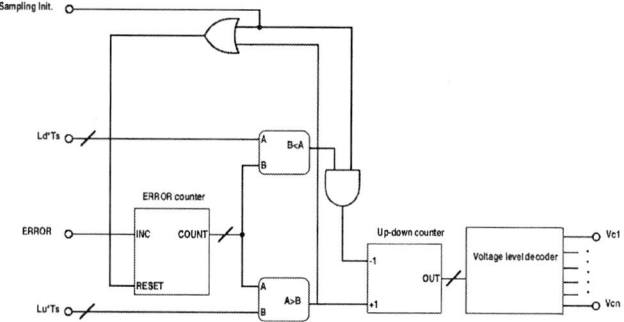

Fig. 5 RTL circuit of our control block.

down counter and reset the error counter. When we receive the next sampling initialization signal, we decrement the up-down counter if the error register is less than L_D*T_S and reset the error counter. At any time, we convert the output of the up-down counter to a format readable by the switch transistors (or by a DC-DC converter if we use that instead).

We depicted in Figure 5 an abstraction of our RTL logic for the control block. As we can see, most of the area of the control block is made of the error and the up-down counter, We computed the overall energy of our control block using Synopsys Design Compiler (see section IV for more details about our experimental setup). The power consumption of the control block was less than 300nW, which is an order of magnitude lower than the power consumption of a 32-bit adder (test circuit we used). Furthermore, the control block can be shared amongst several logic blocks sharing the same voltage level and the logic block will usually be bigger than a simple adder. In the rest of our experiments, we ignored the power consumption of the control block.

IV. EXPERIMENTAL SETUP

We simulated a 32-bit carry look ahead adder as a test logic block in the 28/32nm node, although the proposed approach was targeting bigger blocks. We used mixed-signal simulation, implementing the control block in Verilog and the logic block in SPICE (using Synopsys Interoperable PDK transistor models). We used Synopsys CustomSim XA for analog simulation with speed/accuracy tradeoff (set_sim_level) level 4, and Synopsys VCS for digital simulation. The nominal supply voltage for the transistors was

Table 1. Experimental Parameters

Parameter	Value
Corner tested	FF,TT,SS
Temperature range	-25°C , 25°C, 60°C
Sample period (T_S)	3000 cycles
Maximum error rate	From 0.1% up to 5%
$D=|L_U-L_D|$ (range of the error rates where V_{dd} stays constant)	0.1%

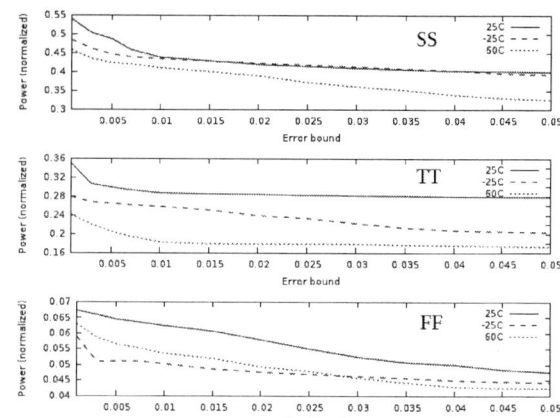

Fig. 6. Experiment results normalized to a single supply voltage source.

1.05V.

We simulated 45K cycles, adding every cycle two different frames of fixed point EEG data as the inputs. Other possible applications of our technique would be IoT/wearable devices. We chose a 200MHz clock frequency. It corresponds to null slack time when we use the critical path of the adder in the worst case (SS corner@-25°C).

As shown in Table I, we chose a relatively short sample period to speed up testing our technique on a maximum error rate of 0.1%. This corresponds to 3 errors in one sample period.

V. RESULTS

A. Comparaison to single supply voltage source

Without the use of any kind of AVS and without allowing any error, the designer would have to set the supply voltage of our logic block to 1.05V (lowering the voltage below this level causes errors at some process corner for a certain range of temperatures). In this section, we compare our technique to that case. Figure 6 shows that in the average case (TT@25°C), our technique which takes into account computational load as well as process and temperature variations achieves 72%

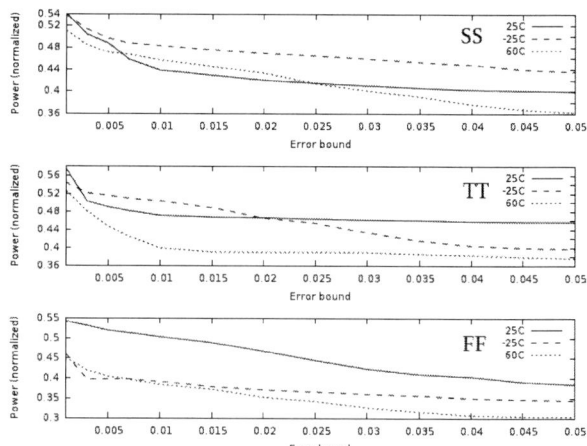

Fig. 7. Experimental results normalized to a multiple voltage supply adapting only to process and temperature variations.

power reduction. In the best case (FF@25°C), it achieves 93.5% power reduction, and in the worst case (SS@25°C), it achieves 46% power reduction. Note that for a given process corner, the different lines can cross since each result was normalized to a different result (i.e., the result at its own temperature) at 1.05V.

Of course, this power reduction comes with all the other costs of having multiple supply lines in the chip, so a more fair comparison would be to compare our technique to a logic block with multiple voltage sources.

B. Comparaison to multiple, static, voltage sources

We compare in Figure 7 our technique to a previous technique taking into account process and temperature variations but not computational load. This corresponds to AVS techniques using a critical path replica as an error detecting circuit. Note that in our comparison we ignore the power consumption overhead of the previous technique, which can only worsen our results.

We see here less energy savings, since the baseline already takes into account a part of the variations. However, we see that even with only 0.1% of error, we can reduce the power consumption by 42% and with 5% of error, we can reduce the power by 54%, for the TT corner at 25°C.

C. Effective and theoretical error rate

We plotted in Figure 8 the difference between the error rate upper bound defined in our design and the actual error rate (measurement was started after the first 10K cycles). We see that the difference is up to 42% (with a mean of 29%) for a low error bound. This is because at this range, small number of errors can make a big difference. When the error bound increases, the discrepancies between the actual number of errors and the bound becomes lower than 10%. Also, notice that most of the time, the actual error rate is lower than our bound: it means that although we loose power since we could slightly increase the number error, our bound is respected.

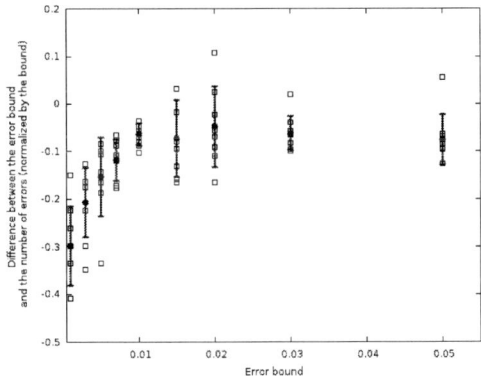

Fig. 8. Mean and standard deviation of the difference between the actual number of errors measured in the simulation and the error bound for several process corners-temperatures.

VI. CONCLUSION

This paper introduces a new approach to dynamic adaptation of the supply voltage to variations and change of computational load. We use a current sensing completion detection technique to track timing errors. Compared to other *critical path replica type* techniques, this allows us to better track local process/temperature variations and dynamically modify the supply voltage according to the change of computational load at a fixed frequency. With the technique, we achieve 42% power reduction by allowing only 0.1% errors for a 32-bit adder, and 54% power reduction by allowing 5% errors. This approach would be parcularly usefull in embedded DSP applications, where energy battery can become rapidly scarce.

ACKNOLEDGMENTS

This work was supported by Samsung Research Funding Center of Samsung Electronics under Project Number SRFC-IT1501-08.

REFERENCES

[1] Dreslinski, Ronald G., et al. "Near-threshold computing: Reclaiming moore's law through energy efficient integrated circuits." Proceedings of the IEEE 98.2 (2010): 253-266.

[2] R. Swanson and J. Meindl, B Ion-implanted complementary MOS transistors in low-voltage circuits, [IEEE J. Solid-StateCircuits, vol. 7, no. 2, pp. 146–153, 1972 Borkar, Shekhar, et al.

[3] "Parameter variations and impact on circuits and microarchitecture." Proceedings of the 40th annual Design Automation Conference. ACM, 2003.

[4] Calhoun, Benton H., Alice Wang, and Anantha Chandrakasan. "Modeling and sizing for minimum energy operation in subthreshold circuits." Solid-State Circuits, IEEE Journal of 40.9 (2005): 1778-1786.

[5] Drake, Alan, et al. "A distributed critical-path timing monitor for a 65nm high-performance microprocessor." Solid-State Circuits Conference, 2007. ISSCC 2007. Digest of Technical Papers. IEEE Internationa, 2007.

[6] Ernst, Dan, et al. "Razor: A low-power pipeline based on circuit-level timing speculation." *Microarchitecture, 2003. MICRO-36. Proceedings. 36th Annual IEEE/ACM International Symposium on.* IEEE, 2003.

[7] Crop, Joseph, Robert Pawlowski, and Patrick Chiang. "Regaining throughput using completion detection for error-resilient, near-threshold logic." DAC, 2012 49th ACM/EDAC/IEEE. IEEE, 2012.

[8] Burd, Thomas D., and Robert W. Brodersen. "Design issues for dynamic voltage scaling." Proceedings of the 2000 international symposium on Low power electronics and design. ACM, 2000.

[9] Kao, James T., Masayuki Miyazaki, and Anantha P. Chandrakasan. "A 175-mV multiply-accumulate unit using an adaptive supply voltage and body bias architecture." Solid-State Circuits, IEEE Journal of 37.11 (2002): 1545-1554.

[10] Kumar, Ranjith, and Volkan Kursun. "Temperature-adaptive voltage scaling for enhanced energy efficiency in subthreshold memory arrays." *Microelectronics Journal* 40.6 (2009): 1013-1025.

[11] Kahng, Andrew B., et al. "Slack redistribution for graceful degradation under voltage overscaling." Design Automation Conference (ASP-DAC), 2010 15th Asia and South Pacific. IEEE, 2010.

[12] Chen, Yu-Guang, et al. "Critical Path Monitor Enabled Dynamic Voltage Scaling for Graceful Degradation in Sub-Threshold Designs." Proceedings of the The 51st Annual Design Automation Conference on Design Automation Conference. ACM, 2014.

Physical Design and Mask Optimization for Directed Self-Assembly Lithography (DSAL)

Seongbo Shim[‡] and Youngsoo Shin[§]

‡ Samsung Electronics, Hwasung 445-330, Korea
§ Department of Electrical Engineering, KAIST, Daejeon 305-701, Korea

Abstract—**In DSAL, contact holes are indirectly formed through guide patterns (GPs). Thus the integrity of GPs is very important, in particular when GP shape is large and complex. The limitation in GP shape calls for careful consideration in physical design stage. In mask optimization, synthesizing ideal GP shape and verifying whether synthesized GP is correct are important but difficult problem. Some challenges in physical design and mask optimization are reviewed in this paper with possible solutions.**

I. INTRODUCTION

It has recently been observed that the scaling of devices is approaching fundamental (i.e. atomic scale) and economic (i.e. cost per fabrication facility) limits, in large part because traditional lithography is facing substantial challenges for printing fine features while maintaining a reasonable cost. Researchers are actively searching for alternative patterning approaches for the next generation lithography. Potential solutions such as extreme ultraviolet lithography, electron beam lithography, and multiple patterning lithography have attracted much attention from the lithography community.

However, they all have practical limitations, e.g. extremely high cost or low throughput. Directed self-assembly lithography (DSAL) stands out due to its low cost and high throughput as well as good patterning resolution for sub-20nm features; it is thus considered as a most promising patterning solution for contact holes and vias in technology node of 7nm and below.

In DSAL, contacts (or vias) that are physically close are clustered as shown in Fig. 1(a), which are then patterned together by DSAL process through one guide pattern (GP) as shown in Fig. 1(b). The DSAL process consists of two steps: optical lithography to form GP on a wafer, and DSA to form contacts (or vias) within each GP. A GP, which is a key component of DSAL, poses a few challenges. As a cluster contains more contacts, the corresponding GP is more sensitive to lithography variations and is more likely to cause a pattern failure during DSA process; for example, DSA patterns from three contact GP are more irregular than those from two contact GP (see Fig. 1(b)). The limitation in contact topologies should be taken into account in layout design. In mask synthesis, challenge lies on runtime since a set of GP images for a given contact layout can only be synthesized through lengthy simulations. Since GP is created by optical lithography, GP may have errors due to lithography variations.

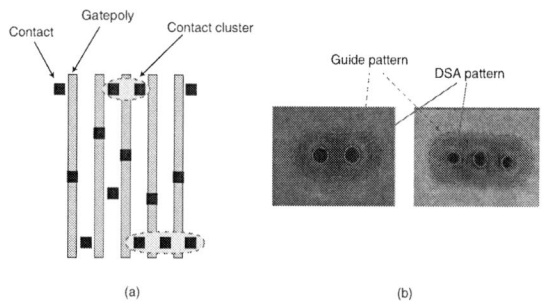

Fig. 1. (a) Contact clusters and (b) SEM images of corresponding GP and DSA pattern on a wafer.

Therefore, once GP images are obtained, they should be verified to see whether target contacts are correctly formed, which is also a difficult problem.

This paper presents the basics of DSAL technology together with current issues and a few state-of-the-art in design optimization and mask synthesis as follows:

- **Overview of DSAL technology:** Patterning mechanism and flow of DSAL is briefly introduced and compared with those of standard optical lithography in Section II, which will also cover recent development issues that affect circuit design.

- **DSAL-Aware Physical Design Optimization:** In Section III, we will present standard cell layout optimization and post-placement optimization to avoid undesired GP with high defect probability within cell region and in between adjacent cells.

- **DSAL Mask Synthesis:** Conventional mask synthesis and verification (for optical lithography) are obsolete in DSAL. New problems in mask synthesis and verification will be introduced and addressed in Section IV. Inverse DSA problems is to synthesize ideal GPs for a given contact layout, but it takes too much time due to lengthy simulation. Verification of GP also requires a number of simulations, so alternative is required to handle GPs on full-chip layout.

978-1-4673-9141-2/15 $31.00 © 2015 IEEE

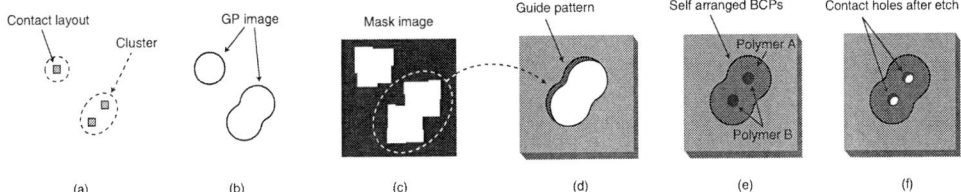

Fig. 2. Mask synthesis: (a) contact layout and clusters, (b) GPs, and (c) mask image after OPC; DSA process: (d) a GP created by optical lithography, (e) a GP filled with BCPs, and (f) contact holes after polymer B is etched away.

Fig. 3. Examples of DSA defect: (a) too large contact may cause electrical short, (b) actual contact is not open, and (c) unexpected contact appears.

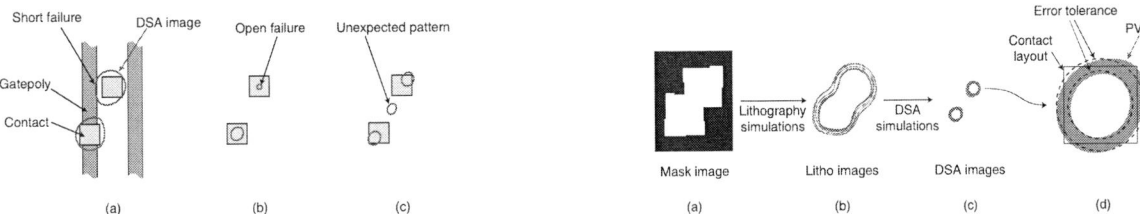

Fig. 4. (a) Mask image of a GP, (b) expected shapes of GP after optical lithography (lithography images) under various conditions, (c) expected shapes of contacts (DSA images) after DSA process, and (d) PVB of a contact with error tolerance.

II. PRELIMINARIES

A. DSAL process

Block copolymers (BCPs) are unique soft materials that assemble by themselves through microphase separation into various periodic nanostructures such as cylinders, spheres and lamellas [1] [1]. The feature size of these nanostructures is dependent on the length of the block copolymers and therefore not limited by the same factors that limit optical lithography such as ultraviolet light wavelength. In addition, the self-assembly could be controlled by a simple thermal annealing process, which significantly reduces the cost and improves the throughput. Among all the varieties of nanostructures, the cylindrical self-assembled patterns are especially suitable for patterning contacts and vias in integrated circuits (ICs).

The standard procedure of DSAL is illustrated in Fig. 2.

- **Mask synthesis:** Neighboring contacts in a layout are clustered with respect to their physical distance [2] (Fig. 2(a)); a GP is obtained for each cluster through the process named inverse DSA or GP synthesis (Fig. 2(b)); a mask image is then synthesized through OPC with GPs as targets (Fig. 2(c)).

- **DSA process:** GPs are created by the optical lithography process such as 193nm immersion lithography, which has enough pitch resolution to pattern the sparsely located GPs (Fig. 2(d)); each GP (a large hole) is filled with BCPs, which are self arranged due to forces between the polymers and the GP (Fig. 2(e)); one of the polymers (polymer B) is then etched away leaving small contact (or via) holes (Fig. 2(f)).

B. DSA Defect due to Lithography Variations

Even though an ideal GP and its mask image are synthesized (Fig. 2(b) and (c)), GP may have errors on a wafer due to the lithography variations, which may cause a pattern failure, called DSA defect. Some defect examples are shown in Fig. 3: too large contact that may cause electrical short, target contact is not open due to too small contact image on a wafer, and unexpected contact appears during DSA process. Since contacts are patterned through two independent steps as above, small variations on some critical boundaries of GP may result in huge interference on the final DSA image.

DSA defect can be predicted through repeated lithography and DSA simulation. As illustrated in Fig. 4, a lithography simulation with a mask image of a GP yields an expected GP shape on a wafer, called litho image. To account for lithography variations, the simulation is in fact repeated while lithography parameters (e.g. scanner focus, exposure energy, and mask manufacturing error) are varied [3]; the result is a set of litho images. Each litho image is then submitted to a DSA simulator [4], which outputs an expected shape of a contact (or contacts), called DSA image. Each contact, in the end, is associated with multiple DSA images as shown in Fig. 4(c), and we can verify whether DSA defect occurs by examining the size and location of the DSA images.

C. DSA Defect Probability

The probability that DSA defect occurs, called DSA defect probability (or simply defect probability), is proposed to quantitatively measure how sensitive a GP is to the lithography variations [5]. The region bounded by the outermost- and innermost-contours of the multiple DSA images is called DSA process variation band (PVB) as illustrated in Fig. 4(d). The size of DSA image mainly determines the defect, and PVB indicates how much that size varies. So, defect probability can be defined by comparing the PVB area (red region) with some target region, called error tolerance as shown in Fig. 4(d) [5]. If

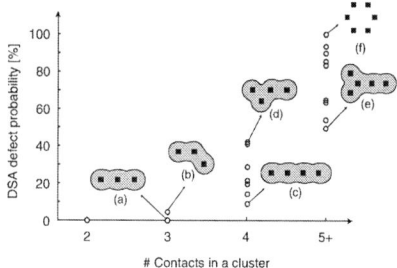

Fig. 5. DSA defect probability for various contact clusters.

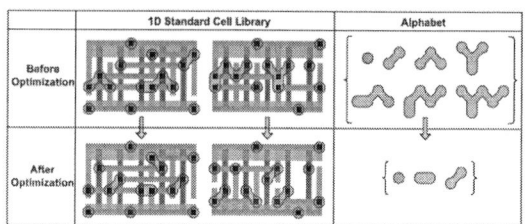

Fig. 6. Layouts and GPs for a standard cell before and after optimization [6].

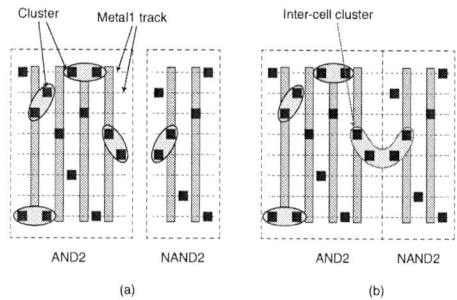

Fig. 7. Contact clusters (a) before- and (b) after-placement.

whole PVB is contained within the target region, defect is not expected (i.e. zero defect probability). Otherwise, we define DSA defect probability by area ratio of PVB beyond tolerance to PVB area. The defect probability of a contact cluster (or GP) is assumed to be the maximum defect probability of its member contacts.

Experiments using 7-nm technology indicate that the defect probability increases as a cluster contains more contacts as shown in Fig. 5 [5]. There is a substantial variation of defect probability even for the clusters of the same number of contacts. Cluster (a), (c), and (e) have lower defect probability. This is because of symmetry that GP has, which makes the cluster less sensitive to lithography variations because BCPs tend to be aligned periodically for lower energy. Due to the absence of symmetry, cluster (b) and (d) have high defect probability. Cluster (f) cannot be patterned by DSAL since corresponding GP is not synthesized, so its defect probability is 100%.

III. DSAL-AWARE PHYSICAL DESIGN OPTIMIZATION

As we discussed in Section II, the DSA defect probability usually increases as a cluster contains more contacts because corresponding GP contour becomes more complex. To avoid such undesired clusters, contact layout is carefully modified in physical design stages.

An alphabet approach [6] has been proposed to optimize standard cell layout for this, in a way that such large clusters are split into smaller ones by relocating some contacts as shown in Fig. 6. The approach is to use a minimal set of alphabet, which are associated with allowable contact clusters (or GPs), to pattern all contact configurations on integrated circuits. This method successfully avoids all large clusters within a cell region.

However, when two cells are located side by side (e.g. during placement), large clusters can still form across the cells as illustrated in Fig. 7. Verifying the correctness of all inter-cell clusters is very important but difficult. Applying lithography- and DSA-simulations on whole layout may be considered as a possibility, but impractical amount of time does not allow this approach.

Key ideas of our approach is to determine defect probabilities of all cell pairs in advance, when two cells of each pair are placed side by side (Section III-A); and after standard placement, we perform post-placement optimizations with the

goal of minimizing the use of whitespace, which is inserted in between a cell pair whose defect probability is unacceptably high (Section III-B).

A. Fast Identification of Defect Probability

Fortunately, DSA process is localized within a GP [7], which implies that the identical clusters are associated with the same GP. And there are many identical clusters, which have identical contact topology (the number of contacts, distances between adjacent contacts, and the angle between lines connecting the contacts) because Gridded Design Rules (GDR) limits diversity of contact topology. So, the number of unique (inter-cell) clusters is quite manageable, and they are submitted to defect probability computation. The defect probability of a cell pair (see Fig. 7(b)) is now represented by the maximum defect probability of inter-cell clusters.

In the experiments using a 7-nm synthetic library, in which the minimum contacted-poly pitch (CPP) is 44nm, size of contact is 22nm, and metal 1 track pitch is 36nm, we identify about 40000 inter-cell clusters, but unique clusters are only 19. Defect probability calculation takes about 50 hours, which however is required only once. About 35% of pairs have zero defect probability: 17% pairs have no inter-cell clusters; 11% have clusters of two contacts, which have zero defect probability as shown in Fig. 5; the remaining 7% have clusters of three contacts that are linearly aligned (cluster (a) in Fig. 5), which also have 0% defect.

B. Post-Placement Optimization

Once standard placement has been performed, if there is a cell pair whose defect probability exceeds the threshold (usually 0%), a whitespace needs to be inserted in-between so that no inter-cell cluster can form. The goal of post-placement optimization is to minimally perturb the placement

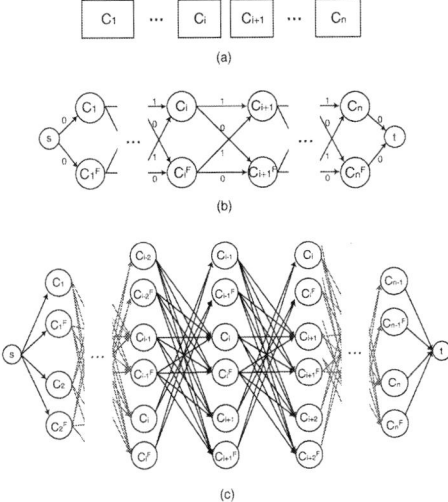

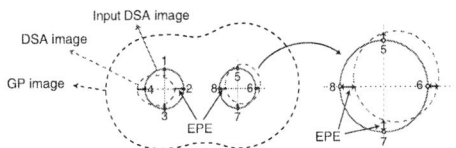

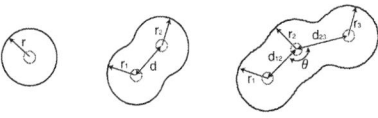

Fig. 8. (a) A placement row of n cells, (b) graph modeling to determine optimal cell flipping, and (c) graph modeling to determine optimal cell swapping and flipping.

Fig. 9. GP image and its DSA image; each contact of input DSA image has 4 EPEs at certain measurement points.

Fig. 10. Geometry parameters of GP image.

so that the amount of whitespace we insert is minimized. Our optimization has two options: (1) flip some cells (along y-axis) and (2) swap some adjacent cells as well as flip some.

Our problem is modeled as the shortest path problem in a directed acyclic graph (DAG) with nonnegative weights. For a given cell row as shown in Fig. 8(a), each position in the row (i.e. each cell) corresponds to two vertices as shown in Fig. 8(b), where superscript F indicates that the cell is flipped. If two adjacent cells can be swapped, each position now corresponds to 6 vertices since three cells can be located at one position; note that the positions at the end of the row corresponds to 4 vertices. Two adjacent cells have 4 edges, which represents 4 different paring configurations of the two cells. If a paring configuration requires a whitespace due to high defect probability, corresponding edge is associated with weight of 1; otherwise weight is 0. The shortest path from s to t is a list of vertices, which denotes a single row placement that requires smallest number of whitespace.

We demonstrate our method using a few test circuits from Open Cores [8]. Cell flipping alone achieves 6% increase of placement density with marginal increase of wirelength (2%). Cell swapping and flipping allows 11% increase of placement density, which comes at the cost of 9% increase of wirelength. The same comparison is also made when the threshold increases to 5% and 10%. Placement density generally increases with increasing threshold because we need whitespaces in-between less number of cell pairs. Since our optimizations are applied to less number of cell pairs, the benefit clearly decreases.

IV. DSAL MASK OPTIMIZATION

DSA simulation [4] predicts DSA image for a GP, which therefore can be employed to synthesize GP for a given contact cluster (i.e. inverse DSA) or to verify DSA defect. However, the extremely low efficiency makes it impossible to be adopted

in the full-chip level implementation. For example, 10000 GPs on $100\mu m \times 100\mu m$ layout, verification is estimated to reach more than 6 months, if 27 lithography settings are considered [9].

In this section, we introduce and address two new problems in DSAL mask optimization: (1) inverse DSA problem, which is to synthesize perfect GP image for a contact cluster (Fig. 2(b)) and (2) fast verification problem of GP on a full-chip level.

A. Inverse DSA

There have been a few studies for the opposite process of inverse DSA, e.g. form various shapes of GP images on a wafer and check how contacts are formed for each GP [10], [11]. But inverse DSA has never been studied before. We propose an iterative solution, in which a GP image is progressively refined while the resulting DSA image, obtained through DSA simulation, is assessed against a target DSA image. A GP image is defined as a function of a few geometry parameters; it is refined by using sensitivity matrix, which contains the extent of how sensitive the DSA image is to each parameter change.

The algorithm receives one contact cluster in DSA image (input DSA image in Fig. 9), and returns corresponding perfect GP image. An initial GP image is constructed by placing circles (each is a concentric circle for a circle in DSA image with radius being set to the length of one BCP string), taking only the boundary arc of merged circles, and smoothing the boundary (see GP image in Fig. 9). A DSA simulation is performed on the initial GP image to obtain DSA image, which is, in most cases, different from input DSA image. The edge placement errors (EPEs) between the two DSA images are measured at certain points (four for each circle), and they are arranged as a vector $\mathbf{e} = (e_1, e_2, ..., e_m)$ (see Fig. 9).

Our goal is to refine the GP image so that resulting DSA image resembles the input DSA image as much as possible; this is achieved through iteration. To guide the refinement, GP image is defined as a function of geometry parameters, i.e.

$$\mathcal{G} = f(\mathbf{g}) = f(g_1, g_2, g_3, \cdots, g_n). \qquad (1)$$

The parameters include the radius of each constituent circle, the center-to-center distance between adjacent circles, and the

978-1-4673-9141-2/15 $31.00 © 2015 IEEE

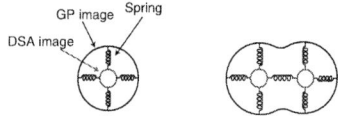

Fig. 11. Compact DSA model using springs [12], [13].

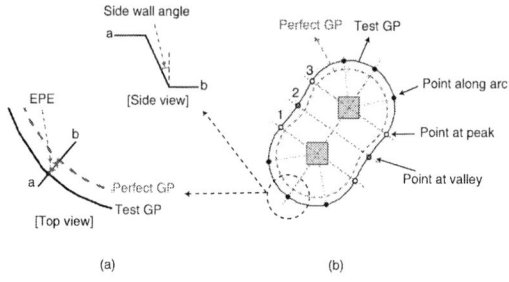

Fig. 12. Parameterization of two contact GP: (a) EPE and side wall angle and (b) measurement points.

angle between adjacent center-to-center lines (see Fig. 10). We perturb each parameter to obtain a new GP image, perform a DSA simulation that returns corresponding DSA image, and assess its EPEs; a set of EPE changes allows us to introduce the sensitivity matrix:

$$
\mathbf{M} = \begin{bmatrix} \frac{\partial e_1}{\partial g_1} & \frac{\partial e_1}{\partial g_2} & \cdots & \frac{\partial e_1}{\partial g_n} \\ \frac{\partial e_2}{\partial g_1} & \frac{\partial e_2}{\partial g_2} & \cdots & \frac{\partial e_2}{\partial g_n} \\ \cdots & \cdots & \cdots & \cdots \\ \frac{\partial e_m}{\partial g_1} & \frac{\partial e_m}{\partial g_2} & \cdots & \frac{\partial e_m}{\partial g_n} \end{bmatrix}. \tag{2}
$$

It can be shown that the following holds:

$$
\mathbf{M} \times \Delta \mathbf{g}^{\mathrm{T}} = \Delta \mathbf{e}^{\mathrm{T}}, \tag{3}
$$

where $\Delta \mathbf{e}$ is EPE variations due to some extents of the parameter changes ($\Delta \mathbf{g}$). Rearranging (3) yields

$$
\Delta \mathbf{g}^{\mathrm{T}} = \mathbf{M}^{-1} \times \mathbf{e}^{\mathrm{T}}, \tag{4}
$$

where $\Delta \mathbf{g}$ now indicates how much we should adjust each parameter to compensate current EPEs to obtain a new GP image[1]. DSA simulation and EPE measurement then follow. If the largest EPE does not exceed a certain threshold, iteration stops; otherwise iteration continues until user defined maximum iterations are reached.

This algorithm synthesizes a GP image for one cluster, so it has to be applied to all clusters (of contacts and vias) in a layout. Fortunately, there are many identical clusters and DSA process is localized within a GP image, so we classify all contact clusters in a layout by examining contact topology and apply inverse DSA to a few unique clusters.

We implement our method in MATLAB and C++, and DSA simulator that we used is based on self consistent field theory [4]. To compare with the state-of-the-art of DSAL mask synthesis [14], we combine our inverse DSA method with conventional OPC tool, which receives resulted perfect GP as a target. Test via layout with $44 \mu m \times 44 \mu m$ size is prepared by synthesizing a circuit from Open Cores [8] using 15-nm NanGate library [15] and layout was scaled down so that the layouts follow the design rules of 7-nm technology. The layout contains 1263 clusters, only 14 (or about 1%) required inverse DSA in our approach while all clusters are individually concerned in the state-of-the-art approach. It takes 25 hours (using 42 cpu cores) [14] to obtain DSAL mask in the state-of-the-art, but our method finishes it in 1 hours (using 42 cpu cores), which is understandable consequence of applying lengthy inverse DSA to only a few unique clusters in our approach.

[1]If $\mathbf{M}$ is not square, $\mathbf{g}^{\mathrm{T}} - (\mathbf{M}^{\mathrm{T}} \times \mathbf{M})^{-1} \times \mathbf{M}^{\mathrm{T}} \times \mathbf{e}^{\mathrm{T}}$ replaces $\mathbf{g}^{\mathrm{T}}$ or singular value decomposition (SVD) is applied.

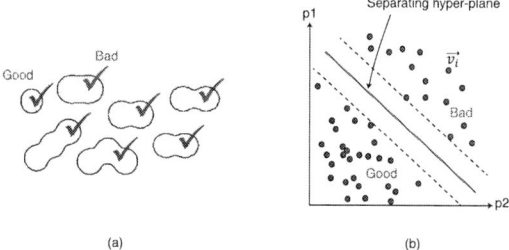

Fig. 13. Construction of decision function: (a) DSA simulations on test GPs and (b) identifying hyper-plane that separates good and bad GPs.

B. DSA Verification

Once GP and mask images are obtained, we should verify whether the mask image results in DSA defect due to lithography variations. This can be straightforwardly performed by repeated lithography and DSA simulation with various lithography settings, which is however not practical due to huge runtime.

One of the alternative approaches is to make fast model instead of the rigorous simulator. In [12], [13], force between contacts and GP is modeled as springs as shown in Fig. 11; the spring can be compressed and stretched but has an equilibrium position where the force of the spring is zero, which denotes the equilibrium state of BCPs after annealing. This method can predict center locations of contacts with reasonable errors, but a pattern failure is not correctly predicted because size variation of DSA image is not taken into account.

On the other hands, machine learning (ML) has demonstrated its effectiveness and efficiency for GP verification problem [9], [16], [17]. This approach is to construct a decision function of n variables, which correspond to geometry parameters of GP (e.g. edge placement error (EPE) and side wall angle at various points along GP contour as shown in Fig. 12 [9]); the function is evaluated for each GP in a given test layout and decides whether GP can faithfully produce desired contacts (good) or not (bad).

The function type depends on machine learning algorithm we use. In artificial neural network (ANN), a set of weight values in hidden layers plays a role of a decision function [16], [17], and in support vector machine (SVM), hyper-plane in n-dimensional space is used as a decision function [9]. For example, a decision function of SVM is determined as follows:

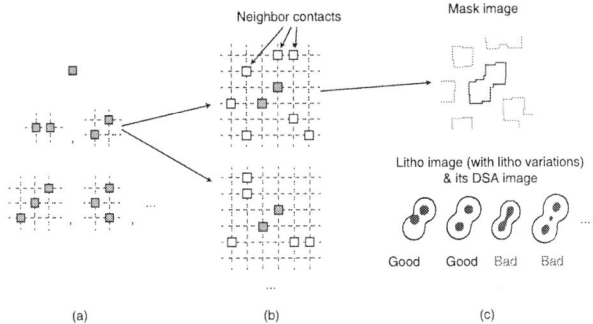

Fig. 14. Test GPs: (a) generating all possible contact topologies, (b) introducing random neighbor contacts, and (c) mask synthesis, litho- and DSA-simulation, and assessing GPs.

a set of test GPs is prepared (Fig. 13(a)); after DSA simulations, each GP is marked good or bad in n-dimensional space; the hyper-plane that separates good marks and bad marks is determined through machine learning process (Fig. 13(b)), which corresponds to our decision function.

The accuracy and coverage of a decision function is affected by how comprehensive test GPs are. Introducing random noise to a perfect GP has been tried to generate various test GPs [16], [17], but this may yield many unrealistic GPs, e.g. too curvy and fine GP contour that lithography cannot support. To prepare more comprehensive and realistic test GPs, we explore all possible contact topologies under GDR, and generate corresponding deformed GPs (as well as original GP) by performing lithography simulations with various lithography settings, as shown in Fig. 14.

How effectively geometry parameters are chosen is also important to construct stable model. A few popular parameters from image processing and computer vision, e.g. histogram of oriented gradients, have been employed in [16], [17], but a number of parameters may cause over fitting issue during constructing decision function. We therefore propose to analyze the parameters and try to remove redundant ones, which can be performed through principal component analysis (PCA). For example, we first measure more than 50 parameters (e.g. EPEs along a GP contour) so a GP is marked in 50-dimension space, but we can extract only 12 dominant and orthogonal new parameters, which are obtained by a linear combination of the initial 50 parameters, implying that a GP is now marked in 12-dimension space. Our experiment indicates that this reduction of parameter space increase fitting error (by 11%) when constructing a decision function, but increases prediction accuracy for new (unknown) GPs (by 27%).

To apply a decision function to new GPs, the new GPs should be parameterized using the same number of parameters used in function construction. This implies that a decision function should be constructed for each GP type (e.g. two contact and three contact GPs), and it is not possible to apply a decision function to new GP type, which limits coverage of this approach. A single global model that can be applied to all GP types has been proposed in [9], which is made possible by rearranging the parameters for different GP types and unifying them.

V. SUMMARY

We have presented some new challenges posed by DSAL in physical design and mask optimization. To avoid lithography-induced DSA defect, standard cell layout and its placement have been optimized, in which DSA defect probability is proposed and its fast identification is realized by grouping identical contact topologies. In mask optimization, inverse DSA problem has been addressed for the first time, and machine learning has been applied to fast GP verification.

REFERENCES

[1] M. Muramatsu, M. Iwashita, T. Kitano et al., "Nanopatterning of diblock copolymer directed self-assembly lithography with wet development," Journal of Micro/Nanolithography, MEMS, and MOEMS, vol. 11, no. 3, pp. 1–6, Nov. 2012.

[2] H. Yi, X. Bao, R. Tiberio, and P. Wong, "Design strategy of small topographical guiding templates for sub-15nm integrated circuits contact hole patterns using block copolymer directed self assembly," in Proc. SPIE, Mar. 2013, pp. 1–9.

[3] L. Liebmann, S. Mansfield, G. Han, J. Culp, J. Hibbeler, and R. Tsai, "Reducing DfM to practice: the lithography manufacturability assessor," in Proc. SPIE, Feb. 2006, pp. 786–798.

[4] N. Laachi, K. T. Delaney, B. Kim et al., "Self-consistent field theory investigation of directed self-assembly in cylindrical confinement," Journal of Polymer Science: Part B Polymer Physics, vol. 53, no. 2, pp. 142–153, Jan. 2015.

[5] S. Shim, W. Chung, and Y. Shin, "Defect probability of directed self-assembly lithography: fast identification and post-placement optimization," in Proc. Int. Conf. on Computer Aided Design, Nov. 2015, to be presented.

[6] Y. Du, D. Guo, M. Wong, H. Yi, H. Wong, H. Zhang, and Q. Ma, "Block copolymer directed self-assembly (dsa) aware contact layer optimization for 10nm 1D standard cell library," in Proc. Int. Conf. on Computer Aided Design, Nov. 2013, pp. 186–193.

[7] H. Yi, A. Latypov, and P. Wong, "Computational simulation of block copolymer directed self-assembly in small topographical guiding templates," in Proc. SPIE, Mar. 2013, pp. 1–7.

[8] "Opencores," http://www.opencores.org/.

[9] S. Shim, W. Chung, and Y. Shin, "Synthesis of lithography test patterns through topology-oriented pattern extraction and classification," in Proc. SPIE, Mar. 2014, pp. 1–10.

[10] H. Yi, Y. Bao, J. Zhang et al., "Flexible control of block copolymer directed self-assembly using small, topographical templates: potential lithography solution for integrated circuit contact hole patterning," Advanced Materials, vol. 14, no. 23, pp. 3107–3114, Jul. 2012.

[11] Y. Seino, H. Yonemitsu, H. Sato, M. Kanno, H. Kato, K. Kobayashi, A. Kawanishi, T. Azuma, S. Nagahara, T. Kitano, and T. Toshima, "Contact hole shrink process using graphoepitaxial directed self-assembly lithography," Journal of Micro/Nanolithography, MEMS, and MOEMS, vol. 12, no. 3, pp. 1–6, Dec. 2013.

[12] J. Torres, S. Kyohei, D. Fryer, Y. Granik, Y. Ma, P. Krasnova, G. Fenger, S. Nagahara, S. Kawakami, B. Rathsack, G. Khaira, J. Pablo, and J. Ryckaert, "Physical verification and manufacturing of contact/via layers using grapho-epitaxy DSA process," in Proc. SPIE, Mar. 2014, pp. 1–8.

[13] G. Fenger, A. Burbine, J. Torres, Y. Ma, Y. Granik, P. Krasnova, G. Vandenberghe, R. Gronheid, and J. Bekaert, "Calibration and application of a dsa compact model for grapho-epitaxy hole processing using contour-based metrology," in Proc. SPIE, Mar. 2014, pp. 1–12.

[14] W. Wang, L. Azat, Y. Zou, and T. Coskun, "A full-chip DSA correction framework," in Proc. SPIE, Mar. 2014, pp. 1–11.

[15] "Nangate 15nm open cell library," http://www.nangate.com/.

[16] Z. Xiao, Y. Du, H. Tian, and M. Wong, "Directed self-assembly (DSA) template pattern verification," in Proc. Design Automation Conf., Jun. 2014, pp. 1–6.

[17] Z. Xiao, Y. Du, M. Wong, H. Yi, H. Wong, and H. Zhang, "Contact pitch and location prediction for directed self-assembly template verification," in Proc. Asia South Pacific Design Automation Conf., Jan. 2015, pp. 644–651.

978-1-4673-9141-2/15 $31.00 © 2015 IEEE

Qualifying Non-Volatile Register Files for Embedded Systems through Compiler-directed Write Minimization and Balancing

Chengmo Yang
Electrical and Computer Engineering
University of Delaware, Newark, DE 19716
Email: chengmo@udel.edu

Maria Ruiz Varela
Computer and Information Sciences
University of Delaware, Newark, DE 19716
Email: mruizv@udel.edu

Abstract—**Recent research shows non-volatile flip-flops can be attached to register files to hold computation state thus enabling fast recovery upon power failure. However, the endurance limitation of NVM cells challenges their usage for holding register values that are frequently updated during program execution. To extend the lifetime of non-volatile register files, we propose two compiler-directed optimizations. First, through analyzing the register access patterns in frequently executed loops, a minimum set of registers is identified to be periodically written to NVM cells, thus minimizing the total number of writes to the NVM register file. Meanwhile, the register mapping is also adjusted to enable an efficient dynamic register rotation to further balance the writes to different NVM registers. Experimental studies show that the proposed two techniques can significantly extend the lifetime of non-volatile registers, thus qualifying them for various embedded systems.**

I. INTRODUCTION

Current embedded systems are facing energy and reliability challenges since their computation is confined to a limited or unstable power supply. Although sometimes self-powered devices may be able to derive energy from external sources such as solar or wind, they suffer from power failures due to the instability of the power source. As the computation state is not maintained when the power is off, not only may critical data get lost, but a fair amount of energy will also be wasted to restart the execution from scratch.

A very promising solution to overcome these energy and reliability difficulties is to deploy non-volatile memories (NVM) in embedded systems. Flash Memory [1] has been widely used in many portable electronics as secondary storage, while Phase Change Memory (PCM) [2], Spin-Transfer Torque Random Access Memory (STT-RAM) [3], [4], Magnetic Random Access Memory (MRAM) [5], and Ferroelectric Random Access Memory (FeRAM) [6], [7], are considered as promising candidates to replace main memory, caches, and even register files. These non-volatile devices offer a set of advantages, such as negligible standby power, high resilience to soft errors, as well as the ability to maintain computation states across interrupted executions.

Yet the deployment of NVM storage devices is challenging, primarily because NVMs are usually constrained by limited write endurance, high write energy, and slow write performance compared to read operations. These challenges are particularly difficult to overcome when NVMs are deployed at the top level of the conventional memory hierarchy, that is, used as register files. Among all NVMs, FeRAM has the highest access speed. Recent studies [8], [9] have proposed to attach a FeRAM-based flip-flop to each register, so as to not only reduce energy consumption but also enable the system to quickly recover from sudden

power failures. Although the ferroelectric material can sustain a much higher number of writes (10^{12}) than Flash (10^5) or PCM (10^7), it still suffers from early wear out because registers are updated extremely frequently, almost once per instruction. More precisely, for a system with an expected lifetime of 10 years and expected usage of 10 hours per day, a FeRAM register can be updated only $10^{12}/3600/10/365/10 = 76104$ times per second, which is far less than the clock speed of most embedded systems.

Previous NVM research mainly focuses on extending the lifetime of NVM cache or memory [10], [11] whose accesses are managed by a hardware controller and/or the operating system. Those solutions cannot be directly adopted at the register level since register access patterns are directly determined by the compiler. In this paper, we propose two compiler-directed optimizations to collaboratively extend the lifetime of FeRAM-based register files. Through analyzing the register access patterns in frequently executed loops, the first technique identifies a minimum set of registers that need to be written to the attached NVM flip-flops. Meanwhile, to avoid the undesired case of a small set of registers dominating the lifetime of the entire NVM register file, we propose another technique to balance the number of writes to each individual register, achieved through an efficient, low-cost register rotation at runtime. Experimental studies on embedded benchmarks show that the proposed two techniques can extend the lifetime of non-volatile registers by up to 10 times, thus qualifying them for various embedded systems.

The rest of this paper is organized as follows. Section II briefly reviews non-volatile registers and techniques to extend the life of non-volatile devices. Sections III and IV respectively present the proposed register write minimization and register mapping rotation techniques. Section V verifies the efficacy of the technique, while Section VI concludes the paper.

II. BACKGROUND AND RELATED WORK

A. Non-volatile memories

Non-volatile memory (NVM) technologies are revolutionizing the way we access and manipulate information. Their non-volatility allows computation states to be maintained across power interrupts, thus supporting long running computations even with unreliable power sources. This advantage, in conjunction with their high density, negligible standby power, and resilience to soft errors, is driving their usage at various layers of the memory hierarchy, for example, Fe-RAM as registers [12], [13], STT-RAM as on-chip memory [3], [4], and PCM as off-chip memory [2], [11]. However, two drawbacks of NVMs are preventing them from fully replacing traditional SRAM and DRAM, namely, *limited write endurance* and *expensive writes* compared to read operations. To overcome these challenges, previous research aims either to reduce the number of writes by eliminating unnecessary writes [2], [14], or to even out write

---

This work was partially supported by NSF grant #1253733.

978-1-4673-9141-2/15 $31.00 © 2015 IEEE

distributions by periodically swapping the positions of hot and cold data [11], [15].

B. Non-volatile register files

Among the different NVM devices, FeRAM has the highest access speed, making it a viable technology for implementing non-volatile registers. FeRAM [6], [8], [7] uses a ferroelectric layer to achieve non-volatility, while 0's and 1's are stored as one of two possible electric polarizations. As FeRAM is immune to electric fields and radiation, it is very power efficient and highly reliable.

Previous work has demonstrated different hardware implementations for using FeRAM in embedded systems. Ferroelectric flip-flops (FeFFs) designed to save energy and protect the system against power failures are illustrated in [8], [9]. FeRAM-based non-volatile processor (NVP) designs are shown in [12], [13]. Another set of work focused on reducing the area overhead and backup time of NVPs [16], [17]. However, same as the other non-volatile memories, FeRAM also suffers from limited write endurance and high write latency and energy. These drawbacks are extremely crucial to FeRAM when it is used to hold register values that are frequently updated during program execution. To extend FeRAM lifetime, several architecture and circuit level solutions have been proposed. The work in [6] extends FeRAM lifetime by eliminating unnecessary programming cycles through a compare-and-write optimization. In [18], a hybrid flip-flop structure is presented, wherein a persistent counter is used to enhance the endurance of non-volatile flip-flops.

Yet solutions at the architecture and circuit levels are unable to eliminate the inefficiency and imbalance in register access patterns. Similarly, most of the write optimizations developed for NVM cache and memory cannot be directly adopted at the register level. This is because the accesses in cache and memory are managed by a hardware controller and/or the operating system, while the register access patterns are directly determined by the compiler. In contrast, the work presented in this paper exploits application information regarding register access patterns, so as to backup only those values that are necessary to restart the application in case of a power failure.

C. Register file checkpointing and rotation

Checkpointing can be used to accelerate recovery in case of execution failures. In [19], a self-checkpointing register file is presented. Each register cell is extended to include a bit of non-volatile storage to support checkpointing. The work in [20] proposes a redundant execution-based approach for fault detection and recovery at the register space. By analyzing the control and data flow of application hotspots, it is able to select a minimum set of registers for checkpointing. A deterministic register rotation technique is proposed in [21] to balance the accesses to individual registers. The goal of that work is to reduce heat consumption, and their remapping algorithm focuses on preserving loop-carried dependences. In contrast, our work aims at evenly distributing the writes to different non-volatile registers so as to extend their lifetime.

III. REGISTER WRITE MINIMIZATION

A. Technical Motivation

In state-of-the-art non-volatile processors, a non-volatile register file (NVRF) can be used in two different settings, as shown in Fig. 1. In the first setting, a NVRF is used exclusively and the processor directly reads from and writes to the NVRF. Every register write request issued by the processor needs to be committed to the NVRF, implying that the total number of writes

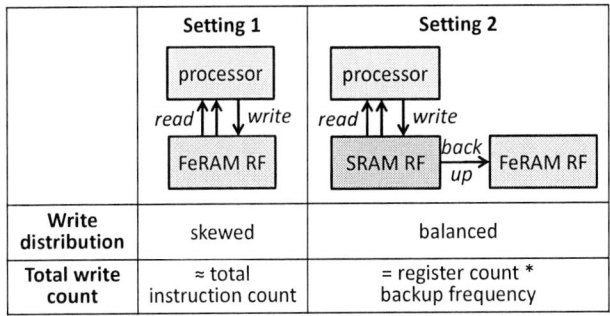

Fig. 1. Two utilizations of FeRAM-based non-volatile register file

to NVRF equals the total number of executed instructions with register destinations. Clearly, this value is more-or-less fixed for a given application. Furthermore, the write distribution across registers is highly imbalanced, since the compiler produces asymmetric write patterns to different registers.

In the second setting, a non-volatile flip-flop is attached to each SRAM-based register. The SRAM register file serves read and write requests from the processor, while all the register values are periodically backed up to the NVRF. In this setting, all non-volatile registers are written at the same frequency, thus the write distribution in NVRF is balanced. The total number of write operations to NVRF is the product of the total number of registers and the backup frequency.

Clearly, the second setting offers more flexibility than the first since the backup frequency can be adjusted to reduce the number of writes to NVRF. However, there still exist **two types of inefficiency** in backing up register values. First, the approach backs up all register values to NVRF regardless of their access patterns. Yet a quick examination indicates that not all register values need to be written to NVRF. Upon a power failure or other unexpected faults, only the *live variables*[1] are needed to restart the computation. The second type of inefficiency lies in the fact that not all the live variables need to be backed up at the same frequency. As some live variables may not be updated within a frequently executed loop, they only need to be backed up once, prior to entering the loop. In light of these observations, we propose to leverage the application information regarding register access patterns, which can be collected at compile time, so as to minimize the total number of writes to NVRF.

B. Access Pattern-based Register Backup

A widely-recognized program characteristic is locality: 90% of the execution time is spent on loops that constitute only 10% of the code size. This property has been intensely exploited, primarily in application-specific embedded systems, to achieve performance, power, and reliability optimizations [20]. In this paper, we will demonstrate that such information can also be exploited to identify the minimum set of register values to be written to a non-volatile register file.

Most execution hotspots are composed of loops that have highly regular control and data flows and generate repetitive access patterns to registers. Based on their access patterns, registers can be classified into four types according to two criteria: (1) Whether a register is *assigned*, that is, updated within the loop body, and (2) whether a register is *live* at loop entry. A concrete loop example and its register classification are shown in Fig. 2 and Table I, respectively. This is the hottest loop of the

---

[1]At a given time point, a register is *live* if the very first subsequent access to that register is a read operation.

978-1-4673-9141-2/15 $31.00 © 2015 IEEE

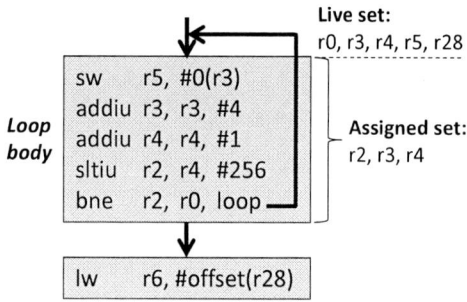

Fig. 2.   Example: hottest loop of string search

TABLE I.   REGISTER CLASSIFICATION OF THE LOOP EXAMPLE

	Assigned	Not assigned
Live	r3, r4	r0, r5, r28
Not live	r2	r6

string search benchmark [22] that accounts for 38% of execution time.

The figure shows a loop body composed of five instructions, as well as one instruction that will be executed when exiting the loop. Register $r2$ is *not live* at loop entry, since the first access to $r2$ in the loop is a write in *sltiu*. Accordingly, it is unnecessary to back up $r2$ to NVRF, even if the loop has to be re-executed due to a failure. In contrast, register $r3$ is *live* at the entry of the loop, since the first access to $r3$ in the loop is a read in *sw*. Therefore, the value of $r3$ is needed for re-executing the loop and hence should be written to NVRF. While *live* registers are necessary for re-executing the loop, some of them do not need to be backed up multiple times during loop execution. An illustrative example is register $r5$ in Fig. 2. As $r5$ does not serve as a destination register, its value will not change during loop execution. Similarly, although $r28$ is a live register because it is read right after exiting the loop, it is not accessed within the loop body. Therefore, both $r5$ and $r28$ need to be backed up only once.

Overall, the loop example illustrates the minimum register backup requirements, determined by the register access patterns within the loop body. Registers that are:

- *Not live* at the loop entry, such as $r2$ and $r6$, do not need to be backed up.
- *Live* at the loop entry but *not assigned* in loop body, such as $r0$, $r5$, and $r28$, need to be written to NVRF only once, at the loop entry.
- Both *live* at the loop entry and *assigned* in loop body, such as $r3$ and $r4$, need to be written to NVRF multiple times, at a frequency selected by the user.

Assume the loop is executed n iterations in total, there are i *live but not assigned* registers as well as j *live and assigned* registers that are written to NVRF every k iterations. During loop execution, the total number of write operations performed to NVRF is:

$$\text{Total NVRF writes} = i + j \times n/k \qquad (1)$$

In comparison, the traditional approach described in setting 2 of Fig. 1 needs to back up all registers every k iterations. For a system with N general purpose registers, the proposed backup strategy is able to reduce NVRF writes by a rate of $[1 - (i \times k/n + j)/N]$.

C. Backup Position Selection

The criteria outlined above can be used to determine the minimum register backup requirements for a given loop. However,

TABLE II.   VARIABLES USED IN EQUATIONS (2),(3),AND (4)

$Live\_in(B)$	– variables live at the entry of basic block B
$Live\_out(B)$	– variables live at the exit of basic block B
$Assigned(B)$	– variables written in basic block B
$Def(B)$	– variables written before read in basic block B
$Use(B)$	– variables read before written in basic block B

an application may contain more than one frequently executed loop, and these loops may be nested. As a result, there may exist multiple possible backup positions in the code. The best positions for writing register values to NVRF will be selected based on the following three considerations.

First, to maximize the benefit attainable dynamically while minimizing the overhead of static analysis, a loop is considered as one possible candidate for analysis only if it accounts for more than 10% of program execution time.

Second, a loop is considered as one possible candidate only if it allows a backup frequency that is smaller than the maximum backup frequency constrained by the expected lifetime of the NVRF. As a concrete example, assuming FeRAM write endurance of 10^{12}, system lifetime of 10 years, per-day usage of 10 hours, and clock frequency of $1GHz$, the maximum frequency for updating a non-volatile register is $10^{12}/10/365/10/3600 = 76104$ times per second, which is approximately once per 13,000 clock cycles. Since registers have to be backed up at least once at the loop entry, this implies that a possible loop candidate should include more than 30,000 dynamic instructions that will be executed before exiting the loop.

Finally, while loops with non-overlapping code body can be analyzed independently, nested loops have to be considered together for adjusting the backup frequency. The outer loop allows a lower backup frequency to reduce the backup cost, and hence is preferable when the expected probability of power failure is low. In contrast, inner loops allow a higher backup frequency that lets computation progress in finer steps, and hence is preferable when the expected failure probability is high. Another consideration is that since the inner loop is contained within the outer loop, it usually has fewer *assigned* registers than the outer loop. Therefore the cost per backup, in terms of the number of registers to write, may be smaller in the inner loop.

D. Software and Hardware Support

Equation (1) shows that the total number of NVRF writes is determined by the number of the *assigned* registers and the *live* registers within a given loop. These two sets of register can be identified by the compiler, through control and data flow analysis [23]. For a loop composed of multiple basic blocks[2], the following equations can be used, with the variable definitions outlined in Table II. More detailed explanation of these equations can be found in [23].

$$Assigned(loop) = \bigcup_{B \in loop} Assigned(B) \qquad (2)$$
$$Live\_in(B) = Use(B) \cup \{Live\_out(B) - Def(B)\} \qquad (3)$$
$$Live\_out(B) = \bigcup_{S \in succ(B)} Live\_in(S) \qquad (4)$$

The statically extracted information will be embedded into the application at the end of the static analysis phase. For each hotspot, two pieces of information will be delivered to the runtime system: The address of the selected backup position, and a "backup enable" bit vector with each bit indicating whether a register is selected for backup. At runtime, a switch between

---

[2]A basic block is a linear sequence of instructions with single entry and single exit points.

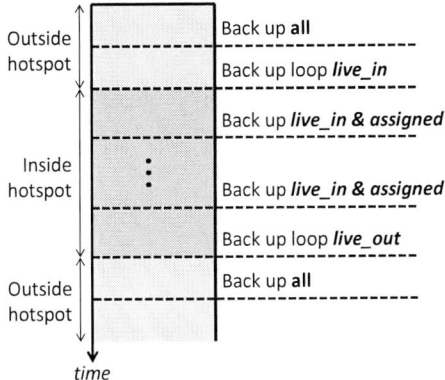

Fig. 3. Register backup strategies inside and outside a hotspot

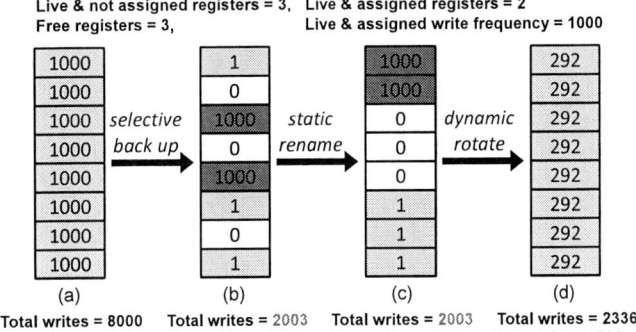

Fig. 4. Balancing register writes through rotations

TABLE III. REGISTER WRITE COUNTS AS A RESULT OF ROTATION.

	No rotation	Rotate live&assigned	Rotate all live
Not live [N-i-j]	0	$\frac{m*j}{N-i}$	$\frac{m*j+i}{N} \times \frac{N-i+2}{N-i+1}$
Live&assigned [j]	m	$\frac{m*j}{N-i}$	$\frac{m*j+i}{N} \times \frac{N-i+2}{N-i+1}$
Live&not-assigned [i]	1	1	$\frac{m*j+i}{N} \times \frac{N-i+2}{N-i+1}$

register backup strategies takes place when entering or exiting an execution hotspot. This process is shown in Fig. 3. When the execution is outside any hotspot, the traditional strategy that backs up all registers to NVRF is adopted. Within a hotspot, the proposed technique can be applied to minimize the writes to NVRF. Only those *live* and *assigned* registers need to be backed up. Additionally, right before entering a hotspot and right after exiting it, the *live_in* and *live_out* registers need to be backed up, respectively, so as to achieve a smooth transition between the two mechanisms.

To switch between different strategies, a store instruction is inserted prior to the entrance to the application hotspot, in order to set up the "backup enable" register. Similarly, another store instruction is executed when exiting the hotspot to set all the bits in the "backup enable" register to 1.

Finally, to support the arbitrary selection of backup frequency, a saturating hardware counter can be used to record the desired frequency, for example, once per M instructions. During application execution, the counter value is set to M in three conditions: (1) At the beginning of execution, (2) upon entering a hotspot, and (3) upon exiting a hotspot. The counter value decrements by 1 upon completing an instruction. Outside execution hotspots, registers are written to NVRF when the counter value reaches 0. Yet within a hotspot, the selected register set is written to NVRF when the counter is 0 AND the program counter reaches the pre-selected backup position. After performing a backup, the counter will be reset to M.

IV. REGISTER WRITE BALANCING

Through identifying the necessary conditions for backing up register values, the previous approach effectively minimizes the total number of writes to NVRF, thus minimizing the performance and energy overhead in backing up register values. However, the write distribution is still imbalanced. As shown in Fig. 4(b), selectively backing up a subset of registers does not reduce the maximum number of writes to a non-volatile register. As the lifetime of NVRF is usually determined by the maximum number of writes to an individual non-volatile register, we propose to balance writes across different non-volatile registers through a *deterministic register rotation* technique.

A. Functional Overview

While rotating the register mapping is desirable for balancing the writes to individual non-volatile registers, a critical consideration is the complexity of the rotation algorithm. Since the mapping between the SRAM registers and the FeRAM registers is not fixed, it is necessary to record this mapping, so that upon

a power failure, the mapping can be quickly identified and the necessary values can be correctly restored. However, this task should be accomplished without using a hardware mapping table, as it imposes notable amount of hardware complexity, energy consumption and performance overhead. Instead, it is preferable to establish a regular, pre-determined mapping between the two sets of registers, enabling register indices to be computed based on loop iterations.

To simplify the runtime control logic for rotating registers, the proposed technique places registers of the same category in contiguous positions. This can be achieved through inserting extra move instructions right before entering a hotspot. The regularity embedded in register names will enable a highly regular dynamic rotation process. An illustrative example is shown in Fig. 4. The example has 3 *live&not-assigned* registers, 2 *live&assigned* registers, and 3 *free* registers. As a result of static renaming, registers of the same type are placed in contiguous positions. Then, after dynamic rotation, writes to NVRF are balanced, and each register only needs to sustain 30% of the original writes.

B. Deterministic Register Rotation

As different backup strategies are used for different registers, they require different rotation strategies as well. A *live&assigned* register is written multiple times during loop execution. Accordingly, each time it is written, it can be rotated to an empty position. In contrast, a *live&not-assigned* register is written only once per loop. However, since it is live, prior to its overwritten, if any, its value should be migrated to another position.

Assume there are i *live&not-assigned* registers as well as j *live&assigned* registers that are written m times to NVRF. Therefore for a NVRF of N registers, N-i-j registers are free for rotation. If they are used for rotating those *live&assigned* registers, each of them will be written $m \times j/(N-i)$ times. Yet these N-i registers still dominate the write count of the entire NVRF since the other i registers are written only once. Overall, the write counts to the different types of registers before and after this rotation are summarized in Table III.

As the profiling results in the next section will show, a hotspot usually has more *live&not-assigned* registers than *live&assigned* registers. To fully exploit rotation capability, it is preferable to migrate the *live&not-assigned* registers as well, however, at a reduced frequency. Specifically, the proposed rotation algorithm rotates one *live&not-assigned* register when all the free registers

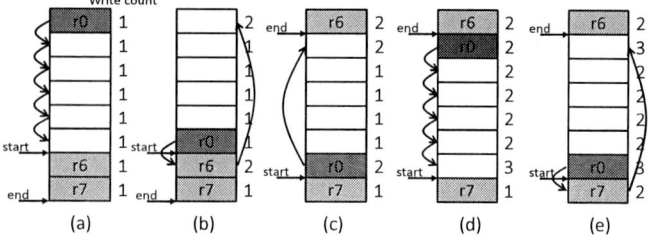

Fig. 5.   Rotation of $r0$ (live&assigned) and $r6$, $r7$ (live&not-assigned)

in NVRF have been consumed. To illustrate this process, a concrete example is shown in Fig. 5.

Fig. 5(a) shows that $r0$, a *live&assigned* register, is rotated 5 times. As a result, all the free registers are written once. At that point, one of the *live&not-assigned* register, namely, $r6$, is migrated to the position after $r7$, allowing $r0$ to be written to the original position of $r6$. This is shown in 5(b). As a result, both the *start* and the *end* of the *live&not-assigned* region are shifted by one position, shown in 5(c). The rotation of $r0$ continues for 6 times in 5(c) and 5(d). Then, the other *live&not-assigned* register, $r7$, is migrated, shown in 5(e). The rotation process repeats in this way. Clearly, the contiguity in the register positions is preserved throughout the rotation process. Such high regularity enables the register mapping to be easily computed during execution. Specifically, the position of each register can be computed according to the rotation count and the *start* and the *end* pointers that separate the *live&not-assigned* region and the rotation region, while the positions of the *start* and the *end* will be incremented by 1 upon every N-i+1 writes to NVRF, with N denoting the size of NVRF and i the number of *live&not-assigned* registers.

Overall, the proposed rotation algorithm migrates one *live&not-assigned* register after N-i+1 writes to NVRF. Therefore the overhead, in terms of the extra NVRF write count, for migrating *live&not-assigned* registers is $1/(N$-i+1$)$. As the rotation process iterates, all registers in the NVRF will be written at the same frequency, as shown in the last column of Table III.

V. EXPERIMENTAL EVALUATION

A. Methodology

To evaluate the effectiveness of the proposed technique in reducing and balancing the writes to non-volatile registers, we conducted a set of experimental studies on Mediabench [24] and Mibench [22] programs. We first extend the Simplescalar toolset [25] to extract useful program information, including instruction execution counts, branch targets, destination and source operands, and instruction execution sequence. Then hotspots for each application are selected based on the instruction execution counts. Subsequently, control and data flow analysis [23] is performed for each identified hotspot. The *assigned* set and the *live* set of registers are identified using Equations (2)-(4). Based on these values, the best backup positions throughout program execution are manually selected using the approach outlined in Section III-C. Finally, the statically extracted register backup information is delivered to the runtime simulator, which performs both selective backup and register rotation to reduce and balance the writes to NVRF.

B. Results

1) Static Analysis Results: Table IV reports the collected profiling results. For each benchmark, five values are reported: hotspot count, the amount of execution time covered by hotspots,

TABLE IV.   STATIC ANALYSIS RESULTS

benchmarks	hotspots	time%	BBs	live∪assigned	live∪(assigned)'
adpcm_dec	1	99.8%	18	7	8
crc	1	99.8%	4	1	6
epic	1	77.3%	9	2	8
gsm_enc	2	66.1%	5	2	6
			3	2	3
jpeg_dec	2	41.8%	4	3	10
			6	2	8
string search	1	38.4%	1	2	7
susan	1	97.9%	11	2	6

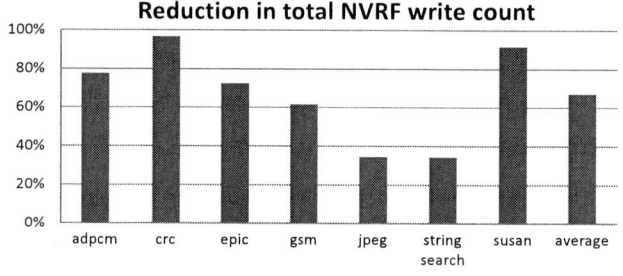

Fig. 6.   Reduction in the total number of NVRF writes

the number of basic blocks in each hotspot, the number of *live* and *assigned* registers, as well as the number of live but not assigned registers. These benchmarks constitute our study set as their hotspots occupy the largest fraction of the total execution time among all the programs we have tested.

As can be seen, all benchmarks display high potential for reducing the total number of writes to NVRF. Seven out of 8 benchmarks require only three or fewer registers (out of 32) to be written to NVRF during loop execution. The only exception is *adpcm*, which contains seven *live* and *assigned* registers in the loop body. The number of *live* but not *assigned* registers to be written to NVRF at loop entry and exit ranges from 3 to 10.

2) Write Reduction Results: Reducing the total number of NVRF writes is critical to the performance and energy consumption of NVRF. In our experiments, the frequency for backing up registers is set to once per 100,000 instructions. Outside any hotspot, the entire set of registers are written to NVRF every time after executing 100,000 instructions. Within a hotspot, only the selected set of registers are written to NVRF, every time after finishing 100,000 instructions AND when the PC reaches the pre-selected backup position. Meanwhile, the live registers are written to NVRF at both loop entry and exit.

Fig. 6 shows the reduction in the total number of NVRF writes, normalized to a baseline scheme that backs up all 32 registers every 100,000 instructions. On average, the proposed scheme can reduce 67% of writes to the NVRF. Over 90% reduction can be observed for *crc* and *susan*, whose hotspots cover 99% of execution time, and only a very small number of registers need to be backed up in hotspots. In comparison, *jpeg* and *string search* only achieve 35% reductions, primarily because the execution time spent in hotspots is only around 40%. Overall, these results confirm that the proposed selective backup strategy significantly reduces the total number of writes to NVRF and hence improves its performance and energy efficiency.

3) Write Balancing Results: The lifetime of a NVRF is constrained by the maximum number of writes to a non-volatile register. To evaluate the effectiveness of the proposed register rotation algorithm, we implement it in the simulator and report the maximum write count. The reduction in the maximum number of NVRF writes achieved by the rotation algorithm is

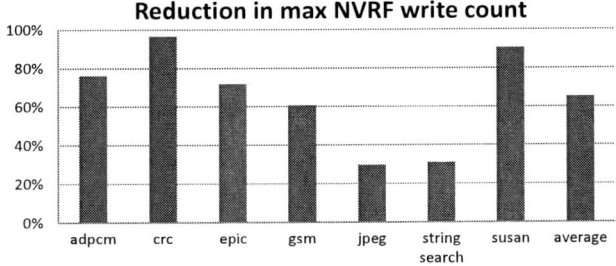

Fig. 7. Reduction in the maximum number of writes

shown in Fig. 7. As expected, Fig. 7 is highly similar to Fig. 6. For each individual benchmark, the reduction in maximum writes is slightly smaller than the reduction in total writes. This is because extra NVRF write operations are needed for migrating those *live&not-assigned* registers. On average, a 65% reduction is achieved, confirming the effectiveness of the proposed rotation algorithm in reducing the maximum number of writes to the NVRF and hence improving NVRF lifetime.

VI. CONCLUSIONS

In this paper, we have presented two compiler-directed optimizations aimed at improving the write efficiency and extending the lifetime of non-volatile register files (NVRF). The first, a write reduction technique, selects a minimum set of registers for backup based on register access patterns of frequently executed loops. It effectively minimizes the total number of writes to NVRF. The second, a write balancing technique, implements a deterministic rotation algorithm that effectively reduces the maximum number of writes to individual non-volatile registers. The experiments show that the selective backup technique reduces 67% NVRF writes on average, with some benchmarks achieving over 90% reduction. The register rotation technique achieves an average of 65% reduction in the maximum write count, confirming its effectiveness in extending NVRF lifetime. Such drastic reductions increase the performance, energy efficiency, and lifetime of FeRAM, thus qualifying NVRFs for adoption in various embedded systems.

REFERENCES

[1] L. P. Chang and T. W. Kuo, "Efficient management for large-scale flash-memory storage systems with resource conservation," *ACM Transactions on Storage*, vol. 1, no. 4, July 2005.

[2] B. Lee, P. Zhou, J. Yang, Y. Zhang, B. Zhao, E. Ipek, O. Mutlu, and D. Burger, "Phase-change technology and the future of main memory," *IEEE Micro*, vol. 30, no. 1, January 2010.

[3] P. Zhou, B. Zhao, J. Yang, and Y. Zhang, "Energy reduction for STT-RAM using early write termination," in *International Conference on Computer-Aided Design (ICCAD)*, 2009, pp. 264–268.

[4] G. Sun, X. Dong, Y. Chen, and Y. Xie, "An energy-efficient 3D stacked STT-RAM cache architecture for CMPs," in *Emerging Memory Technologies*. Springer, 2014.

[5] X. Dong, X. Wu, G. Sun, Y. Xie, H. Li, and Y. Chen, "Circuit and microarchitecture evaluation of 3D stacking magnetic RAM (MRAM) as a universal memory replacement," in *Design Automation Conference (DAC)*, 2008, pp. 554–559.

[6] J. Wang, Y. Liu, H. Yang, and H. Wang, "A compare-and-write ferroelectric nonvolatile flip-flop for energy-harvesting applications," in *International Conference on Green Circuits and Systems (ICGCS)*, 2010, pp. 646–650.

[7] H. Nakamoto, D. Yamazaki, T. Yamamoto, H. Kurata, S. Yamada, K. Mukaida, T. Ninomiya, T. Ohkawa, S. Masui, and K. Gotoh, "A passive UHF RF identification CMOS tag IC using ferroelectric RAM in 0.35um technology," *IEEE Journal of Solid-State Circuits*, vol. 42, no. 1, July 2007.

[8] M. Zwerg, A. Baumann, R. Kuhn, M. Arnold, R. Nerlich, M. Herzog, R. Ledwa, C. Sichert, V. Rzehak, P. Thanigai, and B. Eversmann, "An 82ua/mhz microcontroller with embedded FeRAM for energy-harvesting applications," in *International Solid-State Circuits Conference (ISSCC)*, 2011, pp. 334–336.

[9] S. Masui, W. Yokozeki, M. Oura, T. Ninomiya, K. Mukaida, Y. Takayama, and T. Teramoto, "Design and applications of ferroelectric nonvolatile SRAM and flip-flop with unlimited read/program cycles and stable recall," in *Custom Integrated Circuits Conference (CICC)*, 2003, pp. 403–406.

[10] Q. Li, L. Jiang, Y. Zhang, Y. He, and C. J. Xue, "Compiler directed write-mode selection for high performance low power volatile PCM," in *International Conference on Languages, Compilers and Tools for Embedded Systems (LCTES)*, 2013, pp. 101–110.

[11] H. A. Khouzani, Y. Xue, and C. Yang, "Prolonging PCM lifetime through energy-efficient, segment-aware, and wear-resistant page allocation," in *International Symposium on Low Power Electronics and Design (ISLPED)*, 2014, pp. 327–330.

[12] Y. Liu, H. Yang, Y. Wang, C. Wang, X. Sheng, S. Li, D. Zhang, and Y. Sun, "Ferroelectric nonvolatile processor design, optimization, and application," in *Emerging Memory Technologies*. Springer, 2014.

[13] Y. Wang, Y. Liu, S. Li, D. Zhang, B. Zhao, M.-F. Chiang, Y. Yan, B. Sai, and H. Yang, "A 3us wake-up time nonvolatile processor based on ferroelectric flip-flops," in *European Conference on Solid-State Circuits (ESSCIRC)*, 2012, pp. 149–152.

[14] H. A. Khouzani, C. Yang, and J. Hu, "Improving performance and lifetime of DRAM-PCM hybrid main memory through a proactive page allocation strategy," in *Asia and South Pacific Design Automation Conference (ASP-DAC)*, 2015, pp. 508–513.

[15] M. Zhao, Y. Xue, C. Yang, and C. Xue, "Minimizing MLC PCM write energy for free through profiling-based state remapping," in *Asia and South Pacific Design Automation Conference (ASP-DAC)*, 2015, pp. 502–507.

[16] Y. Wang, Y. Liu, Y. Liu, D. Zhang, S. Li, B. Sai, M.-F. Chiang, and H. Yang, "A compression-based area-efficient recovery architecture for nonvolatile processors," in *Design, Automation, and Test in Europe Conference Exhibition (DATE)*, 2012, pp. 1519–1524.

[17] Y. Wang, Y. Liu, S. Li, X. Sheng, D. Zhang, M.-F. Chiang, B. Sai, X. Hu, and H. Yang, "PaCC: A parallel compare and compress codec for area reduction in nonvolatile processors," *IEEE Transactions on Very Large Scale Integration (VLSI) Systems*, vol. 22, no. 7, July 2013.

[18] Rohm, "Nikkei electronics asia: Rohm develops non-volatile register slashes dissipation," http://techon.nikkeibp.co.jp/article/HONSHI/20080729/155646/, 2008.

[19] E. Lundgren, N. Navale, and N. Carter, "A magnetoelectronic self-checkpointing register file," in *International Conference on Electromagnetics in Advanced Applications (ICEAA)*, 2007, pp. 609–612.

[20] H. Chen and C. Yang, "Fault detection and recovery efficiency co-optimization through compile-time analysis and runtime adaptation," in *International Conference on Compilers, Architecture and Synthesis for Embedded Systems (CASES)*, 2013, pp. 1–10.

[21] C. Yang and A. Orailoglu, "Processor reliability enhancement through compiler-directed register file peak temperature reduction," in *International Conference on Dependable Systems Networks (DSN)*, 2009, pp. 468–477.

[22] M. Guthaus, J. Ringenberg, D. Ernst, T. Austin, T. Mudge, and R. Brown, "Mibench: A free, commercially representative embedded benchmark suite," in *International Workshop on Workload Characterization (WWC)*, 2001, pp. 3–14.

[23] A. V. Aho, M. S. Lam, R. Sethi, and J. D. Ullman, *Compilers: Principles, Techniques, and Tools*. Pearson/Addison Wesley, 2007.

[24] C. Lee, M. Potkonjak, and W. H. Mangione-Smith, "Mediabench: A tool for evaluating and synthesizing multimedia and communicatons systems," in *International Symposium on Microarchitecture (MICRO)*, 1997, pp. 330–335.

[25] T. Austin, E. Larson, and D. Ernst, "Simplescalar: An infrastructure for computer system modeling," *IEEE Computer*, vol. 35, no. 2, Feb. 2002.

Non-Volatile Memories in FPGAs: Exploiting Logic Similarity to Accelerate Reconfiguration and Increase Programming Cycles

Yuan Xue[1]   Patrick Cronin[1]   Chengmo Yang[1]   Jingtong Hu[2]

[1]Department of Electrical and Computer Eng.
University of Delaware
Newark, DE 19716 USA
{xueyuan, ptrick, chengmo}@udel.edu

[2]School of Electrical and Computer Eng.
Oklahoma State University
Stillwater, OK 74078 USA
jthu@okstate.edu

Abstract—**Non-volatile memory (NVM) technologies have been known for their advantages of large capacity, low energy consumption, high error-resistance, and near-zero power-on delay. It is expected that they will replace traditional SRAM as FPGA reconfigurable blocks. While NVMs promise FPGAs with more reconfigurable resources, lower power consumption, and higher resilience to power interruptions, they also impose two new design challenges: the slow write performance of NVMs may degrade FPGA reconfiguration speed, while their limited write endurance constrains FPGA programming cycles. To overcome these challenges, we propose a similarity driven approach to reduce reconfiguration cost in NVM-based FPGAs. When synthesizing a new design, its similarity to the design currently on the FPGA is characterized by taking both LUT contents and CLB-level topology into consideration. The reconfiguration cost minimization problem is formulated as a bipartite graph matching problem and solved optimally. Experiments on standard circuit benchmarks show that the proposed algorithms eliminate more than 57.4% of NVM writes during the reconfiguration process, thus effectively improving performance and endurance of NVM-based FPGAs.**

Keywords—*non-volatile memory, FPGA, reconfiguration, logic similarity*

I. Introduction

Non-volatile Memory (NVM) devices such as Phase Change Memory (PCM) and Spin Transfer Torque Magnetic RAM (STT-MRAM) have been known for their advantages of large capacity, low energy consumption, high error-resistance, and near-zero power-on delay. They are promising candidates for replacing traditional dynamic and static RAMs as off-chip and on-chip memories, and even as reconfigurable blocks in Field Programmable Gate Arrays (FPGAs).

In current FPGA design, all the reconfigurable components are implemented in SRAM. Configuration information is stored in off-chip Flash memory, and loaded to on-chip SRAM when the FPGA is powered on. While the fast write/read speed of SRAM accelerates FPGA reconfiguration and performance, it unfortunately also leads to four major drawbacks. First, the large standby current of SRAM makes the FPGA suffer from high leakage power consumption which dominates total power dissipation [1]. Second, the low scalability of SRAM limits the amount of on-chip reconfigurable resources in the FPGA, which in turn constrains the size and complexity of FPGA-implementable designs. Third, SRAM is vulnerable to radiation and particles, leading to not only transient data errors but also permanent logic malfunctions when errors occur in configuration data [2]. Finally, since SRAM is volatile, upon a power outage, all the on-chip configuration information gets lost and needs to

be reloaded from off-chip storage when the power is back on. This long recovery time is unacceptable in critical application domains such as military.

A promising approach to overcome the limitations of SRAM is to use NVMs as FPGA on-chip storage, such as Flash-based [3], PCM-based [1], [4], and MRAM-based FPGAs [5]. These NVM-based FPGAs are expected to contain more logic and storage cells on a single FPGA chip, which consumes much less power and is more resilient to power interruptions than SRAM-based FPGAs. However, the characteristics of NVMs pose two new challenges to FPGA design. First, the slow write performance of NVMs makes reconfiguration time non-trivial. Second, the limited write endurance of NVMs constrains FPGA programming cycles. Frequently reconfiguring the FPGA may wear some building blocks heavily and cause stuck-at faults [6]. These obstacles must be overcome in order to qualify NVM-based FPGAs for industrial needs.

One effective technique to mitigate the adverse impact of NVM writes is *redundant write elimination* [7]. Before programming a block, the original content is read out and compared with the new content, and the identical part is excluded from programming. This technique was not adopted in SRAM-based FPGAs since SRAM read and write operations have similar speed and overhead. However, in NVMs, write operations are usually much slower and more costly than read operations [8]. Therefore, adopting a *read-before-write* strategy is effective to reduce reconfiguration cost, with the amount of reduction proportional to the number of NVM writes eliminated. In light of this observation, we propose to exploit the flexibilities inherent in placing lookup tables (LUTs), so as to maximize the logic similarity between two designs and minimize the cost for reconfiguring one to the other. The main contributions of this paper include:

- Two types of similarity based metrics to evaluate LUT content similarity and CLB-level topology similarity.
- Formulation of NVM write minimization as a *weighted bipartite graph matching* problem, and generation of a *min-reconfig-cost* (MRC) placement with a polynomial-time optimal algorithm.
- Incorporation of the obtained MRC placement into the Verilog-to-Routing (VTR) CAD tool [9] to enable the co-optimization of reconfiguration cost and traditional area, routing, and delay goals.

The rest of this paper is organized as follows. Section II reviews background knowledge of NVM-based FPGAs and related works. Section III outlines the technical motivation and challenges. Section IV describes the proposed similarity model and the reconfiguration optimization algorithm. Section V presents the experimental results while Section VI concludes this work.

978-1-4673-9141-2/15 $31.00 © 2015 IEEE

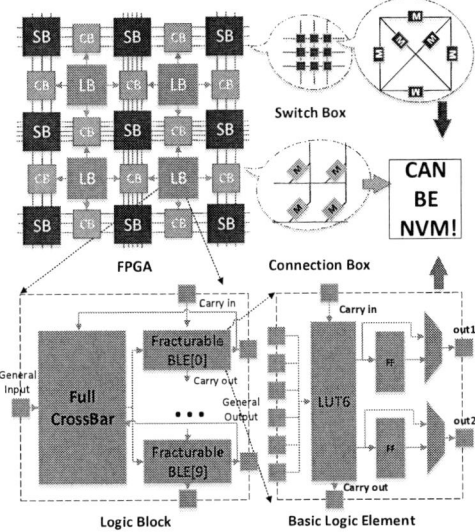

Fig. 1. Representative NVM-based FPGA Architecture

II. TECHNICAL BACKGROUND

A. NVM-based FPGAs

One representative FPGA architecture is presented in Fig. 1, which consists of three types of configurable blocks: configurable logic blocks (CLB/LB), switch boxes (SB), and connection blocks (CB). Each CLB includes 10 Basic Logic Elements (BLEs), connected with a full crossbar that implements arbitrary connections across BLEs. In this architecture, the SBs, CBs, BLEs, crossbars, flip-flops and multiplexors can be implemented with NVM cells.

Various types of NVM-based FPGAs architectures have been developed or proposed recently. Flash memory is one of the most mature NVM technologies. Commercial Flash-based FPGAs have been developed by Actel [10] and Lattice [11]. However, as Flash suffers from a quite short endurance (10^5 cycles), the programming cycles of Flash-based FPGAs is quite limited. Another shortcoming is its requirement for a time-consuming erase operation before programming a block, thus slowing down the reconfiguration process in Flash-based FPGAs. In comparison, STT-MRAM uses magnetic storage elements. An FPGA using STT-MRAM as basic memory cells was proposed in [12]. In [13], STT-MRAM was used to implement LUTs. STT-MRAM delivers better scalability than SRAM, but suffers from slower performance of read and write operations, which respectively constrain the shortest clock period and reconfiguration speed in STT-MRAM based FPGAs. The third type of NVM-based FPGA uses PCM, a resistive memory regarded as one of the most promising NVMs. Researchers have proposed several PCM-based FPGA architectures: the work in [1] proposed several storage organization and management schemes for replacing SRAM with PCM, the work in [4] proposed a 3D architecture for PCM-based FPGA, while the work in [14] demonstrated the feasibility of using PCM in SBs and LUTs. Other types of NVM-based FPGAs include MRAM-based [5] and Resistance-change memories (RRAM) based [15].

B. Reconfiguration Optimizations

The conventional goal of reconfiguration optimization is to reuse parts of existing blocks on the FPGA to reduce the configuration size of a new design. Existing techniques can be grouped in two categories. One type of techniques exploit *partial reconfiguration* (PR) offered in many commercial FPGA products

to support hardware reuse across different tasks [16]. Several techniques [17], [18], [19] have been proposed to perform design partitioning and/or scheduling based on coarse-grained functional information.

Another type of technique aims at maximizing similarity between two designs to reduce the reconfiguration cost. Design similarity can be checked either at the register-transfer level [20] or the gate level [21], [22]. At the higher level, the goal is to reuse entire logic components such as ALUs [20]. In comparison, gate-level approaches try to find identical LUTs for reuse. In [22], a logic remapper is proposed to detect topological similarity between two designs. In [21], similarity between two nodes is defined as the weighted sum of the position similarities between them and their adjacent nodes. Since the target platform is an SRAM-based FPGA, these works abstract LUTs as nodes without taking the similarity between LUT contents into consideration. In other words, two LUTs are considered "similar" only if they are 100% identical. Another difference is that [21] extracts design topology as an undirected graph, while the proposed work extracts topology as a directed graph to differentiate incoming and outgoing interconnects.

To summarize, most of of previous reconfiguration optimizations target SRAM-based FPGAs. They model reconfiguration overhead as the number of LUTs and/or SBs to be programmed when implementing a new design. In contrast, the proposed technique targets NVM-based FPGAs, and models reconfiguration overhead as the number of NVM cells to be programmed, which is at a much finer granularity than existing approaches.

III. TECHNICAL MOTIVATION

While NVMs outperform SRAM in many aspects such as high density, near-zero power-on delay, and superior energy efficiency [1], their long write latency makes the reconfiguration time in NVM-based FPGA non-trivial. Taking Phase Change Memory (PCM) as an example, ITRS reports [6] indicate that its density is 35 times larger than SRAM, and its programming cycles is 10^4 times larger than Flash. However, the write speed of PCM is 500 times slower than SRAM. This will become the bottleneck in reconfiguration when PCM replaces SRAM as FPGA on-chip storage. What is more, in PCM, write is more than 8 times slower than read. This feature motivates the adoption of a *read-before-write* strategy: by reading out the original content of a reconfigurable component before programming it, the identical part in the new content can be eliminated from programming. Since read is much cheaper and faster than write, this strategy is effective in reducing reconfiguration overhead, eliminating unnecessary NVM writes, and increasing FPGA programming cycles.

It is not only the technology characteristics that dictate the need for reducing reconfiguration overhead. From the application's perspective, frequent reconfiguration is commonly required in a number of scenarios. First, limited by the size of on-chip storage, sometimes the FPGA is not large enough to hold multiple hardware accelerators simultaneously. It is therefore necessary to quickly switch between different hardware accelerators so that the application can quickly switch from one task to another. Second, many safety-critical applications use FPGAs as backup hardware resources. If an active unit fails, it is necessary to reconfigure the FPGA to conduct the functionality of the failed unit. Reconfiguration latency therefore directly determines how quick the recovery process can be. Another possible scenario is incremental design and implementation. Designers usually rely on FPGA implementations to test and refine their product. By

978-1-4673-9141-2/15 $31.00 © 2015 IEEE

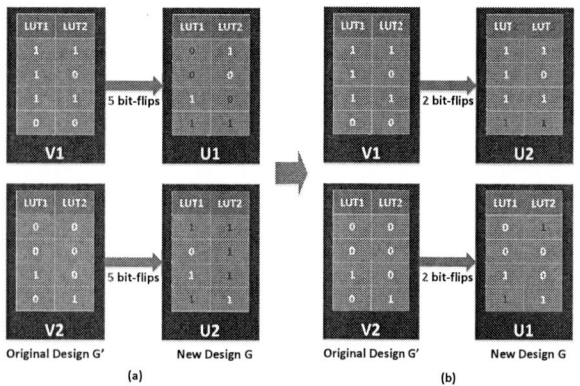

Fig. 2. Content similarity. (a) shows the original cost (10 bit-flips) for configuring $U1$ and $U2$. (b) shows by switching $U1$ and $U2$ and $LUT1$ and $LUT2$ in $U2$, the cost is reduced to 4 bit-flips.

Algorithm 1 Content Similarity Computation

Input: All LUTs of each CLB in G and each BLK in G'
Output: Matrix C
1: **for all** $CLB(i)$ in G **do**
2:     **for all** $BLK(j)$ in G' **do**
3:         **for** LUT $x = 1$ to l in $CLB(i)$ **do**
4:             **for** LUT $y = 1$ to l in $BLK(j)$ **do**
5:                 LUT-level similarity $W[x][y] \Leftarrow p/2^k$;
6:             **end for**
7:         **end for**
8:         Best matching score $Sum_{max} \Leftarrow$ **KM** $(W[l][l])$;
9:         CLB-level similarity $C_{i,j} \Leftarrow Sum_{max}/l$;
10:     **end for**
11: **end for**

---

exploiting the high similarity between two consecutive implementations, the reconfiguration overhead could be largely saved.

To summarize, both the technology characteristics and the application needs motivate the proposal of a framework to speed up reconfiguration and reduce the number of writes in NVM-based FPGAs. We achieve this goal by characterizing the logic similarity between two designs at the bit level, and minimizing the bit-flips when reconfiguring one to implement the other. More specifically, our previous work [23] considered the reconfiguration cost in NVM-based FPGAs and exploited different placement approaches to reduce such cost. However, that work only considered LUT contents and could only reduce CLB reconfiguration cost. On the other hand, as Fig. 1 shows, the majority portion of reconfigurable resources in the FPGA are SBs. Therefore, in this work we propose several similarity metrics that take not only LUT contents but also CLB-level topology into consideration, with the goal of reducing the overall reconfiguration cost of both CLBs and SBs.

IV. PROPOSED RECONFIGURATION OPTIMIZATIONS

To reduce the total number of NVM cells to be programmed during FPGA (re)configuration, the proposed framework identifies a *min-reconfig-cost* (MRC) mapping between two designs (or two tasks in one design), one to be configured and one currently on the FPGA. This section describes the proposed approaches for modeling similarities between two designs, identifying the MRC mapping, and integrating the MRC mapping into the VTR synthesis flow.

A. Design Similarity Characterization

Assuming that G is the design to be configured and G' is the design already on the FPGA, the goal of the proposed work is to minimize the cost for configuring G by increasing its similarity with G'. This goal will be achieved by exploiting both the flexibility in determining CLB positions and the flexibility in changing LUT positions within a CLB. Fig. 2 shows an example illustrating these two types of flexibilities. There are four CLBs, with $V1$ and $V2$ belonging to the old design G' and $U1$ and $U2$ to the new design G. Each CLB has two LUTs. Fig. 2(a) shows that by directly mapping $U1$ to $V1$ and $U2$ to $V2$, the reconfiguration cost is 10 bit-flips. In comparison, the two types of flexibilities imply that both $V1$ and $V2$ can be reconfigured to either $U1$ or $U2$, and the LUT positions in $U1$ and $U2$ can be permuted as well. Fig. 2(b) shows that by mapping $U1$ to $V2$ and $U2$ to $V1$ and switching LUT1 and LUT2 in $U2$, the reconfiguration cost is reduced to 4 bit-flips.

Exploiting the flexibilities outlined above, the proposed approach characterizes similarity between two designs combining two factors: *CLB content similarity* and *CLB-level topology similarity*. More specifically, the approach takes two sets of inputs: *post-placement* netlist of design G', as well as *pre-placement* netlist of design G. Using $CLB(i)$ to denote a CLB in G and $BLK(j)$ a CLB in G', three matrices will be computed:

- Matrix C: $C_{i,j}$ measuring *content similarity* between $CLB(i)$ and $BLK(j)$;
- Matrix T: $T_{i,j}$ measuring *topological similarity* between $CLB(i)$ and $BLK(j)$.
- Matrix S: $S_{i,j}$ measuring *similarity index*, a weighted sum of $C_{i,j}$ and $T_{i,j}$.

Each matrix has m rows and n columns, with m and n respectively denoting the number of CLBs in G and G'. Approaches for computing these three matrices are presented below.

1) CLB Content Similarity: Content similarity represents the cost for reconfiguring $CLB(i)$ to $BLK(j)$. The higher the similarity, the less the reconfiguration cost.

Given two LUTs, computing their content similarity is quite straightforward. For two k-input LUTs, if p bits are identical[1], their similarity can be measured by $p/2^k$. As a concrete example, in Fig. 2(a), the content of $LUT1$ in $V1$ is 1110[2] and the content of $LUT1$ in $U1$ is 0011. Since these two LUTs only have one identical bit, their similarity is $1/2^2 = 0.25$.

Computation of CLB-level content similarity depends not only on LUT contents but also on the positions of LUTs within a CLB. For a CLB with l LUTs, there are a total number of $l!$ permutations of LUT positions, and $C_{i,j}$ should represent the highest similarity score among these cases when mapping $CLB(i)$ to $BLK(j)$. Instead of enumerating all these possibilities, the proposed approach first computes the similarity between each LUT in $CLB(i)$ and each LUT in $BLK(j)$. Algorithm 1 shows the details. An $l \times l$ array W is used to record similarity scores between all the LUTs in $CLB(i)$ and $BLK(j)$. The similarity score between LUT x in $CLB(i)$ and LUT y in $BLK(j)$ is computed at line 5. Subsequently, a bipartite graph is constructed between the two sets of LUTs, with the LUT similarity scores represented as weights of edges. The best matching between the LUTs can be obtained with the classical *Kuhn-Munkres* (KM) algorithm [24], which computes the highest matching score Sum_{max} based on Matrix W at line 8. Then this score is normalized (i.e., divided by l) to obtain the CLB-level similarity score between $CLB(i)$ and $BLK(j)$ at line 9.

---

[1]Sometimes the inputs of a LUT are not fully used. Therefore certain entries in a LUT are *don't cares*, whose configuration cost will be counted as 0.

[2]1110 represents a 2-input LUT of outputs 1, 1, 1, and 0 (a NAND gate).

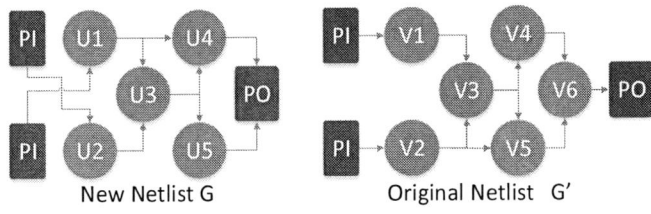

New Netlist G   Original Netlist G'

Fig. 3.   CLB-level topologies.

Since the complexity of the KM algorithm is $O(l^3)$, the overall complexity of Algorithm 1 is $O(l^3mn)$.

As a concrete example, consider the computation of CLB content similarity between $U1$ and $V2$ in Fig. 2. Algorithm 1 first computes LUT content similarity $W[2][2] = \{0.75, 0.75, 0.25, 0.75\}$. Then it calls the KM algorithm, which identifies the best matching {LUT1→LUT1, LUT2→LUT2} and the corresponding score Sum_{max}=1.5. Finally, the CLB content similarity score $C_{1,2}$=1.5/2=0.75.

2) Topological Similarity: Topology similarity represents the position relation between two CLBs. As CLBs are connected, such similarity depends also on how similar their neighbors are. In light of this observation, we develop an iterative algorithm to compute topological similarity, with the kth iteration depending on the result obtained in the $(k-1)^{th}$ iteration. Specifically, at each iteration k, the algorithm uses Equation (1) to compute the similarity score $T_{i,j}$. The first part of the equation accounts for self-influence, computed during the previous iteration $k-1$. The second part accounts for neighbor-influence, including both the *in* sets (nodes coming to $CLB(i)$ and $BLK(j)$) and *out* sets (output nodes of $CLB(i)$ and $BLK(j)$). The two parts are summed up with a weight α that determines their relative importance. A higher value of α implies self-influence is more important than neighbor-influence, and vice versa. In our computation, α is set to 0.75.

$$T_{i,j}^k = \alpha T_{i,j}^{k-1} + (1-\alpha) \frac{1}{|in(i)| + |out(i)|}$$
$$(max(\sum_{u \in in(i), v \in in(j)} T_{u,v}^{k-1}) + max(\sum_{u \in out(i), v \in out(j)} T_{u,v}^{k-1})) \quad (1)$$

In Equation (1), the *in* and *out* sets characterize the local topology of $CLB(i)$ and $BLK(j)$. To compute the weight of the *in* set, the best mapping from $in(i)$ to $in(j)$ is determined using the KM algorithm [24]. If $in(i)$ is larger than $in(j)$, the best mapping from $in(j)$ to $in(i)$ is determined instead. Weight of the *out* set can be computed in the same way. Then these two weights are added and then divided by the total number of neighbors of $CLB(i)$, represented as $|in(i)| + |out(i)|$. As a concrete example, Fig. 3 shows two designs with their CLBs, primary I/Os, and CLB-level topologies. The new design G contains five CLBs, $U1$ through $U5$, to be mapped the CLBs in the old design G', $V1$ through $V6$. Consider the computation of $T_{3,4}$, the topological similarity between $U3$ and $V4$. Fig. 3 shows that $in(U3) = \{U1, U2\}$, $out(U3) = \{U4, U5\}$, $in(V4) = \{V3\}$, and $out(V4) = \{V6\}$. The neighboring-influence score is the sum of the highest mapping scores from $in(V4)$ to $in(U3)$ and from $in(V4)$ to $out(U3)$.

To initiate the computation of the T matrix, we compute $T_{i,j}^0$ considering the difference between $CLB(i)$ and $BLK(j)$ in their numbers of incoming and outgoing neighbors. This is shown in Equation (2). Clearly, if $CLB(i)$ and $BLK(j)$ has the same number of incoming neighbors, the first fraction in Equation (2) is 1. When both $in(i)$ and $in(j)$ are empty (no neighbor), the

---

Algorithm 2 Topological Similarity Computation

Inputs: Netlists of G and G', Termination threshold V_{end}, Max loop iterations N_{max}
Output: Matrix T
1: Initialize $T_{i,j}^0$ with Equation (2)
2: **for** $k = 1$ to N_{max} **do**
3:   $\delta \Leftarrow 0$
4:   **for all** $CLB(i)$ in G **do**
5:     **for all** $BLK(j)$ in G' **do**
6:       **if** $|out(i)| + |in(i)| > 0$ **then**
7:         Compute $T_{i,j}^k$ with Equation (1)
8:         $\delta \Leftarrow \delta + |T_{i,j}^k - T_{i,j}^{k-1}|$
9:       **else**
10:        $T_{i,j}^k \Leftarrow T_{i,j}^{k-1}$
11:      **end if**
12:    **end for**
13: **end for**
14: **if** $\delta < V_{end}$ **then**
15:    break
16: **end if**
17: **end for**

---

fraction is also set to 1. On the other hand, if only one of $in(i)$ and $in(j)$ is empty, the first fraction is set to 0. For $U3$ and $V4$ in Fig. 3, we have $|in(U3)|$=2, $|in(V4)|$=1, $|out(U3)|$=2, and $|out(V4)|$=1. Therefore the initial topological similarity score between $U3$ and $V4$ is $T_{3,4}^0 = (\frac{1}{2} + \frac{1}{2})/2$=0.5.

$$T_{i,j}^0 = \left[\frac{min(|in(i)|, |in(j)|)}{max(|in(i)|, |in(j)|)} + \frac{min(|out(i)|, |out(j)|)}{max(|out(i)|, |out(j)|)} \right] / 2 \quad (2)$$

Algorithm 2 shows the proposed topological similarity computing algorithm. Equations (2) and (1) are used to initialize $T_{i,j}^0$ at line 1 and compute $T_{i,j}^k$ at line 7, respectively. A variable δ is used to keep track of the sum of the differences between two consecutive iterations. The algorithm terminates when reaching an upper-bound of iterations N_{max}, or when δ is less than a given threshold V_{end}.

3) Similarity Index: Once the two matrices C and T are computed, they can be combined to form the similarity index matrix S with a weight β:

$$S = \beta C + (1-\beta)T \quad (3)$$

The value of β should be determined based on the actual reconfiguration overhead of CLBs and switch boxes. For the FPGA platform used in our experiments, the ratio of actual configuration bits needed for CLBs and SBs is around 4:1. Therefore a higher weight is given to content similarity and β is set to 0.8.

B. Optimal Placement Identification

Once the similarity index matrix S is obtained, the algorithm starts to search for the *min-reconfig-cost* (MRC) placement of the new design. One key observation is that this problem can be converted to a *weighted bipartite matching* problem. CLBs of the two designs can be represented as the two sets of nodes in the bipartite graph, while $S_{i,j}$ is the weight of the edge from node i to node j. The graph includes $m + n$ nodes and $m \times n$ edges, with m and n respectively representing the number of CLBs in the new design G and the number of BLKs in old design.

Given the bipartite graph model, the problem of maximizing similarity between the two designs is equivalent to the problem of finding a one-to-one mapping that maximizes the sum of the weights of the matched edges. This problem can be solved

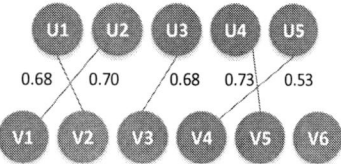

Fig. 4. Similarity based optimal bipartite graph matching

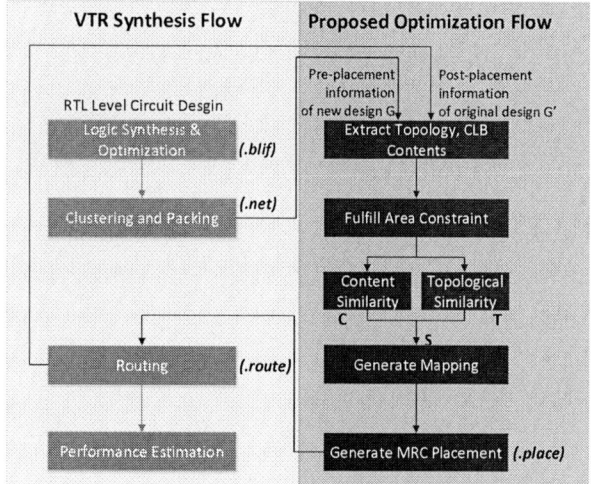

Fig. 5. Proposed reconfiguration optimization design flow

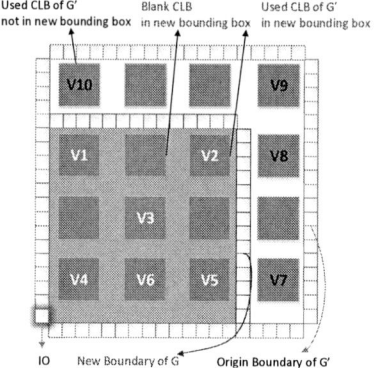

Fig. 6. Areas of new design G and old design G' (larger than G). The proposed algorithm maps G to its original area (shown in gray) including 6 CLBs of G' (V1 through V6) and 3 blank blocks.

TABLE I. BENCHMARK CIRCUITS

No	Benchmark	CLB#	LUT#
1	tseng	105	1046
2	ex5p	107	1064
3	diffeq	150	1494
4	alu4	153	1522
5	seq	175	1750
6	s298	194	1930
7	frisc	356	3539
8	spla	369	3690
9	ex1010	460	4598

optimally in $O(m^2 n)$ time with the classical *Kuhn-Munkres* (KM) algorithm [24]. As a concrete example, the bipartite graph model and the best mapping between the two designs in Fig. 3 is shown in Fig. 4. Given this mapping and the coordinates of the CLBs in the old design G', positions of each CLB in the new design G can be determined.

C. Implementation in VTR

The MRC placement identified in the last subsection should be integrated into the VTR CAD tool [9], so that the NVM reconfiguration overhead can be considered together with traditional design constraints such as area, timing, and routability. The augmented VTR synthesis flow is shown in Fig. 5.

1) Fulfilling Area Constraint: When determining the best positions of the CLBs in the new design G, the first consideration is the set of blocks a CLB could be mapped to. The proposed algorithm will not map G to the old design G' directly. Instead, it will map G to blocks within its given area. Specifically, before placement, VTR specifies an area constraint for G as well as the positions of boundary boxes. A concrete example is given in Fig. 6, which shows the boundaries of both designs. As can be seen, the old design G' is larger than the new design G, and only six out of the 10 CLBs of G' are included in the area specified for G (shown in gray). Moreover, within that area, a set of blocks are not used in G'. The proposed algorithm will map a CLB in G to either one of the six CLBs of G' (V1 through V6) or one blank block.

2) Replacing Original VTR Placement Procedure: Under the given area constraint, the proposed similarity maximization algorithm is applied. As Fig. 5 shows, the *content similarity* matrix C, the *topological similarity* matrix T, and the *similarity index* matrix S are generated with the algorithms presented before. Then, a bipartite graph is constructed based on S, and the similarity maximization problem is solved with the Kuhn-Munkres algorithm [24]. Based on the obtained G to G' mapping

and the post-placement information of G' extracted by VTR, the MRC placement of the new design G is generated. Finally, this placement is sent to VTR [9], which performs routing and applies its traditional timing and congestion optimizations.

V. EXPERIMENTAL EVALUATION

A. Experimental Setup

The experimental FPGA architecture is *k06n10* (based on Altera Stratix IV) with a 45nm fabrication library and an area-delay model offered by the VTR 7.0 CAD tool [9]. Each CLB consists of 10 BLEs, while each BLE contains one 4-input LUT and two flip-flops, as shown in Fig. 1. The proposed techniques are implemented in C++ and incorporated into the VTR synthesis flow, as shown in Fig. 5. Experiments are performed on standard MCNC benchmark circuits [25] to make sure that the obtained results are representative and convincing. Table I shows the 9 benchmarks used in the experiments, sorted in the ascending order of their sizes.

Eight evaluation cases have been performed following a "$Design_{new} \rightarrow Design_{old}$" pattern. All the mapping pairs are shown in Table II. In each case we compare the proposed MRC placement with the original placement result generated by VTR. The results are shown in three metrics: *reconfiguration cost*, *critical path delay*, and *total wire length*. Note that area data are not reported since the proposed algorithm fulfills the given area constraint and imposes no overhead.

B. Results

The first set of results in Table II shows the reconfiguration costs, collected by counting the number of bit-flips of all CLBs and SBs used to implement $Design_{new}$ in the FPGA. The fewer the bit-flips, the better the performance and the lower the overhead. Results in Table II confirm that the proposed reconfiguration optimization is very effective, as it reduces 57.4% bit-flips on average.

TABLE II.   RECONFIGURATION COST IN BIT-FLIPS AND DESIGN PERFORMANCE

No	Benchmark Mapping	Reconfiguration Cost			Critical Path (ns)			Wire Length		
		Baseline	Proposed	Reduction	Baseline	Proposed	Overhead	Baseline	Proposed	Overhead
1	tseng→ex5p	17260	8188	52.6%	6.24	7.28	16.6%	8646	10640	23.1%
2	ex5p→diffeq	19352	9652	50.1%	5.11	5.18	1.4%	13255	13294	0.3%
3	diffeq→alu4	23763	10102	57.5%	6.43	7.32	14.0%	13119	16176	23.3%
4	alu4→seq	24588	9233	62.4%	4.61	4.76	3.2%	13243	15314	15.6%
5	seq→s298	28379	10502	63.0%	4.88	4.97	1.9%	20310	24209	19.2%
6	s298→frisc	34032	11782	65.4%	8.83	8.84	0.1%	12327	15637	26.9%
7	frisc→spla	62763	30891	50.8%	8.08	8.98	11.1%	40732	54071	32.7%
8	spla→ex1010	71143	30343	57.3%	6.15	7.06	14.8%	49892	64434	29.1%
	Average			**57.4%**			**7.9%**			**21.3%**

Besides reconfiguration cost, routing cost and critical path delay are also important design metrics. These data are collected at the post-routing stage for the "Baseline" and "Proposed" schemes. As the baseline VTR placement is a timing-driven scheme that optimizes timing and wire length, it is expected for the proposed scheme to have slightly longer critical paths and wire lengths. As Table II shows, the proposed scheme slightly degrades the critical path by 7.9% and increases the wire length by 21.3% on average. Such overhead is insignificant compared to the 57.4% reduction in reconfiguration cost.

Overall, the results show that the proposed scheme is able to significantly reduce the cost for reconfiguring a design on NVM-based FPGA while at the same time fulfilling area constraints with acceptable delay overhead.

VI. CONCLUSIONS

This paper has proposed a similarity-based reconfiguration optimization framework for NVM-based FPGAs, aiming to mitigate the adverse impact on reconfiguration caused by NVM writes. Several similarity metrics have been proposed to characterize LUB content similarity and CLB-level topology similarity between two designs, and an algorithm has been developed to generate a placement that minimizes reconfiguration cost. The proposed optimizations have been incorporated into the VTR synthesis flow. Experimental results show that our technique is able to reduce the number of bit-flips by 57.4% when configuring a new design on NVM-based FPGA, and introduce no area overhead and acceptable performance overhead in wire length and critical path delay.

REFERENCES

[1] K. Huang, Y. Ha, R. Zhao, A. Kumar, and Y. Lian, "A low active leakage and high reliability phase change memory (PCM) based non-volatile FPGA storage element," *IEEE Transactions on Circuits and Systems*, vol. 61, Aug. 2014.

[2] H. Asadi and M. Tahoori, "Analytical techniques for soft error rate modeling and mitigation of FPGA-based designs," *IEEE Transactions on Very Large Scale Integration Systems*, vol. 15, Dec. 2007.

[3] K. J. Han, N. Chan, S. Kim, B. Leung, V. Hecht, and B. Cronquist, "A novel flash-based FPGA technology with deep trench isolation," in *Non-Volatile Semiconductor Memory Workshop (NVSMW)*, 2007, pp. 32–33.

[4] Y. Chen, J. Zhao, and Y. Xie, "3D-NonFAR: Three-dimensional non-volatile FPGA architecture using phase change memory," in *International Symposium on Low-Power Electronics and Design (ISLPED)*, 2010, pp. 55–60.

[5] W. Zhao, E. Belhaire, V. Javerliac, C. Chappert, and B. Dieny, "Evaluation of a non-volatile FPGA based on MRAM technology," in *International Conference on Integrated Circuit Design and Technology (ICICDT)*, 2006, pp. 1–4.

[6] *International Technology Roadmap for Semiconductors*, 2013.

[7] A. P. Ferreira, M. Zhou, S. Bock, B. Childers, R. Melhem, and D. Mosse, "Increasing PCM main memory lifetime," in *Design, Automation and Test in Europe (DATE)*, 2010, pp. 914–919.

[8] M. K. Qureshi, V. Srinivasan, and J. A. Rivers, "Scalable high performance main memory system using phase-change memory technology," in *International Symposium on Computer Architecture (ISCA)*, 2009, pp. 24–33.

[9] J. Luu, J. Goeders, M. Wainberg, A. Somerville, T. Yu, K. Nasartschuk, M. Nasr, S. Wang, T. Liu, N. Ahmed, K. B. Kent, J. Anderson, J. Rose, and V. Betz, "VTR 7.0: Next generation architecture and CAD system for FPGAs," *ACM Transactions on Reconfigurable Technology and Systems*, vol. 7, Jun. 2014.

[10] *Proasic3 flash family fpgas handbook*, Actel, 2011.

[11] *Lattice xp2 family handbook*, Lattice Semiconductor, 2010.

[12] W. Zhao, E. Belhaire, C. Chappert, and P. Mazoyer, "Spin transfer torque (STT)-MRAM–based runtime reconfiguration FPGA circuit," *ACM Transactions on Embedded Computing Systems*, vol. 9, Oct. 2009.

[13] S. Paul, S. Mukhopadhyay, and S. Bhunia, "A circuit and architecture codesign approach for a hybrid CMOS STTRAM nonvolatile FPGA," *IEEE Transactions on Nanotechnology*, vol. 10, Feb. 2011.

[14] D. Choi, K. Choi, and J. D. Villasenor, "New non-volatile memory structures for FPGA architectures," *IEEE Transactions on Very Large Scale Integration (VLSI) Systems*, vol. 16, Jul. 2008.

[15] P. Gaillardon, D. Sacchetto, G. Beneventi, M. Ben Jamaa, L. Perniola, F. Clermidy, I. O'Connor, and G. De Micheli, "Design and architectural assessment of 3-D resistive memory technologies in FPGAs," *IEEE Transactions on Nanotechnology*, vol. 12, Jan. 2013.

[16] *Partial Reconfiguration User Guide*, Xilinx, March 2011.

[17] H. Tan and R. DeMara, "A physical resource management approach to minimizing FPGA partial reconfiguration overhead," in *International Conference on Reconfigurable Computing and FPGAs (ReConFig)*, 2006, pp. 1–5.

[18] R. He, Y. Ma, K. Zhao, and J. Bian, "ISBA: an independent set-based algorithm for automated partial reconfiguration module generation," in *International Conference on Computer-Aided Design (ICCAD)*, 2012, pp. 500–507.

[19] S. Hansen, D. Koch, and J. Torresen, "High speed partial run-time reconfiguration using enhanced ICAP hard macro," in *International Symposium on Parallel and Distributed Processing Workshops and Phd Forum (IPDPSW)*, 2011, pp. 174–180.

[20] Z. Yang and G. Choi, "Reconfigurable multi-functioning logic structures: a case study of MMX/floating-point unit design," in *Computer Society Workshop On VLSI (ISVLSI)*, 1999, pp. 76–81.

[21] X. Shi, D. Zeng, Y. Hu, G. Lin, and O. Zaiane, "Enhancement of incremental design for FPGAs using circuit similarity," in *International Symposium on Quality Electronic Design (ISQED)*, 2011, pp. 1–8.

[22] M. Rullmann and R. Merker, "A reconfiguration aware circuit mapper for FPGAs," in *International Symposium on Parallel and Distributed Processing (IPDPS)*, 2007, pp. 1–8.

[23] Y. Xue, P. Cronin, C. Yang, and J. Hu, "Fine-tuning CLB placement to speed up reconfigurations in NVM-based FPGAs," in *International Conference on Field Programmable Logic and Applications (FPL)*, 2015, pp. 1–8.

[24] H. W. Kuhn, "The Hungarian Method for the Assignment Problem," in *Naval Research Logistics Quarterly (NRLQ)*, 1955, pp. 29–47.

[25] F. Brglez, D. Bryan, and K. Kozminski, "Combinational profiles of sequential benchmark circuits," in *International Symposium on Circuits and Systems (ISCAS)*, 1989, pp. 1929–1934.

978-1-4673-9141-2/15 $31.00 © 2015 IEEE

Prefetch-based Dynamic Row Buffer Management for LPDDR2-NVM Devices

Jaehyun Park
Dept. of EECS
Seoul National University
cyrano06@snu.ac.kr

Donghwa Shin
Dept. of Computer Engineering
Yeungnam University
donghwashin@yu.ac.kr

Hyung Gyu Lee
School of CCE
Daegu University
hglee@daegu.ac.kr

Abstract—**LPDDR2-NVM has been announced as an industry standard to efficiently interface with non-volatile memory devices such as phase change memory (PCM). This standard interface has been adopted in most commercial PCM devices. In this paper, we devise a prefetch-based dynamic row buffer management that targets the LPDDR2-NVM devices for enhancing performance with almost negligible implementation overhead. Our extensive simulations with timing parameters from the industry's commercial PCM devices demonstrate that the proposed method enhances the performance of memory systems up to 11.3% when compared with the static optimum configuration with fairly low-cost overheads.**

I. INTRODUCTION

Phase change memory (PCM) is emerging as a next-generation non-volatile memory device that is expected to replace conventional DRAM due to its ability to scale very deep down into the low nanometer regime and its low power consumption with non-volatility [1]. However, there are two major drawbacks in adopting this PCM technology in current memory system architecture: low write performance and limited long-term endurance. Various architectural techniques for mitigating these drawbacks have been proposed, maximally exploiting the benefits of the PCM [2].

Though the proposed techniques have contributed to enhance the performance and energy consumption of PCM memory systems, most approaches have considered little or overlooked the realistic characteristics of industry-announced PCM devices. The performance observed in industry-announced PCM devices and interfaces are such examples. Most literature has tried to reduce the number of memory access or to optimize the internal architecture of PCM cell arrays in that the internal architecture and interface to the memory controller of the PCMs are very similar to DRAM [3], [4]. However, the PCM's internal architectures and interfaces considered in the industry are different than those of conventional DRAM. For example, it has been announced that the LPDDR2-NVM standard interface will be used to come up with many different characteristics of newly announced non-volatile memory devices including PCMs [5]. Major PCM manufacturers have announced their PCM prototypes to be compatible with this LPDDR2-NVM standard interface [6], [7]. Although this standard interface inherits many common features from conventional double data rate (DDR) interfaces of DRAM, many distinctive features

such as a three-phase addressing mechanism, different row buffer and bank architecture, and the support for asymmetric read and write operations using overlay windows are included as well. Detailed information about this standard interface is described in a later section.

Among several distinctive features of LPDDR2-NVM, the row buffer architecture has been significantly revised from conventional DRAM architecture in terms of the number of row buffers, the unit size of a single row buffer, and the buffer management policy. The LPDDR2-NVM interface defines 4 or 8 pairs of row buffers, and each pair consists of one row address buffer (RAB) and row data buffer (RDB). These multiple row buffers are arbitrarily selected by the memory controller, regardless of the physical memory address in the LPDDR2-NVM interface. The bank address (BA), signals which are originally coupled to the physical memory address in the DRAM, are used to select an adequate row buffer. In addition, the unit size of a single RDB is pretty much smaller, 32 bytes in typical case, than the unit size of a single row buffer of DRAM. All of the above-mentioned differences indicate that we have more flexibility to control LPDDR2-NVM's row buffers, and that the sophisticated control mechanism of these row buffers is desirable to maximize the performance of the memory systems.

In this paper, we investigate the performance of the memory systems affected by the row buffer architecture in the PCM device supporting an LPDDR2-NVM interface. And then we devise prefetch-based dynamic row buffer management that mimics a dynamic reconfiguration of row buffer architecture, the unit size of single row buffer and the number of row buffer, by utilizing a prefetch considering the memory access patterns without physically changing its internal architecture. The proposed scheme efficiently traces and adaptively controls the number of prefetched rows depending on the memory access characteristics. Our trace-driven simulations using real workloads and practical timing parameters extracted from industries' PCM prototypes demonstrate that the proposed row buffer architecture considering an LPDDR2-NVM industry standard enhances system-level memory performance up to 11.3% compared to the static optimum row buffer architecture under the same cost (area) restrictions.

The rest of the paper is organized as follows. Section II metions the background of this work, including a brief introduction of the LPDDR2-NVM interface and related work. The motivation for this work is provided with an example in Section III, and then we introduce prefetch-based dynamic

---

H.G. Lee is a corresponding author

978-1-4673-9141-2/15 $31.00 © 2015 IEEE

row buffer management in Section IV. Section V evaluates the proposed scheme. Finally, Section VI draws conclusions.

II. BACKGROUND

A. LPDDR2-NVM interface

The representative features of the LPDDR2-NVM interface compared to conventional DDR interfaces are:

- Three-phase addressing mechanism for supporting the larger sizes of memory devices (up to 32 Gb).

- No multi-bank architecture.

- RABs and RDBs are not coupled with the physically accessed memory addresses.

- Smaller size of single RDB (typically 32 bytes) than that of conventional DRAM.

- Indirect write operations via overlay window.

- Dual operation: read operation while performing cell programs in other partition.

Fig. 1 shows the internal structure of an LPDDR2-NVM compatible PCM device and its interface. The LPDDR2-NVM specifies 10 bits of CA pins, and they are used with a DDR architecture even for the address phase. This indicates that the memory controller transfers up to 20 bits of command or address data for each memory clock cycle. In addition, a three-phase addressing mechanism is used to support a larger size of memory devices than conventional DRAM which originally uses a two-phase address mechanism. Three-phase addressing consists of preactive, activate, and read/write phases. LPDDR2-NVM also supports multiple pairs (4 or 8) of RAB and RDB. The BA signals, which are originally used to select a bank in conventional DRAM, are used to select a row buffer. Note that these BA signals are only intended to select a row buffer, not a physical bank address of the memory array [5]. In each phase, the memory controller selects a proper RAB and/or RDB by controlling these BA signals, regardless of the physically-accessed memory address.

PCM shows a relatively long cell program time because of its operating principle. This long cell program time may also affect read performance if a read request arrives during the cell program operation. Similar to the multi-bank architecture in traditional DRAM devices, multi-partition architecture and parallel operation in LPDDR2-NVM alleviate read performance degradation. Parallel operations can read the data in a partition while another partition is being programmed. This increases the performance of LPDDR2-NVM significantly. However parallel cell program operations are not allowed.

Another distinctive feature of LPDDR2-NVM is asymmetric read and write operations. The process of the read operation is very similar to conventional DRAM except for three-phase addressing and row buffer management. However, write operation, strictly speaking non-volatile cell programming is completely different from the conventional DRAMs. Write operations are done indirectly through the special registers, called overlay window. A single write operation from the processor requires several overlay window access to complete the cell program. The overlay window consists of several

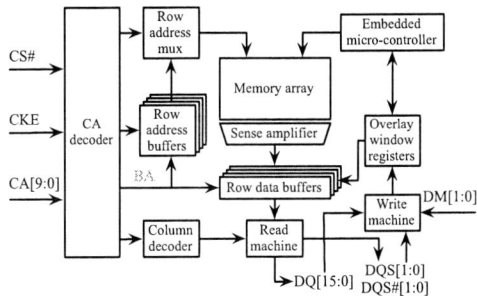

Fig. 1. Functional block diagram of JEDEC LPDDR2-NVM standard-compatible PCM device.

memory-mapped registers such as a command address register, a command code register, a command execution register, and program buffers to properly control the LPDDR2-NVM devices and program the memory array.

B. Related work

Many system-level approaches have been conducted to enhance the performance and energy consumption of PCM memory systems. However, only a few studies have researched on optimizing the row buffer architecture and its management. The row buffer architecture of the PCM is optimized under the assumption that it is similar to that of conventional DRAM [4]. They use multiple small size row buffers so that the energy consumption on a small size row buffer decreases significantly by reducing the number of required sense amplifiers. Performance is also improved through reorganizing the row buffer architecture. Hot and cold row separation has been considered in [8]. By caching the hot rows in DRAM, this approach tries to decrease the number of row misses. However, this is a type of system-level technique and does not care about the row buffer architecture itself. They just exploit the locality information of memory access. However, their architecture or proposed techniques do not consider the asymmetric read and write characteristics of PCM, where the write operation is performed only through overlay window access in industry PCM prototypes. In addition, both approaches assume a similar internal buffer architecture to that of DRAM.

There have been many approaches to optimizing memory requests by reordering their sequences to consider the activated rows at the memory controller [9], [10]. It changes the order of memory requests at the memory controller by using the reordering buffer and returns the results to the processor as ordered. Though their contributions are significant, we do not directly compare these techniques with ours, because these types of approaches are almost orthogonal to our approach, which is focusing on optimizing the configuration of row buffers residing inside of the memory devices. In addition, conventional reordering schemes are effective only if the command queue of the memory controller has multiple memory requests while our proposed dynamic RDB management is activated if the command queue has only a single (or zero) request.

To the best of our knowledge, this is the first practical work that analyzes row buffer architecture and its efficient management by considering the LPDDR2-NVM industry standard interface.

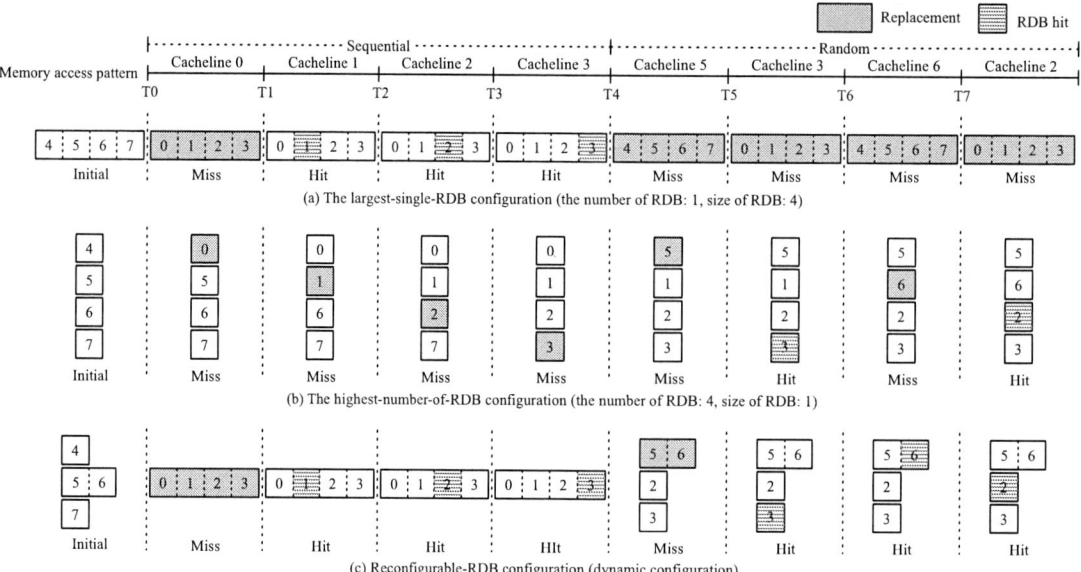

Fig. 2. RDB hit ratio varying on row buffer architecture and memory access patterns.

III. MOTIVATION

As described earlier, the LPDDR2-NVM standard provides more flexibility in designing and managing the row buffer architecture than conventional DRAM architecture.This flexibility enables us to design various row buffer management schemes considering the access patterns of the applications.

While designing the row buffer architecture, determining the unit size of the single RDB and the number of RDBs are as important as determining the size of entire RDBs. Fig. 2 shows a simple motivational example of this work. We compare the number of RDB hits on the three different configurations: (a) The largest-single-RDB configuration, (b) The highest-number-of-RDB configuration, and (c) Reconfigurable-RDB configuration. In the figure, the box with a thick solid line signifies one physical RDB, and one physical RDB consists of one (or more) basic units: the box with a dotted line. The size of one basic unit is equal to the size of one cacheline. The largest-single-RDB configuration has only one RDB, which consists of 4 basic units, while the highest-number-of-RDB configuration has four RDBs, where each RDB has just one basic unit. The example patterns of memory access are presented on the top of the figure. The first half of the memory access pattern is sequential, while the second half of the pattern is random. A grayed-box and horizontally-lined-box represents an RDB miss and an RDB hit, respectively. For a fair comparison, all three configurations start with the same initial state. Cachelines 4, 5, 6, and 7 are stored in the RDBs.

In the largest-single-RDB configuration, the request of Cacheline 0 incurs an RDB miss at time T0. This RDB miss evicts all cachelines in the RDB, and Cachelines 0 to 3 are fetched from the NVM array as shown in Fig. 2(a). Since the next three memory access are sequential, all three requests incur RDB hits. However, the remaining memory access pattern from time T4 to T7 incurs RDB misses again, because this configuration has only one RDB. In total, 5 RDB misses and 3 RDB hits occur during the memory access.

In contrast, the highest-number-of-RDB configuration efficiently handles a random memory access pattern efficiently because one RDB has only one cacheline. However, it is very weak for a sequential memory access pattern. The requests of Cachelines 0 to 3 continuously incur RDB misses from time T0 to T3, as shown in Fig. 2(b). In total, we observe 6 RDB misses and 2 RDB hits.

Both the largest-single-RDB configuration and the highest-number-of-RDB configuration provide limited capability to the given memory access pattern in the motivational example. Each configuration has advantages and disadvantages according to the characteristics of memory access patterns. In real applications, this memory access pattern may vary on application changes as well as time changes, even in the same application. Thus, changing the RDB configuration dynamically increases the number of RDB hits as shown in Fig. 2(c). It reduces the number of RDB misses by reconfiguring the row buffer from 3 RDBs to one RDB with 4 basic units to hold four consecutive cachelines at T0, which results in 3 consecutive RDB hits. Requesting Cacheline 5 at time T4, the dynamic-RDB reconfiguration modifies the configuration of one to RDB with two basic units. This reconfiguration is kept until time T7. In total, 2 RDB misses and 6 RDB hits occur in this dynamic configuration. Surely this dynamic RDB reconfiguration shows the best RDB hit ratio with the information of an incoming memory access pattern. It is important to predict the characteristics of memory access accurately because this adaptive RDB configuration with incorrect information is also possible to incur more RDB misses. In the later section, we devise a history-based prediction method to figure out the characteristics of incoming memory access and then use it for devising prefetch-based dynamic row buffer reconfiguration methods targeting the LPDDR2-NVM standard.

IV. DYNAMIC ROW BUFFER MANAGEMENT POLICY

Reconfigurable RDB architecture must be an attractive solution for further increasing the RDB hit rate. However, beside

978-1-4673-9141-2/15 $31.00 © 2015 IEEE

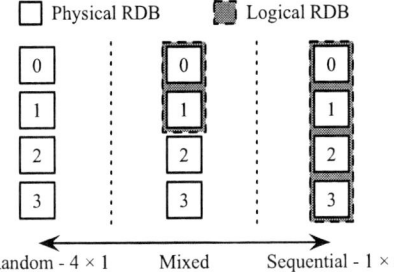

Fig. 3. The concept of a prefetch-based dynamic RDB reconfiguration.

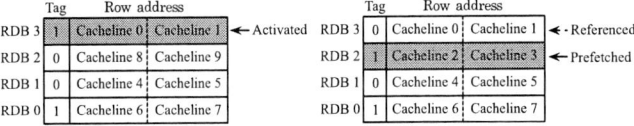

Fig. 4. Operation of the tagged row buffer prefetch.

the prediction policy, physically implementing reconfigurable RDB hardware architecture in a real memory system requires severe overhead. This section mainly describes a way to mimic the dynamic reconfiguration of RDB without hardware modification in the LPDDR2-NVM specification by utilizing an arbitrarily-selectable RDB architecture and prefetch method.

A. Prefetch-based dynamic RDB management

A prefetch technique has been used in traditional cache memory architecture to maximally utilize the limited size of cache memory. Similarly, we exploit a row buffer prefetch technique to mimic the reconfiguration of the RDB architecture without physically modifying the existing RDB architecture. Fig. 3 simply presents a concept of a prefetch-based dynamic RDB reconfiguration. Physically, the row buffer consists of 4 RDBs, where each RDB is not coupled with the physical address but is selected by the memory controller using BA pin signals. When there is a RDB miss, one RDB is allocated to serve this request. In addition to this basic operation, we propose to allocate more RDBs for prefetching the consecutive row data if the next request is expected to be sequential. This prefetch operation acts like allocating more than two RDBs for one memory request, which eventually creates a similar effect of increasing the size of a single RDB. Depending on the number of RDBs used for prefetching, the size of a single RDB (not the physical size but the logical size) can be varied, and this is the basic principle of our proposed dynamic RDB management. The number of RDBs used for prefetching is increased if the access patterns are expected to be strongly sequential, while it is decreased when expecting random access patterns.

All necessary additional control logic is implemented in the memory controller, and there is no need to modify the RDB architecture of the PCM devices. The memory controller issues a row buffer prefetch command to the command queue if the memory controller predicts that the characteristics of the incoming memory access pattern is sequential. In our scheme, we do not consider scheduling between the memory access from the processor, because this is orthogonal with our proposed scheme as described in II. In the case of a conflict between the processor's request and prefetch request, we give a higher priority to the prefetch request than the normal processor's request. The latency of the prefetch request is lower than a normal memory request because the prefetch does not need to execute the read/write phase (the last phase of the operation) of three-phase addressing. Though this may slightly and temporarily increase the response time of the coming memory request, we found that the benefit of this

policy is greater than the temporal response time degradation if the prediction is accurate.

In our proposed dynamic RDB management, performance enhancement depends heavily on the accuracy of the prediction, because a prediction failure may be responsible for performance degradation. In the latter two subsections, we propose a simple but efficient system-level row buffer prediction policy for the dynamic RDB reconfiguration targeting LPDDR2-NVM.

B. Tagged row buffer prefetch

We first devise a simple tagged row buffer prefetch, T_{pre}, that is similar to a tagged prefetch in a cache [11]. Fig. 4 shows its simple operation. It uses one tag bit for issuing a prefetch command. The controller sets the corresponding tag bit when the RDB is first activated. When this RDB is referenced, this bit is cleared. Then, the prefetch is requested to fetch the consecutive data to another RDB. In this case, T_{pre} evicts one RDB to execute a prefetch operation. The victim selection policy is the same as when a RDB miss occurs, LRU.

T_{pre} issues a prefetch command if the RDB is referenced more than twice, and will work efficiently if the access patterns are almost sequential. This assumption is justified in that the size of a single RDB is always larger than the size of the cacheline.

The prefetch is requested only for read access, because write memory access is translated into several overlay window access in LPDDR2-NVM. The address of the overlay window does not change according to the address of the write memory access. This address translation makes the write memory access a random memory access, even though it is a part of a sequential memory access pattern. Moreover, the overlay window access is executed seamlessly without idle time, so the row buffer prefetch does not reduce memory access time.

C. Multiple row buffers prefetch

There are two reasons for potential performance degradation in the proposed dynamic RDB reconfiguration using a prefetch. First the unnecessary row buffer prefetch caused by a missed prediction may degrade the memory access time. In addition, a regular memory request can be delayed until the row buffer prefetch is carried out, if the request arrives during a prefetch operation. In order to minimize these types of performance degradations, we devise a sophisticated row buffer prefetching algorithm using a two-bit saturating counter. The saturating counter tracks the recent activity of RDB as shown in Fig. 5. It initially starts from the weak random mode (WEAK RAND). When an RDB hit occurs, it indicates that the incoming memory access request will be highly likely to be a sequential memory access. Then it promotes the prefetching mode to a weak sequential mode (WEAK SEQ) from WEAK

978-1-4673-9141-2/15 $31.00 © 2015 IEEE

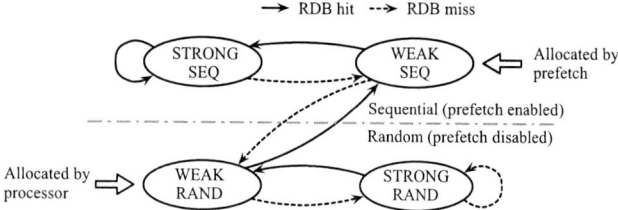

Fig. 5. Prefetch mode transition using two-bit saturating counter to predict incoming memory access.

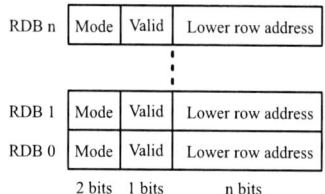

Fig. 6. Row data buffer tracking table.

RAND. On the other hand, if an RDB miss occurs, it implies that the next coming memory access will be highly likely yo be a random memory access. It then demotes the prefetching mode to a strong random mode (STRONG RAND) from WEAK RAND.

For implementing this prediction idea with minimal cost, it is possible to use a single global saturating counter for all RDBs. Using a single global saturating counter may reduce the unnecessary row buffer prefetches. However, at the same time, it may reduce the necessary prefetches, called good prefetch. The reason is that the global saturating counter changes its prefetch mode quickly. In addition, using a single global counter may fail to predict the characteristic of the next coming memory requests if one group of sequential requests are mixed with the other group of sequential requests (or random access patterns) within a short time window. In this case, the prefetch mode will be stayed at the random mode, even though most memory patterns are sequential.

In order to prevent non-desirable predictions, we propose a multiple row buffer prefetch, M_{pre}, which uses multiple saturating counters. Basically, each RDB has one saturating counter, and multiple saturating counters are able to track several groups of sequential memory access patterns simultaneously. When an RDB hit occurs, M_{pre} promotes the prefetch mode of the corresponding RDB only, and it does not demote the prefetch modes of other RDBs. On the other hand, M_{pre} demotes the prefetch modes of all RDBs at once if an RDB miss occurs. The RDB miss highly likely implies that one of the sequential access patterns is over.

Basically, the initial prefetch mode starts from WEAK RAND if memory access is requested from the processor. The memory access requested from the processor implies that it misses all RDBs. However, we set the initial prefetch mode to WEAK SEQ if the memory access is requested from the memory controller to execute the prefetch operation.

Fig. 6 shows the RDB tracking table that includes two-bit saturating counter bits. For determining the RDB hit, the memory controller also keeps the address information of the

TABLE I. SIMULATED SYSTEM CONFIGURATION DETAILS.

System	8 cores in-order processor
Processor	UltraSPARC-III+, 2 GHz
L1 cache (Private)	I/D-cache 32 KB, 4 way 64 B block
L2 cache (Shared)	2 MB, 4-way 64 B block
PCM main memory	1 GB, LPDDR2-800, 64-bit wide
Preactive to Activate (t_{RP})	$3\ t_{CK}$[1]
Activate to Read/Write (t_{RCD})	120 ns
Read latency (RL)	$6\ t_{CK}$[1]
Write latency (WL)	$3\ t_{CK}$[1]
Cell program time ($t_{program}$)	150 ns
The number of partitions	16

[1]t_{CK} is a memory clock cycle (2.5 ns at LPDDR2-800)

data stored in each RDB.

V. EXPERIMENTAL RESULTS

A. Evaluation setup

We evaluate the memory access time of the proposed methodology using a cycle-accurate trace-driven SystemC simulator. The traces have been extracted from Simics full-system simulator [12] with processor clock cycle information. We simulate an 8-core in-order processor system with a shared last level cache. The timing parameters of PCM are extracted from the JEDEC LPDDR2-NVM standard and an industrial prototype [13]. The details of the simulation setup are summarized in Table I. Thirteen multi-threaded benchmarks from the PARSEC benchmark suite [14] are selected.

Based on this setup, we compare the memory access time of the static optimum RDB configuration, T_{pre} and M_{pre}. The size of the entire row buffer is the same for all three configurations for a fair comparison. The static optimum RDB configuration uses a certain RDB configuration that has minimum memory access time among all possible RDB configurations: 2×512, 4×256 or 8×128 while T_{pre} and M_{pre} use the 8×128 configuration.

Our experiments assume that the PCM device supports dual operation of LPDDR2-NVM, which means that read operations are possible during cell program operation unless the accessed partition is the same. However, the read latency increases when the processor tries to read data from the same partition where the cell program is operating.

B. Performance comparison

We first analyze the RDB hit ratios of all three configurations as shown in Table II. We present the RDB hit ratio of regular read access, r_{RD}, and overlay window access, r_{OW}, separately. In all experiments, r_{OW} is higher than r_{RD} because the access of the overlay window for the write operation shows higher spatial and temporal locality than regular read operations. T_{pre} shows higher r_{RD} than that of the static optimum RDB configuration because T_{pre} aggressively prefetches the data to the row buffer in advance based on the prediction. However, this simple and aggressive row buffer prefetch decreases the r_{OW} of T_{pre} because the prefetch often requires to evict the data that is expected to be referenced in the near future from the row buffer. On the other hand, M_{pre} prefetches the data to the row buffer more carefully than T_{pre} based on the access history and this results in reducing the total number of row buffer prefetches. In several

TABLE II. COMPARISON OF THE HIT RATIO.

Application	Static optimum		T_{pre}		M_{pre}	
	r_{RD}	r_{OW}	r_{RD}	r_{OW}	r_{RD}	r_{OW}
blackscholes	16.8	96.3	29.0	85.9	27.1	96.2
bodytrack	35.7	81.8	63.2	78.5	61.1	81.1
canneal	0.6	74.9	0.5	63.0	0.7	75.0
dedup	30.6	82.5	39.8	77.9	42.2	82.0
facesim	56.6	68.0	73.7	72.8	74.3	74.6
ferret	58.2	82.6	62.0	84.2	69.8	85.0
fluidanimate	18.0	80.2	18.3	74.9	19.5	80.1
freqmine	52.5	78.2	56.7	81.9	61.5	83.5
raytrace	48.3	54.7	53.5	62.3	47.4	70.2
streamcluster	55.3	66.6	60.9	76.9	58.2	81.1
swaptions	67.4	71.5	88.1	74.8	86.9	75.2
vips	35.0	83.7	57.7	81.7	57.3	83.3
x264	33.5	82.0	48.7	76.7	50.7	81.3

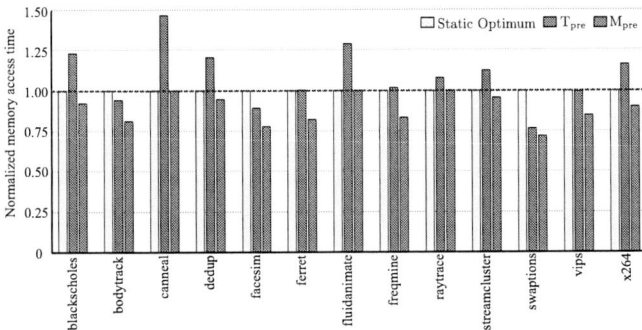

Fig. 7. Comparison of memory access times normalized to the static optimum RDB configuration (lower is better).

applications such as *bodytrack, raytrace, streamcluster,* and *swaptions*, we find that this results in lowering r_{RD} than that of T_{pre}. However, the r_{OW} of M_{pre} is always higher than that of T_{pre} in all applications because of the increased the accuracy of prediction in M_{pre}.

Finally, we compare the memory access time of three configurations as shown in Fig. 7. For easy comparison, the memory access times of two other configurations are normalized to the static RDB configuration. In spite of the prediction, T_{pre} mostly shows higher memory access time than the static optimum configuration, except for *bodytrack, facesim, swaptions,* and *vips*. In fact, the row buffer prefetch makes the effective size of single RDB larger, and this may reduce memory access time for the next consecutive memory access if the prediction is correct. However, T_{pre} uses a simple and aggressive prediction heuristic, and this incurs frequent and unnecessary row buffer prefetches frequently. T_{pre} increases the memory access time by almost 9.0% on average.

As described, M_{pre} efficiently predicts the coming memory request patterns and minimizes unnecessary row buffer prefetches. We successfully confirm that the proposed M_{pre} reduces memory access times in most applications, except *fluidanimate* and *raytrace*. The average memory access time reduction of M_{pre} is 11.3%.

VI. CONCLUSIONS

The performance of the memory system largely depends on the row buffer configuration as well as the memory access patterns in LPDDR2-NVM compatible memory devices. In this paper, we have shown the limitation of the static row buffer architecture and proposed a prefetch-based dynamic row buffer management that mimics a reconfigurable row buffer architecture. We confirm that the proposed row buffer management efficiently enhances the row hit rate that finally results in reducing memory access time. Our trace-driven simulations uses real memory traces extracted from a full system simulator, which demonstrates that the proposed row buffer management enhances the memory access time by 11.3% on average, even when compared with a static optimum RDB configuration.

ACKNOWLEDGMENT

This research was supported by the Basic Science Research Program through the National Research Foundation of Korea (NRF), funded by the Ministry of Education (NRF-2013R1A1A2063350), NRF Grant 2015R1C1A1A02004746, the 2015 Yeungnam University Research Grant, and Brain Korea 21 Plus Project, Department of Computer Engineering, Yeungnam University.

REFERENCES

[1] S. Raoux *et al.*, "Phase-change random access memory: A scalable technology," *IBM Journal of Research and Development*, vol. 52, no. 4/5, pp. 465–479, 2008.

[2] O. Zilberberg, S. Weiss, and S. Toledo, "Phase-change memory: an architectural perspective," *ACM Computing Surveys*, vol. 45, no. 3, p. 29, 2013.

[3] M. K. Qureshi, V. Srinivasan, and J. A. Rivers, "Scalable high performance main memory system using phase-change memory technology," in *Proceedings of the 36th annual international symposium on computer architecture*, 2009, pp. 24–33.

[4] B. C. Lee, E. Ipek, O. Mutlu, and D. Burger, "Architecting phase change memory as a scalable dram alternative," in *Proceedings of the 36th annual international symposium on computer architecture*, 2009, pp. 2–13.

[5] JEDEC, "JESD209-F," 2013.

[6] P. Clarke, "Samsung preps 8-Gbit phase-change memory," Nov. 2011. [Online]. Available: http://www.eetimes.com/document.asp?doc_id=1266495

[7] R. Krishnamurthy, "First volume production phase change memory by micron," May 2013. [Online]. Available: http://www.chipworks.com/en/technical-competitive-analysis/resources/blog/first-volume-production-phase-change-memory-by-micron/

[8] H. Yoon *et al.*, "DynRBLA: A high-performance and energy-efficient row buffer locality-aware caching policy for hybrid memories," *SAFARI Technical Report No. 2011-005*, 2011.

[9] C.-Y. Lai, G.-Y. Pan, H.-K. Kuo, and J.-Y. Jou, "A read-write aware dram scheduling for power reduction in multi-core systems." in *19th Asia and South Pacific Design Automation Conference*, 2014, pp. 604–609.

[10] Y.-J. Lin, C.-L. Yang, J.-W. Huang, T.-J. Lin, C.-W. Hsueh, and N. Chang, "System-level performance and power optimization for MPSoC: A memory access-aware approach," *ACM Transactions on Embedded Computing Systems*, vol. 14, no. 1, p. 8, 2015.

[11] V. Srinivasan, E. S. Davidson, and G. S. Tyson, "A prefetch taxonomy," *IEEE Transactions on Computers*, vol. 53, no. 2, pp. 126–140, 2004.

[12] P. S. Magnusson *et al.*, "Simics: A full system simulation platform," *Computer*, vol. 35, no. 2, pp. 50–58, 2002.

[13] Y. Choi *et al.*, "A 20nm 1.8V 8Gb PRAM with 40MB/s program bandwidth," in *Proceedings of the 2012 IEEE International Solid-State Circuits Conference Digest of Technical Papers*, 2012, pp. 46–48.

[14] C. Bienia, S. Kumar, J. P. Singh, and K. Li, "The PARSEC benchmark suite: Characterization and architectural implications," in *Proceedings of the 17th international conference on Parallel architectures and compilation techniques*, 2008, pp. 72–81.

Design Space Exploration of Row Buffer Architecture for Phase Change Memory with LPDDR2-NVM interface

Jaehyun Park
Dept. of EECS
Seoul National University
cyrano06@snu.ac.kr

Donghwa Shin
Dept. of Computer Engineering
Yeungnam University
donghwashin@yu.ac.kr

Hyung Gyu Lee
School of CCE
Daegu University
hglee@daegu.ac.kr

Abstract—**Phase change memory (PCM) is an attractive candidate for the future memory, but it still has several limitations to overcome such as write latency and long-term endurance. A large body of literature has been dedicated to solving these problems. However, almost all of the previous studies did not consider an important practical aspect of the PCM – an interface. The LPDDR2-NVM standard interface recently introduced by JEDEC is widely adopted by the manufacturers of commercial PCM these days. The LPDDR2-NVM standard allows a more flexible use of row buffers compared to the conventional DRAM interface. In this paper, we explore the design space of row buffer architecture in the PCM with LPDDR2-NVM interface. The effect of row buffer architecture on memory performance is investigated in terms of unit size and number of RDBs, and its management policy. We use the timing parameters from industry prototype PCM and analyze the result from the perspective of Pareto's optimum. The experimental results show that a properly-designed row buffer architecture enhances system-level performance up to 44.2% even at the same cost.**

I. INTRODUCTION

Among the candidates of the next-generation memory, phase change memory (PCM) is often regarded as one of the device that is closest to commercialization. It is a type of non-volatile random-access memory that is expected to replace DRAM and Flash memory simultaneously. Different from the older device that depends on thermal effects to change the state, the recent one only with change of the coordination state (interfacial phase change memory, IPCM) raises its potential as a next-generation memory device. It can be scaled down to a very small size, and low power consumption is also an important attractive feature [1].

To prepare for commercialization, we should overcome several drawbacks of PCM from the perspective of system design. First, PCM in general has longer write latency - low write performance - due to the physical limitations of the device. It requires some time to change the state. The next drawback is also physical. The possible number of phase changes - the number of memory access - is limited, and is generally much less than that of DRAM. This causes a long-term reliability problem in memory systems. Various attempts have been made to mitigate the effect of these drawbacks

H.G. Lee is a corresponding author

from the perspective of system architecture and control. An internal architecture optimization of a PCM cell array has been presented [2], [3]. Another important architectural research direction is to reduce the number of write operations to solve the write latency and endurance problems [4]

There is an important point to be considered to devise a system- and architectural-level solution to solve such a problem. All the memory devices should be connected to the system through the same interface standard. The available feature of the interface standard decisively limits the available design space of memory systems. Therefore, we should be informed about the trend of system design from industry related to the interface standard. After intensive survey on memory system design that includes PCM, we found that major PCM manufacturers have announced their PCM prototypes to be compatible with LPDDR2-NVM [5], [6].

The LPDDR2-NVM standard is a memory interface standard for non-volatile memory devices that is defined by JEDEC JESD209-2 standard [7]. The LPDDR2-NVM standard shares various features with the LPDDR2-DRAM standard; whereas the LPDDR2 standard is basically introduced to replace the LPDDR1 standard with more advanced features such as faster speed, higher density, less power, lower voltage, fewer pins, etc. The LPDDR2-NVM is designed to be compatible to LPDDR2-DRAM with similar system assumptions such as random access and buffered write. It is generally regarded that a good architectural design of an LPDDR2-DRAM controller will yield good results with LPDDR2-NVM [8]. Accordingly, previous studies on the PCM controller design have been performed based on DRAM controllers [3], [9].

However, LPDDR2-NVM also has unignorable distinctive features at the same time [8]. For instance, LPDDR2-NVM conducts addressing with three phases in addition of the "Pre-activate" command as presented in Fig. 1. Based on this feature of LPDDR2-NVM, a performance enhancement method has been introduced, which is called the address phase skipping technique [10]. Another important feature is that LPDDR2-NVM provides a flexible use of a row buffer with various kinds of allocation mechanisms. By using bank address (BA) signals, we can arbitrarily select the row buffer to be used for the memory access. The design of the command queue should be selected by the designer, and the write operation based on an overlay window also has a different feature.

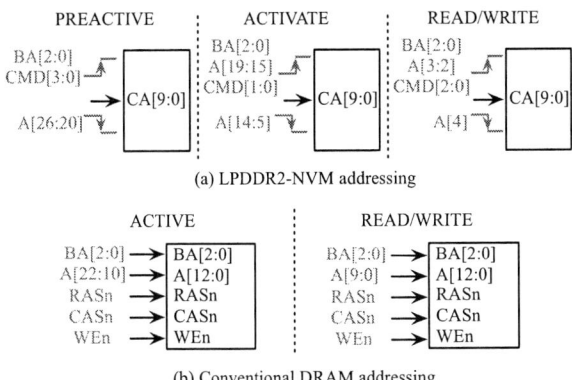

(a) LPDDR2-NVM addressing

(b) Conventional DRAM addressing

Fig. 1. Comparison of LPDDR2-NVM and the conventional DRAM in a 1Gb device with a 16-bit data width.

The LPDDR2-NVM supports multiple row address buffers (RABs) and row data buffers (RDBs). The designer can select the number of buffers and its size. The BA signals are originally coupled with the memory address in DRAM, but the LPDDR2-NVM standard explicitly notifies that these BA signals are only designed to select a row buffer, not a physical bank address of the memory array [7]. It means that the controller should provide a buffer allocation mechanism with the LPDDR2-NVM standard, but a more flexible management of the row buffer is possible at the same time. We focus on evaluating the effect of the row buffer design and control of system performance in this work.

In this paper, we explore the design space of the row buffer setup in the LPDDR2-NVM interface standard with PCM. We employ a trace-driven simulator to evaluate the effects of the changing row buffer architecture. The timing parameters of PCM are extracted from the JEDEC LPDDR2-NVM standard and the industrial prototype [11], [10], and various kinds of traces from widely-used benchmarks are used for the simulation. We change the row buffer setup and evaluate system-level performance as follows:

- We evaluate a variation in performance according to different row buffer mapping policies – a fully-associative mapping policy and a direct-mapped policy.

- We evaluate a variation in performance according to different unit size and the number of RDBs – 2 to 64 RDBs with 32 to 4,096 byte unit sizes.

Then, we analyze the performance and cost of the row buffer setup by considering the total size of the RDB. The simulation result shows that the system-level memory performance can vary more than 40% according to the row buffer setup for the same workload and system environment.

The rest of the paper is organized as follows. Section II introduces the related work. We explore the design space of the row buffer setup in Section III, and then we discuss the design issues including cost in Section IV. Section V concludes this work with the key insights gathered from the experimental results.

II. RELATED WORK

Relatively less research has been conducted about optimizing the row buffer optimization when compared to many system-level approaches developed to enhance the performance and energy consumption of the PCM memory system.

Lee et al. analyze the row buffer architecture under the assumption that the baseline buffer architecture of the PCM is similar to that of conventional DRAM [3]. Instead of using a single 2-KB big size buffer, they propose to reorganize the row buffer with multiple small-sized row buffers under the premise that energy consumption in the small size row buffer can be significantly enhanced by reducing the number of sense amplifiers. At the same time, reorganizing the buffer affects read and write performance. However, their analysis did not consider that the write operation is performed only through the overlay window access in the industry PCM prototypes. We found that write operations using overlay window use the RDB in a different way.

Yoon et al. make separately consideration for hot and cold rows, where a hot row is cached in DRAM [9]. This approach decreases the number of hot row misses, which finally leads to performance and energy enhancement. However, their approach is a type of system-level technique that does not care about the row buffer architecture and its optimization. They just exploit the locality information of row buffers. Their analysis is also assumes a similar internal buffer architecture in DRAM.

Li et al. consider the LPDDR2-NVM interface, but they just utilize access protocol of the LPDDR2-NVM to design a photonic-channel based memory communication infrastructure for PCM [12]. To the best of our knowledge, this is the first work that analyzes the row data buffer architecture and its management by considering practical row buffer architecture currently used and defined in most commercial PCM devices and the LPDDR2-NVM industry standard interface.

III. ROW BUFFER ARCHITECTURE AND MANAGEMENT POLICY

Like a cache in the traditional memory hierarchy, the row buffer within a memory device plays an important role for enhancing the performance of the memory system. It is even more important in PCM devices because of the relatively long latency of the PCM compared to conventional DRAM. This section mainly investigates the effects of row buffer architecture in the PCM by exploring the design space of the PCM row buffer in terms of the system-level performance.

A. Evaluation setup

We evaluate the performance variation according to the row buffer architecture and management policy. We employ a trace-driven simulator to evaluate the effects of the changing row buffer architecture. The traces have been extracted with a processor clock cycle information from the Simics full-system simulator [13]. We simulate an 8-core in-order processor system that operates at 2 GHz clock frequency with shared last level cache. The main memory system has a 64-bit bus with four LPDDR2-NVM compatible PCM chips. The unit size of RDB and the number of RDBs vary depending on the

978-1-4673-9141-2/15 $31.00 © 2015 IEEE   105

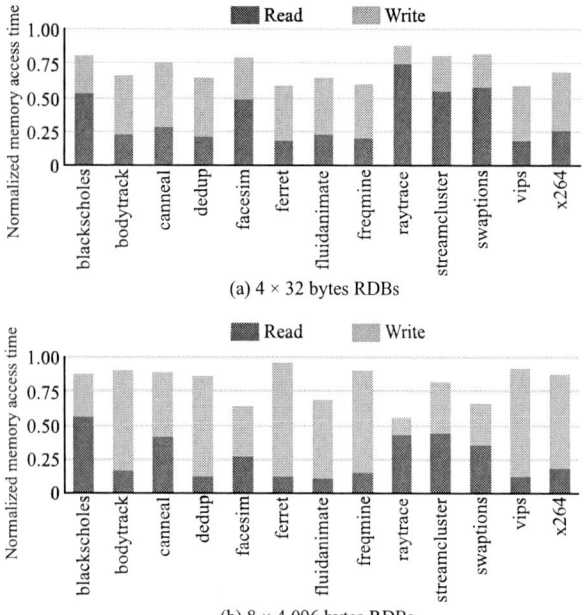

(a) 4 × 32 bytes RDBs

(b) 8 × 4,096 bytes RDBs

Fig. 2. Memory access time of a fully-associative mapping policy normalized to a direct-mapped policy (the lower is better).

row buffer configuration. The timing parameters of PCM are extracted from the JEDEC LPDDR2-NVM standard and the industrial prototype [11]. The details of the simulation setup are summarized in Table I.

Thirteen multi-threaded benchmarks from the PARSEC benchmark suite [14] are selected. Table II summarizes the characteristics of each benchmark in terms of the ratio of read operations normalized to the write operations and the frequency of memory access. Based on this setup, we intensively evaluate the memory system performance by changing the mapping policy, the unit size and the number of RDBs.

B. Row buffer mapping

As described in Section I, the memory controller selects a row buffer arbitrarily in the LPDDR2-NVM like a fully-associative cache while the row buffer selection in conventional DRAM is fixed by the internal architecture of DRAM similar to a directed-mapped cache. This architecture in the LPDDR2-NVM enables us to choose a proper row buffer management

TABLE I.   SIMULATED SYSTEM CONFIGURATION DETAILS.

System	8 cores in-order processor
Processor	UltraSPARC-III+, 2 GHz
L1 cache (Private)	I- and D-cache: 32 KB, 4 way 64 B block
L2 cache (Shared)	2 MB, 4-way 64 B block
PCM main memory	1 GB, LPDDR2-800, 64-bit wide
Preactive to Activate (t_{RP})	3 t_{CK}[1]
Activate to Read/Write (t_{RCD})	120 ns
Read latency (RL)	6 t_{CK}[1]
Write latency (WL)	3 t_{CK}[1]
Cell program time ($t_{program}$)	150 ns

[1]t_{CK} is a memory clock cycle (2.5 ns at LPDDR2-800)

TABLE II.   MEMORY ACCESS CHARACTERISTICS OF THE BENCHMARK APPLICATIONS.

Application	R/W ratio	Memory Access / CPU cycles
blackscholes	5.96	0.00023
bodytrack	5.96	0.00083
canneal	1.84	0.01262
dedup	1.39	0.00474
facesim	6.58	0.00022
ferret	1.52	0.00091
fluidanimate	1.39	0.00474
freqmine	1.55	0.00163
raytrace	27.54	0.00027
streamcluster	8.24	0.00015
swaptions	11.11	0.00058
vips	1.32	0.00164
x264	1.78	0.00232

policy. The difference of management policy causes a different RDB hit ratio during read and write operations and this, in turn, leads to different memory access time.

Fig. 2 shows the comparison of the memory access time of a fully-associative mapping policy with a direct-mapped policy. We evaluate two configurations of RDBs in each chip: 4×32 bytes RDBs and 8×4,096 bytes RDBs. The memory access time of the fully-associative policy is normalized to that of the direct-mapped policy. Most commercial PCMs compatible with an LPDDR2-NVM interface has 4×32 bytes RDBs, while the 8×4,096 bytes RDBs is the largest capable RDB configuration in the LPDDR2-NVM standard.

Overall, the fully-associate policy outperforms the direct-mapped policy in all applications and configurations. In LPDDR2-NVM compatible PCMs, one write operation requires several consecutive write access to control registers located in the overlay window where the row address of control registers are close to each other. Therefore, in the direct-mapped policy, only one RDB is allocated to the entire address space of the overlay window, and this RDB is continuously selected as a victim. This happens even though other RDBs are available, like a conflict miss in a cache.

On the other hand, the conflict miss seldom happens in the fully-associative mapping policy, because several RDBs are allocated for the address space of the overlay window. This indicates almost no conflict miss during a write operation, and only a capacity miss happens in a fully-associative mapping policy. The performance gaps between two management policies reduces when the unit size of RDB and the number of RDBs are increased, as shown in Fig. 2(b). One 4-KB RDB can hold the entire space of the overlay window in the both policies. We do not explore the victim selection policy of the row buffer management because we found very small variations, even when we changed the victim selection policy to round robin, least recently used (LRU), and other conventional ones. Therefore, simple LRU is used for the remaining experiments unless otherwise stated.

C. Unit size of RDB

We evaluate the effect of the unit size of RDB on the memory access time, as shown in Fig. 3. The plots with

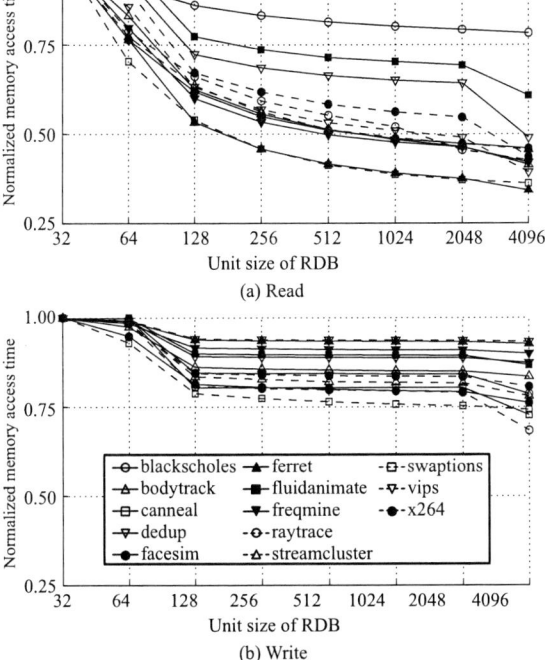

Fig. 3. Memory access time depending on unit size of RDB. The number of RDBs is set to four. The value of each memory access time is normalized to the memory access time of the 4×32-bytes RDBs configuration.

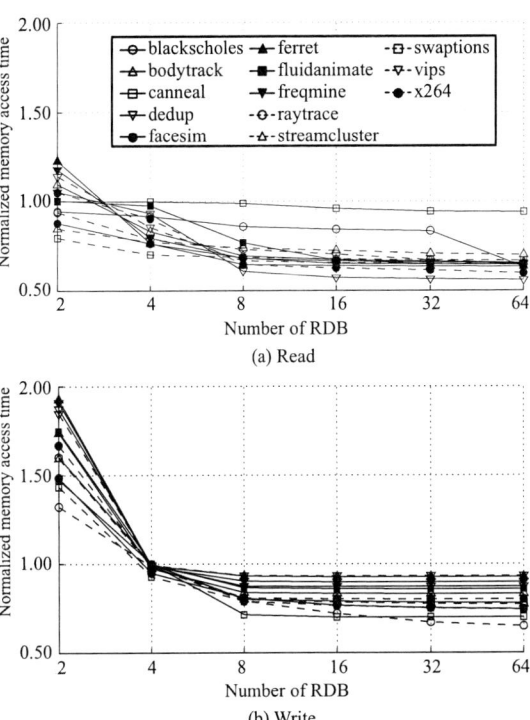

Fig. 4. Memory access time depending on the number of RDBs. The unit size of RDB is fixed to 64 bytes for all configurations. The value of each memory access time is normalized to the memory access time of the 4×32-bytes RDBs configuration.

different markers and lines represent the results from different traces. The unit size of each RDB varies from 32 to 4,096 bytes, while the number of RDBs is 4 for all configurations. We normalize the memory access time of each configuration to the baseline configuration (4×32 bytes RDBs).

The comparison result shows that the effect of the unit size of RDB varies according to the behavior of memory access. For instance, *canneal*: a square with a solid line is almost not affected by the size of the RDB. The reason is because increasing the unit size of RDB does not increase the hit ratio in RDBs, especially for *canneal*, because the access patterns of the *canneal* trace are mostly random. The read performance is more sensitive than the write performance on the unit size of RDB because single write operation requires several consecutive overlay window access regardless of the traces.

Another observation is that the effect of the unit size increment of the RDB becomes limited after 512 bytes in most cases. Even in the most sensitive trace, *raytrace* (a circle with a dashed line), a change of unit size of RDB decreases the memory access time by just 9.3% when we change the unit size from 1,024 bytes to 4,096 bytes. The memory access time decreases significantly when the unit size of RDB changes from 2,048 to 4,096 bytes in some traces. We have noticed that accessing the overlay window causes frequent RDB misses in those traces when the unit size of RDB is less than 4,096 bytes. However, the number of misses decreases when the unit size of RDB is the same as the size of the overlay window. The experimental results imply that there are many possibilities to optimize the row buffer space if we adequately consider the behaviors of the memory access.

D. The number of RDBs

We conduct another design space exploration by changing the number of RDBs as shown in Fig. 4. Instead of increasing the unit size of RDB, we change the number of RDBs from 2 to 64. In this analysis, the unit size of RDB is fixed to 64 bytes for all configurations. We normalize the memory access time of each configuration to the baseline configuration (4×32 bytes RDBs).

As in the case of the results of design space exploration, which varies according to the unit size of RDB, the number of RDBs also affects the memory system performance, depending on the applications' memory access behavior. Among the applications, *dedup, ferret*, and *freqmine* are more sensitive than *blackscholes, facesim, streamcluster*, and *racetrace* applications because their read/write ratios are lower than others. As described already, more write operations occupy more RDBs, which means the RDBs occupied by the read operations are evicted frequently. In our analysis, we have found that a single write operation occupies up to 3 RDBs.

For write operation in all applications, memory access time significantly increases when the number of RDBs are smaller than 4 as shown in Fig. 4(b). It is mainly due to the heavy row buffer misses caused due to the lack of RDBs allocated for the overlay window.

Overall, changing the number of RDBs has relatively less effects on the memory access time compared to changing the unit size of the RDB. We analyze that the number of RDBs are not enough to capture the frequently accessed rows even after increasing the number of RDBs.

978-1-4673-9141-2/15 $31.00 © 2015 IEEE

IV. DISCUSSION

This section mainly discusses the practical design issues based on the results of the design space explorations mentioned in the previous section. We then discuss the design overhead of the row buffer, which is followed by a tradeoff between the design overhead and performance enhancement.

A. Row buffer design overhead

Additional resources are required to efficiently utilize multiple RDBs specified in the LPDDR2-NVM interface. We categorize the design overhead for RDB management as follows:

a) Control logic: Unlike the conventional DRAM interface, the LPDDR2-NVM standard allows a more flexible use of RDBs such as fully-associative row buffer management. However, a more flexible use of RDBs requires a more complex control logic including generation of proper BA signals in the memory controller, whereas the conventional DRAM controller generates BA signals by simply decoding the part of the accessed address. For instance, we need to maintain the access history of RDBs to select a victim RDB in compliance with the replacement policy to support a fully-associative row buffer architecture. It incurs extra area cost and power consumption. The extra area is regarded as cost overhead, even though the extra power consumption can be compensated for by the power savings derived from performance enhancement.

b) Extra area for bigger RDBs: We investigated the effect of different unit sizes of RDBs and the number of RDBs in terms of the PCM performance, as is presented in Section III. Increasing the unit size of RDB or the number of RDBs also requires extra area, and this incurs extra cost for silicon and power consumption. In addition, we may need extra signals for BA when we increase the number of RDBs to more than 8. However, this overhead can be minimized by utilizing the unused CA signals.

B. Tradeoff between design overhead and performance enhancement

We explore the tradeoffs between the design overhead and performance enhancement by taking into account the design overheads as discussed in the previous section. We found that the energy overhead for the control logic and extra bits are almost negligible compared to the energy gain due to the time savings. Therefore, in this section, we focus on the relationship between the area overhead of the silicon and performance enhancement.

We represent the total size of the RDB space in bytes to present the silicon area overhead, where the complexity of the control logic is also expected to be proportional to the unit size of RDB and the number of RDBs in general. Since the silicon area and performance enhancement are not exchangeable, we derive a Pareto optimum at the same area cost (the unit size of RDB and the number of RDBs), which implies the maximum performance gain at a given cost.

Fig. 5 shows variations of the memory access time at different RDB configurations under the same cost for three applications. Due to the lack of space, we only plot three applications, which are the representative applications for random, sequential, and balanced random and sequential behaviors. We

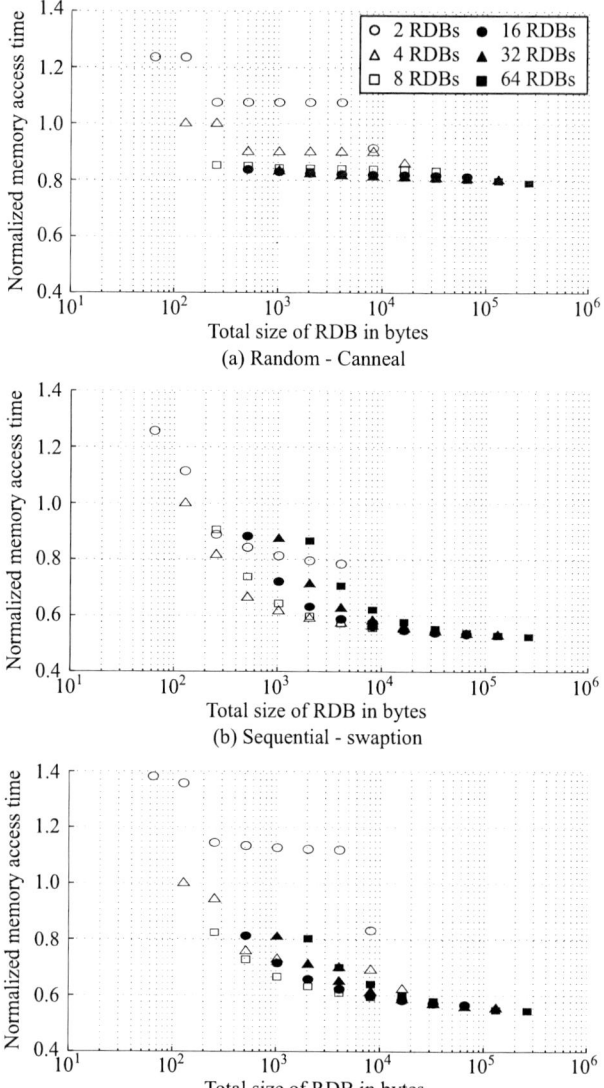

Fig. 5. Normalized memory access time at different RDB configurations.

normalize the memory access time of each benchmark with different unit sizes and number of RDBs to the configuration with 4×32 bytes RDBs. An LRU replacement policy is used due to fully associative policy.

The result presented in Fig. 5 clearly shows Pareto optima in RDB configurations under the same area cost restriction. For instance, at 512 bytes of RDBs, the 16×32 bytes configuration, the 4×128 bytes configuration, and the 8×64 bytes configuration are Pareto optimum for *canneal* (Fig. 5 (a)), *swaption* (Fig. 5 (b)), and *x264* (Fig. 5 (c)), respectively. This analysis clearly shows that the performance of the memory system varies depending on the design of the row buffer architecture, even when the cost is the same. In our evaluation, we observed that a properly designed row buffer architecture enhances memory performance up to 44.2%.

Fig. 6 shows extra gain in performance per additional byte at different traces. The closed inverted triangle in the solid line in Fig. 6 represents the Pareto optima for row buffer configurations at the same area cost. The circle with the dashed line represents extra gain per additional byte of each

978-1-4673-9141-2/15 $31.00 © 2015 IEEE

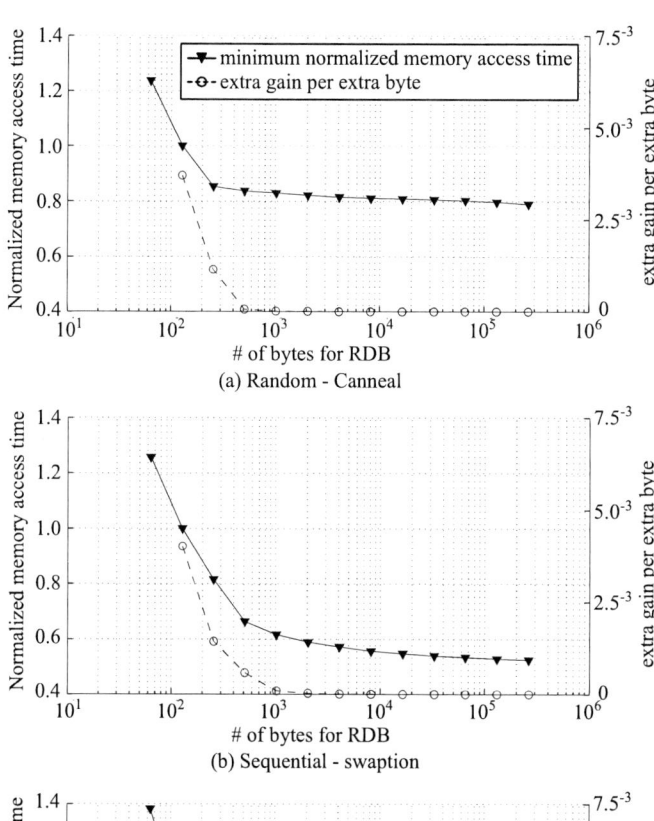

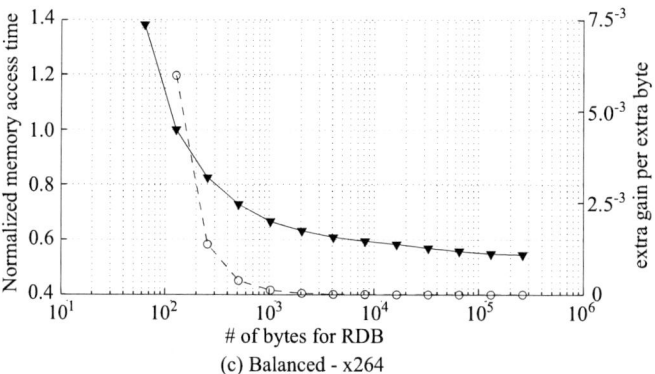

Fig. 6. Optimal memory access time with pareto optimum of RDB configurations under the same area cost and extra gain per extra bits.

total bytes. It shows the point that the additional byte do not significantly affect the performance.

V. CONCLUSIONS

Because the row buffer architecture in the PCM is designed differently than the conventional DRAM interface, we need to devise a different way to utilize the interface efficiently. In this paper, we have conducted an extensive design space exploration to evaluate the effect of the unique features of LPDDR2-NVM interface on the whole system performance. We have obtained several insights about the design of row buffer architecture through this work.

First, the fully-associative row buffer gives better results when compared to the direct-mapped row buffer that has been used in the conventional DRAM interface. The fully-associative one attains 29% performance enhancement on average with a 4×32 bytes row buffer architecture. Next, to varying degrees of memory access traces, changing the unit

size of RDB has more effect on the memory access time than changing the number of RDBs. In general, 4 to 8 RDBs show the best results for benchmark memory access traces with a given size of row buffers. Finally, we observed that a properly-designed row buffer architecture enhances the system-level performance up to 44.2%, even for the same cost.

ACKNOWLEDGMENT

This research was supported by the Basic Science Research Program through the National Research Foundation of Korea (NRF), funded by the Ministry of Education (NRF-2013R1A1A2063350), NRF Grant 2015R1C1A1A02004746, the 2015 Yeungnam University Research Grant, and Brain Korea 21 Plus Project, Department of Computer Engineering, Yeungnam University.

REFERENCES

[1] S. Raoux et al., "Phase-change random access memory: A scalable technology," IBM Journal of Research and Development, vol. 52, no. 4/5, pp. 465–479, 2008.

[2] O. Zilberberg, S. Weiss, and S. Toledo, "Phase-change memory: an architectural perspective," ACM Computing Surveys, vol. 45, no. 3, p. 29, 2013.

[3] B. C. Lee, E. Ipek, O. Mutlu, and D. Burger, "Architecting phase change memory as a scalable dram alternative," in Proceedings of the 36th annual international symposium on computer architecture, 2009, pp. 2–13.

[4] M. K. Qureshi, V. Srinivasan, and J. A. Rivers, "Scalable high performance main memory system using phase-change memory technology," in Proceedings of the 36th annual international symposium on computer architecture, 2009, pp. 24–33.

[5] P. Clarke. (2011, Nov.) Samsung preps 8-Gbit phase-change memory. [Online]. Available: http://www.eetimes.com/document.asp?doc_id=1266495

[6] R. Krishnamurthy. (2013, May) First volume production phase change memory by micron. [Online]. Available: http://www.chipworks.com/en/technical-competitive-analysis/resources/blog/first-volume-production-phase-change-memory-by-micron/

[7] JEDEC, "JESD209-F," 2013.

[8] M. Greenberg. (2009, August) Modifying a dram controller to support lpddr2-nvm. [Online]. Available: http://www.flashmemorysummit.com/English/Collaterals/Proceedings/2009/20090812_S106_Greenberg.pdf

[9] H. Yoon et al., "DynRBLA: A high-performance and energy-efficient row buffer locality-aware caching policy for hybrid memories," SAFARI Technical Report No. 2011-005, 2011.

[10] J. Park, D. Shin, N. Chang, and H. G. Lee, "Accelerating Memory Access with Address Phase Skipping in LPDDR2-NVM," Journal of Semiconductor Technology and Science, vol. 14, no. 6, pp. 741–748, 2014.

[11] Y. Choi et al., "A 20nm 1.8V 8Gb PRAM with 40MB/s program bandwidth," in Proceedings of the 2012 IEEE International Solid-State Circuits Conference Digest of Technical Papers, 2012, pp. 46–48.

[12] Z. Li, R. Zhou, and T. Li, "Exploring high-performance and energy proportional interface for phase change memory systems," in 2013 IEEE 19th International Symposium on High Performance Computer Architecture, 2013, pp. 210–221.

[13] P. S. Magnusson et al., "Simics: A full system simulation platform," Computer, vol. 35, no. 2, pp. 50–58, 2002.

[14] C. Bienia, S. Kumar, J. P. Singh, and K. Li, "The PARSEC benchmark suite: Characterization and architectural implications," in Proceedings of the 17th international conference on Parallel architectures and compilation techniques, 2008, pp. 72–81.

978-1-4673-9141-2/15 $31.00 © 2015 IEEE

Architecture Exploration of 3D FPGA to minimize internal layer connection

Motoki Amagasaki, Yuto Takeuchi, Qian Zhao, Masahiro Iida, Morihiro Kuga, Toshinori Sueyoshi

Graduate School of Science and Technology, Kumamoto University

2-39-1 Kurokami, Chuo-ku, Kumamoto 860-8555, Japan

Email: {amagasaki, iida, kuga, sueyoshi}@cs.kumamoto-u.ac.jp, {takeuchi, cho}@arch.cs.kumamoto-u.ac.jp

Abstract—A three-dimensional (3D) integration based on wafer-to-wafer bonding using through-silicon vias (TSVs) has been developed for the fabrication of new 3D large-scale integrated chips. To balance between cost and performance, and to explore 3D field-programmable gate array (FPGA) with realistic 3D integration processes, we propose spatially distributed and functionally distributed types of 3D FPGA architectures. The functionally distributed architecture consists of two wafers, a logic layer and a routing layer, and is stacked by a face-down process technology. Since vertical wires pass through microbumps, no TSVs are needed. In contrast, the spatially distributed architecture is divided into multiple layers with the same structure, unlike in the functionally distributed type. This architecture can be expanded to more than two layers by stacking multiples of the same die. The goal of this paper is to elucidate the advantages and disadvantages of these two types of 3D FPGAs. According to our evaluation, when only two layers are used, the functionally distributed architecture is more effective. When higher performance is achieved by using more than two layers, the spatially distributed architecture achieves better performance.

I. INTRODUCTION

For the past 40 years, Moore's law has been correct in predicting that the number of transistors that can be placed inexpensively on an integrated circuit (IC) will double approximately every 18 months. With the scaling down of device dimensions on semiconductor wafers, the speeds of integrated circuits have been improved greatly. However, as process shrinking has proceeded into scales much smaller than micrometers, the problems of circuit delay and power leakage have become critical. This is especially true for field-programmable gate arrays (FPGAs), where the routing resources account for the majority of chip area and circuit-delay performance. Employing three-dimensional (3D) integration technologies to vertically stack several silicon dies is considered as one way to solve this problem.

Conventional 3D FPGAs are classified into spatially distributed types and functionally distributed types on the basis of the distribution of die stacking. Spatially distributed 3D FPGAs are realized by stacking a set of similar silicon dies. Such 3D FPGAs employ a number of through-silicon vias (TSVs) in 3D switch boxes (SBs) to ensure high routability [1][2][3][4]. The relation between TSVs and SBs is such that when the channel width (CW) is 50, 100 inter-layer connections will be necessary in each SB. In light of the size of microbumps and TSVs, such architectures will be infeasible to scale down in the near future due to the area overhead. In contrast, functionally

distributed types specialize each layer to one function. Existing FPGAs [5] [6] have a structure in which logic circuits and configuration memory bits are placed on different layers. In this type, circuits on each layer are optimized separately. However, the connections between layers are specialized to each design, and so the generalization of connections is difficult. Therefore, although 3D stacking technology is very attractive, effective 3D FPGA architectures with good cost and performance are yet to be introduced.

The development of computer-aided design (CAD) tools for 3D FPGA is also crucial. The design flow differences between 2D and 3D FPGAs occur in processes after logic clustering. First, logic blocks (LBs) are distributed into several layers. Then, intra- and inter-layer placement and routing are performed. In [7], TPR, an open-source 3D FPGA placement and routing tool, was developed to handle second-step processes. TPR is based on VPR 4.0. This tool was published over a decade ago, however, and therefore does not support some recent FPGA architectures. A design flow to explore minimal vertical connections is proposed in [2], but the details of their novel tools are not mentioned.

To balance cost and performance, and to explore 3D FPGA architectures with realistic 3D LSI processes, we present two facile 3D FPGA architectures to distinguish between the features of functionally distributed and spatially distributed approaches. **(1) Functionally distributed approach**: This FPGA consists of two wafers (a logic layer and a routing layer) and mitigates the side effects of vertical wires[8]. Since vertical wires pass through microbumps by using face-down stacking, no TSVs are needed. By dividing routing resources into two layers, a smaller tile can be achieved. Smaller tile sizes result in shorter routing wires and faster signal transport, which improves routing performance. **(2) Spatially distributed approach**: This FPGA is divided into multiple layers that have the same structure, unlike in the functionally distributed type. This architecture can be expanded to more than two layers easily. To decrease total number of vertical connections, vertical wire between layers is limited to outputs of logic cluster.

In this paper, we developed 3D FPGA computer-aided design (CAD) tools and used these to explore the architectures, comparing the two 3D FPGA architectures in terms of area and delay performance. The goal of this paper is to elucidate the advantages and disadvantages of these two types of 3D FPGAs. The rest of the paper is organized as follows. Section 2 discusses the proposed 3D routing architecture. Section 3 describes the design flow and CAD tools. Section 4 describes

978-1-4673-9141-2/15 $31.00 © 2015 IEEE

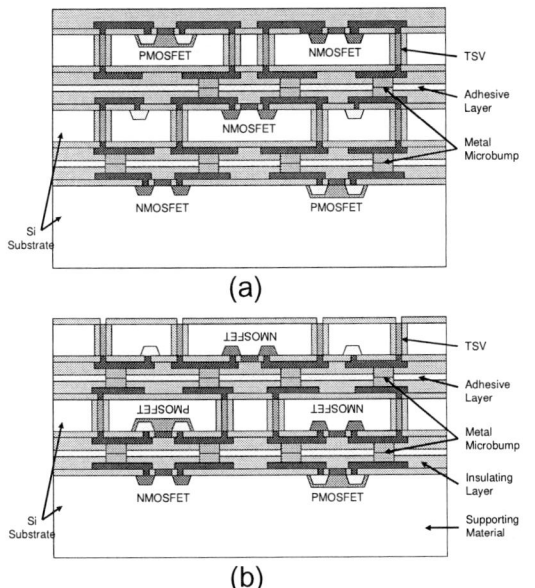

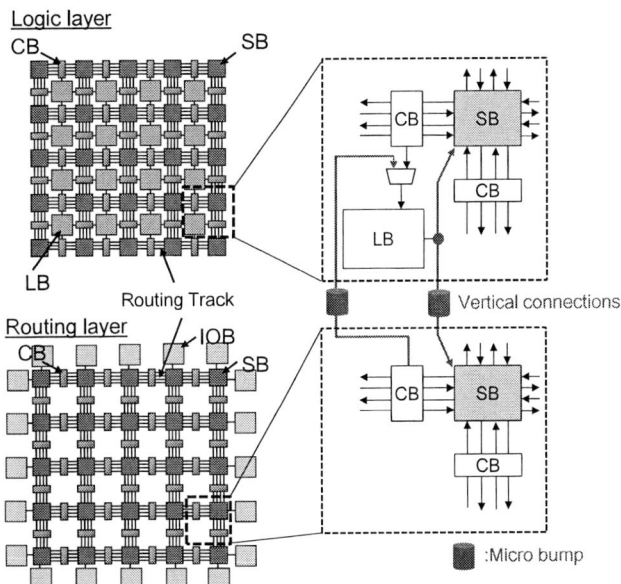

Fig. 1. Cross-sectional structure of 3D LSI architecture: (a) face-up stacking; (b) face-down stacking.

Fig. 3. Functionally distributed 3D architecture with face-down stacking.

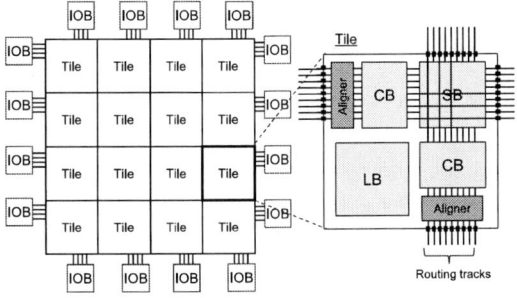

Fig. 2. Completely homogeneous routing architecture.

the evaluation process and results. Conclusions and directions for future research are given in Section 5.

II. PROPOSED 3D FPGAs

A. 3D Integration and Basic Tile Structure

Figure 1 shows the cross-sectional structure of a 3D LSI circuit fabricated by Koyanagi's 3D integration technology [9]. The thinned upper layers are stacked face-up onto a thick LSI wafer that is face-up (Fig. 1(a)). Figure 1(b) shows the arrangement when the thinned upper layers are stacked face-down onto the thick LSI wafer. If the number of layers is limited to two, no TSVs are needed.

It is important to balance cost and performance when deciding on a 3D FPGA architecture. To this end, we consider the homogeneous (uniform) tile structure[10]. To treat various array sizes in a similar manner, it is important to simplify the tile structure. In this paper, the proposed 3D FPGAs are based on a homogeneous tile structure (Fig. 2). Most FPGAs use complex structures to achieve high programmability, but such structures complicate design. We have been studying simple and regular FPGA structures [10] like that shown in Fig. 2 with the aim of increasing programmability. Unlike traditional

island-style FPGA architectures, in this type of architecture all tiles have the same structure, which is composed of several tile types. We also implement aligners to simplify the connections of wire segments.

Figure 2 shows a diagram of the homogeneous FPGA architecture. The most distinctive feature of this architecture is that it is composed of only a single type of tile and input/output block (IOB). In other words, all switch boxes (SBs) and circuit blocks (CBs) have the same structure, unlike the traditional island-style architecture. To simplify the task of designing circuits, the connections between tiles are kept simple, and are formed by only the routing track.

B. Functionally Distributed 3D FPGA Architecture

Figure 3 shows details of the structures in the functionally distributed 3D FPGA. This architecture has been enhanced previously proposed[8]. There are two layers in the architecture examined in the current research: a logic layer and a routing layer. The tiles on the logic layer have an LB and a small part of the routing resources; the tiles on the routing layer have only routing resources. The difference between conventional and proposed 3D SBs is that the 3D connections are made at the LB input and output pins. The number of inter-layer connections within one tile is equal to the total number of LB input and output pins. The number of LB inputs I is determined by the following formula [11].

$$I = \frac{K}{2} \times (N + 1) \qquad (1)$$

Here, K is the number of logic cell inputs, and N is the cluster size. The number of vertical wires per one tile are $I + N$. For example, when $K = 6$ and $N = 4$, the numbers of LB inputs and outputs are 15 and 4, respectively. Therefore, the number of vertical connections per tile is 19.

We next discuss a method for determining the minimum width of the routing track channel for the two layers. To find

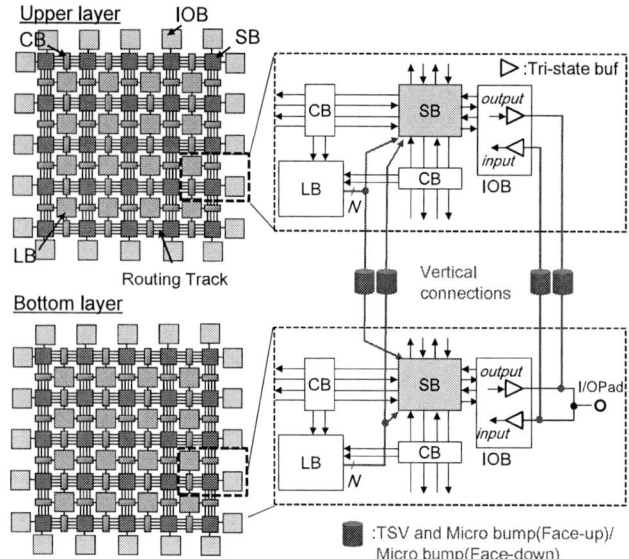

Fig. 4. Proposed 3D routing structure.

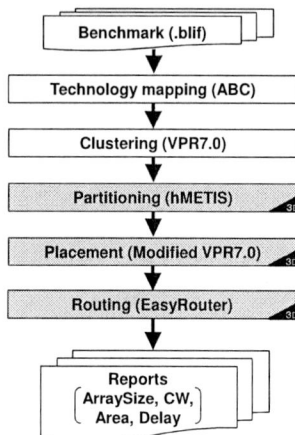

Fig. 5. Design flow of proposed 3D FPGA.

this, we first set the initial CW of the logic layer to 1.5 times the number of LB input pins. Then, the areas used by the small part of connection blocks (CBs) and SB on logic layer can be calculated as the routing area of the logic layer; the tile area of the logic layer is the sum of the logic and small routing resource to connect neighbor LBs. The CW of the routing layer is calculated by allocating the size of the routing area as the size of the logic layer tile area. We next perform routing. If the routing is successful, then the next trail of the logic layer will have its CW set to half the current one. If routing fails, the CW of the next trail of the logic layer will be set to twice the current one. This process is repeated until the minimum CW that can lead to successful routing is found.

By dividing routing resources into two layers, we can achieve a smaller tile. Smaller tiles allow shorter routing wires and thereby enable faster signal transport, which improves the routing performance. The router can choose a network route on the logic layer or the routing layer. Although the number of layers in this type of approach is limited to two, when face-down stacking is used, no TSVs are necessary.

C. Spatially Distributed 3D FPGA Architecture

Figure 4 shows details of the structures in the spatially distributed 3D FPGA. There are two layers in the architecture examined in this example: an upper layer and a lower layer. Both layers have the same structure, which allows using a uniform mask set. The IOBs are connected to the two layers by vertical wires. One difference between this structure and that in our functionally distributed 3D FPGA (Fig. 3) is that vertical wires for LB inputs have been eliminated. The number of inter-layer connections within each tile is equal to twice the number of LB output pins. The total number of vertical wire connections, T_{VC}, is determined by the following formula.

$$T_{VC} = (Arraysize)^2 \times 2N + 4 \cdot Arraysize \times 2C_{IO} \quad (2)$$

In this formula, $Arraysize$ is the side length of the FPGA array, N is the cluster size, and C_{IO} is the IO capacity.

Compared with the functionally distributed architecture, which requires $\frac{LUTinputs}{2} \times (N+1) + N$ vertical connections per tile, this architecture reduces the required number of inter-layer connections. In addition, this architecture can be expanded to more than two layers by stacking dies of the same type. When the number of layers are 3 and 4, the maximum number of TSVs per one tile are $4N$ and $8N$, respectively.

III. DESIGN FLOW AND CAD TOOLS

The design of functionally distributed architectures can use the same CAD tools as are used for the design of traditional 2D FPGAs. In this section, we introduce a design flow and CAD tools that can be used for designing spatially distributed architectures. We developed the 3D FPGA design flow (Fig. 5) by using VTR 7.0 [12], which is the most current version as of this writing. As is done with 2D FPGAs, the circuit (in 'BLIF' format) is first technology-mapped with ABC [13] and then clustered with AAPack, which is included as part of VPR 7.0.

Next, the clustered netlist is partitioned by using hMETIS [14], which can efficiently group LBs into a specified number of layers of similar size and with a minimal number of interconnections. These constraints are important for 3D FPGA partitioning. Allocating a similar number of LBs on all layers ensures that the resources on each layer can be fully used. Simultaneously, in order to minimize the overhead from TSVs, partitions with fewer interconnections are preferable. We wrote a script that can generate hypergraphs of LBs from the clustered netlist. The hMETIS tool is used to perform partitioning on the hypergraph generated by the script. Finally, layer allocation information is added back to the clustered netlist as LB attributes.

We created a 3D placement tool that uses VPR 7.0 placer as a base. The tool operates as follows. First, the layer allocation information is read from the clustered and partitioned netlist. Then, a conventional placement by simulated annealing is performed. During the placement process, logic modules are freely swapped within each layer. The algorithm used is the bounding box(BB) method, which focuses on minimizing the bounding-box wire length of the circuit(Fig. 6). In order to

978-1-4673-9141-2/15 $31.00 © 2015 IEEE

TABLE I.   Target architectures parameters.

Name	Value	Name	Value
LUT size	6	IO capacity	10
Cluster size	8	TSV area	$96\mu m^2$
Fc_{in}	0.5	TSV delay	2.2 ps

evaluate the BBcost of a net across multiple layers, we calculate bb_x (BB distance on x direction) and bb_y (BB distance on y direction) of each layer, and then use the maximum bb_x and bb_y of all layers in the final BBcost. The vertical connection cost is not considered yet in the placement for two reasons: (1) A typical 3D-FPGA CAD flow processes the vertical allocation of logic modules in the partitioning step. During partition step, inter-layer-connections are minimized with algorithms like hMETIS. For the placement step, logic modules are placed within each layers, therefore, vertical connection cost is not necessary to be considered. (2) When the vertical connection cost is comparable with horizontal connection cost, we have to build a cost model for the partitioning algorithm. However, in this approach, the delay of the TSV is small compare to horizontal wire delay, and the number of layers is small, the vertical connection cost is negligible during partitioning and placement steps. However, the final delay derived from the router does certainly include the TSV delay. In order to handle 3D-FPGAs with different 3D-VLSI processes and more layers, we will improve 3D partitioning that considering vertical cost in the future work. We also plan to implement timing-driven algorithms.

Finally, routing was performed with our novel tool, the EasyRouter [15]. EasyRouter implements a pathfinder routing algorithm similar to that in VPR; however, EasyRouter simplifies the implementation of new FPGA architectures with various routing topologies. In addition, EasyRouter combined with VLSI CAD can provide highly accurate reports on area and critical path delays for FPGA designs that are based on standard cells.

Routing is performed in two main steps. First, the router explores the minimum channel-width for each circuit. Next, the CW is fixed at 1.2 times the CW of the circuit with the highest minimum CW (i.e., 1.2-fold the maximum width), and the results are evaluated. This method ensures that all circuits are fairly evaluated on the same device with sufficient resources.

IV. Evaluation

This section compares the architectures of a functionally distributed 3D FPGA (type 1) with that of the proposed spatially distributed 3D FPGA (type 2), evaluating the area, the critical path delay, and the area delay product. An island-style 2D FPGA(2D_Island) and 2D homogeneous FPGA(2D_Homo) are used as the baseline for evaluations. An analysis of the proposed 3D FPGAs from evaluation results is also given in this section.

A. Evaluation Conditions

The parameter values for the target architectures are listed in Table I. All implemented 3D FPGA architectures are homogeneous FPGAs with a lookup-table (LUT) size of 6 and a cluster size of 8. For the routing architecture, we used a

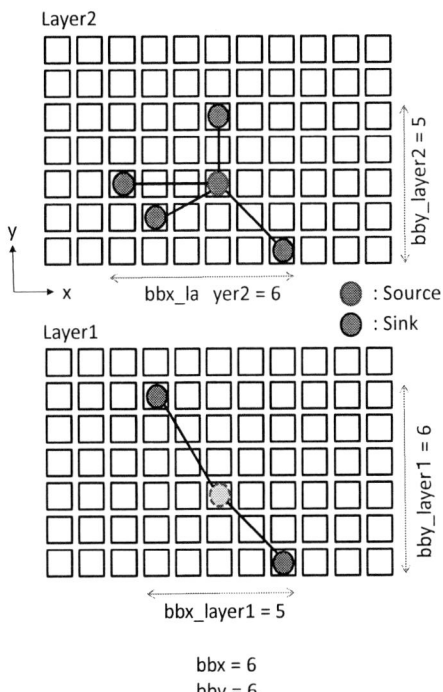

Fig. 6.   Calculation of BBcost.

Wilton-type SB with $Fs = 3$. The Fc_{in} parameter was set to 0.5, which means that half of the tracks in the routing channel are connected to an LB input through a CB. We set the area of each TSV to $96 \mu m^2$[1] and the delay to 2.2 ps, taking these values from the report in [9]. The type 1 3D FPGA has only two layers, which is denoted by type 1_L2. The type 2 3D FPGAs were implemented with from 2 to 4 layers, and these are denoted by type 2_L2, type 2_L3 and type 2_L4 for 2, 3, and 4 layers, respectively. The face-up stacked architectures are marked as "(face-up)", and the face-down stacked architectures are noted as "(face-down)".

The 20 largest MCNC benchmarks were used as the evaluation test suite. The device was designed to use 65-nm CMOS technology. The tile was synthesized with a Synopsys Design Compiler F-2011.09-SP2. For the area calculation, tile areas of all architectures were from synthesized results. The total area was calculated by multiplying the number of tiles with the area of one tile. For the delay evaluation, all CBs and SBs are considered to be composed of 2-to-1 multiplexers (MUXes) for both area and delay. The MUXes' physical parameters are incorporated by referenced to the standard cell library.

B. Preparation: Partitioning Results

The partitioning is performed by hMETIS with UBfactor set at 0.01, which allows an imbalance of up to 1% between the partitions in order to minimize inter-layer connections. All circuits are partitioned into layers of similar sizes. Because it is necessary to set the array size of all FPGA layers to the size of the largest partition, a good partitioning balance ensures that the resources on each layer can be fully used.

---

[1] In [9], four poly-Si TSVs with a size of $2 \times 12 \ \mu m^2$ are connected in parallel in one vertical interconnection.

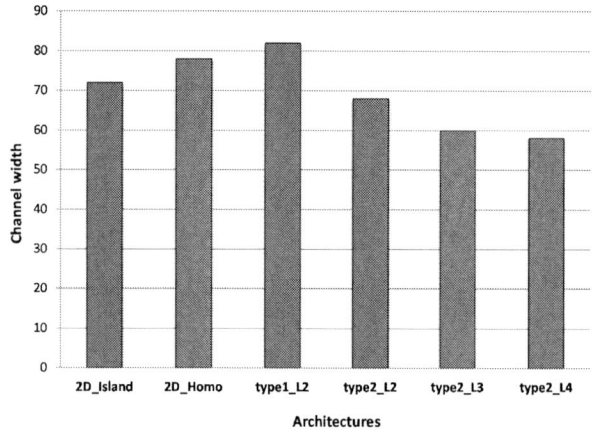

Fig. 7.   Channel width.

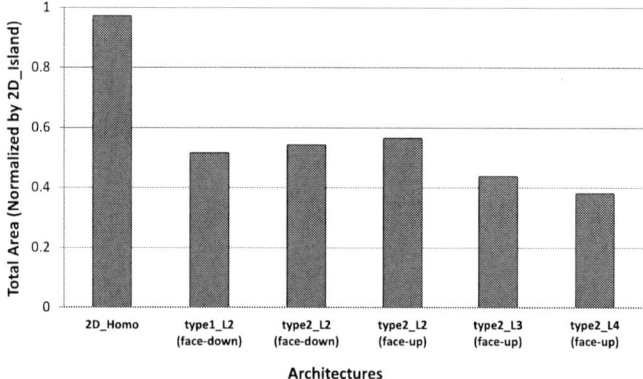

Fig. 8.   Total area per layer.

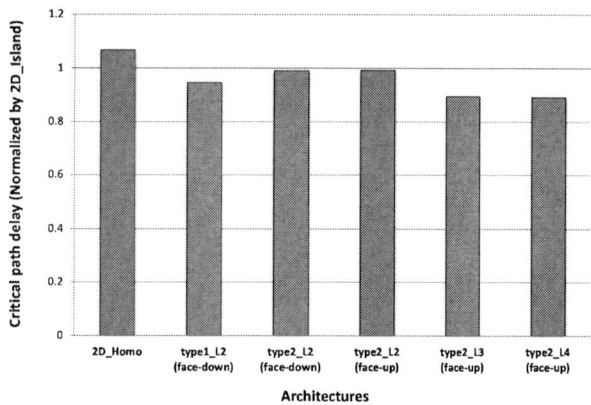

Fig. 9.   Critical path delay.

C. CW Exploration Results

Figure 7 shows the results of CW exploration for each architecture for the 2D and 3D FPGAs. According to the CW exploration results, the fixed CW result for each architecture was taken as 1.2 times the largest of explored CW of all benchmarks.

As reported in [10], the 2D_Homo layout uses slightly more channels than the 2D_Island layout. The CW of the routing layer of the type 1 3D FPGA was determined by examining the size of the logic layer tile. A CW of 82 (the total CW from the logic and routing layers) was the minimum CW of this architecture, which was sufficient to implement all circuits for this evaluation. The type 2 architectures permit narrower CWs as the number of layers increases. This is because the 3D allocation of LBs brings some logics closer together than in the 2D allocation, which reduces routing congestion. The type 2 3D FPGAs with 2, 3, and 4 layers have, respectively, 5.6%, 16.7%, and 19.4% narrower CWs, on average, than the 2D_Island does.

D. Area Results

The results of comparing chip areas between type 1 and type 2 3D FPGAs are shown in Fig. 8. We first implemented all benchmarks on each architecture and normalized the area results by the area of the 2D_Island FPGA. All results shown

in Fig. 8 are an average of the normalized results for the corresponding architecture. We evaluated face-down and face-up 3D stacking methods for 3D FPGAs. The face-up counts include TSV area in the total area, and the face-down counts do not (more specifically, the area used by TSVs is 0 for face-down stacking).

First, we compare two-layer 3D FPGAs. The face-down stacked type 1_L2 (face-down) and type 2_L2 (face-down) reduce area by about 48.2% and 45.6% from the area of 2D_Island. In contrast, the reduction from the face-up stacked type 2_L2(face-up) is about 43.4%. Next, we compare type 2 3D FPGAs with different numbers of layers. We can see the trend of area reduction from the results of type 2 3D FPGAs having from 2 to 4 layers. Relative to the area of 2D_Island, the type 2_L2 (face-up), type 2_L3 (face-up), and type 2_L4 (face-up) designs reduce area by 43.4%, 56.1%, and 61.7%, respectively. The type 2_L4 (face-up) design offers the best performance of all examined architectures.

To summarize, for a two-layer 3D FPGA, type 1 with face-down stacking has the best performance on area. However, when implementing 3D FPGAs with more than two layers, where face-up stacking is necessary, the type 2 architecture offers a much smaller area.

E. Delay Results

The results of testing the type 1 and type 2 architectures for critical path delay are shown in Fig. 9, where the delays are normalized by the critical path delay of the 2D_Island architecture. Delay results are not significantly different between (all are within 10% of one another). This is because the MCNC benchmark circuits are not very large, and so the critical path delay is mainly from the LB rather than from routing. However, the results still show some trends. The type 1_L2 (face-down) has 5.3% better delay performance than the 2D_Island FPGA. The improvement is a result of the type 1 3D FPGA having more routing channels on the routing layer. When comparing the type 2_L2 (face-down) and type 2_L2 (face-up) designs, we can see that the TSV overhead in the critical path delay is very small. The TSV delay overhead is caused mainly by increased routing-wire delay due to the larger tile sizes and the delays of the TSVs themselves.

For the type 2 3D FPGA, the critical path delay was

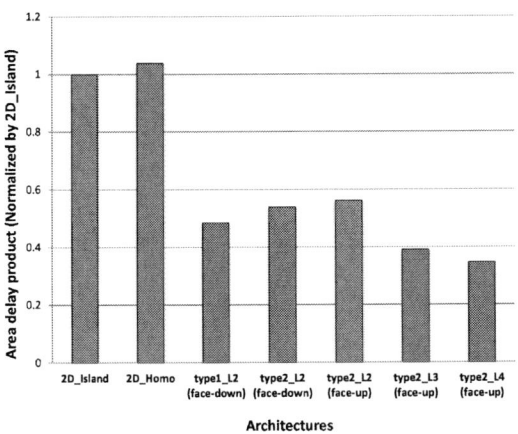

Fig. 10.   Area delay product.

mainly affected by two factors. More LBs are allocated closer together as more layers are stacked, which improves the delay. Opposing this, the delay from MUXes of the SBs increase when stacking more layers. In the proposed type 2 3D FPGA, all LB outputs are connected to an SB of tiles at the same coordinate across all layers. When the number of MUX inputs per SB is m, each SB will have $(m-1)$ 2-to-1 MUXes with logic depth $log_2(m)$. As a result, each type 2 SB MUX has more levels in its logic depth, which lengthens the total routing path. The performance of the type 2_L3 (face-up) architecture is similar to that of the type 2_L4 (face-up) architecture, which is because the increased SB MUX depth offsets the advantages of 3D allocation.

Additionally, partitioning and placement algorithms affect delay performance. In future work, we plan to develop a timing-driven partitioning and placement algorithm for the proposed 3D FPGA.

F. Area Delay Product Results

Finally, the area delay product results are shown in Fig. 10. For the face-down stacking method, the type 1_L2 (face-down) design performs 51.5% better than 2D_Island. For the face-up stacking method, the 4-layer type 2_L4 (face-up) performs 65.2% better than 2D_Island. When limiting the analysis to two layers, the functionally distributed type of architecture (type 1) is most effective. However, if we prioritize performance and use more than two layers, the spatially distributed architecture (type 2) offers better performance.

V. CONCLUSION

In this paper we proposed a functionally distributed type and a spatially distributed type of 3D FPGA architecture to allow simple die stacking. According to the evaluation results, the functionally distributed type is more effective when limiting designs to two layers, but the spatially distributed architecture with more than two layers is a better choice when performance is prioritized. In this research, we used relatively large TSVs. However, since more smaller TSVs are been developed in recent researches, functionally distributed type is more effective using these smaller TSVs. In future work, we are planning to stack multiple functionally distributed FPGAs

with the face-up method and propose inter-layer high-speed communications that use TSV serial connections [16]. We also intend to improve the CAD toolsets to support algorithms that consider power consumption and timing in order to achieve better power and delay performance. At the same time, to show the effectiveness of 3D technology, it is necessary to evaluate using more larger circuits, such as VTR benchmark sets[12].

REFERENCES

[1]   M. J. Alexander, J. P. Cohoon, J. L. Colflesh, J. Karro, G. Robins, and C. Science, "Three-Dimensional Field-Programmable Gate Arrays," Proc. of the Eighth Annual IEEE International ASIC Conference and Exhibit, pp.253-256, Sept. 1995.

[2]   A. Gayasen, V. Narayanan, M. Kandemir, and A. Rahman, "Designing a 3-D FPGA: Switch Box Architecture and Thermal Issues," IEEE Trans. on VLSI Systems, Vol.16, Issue.7, pp.882-893, July 2008.

[3]   K. Siozios, A. Barzas and D. Soudris, "Architecture-Level Exploration of Alternative Interconnection Schemes Targeting 3D FPGAs: A Software Supported Methodology," International Journal of Reconfigurable Computing, Vol. 2008, Article ID 764942, 18 pages, 2008

[4]   S. A. Razavi, M. S. Zamani and K. Bazargan, "A Tileable Switch Module Architecture for Homogeneous 3D FPGAs," Proc. of IEEE International Conference on 3D System Integration(3DIC), pp.1-4, Sep. 2009.

[5]   T. Naoto, T. Ishida, T. Onoduka, M. Nishigoori, T. Nakayama, Y. Ueno, Y. Ishimoto, A. Suzuki, W. Chung, R. Madurawe, S. Wu, S. Ikeda and H. Oyamatsu, "World's first monolithic 3D-FPGA with TFT SRAM over 90nm 9 layer Cu CMOS," Proc. of Symposium on VLSI Technology(VLSIT), pp.219-220, June 2010.

[6]   M. Lin, A. E. Gamal, Y. Lu and S. Wong, "Performance Benefits of Monolithically Stacked 3D-FPGA," IEEE Trans. on CAD of Integrated Circuits and Systems, Vol.26, Issue.2, pp.216-229, Feb. 2007.

[7]   C. Ababei, H. Mogal, and K. Bazargan, "Three-dimensional Place and Route for FPGAs," IEEE Trans. on CAD of Integrated Circuits and Systems, Vol.25, Issue 6, pp.1132-1140, June 2006.

[8]   T. Hamada, Q. Zhao, M. Amagasaki, M. Iida, M. Kuga and T. Sueyoshi, 'Three-Dimensional Stacking FPGA Architecture Using Face-to-Face Integration," Proc. of International Conference on Very Large Scale Integration (VLSI-SoC) , pp.196-201, Oct. 2013.

[9]   M. Koyanagi, T. Fukushima, and T. Tanaka, "High-Density Through Silicon Vias for 3-D LSIs," IEEE Journals & Magazines, Vol. 97, Issue 1, pp.49-59, Jan. 2009.

[10] K. Inoue, H. Yosho, M. Amagasaki, M. Iida, and T. Sueyoshi, "An Easily Testable Routing Architecture And Effecient Test Technique," Proc. of International Conference on Field-programmable Logic and Applications (FPL) , pp.291-294, Aug. 2011.

[11] E. Ahmed and J. Rose, "The Effect of LUT and Cluster Size on Deep-Submicron FPGA Performance and Density," Proc. of International Symposium on Field-Programmable Gate Arrays(FPGA), pp.3-12, Feb. 2000.

[12] J. Luu, J. Goeders, M. Wainberg, A. Somerville, T. Yu, K. Nasartschuk, M. Nasr, S. Wang, T. Liu, N. Ahmed, K. B. Kent, J. Anderson, J. Rose and V. Betz, "VTR 7.0: Next Generation Architecture and CAD System for FPGAs," ACM Transactions on Reconfigurable Technology and Systems (TRETS), Vol. 7, No. 2, pp. 6:1 - 6:30, June 2014.

[13] A. Mishchenko et al., "ABC: A System for Sequential Synthesis and Verification," http://www.eecs.berkeley.edu/ alanmi/abc/, 2009.

[14] N. Selvakkumaran and G. Karypis, "Multi-Objective Hypergraph Partitioning Algorithms for Cut and Maximum Subdomain Degree Minimization," IEEE Trans. on CAD of Integrated Circuits and Systems, Vol.25, Issue 3, pp.504-517, Feb. 2006.

[15] Q. Zhao, K. Inoue, M. Amagasaki, M. Iida, and T. Sueyoshi, "FPGA Design Framework Combined with Commercial VLSI CAD," IEICE Trans. on Information and Systems, Vol.E96-D, No.8, pp.1602-1612, Aug. 2013.

[16] T.Kajiwara, Q. Zhao, M. Amagasaki, M. Iida, M. Kuga and T. Sueyoshi, "A Novel Three-Dimensional FPGA Architecture with High-speed Serial Communication Links," Proc. of nternational Conference on Field Programmable Technology(ICFPT), Dec. 2015.

978-1-4673-9141-2/15 $31.00 © 2015 IEEE

Compact Interconnect Approach for Networks of Neural Cliques Using 3D Technology

Bartosz Boguslawski*, Hossam Sarhan*, Frédéric Heitzmann*, Fabrice Seguin[†], Sebastien Thuries*,
Olivier Billoint* and Fabien Clermidy*
*Univ. Grenoble Alpes, F-38000 Grenoble, France.
CEA, LETI, MINATEC Campus, F-38054 Grenoble, France
Email: {name.surname}@cea.fr
[†]TELECOM Bretagne, Electronics Department, 29238 Brest, France
Email: fabrice.seguin@telecom-bretagne.eu

Abstract—Thanks to their brain-like properties, neural networks outperform traditional algorithms in certain group of applications. However, since they are wire-dominated systems, their hardware implementation poses numerous challenges as high latency and energy consumption. The recent technological improvements allow for stacking few dies one on another and designing 3D electronic circuits. This creates opportunities for 3D efficient implementations of neural networks targeting high-performance applications. This work explores the gains of 3D technology for neural networks relying on neural cliques. A general study shows up to 55% reduction in terms of total interconnect length and interconnect power consumption, and 74% reduction of the maximal interconnect delay. The proposed approach is validated with a power management applicative test-case. We demonstrate that, in this scenario, the 3D architecture reduces interconnect length and power by 35% and the maximal delay by 57%, compared to 2D.

I. INTRODUCTION

The brain's capabilities of learning, processing a large amount of information, and taking decision have always interested scientists from various domains. More specifically, engineers are interested in the brain because of its a) distributed structure enabling massive parallelism [1], b) capability to deal with unreliability and uncertainty [1], c) energy efficiency [2] and d) the ease of solving associative tasks so difficult to solve with traditional algorithmic approaches [3]. Furthermore, the lack of distinction between calculation and memory blocks inspires scientists searching for new, different from Von Neumann, architectures [3]. However, the aforementioned characteristics come with a huge number of neurons (elementary processing units) and synapses (connections allowing communication between the neurons). In fact, the capabilities of brain are the sum of simple, but numerous, operators.

Due to the number of elements that are interconnected, neural networks are wire-dominated systems, *i.e.* this entails challenges in hardware implementation - high latency, energy consumption and large memory requirements among others. One way to overcome these issues is to use a shared, multiplexed communication medium as bus or Network on Chip (NoC). Nevertheless, this approach comes with performance reduction and strongly limits the application field. To obtain high-performance neural networks, physical connections are needed for all the synapses. That is why first neural networks in 3D technology are arising. 3D technology relies on stacking few dies one on another and communicating between them over Through-Silicon-VIAs (TSVs). This allows to overcome the limitations of 2D circuits and provide compact connections for all the synapses. In [4] a 3D Spiking Neural Network (SNN) based accelerator is proposed. The authors report 52% energy savings and 64% bandwidth improvement. The authors of [5] study a two-layer neural network for objects recognition in a video stream and demonstrate that the total connections' length is reduced three times. The work [6] shows that 3D technology can also be used to design spiking neurons (TSV-neuron) by exploiting the TSV's capacitance.

Recently Gripon and Berrou proposed a new family of neural networks [7]. It relies on a neural clique which is an assembly of neurons representing an information stored in the neural network. The elementary part of the information is called message and is associated with the clique. The network is able to store a large number of cliques-messages and can be used either to check if a given information is known by the network (an exemplary application is an intrusion detection system) or an associative memory. The principle of the associative memory is that the retrieval of the message from the memory is accomplished presenting a part (possibly small and even partly incorrect) of it to the memory. Then, the memory outputs the remaining part of the message. Associative memories are used for example in processing units' caches [8] or routers [9]. This type of neural networks showed a huge gain in performance compared to state-of-the-art neural networks [7].

The clique that represents a piece of information stored in the network contains more information than the necessary minimum [10]. That is why it allows the retrieval based on partial and/or noisy information. Consequently, the hardware implementation is wire-dominated. The long wires impact the energy consumption and time response since all the neurons in the clique have to exchange some signals between them. For that reason, it is interesting to organize the neurons in 3D so that they create 3D cliques with shorter connections, and therefore lower energy consumption and time response. By this time, 3D technology has been analyzed for neural cliques in [11]. Contrary to this work, [11] is a digital NoC-based prototype architecture synthesized on an FPGA platform where the use of 3D NoC is analyzed. Even if [11] targets different applications from this work, the authors report that the 3D technology offers the best performances for their architecture

978-1-4673-9141-2/15 $31.00 © 2015 IEEE

which is an additional motivation for the current work.

The paper is organized as follows. Section II outlines the theory and implementation of the considered neural networks and introduces the 3D technology. Section III explains how the 3D technology is used to create 3D neural cliques. Later, the model used for the simulations is outlined. Section IV gives simulation results for a general study. Finally, Section V gives simulation results for a practical application.

II. BACKGROUND

A. Networks of neural cliques

1) Message definition: Throughout this work, it is considered that associative memories store *messages* m that they are later capable of retrieving given a sufficiently large part of their content. In order to ease the readability of this document, and without loss of generality, it is considered that a message consists of C sub-messages or segments. Each segment is a value in range from 0 to $\ell - 1$. An exemplary message, for $C = 4$ and $\ell = 4$, is $m = \{2, 0, 2, 3\}$.

2) Network structure: In order to store messages, a network that consists of binary neurons called *fanals* and binary connections is used. The authors of [7], use the term *fanal* (which means lantern or beacon) instead of neuron for two reasons: a) at a given moment, in normal conditions, only one fanal within a group of them can be active and b) for biological inspirations, fanals do not represent neurons but microcolumns [12]. Fig. 1a represents the general structure of the network and the notation. All of the n fanals are organized in C disjoint groups called *clusters*. Fanals belonging to specific clusters are represented with different shapes. Each cluster groups $\ell = n/C$ fanals. Note that the number of clusters is equal to the number of segments and the number of fanals within each cluster is equal to the number of values possible on each segment. The connections (synapses) are allowed only between fanals belonging to different clusters. Contrary to classical neural networks the connections do not have weights, the connection exists or not. Hence, the weight (or adjacency) matrix of such a network consists of values $\{0, 1\}$ where 1 indicates the connection between two fanals, and 0 the lack of connection. In this paper, the connection w between two fanals is identified by their coordinates (i, y) and (i', y') where the first one is the row number and the second one the column number of the fanal on an xy plane. Fig. 1b shows an example of a network with $C = 4$ and $\ell = 4$.

3) Message storing and retrieval procedures: To store a message m in the network, each of its C segments is associated with a distinct cluster, and more precisely with a unique fanal in its cluster (the one which index corresponds to the value of the segment). Then, this subset of fanals is fully interconnected forming a *clique* representing the message in the network. This term is also used in neurobiology to describe such groupings of neurons [13]. When a new message shares the same connection as an already stored message, this connection remains unchanged. Therefore, the result of the storing procedure is independent of the order in which messages are presented to the network [7]. In Fig. 1b, message $m = \{2, 0, 2, 3\}$ is stored in the network.

As a result of the strong correlation brought by the connections of the clique embodying a message in the network, it

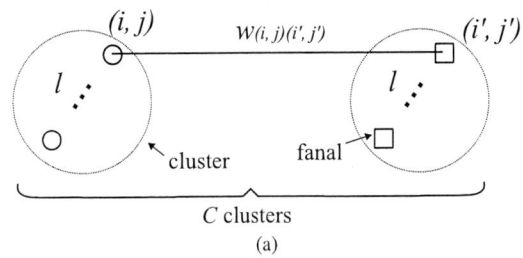

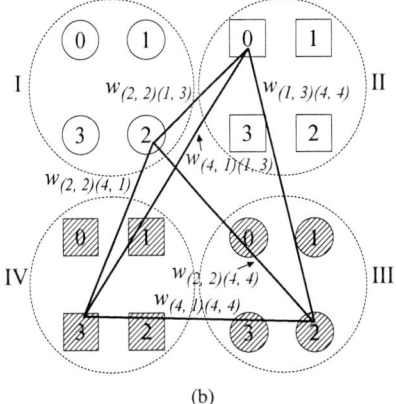

Fig. 1. (a) The network general structure and notation. Different shapes (circles, squares) represent fanals belonging to different clusters. (b) Exemplary network with $C = 4$, $\ell = 4$ and message $m = \{2, 0, 2, 3\}$ stored. The numbers inside each fanal correspond to its index.

is possible to retrieve the stored message based on partial or noisy information put into the network. The retrieval procedure is described in detail in [14]. From the interconnect point of view it is important that during the retrieval all the fanals belonging to the clique have to exchange some signals and the performance of the whole system is limited by the longest wire.

4) Physical implementation: A high-peformance implementation of neural cliques with physical connections for all the synapses has been first proposed in [15]. Because of their characteristics these circuits were used to target first applications [16][17]. The network circuit is built of two types of blocks: the fanal circuit and the synapse circuit. The functioning of these circuits is described in detail in [15]. The signals between the clusters are exchanged thanks to the synapses. Consequently, each fanal circuit has a number of synapse circuits attached to itself. However, an important wiring is needed to provide the communication between all these circuits. In this work a compact 3D interconnect to provide a communication structure for these circuits is studied.

B. 3D technology

As technology node advances, the impact of wiring interconnects parasitics, *i.e.* resistance and capacitance increases. 3D stacking technology affords an effective technique to reduce the wiring length. 3D technology varies according to the fabrication techniques. In this work we will focus on 3D technology using TSVs as vertical interconnections, however the proposed approach is valid using other 3D technologies.

3D TSVs are used to interconnect between stacked dies. The pitch value of TSVs varies depending on the fabrication

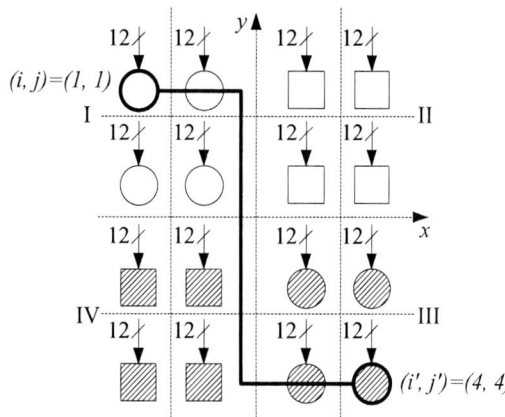

Fig. 2. 2D network example. Different shapes (circles, squares) represent fanals belonging to different clusters. Each fanal has 12 synapses to connect to other fanals. The thick line represents the wire necessary to connect two fanals.

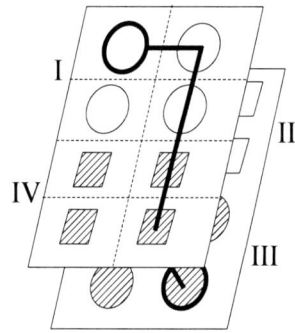

Fig. 3. 3D network (folded network from Fig. 2). Different shapes (circles, squares) represent fanals belonging to different clusters. The thick line represents the wire necessary to connect two fanals.

process. Reference [18] shows a TSV pitch of $1.2\mu m$. The main advantage of 3D technology is to replace the long horizontal wires, which introduce large parasitics, by such vertical connections, *i.e.* TSVs with lower parasitics. Proper partitioning is needed to achieve such goal. In the next sections, we will show how to partition a 2D neural clique network to create a 3D one, then we will introduce the methodology used to evaluate such 3D neural cliques.

III. 3D NEURAL CLIQUES

A. Methodology

Since the clique is created by interconnecting the fanals from distinct clusters, the wires span all over the network. Moreover, the performance of the system depends on the longest wire in the clique since all the fanals activated in the retrieval process exchange the signals through their connections. Therefore, in such a wire-dominated structure, it is beneficial to reduce the length of the connections in the clique to reduce the delays and the energy consumption due to the signals exchanged between the fanals. This can be obtained by folding the network. Fig. 2 shows an exemplary network. The network is made of four clusters of four fanals each. Fanals belonging to specific clusters are represented with different shapes. Each fanal has 12 synapses to provide all the possible connections. For the simplicity, the length of the wires is measured with the

number of hops between the fanals in terms of the Manhattan distance. In the presented example the fanal $(i, j) = (1, 1)$ in the cluster I is connected with the fanal $(i', j') = (4, 4)$ in the cluster III. This connection represents the longest possible distance that equals six (the number of dotted lines that have to be crossed). Fig. 3 shows the same network after folding. One can see that the clusters II and III are moved to another layer and put below the clusters I and IV. To realize the connection from Fig. 2 a wire of length four is used. The connection to the layer below is ensured by the TSV. Depending on the size and arrangement of the clusters the folding can be done either in x or y direction.

B. Simulation model

To analyze the gains introduced by using 3D technology the lengths of all the possible connections have to be calculated.

First, the elementary distance between two fanals has to be calculated. This distance depends on the space occupied by the fanal and all its synapses. The area A_{f+s} occupied by one fanal and all its synapses is calculated as:

$$A_{f+s} = A_{fanal} + N_s A_{synapse} \qquad (1)$$

where A_{fanal} is the area of the fanal, $A_{synapse}$ is the area of the synapse, N_s is the number of the synapses connected to one fanal and is calculated as:

$$N_s = (C - 1)\ell. \qquad (2)$$

In accordance with the aforementioned model each fanal can be connected to any fanal in a different cluster.

Second, the Manhattan distance $d_{(i,j)(i',j')}$ between two fanals with coordinates (i, j) and (i', j') is (*cf.* Fig. 2):

$$d_{(i,j)(i',j')} = |i - i'|\sqrt{A_{f+s}} + |j - j'|\sqrt{A_{f+s}}. \qquad (3)$$

Knowing the distance, the unitary resistance and capacitance for the targeted technology, one can obtain the RC delay τ as:

$$\tau_{2D} = r_{per\,\mu m}\, c_{per\,\mu m}\, d^2_{(i,j)(i',j')} \qquad (4)$$

where $r_{per\,\mu m}$ and $c_{per\,\mu m}$ are the resistance and capacitance per unit length. In case of a 3D circuit, the resistance and capacitance of the TSV are added to the RC delay:

$$\tau_{3D} = \tau_{2D} + r_{TSV}\, c_{TSV}. \qquad (5)$$

IV. GENERAL STUDY RESULTS

To evaluate the proposed 3D architecture, the gains in terms of total wire length d_{total} and maximal RC delay τ_{max} obtained by using the 3D technology are explored for different configurations of the network. This includes scaling the number of clusters C and the cluster's size ℓ. The results are based on physical implementation using 65nm technology and 3D technology with TSV of resistance and capacitance equal to $2m\Omega$ and 5fF, respectively. Using our technology specifications, each TSV is equivalent to horizontal wire length of $0.06\mu m$. The equivalent horizontal wire length of TSVs represents less than 1% of the total wire length in the 3D worst simulated case. These values have been included in the 3D results to count the TSV vertical effects.

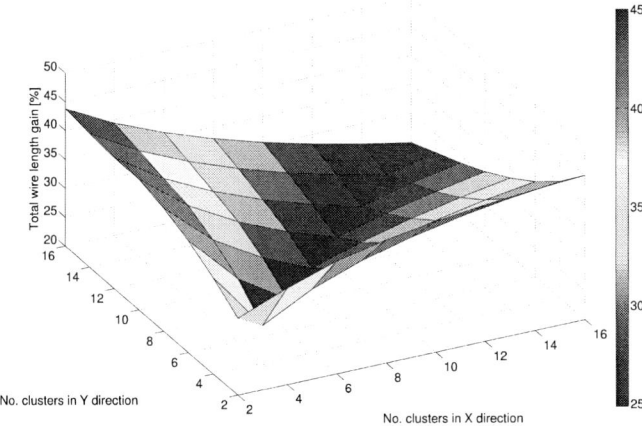

Fig. 4. Total wire length gain compared to 2D in function of the number of clusters in each direction. Square cluster of size two by two is used. The numbers of clusters are given for 2D. For 3D the network is split in two equal parts in such a way to cut its longest dimension.

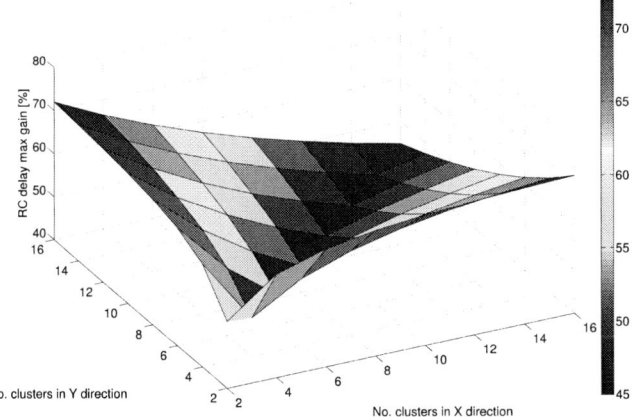

Fig. 5. Maximal RC delay gain compared to 2D in function of the number of clusters in each direction. Square cluster of size two by two is used. The numbers of clusters are given for 2D. For 3D the network is split in two equal parts in such a way to cut its longest dimension.

In the beginning, to facilitate the study, the size of the cluster ℓ is fixed to four (the cluster is square - two by two fanals) and the number of clusters C is scaled. Fig. 4 shows the gain of the 3D technology in terms of total wire length compared to the conventional 2D circuit. The gains reach 45% when the network is strongly rectangular. For a given network size it is therefore better to organize the clusters in a rectangle. For instance, for C=16, the gain is 41% when the network is organized in eight by two clusters compared to 27% for four by four clusters. The square networks are clearly distinguished on the surface by their lower gains. Note that it is possible to use few rectangular networks to organize them in a square. Similar trends can be observed in Fig. 5 that shows the gains in terms of maximal RC delay τ_{max}. In this case the maximal gains reach 72%. It is important to note that the time response of the neural cliques is determined by the longest path in the clique.

Fig. 4 and Fig. 5 show that by increasing the number of clusters equally in both x and y directions (square network), the 3D gain is higher for smaller numbers of clusters (e.g. for total wire length 33% for 2x2) than bigger ones (25% for

8x8, 24% for 16x16). The reason is that 3D cut partitioning is done in only one direction and, consequently, the 3D gain is larger in that direction. Therefore, in case of increasing number of clusters equally in both directions, the effect of long interconnects will not be reduced in one of the two directions which will reduce the overall 3D gain.

Fig. 6a shows the total wire length gain. The cluster size is kept the same as in Fig. 4 and Fig. 5. The red curve shows the evolution of the total wire length gain when the number of clusters in one direction is fixed to two and the network is scaled in another direction. The gain increases quickly for smaller numbers of fanals, then it saturates. It reaches 90% of the maximal value for n=112 fanals which corresponds to C=28 clusters. Blue crosses show the gains when the number of clusters is scaled in both directions, resulting in a network that is less rectangular. The obtained gains are never bigger than when scaling in only one direction. Therefore, it is more beneficial to scale the network in one direction. The black curve presents the gain when C is fixed to four (two by two clusters) and the number of fanals in the cluster ℓ is scaled. Now the gains are higher than in the former case for the same total number of fanals. Again, the same trend is observed. After the fast increase in the gain for smaller numbers of fanals, there comes a saturation. 90% of the maximal gain is reached for n=128 fanals which corresponds to ℓ=32. Comparing these two curves leads to the conclusion that for a given total number of fanals n from the 3D point of view it is more beneficial to have bigger clusters. This is consistent with [7] where authors state that from the storage capacity point of view for a given total number of fanals n it is better to have bigger clusters since the density of the connections established in the network grows slower.

Fig. 6b shows the similar analysis for the gains in terms of maximal RC delay τ_{max}. The gains, reaching 74%, are bigger than for total wire length. There is no difference between the maximal RC delay when increasing the number of clusters C or the number of fanals per cluster ℓ because in both cases the length of the longest connection is the same. Similarly, it is beneficial to scale the network only in one direction.

To give more insight in the gains represented by the blue crosses (when the number of clusters is increased in both x and y directions), Table I shows the gains obtained for total wire length (cf. Fig. 6a). The table gives the number of clusters in each direction (x or y) and the corresponding gain compared to 2D. For instance, when x=2 and the clusters are added only in y direction, the gain compared to 2D increases from 33 to 45%. If a network of 16 clusters is considered, for two clusters in x direction and eight clusters in y direction, one obtains 41% gain whereas for four clusters in x and y direction one obtains only 27%. This shows once again, that it is more beneficial to organize the clusters in a rectangle rather than in a square. Similar analysis is shown in Table II for the maximum RC delay.

Additionally, the power of the interconnects is directly proportional to the total wire length d_{total}:

$$P_{interconnects} \propto d_{total}. \qquad (6)$$

Consequently, in case of reducing the total wire length d_{total} by 55%, as shown in Fig. 6a, the power of the interconnects is reduced by the same percentage.

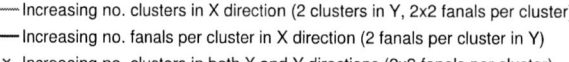

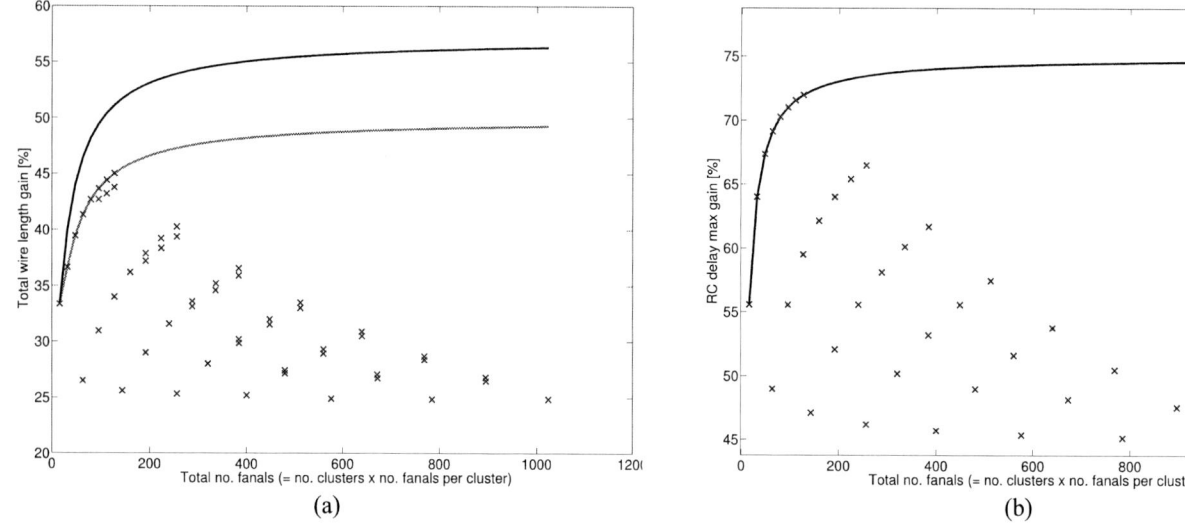

Fig. 6. (a) Total wire length gain (b) maximal RC delay gain in function of the total number of fanals n for different network dimensions.

TABLE I. TOTAL WIRE LENGTH GAIN IN PERCENTAGE COMPARED TO 2D FOR A GIVEN NUMBER OF CLUSTERS IN x AND y DIRECTION

y \ x	2	4	6	8	10	12	14	16
2	33	37	40	41	43	44	44	45
4	37	27	31	34	36	38	39	40
6	40	31	26	29	32	34	35	37
8	41	34	29	25	28	30	32	34
10	43	36	32	28	25	27	29	31
12	44	38	34	30	27	25	27	29
14	44	39	35	32	29	27	25	27
16	45	40	37	34	31	29	27	25

TABLE II. MAXIMAL RC DELAY GAIN IN PERCENTAGE COMPARED TO 2D FOR A GIVEN NUMBER OF CLUSTERS IN x AND y DIRECTION

y \ x	2	4	6	8	10	12	14	16
2	56	64	67	69	70	71	72	72
4	64	49	56	60	62	64	65	67
6	67	56	47	52	56	58	60	62
8	69	60	52	46	50	53	56	58
10	70	62	56	50	46	49	52	54
12	71	64	58	53	49	45	48	51
14	72	65	60	56	52	48	45	48
16	72	67	62	58	54	51	48	45

V. CASE STUDY SIMULATION RESULTS

In this section, a real-world test-case is used to obtain the dimensions of the neural cliques and explore the gains of using 3D technology. In the considered application the time response of neural cliques is of first importance. Therefore, high-performance communication structure is essential.

The considered test-case is a neural cliques-based power management controller for a Multiprocessor-System-on-Chip (MPSoC) firstly proposed in [16]. An MPSoC is built of multiple Processing Elements (PE) that can work in parallel. Each PE or set of PEs form a Voltage/Frequency Island (VFI), *i.e.* they work within the same power domain. The supply voltage V_{dd} and frequency f are set by dedicated switching circuits allowing Dynamic Voltage and Frequency Scaling (DVFS). By decreasing the speed of the PEs with lower performance requirements the energy consumption is reduced. The controller decides on (V_{dd}, f), or power modes, based on latency constraint, current workload, and temperature among others.

Nowadays, DVFS switching circuits allow switching between two different power modes in time of the order of tens of nanoseconds [19][20][21]. It has been shown in [17] that

providing a controller with time response of the same order of magnitude as DVFS switching circuits, allows for 60% energy savings compared to 38% in case of a controller with time response of some μs. That is why neural cliques with high-performance connections for all the synapses need to be used so that the power management controller does not limit the power management reactivity.

Here, MAGALI MPSoC platform is considered [22]. The application mapped on the platform is the LTE receiver [23]. Six PEs are used in the test-case, each PE can choose from 256 frequencies. The speed of each processor is determined by a global latency constraint and an operating mode offering different data rates. The global latency constraint has 51 possible values, there are 5 different operating modes. Based on the optimization problem defined in [17], the frequencies corresponding to each of the possible latency and operating mode combination are obtained. Then, neural cliques are used to store messages containing all the latency and operating mode combinations and the associated six frequencies. During the system operation, latency and operating mode are input to the network and the corresponding frequencies are retrieved. These frequencies are applied to PEs by the DVFS actuators (the DVFS actuator is able to adjust the necessary V_{dd} upon

TABLE III. THE GAINS OBTAINED FOR 3D NEURAL CLIQUES USED AS POWER MANAGEMENT CONTROLLER. THE RESULTS ARE NORMALIZED TO 2D CIRCUIT

	2D	3D		Gain	
		Case 1	Case 2	Case 1	Case 2
Total wire length	1	0.65	0.83	35%	17%
RC delay max	1	0.43	0.69	57%	31%

the given frequency). Additionally, in each message a global estimation of the energy consumed by all the PEs is included. Thanks to that, when the energy consumption is the main constraint (*e.g.* low battery level), the maximum affordable energy is used as the input to the network and the frequencies and the corresponding latency are retrieved. This kind of flexible controller is of high interest in low-power systems.

The aforementioned paremeters of the application allow to obtain the dimensions of the network of neural cliques. The network is made of six clusters of 256 fanals to store all the possible frequency values, one cluster of 255 fanals to store all the possible latency and operating mode values (51 latencies times 5 operating modes), and one cluster of 255 fanals to store all the corresponding energies. For 2D, the network is organized in four clusters in x direction, two clusters in y direction. In each cluster 16 fanals are placed in each direction. For 3D, there are two possible cases: 1) network is cut in x direction (two clusters in each direction on each die), 2) network is cut in y direction (four clusters in x direction and one cluster in y direction on each die).

The gains in terms of total wire length and RC maximum delay compared to the conventional 2D circuit are summarized in Table III. Case 1 gives better results. In this application, using 3D technology allows for 35% total wire length reduction and, consequently, the same reduction in terms of the power of the interconnects. Furthermore, the maximum RC delay is reduced by 57%.

As TSV insertion adds additional area for creating the vertical connections, area overhead analysis should be done. In the test case, the number of used TSVs is 1048576. The TSV pitch is $1.2 \mu m$ [18], consequently TSVs need area of $1.5 mm^2$ which represents 5% of the total chip area. Advanced 3D technologies with smaller pitch can be used to reduce the area overhead. For example, 3D sequential integration [24] can reduce TSVs area overhead to 1.4%.

VI. CONCLUSION

In this paper the gains of using 3D technology for a high-performance implementation of networks of neural cliques are explored. The hardware implementation of neural clique is wire-dominated. The paper shows that using 3D technology allows important gains in terms of total interconnect length, power and delay. It is also shown that dimensioning the network in a way to obtain the highest storage capacity is consistent with dimensioning the network in a way to obtain the maximal gains coming from 3D technology. This means that optimizing the gains coming from the theoretical model and hardware 3D implementation is not contradictory. Furthermore, an applicative test-case of an MPSoC power management controller validates the gains in a real-world

application. This application requires fast time response, and therefore high-performance interconnect is essential.

REFERENCES

[1] R. Rojas, *Neural Networks: A Systematic Introduction*. New York, NY, USA: Springer-Verlag New York, Inc., 1996.

[2] B. Sengupta and M. B. Stemmler, "Power consumption during neuronal computation," *Proceedings of the IEEE*, vol. 102, no. 5, pp. 738–750, 2014.

[3] B. Whitworth, "Some implications of comparing brain and computer processing." in *HICSS*. IEEE Computer Society, 2008, pp. 1–10.

[4] B. Belhadj, A. Valentian, P. Vivet, M. Duranton, L. He, and O. Temam, "The improbable but highly appropriate marriage of 3D stacking and neuromorphic accelerators," in *Proc. CASES'14*, 2014, pp. 1–9.

[5] F. Clermidy *et al.*, "Advanced technologies for brain-inspired computing," in *Proc. ASP-DAC'14*, 2014.

[6] A. Joubert, M. Duranton, B. Belhadj, O. Temam, and R. Heliot, "Capacitance of TSVs in 3D stacked chips a problem? Not for neuromorphic systems," in *Proc. DAC'12, WACI session*, 2012, pp. 1260–1261.

[7] V. Gripon and C. Berrou, "Sparse neural networks with large learning diversity," *IEEE Trans. Neural Netw.*, vol. 22, no. 7, pp. 1087–1096, Jul. 2011.

[8] N. P. Jouppi, "Improving direct-mapped cache performance by the addition of a small fully-associative cache and prefetch buffers," in *ISCA '90*, 1990, pp. 364–373.

[9] N. Huang *et al.*, "Design of multi-field IPv6 packet classifiers using ternary CAMs," in *Proc. IEEE GLOBECOM*, vol. 3, 2001, pp. 1877–1881.

[10] V. Gripon and C. Berrou, "A simple and efficient way to store many messages using neural cliques," in *Proc. IEEE CCMB*, 2011, pp. 54–58.

[11] J.-P. Diguet *et al.*, "Scalable NoC-based architecture of neural coding for new efficient associative memories." in *Proc. CODES+ISSS'13*. IEEE, 2013, pp. 1–9.

[12] B. K. Aliabadi *et al.*, "Storing sparse messages in networks of neural cliques," *IEEE Trans. Neural Netw. Learn. Syst.*, vol. 25, no. 5, pp. 980 – 989, 2014.

[13] L. Lin, R. Osan, and J. Z. Tsien, "Organizing principles of real-time memory encoding: Neural clique assemblies and universal neural codes," *Trends Neurosci.*, vol. 29, no. 1, pp. 48–57, Jan. 2006.

[14] B. Boguslawski, V. Gripon, F. Seguin, and F. Heitzmann, "Huffman coding for storing non-uniformly distributed messages in networks of neural cliques," in *Proc. AAAI*, July 2014, pp. 262–268.

[15] B. Larras, C. Lahuec, M. Arzel, and F. Seguin, "Analog implementation of encoded neural networks," in *Proc. ISCAS'13*, 2013, pp. 1612 – 1615.

[16] B. Larras, B. Boguslawski, C. Lahuec, M. Arzel, F. Seguin, and F. Heitzmann, "Analog encoded neural network for power management in MPSoC," in *Proc. NEWCAS'13*, 2013, pp. 1–4.

[17] B. Larras, B. Boguslawski, C. Lahuec, M. Arzel, F. Seguin, and F. Heitzmann, "Analog encoded neural network for power management in MPSoC," *AICSP*, vol. 81, no. 3, pp. 595–605, 2014.

[18] A. Gutierrez *et al.*, "Integrated 3D-stacked server designs for increasing physical density of key-value stores," in *Proc. ASPLOS*. ACM, 2014, pp. 485–498.

[19] W. Kim, M. S. Gupta, G.-Y. Wei, and D. Brooks, "System level analysis of fast, per-core DVFS using on-chip switching regulators." in *HPCA*. IEEE Computer Society, 2008, pp. 123–134.

[20] D. Truong *et al.*, "A 167-processor computational platform in 65 nm CMOS," *IEEE J. Solid-State Circuits*, 2009.

[21] E. Beigné *et al.*, "An asynchronous power aware and adaptive NoC based circuit," *Solid-State Circuits, IEEE Journal of*, vol. 44, no. 4, pp. 1167–1177, 2009.

[22] F. Clermidy *et al.*, "A 477mW NoC-based digital baseband for mimo 4G SDR." in *ISSCC*. IEEE, 2010, pp. 278–279.

[23] F. Clermidy, R. Lemaire, X. Popon, D. Ktenas, and Y. Thonnart, "An open and reconfigurable platform for 4G telecommunication: Concepts and application." in *DSD*. IEEE Computer Society, 2009, pp. 449–456.

[24] P. Batude *et al.*, "3D sequential integration opportunities and technology optimization," in *IEEE IITC/AMC*, May 2014, pp. 373–376.

A Thermal Estimation Model for 3D IC Using Liquid Cooled Microchannels and Thermal TSVs

Surajit Kr Roy[1], Supriyo Mandal[2], Chandan Giri[3] and Hafizur Rahman[4]

Department of Information Technology

Indian Institute of Engineering Science and Technology,

Shibpur, Howrah 711103, WB, INDIA

email:{ [1]suraroy, [2]supriyo.mandal2013, [3]chandangiri}@gmail.com, [4]rahaman_h@yahoo.co.in

Abstract—**3D integrated circuit (3D IC) is becoming challenging for increasing power density and design complexity. Due to vertical integration heat dissipation in 3D IC is increased that creates hotspots on chip and hence temperature of the chip is very serious issue. Traditional fan-based cooling technique is insufficient for 3D ICs. Hence inter-die integrated microchannel cooling technique and dummy thermal Through-Silicon-Vias (TSVs) based techniques are used for controlling thermal behaviour of the chip. In this paper, we have proposed two temperature estimation models for 3D IC based on liquid cooled microchannels and thermal TSVs. Experimental results show that thermal simulation using our approach provide same result compared to [1]. For microfluidic cooling, our proposed thermal simulator is much more accurate than proposed model [2] and less than 0.5% error respect to simulation result with COMSOL thermal simulator using the same experimental environment of [2].**

Keywords: 3D SoC, COMSOL, TSV, 3D-LCM, 3D-DTT.

I. INTRODUCTION

With the help of interconnection along vertical dimension, 3D integrated circuits (3D ICs) have become a promising solution for process scaling and heterogeneous system integration [3]. In a 3D IC multiple device-layers (tiers) are vertically stacked one above other and are interconnected by Through-Silicon-Vias (TSVs). 3D IC provides numerous potential benefits over traditional 2D ICs [4]. Despite some advantages like reduced latency, reduced power consumption, smaller footprint and heterogeneous integration, chip overheating in 3D IC is a major challenge because multiple stacked device layers increase the power density, resulting rise in temperature of 3D IC.

To remove the excessive heat generated within a chip traditional fan-based cooling technique [5] is not sufficient due to its limited heat removal capacity. Different heat sink models like AIR-SINK, OIL-SILICON model are proposed. These type of methods are not applicable when 3D ICs contain two or maximum three layer. Due to increasing number of layers and circuit design complexity, these type of models are not applicable and microchannel based liquid cooling technique is an efficient technique to keep temperature of 3D ICs under control. Due to the small size of the microchannels, the heat transfer coefficient is very high. In early work a microchannel heat sink is used containing $50\mu m$ wide and $302\mu m$ deep

parallel micro flow passages, very small thermal resistance $9*10^{-6}$ Ω for a pumping power 1.84 Watt.

The thermal resistance of a microchannel is evaluated under an analytic model in [6]. There are many comparisons of conventional microchannel heat sink and manifold microchannel [7]. Gillot et al. [8] calculated thermal performance of microchannel heat sink for multichip power module. Thermal resistance of microchannel is so important to reduce temperature of proposed model. Thermal resistance of a liquid cooled microchannel can be computed as presented in [9]. In [10] the authors have calculated thermal resistance of liquid cooled microchannel on the basis of the properties of working fluid and liquid flow rate. Channel dimensions can be optimized to minimize the overall thermal resistance. The overall thermal resistance can be reduced by increasing the pumping power. Another way to reduce thermal resistance is to increase heat transfer area. As a liquid cooling water, laminate liquid is used in [10].

Through-Silicon-Vias(TSVs) can be used for electrical connection are known as signal TSVs and also can be used for heat dissipation are considered as thermal TSVs. These thermal TSVs have no functional significant, reduce chip temperature for its high thermal conductivity. It provides more uniform thermal profile. Recent device modelings [11]show that due to the existence of liner for isolation, TSVs work quite similarly to the nonlinear MOS-capacitance (MOSCAP) under different signal voltages and operating frequencies. A nonlinear programming-based algorithm is developed and solved for an optimal thermal TSV insertion to minimize the clock skew. Several approaches are also used to remove excessive heat of 3D IC using dummy thermal through-silicon-vias [12], [13]. Many works are proposed based on thermal TSVs planing by using iterative thermal conductivity modification and nonlinear programming problem [12].

Different simulators are used to estimate the temperature of a 3D IC. Numerical multiphysics simulators such as [14] are also accurate to measure temperature of 3D IC.

The main contributions of this paper are as follows:

(i) Previously proposed thermal models are used to evaluate temperature of 3D ICs. Our main target is to design thermal estimation models to estimate temperature of 3D ICs. Here two thermal estimation models, one is 3D-IC Using Liquid-

978-1-4673-9141-2/15 $31.00 © 2015 IEEE

Cooled Microchannels (3D-LCM) and other is Thermal Model Using Dummy Thermal TSVs (3D-DTT) are proposed to provide temperature estimation for 3D IC and compared with previously proposed model [1] .

(ii) 3D-LCM is investigated and compared with 3D-ICE [2], COMSOL [2] and previously proposed model [2] using same experimental environment of [2] .

The rest of the paper is organized as follows: Section II provides an overview of target Architecture Of 3D-IC using Liquid-Cooled Microchannels(3D-LCM). Section III presents Thermal Model Using Dummy Thermal TSVs(3D-DTT). Section IV shows experimental result of proposed thermal models for 3D SIC. Finally Section V concludes this paper.

II. TARGET ARCHITECTURE OF 3D-IC USING LIQUID-COOLED MICROCHANNELS(3D-LCM)

The proposed thermal model contains a 3D IC with N number of dies and microchannels are placed between two dies, also at above of top most die and below of bottom most die. The microchannels have liquid inlets and outlets which are connected to the micropump for liquid circulation. Flow rate of liquid depends on the pumping power of micropump. Heat is absorbed by the liquid of microchannels when the cooling liquid passes through the upper and bottom layer dies. Total amount of temperature absorbed by this liquid totally depends on thermal resistance of die and liquid cooled microchannels, power consumption of die. Obviously temperature of liquid outlet is higher than liquid inlet. There is a mechanism outside the chip that converts the heated liquid to initial temperature of liquid. Fluorochemical Inert Liquid (FC 43) is used in microchannel at 10^0C temperature for its high boiling point (174^0C), electric strength, non flammability, thermal stability and low temperature fluidity. Fig.1 presents a 3D IC model containing two dies and liquid cooled microchannels are placed between two dies.

Fig. 1.   **3D IC with microchannels**

A chip contains multiple dies and a die contains multiple blocks. In a die one or more than one block may active, then we assume that total die is active. If no block is active, then total die is inactive. Here we assume power consumption based on total die not for particular block or blocks and maximum power consumption of a die is considered. So we do not consider lateral heat dissipation effect between blocks. Here only vertical heat flow of a die is assumed since lateral heat flow is negligible due to very small thickness (in the range of tens of microns) of each die. The heat dissipation path is from die to liquid cooled microchannels through the bonding interface.

The thermal model has considered the well known duality between heat transfer and electrical phenomena. Here heat flow is considered as a current passing through a thermal resistance and the temperature difference analogous to voltage.

Fig.2 shows the thermal resistive model, where die0 is the bottom most die and its thermal resistance is denoted as R_0 and die(N-1) is the top most die of 3D IC and its thermal resistance is denoted as R_{N-1}. Power consumption and thermal resistance of the dies from bottom to top are P_0 , P_1 , P_2 ,......., P_{N-1} and R_0 , R_1 , R_2 ,......., R_{N-1} respectively. Thermal resistance of microchannels with cooling liquid and resistance of bonding interface are considered as R_L, R_X respectively.

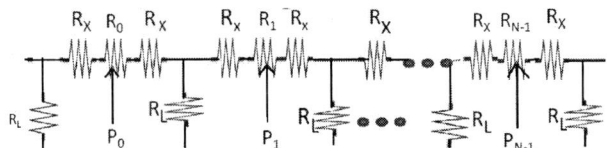

Fig. 2.   **Thermal Resistive Model**

Now thermal resistance of liquid cooled microchannel ($R_{channel}$) can be found by equation 1 as obtained from [10] [15]:

$$\mathbf{R_{channel} = R_{cond} + R_{conv} + R_{heat}} \qquad (1)$$

R_{cond} is the thermal resistance of heat sink area of a chip. It is determined by the silicon thermal conductivity k_{si}, the substrate thickness t, and the heat sink area A_{chip} and represented as:

$$\mathbf{R_{cond}} = \frac{t}{k_{si} A_{chip}} \qquad (2)$$

which is typically very small and negligible.

$$\mathbf{R_{conv}} = \frac{1}{nh A_1} \qquad (3)$$

R_{conv} is the thermal resistance of cooling area of microchannel. It is calculated by overall cooling area of the channels, i.e, $A_1 = 2(W_c+H_c) \times L_c$ where L_c, H_c, W_c are respectively length, height, width of microchannel, convective heat transfer coefficient h and number of microchannel n. R_{heat} is resistance of cooling liquid where

$$\mathbf{R_{heat}} = \frac{1}{nc_p \rho v_m A_2} \qquad (4)$$

Here c_p, ρ, v_m, A_2 are specific heat, density, maximum flow rate, area of cooling liquid respectively. Here $A_2 = (H_C \times W_C)$.
So thermal resistance of liquid cooled microchannels is represented as:

$$R_L = R_{channel} = \frac{1}{n}(\frac{1}{h A_1} + \frac{1}{c_p \rho v_m A_2}) \qquad (5)$$

A. Temperature Evaluation When single die active

In a 3D IC a single die or a number of dies are tested at any time. First we evaluate the temperature for 3D IC when a single die is tested. A 3D IC model and corresponding thermal resistive model of 3D IC is shown in Fig. 3(a) and Fig.

3(b) respectively. Blue, black and gray color indicate liquid, bonding interface and die respectively. Suppose only single die means die0 is tested. R_0 is thermal resistance of die0. Suppose n_1 number of microchannels are used below layer of die0 and n_2 number of microchannels are used above layer of die0. Thermal resistance of liquid cooled microchannels at lower and upper portion of die0 are respectively $R_{L0} = \frac{1}{n_1}(\frac{1}{hA_1} + \frac{1}{c_p\rho v_m A_2})$ and $R_{L1} = \frac{1}{n_2}(\frac{1}{hA_1} + \frac{1}{c_p\rho v_m A_2})$ which are estimated from eqn (5). It is assumed that heat is generated at the middle of a die and heat flows towards the microchannels above and bellow of that die. In Fig. 3(b) power P_0 is applied on the middle point(**A**) of die0 with thermal resistance R_0.

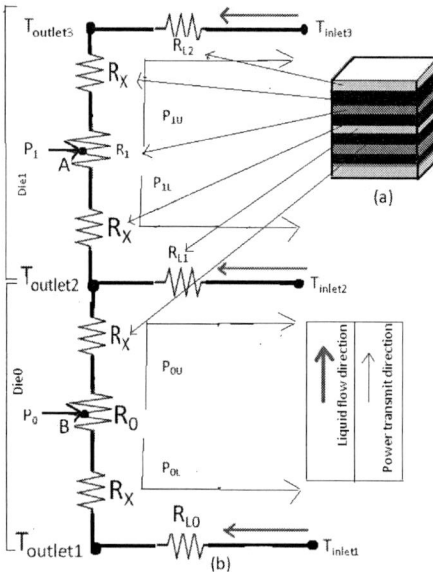

Fig. 3. **(a) 3D IC Model (b) Thermal Resistive Model of 3D IC**

Using current division rule the power consumption P_0 of die0 is divided as P_{0U} and P_{0L} such that $P_0 = P_{0U} + P_{0L}$. Here P_{0U} power is transmitted through upper portion of die0 that is thermal resistance $R_0/2$ and n_2 number of microchannels. P_{0L} is transmitted through rest $R_0/2$ thermal resistance and n_1 number of microchannels. T_{inlet} and T_{outlet} are inlet and outlet temperature of liquid respectively.

Power P_{0U} and P_{0L} transmit towards the upper portion and lower portion of die0 respectively where

$$P_{0U} = \frac{R_X + R_0/2 + R_{L0}}{2R_X + R_0 + R_{LO} + R_{L1}} \qquad (6)$$

$$P_{0L} = \frac{R_X + R_0/2 + R_{L1}}{2R_X + R_0 + R_{LO} + R_{L1}} \qquad (7)$$

Now, we calculate temperature of die0. Outlet temperature of n_1 number of microchannels is presented as:

$$T_{outlet1} = P_{0L} * R_{L0} + T_{inlet1} \qquad (8)$$

Due to the liquid cooled microchannels below die0, temperature of die0 is

$$T_0' = P_{0L}(R_X + R_0/2) + T_{outlet1} \qquad (9)$$

Outlet temperature of n_2 number of microchannels is presented as:

$$T_{outlet2} = P_{0U} * R_{L1} + T_{inlet2} \qquad (10)$$

Due to the liquid cooled microchannels above die0, temperature of die0 is

$$T_0'' = P_{0U}(R_X + R_0/2) + T_{outlet2} \qquad (11)$$

Here $n_1 \neq n_2$ so $T_{outlet1}$ and $T_{outlet2}$ will be different. Hence T_0' and T_0'' will not be same and temperature of die0 depends on maximum of T_0' and T_0''.

Now temperature of die0 is represented as:

$$T_0 = MAX(T_0', T_0'') \qquad (12)$$

$$T_0 = Max \begin{cases} P_{0L}(R_X + R_0/2 + R_{LO}) + T_{inlet1}, \\ P_{0U}(R_X + R_0/2 + R_{L1}) + T_{inlet2} \end{cases} \qquad (13)$$

If $n_1 = n_2$ then obviously $P_{0L} = P_{0U}$. Hence

$$T_0 = P_{0L}(R_X + R_0/2 + R_{LO}) + T_{inlet1} \qquad (14)$$

or

$$T_0 = P_{0U}(R_X + R_0/2 + R_{L1}) + T_{inlet2} \qquad (15)$$

So temperature of 3D IC will be T_0 as a single die of the 3D IC is tested.

B. Temperature Evaluation When Two Consecutive Dies Active

Suppose two consecutive dies means in Fig. 3 die0 and die1 are tested. Thermal resistance of die1 is R_1 and n_3 number of microchannels are used above layer of die1. Power P_1 is applied on middle point (**B**) of Die1. Here we assume a node **B** is on the middle point of R_1. Suppose P_{1U} power is transmitted through thermal resistance $R_1/2$ and n_3 number of microchannels and rest power P_{1L} of P_1 is transmitted through rest thermal resistance $R_1/2$ and n_2 number of microchannels. When two consecutive dies are tested then total $(P_{1L} + P_{0U})$ power transmits through n_2 number of microchannels means thermal resistance R_{L1}. It is assumed that negligible power of P_{1L} transmits through R_0 resistance.

Now temperature of die0 is represented as:

$$T_0 = Max \begin{cases} P_{0U}(R_X + R_0/2) + (P_{0U} + P_{1L})*R_{L1} + T_{inlet2}, \\ P_{0L}(R_X + R_0/2 + R_{L0}) + T_{inlet1} \end{cases} \qquad (16)$$

and temperature of die1

$$T_1 = Max \begin{cases} P_{1L}(R_X + R_1/2) + (P_{0U} + P_{1L})*R_{L1} + T_{inlet2}, \\ P_{1U}(R_X + R_1/2 + R_{L2}) + T_{inlet3} \end{cases} \qquad (17)$$

Using this formula we also can calculate temperature of 3D ICs when more than two consecutive dies are tested. Suppose a 3D IC model contains N number of dies and all dies are tested at the same time. Now generalized formula to evaluate temperature of ith die is represented as:

$$T_i = Max \begin{cases} P_{iU}*(R_X + R_i/2) + (P_{(i+1)L} + P_{iU})*R_{L(i+1)} + T_{inlet}, \\ P_{iL}*(R_X + R_i/2) + (P_{iL} + P_{(i-1)U})*R_{Li} + T_{inlet} \end{cases} \qquad (18)$$

Here thermal resistance of ith die is R_i. Power P_i is applied on the middle point of ith die. Thermal resistance of microchannel of upper layer and bottom layer of ith die are $R_{L(i+1)}$ and R_{Li} respectively. Using current division rule P_{iU} power of P_i transmits through upper portion of middle point of ith die and thermal resistance $R_{L(i+1)}$. Rest P_{iL} power of P_i transmits through bottom portion of ith die and thermal resistance R_{Li}. $P_{(i+1)}$ power is applied on the middle point of upper die of ith die. According current division rule $P_{(i+1)L}$ transmits through bottom portion of middle point of $(i+1)^{th}$ die and $R_{L(i+1)}$ resistance. $P_{(i-1)U}$ power of bottom die of ith die transmits through R_{Li} resistance. Assume power consumption of bottom die of ith die is $P_{(i-1)}$.
When i^{th} die is top most die then

$$T_i = Max \begin{cases} P_{iU}*(R_X+R_i/2+R_{L(i+1)})+T_{inlet}, \\ P_{iL}*(R_X+R_i/2)+(P_{iL}+P_{(i-1)U})*R_{Li}+T_{inlet} \end{cases}$$
(19)

When i^{th} die is bottom most die then

$$T_i = Max \begin{cases} P_{iU}*(R_X+R_i/2)+(P_{(i+1)L}+P_{iU})*R_{L(i+1)}+T_{inlet}, \\ P_{iL}*(R_X+R_i/2+R_{Li})+T_{inlet} \end{cases}$$
(20)

Now, estimated temperature with respect to our proposed model= MAX $(T_i, T_{i+1}.....,T_{N-1})$ where $0\leq i \leq$ N-1.

III. PROPOSED THERMAL MODEL USING DUMMY THERMAL TSVs(3D-DTT)

The proposed thermal model assumes a 3D IC having N number of hard dies. Signal TSVs make the connections between Back End Of Line (BEOL) layers of adjacent tiers but are not connected to the heat-sink. Here signal TSVs are not considered. Only dummy thermal TSVs are used for heat dissipation of proposed model. Here heat sink is connected to the top most die of the chip. Dummy thermal TSVs are inserted between top most die and heat sink. Heat of chip transfers through dummy thermal TSVs due to its high thermal conductivity and heat sink. Thus thermal TSVs are used to reduce temperature of the chip.

Fig. 4(a) presents 3D IC containing four dies and dummy thermal TSVs based on heat flow is considered as a current passing through a thermal resistance and temperature difference analogous to voltage. Fig. 4(b) shows the thermal resistive model containing N number of dies respectively where die0 is the bottom most die and die(N-1) is the top most die of 3D IC. Power consumption and thermal resistance of the dies from bottom to top are P_0 , P_1 , P_2 ,......., P_{N-1} and R_0 , R_1 , R_2 ,......., R_{N-1} respectively.

Aluminium alloys 6061 and 6063 are commonly used as heat sink materials. These materials have the high thermal conductivity values of 166 and 237 W/mK , respectively. These thermal conductivity values depend on the temper of the alloy. Aluminium alloys 6063 are used on the top of die4 as an effective heat sink. Si_3N_4 is assumed to be the linear material and copper is used as the filling material of TSV. Thermal conductivity of copper material of thermal TSV and silicon material are 400 W/mK and 150 W/mK [16]

respectively. T_{amb} is the ambient temperature of environment and considered as room temperature. T_H is thermal effect on heat sink due to power consumption of N number of dies and T_{amb}. R_{0TSV}, R_{1TSV},...... $R_{(N-3)TSV}$, $R_{(N-2)TSV}$ and $R_{(N-1)TSV}$ are thermal resistance of thermal TSVs between die0 and die1, die1 and die2, die(N-3) and die(N-2), die(N-2) and die(N-1), die(N-1) and heat sink respectively. Here thermal TSVs are used between die0 and heat sink thats why $R_{0TSV}=R_{1TSV}=......$ $R_{(N-3)TSV}= R_{(N-2)TSV}=R_{(N-1)TSV}$. R_H is thermal resistance of heat sink.

Here only vertical heat flow is assumed since lateral heat flow is negligible due to very small thickness (in the range of tens of microns) of each die. The heat dissipation path is from die to heat sink through thermal TSVs.

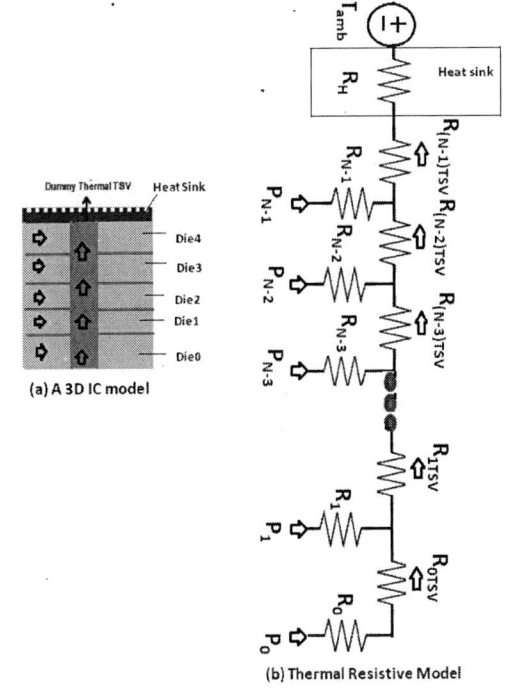

Fig. 4.   **3D IC Model and Thermal Resistive Circuit Model**

A. Thermal Resistance Of Dummy Thermal TSVs

The vertical thermal resistance R of a dummy thermal TSV can be computed :

$$R = \frac{L}{K_{TSV} A_{tsv}}$$
(21)

where L is the TSV length in the vertical direction, $A_{tsv} = \pi r^2$ is the cross section area of the TSV cylinder in the vertical direction and K_{TSV} is the thermal conductivity of the TSV material. For a TSV array or farms, the vertical resistance can be computed as:

$$R_{arraytsv} = \frac{L}{K_{eff} A_{arraytsv}}$$
(22)

where K_{eff} is the effective thermal conductivity of a TSV array, L is the vertical length of TSV array and $A_{arraytsv}$ is the total cross sectional area of TSV array. Here

$$K_{eff} = \eta * K_{TSV} + (1 - \eta) * K_0 \qquad (23)$$

where K_{TSV} and K_0 are the thermal conductivity of thermal TSVs inserted material(copper metal is used for all metallizations in the IC) and surrounding material of thermal TSVs respectively and η is thermal TSV density.

B. Temperature Evaluation When Single Die Active

During testing of 3D ICs dies are tested in sequentially or parallely based on test scheduling algorithm. Hence a single die or a number of dies are tested at any time. First we evaluate the temperature for 3D IC when a single die is tested. In Fig. 4(b) a single die die0 is tested means only P_0 power is transmitted through thermal TSVs and heat sink. The thermal model has considered the well known duality between heat transfer and electric phenomena. Heat flow is considered as a current passing through a thermal resistance, temperature difference analogous to voltage. So due to P_0 power, thermal effects on heat sink is represented as according kirchhoff's voltage law

$$T_H = R_H * P_0 + T_{amb} \qquad (24)$$

Temperature of die0 is represented as

$$T_0 = P_0 R_0 + P_0 (R_{0TSV} + R_{1TSV} + R_{2TSV} + R_{3TSV} + R_{4TSV}) + T_H \qquad (25)$$

C. Temperature Evaluation When Two Consecutive Dies Active

Suppose two consecutive dies means in Fig. 4(b) die0 and die1 are tested. So total (P_0+P_1) power is transmitted through thermal TSVs and heat sink. So due to (P_0+P_1) power, thermal effects on heat sink is represented as

$$T_H = R_H(P_0 + P_1) + T_{amb} \qquad (26)$$

Temperature of die0 is represented as

$$T_0 = P_0 R_0 + P_0 (R_{0TSV} + R_{1TSV} + R_{2TSV} + R_{3TSV} + R_{4TSV}) + P_1(R_{1TSV} + R_{2TSV} + R_{3TSV} + R_{4TSV}) + T_H \qquad (27)$$

Temperature of die1 is represented as

$$T_1 = P_1 R_1 + (P_0 + P_1)(R_{1TSV} + R_{2TSV} + R_{3TSV} + R_{4TSV}) + T_H \qquad (28)$$

A generalized formula is proposed for N number of die of a 3D ICs model:

$$T_i = P_i R_i + \sum_{i}^{N-1} R_{iTSV} \left(\sum_{i=0}^{i} P_i \right) + R_H \sum_{i=0}^{N-1} P_i + T_{amb} \qquad (29)$$

where T_i, P_i, R_i are temperature, power and resistance of ith die respectively.

IV. EXPERIMENTAL RESULT

Experimental results are obtained for two different thermal estimation model. For simulation purpose two different simulators are implemented in C language, GCC compiler and run on an Intel Core i3 processor having 4GB RAM in Linux operating system.

In our experiment case a 3D-IC with two microchannel layers in between of three active layers is designed. We use IBM-PLACE 2.0 circuits with placement information as the benchmark [17]. IBM-PLACE 2.0 benchmarks were converted from the original ISPD98 IBM netlists. For each test case, three circuits from IBM-01 to IBM-10 circuits are selected, each circuit corresponds to one 3D IC layer. Power information value is taken from [1].

The width and length of chip dimension is considered as 9 mm both. The microchannel width = 100 μm and height = 200 μm [1] and the pressure drop is Δp = 600 KPa [1]. The pumping power is obtained as $Q_{PUMP} = \sum_{n=1}^{N} f_n \Delta p$ where N is the number of microchannels. Parameters Δp and f_n are the pressure drop and fluid flow rate of nth microchannel respectively. Usually the microchannels are assumed to have the same Δp_n and f_n.

Estimated temperature from proposed thermal models 3D-LCM and 3D-DTT are compared with [1] and shown in TABLE I. We use same environment as [1]. In this comparison table, Q_{CHIP} (in Watt) denotes the total dissipated power of each benchmark. For each design method, N_{MC} is the number of channels used by this design, N_{TSV} is the percentage of total chip area occupied thermal TSVs. T_{MAX} in °C is the peak temperature of the resultant thermal profile, Q_{PUMP} indicates the pumping power. Both of our proposed models give same result compare to [1].

TABLE I
COMPARISON OF DIFFERENT 3D-IC COOLING APPROACHES

Benchmark [17]	Q_{CHIP} [1]	Microchannels only [1]			3D-LCM [our]	Thermal TSVs [1]		3D-DTT [our]
		N_{MC}	Q_{PUMP}	T_{MAX}	T_{MAX}	N_{TSV}(%)	T_{MAX}	T_{MAX}
IBM-01	182.35	12	2.93	84.37	84.43	5.27	87.02	87.35
IBM-02	203.75	20	5.06	84.36	84.32	6.03	93.48	93.35
IBM-03	220.80	33	8.52	81.36	80.25	5.29	99.63	97.59
IBM-04	241.79	36	9.32	82.96	82.09	5.78	106.95	104.35
IBM-05	254.60	40	10.38	84.58	83.65	5.56	110.99	109.30
IBM-06	275.68	52	13.58	84.64	83.73	5.43	118.23	115.95
IBM-07	292.08	53	13.84	84.56	83.69	5.58	123.65	119.37
IBM-08	308.64	68	17.83	84.18	83.43	5.17	129.56	123.83
IBM-09	332.46	81	21.29	84.93	83.85	5.41	137.42	129.71
IBM-10	344.91	84	22.11	80.09	78.93	5.25	141.74	132.08

Fig.5 shows the variation of T_{MAX} with N_{TSV}(%). For this experiment we choose three circuits from IBM05 [17]. All other environment is same as [1]. From this graphical presentation it is cleared that T_{MAX} decreases with increasing of N_{TSV}(%) where Q_{CHIP} is fixed.

Fig.6 shows the variation of T_{MAX} with N_{MC}. For this experiment we choose three circuits from IBM-05 [17]. All other environment is same as [1]. From this graphical presentation it is cleared that T_{MAX} decreases with increasing of N_{MC} where Q_{CHIP} = 254.60 W and pumping power is 10.38 W.

978-1-4673-9141-2/15 $31.00 © 2015 IEEE

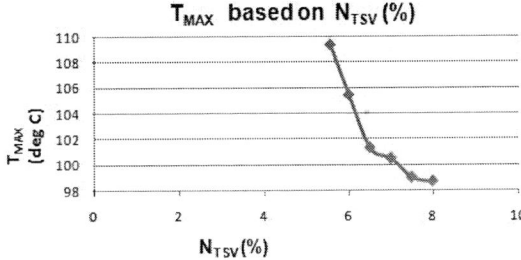

Fig. 5. Variation of T_{MAX} with $N_{TSV}(\%)$

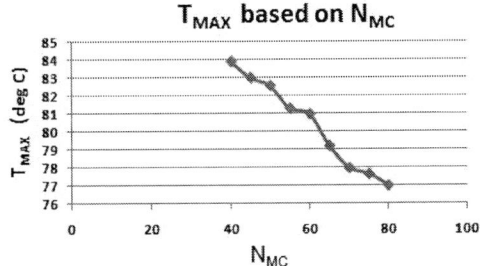

Fig. 6. Variation of T_{MAX} with N_{MC}

In this section, we also verify the accuracy of our proposed thermal simulator of 3D-LCM. The chip size is 1.8×1.8 mm^2 with 12 μm BEOL layers and 2 μm device layers. The thickness of bulk layer is 48 μm for the top tier and 148 μm for the bottom tier. Ambient temperature is 300 K. Microchannel size and channel wall width are 100×100 μm^2 and 100 μm respectively. In the absence of physical prototypes, our thermal simulator named as 3D-LCM are compared against COMSOL [2], 3D ICE 2.1 [2] using same experimental environment of [2] and proposed model [2]. Two layer 3D circuit is used and the microfluidic channels are used between two circuit layers. The maximum temperature of the top device layer is reported in TABLE II with their corresponding percentage of errors enclosed in brackets. It is shown that our model 3D-LCM is very accurate with a percentage error of less than 0.5% with COMSOL where 3D-ICE exhibits larger errors. Our proposed model gives more accurate result respect to [2]. It can be seen that temperature distribution of 3D-LCM agrees very well with COMSOL where result of 3D-ICE deviates from COMSOL significantly.

V. CONCLUSION

In this paper, two different thermal estimation models are proposed. These two models have capacity for estimating thermal effect of liquid cooled microchannels and thermal TSVs in fine granularity. Our proposed two models show same result with respect to [1]. An accurate steady state thermal simulator has been presented for liquid cooled microchannels based 3D IC. It is shown that our model 3D-LCM is very accurate with a percentage error of less than 0.5% for COMSOL where 3D-ICE exhibits larger errors. Our proposed model gives more accurate result with respect to [2]. It can be seen that temperature distribution of 3D-LCM agrees very well with COMSOL where result of 3D-ICE deviates from COMSOL significantly.

REFERENCES

[1] B. Shi, A. Srivastava, and A. Bar-Cohen, "Hybrid 3D IC cooling system using micro-fluidic cooling and thermal TSVs," in *2012 IEEE Computer Society Annual Symposium on VLSI (ISVLSI)*, Aug 2012, pp. 33–38.

[2] H. Qian, H. Liang, C.-H. Chang, W. Zhang, and H. Yu, "Thermal simulator of 3D-IC with modeling of anisotropic TSV conductance and microchannel entrance effects," in *2013 18th Asia and South Pacific on Design Automation Conference (ASP-DAC)*, Jan 2013, pp. 485–490.

[3] E. Marinissen and Y. Zorian, "Testing 3D chips containing Through-Silicon Vias," in *International Test Conference*, Nov 2008, pp. 1–11.

[4] W. Davis, J. Wilson, S. Mick, J. Xu, H. Hua, C. Mineo, A. Sule, M. Steer, and P. Franzon, "Demystifying 3D ICs: the pros and cons of going vertical," *Design Test of Computers, IEEE*, vol. 22, no. 6, pp. 498–510, Nov 2005.

[5] J. L. Ayala, A. Sridhar, and D. Cuesta, "Thermal modeling and analysis of 3D multi-processor chips," *"Integration, the VLSI Journal "*, vol. 43, no. 4, pp. 327 – 341, 2010.

[6] A. Weisberg, H. H. Bau, and J. Zemel, "Analysis of microchannels for integrated cooling," *International Journal of Heat and Mass Transfer*, vol. 35, no. 10, pp. 2465 – 2474, 1992.

[7] G. Harpole and J. Eninger, "Micro-channel heat exchanger optimization," in *Seventh Annual IEEE, Semiconductor Thermal Measurement and Management Symposium, 1991*, Feb 1991, pp. 59–63.

[8] C. Gillot, C. Schaeffer, and A. Bricard, "Integrated micro heat sink for power multichip module," *IEEE Transactions on Industry Applications,*, vol. 36, no. 1, pp. 217–221, Jan 2000.

[9] D. Liu and S. V. Garimella, "Analysis and optimization of the thermal performance of microchannel heat sinks," *International Journal of Numerical Methods for Heat and amp, Fluid Flow*, vol. 15, no. 1, pp. 7–26, 2005.

[10] Z. Feng and P. Li, "Fast thermal analysis on GPU for 3D ICs with integrated microchannel cooling," *IEEE Transactions on Very Large Scale Integration (VLSI) Systems*, vol. 21, no. 8, pp. 1526–1539, Aug 2013.

[11] T. Bandyopadhyay, R. Chatterjee, D. Chung, M. Swaminathan, and R. Tummala, "Electrical modeling of through silicon and package vias," in *IEEE International Conference on 3D System Integration, 3DIC 2009.*, Sept 2009, pp. 1–8.

[12] J. Cong and Y. Zhang, "Thermal via planning for 3D ICs," in *ICCAD-2005. IEEE/ACM International Conference on Computer-Aided Design, 2005.*, Nov 2005, pp. 745–752.

[13] B. Goplen, "Thermal via placement in 3D ICs," in *Proceedings of the International Symposium on Physical Design*, 2005, pp. 167–174.

[14] ANSYS Multiphysics . Available: http://www.ansys.com.

[15] X. Wei and Y. Joshi, "Optimization study of stacked micro-channel heat sinks for micro-electronic cooling," in *The Eighth Intersociety Conference on Thermal and Thermomechanical Phenomena in Electronic Systems (ITHERM '02)*, 2002, pp. 441–448.

[16] Y. Shang, C. Zhang, H. Yu, C. S. Tan, X. Zhao, and S. K. Lim, "Thermal-reliable 3D clock-tree synthesis considering nonlinear electrical-thermal-coupled TSV model," in *18th Asia and South Pacific Design Automation Conference (ASP-DAC), 2013*, Jan 2013, pp. 693–698.

[17] IBM-place 2.0 benchmark. In http://er.cs.ucla.edu/benchmarks/ibmplace2/.

TABLE II
STEADY STATE TEMPERATURE(K)COMPARISON AMONG COMSOL, 3D-ICE, [2] AND 3D-LCM

Power Density	COMSOL [2]	3D-ICE [2]	Proposed Model [2]	3D-LCM(Ours)
50	314.3	317.4	314.0(0.1%)	314.5(0.1%)
100	328.6	334.8	328.0(0.2%)	327.2(.42%)
150	342.9	352.2	341.9(0.3%)	343.0(.03%)
200	357.2	369.6	355.8(0.4%)	358.1(.25%)
250	371.5	387.0	369.8(0.5%))	372.8(.35%)

978-1-4673-9141-2/15 $31.00 © 2015 IEEE

An Equilibrium Partitioning Method for Multicast Traffic in 3D NoC Architecture

Lin Wei*, Lei Zhou†

*†School of Information Engineering, Yangzhou University, Yangzhou, China
*Email: elitals@163.com
†Email: tomcat800607@126.com

Abstract—Recently, three-dimensional Network-on-Chip (3D NoC) has been a promising transmission architecture for the communication requirement in Multiprocessor system-on-chips (MPSoCs) due to its capability of providing better scalability and higher throughput. The mode of information transmission in the network has significant impact on the overall system performance. Multicast routing can take advantage of the inherent parallelization offered by the MPSoCs to communicate the messages efficiently. However, the transmission latency is a bottleneck that has become a barrier to improve the performance. In this paper, to reduce the total transmission latency in the network, we propose an equilibrium partitioning method, which can obtain a tradeoff between the startup latency and the network latency to reduce the total transmission latency. The experiment results demonstrate that the proposed method can reduce the average latency by 18.3% at most than the previous recursive partitioning (RP) method, when the architecture is $16 \times 16 \times 3$ mesh and the packet size is 15 flits. In addition, the scalability of the equilibrium partitioning method is better than the contrastive method.

Index Terms—Three-dimensional Network-on-Chip (3D NoC), multicast, partitioning method, multicast transmission latency.

I. INTRODUCTION

With the development of the integrated circuit technique, the scale and complexity of the multiprocessor systems-on-chips (MPSoCs) is increasing with a greater speed than that described by the Moore's law [1], [2], [3]. Since the insufficiency of the traditional global interconnection, networks-on-chip (NoC) has been developed as a promising solution for the communication requirement in MPSoCs [4], [5], [6]. However, the performance of two-dimensional (2D) NoC is limited by the chip area and long latency [7]. To overcome the the design bottleneck in 2D NoC, 3D NoC has emerged for its capability of providing lower global interconnect length, better scalability, higher throughput and higher packing density, which increases the performance in system [8]. Since the temperature information sharing is important for temperature controlling in real time to achieve efficient and stable performance in 3D NoC, it is necessary to share the regional temperature of each node in the system efficiently. That means the regional temperature should be transferred from one source node to all other destinations. Therefore, multicast routing is implemented in this field to achieve more parallelism with efficient and fast communication between IP cores [9].

Multicast routing algorithm can be broadly classified into three categories: unicast-based, tree-based [2], [10], [11], [12]

and path-based [1], [5], [13], [14]. Unicast-based multicast is the simplest one, however, it causes a significant amount of traffic and high overhead [6], [8]. Tree-based approaches can generate the shortest travel path for the source node, however, branches of the tree may compete with each other to increase network contention and deadlocks [3], [15]. Thus, path-based multicast has been proposed as a solution to overcome the disadvantages of unicast-based approaches and tree-based approaches, such as dual-partitioning(DP), multi-partitioning (MP), fix-partitioning (FP) and column-partitioning (CP) [9]. In a path-based mulicast approach, the source node prepares a message to traverse the predetermined path to form Hamiltonian paths. Therefore, path-based approach can overcome the disadvantage of tree-based approach by forbidding branching to different routers, which may cause deadlocks in 3D NoC [15]. Path-based multicast approach has been used frequently in 3D NoC due to the better performance of deadlock-free.

However, the performance of path-based approach may not good due to high network latency, when the path in the network is long. To reduce the length of the multicast path, partitioning methods can be used by dividing destinations in several disjoint sets to reduce the network latency [8], [9]. For instance, Daneshtalab *et.al.* [16] introduced four path-based multicast partitioning methods including two-block partitioning (TBP), multi-block partitioning (MBP), vertical-block partitioning (VBP), and hybrid partitioning (HP). In these methods, each message in the subset is transmitted in the optimization Hamilton path to avoid deadlocks effectively. However, due to the uncertainty of partitioning numbers, utilizing these methods may affect the total latency, including the startup latency and the network latency. On the one hand, the more partitions increase the number of packets and shorten the shared path, which will increase the startup latency. On the other hand, the fewer partitions increase the length of the transmission path, which will increase the network latency. To the best knowledge of authors, synthesizing the startup latency and the network latency to reduce the total latency in 3D network is still rarely investigated in these existing methods. Only recently in [8], Ebrahimi *et.al.* proposed a recursive partitioning (RP) method which recursively partitions the network until all partitions contain a comparable number of switches. The RP method outperforms other partitioning methods including TBP, VBP, HP due to its restriction for the longest path in each subnetwork. However, the reference

978-1-4673-9141-2/15 $31.00 © 2015 IEEE

value indicating the number of switches in each partition is fixed, so that it cannot be adjusted dynamically according to the location of the source node to reduce the total latency.

Motivated by this observation, in this paper, the Hamiltonian Path strategy is used due to its advantages of deadlock-free, and an equilibrium partitioning (EP) method based on the 3D mesh architecture is proposed, which can obtain a tradeoff between the startup latency and the network latency. The optimal partitioning number can be determined by this this partitioning method for the current source node, and then, the corresponding areas are divided according to the location of the source node to make the total latency as small as possible.

The rest of this paper is organized as follows. Section II presents the equilibrium partitioning method and the example. In Section III, the experimental results are shown and discussed. The conclusions are provided in Section IV.

II. THE PROPOSED PARTITIONING METHOD

A. The Equilibrium Partitioning Method

To simplify the analysis, in this paper, it is assumed that the network is 3-stage pipeline architecture which is composed of three stages: request, response and transfer, i.e., the latency for one hop is 3 cycles, the most ideal time which is required to complete a multicast transmission can be given as follows [8]

$$\text{Total Latency (cycles)} = \underbrace{n \times L}_{\text{Startup latency}} + \overbrace{\left\lceil \frac{N}{n} \right\rceil \times 3}^{\text{Network latency}}, \quad (1)$$

where n and N are the numbers of partitions and all nodes in the system, respectively, and L is the number of flits per message, in which the unit is the cycle. To achieve the minimum total latency, it is important to determine the number of partitions reasonably.

1) Adjustment of Label Assignment: Consider the generic 3D NoC architecture (each node has one IP core and one router), as shown in Fig. 1(a), which denotes a $4 \times 4 \times 3$ mesh network. In 3D $a \times b \times c$ mesh network, each node has the corresponding triple coordinate (x,y,z) to represent its position in the network. In order to partition the network expediently and avoid deadlocks effectively, all the nodes in the same layer with the source node are numbered according to the ordered triple coordinates of them. The following equation describes the common label assignment:

$$L(x,y) = \begin{cases} a \times y + x & \text{where } y \text{ is even} \\ a \times y + a - x - 1 & \text{where } y \text{ is odd} \end{cases} \quad (2)$$

Since only the nodes in the same layer with the source node are numbered, the messages may not be transmitted to the partitions through the other layers when the source node is located in the boundary of the y axis, i.e., $y = 0$ ($x \neq a - 1$) or $y = b - 1$ ($x \neq a - 1$). In these situations, the method of

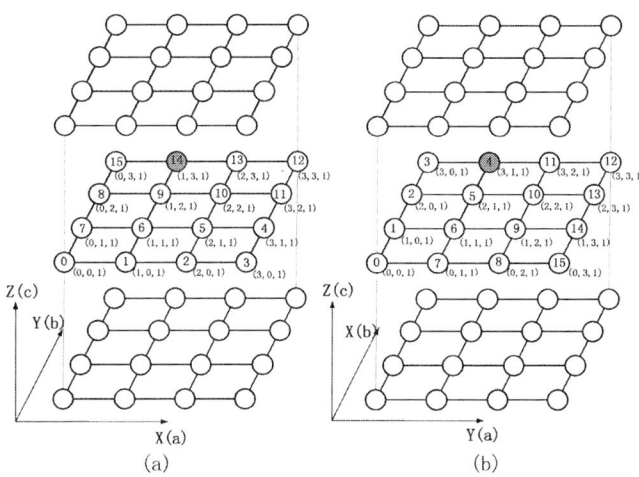

Fig. 1. (a) $4 \times 4 \times 3$ 3D NoC, (b) The adjustment of assigning the labels.

assigning the labels should be regulated as

$L(x,y) = b \times x + y$,

where ($y = 0, b - 1$ and $x \neq a - 1$) and x is even

(expect $x, y = 0$);

$L(x,y) = b \times y + b - y - 1$,

where ($y = 0, b - 1$ and $x \neq a - 1$) and x is even   (3)

(expect $x, y = 0$);

$L(x,y) = a \times y + x$,

where (($y \neq 0, b - 1$) or ($y = 0, b - 1$ and $x = a - 1$)

and y is even) or ($x, y = 0$ and y is odd);

$L(x,y) = a \times y + a - x - 1$,

where (($y \neq 0, b - 1$) or ($y = 0, b - 1$ and $x = a - 1$)

and y is odd) or ($x, y = 0$ and y is even).

Fig. 1(b) describes an example of the adjustment, which converts the original label of the source node 14 to 4, i.e., making the source nodes not in the boundary of y axis in the network by exchanging the X axis and Y axis.

2) Partitioning Method: At first, partitions formed after employing the TBP method can be a high set (D_H) and a low set (D_L). The D_H set contains the nodes labeled in ascending order in the layer where the source node is present, and all nodes in higher layers. The D_L set contains the rest in the system.

In most cases, the longest path is generated from the larger set, therefore, several parameters of the larger set should be calculated firstly. Assuming that the number of the nodes in the larger set is N_1. By calculating the ideal minimum value for (1), the optimal number of partitions for the set can be obtained as $n_1 \approx \sqrt{\frac{3N_1}{L}}$, which is rounded to the nearest integer. In order to enlarge the choice scope of partition numbers, partitions are divided by column, hence, if n_1 is greater than or equal to the maximum columns in the X axis, i.e., $n_1 \geq x_{\max} + 1$, the n_1 value should be adjusted

978-1-4673-9141-2/15 $31.00 © 2015 IEEE   129

to $x_{\max} + 1$, and the ideal largest and smallest number of the nodes in each partition of the larger set can be computed as $N_{\text{large-average}(N_1)} = \lceil \frac{N_1}{n_1} \rceil$ and $N_{\text{small-average}(N_1)} = \lfloor \frac{N_1}{n_1} \rfloor$, respectively. Note that this $x_{\max}$ is the one which has been changed according to the location of the source node by the adjustment of label assignment.

In order to avoid deadlocks, the partitions should be partitioned in specified orders to make partitions not intersect with each other. If the set which is going to be divided is D_H, the first column of the first partition is divided up starting from the column, where the node label in the last row is the smallest. In contrast, it is divided up starting from the column where the node label in the first row is the largest, if the set is D_L. Then, the partition is extended in the vertical direction. If the corresponding column in the vertical direction has already been divided, the number of columns in the partition of the current layer should be increased in accordance with established direction until the corresponding column in the vertical direction can be divided up, and then the partition being extended sequentially in the vertical direction until the last layer. Assuming that the number of columns in the j_{th} partition in the i_{th} layer is α_i^j in the set, the number of the nodes in one column of the j_{th} partition in the i_{th} layer is C_i^j, and the z-coordinate of the source node layer is s. At this point, the number of the nodes in the j_{th} partition can be expressed as $C^j = \alpha_s^j C_s^j + \alpha_{s+1}^j C_{s+1}^j + \cdots + \alpha_{z_{\max}}^j C_{z_{\max}}^j$, $(C_{s+1}^j = C_{s+2}^j = \cdots = C_{z_{\max}}^j = y_{\max} + 1, \alpha_{s+1}^1 = \alpha_{s+2}^1 = \cdots = \alpha_{z_{\max}}^1 = 1, \alpha_{z_{\max}}^j = 1)$ for D_H, or $C^j = \alpha_s^j C_s^j + \alpha_{s-1}^j C_{s-1}^j + \cdots + \alpha_0^j C_0^j$, $(C_{s-1}^j = C_{s-2}^j = \cdots = C_0^j = y_{\max} + 1, \alpha_{s-1}^1 = \alpha_{s-2}^1 = \cdots = \alpha_0^1 = 1, \alpha_0^j = 1)$, $(j = 1, 2, \cdots, n_1 - 1)$ for D_L. And the number of columns in the j_{th} partition in other layers in the set can be computed as

$$\alpha_s^j = 1 + \alpha_{s+1(s-1)}^{j-1} + \alpha_{s+1(s-1)}^{j-2} + \cdots + \alpha_{s+1(s-1)}^{1} - \alpha_s^{j-1} - \alpha_s^{j-2} - \cdots - \alpha_s^1;$$

$$\alpha_{s+1(s-1)}^j = 1 + \alpha_{s+2(s-2)}^{j-1} + \alpha_{s+2(s-2)}^{j-2} + \cdots + \alpha_{s+2(s-2)}^{1} - \alpha_{s+1(s-1)}^{j-1} - \alpha_{s+1(s-1)}^{j-2} - \cdots - \alpha_{s+1(s-1)}^{1};$$

$$\vdots$$

$$\alpha_{z_{\max}-1(1)}^j = 1 + \alpha_{z_{\max}(0)}^{j-1} + \alpha_{z_{\max}(0)}^{j-2} + \cdots + \alpha_{z_{\max}(0)}^{1} - \alpha_{z_{\max}-1(1)}^{j-1} - \alpha_{z_{\max}-1(1)}^{j-2} - \cdots - \alpha_{z_{\max}-1(1)}^{1}.$$

(4)

Subsequently, C^j and $N_{\text{small-average}(N_1)}$ are weighed to determine whether the number of columns in the last layer ($\alpha_{z_{\max}}^j$ or α_0^j) of this set should be increased. If C^j is lower than the ideal largest number of the nodes $N_{\text{small-average}(N_1)}$, the number of columns in the last layer of this set will be increased until C^j is closest to $N_{\text{small-average}(N_1)}$ or $N_{\text{large-average}(N_1)}$ according to the judgment algorithm, as shown in Algorithm 1. This partitioning strategy is repeated until the $(n_1 - 1)_{\text{th}}$ partition is completed, and the rest of the nodes in the set are divided in the $n_{1_{\text{th}}}$ partition. The number

---

Algorithm 1 Determine $\alpha_{\text{optimal}}()$

Input: $\alpha_s^j, \ \alpha_{s+1(s-1)}^j, \ \cdots, \ \alpha_{z_{\max}-1(1)}^j, \ C_s^j, \ y_{\max}, \ N_{\text{small-average}(N_1)(N_2)}, N_{\text{large-average}(N_1)(N_2)};$

Output: $\alpha_{\text{optimal}};$

1: $\alpha \Leftarrow 1; \alpha_{\text{optimal}} \Leftarrow \alpha;$
2: **while** $\alpha_s^j \cdot C_s^j + \alpha_{s+1(s-1)}^j \cdot (y_{\max} + 1) + \cdots + \alpha_{z_{\max}-1(1)}^j \cdot (y_{\max} + 1) + \alpha \cdot (y_{\max} + 1) \leq N_{\text{small-average}(N_1)(N_2)}$ **do**
3: $\quad \alpha_{\text{optimal}} \Leftarrow \alpha;$
   $\quad \beta \Leftarrow \alpha_s^j \cdot C_s^j + \alpha_{s+1(s-1)}^j \cdot (y_{\max} + 1) + \cdots + \alpha_{z_{\max}-1(1)}^j \cdot (y_{\max} + 1) + \alpha \cdot (y_{\max} + 1);$
4: $\quad$ **if** $(\beta < N_{\text{small-average}(N_1)(N_2)})$ and $(\beta + (y_{\max} + 1) \geq N_{\text{large-average}(N_1)(N_2)})$ **then**
5: $\quad\quad$ **if** $|\beta - N_{\text{small-average}(N_1)(N_2)}| > |\beta + (y_{\max} + 1) - N_{\text{large-average}(N_1)(N_2)}|$ **then**
6: $\quad\quad\quad \alpha_{\text{optimal}} \Leftarrow \alpha + 1;$
7: $\quad\quad$ **else**
8: $\quad\quad\quad \alpha_{\text{optimal}} \Leftarrow \alpha;$
9: $\quad\quad$ **end if**
10: $\quad$ **end if**
11: $\quad \alpha \Leftarrow \alpha + 1;$
12: $\quad$ **return** α_{optimal}
13: **end while**

---

of columns in the $n_{1_{\text{th}}}$ partition can be expressed as

$$\alpha_s^{n_1} = x_{\max} + 1 - \alpha_s^1 - \alpha_s^2 - \cdots - \alpha_s^{n_1-1};$$

$$\alpha_{s+1(s-1)}^{n_1} = x_{\max} + 1 - \alpha_{s+1(s-1)}^1 - \alpha_{s+1(s-1)}^2 - \cdots - \alpha_{s+1(s-1)}^{n_1-1};$$

$$\vdots$$

$$\alpha_{z_{\max}(0)}^{n_1} = x_{\max} + 1 - \alpha_{z_{\max}(0)}^1 - \alpha_{z_{\max}(0)}^2 - \cdots - \alpha_{z_{\max}(0)}^{n_1-1}.$$

(5)

Then, the larger set will be divided into n_1 partitions.

Since the longest path is generated from the larger set, the limits of the number of nodes for each partition in the smaller set should be determined according to $N_{\text{large-average}(N_1)}$ of the larger set. If the total path length in the smaller set, which is equal to the number of nodes in the smaller set, i.e., N_2, is not higher than the longest transmission path length in the larger set, i.e., $N_{\text{large-average}(N_1)} + \max(x, x_{\max} - x)$, in which x is the x-coordinate of the source node, the smaller set should not be partitioned. If this condition is not satisfied, to ensure the longest path length in the smaller set is less than that in the larger set, the optimal number of partitions in this set, i.e., n_2, can be given as $n_2 = \lceil \frac{N_2}{N_{\text{large-average}(N_1)}} \rceil$, and the ideal largest and smallest numbers of the nodes in each partition are $N_{\text{large-average}(N_2)} = \lceil \frac{N_2}{n_2} \rceil$ and $N_{\text{small-average}(N_2)} = \lfloor \frac{N_2}{n_2} \rfloor$, respectively. The partitioning strategy in the smaller set is same as that in the larger set. Finally, all partitions of the network has been completed.

In the proposed equilibrium partitioning (EP) method, the nodes in the source node layer are assigned the labels by

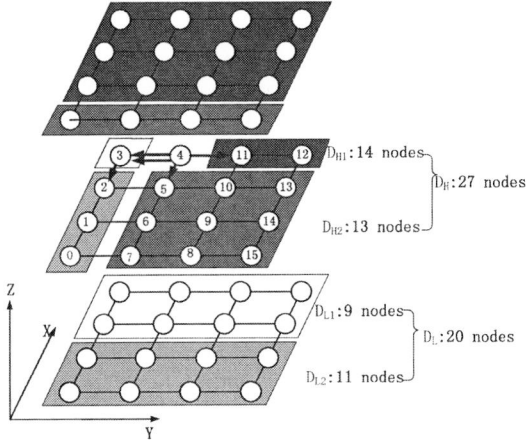

Fig. 2. The equilibrium partitioning (EP) method where the coordinate and label of the source node are (1,3,1) and 4, respectively.

Algorithm 2 The Equilibrium Partitioning

Input: $a \times b \times c$ network; The source node (x_s,y_s,z_s) and its label; Other nodes labels on the source layer; The Z value of all nodes; Flit per message (L); $x_{\max}$; $y_{\max}$; $z_{\max}$;

Output: High destination set(D_{H1}, D_{H2},...); Low destination set(D_{L1}, D_{L1},...);

1: $D_H \Leftarrow \{(x,y,z)|L(x,y) > L(x_s,y_s) \text{ and } z > z_s\}$;
   $D_L \Leftarrow \{(x,y,z)|L(x,y) < L(x_s,y_s) \text{ and } z < z_s\}$.
2: Find the larger set (D_{large}) and the smaller set (D_{small}) between D_H and D_L.
3: Get the number of the nodes in the D_{large} and D_{small}: N_1 and N_2.
4: Compute n_1, $N_{\text{large}-\text{average}(N_1)}$ and $N_{\text{small}-\text{average}(N_1)}$.
5: **if** $N_2 \leq N_{\text{large}-\text{average}(N_1)} + \max(x_s, x_{\max} - x_s)$ **then**
6:     $n_2 \Leftarrow 1$; $D_{H(L)_{n_2}} \Leftarrow D_{\text{small}}$.
7: **else**
8:     Compute n_2, $N_{\text{large}-\text{average}(N_2)}$, $N_{\text{small}-\text{average}(N_2)}$.
9: **end if**
10: **while** $n_1 \neq 1$ or $n_2 \neq 1$ **do**
11:    Find the starting columns.
12:    $\alpha^1_{s+1(s-1)} = \alpha^1_{s+2(s-2)} = \cdots = \alpha^1_{z_{\max}-1(1)} \Leftarrow 1$;
      $\alpha^1_{z_{\max}(0)} \Leftarrow$ Determine $\alpha_{\text{optimal}}()$.
13:    **for** $j \leftarrow 2$ to $n_1 - 1$ **do**
14:       Compute $\alpha^j_s, \cdots, \alpha^j_{z_{\max}-1(1)}$;
         $\alpha^j_{z_{\max}(0)} \Leftarrow$ Determine $\alpha_{\text{optimal}}()$.
15:    **end for**
16:    Divide the corresponding columns in each partition: D_{H_1}, D_{H_2}, $\cdots$, $D_{H_{n_1-1(n_2-1)}}$ or D_{L_1}, D_{L_2}, $\cdots$, $D_{L_{n_1-1(n_2-1)}}$.
17:    Divide the rest columns in the $D_{H_{n_1(n_2)}}$ or $D_{L_{n_1(n_2)}}$ partition.
18:    **if** (All partitions has been completed) **then**
19:       Break.
20:    **end if**
21: **end while**

the adjustment of label assignment, on this basis, all nodes in the network are divided in disjoint partitions at column level. Therefore, for the subsets which are in the source node layer, the Hamiltonian path strategy is used for each partition in this layer to avoid deadlocks. Besides, for the subsets in the other layers of each partition, the message is transmitted following the YXZ path, i.e., the priorities of the transmission channels in y direction are higher than those in x direction, which are higher than those in z direction. Therefore, no cyclic dependency will be formed in this context, and all partitions are disjoint. Then the network is free of deadlocks.

According to the above analysis, the partitioning procedure which is shown in Algorithm 2, can be expressed as follows:

1. Apply the TBP method in the network upon the basis of the adjustment of label assignment.

2. Find the larger set and smaller set between the two sets which are formed after employing the TBP method.

3. Obtain the optimal number of partitions, the ideal largest and smallest numbers of the nodes in each partition for the larger set.

4. Determine whether the smaller set should be partitioned. Obtain the optimal number of partitions, the ideal largest and smallest numbers of the nodes in each partition for the smaller set.

5. For the set which should be partitioned: Seek the starting columns in each layer, and compute the number of columns in each layer for the partition by (4). Determine the number of columns in the last layer of this partition by Algorihtm 1. Divide the corresponding columns in each partition.

6. Repeat step 5 until the $(n_1 - 1)_{\text{th}}$ or $(n_2 - 1)_{\text{th}}$ partition is completed, and the rest of the nodes in the set are divided in the $n_{1_{\text{th}}}$ or $n_{2_{\text{th}}}$ partition.

B. Example

An example of the EP method is shown in Fig. 2, where the architecture is $4 \times 4 \times 3$ 3D mesh, and the coordinate, label of the source node are (1,3,1) and 4, respectively. Firstly, TBP method is applied in the network upon the basis of the

adjustment of label assignment. The larger set is D_H, in which there are 27 nodes, i.e., $N_1 = 27$, and the smaller set is D_L, in which there are 20 nodes, i.e., $N_2 = 20$. Assuming that the number of flits per message (L) is 15, hence, the optimal number of partitions for D_H is $n_1 = 2$, and the ideal largest and smallest numbers of the nodes in each partition for D_H are $N_{\text{large}-\text{average}(N_1)} = 14$, $N_{\text{small}-\text{average}(N_1)} = 13$, respectively. Since the total path length in the smaller set is larger than the longest transmission path in the larger set, i.e., $N_2 > N_{\text{large}-\text{average}(N_1)} + \max(0,3)$, the smaller set should be partitioned. The optimal number of partitions for D_L is $n_2 = 2$, and the ideal largest and smallest numbers of the nodes in each partition for D_L are $N_{\text{large}-\text{average}(N_2)} = N_{\text{small}-\text{average}(N_2)} = 10$. In D_H set, the columns are divided in each partition in descending order based on the nodes label in the last row. Therefore, D_{H1} starts from the column where $x = 3$ in the 1_{th} layer, and expands the columns to the upper layer vertically, i.e., $\alpha^1_1 = 1$. The optimal number of

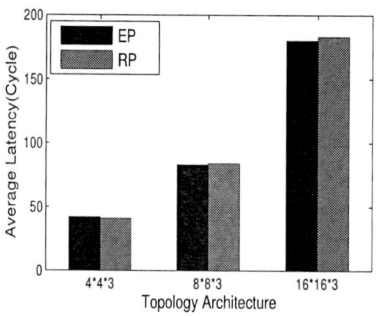

(a) Average latency with 1 flit.

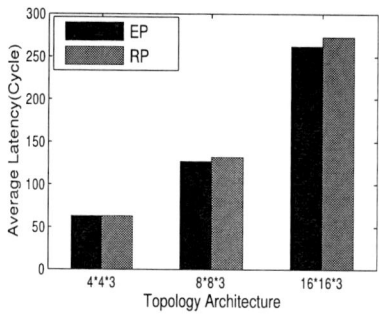

(b) Average latency with 5 flits.

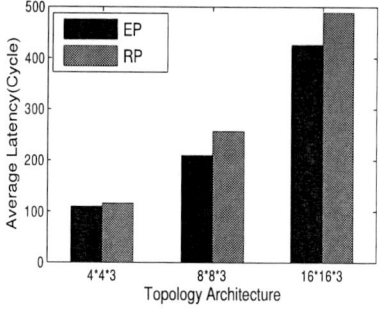

(c) Average latency with 15 flits.

Fig. 3. Average latencies of three topological architectures with 0.1 injection rate when the packet size is (a) 1fit, (b) 5flits, (c) 15flits.

the columns at the top is 3 according to Algorithm 1, i.e., $\alpha_2^1 = 3$. Then, three columns in the 2_{th} layer and one column in 1_{th} layer are divided in D_{H_1} starting from the column where $x = 3$. D_{H2} starts from the column where $x = 2$ in the 1_{th} layer. However, the columns where $x = 2, 1$ in the 2_{th} layer have been divided in D_{H1}, therefore, the number of the columns in the 1_{th} layer for D_{H2} should be increased to 3 by (4), i.e., $\alpha_1^2 = 1 + \alpha_2^1 - \alpha_1^1 = 3$. Then, D_{H2} expands the columns to the upper layer vertically, i.e., $\alpha_2^2 = 1$. At this point, the larger set D_H has been divided into two partitions completely. Similar to the steps in D_H, two partitions in D_L are formed as shown in Fig. 2. Note that the columns in D_L set are divided in each partition in ascending order based on the nodes label in the first row. In Fig. 2, transmission paths of the startup messages from the source node to each partition are represented by using solid arrows.

III. EXPERIMENTAL EVALUATION

To assess the efficiency of the proposed EP method, the recursive partitioning (RP) method in [8] is also implemented in this work for comparison. For the performance metric, the multicast latency can be defined as the number of cycles between the initiation of the multicast message operation and the time when the tail of the multicast message reaches all the destinations [5]. Table 1 shows the configuration of the 3D NoC in this experiment.

TABLE I
THE CONFIGURATION OF THE 3D NOC.

Topology	$4 \times 4 \times 3$, $8 \times 8 \times 3$, $16 \times 16 \times 3$
Pipeline architecture	3-stage
Packet size	1 flit, 5 flits, 15 flits
Switching	Wormhole
Number of destinations	8 destinations, 16 destinations

In first set of simulations, we assume the traffic profile as uniform random traffic that each node generates packets to each destination at equal probability, and the message injection rate is a constant which is 0.1. In order to compare average communication latencies under varied conditions in networks, three mesh sizes of $4 \times 4 \times 3$, $8 \times 8 \times 3$, $16 \times 16 \times 3$ and three packet sizes including 1 flit, 5 flits, 15 flits have been considered.

The average latencies under various network sizes and packet sizes are shown in Fig. 3. It is obvious that the differences between the average latencies achieved by the proposed EP method and the RP method in three topological architectures are small, when the packet size is small as 1 flit. However, as the packet size increases, the differences between the average latencies of these two methods are also increasing. That means the EP method has better performance than the RP method, especially when the sizes of injected packets are long. Comparing the latencies of these two methods in three architectures when the packet size is the same, it can be seen that the proposed EP method outperforms the RP method obviously in larger networks, and the biggest difference between the average latencies of these two methods is 63 cycles when the architecture is $16 \times 16 \times 3$ and the packet size is 15 flits, i.e., the average latency of the proposed EP method is 18.3% smaller than the RP method. That means the scalability of the proposed EP method is better than RP method.

To compare the performances of the RP method and the EP method under different loads, this set of simula-

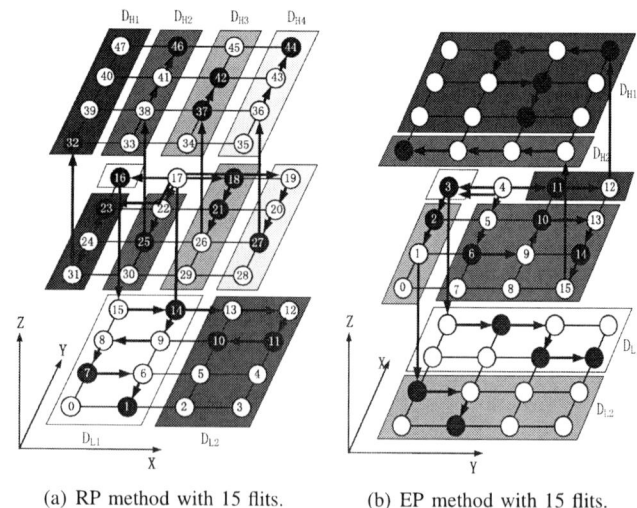

(a) RP method with 15 flits.　　(b) EP method with 15 flits.

Fig. 4. The partitioning results achieved by the RP method and the proposed EP method with 16 destinations.

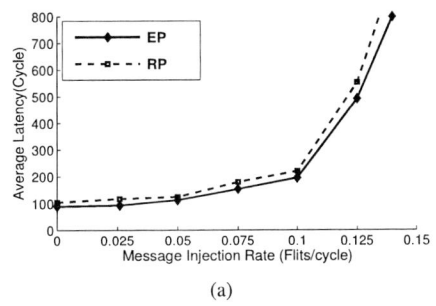

(a)

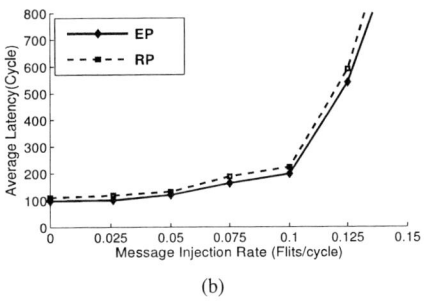

(b)

Fig. 5. Performance under different loads in $4 \times 4 \times 3$ 3D mesh with (a) 8 destinations (b) 16 destinations.

tion was performed in $4 \times 4 \times 3$ 3D mesh for a random multicast traffic profile, the PEs generate 15 flits data messages and inject them into network. In the multicast traffic, each core sends a message to a set of destinations. For simplicity, we choose two fixed destination sets to implement the contrast, i.e., $\{10,14,21,23,27,32,42,46\}$ and $\{1,7,10,11,14,16,18,21,23,25,27,32,37,42,44,46\}$, in which there are 8 destinations and 16 destinations, respectively. Figs. 4(a)(b) show the partitioning results achieved by the RP method and the EP method with 16 destinations in the above example, where the source node is 17. The transmission paths from the source node to destinations are represented by using solid arrows in these figures. The average latency for the RP and EP methods as a function of the message injection rate with multicast messages destined to 8 and 16 destinations, is shown in Fig. 5. As observed, the proposed EP method leads to the lower latency in comparison with the RP method even at high traffic loads or with a large number of destinations. The foremost reason for this performance gain is due to the tradeoff between the startup latency and the network latency in the EP method, which can ease restrictions between the startup latency and the network latency according to the actual network parameters to minimize the total latency.

IV. CONCLUSION

In this paper, from partitioning the network optimally prospective, we propose an equilibrium partitioning method based on the 3D mesh architecture which can obtain a tradeoff between the startup latency and the network latency to reduce the total latency, and guarantee that the network is free of deadlocks. We demonstrate that the proposed equilibrium partitioning method provides better performance than the previous RP method. And as the scale of the network and the size of the packet increase, the performance in the total transmission latency of the network improves by using the proposed equilibrium partitioning method.

ACKNOWLEDGMENT

This work is supported by the National Natural Science Foundation of China (Grant No. 61376025), the Natural Science Foundation of Colleges and Universities of China's Jiangsu Province (Grant No. 13KJB510039) and 2014 Post-Graduate Innovation Project Foundation of China's Jiangsu Province (Grant No. SJZZ_0182).

REFERENCES

[1] P. Bahrebar and D. Stroobandt, "Adaptive routing in MPSoCs using an efficient path-based method," in *Proc. Int. Conf. SoC Des. (ISOCC)*, Nov. 2013, pp. 31–34.

[2] N. Jerger, L. shiuan Peh, and M. Lipasti, "Virtual circuit tree multicasting: A case for on-chip hardware multicast support," in *Proc. ISCA '08 35th Int. Symp. Comput. Arch.*, Jun. 2008, pp. 229–240.

[3] M. Ebrahimi, M. Daneshtalab, P. Liljeberg, and H. Tenhunen, "Hamum - a novel routing protocol for unicast and multicast traffic in MPSoCs," in *Proc. 22nd Euro. Int. Conf. Parallel, Distrib., and Netw.-Based*, Feb. 2014, pp. 525–532.

[4] P. Bahrebar and D. Stroobandt, "The hamiltonian-based odd-even turn model for adaptive routing in interconnection networks," in *Proc. Int. Conf. Reconfigurable Computing and FPGAs (ReConFig)*, Dec. 2013, pp. 1–6.

[5] M. Daneshtalab, M. Ebrahimi, S. Mohammadi, and A. Afzali-Kusha, "Low-distance path-based multicast routing algorithm for network-on-chips," *Computers and Digital Techniques, IET*, vol. 3, no. 5, pp. 430–442, Sep. 2009.

[6] M. Ebrahimi, X. Chang, M. Daneshtalab, and J. Plosila, "Dyxyz: Fully adaptive routing algorithm for 3D NoCs," in *Proc. 21st Euro. Int. Conf. Parallel, Distrib. and Netw.-Based Processing*, 2013, pp. 499–503.

[7] M. Zhu, J. Lee, and K. Choi, "An adaptive routing algorithm for 3D mesh NoC with limited vertical bandwidth," in *Proc. IEEE/IFIP 20th Int. Conf. VLSI and Syst.-on-Chip (VLSI-SoC)*, Oct. 2012, pp. 18–23.

[8] M. Ebrahimi, M. Daneshtalab, P. Liljeberg, and J. Plosila, "Path-based partitioning methods for 3D networks-on-chip with minimal adaptive routing," *IEEE Trans. Computers*, vol. 63, no. 3, pp. 718–733, 2014.

[9] N. Meena, H. Kapoor, and S. Chakraborty, "A new recursive partitioning multicast routing algorithm for 3D network-on-chip," in *Proc. 18th Int. Symp. VLSI Des. and Test*, Jul. 2014, pp. 1–6.

[10] L. Wang, Y. Jin, H. Kim, and E. J. Kim, "Recursive partitioning multicast: a bandwidth-efficient routing for networks-on-chip," in *Proc. 3rd ACM/IEEE Int. Symp. Netw.-on-chip*, May. 2009, pp. 64–73.

[11] W. Hu, Z. Lu, A. Jantsch, and H. Liu, "Power-efficient tree-based multicast support for networks-on-chip," in *Proc. 16th Asia and South Pacific Des. Autom. Conf. (ASP-DAC)*, Jan. 2011, pp. 363–368.

[12] D. Kumar, W. Najjar, and P. Srimani, "A new adaptive hardware tree-based multicast routing in k-ary n-cubes," *IEEE Tran. Computers*, vol. 50, no. 7, pp. 647–659, Jul. 2001.

[13] E. Alceu Carara and F. Moraes, "Deadlock-free multicast routing algorithm for wormhole-switched mesh networks-on-chip," in *Proc. ISVLSI '08 IEEE Comput. Society Annual Symp. VLSI*, Apr. 2008, pp. 341–346.

[14] M. Ebrahimi, M. Daneshtalab, P. Liljeberg, and H. Tenhunen, "An adaptive unicast/multicast routing algorithm for mpsocs," in *Proc. 12th Euro. Conf. Digital Sys. Design, Arch., Methods and Tools*, Aug. 2009, pp. 203–206.

[15] K.-J. Lee, C.-Y. Chang, and H.-Y. Yang, "An efficient deadlock-free multicast routing algorithm for mesh-based networks-on-chip," in *Proc. Int. Symp. VLSI Des., Autom., and Test (VLSI-DAT)*, Apr. 2013, pp. 1–4.

[16] M. Ebrahimi, M. Daneshtalab, P. Liljeberg, and H. Tenhunen, "Partitioning methods for unicast/multicast traffic in 3d noc architecture," in *Proc. 16th IEEE Int. Symp. Des. and Diagnostics of Electron. Circuits and Syst. (DDECS)*, Apr. 2010, pp. 127–132.

978-1-4673-9141-2/15 $31.00 © 2015 IEEE

An Integrated SoC for Science Data Processing in Next-Generation Space Flight Instruments Avionics

Xabier Iturbe[1,2,3], Didier Keymeulen[2], Emre Ozer[3], Patrick Yiu[1], Daniel Berisford[2], Kevin Hand[2] and Robert Carlson[2]

[1] California Institute of Technology, Pasadena CA, USA; Email: xabier.iturbe@jpl.nasa.gov
[2] NASA Jet Propulsion Laboratory, California Institute of Technology, Pasadena CA, USA
[3] ARM R&D, Cambridge, UK

Abstract— We present here an integrated SoC platform, called APEX-SoC, that is aimed at speeding-up the design of next-generation space flight instruments avionics by providing a convenient infrastructure for hardware and software based science data processing. We use a case-study drawn from the JPL Compositional Infrared Imaging Spectrometer (CIRIS) to illustrate the process of integrating instrument-dependent data acquisition and processing stages in this platform. In order to enable the use of APEX-SoC-based instruments in deep space missions, the platform implements Radiation Hardening By Design (RHBD) techniques and offers support for instantiating multiple processing stages that can be used at runtime to increase reliability or performance, based on the requirements of the mission at each particular stage. Finally, in the specific case of CIRIS, the data processing includes a stage to cope with radiation affecting the instrument photo-detector.

Keywords—SoC, RHBD, Avionics, Instrumentation

I. INTRODUCTION

Current space instrument payload systems include several components, typically FPGA+CPU [1][2], that need to be accommodated in a board. Besides, the technology used in these systems, especially CPU technology (e.g., RAD750), lags state-of-the-art terrestrial technology (e.g., ARM) by several generations. Commercial applications have already started the transition towards integrating a complete system composed of programmable logic and ARM processor into a single chip (e.g., Xilinx SoC). While this is extremely convenient in space applications, due to the reduction in power consumption, board space (weight, volume) and system cost, currently there are no flight-qualified SoC devices available yet [3][4][5].

In view of a potential radiation-hardened SoC part that might be ready to fly in the near to mid future, JPL and ARM have partnered together to develop a SoC platform to be used as a research vehicle for powering next-generation flight instruments. Presently this platform, called APEX-SoC, is being prototyped using a commercial Xilinx Zynq device. This initiative is also aligned with NASA's plans to use Zynq devices in precursor CubeSats operating in deep space [6][7].

The APEX-SoC platform has been designed to be generic enough to be used with a wide range of instruments avionics and to meet the processing demands of the payloads expected in the coming years. In order to illustrate the APEX-SoC, this paper takes as a case-study JPL's CIRIS prototype spectrometer, which is based on a commercial COTS instrument which is proposed to be used in future NASA missions [8]. Despite it is not covered in this paper, the APEX-SoC has been also used with the JPL Tunable Laser Spectrometer (TLS) [9].

The APEX-SoC implements a parallel and pipelined architecture that can be used to increase either performance or reliability. Besides, it also implements a number of RHBD features to address its use in harsh space environments. In the specific case of the CIRIS instrument, the APEX-SoC integrates a data processing to detect and remove outlier noise provoked by radiation hits in the instrument's photo-detector without compromising the science returns. This technique has been experimentally proven to be effective to cope with radiation similar to that existing in the Jupiter's moon Europa [10][11], which deteriorated the spectroscopy data collected by NASA's previous-generation Galileo Near-Infrared Mapping Spectrometer (NIMS) a decade ago [12].

The remainder of this paper is organized as follows. Section II introduces the CIRIS instrument and its functioning. Then, section III describes the general architecture of the APEX-SoC platform and the RHBD features included on it. Section IV discusses the implementation, performance and radiation-tolerance results measured so far in the APEX-SoC based CIRIS controller. Finally, section V presents the conclusions of the paper.

II. THE JPL CIRIS SPECTROMETER

CIRIS is one of the new generation JPL instrument prototypes. It is based on a commercial COTS instrument [8] and is proposed to search for life indicators in our solar system, such as in Jupiter's moon Europa [13]. CIRIS is a small, rugged and lightweight Fourier Transform Spectrometer (FTS) with a high Signal-to-Noise Ratio (SNR) in the near-IR to thermal-IR region (2-12 μm), where the strongest and most diagnostic vibrational bands of the compounds of interest in Europa are found (e.g., 'CHNOPS' functional groups).

The major structural novelty introduced by CIRIS is the constant-velocity rotating refractor it uses to vary the optical path difference of the two rays in which incoming light is divided by a beam splitter at the entrance of the instrument

978-1-4673-9141-2/15 $31.00 © 2015 IEEE

(red and green rays in Fig. 1). The rotating refractor replaces the linearly moving mirrors typically used in related spectrometers, such as Michelson interferometers, which are more prone to suffering from vibration-induced motion errors. The reflected rays in the rotating refractor recombine after travelling through the instrument, resulting in a fringe interference light pattern (interferogram) that is measured with a photo-detector (purple ray in Fig. 1). Based on Snell's law, the light rights travel the same distance through CIRIS optics when they are incident on the rotating refractor at 45°. The Zero Path Difference (ZPD) positions occur when the refractor is parallel or perpendicular to the beam splitter, that is, four times over the course of a revolution. The regions where the optical interference between the input light rays can be measured are located at approximately 16° arcs around each of the four ZPD positions. A Faulhaber E2-360I optical incremental encoder mounted on the servomotor that drives the refractor's rotation is used on the ground CIRIS prototype to identify these regions and also provides feedback to the PID that controls the rotation speed of the refractor. As the CIRIS refractor performs 6.5 revolutions per second, the interferograms span over a period of 13.6 ms every 24.8 ms. For in-situ applications, CIRIS is equipped with three servomotors to move the focusing lens on the collected material samples to analyze.

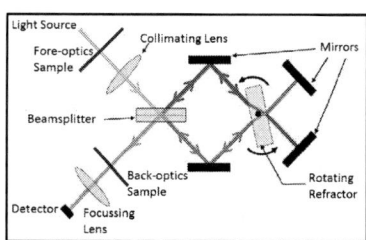 

Fig. 1. CIRIS spectrometer

The optics and functioning of CIRIS result in an interferogram with the high-amplitude values assembled in a narrow central burst, and small-amplitude values spanning the vast majority of the tail positions and carrying the spectral resolution information. This can be seen in Fig. 2. Note the central peak at 0 μs associated with the ZPD position where the light rays are recombined in phase. This shape of the CIRIS interferogram eases the detection (and removal) of radiation hits that induce large current pulses in the instrument's photo-detector (i.e., significantly greater than the standard deviation of the AC interferogram signal). Two radiation hits can be seen in Fig. 2 at -266 μs and -500 μs.

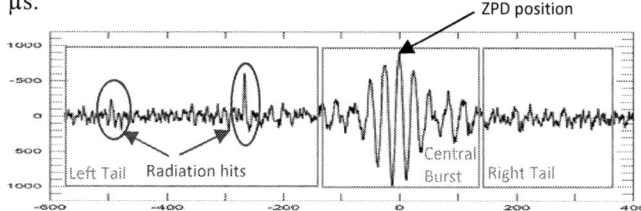

Fig. 2. CIRIS interferogram with radiation hits

The interferogram signal is conditioned, filtered and amplified to ±5V range prior to being digitized at 1 MSPS using an 18-bit resolution ADC. The interferogram samples are then processed via a Fast Fourier Transform (FFT) to produce a spectrum that illustrates the intensity of the wavelengths present in the light beam. This in turn permits to find out the chemical composition of the sample or body under study by looking at the absorption lines in the spectrum. However, spectral leakage (e.g., "picket-fence" effect) and noise are also present in the spectrum due to the limited discretization of the interferograms through time limited digital sampling, and need to be properly handled by the instrument electronics to produce meaningful results [14]. Fig. 3 shows the processed CIRIS uncalibrated spectrum in the laboratory, in which the water and CO_2 spectral features are highlighted.

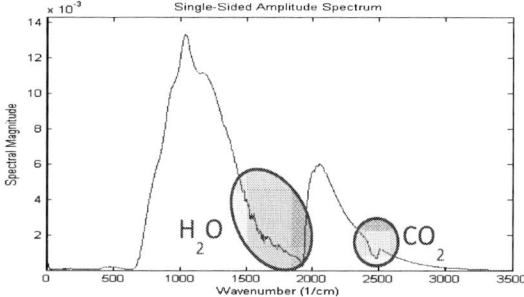

Fig. 3. CIRIS uncalibrated spectrum

The next-generation of CIRIS will be equipped with an array of up to 25 photo-detectors to increase the instrument's spatial resolution and sensitivity in different IR bands.

III.   THE APEX-SoC INSTRUMENTATION CONTROLLER

The APEX-SoC instrumentation controller is prototyped on a ZedBoard mini-ITX board, populated with a Xilinx Zynq 7Z100 SoC device, to which an FPGA Mezzanine Card (FMC) containing the ADC is attached.

Although all the interferogram data delivered by the single photo-detector in the current version of CIRIS can be stored on-chip, an external DDR memory (PL-DDR) is used in advance of the new version of the instrument with multiple photo-detectors, and to enable the use of APEX-SoC with other instruments that process larger amounts of data. The PL-DDR is organized in segments, each of which is read and written by a specific data processing stage. The number and size of the segments can be customized to the specific processing needed by each instrument. The system also includes another DDR memory (PS-DDR) that is solely dedicated to the ARM processor embedded in the Zynq.

In order to make the flight instruments independent and stand-alone subsystems in the spacecraft, a SATA unvolatile Solid State Device (SSD) is used to store large amounts of science results computed by the SoC until a spacecraft-earth downlink communication window is available.

978-1-4673-9141-2/15 $31.00 © 2015 IEEE

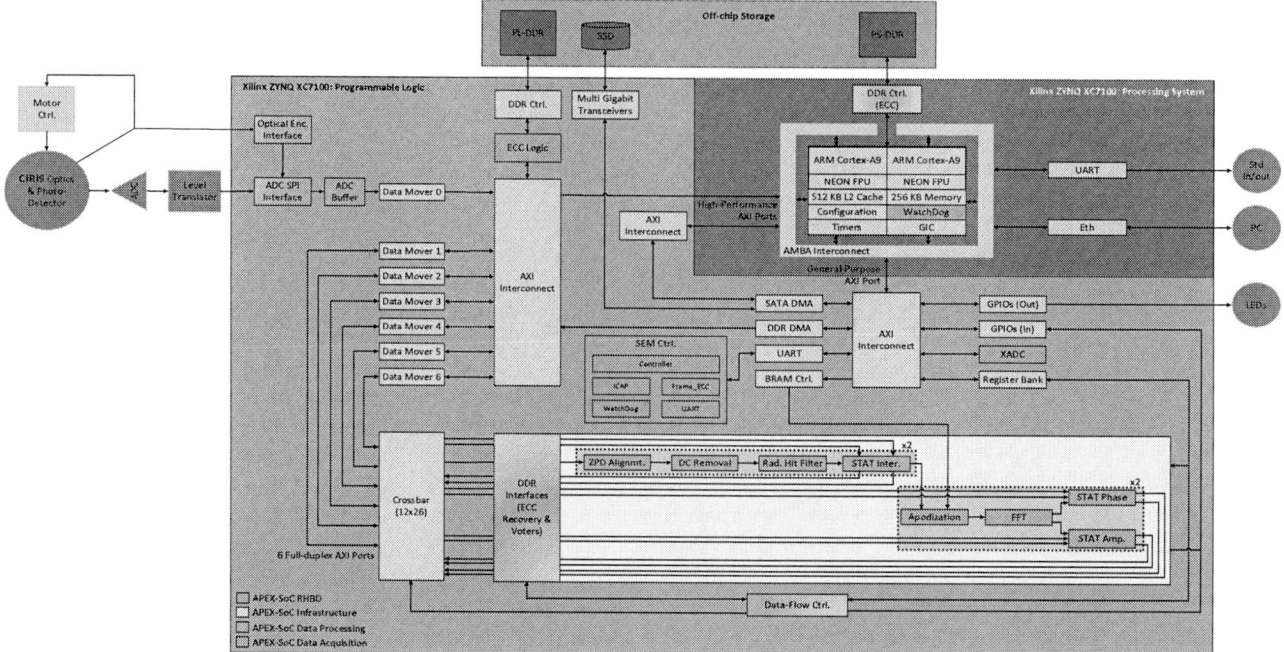

Fig. 4. APEX-SoC block diagram

A. General Architecture of APEX-SoC

The APEX-SoC implements a generic and customizable on-chip infrastructure that gives support to the instrument-dependent acquisition and processing functionality. Despite most of this functionality is usually implemented as custom FPGA logic with standard AXI-Stream interfaces, this logic may be accompanied by software routines to configure and control its functioning and to compute floating-point operations. To deal with this, the APEX-SoC infrastructure (shown in orange color in Fig. 4) includes a fully-functional ARM-based system as well as a customizable multi-ported access to the external DDR and a data-flow controller to ease the implementation of the instrument data processing logic in the FPGA fabric.

The ARM system includes all the peripherals that are typically required by flight science instruments, including: DMA support, GPIOs, Ethernet, SATA, interrupt controller and a memory-mapped register bank to exchange state and configuration data with the FPGA processing logic. On the other hand, process data are exchanged with the FPGA logic through the DMA-accessible PL-DDR memory. In order to speed-up the development of APEX-SoC-based instruments, the ARM processor runs a standard Linux-based operating system, which provides Ethernet protocol to communicate with the spacecraft's main computer and a file system to ease the management of science results stored in the SSD. The Zynq ARM supports also commonly used Real-Time Operating Systems (RTOS) for space, such as VxWorks and more recently Integrity from Green Hills Software.

In order to exploit the full bandwidth delivered by the PL-DDR memory (6.4 GBytes/s), the SoC implements eight 32-bit DDR access ports at 200 MHz using Xilinx-provided AXI-Stream Data Movers, which act as DMA controllers for the FPGA processing logic [15]. One of the DDR ports is dedicated to the instrument data acquisition logic (shown in blue color in Fig. 4), another one is assigned to the ARM DMA, and the remaining six ports are connected to a crossbar that multiplexes them among the instrument data processing stages as requested by the data-flow controller. In effect, the latter data-flow controller schedules the PL-DDR accesses to maximize performance and specifies the memory segment, i.e., address and size, to read/write in each transfer. This controller is based on a Xilinx 8-bit PicoBlaze processor [16] that provides the required flexibility to deal with a wide range of instruments (see Fig. 5).

B. CIRIS-Dependent Data Processing

Fig. 4 shows in green color the CIRIS data processing implemented in the APEX-SoC. This consists of seven stages operating both in time and spectrum domains. Note that most of these stages exchange data using on-chip BRAMs, but some of them (STAT stages) need to store intermediate results in the PL-DDR between executions. A more detailed view of the CIRIS data processing implementation and its integration in the APEX-SoC is shown in Fig. 5.

The first stage prepares the interferogram for the subsequent processing by selecting 8,192 samples centered around the ZPD position. This is done to deal with any temporal shift that might have occurred while sampling the interferogram.

The second stage removes the DC offset in the ZPD aligned interferogram by subtracting its average value, which is computed using a Cumulative Moving Average (CMA).

978-1-4673-9141-2/15 $31.00 © 2015 IEEE

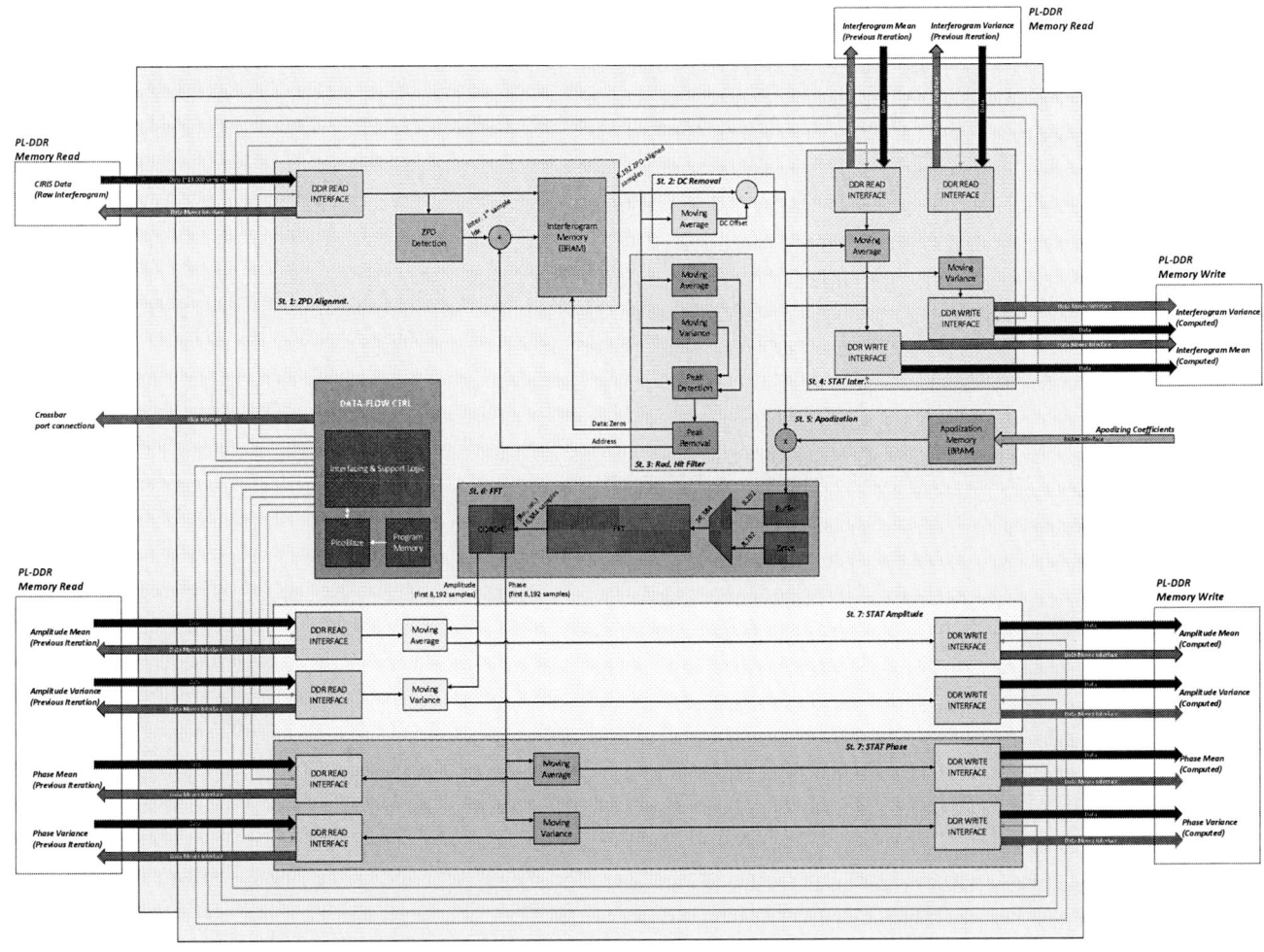

Fig. 5. CIRIS data processing implementation in the APEX-SoC (Green part in Fig. 4)

The third stage implements a radiation hit filter to detect and remove the outlier in the interferogram provoked by radiation striking the CIRIS photo-detector. The radiation pulses at the output of the CIRIS transconductance amplifier, with a bandwidth of 100 kHz, are about 10 μs full width at half maximum and easily recognized in the small-amplitude tail samples using statistics, namely mean and variance (shown by triangles in Fig. 6) [10]. Radiation hit samples are then replaced by zeros without modifying significantly the spectral content of the interferogram [10][17][18]. This property comes from the fact that interferogram points outside the central burst carry mainly resolution information in redundant way. Thus, removing a few points out of 8,192 lead to indistinguishable changes in the spectrum. Moreover, the undetectable radiation hits that are at or below the un-irradiated noise level spread their energy over all wavelengths and therefore average to a constant DC offset in the spectrum. The mean and variance statistics are computed using the Knuth algorithm, which is less prone to loss of precision due to massive cancellation than other related algorithms [19]. In our implementation, updating the mean after processing a new interferogram sample takes 4 clock cycles and 3 additional clock cycles are needed to update the variance.

For the sake of performance, both CMA to determine the DC offset and statistics computation on the small-amplitude tail samples to mitigate the radiation hits are pipelined and performed simultaneously with a latency of 56,144 clock cycles.

The fourth stage (STAT Inter.) computes the variance and performs a CMA on successive interferograms detected around the same ZPD positions with the objective of estimating and increasing the SNR by removing the effect of high frequency and random noise. As in the third stage, the Knuth algorithm is used to calculate these statistics. However, unlike in the radiation hit filter, this stage is fully pipelined as the statistics are independently computed for each interferogram position. The mean and variance results for every interferogram position after each algorithm iteration are stored in dedicated PL-DDR memory segments.

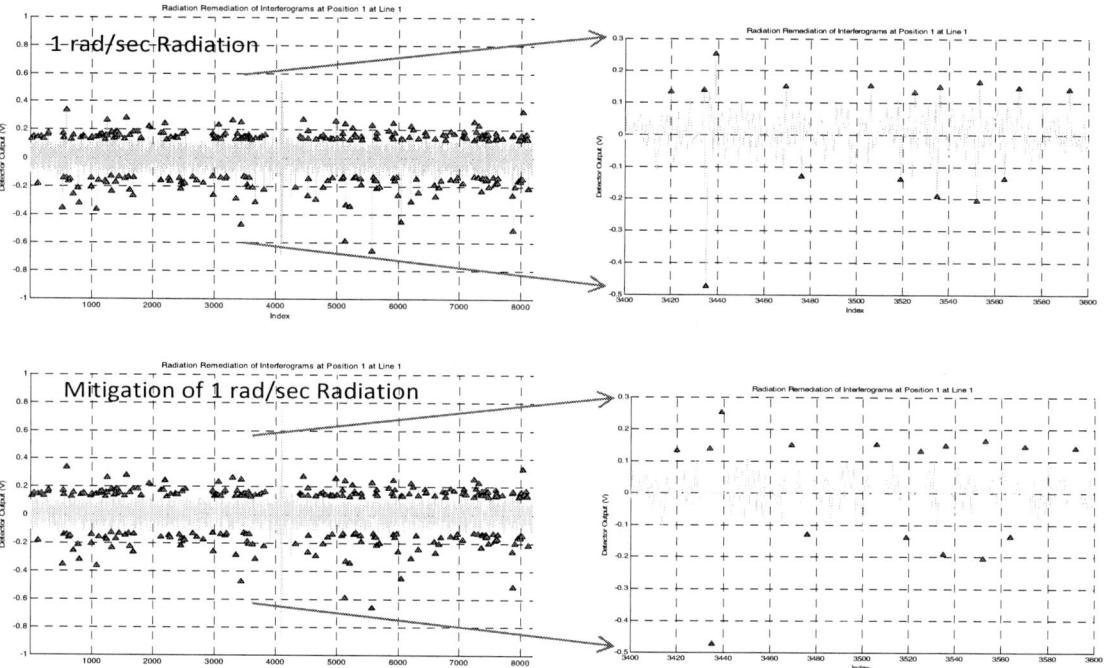

Fig. 6. CIRIS interferogram radiation effects (top) and mitigation (bottom). Right figures shown the zoom over a small interferogram region of the full interferogram shown on the left

The fifth stage apodizes the averaged interferograms at the edges of the sampled regions to minimize the effects of spectral leakage. In order to make the system runtime adaptive, the apodizing coefficients are stored in a BRAM, which is writable by the ARM processor to be able to choose the optimal apodization depending on the instrument (see Fig. 5).

The sixth stage computes the FFT on the interferogram. In light of increasing spectral resolution, this stage adds 4,096 zeros to each of the tails of the interferogram to obtain 8,192 additional interpolated spectrum points in-between the original nonzero-filled spectrum data, that is, 16,384 total spectrum points. This zero padding allows us to reduce the erroneous signal due to the "picket-fence" effect by 36% [20]. In order to obtain the best performance, the FFT stage pipelines several Radix-2 butterfly processing engines, where each engine has its own memory banks to store input and intermediate data. This pipelined implementation allows computing one spectrum data per clock cycle, with a latency of 33,013 clock cycles. A CORDIC logic is then used to translate the real and imaginary representation of the spectrum data used in the internal butterfly processing engines into polar representation, which is more suitable for scientific analysis. This translation is completed within 36 clock cycles.

The seventh and last stage relies on the Knuth algorithm to compute the variance and CMA on the spectrums resulting from the successive interferograms detected around the same ZPD positions. As shown in Fig. 4, both amplitude and phase data are processed in parallel using dedicated logic: STAT Amp. and STAT Phase. As for the fourth stage, the statistics are independently computed for each spectrum position and temporarily stored in dedicated PL-DDR memory segments between algorithm iterations.

Besides the processing stages implemented on the FPGA fabric, CIRIS requires two additional processing stages that involve floating-point operations and hence are executed on the ARM processor, which is equipped with NEON DSP/FPUs. These stages correct the deviations provoked by CIRIS refractor's refractive index variations with wavelength and due to minor dissimilarities in the CIRIS optics between the four ZPD positions. Additionally, the ARM processor executes the CIRIS focusing algorithms, which also involve floating-point operations.

C. RHBD Features

This section lists the RHBD features that have been implemented in the instrumentation APEX-SoC. These are shown in red color in Fig. 4.

The APEX-SoC has a watchdog timer and a recovery mechanism that reboots the ARM processor in the event of an error.

All the storage resources in the APEX-SoC are protected with ECCs. These include off-chip DDR memories and on-chip BRAMs, such as the memory used to store the apodizing coefficients and the program executed by the data-flow controller PicoBlaze. While Xilinx provides an ECC solution for the PS-DDR memory [21], a custom solution to enable the use of ECCs in the PL-DDR memory has been implemented on the FPGA fabric. As shown in Fig. 4, an ECC logic is inserted between the ARM's AXI bus

978-1-4673-9141-2/15 $31.00 © 2015 IEEE

interconnection and the PL-DDR memory controller. This ECC logic implements a hamming (32, 26) code to protect the data words transferred through each of the PL-DDR ports and is pipelined to maximize the performance. It allows detecting and automatically correcting single bit flips (e.g., radiation-induced SEUs) in a PL-DDR data word and detecting, but not correcting, double bit errors. The fact that data are continuously read, processed and written back into the PL-DDR memory reduces the possibility that multiple bit errors are accumulated in the same data word, as the ECC logic corrects every single bit flip that might have occurred while data remains stored in the memory.

Besides, all the finite state machines in the APEX-SoC are implemented using "one-hot" encoding. Hence, if radiation strikes and changes the value of a state flip-flop, the resulting code is not recognized by the state transition logic and the machine is directed to an "illegal" state that signals the ARM processor the error situation.

The APEX-SoC also includes some diagnostic components devoted to detect malfunctions in the CIRIS instrument. The temperature on the Zynq die is continuously monitored using an on-chip sensor (see XADC in Fig. 4) to identify and prevent overheat situations that could lead to the eventual destruction of the chip. In addition to this, excessive noise situations in the power supply are also detected with this sensor. These may indicate that there is a problem with the voltage regulators, power lines in the PCB or even in the spacecraft power subsystem. Besides, the PL-DDR AXI Stream ports are continuously monitored to detect stuck-at situations and errors in memory data transfers. In the specific case of CIRIS, the signals delivered by the optical encoder mounted on the servomotor that drives its internal optics are used to find anomalies in the functioning of the encoder itself, in the PID that controls the refactor's rotation speed or in the servomotor that moves the refractor. In addition, the ADC used to digitize the interferogram signal has a serial SPI interface that is monitored to detect radiation provoked noise and/or malfunctions of the ADC, when the serial data changes are not synchronized with the clock delivered to it.

In order to cope with radiation-induced upsets affecting Zynq's configuration memory [22], the Xilinx Single Event Mitigation (SEM) controller is used [23]. This reads-back configuration data from the Zynq configuration memory using the Internal Configuration Access Port (ICAP) and relies on using the FRAME_ECC block to check the ECCs associated to these data. The SEM controller is able to detect and automatically correct single bit upsets (i.e., SEUs) and detect, but not correct, double bit upsets. In the latter case, the ARM core is signaled, initiating a full reconfiguration of the FPGA fabric from an external flash memory, namely from a QSPI in flight and from a microSD in ground testing.

The APEX-SoC provides support for integrating multiple data processing stages that can be used to process different science data in parallel to increase the performance, or to detect computation errors by comparing their results when they process the same science data. In order to increase the availability of the system, the parallel/redundant data processing stages are spatially isolated in the FPGA fabric, following the Xilinx Isolation Design Flow (IDF) [24]. In light of reducing power consumption, only two redundant data processing stages are used in the APEX-SoC based CIRIS controller (i.e., Dual Modular Redundancy [25][26]). On the other hand, three redundant copies of the same science data are kept in the PL-DDR memory. These are accessed by replicated input and output voters that are coupled to the DMR processing stages, as shown in Fig. 7. Note that data with uncorrectable double-bit errors are not considered in the input voting process. If any output voter detects a result mismatch, a computation error is assumed and the processing of that science dataset is repeated. The data-flow controller coordinates access by the voters to the PL-DDR redundant data segments in a ping-pong fashion, so that voted results do not overwrite source data, in case computation needs to be repeated. Computation errors can occur when radiation affects data registers and/or FPGA configuration [22]. While upsets in data registers cannot be detected by the Xilinx SEM controller, these are automatically corrected when reloading the data to process again. On the other hand, computation errors provoked by FPGA configuration upsets are immediately detected in the DMR scheme when they occur, but the repeated processing is correct only when all configuration upsets are fixed. This usually takes some milliseconds due to the sequential access of the Xilinx SEM controller to FPGA configuration frames, or due to the time needed to fully reconfigure the FPGA, respectively. Although the implemented redundant scheme permits to mask single errors in the voters and provides a good error-detection coverage, false positive errors can occur when radiation affects one of the output voters. Combining the Xilinx SEM controller with circuit redundancy is a very effective solution recommended by Xilinx [27].

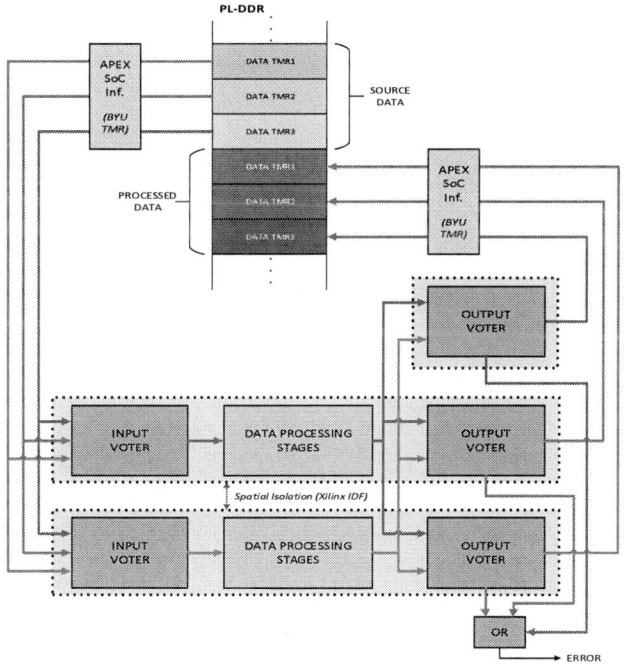

Fig. 7. DMR Scheme used in the APEX-SoC CIRIS controller

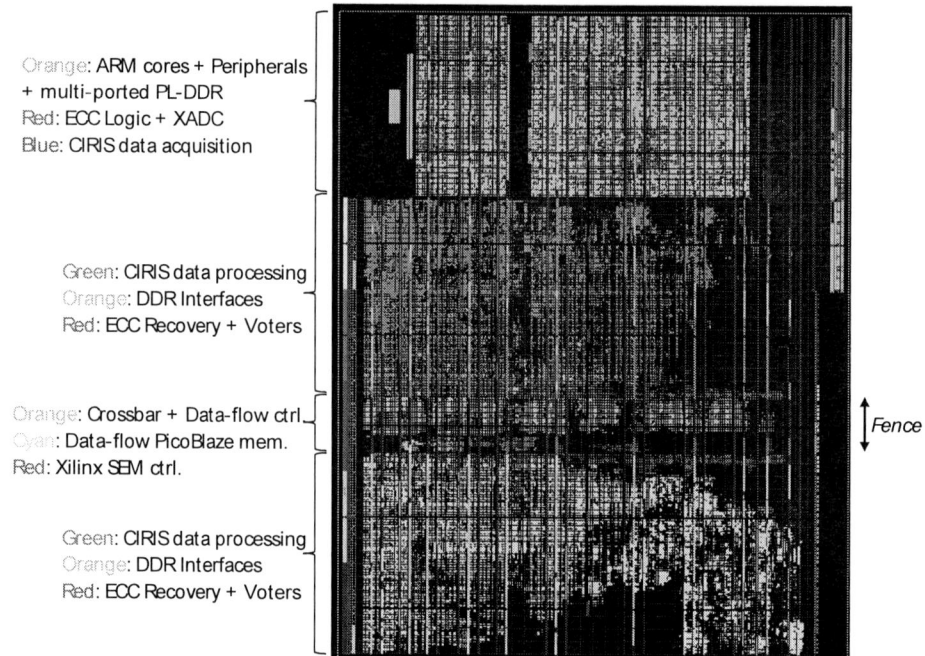

Orange: ARM cores + Peripherals
+ multi-ported PL-DDR
Red: ECC Logic + XADC
Blue: CIRIS data acquisition

Green: CIRIS data processing
Orange: DDR Interfaces
Red: ECC Recovery + Voters

Orange: Crossbar + Data-flow ctrl.
Cyan: Data-flow PicoBlaze mem.
Red: Xilinx SEM ctrl.

Green: CIRIS data processing
Orange: DDR Interfaces
Red: ECC Recovery + Voters

Fence

Fig. 8. APEX-SoC-based CIRIS controller floor-planning

In order to protect selected critical parts in the APEX-SoC infrastructure, we plan to use low-level redundancy techniques, such as the Xilinx TMR tool (if available for the Zynq) [28] or the BYU EDIF tool [29].

In order to increase the reliability, some of the critical parts of the SoC could be implemented using off-chip rad-hard circuits (e.g., watchdog timers, reset, voters) as done partially in [7].

IV. RESULTS

A. Implementation

Table I lists the amount and type of resources required by the APEX-SoC when implemented on a Zynq 7Z100 device. The power consumption estimated by Xilinx Vivado tool is approximately 5W, which is considered to be the power budget for a flight instrument avionics for data processing. The APEX-SoC uses up to 256 MB in the PL-DDR memory to process 25 interferograms simultaneously.

TABLE I.   IMPLEMENTATION RESULTS ON A XILINX ZYNQ 7Z100

Component	LUTs	Flip-flops	DSP48s	BRAMs
Data Acquisition	347	267	-	-
RHBD Features	6,194	3,818	-	14.5
Data Processing	61,874	46,152	414	155
Infrastructure	61,733	49,631	-	395
Total	130,148 (47%)	99,868 (18%)	414 (20%)	564.5 (75%)

Fig. 8 shows the floor-planning of the APEX-SoC-based CIRIS controller. The crossbar and data-flow controller are strategically located in-between the two processing stages, in such a way that they are used as a fence to prevent a single energized particle from corrupting the two redundant stages and the voters coupled to them at the same time. Besides, the

fact that the crossbar and data-flow controller are physically near to the processing stages with which they share most of their connections, leads to short paths and ultimately high usable clock frequency (i.e., 200 MHz). It is also important to note that the small footprint of the voters, in the range of hundreds of LUTs and flip-flops, minimizes the chances of being corrupted by radiation.

B. Performance

When used in the performance mode, the parallel computation delivered by the APEX-SoC permits to process two interferograms every 460 µs, with a latency of 867 µs. This means that a hundred interferograms can be processed within the time span between consecutive ZPD positions (24.8 ms), which is almost four times the processing requirements of the next-generation of CIRIS spectrometer. On the other hand, when used in the reliability mode, the APEX-SoC is still able to fulfill the processing requirement for the next-generation of CIRIS, requiring up to 900 µs to process each interferogram.

C. Robustness Against Radiation

A radiation test was conducted at JPL using a 60Co γ-ray source (1 rad/sec) directed toward the CIRIS photo-detector operating at 77K to reduce detector noise below the radiation hit pulses [10]. The hit rate in this test was approximately 3,400 hits per second, exceeding what is expected in Europa by at least a factor of three. We verified that the radiation hit filter stage implemented in the FPGA data processing reduces the distortion of the line shape and spectrum amplitude provoked by radiation while keeping all other spectral components unaltered. In addition, as shown in Fig. 9, this mitigation considerably increases the instrument SNR by eliminating the noise due to radiation hit pulses.

We have not conducted any specific experiment to test the implemented RHBD features yet, as these are well known and proven to be effective [24]-[27]. Having said this, we do have plans to characterize the APEX-SoC-based CIRIS instrument in a simulated Europa-like thermal and radiation environment when the whole design is finished.

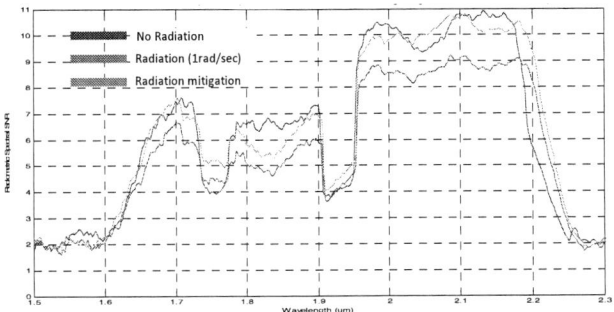

Fig. 9. SNR improvement due to radiation hit mitigation

V. CONCLUSIONS AND FUTURE WORK

This paper has presented the ongoing work at JPL to develop a fault-tolerant SoC to power the next-generation flight science instruments avionics, using the JPL CIRIS instrument as a case-study. The paper has also discussed the RHBD mechanisms and data processing techniques that have been implemented in the APEX-SoC to deal with harsh radiation in space environments. Future work will explore the use of reconfiguration-based fault-tolerance techniques, such as [30], and lockstep ARM processor schemes for space applications [31].

ACKNOWLEDGMENT

The research described in this paper was carried out at the Jet Propulsion Laboratory, California Institute of Technology, under a contract with the National Aeronautics and Space Administration (NASA). Xabier Iturbe is funded by the European Commission's FP7 Marie-Curie International Outgoing Fellowship Program with "Project No. 627579".

REFERENCES

[1] R. C. Wiens et al., "The ChemCam Instrument Suite on the Mars Science Laboratory (MSL) Rover: Body Unit and Combined System Tests", Space Science Reviews, pp. 167-227, Springer, 2012.

[2] D. Petrick, N. Gill, M. Hassouneh, R. Stone, L. Winternitz, L. Thomas, M. Davis, P. Sparacino, T. Flatley, "Adapting the SpaceCube v2.0 Data Processing System for Mission-Unique Application Requirements", in Proc. of the NASA/ESA Conference on Adaptive Hardware and Systems, 2015.

[3] M. Amrbar, F. Irom, S. M. Guertin,G. R. Allen, "Heavy Ion Single Event Effects Measurements of Xilinx Zynq-7000 FPGA", in Proc. of the IEEE Nuclear and Space Radiation Effects Conference, 2015.

[4] L. A. Tambara, F. L. Kastensmidt, N. H. Medina, V. A. P. Aguiar, M. A. G. Silveira, "Heavy Ions Induced Single Event Upsets Testing of the 28 nm Xilinx Zynq-7000 All Programmable SoC", in Proc. of the IEEE Nuclear and Space Radiation Effects Conference, 2015.

[5] D.M. Hiemstra and V. Kirischian, "Single Event Upset Characterization of the Zynq-7000 ARM Cortex-A9 Processor Unit

using Proton Irridiation", in Proc. of the IEEE Nuclear and Space Radiation Effects Conference, 2015.

[6] http://www.nasa.gov/directorates/heo/home/CubeSats_initiative.html

[7] D. Rudolph, C. Wilson, J. Stewart, P. Gauvin, A. George, H. Lam, G. Crum, M. Wirthlin, A. Wilson, A. Stoddard, "CSP: A Multifaceted Hybrid Architecture for Space Computing", in Proc. of the Small Satellite Conference, 2014.

[8] W. Wadsworth and J. P. Dybwad, "Rugged High-Speed Rotary Imaging Fourier Transform Spectrometer for Industrial Use", in Proc. of the International Society for Optics and Photonics Conference on Environmental and Industrial Sensing, pp. 83–88, 2002.

[9] G. Flesch and D. Keymeulen, "Adaptive Control of Tunable Laser Spectrometers for Space Flight Applications", in Proc. of the IEEE Aerospace Conference, 2010.

[10] P. Yiu, X. Iturbe, D. Keymeulen, D. Berisford, K. P. Hand, R. W. Carlson, W. Wadsworth, R. Levy, "Adaptive Controller for a Fourier Transform Spectrometer with Space Applications", in Proc. of the IEEE Aerospace Conference, 2015.

[11] R. Levy, K. P. Hand, R. W. Carlson, W. Wadsworth, J. P. Dybwad, D. Berisford, D. Keymeulen, J. Feldman, "Remote Sensing in Severe Radiation Environments", In proc. of the Sensors Tech Forum, 2011.

[12] R. W. Carlson, P. Weissman, W. Smythe, J. Mahoney, "Near-Infrared Mapping Spectrometer Experiment on Galileo", Springer, 1992.

[13] R. W. Carlson, K. P. Hand, D. Berisford, D. Keymeulen, "The Compositional InfraRed Interferometric Spectrometer (CIRIS) for Assessing the Habitability of Europa", in Proc. of the American Geophysical Union Fall Meeting, 2013.

[14] V. Saptari, "Fourier Transform Spectroscopy Instrumentation Engineering", SPIE Press, vol. 61, 2004.

[15] Xilinx Inc., "AXI DataMover User Guide", PG022, 2014.

[16] K. Chapman, "PicoBlaze for Spartan-6, Virtex-6, 7-Series, Zynq and UltraScale Devices (KCPSM6)", 2014.

[17] W. Herres and J. Gronholz, "Elimination of ThinFilm Infrared Channel Spectra in Fourier Transform Infrared Spectroscopy", Applied Spectroscopy, vol. 3, no. 5, pp 552-553, 1976.

[18] J. Gronholz and W. Herres, "Understanding FT-IT Data Processing", in Instruments & Computers vol. 3, number 10, pp. 1-23, 1985.

[19] D. E. Knuth, "The Art of Computer Programming", vol. 2: Seminumerical Algorithms, Addison-Wesley, 1998.

[20] J. Gronholz and W. Herres, "Understanding FT-IT Data Processing", Instruments & Computers, vol. 3, no. 10, pp. 1-23, 1985.

[21] Xilinx Inc., "Zynq-7000 All Programmable SoC and 7 Series Devices Memory Interface Solutions", UG586, 2014.

[22] F. L. Kastensmidt, L. Carro, R. Reis, "Fault-Tolerance Techniques for SRAM-Based FPGAs", Springer, 2006.

[23] Xilinx Inc., "Soft Error Mitigation Controller", PG036, 2014.

[24] J. D. Corbett, "The Xilinx Isolation Design Flow for Fault-Tolerant Systems", Xilinx Inc., WP412, 2013.

[25] J. Teifel, "Self-Voting Dual-Modular-Redundancy Circuits for Single-Event-Transient Mitigation", IEEE Transactions on Nuclear Science, vol. 55, no. 6, pp. 3435-3439, 2008.

[26] M. Berg, "Fail-Safe Strategies for FPGA Devices Targeted for Critical Applications", in proc. of the Military and Aerospace Programmable Logic Devices Workshop, 2015.

[27] A. Lesea and P. Alfke, "Xilinx FPGAs Overcome the Side Effects of Sub-40 nm Technology", Xilinx WP256, 2011.

[28] http://www.xilinx.com/ise/optional_prod/tmrtool.htm

[29] http://reliability.ee.byu.edu/edif

[30] X. Iturbe, A. Ebrahim, K. Benkrid, C. Hong, T. Arslan, J. Perez, D. Keymeulen, M. D. Santambrogio, "R3TOS-based Autonomous Fault-Tolerant Systems", IEEE Micro, vol. 34, no. 6, pp. 20-30, 2014.

[31] http://www.tcls-arm-for-space.eu

Hardware Architecture and Optimization of Sliding Window Based Pedestrian Detection on FPGA for High Resolution Images by Varying Local Features

Asim Khan, Muhammad Umar Karim Khan, Muhammad Bilal, Chong-Min Kyung

Department of Electrical Engineering
Korea Advanced Institute of Science and Technology (KAIST),
Daejeon, South Korea
Email: {asimkhan, umar, bilalm, kyung}@kaist.ac.kr

Abstract— Pedestrian detection has lately attracted considerable interest from researchers due to many practical applications. However, the low accuracy and high complexity of pedestrian detection has still not enabled its use in successful commercial applications. In this paper, we present insights into the complexity-accuracy relationship of pedestrian detection. We consider the Histogram of Oriented Gradients (HOG) scheme with linear Support Vector Machine (LinSVM) as a benchmark. We describe parallel implementations of various blocks of the pedestrian detection system which are designed for full-HD (1920x1080) resolution. Features are improved by optimal selection of cell size and histogram bins which have been shown to significantly affect the accuracy and complexity of pedestrian detection. It is seen that with a careful choice of these parameters a frame rate of 39.2 fps is achieved with a negligible loss in accuracy which is 16.3x and 3.8x higher than state of the art GPU and FPGA implementations respectively. Moreover 97.14% and 10.2% reduction in energy consumption is observed to process one frame. Finally, features are further enhanced by removing petty gradients in histograms which result in loss of accuracy. This increases the frame rate to 42.7 fps (18x and 4.1x higher) and lowers the energy consumption by 97.34% and 16.4% while improving the accuracy by 2% as compared to state of the art GPU and FPGA implementations respectively.

Keywords— FPGA, Low Power, Object Detection, Real-Time

I. INTRODUCTION

Researchers in industry and academia have been striving for accurate and real-time pedestrian detection (PD) for more than a decade owing to many commercial and military applications. Industries such as surveillance, robotics, and entertainment will be greatly influenced by appropriate application of PD. Advanced driver assistance systems (ADAS) and unmanned ground vehicles (UGV) are merely a distant dream without automated pedestrian detection. The fact that more than 15% of traffic accidents include pedestrians [1] shows the importance of real-time pedestrian detection for the modern society [2].

Amid numerous applications, the search for an accurate yet fast PD algorithm is ongoing. Researchers have shown great interest over the past few years in extracting diverse features

from an image and finding an appropriate classification method to perform robust PD [10-14]. However, the histogram of oriented gradients (HOG) approach has proven to be a groundbreaking effort, and has shown good accuracy in various illumination conditions and multiple textured objects. Inspired from SIFT [5], the authors in their seminal paper [4] present a set of features over a dense grid in a search window. For training and classification they used the linear support vector machine (linSVM). Their work inspired many other researchers and is still used as a benchmark PD scheme.

Although HOG was presented many years back, it is surprising to see that very few efforts have been made for an optimal hardware implementation of HOG. In fact most of the research has been targeting pedestrian detection on a high end CPU or GPU or combination of both [24-29]. Field Programmable Gate Arrays (FPGA) and Application Specific Integrated Circuits (ASIC) often provide better execution speed and energy efficiency as compared to GPUs due to deep pipelined architectures. Furthermore, in many embedded applications, such as surveillance, there are numerous constraints on hardware cost, speed, and power consumption. For such applications, it is more suitable to use task-specific (FPGA, ASIC) rather than general-purpose platforms. Moreover, to meet such constraints, certain parameters of the algorithm need to be tuned and an insight is required into how

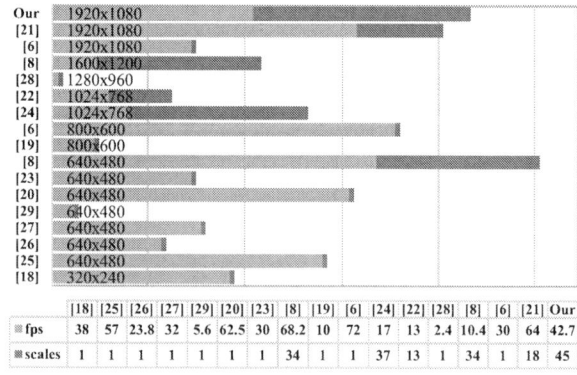

Fig. 1 Comparison of the state of the art in terms of frames per second and number of scales.

	[18]	[25]	[26]	[27]	[29]	[20]	[23]	[8]	[19]	[6]	[24]	[22]	[28]	[8]	[6]	[21]	Our
fps	38	57	23.8	32	5.6	62.5	30	68.2	10	72	17	13	2.4	10.4	30	64	42.7
scales	1	1	1	1	1	1	1	34	1	1	37	13	1	34	1	18	45

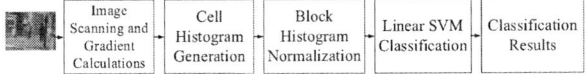

Fig. 2 Block Diagram of HOG Algorithm Flow

the change of parameters of PD affects not only the accuracy but also the hardware complexity.

Efforts have been made in the research community to either improve the accuracy of PD or reduce the hardware complexity of HOG. In [6] and [7], the computational complexity of HOG is reduced with cell-based scanning and simultaneous SVM calculation, authors have presented FPGA and ASIC implementations for full HD resolution; however, their implementations use the parameters as suggested in [4]. In [15], simplified methods are presented for division and square root operations for use in HOG. However, by employing their methods, the accuracy of PD is severely degraded. A multiprocessor system on chip (SoC) based hardware accelerator for HOG feature extraction is described in [16]. In [17], the authors reuse the features in blocks to construct the HOG features of overlapping regions in detection windows and then use interpolation to efficiently compute the HOG features for each window.

In [18], the authors developed an efficient FPGA implementation of HOG to detect traffic signs. In [19], a real-time PD framework is presented which utilizes an FPGA for feature extraction and a GPU for classification. A deep-pipelined single chip FPGA implementation of PD using binary HOG with decision tree classifiers is discussed in [20]. A heterogeneous system is presented in [22] to optimize the power, speed and accuracy. In our study, we are yet to find an effort which analyzes the effects of reducing hardware costs on accuracy of HOG. For real-time PD with power and area constraints, it is imminent to find the set of parameters of HOG that provide the best compromise in terms of computational complexity and accuracy. Recently, a hardware architecture for fixed point HOG implementation has been presented [8] where the bit-width has been optimized to achieve significant improvement in power and throughput. We believe that in addition to bit-width there are other parameters which need to be optimized to provide a holistic understanding of the relationship between accuracy, speed, power, and complexity. Moreover, as shown in Fig. 1, a combination of the number of scales required for maximum accuracy [8]-[9] and throughput for real-time pedestrian detection for full-HD resolution has not been achieved before.

The key contributions of our work can be summarized as follows.

- We present parallel implementation of various blocks of HOG-based PD on an FPGA. Parallel implementation has been used to improve the speed of PD.
- We derive the accuracy, speed, power, and hardware complexity results of HOG-based PD with different choices of cell sizes and number of histogram bins.
- We show that by using the right choice of cell size and number of histogram bins, a significant reduction in power consumption and increase in throughput can be achieved with reasonable accuracy.

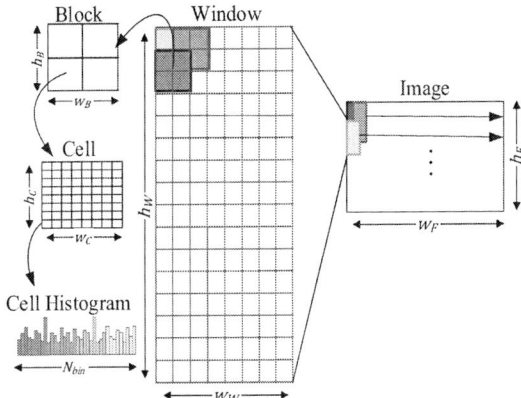

Fig. 3. A depiction of image division for sliding window based object detection. An input image ($w_F \times h_F$) is divided into overlapping windows. The window is divided into overlapping blocks which are further divided into cells. A histogram is generated for every cell.

- Finally features are refined by removing insignificant gradients which results in not only improvement of throughput and power consumption but also accuracy.

The rest of the paper is organized as follows. In Section II, a brief overview of HOG is presented. The proposed hardware implementation is discussed in Section III. In Section IV, the accuracy, speed, power, and hardware complexity results are shown for different choices of parameters and the optimal choice of parameters under given constraints is described. Section V concludes the work.

II. OVERVIEW OF HOG

In this section, we present a brief overview of the HOG algorithm for PD. Although HOG can be used in a part-based PD scheme, we limit our discussion to the rigid HOG as described in the original paper [4]. A block diagram showing functional blocks of the algorithm is shown in Fig. 2 where (w_F, h_F), (w_W, h_W), (w_C, h_C), (w_B, h_B) are the frame, cell and block (width, height) respectively. N_{bin} is the number of histogram bins.

In HOG, a search window is divided into multiple overlapping blocks which are further divided into cells as shown in Fig. 2. The blocks have an overlap of 50%, creating a dense grid over the search window. So a single ($w_W \times h_W$) window has $n_C = w_W/w_C \times h_W/h_C$ cells and $n_B = \left(\frac{w_W - w_C}{w_C} \times \frac{h_W - h_C}{h_C} \right)$ blocks. Gradient features are extracted from these blocks and cells, and are concatenated to create a single feature vector for the whole window.

A filter with coefficients [-1, 0, 1] is applied to the window in horizontal and vertical directions, creating the images G_x and G_y, respectively. These images are used to generate the gradient magnitude image, G_m, and the gradient orientation image, G_o, for each pixel (x, y) as follows.

$$G_m(x, y) = |G(x, y)| = \sqrt{G_x(x, y)^2 + G_y(x, y)^2} \quad (1)$$

$$G_o(x, y) = \tan^{-1} \frac{G_y(x, y)}{G_x(x, y)}, \quad (2)$$

978-1-4673-9141-2/15 $31.00 © 2015 IEEE

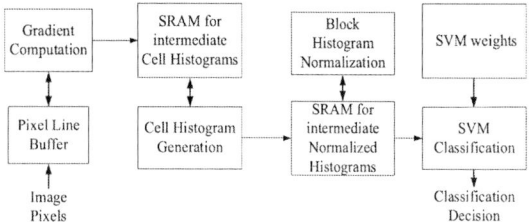

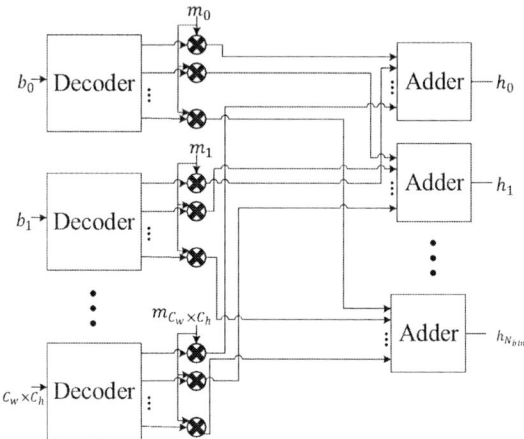

Fig. 4 Block Diagram of Hardware Architecture. Gradient is computed over input pixels stored in Pixel Line buffer, cell histograms are then built using gradients, SRAM is used to store intermediate cell histograms,. Next steps are normalization of histograms generated and finally classification.

The histogram used in the feature accumulates the orientation information of an image. Each histogram has multiple bins, where each bin represents a specific orientation in the interval $[0, \pi)$. The value $G_m(x, y)$ is added to the bin of the histogram which corresponds to $G_o(x, y)$. Such histograms are developed for every cell of the window, as shown in Fig. 3.

The cell histograms belonging to a single block are concatenated to form block histogram of length $M = 4 \times N_{bin}$, where, N_{bin} is the number of bins in each cell histogram. Block histograms are further L2-normalized using (3), and then added to the feature vector. *L2-norm* for an unnormalized feature vector v, is given by

$$x_i^b = \frac{v_i^b}{\sqrt{\|v\|_2^2 + \varepsilon^2}}. \tag{3}$$

where $i = 1, \ldots, M$, $b = 1, \ldots, n_B$, $\|v\|_2^2 = v_1^2 + v_2^2 + \cdots + v_M^2$ and ε is a small constant to avoid division by zero. L2-normalization is performed to improve robustness against illumination changes.

For classification, linSVM is used. From an implementation perspective, a weight vector is obtained after the training stage. During classification, a dot product of the feature extracted from the window and the weight vector is compared against a threshold. If the dot product is greater than the threshold then a pedestrian is identified.

III. HARDWARE ARCHITECTURE

The hardware implementation of HOG presents a unique challenge, which is quite distinct from the software implementation. First, we cannot store and access a complete frame, and read and write from multiple addresses at once as this will require unrealistically large hardware resources. Second, floating point operations are quite costly in hardware, as they use more FPGA area and runs at a lower frequency; therefore, we avoid them in hardware implementation. Finally, the choice of parameters affects hardware complexity significantly compared to software implementation.

Our key objectives in this implementation are to maintain the maximum accuracy and minimum power consumption while performing real time PD by controlling local features. Hardware/memory optimization is done using optimal values of these features. The optimized architecture thus obtained results in a reduced workload and low bandwidth.

The conceptual block diagram of the proposed HOG Accelerator (HOG-Acc) is shown in Fig. 4. In the following we

Fig. 5 Cell Histogram Generation (CHG) Engine: Histogram bins and gradient magnitudes are given as input and cell histograms are generated.

present a description of the major functional blocks shown in Fig. 4.

A. Gradient Computation

To compute the gradient magnitude and orientation, the horizontal and vertical gradient images, i.e., G_x and G_y, need to be generated. Gradient is computed over the 3x3 neighborhood of each pixel; therefore, two line buffers are required to store two consecutive scan lines of the image to maintain a 3x3 neighborhood of every pixel. In order to reduce the computational complexity, the approximations in [10] are used to compute the gradient magnitude and orientation. The hardware utilizes only comparators, shifters and adders, hence reducing the complexity significantly.

Moreover, we argue that using only integer values of gradient magnitude can further improve the accuracy, throughput and power consumption. The details are given in Section IV. The key insight is that by using integer values for gradient magnitudes, we can remove the histogram values which are less significant. The advantages are twofold. 1) It reduces the hardware complexity due to reduced bit width and integer operations. 2) It improves the accuracy because removing these petty gradient magnitudes enhances the feature vector for training and classification.

B. Cell Histogram Generation

We propose a parallel Cell Histogram Generation (CHG) module as shown in Fig. 5. Gradient magnitudes and orientation bins for every $w_c \times h_c$ pixels are given as input to CHG. Decoders and adders are used to build the histogram. Each bin value is given as input to the decoder. Only one output is set to '1' corresponding to the specific bin; gradient magnitude for that bin hence propagates to the input of adder, where all the magnitudes of the same bin are added.

The decoder size is dependent on the number of bits required to represent single bin, i.e; if number of histogram bins increase the size of decoder increases. On the other hand, the cell size ($C_{size} = w_c \times h_c$) affects the number of decoders as the total number of decoders required equals C_{size}. Multipliers required for CHG are dependent on both N_{bin} and C_{size}. Multipliers in each stage depend on N_{bin} while number of stages depends on C_{size}. Finally, the number of adders is equivalent to the N_{bin}

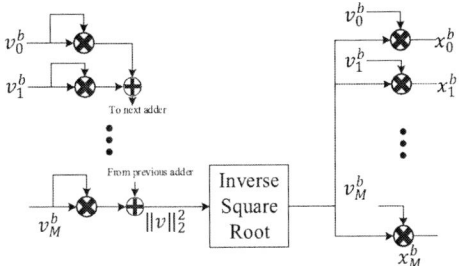

Fig. 6 Block Histogram Normalization (BHN) Engine: Un-normalized Block histograms (concatenated cell histograms) are used as inputs to generate normalized block histograms.

chosen. Adder size, however, varies according to C_{size}. We can clearly see that the complexity of CHG is strongly dependent on N_{bin} and C_{size}.

Since pixels are coming row by row, we have to maintain cell histograms for multiple blocks and windows as each row has multiple windows. Therefore, the gradient magnitudes and orientations computed for every w_C pixels (one cell) are concatenated and stored in memory. Pixels of row index which is a multiple of h_C indicate the completeness of cell. This row is directly stored into registers. At the start of every such row, respective values of previous rows for the particular cell are read into registers from block RAM every clock cycle. As we have considered $w_C = h_C$ the cell completes in horizontal and vertical directions simultaneously. Hence, the number of shift registers required is equivalent to h_C. Each shift register stores magnitudes and bins for w_C pixels. After w_C cycles the data of one cell is completed so it is shifted to the memory, which in turn writes the data for the previous row in the next register.

The resultant cell histogram is given to the next stage for processing. This is done every time the new cell is completed. i.e; when the row index is a multiple of h_C and column index is a multiple of w_C. The cell histograms for multiple windows in a frame are stored in memory while they are shifted to registers for each active window (the window whose cells histograms are completed).

C. Block Histogram Normalization

Cell histograms are maintained in the memory till four neighboring cells are completed and a block is obtained. Note that the memory required to store the cell histograms increases with smaller cells (more cells per row and column) and larger number of histogram bins for every cell (more data per cell). In other words, C_{size} affects the memory locations required while N_{bin} influences the width of each location.

The histogram is normalized using the Block Histogram Normalization Engine (BHN) shown in Fig. 6. Normalization is performed every time a new block is completed. Each histogram value in a block is squared and added. The sum is given as input to inverse square root module which is approximated using "fast inverse square root" algorithm [30]. In summary, logical shifting, subtraction and finally one iteration of Newton's method approximates the inverse square root. Finally, the result of inverse square root is multiplied with each histogram value to generate the normalized block histogram.

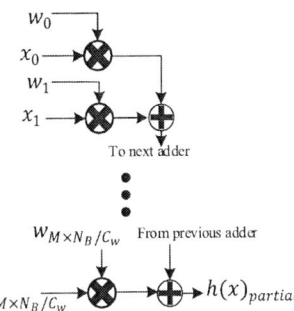

Fig. 7 Partial Classification Engine (PCE), single stage of LinSVM classification to be performed for whole window. Inputs are normalized histograms while output is the partial classification result.

It is seen that the number of multipliers in BHN depends on the size of the block histogram, which is related to number of bins assigned to each cell histogram. Adding a single bin to cell histogram adds eight multipliers to the hardware. The adder size also increases proportionately.

D. SVM Classification

The normalized histograms obtained from the BHN block are again stored in the memory. Once normalized histograms for the whole window are available classification can be performed which can consume a fair amount of memory. Performing classification for the whole window at once also requires a large number of multipliers and adders. The situation gets worse as the feature vector size increases with smaller cell sizes or large number of bins. Therefore, we have opted for partial classification by dividing the classification for the whole window into multiple stages. The hardware shown in Fig. 7 is reused at every stage. The strategy behind reusing the hardware is very straightforward. Since it takes w_C cycles to completely process a cell, we have reused the same hardware over these w_C cycles doing partial classification every N_B/w_C blocks. So the number of partial classification stages is equal to C_w. The results of each stage are accumulated to get the final classification result.

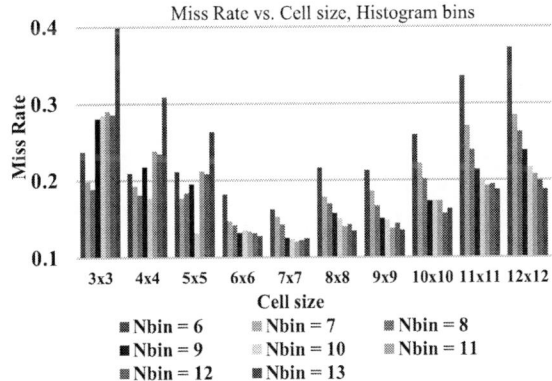

Fig. 8 Accuracy Analysis, miss rate generally reduces with increasing cell sizes and decreasing number of bins

978-1-4673-9141-2/15 $31.00 © 2015 IEEE

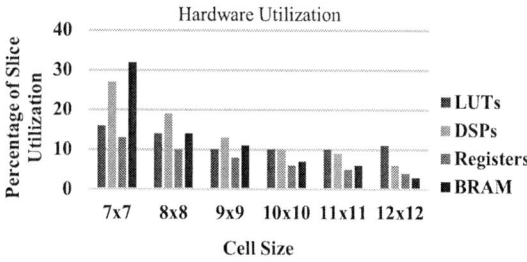

(a) Percentage of slice utilization for multiple cell sizes and fixed N_{bin}=9.

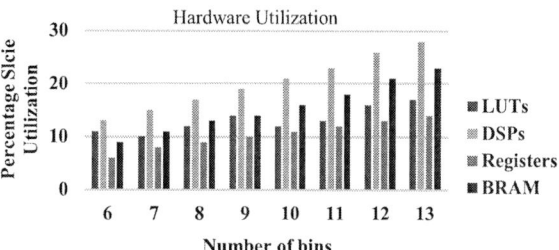

(b) Percentage of slice utilization for multiple histogram bins and fixed C_{size}=8x8

Fig. 9 Hardware Utilization Comparison. Breakdown of usage of multiple slices of Xilinx FPGA with varying cell size and histogram bins. Increasing any one of them results in increased hardware complexity.

The key observation is that the cell size effects the hardware complexity in two ways. First, it has a direct impact on feature vector size. Second, larger the cell size, more cycles will be available to perform classification, thereby, smaller hardware is required for partial classification.

IV. RESULTS AND DISCUSSION

In this section, we evaluate our hardware implementation for multiple cell sizes and histogram bins to obtain optimal set of these parameters. Results are presented for full-HD (1920x1080) resolution videos. Window size is considered to be 64x128. Block size is 2x2 cells, while block and window step size is one cell for both horizontal and vertical directions. Scale factor to rescale images is set to 1.05. This results in 45 scales to be processed per frame. Other parameters depend on the choice of cell size and histogram bins.

Here, we first present our experimental setup then we analyze the effect of different cell sizes and histogram bins on accuracy, throughput and power. Using these results, parameters yielding least power and maximum throughput with negligible loss in accuracy are selected. Finally, using these parameters comparison with the state of the art object detection implementations is presented.

A. Experimental Setup

We have implemented our system on Xilinx Virtex 7 (XC7VX485T) FPGA. There are 75,900 slices, 607,200 Configurable Logic Blocks (CLBs) and 485,760 logic cells in this FPGA. Moreover, 37,080 Kb block RAM and 2,800 DSP slices are present. Image rescaling and window sampling is done for positive and negative images and then sent to HOG-Acc for processing which returns the detection result. Processing 45 scales requires a large amount of memory and pipelined stages so we have utilized the time multiplexing approach of [21]. In addition three HOG Accelerators are employed in parallel to further improve the frame rate. The host software is written using Visual Studio 2012 and Verilog is used for HOG-Acc design. Design is synthesized using Xilinx ISE 14.7 and along with Modelsim 10.2, a hardware/software co-simulation is performed to verify the implementation functionality.

B. Accuracy Analysis

We have used INRIA dataset [31], to evaluate our HOG implementation. There are several other datasets available for pedestrian detection evaluation like Caltech [32], ETH [33], and Daimler [3]. We have, however, restricted our results to INRIA because it provides us with a reasonable variety of images with different poses and backgrounds so these results can be generalized to other datasets and real life scenarios.

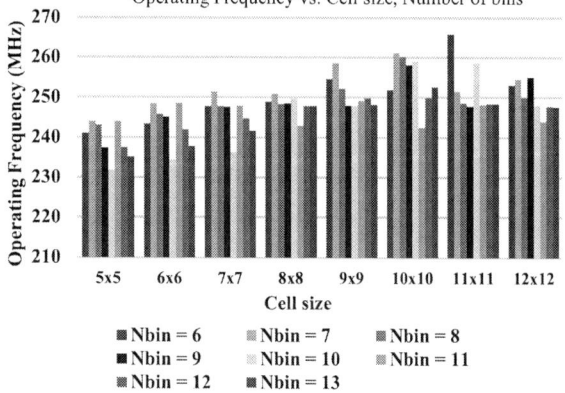

Fig. 10 Throughput Analysis, An increase in throughput is seen for bigger cell sizes and histogram bins.

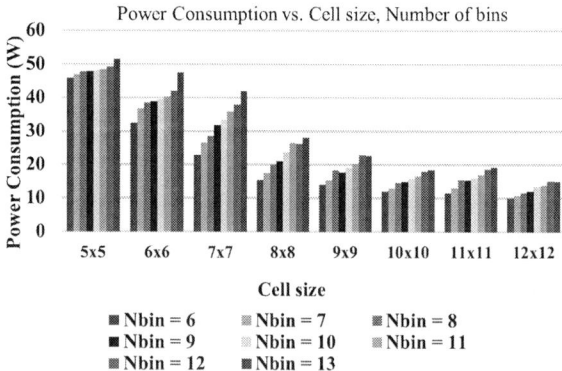

Fig. 11 Power Consumption Analysis, Large cell size and smaller number of histogram bins results in low power consumption.

978-1-4673-9141-2/15 $31.00 © 2015 IEEE

TABLE I
COMPARISON OF PARAMETERS AND THROUGHPUT FOR VARIOUS GPU AND FPGA IMPLEMENTATIONS

	Cell Size	Histogram Bins	Win. Stride	# scales	Resolution	Windows/frame	FPS
			GPU Implementation				
[22], [24]	8x8	9	8	37	1024x768	-	17
[25]	8x8	-	-	>1	640x480	4,096	57
[26]	8x8	8	2	>1	640x480	50,000	23.8
[27]	8x8	9	8	1	640x480	-	32
[28]	8x8	9	4	>1	1280x960	150,000	2.4
[29]	8x8	9	-	>1	640x480	-	5.6
			FPGA Implementation				
[18]	8x8	8	4	>1	320x240	3,615	38
[19]	8x8	9	-	1	800x600	1000	>10
[20]	8x8	8	9	1	640x480	1,540	62.5
[21]	8x8	9	8	18	1920x1080	27,960	64 (estimated)
[22]	8x8	9	8	13	1024x768	20,868	13
[23]	8x8	8	4	>1	640x480	56,466	30 (estimated)
[6]	8x8	9	8	1	800x600	5,580	72
[8]	8x8	9	4	34	640x480	121,210	68.18
[8]	8x8	9	4	34	1600x1200	1,049,886	10.41 (estimated)
HOG_{CONV}	8x8	9	8	45	1920x1080	264,062	32
HOG_{OCB}	9x9	10	8	45	1920x1080	264,062	39.2
HOG_{OCB-RF}	9x9	10	8	45	1920x1080	264,062	42.7

All detection results are collected, and afterwards recall is calculated from number of true positives (TP) and false negatives (FN) as shown in (4).

$$Recall = \frac{TP}{TP + FN} \qquad (4)$$

A false positive per window (FPPW) of 10^{-4} is mostly considered in literature for pedestrian detection results. We also present the *Miss Rate (1- Recall)* results for $FPPW = 10^{-4}$ for multiple cell sizes and number of bins. The results are shown in Fig. 8. It is seen that generally larger histogram bins gives better detection rates. This is obvious, as more histogram bins allow fine division gradient orientations, hence better feature vector for training and classification. On the other hand improvement is seen in detection rates by increasing cell size up to a certain value and it drops increasing cell sizes too much. Smaller cell sizes provide a dense grid of blocks and windows in a frame, therefore, using smaller cells would improve accuracy. However, using too small cell sizes results in degraded performance because there are not enough distinguishing

TABLE II
COMPARISON OF PARAMETERS AND ENERGY CONSUMPTION FOR VARIOUS GPU AND FPGA IMPLEMENTATIONS

	Cell Size	Histogram Bins	Win. Stride	# scales	Resolution	Windows/frame	Power (W)	Energy (J/Frame)
[8] (GPU)	8x8	9	4	34	640x480	121,210	225	17
[8](FPGA)	8x8	9	4	34	640x480	121,210	37	0.54
HOG_{CONV}	8x8	9	8	45	1920x1080	264,062	21	0.656
HOG_{OCB}	9x9	10	8	45	1920x1080	264,062	19	0.485
HOG_{OCB-RF}	9x9	10	8	45	1920x1080	264,062	19.276	0.451

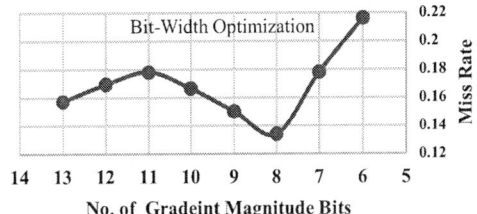

Fig. 12 Variation in miss rate for $(C_{size}, N_{bin}) = (9x9, 10)$.

features within the cells. Minimum miss rate of 12% is achieved at $(C_{size}, N_{bin}) = (7x7, 11)$.

C. Throughput and Power Consumption Analysis

Power consumption and throughput are directly related to the hardware resources used. In the previous section it is seen that the cell size and histogram bins has significant impact on hardware complexity. The effect on different hardware components for different cell sizes and number of bins for a single core is shown in Fig. 9. We see a significant reduction in hardware resources by increasing cell size or reducing number of bins. The reasons being discussed in previous section for independent blocks.

Number of frames processed per second (fps) is dependent on the maximum frequency at which the hardware can operate. In our hardware architecture it is mainly dependent on the size of partial classification engine and the block normalization engine. As discussed in previous section, the complexity of PCE is heavily dependent on C_{size} while that of BHN depends on both C_{size} and N_{bin}. Fig. 10 shows the results. We get the maximum frequency at the point where both PCE and BHN have overall minimal hardware complexity. Specifically, the $(C_{size}, N_{bin}) = (11x11, 6)$.

We have used Xilinx Xpower analyzer (XPA) 14.7 to estimate the deviations in power consumption by varying the parameters. We have simulated the hardware and created 'Value Change Dump' (vcd) files are used to set the toggle rates of all signals. Post place and route results are obtained and are shown in Fig. 11. Power consumption increases by reducing the cell size or increasing the histogram bins. This is fairly understandable due to the fact that both these parameters increase the hardware complexity due to increase in the feature vector size. Minimum power consumption is 9.98 W with $(C_{size}, N_{bin}) = (12x12, 6)$. The maximum cell size and minimum histogram bins as expected.

D. Choice of Parameters

We have seen from the previous analysis that there does not exist a set of parameters which give us best accuracy, power and throughput. Improving the accuracy worsens the power and throughput while maintaining minimum power and maximum throughput severely degrades the accuracy. Similarly, trying to improve throughput may degrade power consumption significantly and vice versa. However, accuracy is changing very slightly at certain regions in Fig. 9. Similarly, there are more than one sets of parameters which give almost the same power consumption. This allows us to select the best of one of these metrics while slightly compromising on another metric. We can achieve best results by selecting $(C_{size}, N_{bin}) = (9x9, 10)$. Further we obtained results for miss rate by changing bit-width for this optimal parameter set. The results are shown in

Fig. 12. Bit-width is hence set to eight bits as it gives maximum accuracy and minimum hardware complexity. Note that this further results in reduced bit-width in all the next blocks.

Parameters optimized for low power and high speed are shown in Table I & II comparing the throughput and energy consumption results with the other state of the art FPGA and GPU implementations. We have presented three results. 1) HOG_{CONV}, which shows the results for conventionally used parameters. We can achieve 32 fps for full-HD while dissipating 0.656 J/frame and 15% miss rate. 2) HOG_{OCB}, presents results for optimized cell size and histogram bins. Frame rate achieved by optimizing cell size and histogram bins is 39.2 fps with energy consumption of 0.484 J/frame while maintaining a miss rate at 15%. Gradient magnitude bit-width is considered to be 13 bits. 3) Finally, HOG_{OCB-RF}, in which features are further refined by removing insignificant gradients, is presented. This results in a frame rate of 42.7 fps while energy consumption is 0.45 J/frame at 13% miss rate.

V. ACKNOWLEDGEMENT

This work was supported by the Center of Integrated Smart Sensors funded by Ministry of Science, ICT & Future Planning as Global Frontier Project (CISS-2013M3A6A6073718).

VI. CONCLUSION

We have presented fully parallel architectures for various modules of pedestrian detection system utilizing Histogram of oriented gradients (HOG). HOG has shown high detection accuracy but the detection speed and power consumption are major bottlenecks for real time embedded applications. We have optimized parameters, cell size and histogram bins, to achieve low power and high throughput while maintaining the detection accuracy. Feature refinement is done to further improve the results. Combination of optimal parameters and our hardware accelerator results in a frame rate of 42.7 fps for full-HD resolution and lowers the energy consumption by 97.34% and 16.4% while improving the accuracy by 2% as compared to state of the art GPU and FPGA implementations respectively. This work can be extended to use multiple cores on a single FPGA or using multiple FPGAs to further increase throughput while an ASIC implementation would greatly reduce the power consumption. It can also be extended to include other features and classifiers or combinations of those to optimize for objects other than pedestrians.

REFERENCES

[1] U. Shankar, "Pedestrian roadway fatalities," Department of Transportation Tech. Rep., 2003.

[2] D. Geronimo, A. M. Lopez, A. D. Sappa, and T. Graf, "Survey on pedestrian detection for advanced driver assistance systems," IEEE Trans. Pattern Analysis and Machine Intelligence, vol. 32, no. 7, pp. 1239–1258, 2010. 1, 2, 10, 16, 18

[3] A. Ess, B. Leibe, K. Schindler, and L. Van Gool, "A mobile vision system for robust multi-person tracking," in Computer Vision and Pattern Recognition (CVPR), IEEE Conf. on, 2008, pp. 1–8.

[4] N. Dalal and B. Triggs, "Histograms of oriented gradients for human detection," in Proc. IEEE Conf. Comput. Vision Pattern Recog., Jun. 2005, vol. 1, pp. 886–893.

[5] D. G. Lowe. Distinctive image features from scale-invariant keypoints. IJCV, 60(2):91–110, 2004

[6] K. Mizuno, Y. Terachi, and K. Takagi, "Architectural study of HOG feature extraction processor for real-time object detection," in Proc. IEEE Workshop Signal Process. Syst., Oct. 2012, pp. 197–202.

[7] K. Takagi et al. A sub-100-milliwatt dual-core HOG accelerator VLSI for real-time multiple object detection. In ICASSP, 2013.

[8] Ma, X.; Najjar, W.A.; Roy-Chowdhury, A.K., "Evaluation and Acceleration of High-Throughput Fixed-Point Object Detectio n on FPGAs,"Circuits and Systems for Video Technology, IEEE Transactions on , vol.PP, no.99, pp.1,1

[9] P. Dollar, C. Wojek, B. Schiele, and P. Perona, "Pedestrian detection: An evaluation of the state of the art," Pattern Analysis and Machine Intelligence, IEEE Trans. on, vol. 34, no. 4, pp. 743–761, 2012.

[10] S. J. Krotosky and M. M. Trivedi, "Person surveillance using visual and infrared imagery," IEEE Trans. Circuits Syst. Video Technol., vol. 18, no. 8, pp. 1096–1105, Oct. 2008.

[11] C. Papageorgiou and T. Poggio, "A trainable system for object detection," Int. J. Comput. Vision, vol. 38, no. 1, pp. 15–33, Jun. 2000.

[12] M. Oren, C. Papageorgiou, P. Sinha, E. Osuna, and T. Poggio, "Pedestrian detection using wavelet templates," in Proc. IEEE Conf. Comp. Vision Pattern Recog., Jun. 1997, pp. 193–199.

[13] G. Lowe, "Distinctive image features from scale-invariant keypoints," Int. J. Comput. Vision, vol. 60, no. 2, pp. 91–110, Nov. 2004.

[14] H. Cheng, N. Zheng, and J. Qin, "Pedestrian detection using sparse Gabor filter and support vector machine," in Proc. IEEE Intell. Veh. Symp., Jun. 2005, pp. 583–587.

[15] P. Y. Chen, C. C. Huang, C. Y. Lien, and Y. H. Tsai, "An Efficient Hardware Implementation of HOG Feature Extraction for Human Detection," IEEE Trans. Intell. Transp. Syst., vol. 15,NO. 2, pp. 656-662, 2014.

[16] S. E. Lee, K. Min, and T. Suh, "Accelerating histograms of oriented gradients descriptor extraction for pedestrian recognition," Comput. Elect. Eng., vol. 39, no. 4, pp. 1043–1048, May 2013.

[17] Y. Pang, Y. Yuan, X. Li, and J. Pan, "Efficient HOG human detection," Signal Process., vol. 91, no. 4, pp. 773–781, Apr. 2011.

[18] M. Hiromoto and R.Miyamoto, "Hardware architecture for high-accuracy real-time pedestrian detection with CoHOG features," in Proc. IEEE ICCVW, 2009, pp. 894–899.

[19] S. Bauer, S. Kohler, K. Doll, and U. Brunsmann, "FPGA-GPU architecture for kernel SVM pedestrian detection," in Computer Vision and Pattern Recognition Workshops (CVPRW), 2010 IEEE Comp. Soc. Conf. on, June 2010, pp. 61–68.

[20] K. Negi, K. Dohi, Y. Shibata, and K. Oguri, "Deep pipelined one-chip FPGA implementation of a real-time image-based human detection algorithm," in Proc. Int. Conf. FPT, Dec. 12–14, 2011, pp. 1–8.

[21] M. Hahnle, F. Saxen, M. Hisung, U. Brunsmann, and K. Doll, "FPGAbased real-time pedestrian detection on high-resolution images," in Computer Vision and Pattern Recognition Workshops (CVPRW), 2013 IEEE Conf. on, June 2013, pp. 629–635.

[22] C. Blair, N. Robertson, and D. Hume, "Characterizing a heterogeneous system for person detection in video using histograms of oriented gradients: Power versus speed versus accuracy," Emerging and Selected Topics in Circuits and Systems, IEEE J. on, vol. 3, no. 2, pp. 236–247, 2013.

[23] R. Kadota, H. Sugano, M. Hiromoto, H. Ochi, R. Miyamoto, and Y. Nakamura, "Hardware architecture for HOG feature extraction," in Intelligent Information Hiding and Multimedia Signal Processing (IIHMSP), 5th Int. Conf. on, 2009, pp. 1330–1333.

[24] OpenCV: http://opencv.org/

[25] P. Sudowe and B. Leibe, "Efficient use of geometric constraints for sliding-window object detection in video," in Int. Conf. on Computer Vision Systems (ICVS'11), 2011

[26] T. Machida and T. Naito, "GPU & CPU cooperative accelerated pedestrian and vehicle detection," in Computer Vision Workshops (ICCV Workshops), IEEE Int. Conf. on, 2011, pp. 506–513.

[27] C. Yan-ping, L. Shao-zi, and L. Xian-ming, "Fast HOG feature computationbased on CUDA," in Computer Science and Automation Engineering (CSAE), IEEE Int. Conf. on, vol. 4, 2011, pp. 748–751.

[28] B. Bilgic, B. K. P. Horn, and I. Masaki, "Fast human detection with cascaded ensembles on the GPU," in Intelligent Vehicles Symp. (IV), 2010 IEEE, 2010, pp. 325–332.

[29] V. Prisacariu and I. Reid, "fastHOG - a real-time GPU implementation of HOG," Department of Engineering Science, Oxford University, Tech. Rep. 2310/09, 2009.

[30] http://en.wikipedia.org/wiki/Fast_inverse_square_root

[31] INRIA Person Dataset. http://pascal.inrialpes.fr/data/human/

[32] P. Dollar, C. Wojek, B. Schiele, and P. Perona, "Pedestrian detection: A benchmark," in Computer Vision and Pattern Recognition (CVPR), IEEE Conf. on, 2009, pp. 304–311.eth

[33] M. Enzweiler and D. Gavrila, "Monocular pedestrian detection: Survey and experiments," Pattern Analysis and Machine Intelligence, IEEE Trans. on, vol. 31, no. 12, pp. 2179–2195, 2009.

978-1-4673-9141-2/15 $31.00 © 2015 IEEE

10Mbps Human Body Communication SoC for BAN

Hyungil Park, Ingi Lim, Sungweon Kang
Human Interface SoC Research Section
Electronics and Telecommunications Research Institute
Daejeon, Rep. of Korea
hipark@etri.re.kr

Whan-woo Kim
Division of Electrical and Computer Engineering
Chungnam National University
Daejeon, Rep. of Korea
wwkim@cnu.ac.kr

Abstract—**A plurality of wearable devices have appeared recently. The requirements of communication method for wearable applications are wearing comfort, easy use with intuitiveness and low power consumption. In this paper, 10Mbps human body communication method which has extremely low power consumption is presented. The differences between HBC in IEEE 802.15.6 and proposed 10Mbps HBC, SoC architecture and implementation, performance of evaluation board and application are discussed.**

Keywords—Human Body Communications (HBC); Frequency Selective Spreader (FS-Spreader); Body Area Network (BAN);Body Area Adverts

I. INTRODUCTION

In recent years, while the needs of the market about wearable computing and the associated are increasing, release of wearable items such as glasses and watches is increasing explosively. Main application fields for them are health-care and fitness. Moreover it is expanding expectations of users according to including SNS, schedule check, alarm and simple games.

BAN standard of IEEE 802.15 which was published in February 2012, may give the communication method for wearable application. Human body communication in BAN included Narrow band, UWB and HBC will be the most suitable method for them.

By the way, the data rate of BAN HBC has rate of 164kbps mandatorily and from 328Kbps to 1.3123Mbps optionally. It gives the significant limitations in various services which is required recently [1].

In this paper, HBC PHY specification with 10Mbps data rate in order to support various multimedia services for wearable application, SoC architecture and implementation, Prototyping, evaluation of function and performance and interesting application are presented.

II. 10MBPS HBC SPECIFICATION

A. Comparison of IEEE 802.15.6 HBC with Proposal

According to the TRD(Technical Requirement Document) of IEEE 802.15.6, BAN aims to support a low complexity, low cost, ultra-low power and highly reliable wireless communication for use in close proximity to, or inside, a human body to satisfy an evolutionary set of entertainment and healthcare products and services. Typical link data rate should be some tens kbps in most cases. However, raw data rate up to 10Mbps is expected in some applications [2].

The comparison of HBC in IEEE 802.15.6 with 10Mbps HBC proposed is shown in Tab. I.

TABLE I.   COMPARISON OF HBC IN BAN WITH PROPOSAL

	IEEE802.15.6	*10Mbps HBC*
Frequency Band	21MHz Centered	32MHz Centered
Spreading	FS-Spreader 16chip Walsh Mod with FSC(8~64bits)	FS-Spreader 64chip Walsh Mod
Data Rate	164Kbps(mandatory) 328Kbps(optional) 656Kbps(optional) 1.3125Mbps(optional)	1Mbps 4Mbps 10Mbps

HBC PHY specification of IEEE 802.15.6 has center frequency of 21MHz and following modulation scheme; Selection of 16 Walsh code by Serial to parallel of 1:4, modulation with 8 to 64 frequency shift code (FSC) according to data rate. The data rate is supported to 164kbps mandatorily and from 328Kbps to 1.3123Mbps optionally.

The detail modulation method of IEEE 802.15.6 HBC through the block diagram of Serial to parallel and FS-Spreader is shown in Fig. 1 [1].

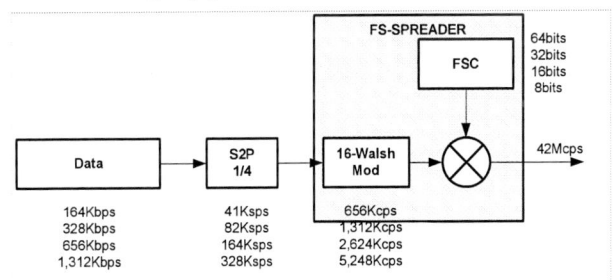

Fig. 1. S2P and FS-Spreader of IEEE 802.15.6 HBC.

Manuscript received May 4, 2015. This work was supported by the Ministry of Science, ICT and Future Planning, Rep. of Korea, under the IT R&D program supervised by the KOREA Evaluation Institute of Industrial Technology (KEIT) (100-385-99).

How to modulate data in 10Mbps HBC proposed is shown in Fig. 2 through the block diagram including scrambler, serial to parallel and FS-Spreader.

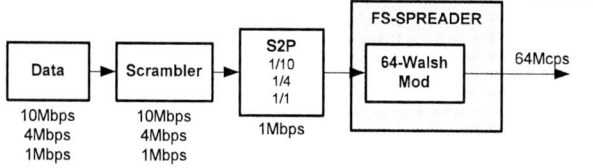

Fig. 2. S2P and FS-Spreader of Proposed 10Mbps HBC.

B. 3 Modes according to Data Rate

Power consumption is the key issue in BAN. HBC uses FSDT(Frequency Selective Digital Transmission, FSBT: Frequency Selective Baseband Transmission) without RF/IF components in order to satisfy the extremely low power consumption. The characteristics of a Walsh code satisfy the requirements of FSDT for human body communication. It is possible to transmit a digital baseband signal through a human body in the selected band without a carrier signal. [3]

There are 3 modes in 10Mbps HBC proposed according to data rate.

- High-Rate Mode: 10Mbps

- Low-Rate Mode: 4Mbps

- High-Gain Mode: 1Mbps

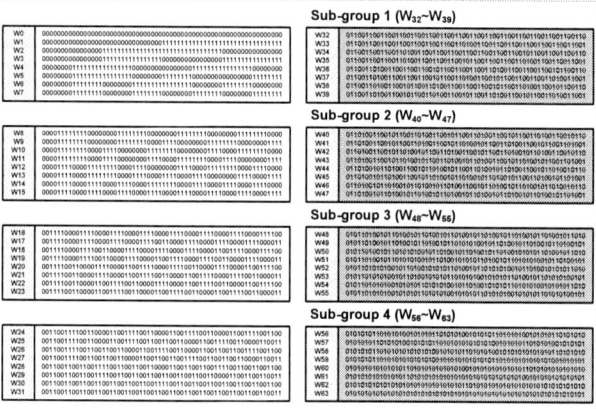

Fig. 3. 8 sub-groups of 64 Walsh Codes.

The transmit symbols of each mode are generated by serial to parallel and FS-Spreader. 64 Walsh code is divided into eight sub-groups as Fig. 3. Each sub-group has eight Walsh codes. Sub-groups from W32 to W63 on the right side of Fig. 3 are selected. Even though basically all sub-groups can be selected depending on channel environments, it is good to avoid to select sub-groups (W0~W31 in Fig. 3) with low frequency band existing a large amount of noise [4]. Header of PHY includes the information about sub-group selection.

First, In the High-Rate Mode 3 sub-groups of 64 Walsh codes are selected such as sub-group 2, 3 and 4 in Fig. 3. 10bits after serial to parallel of 1:10 are divided as three indexing

symbols with 3bits and 1bit symbol. Indexing symbols with 3bits is used to select a Walsh code in sub-group. Three 64 Walsh code is combined by using majority selection and then multiplied by 1bit symbol. The detail operation of FS-Spreader for High-Rate Mode is shown in Fig. 4. The data rate of FS-Spreader Output is 64Mcps finally.

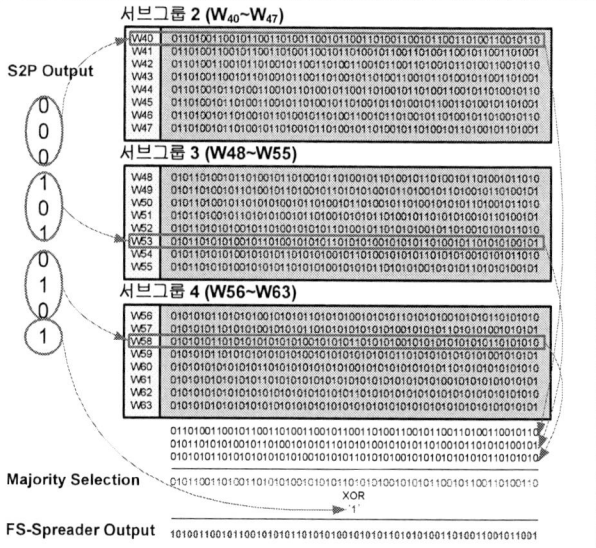

Fig. 4. FS-Spreader Operation for High-Rate Mode.

In the Low-Rate Mode 2 sub-groups are selected in Fig. 3. 4bits after serial to parallel of 1:4 are divided as one indexing symbol with 3bits and 1bit symbol. Indexing symbols with 3bits is used to select a Walsh code in sub-group. Output symbol is generated by selecting one of the two 64 Walsh code by using 1bit symbol.

In the High-Gain Mode Output symbol is generated by logical operation of XOR with the repetition pattern of 0 and 1 (W63 in Fig. 3) with the highest frequency band.

III. PERFORMANCE EVALUATION

Functional floating point model using MATLAB was used to evaluate performance of 10Mbps HBC proposed.

In transmitter maximum 10Mbps data is modulated using 64 Walsh code as Fig. 2. Modulated symbols are transmitted in baseband as binary symbols without DAC. RX receives the signal which is degraded through body channel and added noise [5]. With assumption of perfect timing synchronization, received signals are demodulated through maximum likelihood detection by correlating received signals and Walsh code.

Frequency spectrums are shown in Fig. 5: Baseband TX Signal (Blue Curve), Walsh modulated signal (Green Curve), channel output (Purple Curve) and RX band pass filter output (Red Curve).

978-1-4673-9141-2/15 $31.00 © 2015 IEEE

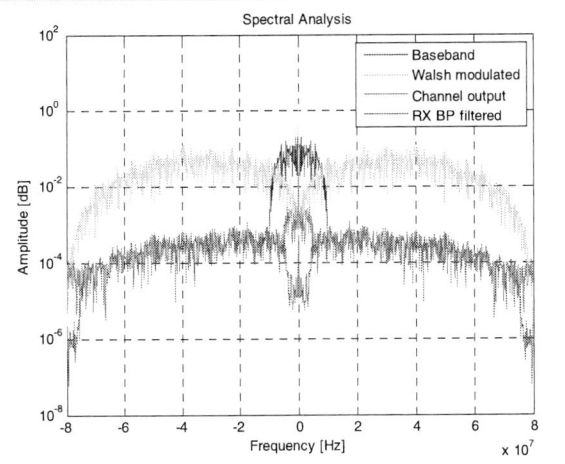

Fig. 5. Frequency Spectrum of TX Signal.

The results of BER performance evaluation by floating point simulation are shown in Fig. 6. The required SNR for achieving the BER of 10^{-6} in 10Mbps data rate ensured 6dB margin. Each curve shows the performance depending on the cutoff frequency of band pass filter in the input stage of receiver which is used to remove noise of low frequency band.

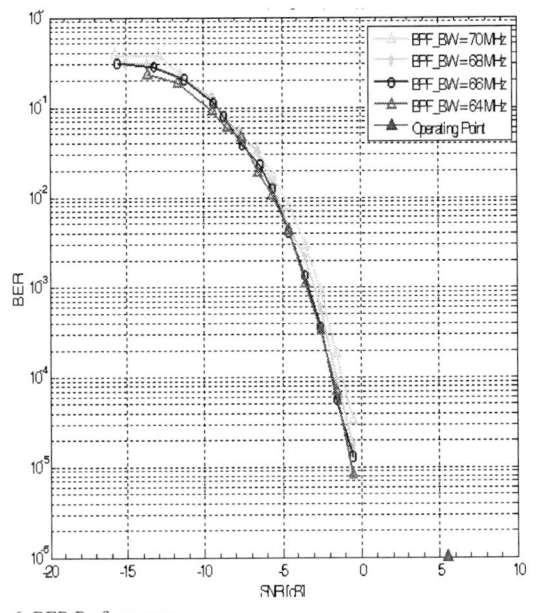

Fig. 6. BER Performance.

IV. IMPLEMENTATION

C. 10Mbps HBC SoC

10Mbps HBC proposed was implemented as SoC using TSMC 0.13um library through front-end and back-end design process. The equivalence gate counts are about 60,000 gates of standard cell and about 350,000 gates for memory to support 10Mbps data rate.

Pre-simulation for function test and Static Timing Analysis (STA) for timing analysis were performed after logic synthesis. Equivalence check was performed before and after each step; logic synthesis, DFT insertion, Layout.

BIST for testing memory block, DFT with boundary scan and full scan were added. Overall fault coverage using automatic test pattern generation (ATPG) was achieved up to 94%.

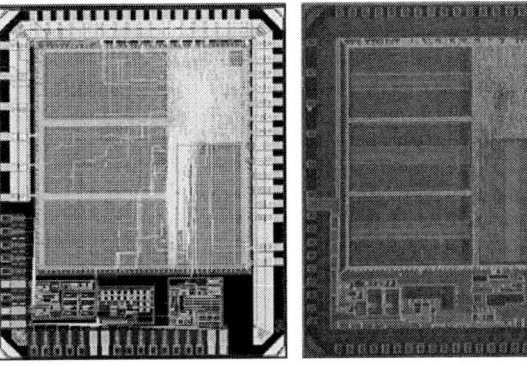

Fig. 7. Layout, Chip Photograph and Package Diagram of 10Mbps HBC.

After layout, post simulation and STA for function and timing verification, DRC, Antenna rule check and LVS check for back-end verification was performed.

Layout and chip photographs of the implemented HBC SoC are shown in Fig. 7. The size of HBC SoC is 2.19mm x 2.66mm. MLF 56 pin package with 7mm x 7mm size was used.

D. HBC SoC Prototype Evaluation Board

USB dongle as prototype board was developed to implement 10Mbps multimedia application. HBC SoC prototype evaluation board is composed of the controller board based of USB as HBC platform and the modem board which is responsible for HBC physical layer.

The modem board to verify HBC SoC is shown in Fig. 8. There are 50 pins connector for the interface between the control processor and modem, 2 pins connector for electrode, V-cut connector probing Analog Front End (AFE) part in the modem board which has 48mm x 25mm size.

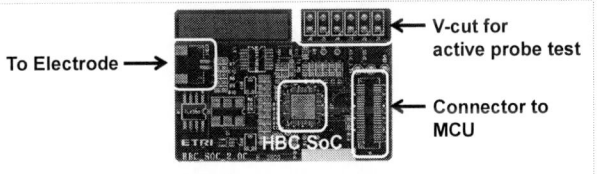

Fig. 8. HBC SoC Test Board.

AT91SAMx256 of Atmel deals with data between HBC modem and Host. The controller board to control the modem board is shown in Fig. 9. It has 70mm x 30mm size.

978-1-4673-9141-2/15 $31.00 © 2015 IEEE

Fig. 9. HBC Controller Board.

There are some interfaces in HBC SoC Prototype Board; I2C to read and write register set in HBC SoC, USART to transfer TX/RX message data, UART for console to control and check ARM7 processor, USB to supply power and connect HOST such as laptop computer, mobile phone, etc.

HBC USB dongle prototype is shown in Fig. 10. Steel electrode for touching and status LED to check operations are presented in USB dongle case.

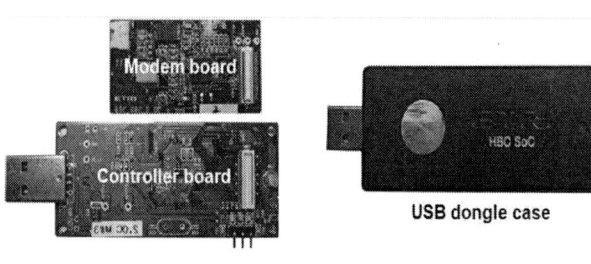

Fig. 10. HBC USB Dongle Prototype.

V. PROTOTYPE EVALUATION

The testbed of 10Mbps HBC SoC to evaluate function and performance is presented in Fig. 11. Four video streams have been transmitted at the same time between 2 laptops through HBC USB dongle prototype and body channel. Each video stream has the average rate of 2.5Mbps. Testbed components are as follows: Two laptops for video streaming server and user terminal, two HBC USB dongle prototypes, touching electrodes using the thumbs of both hands.

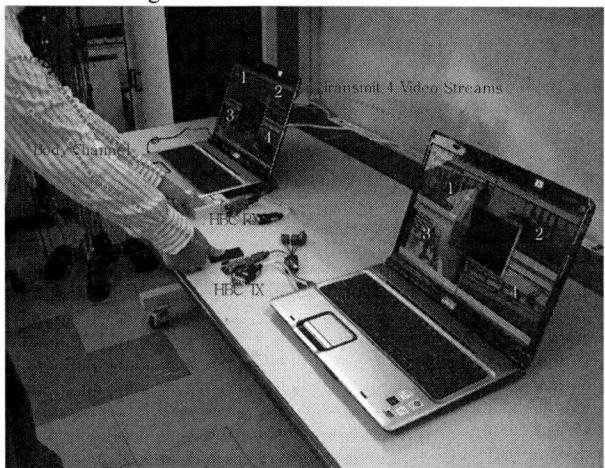

Fig. 11. Transmit 4 Video Streams through body channel.

Power consumption is the key issue in BAN and wearable

applications. So, the power consumption of HBC prototype dongle has been measured under the maximum data rate transmission. The measured results are presented in Tab. II.

Power consumption according to operation voltage are as follows: 7.1mW at AFE(1.2V), 4.4mW at HBC modem core(1.2V), and 11.5mW at modem IO(3.3V). Total power consumption was 23mW. If test IO pins will be removed, the amount of power consumption could be reduced significantly.

TABLE II.   POWER CONSUMPTION

1.2V		3.3V	*Power Consumption*
AFE	**Modem Core**	*Modem IO*	
5.88mA	3.64mA	3.5mA	13mA
7.1mW	4.4mW	11.5mW	23mW

APPLICATOIN

"Body area adverts" patent has been introduced as a promising technology to Newscientist magazine in UK in October 2008 [6]. Body area adverts system as 10Mbps HBC application service was implemented by using dongle prototype. User can download the product information and brochures, multimedia data such as music or movie preview and educational contents. Demonstration of "Body area adverts" application system implemented using HBC USB dongles is presented in Fig. 12.

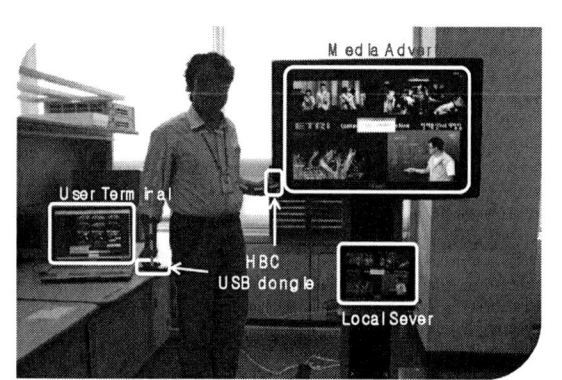

Fig. 12. HBC Media Adverts Service.

VI. CONCLUSION

Meanwhile, the human body communication technique has been developed primarily for use of a simple data transmission such as ID authentication, which requires a data rate of several tens of Kbps ~ Kbps. The system was developed for high-speed application, which can be used only under certain circumstances with some limitation.

Transmitting four video streams and Body area adverts

system using HBC SoC which was implemented by 10Mbps specification proposed was demonstrated.

Based on these results, without any constraints of the environment it can be seen data transfer through body channel is possible in low-speed as well as high-speed. Various services of wearable application could be provided using 10Mbps HBC specification and HBC SoC which is implemented.

REFERENCES

[1] IEEE Std.802.15.6TM-2012, "IEEE Standard for Local and metropolitan area networks – Part 15.6: Wireless Body Area Networks", February 2012.

[2] IEEE P802.15-08-0644-09-0006, "Technical Requirements Document", November 2008.

[3] Hyung-il Park et al., "Human Body Communication System with FSBT", International Symposium on Consumer Electronics, June, 2010

[4] Chang-Hee Hyoung et al., "Transceiver for Human Body Communication Using Frequency Selective Digital Transmission", *ETRI Journal*, vol. 34, no. 2, pp. 216-225, April, 2012.

[5] IEEE P802.15-08-0780-12-0006, "Channel Model for Body Area Network (BAN)", November 2010.

[6] http://www.newscientist.com/article/dn14921-invention-billboards-th

Integrating Wearable Low Power CMOS ECG Acquisition SoC with Decision Making System for WSBN Applications

Manikandan Pandiyan, Geetha Mani, Jovitha Jerome and Natarajan S

Abstract— **The paper aims to present an ultra-low power electrocardiogram (ECG) on chip with integrated fuzzy decision making (FDM) chip for body sensor networks. The proposed device is small in size, wearable, battery life. The proposed device has two designed chips: (1) ECG on Chip and 2) FDM chip. The ECG on chip contains an analog front end circuits, a 12-bit SAR ADC, a QRS detector and relevant control circuitry interfaces. The analog front end circuits accurately senses and digitizes the raw ECG signal, which is then filtered to extract the QRS complex with sampling frequency of 256 Hz. The obtained ECG details are sent to FDM chip for decision making where abnormalities are found and an alert signal is sent to the patient via microcontroller. The patient's ECG data is wirelessly transmitted to mobile phone or PC using ZigBee. The chip was designed and implemented in 0.35μm standard CMOS process. The digital circuits and SRAM operate at 3.3V. The total area of the device is about $6cm^2$ and consumes about 8.5μW. Small size and low power consumption show the effectiveness of the proposed design suitable for wireless wearable ECG monitoring devices.**

Index Terms— **ECG, FDM, HMI, System on Chip.**

I. INTRODUCTION

Cardiovascular disease is one of the main causes of mortality worldwide. Especially, cardiac diseases are the leading cause of death among people aged 60 years or older stated by World Health Organization. These types of diseases can be controlled by having close and potentially continuous medical supervision and are expected to have healthcare costs and medical management needs that are unsustainable for traditional healthcare delivery systems. The effective remedy for this is Personal health monitoring systems. This primarily requires the support of innovative sensor technologies, especially wireless body sensor networks (WBSN) formed with various wearable biomedical sensors. Since the constraints on battery life and form factor is crucial, this type of sensors has very stringent power requirement. To aid low cost ultra-low power design is essential in the development of these sensors. In terms of cost, size and performance SoC implementation is an attractive option.

Cardiac arrhythmias and coronary heart disease and atrial fibrillation constitute significant public health burdens. These afflict nearly 9% of persons over 80 years old, and are prone to have increased stroke risk [1]. Since abnormal ECG episodes often occur sporadically and are easily missed, many patients with cardiac arrhythmia or silent myocardial ischemia remain undiagnosed and untreated, This increases the need for low cost and easy to use wearable wireless ECG sensors with integrated decision making algorithm to alert the personal.

In Recent years [2]–[4] the development of ECG SoC has attracted much attention. 'Smart Vest', a wearable physiological monitoring system is to monitor various physiological parameters such as electrocardiogram (ECG), heart rate, blood pressure, body temperature and galvanic skin response (GSR) have been developed [2]. The geo-location of the wearer with the acquired physiological parameters is transmitted wireless to a remote monitoring station. A wrist worn wearable medical monitoring and alert system (AMON) targets high-risk cardiac/respiratory patients and was developed to monitor physiological parameters such as ECG, heart rate, blood pressure [3]. 'Life Guard' is a wearable physiological monitoring system for space and terrestrial applications to monitor the health status of the astronauts in space is developed [2]. The Georgia Tech, Smart Shirt characterized as a " wearable mother board " does a variety of vital parameters to be incorporated into the vest that can be easily and comfortably worn by the soldiers [4].

Many of ECG SoCs [3]–[4] integrate microcontroller or microprocessor on the use remote gateway The paper talks about an ultra-low power ECG on Chip with Integrated CMOS Fuzzy Decision Making Chip that addresses issues in existing solutions.

This paper is organized as follows. In Section II, the architecture of the proposed wireless ECG SoC is presented, followed by the description of ECG on chip in Section III. Section IV shows chip implementation details of fuzzy decision making and section V discusses measurement results. Conclusion is drawn in Section VI.

II. SYSTEM DESCRIPTION

The proposed health care architecture includes two parts: (a) Main unit and (b) a remote unit as shown in Fig 1. The Main unit contains wearable textile electrodes, designed ECG front-end chip, FDM chip, a controller and a ZigBee transceiver. The remote unit (personal gateway) can be either a mobile phone or a personal computer with an USB interface.

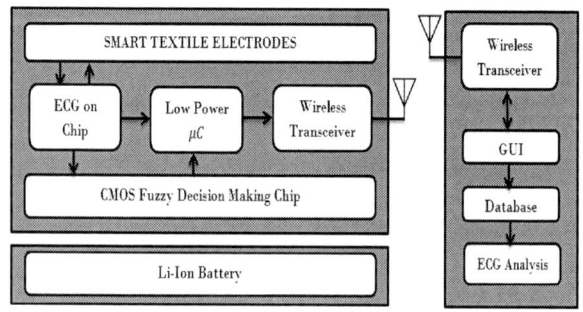

Fig.1. Block Diagram of proposed healthcare architecture

A. Main Unit

The main unit records the ECG from wearable textile electrodes and transfers the data to remote unit wirelessly. The ECG acquisition chip is designed for low power use is

---

Manikandan P was with Dept. of Instrumentation and Control systems Engineering, PSG College of Technology, India (2011-14 B.E. Batch).

Geetha M , Jovitha Jerome and Natarajan S are with Dept. of Instrumentation and Control systems Engineering, PSG College of Technology, India.

*Address for Correspondence – (Email ID: vanajapandi@gmail.com)

978-1-4673-9141-2/15 $31.00 © 2015 IEEE

described in Section III. The ECG acquisition chip consists of: (1) a specially designed textile electrodes for acquiring the ECG; (2) a miniature printed circuit board with ECG front end circuits; (3) Analog to Digital Conversion unit; (4) QRS Detection; and (4) System control unit. The ECG data is buffered using low power microcontroller internal memory to minimize power consumption before sending to the remote unit wirelessly.

The main unit also performs the tasks such as ECG front-end and microprocessor initialization, managing ECG data buffering, and scheduling the ZigBee transceiver. FDM chip is designed to take decisions when necessary, which is explained detail in Section IV. ZigBee protocol (TI CC2420) is used for wireless communication as it offers sufficient data rate at reasonable power consumption. The low power microcontroller (TI MSP430) is used for ZigBee baseband and ECG data management. The device was designed for users with respect to comfort and ease of use. Hence, it does not affect the daily activities of users. In addition, the entire unit is sealed with smart shirt, so the patient can wear and remove it easily.

B. Remote Unit

The remote unit (personal gateway) can be either a mobile phone or a personal computer with an USB interface. The important responsibilities of the remote unit are receiving the data wirelessly, database management, ECG analysis using graphical user interface (GUI) interface and customization. In order to avoid signal interference from other wireless devices, GUI interface has specific authentication to process the received data. In addition, there are several options in the GUI interface for customization.

III. ECG ON CHIP

A. ECG Analog Front-end Amplifier

The ECG front-end amplifier is mainly responsible for noise suppression, signal conditioning and amplification, which consists of two phases as shown in Fig.2. Namely, low noise AFG with band pass function and a programmable gain amplifier (PGA) adopting flip-over-capacitor technique. In the AFG design, two switches (S1 and S2) are integrated in order to avoid very long time for low AFG to settle down when power is applied due to the large resistance by the pseudo-resistors. The speeding up of the AFG is considerably done by a reset signal with an appropriate switching during startup of the system.

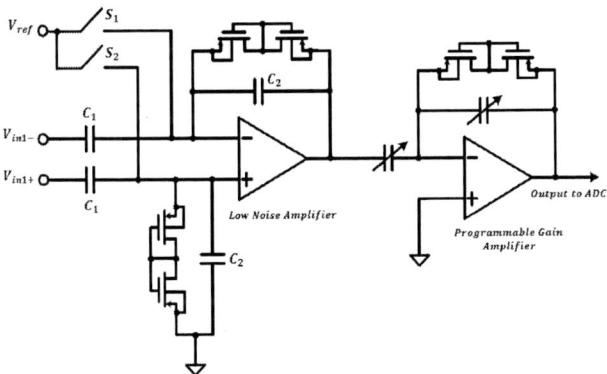

Fig.2. Typical circuit diagram of ECG Front-end low noise amplifier

B. Analog to Digital Converter (ADC)

SAR ADC is chosen for this application because of its moderate accuracy and low power overhead. Fig.3 depicts the architecture of the SAR ADC adopted from literatures. The analog ECG output is driven directly by the preceding buffer stage without the need of additional hold amplifier, and sampled through a bootstrapped switch and held on the capacitive 12 bit DAC. This is then employed by open-loop S/H that achieves fast settling and small offset error at low power cost. The analog signals that were captured are compared with a fixed reference REF and level-shifted by the DAC in a sequence of binary search. The fixed REF largely eliminates the dynamic offset associated with the comparator. The SAR logic and timing sequence are driven by an on-chip crystal oscillator that features both low jitter and low power consumption. The obtained digital codes are level-converted and passed to the SCU and QRS detection modules for further processing.

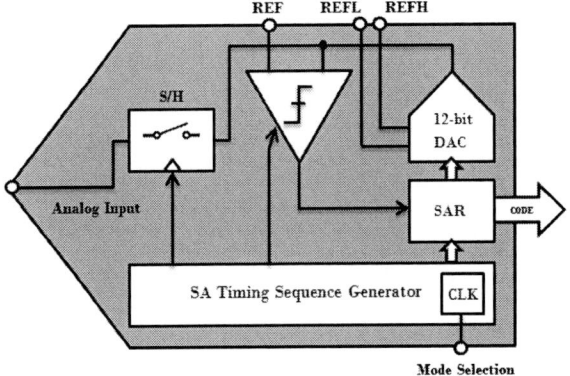

Fig.3. Schematic Architecture of Successive Approximation ADC

C. Hear Rate Calculation and QRS Detection

QRS complexes and estimate R-R intervals are located by morphological filter [4] which is used to remove noise artifacts in the ECG signal and to detect the QRS complex. The morphological filter contains a pair of Opening and Closing operations that suppresses peaks and valleys are implemented using dilation and erosion operators as shown in Fig.4. The effect of wandering baseline drift is removed by subtracting the averaged output of opening and closing operations with the original input. The ECG samples are overloaded into the shift register serially and then added (or subtracted, for erosion) with the equivalent structure element $g(x)$. The outcomes are continuously compared to find the minimum (or maximum, for erosion) using a comparator tree. The impulse noises are further removed to smooth the filtered signal using a serial structure moving average filter. The results are monitored continuously to detect the R-peak by comparing the received signal against an adaptive threshold. The current threshold is updated when a new peak is detected. The R-R interval is measured by counting the number of clocks between R peaks using a binary counter. In addition, the heart rate is also calculated by counting the number R peaks in the last 60s. A parallel-to-serial converter is incorporated for sending out the heart rate through SPI interface.

978-1-4673-9141-2/15 $31.00 © 2015 IEEE

D. System Control Unit and SPI Interface

Interface control signals are generated by System Control Unit for all blocks in the ECG on chip, based on the host CPU commands. Data framing, CPU interrupt generation etc. are done by this system based on state machine control signals. It also includes an asynchronous FIFO with 8 Kb buffers, in-order to interface the chip with different type of host CPUs. Data from ADC and QRS block is continuously written into the FIFO at the sampling frequency. The FIFO write, read controllers generate status signals such as "full", "nearly full", "empty", "nearly empty" based on the FIFO status.

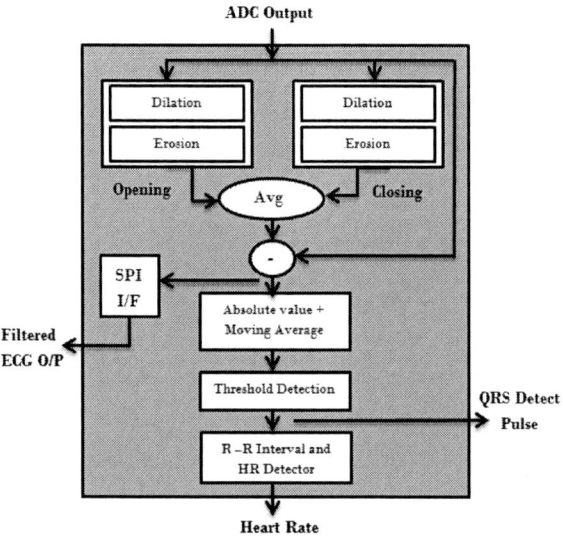

Fig.4. Flow process for QRS Detection and HRV Calculation

External microcontroller is communicated with proposed device with via a duplex SPI slave interface. The data link carries the ECG and QRS codes, along with the internal FIFO status flags. The control vector is delivered by the command link from the external microcontroller to the internal control registers. Additionally, the driving clocks for the I/O shift registers, the clock extraction circuit also extracts the CCU clocks that pace the FIFO reading thread.

IV. DESIGN OF FUZZY DECISION MAKING CHIP

The main functional blocks of FDM chip are described in detail. It consists of three parts: fuzzier, inference engine and defuzzier. In the fuzzier, input variables (non-fuzzy) are mapped to input membership functions. Inference engine is responsible for fuzzy inference depending on the type of inference method. In this work we used Min-Product inference method. Finally, defuzzier is used for converting fuzzy output resulting from inference engine to non-fuzzy values.

A. Fuzzy Interface

This architecture is a two-input one-output fuzzy controller. Each input has three trapezoidal membership functions or linguistic terms abbreviated as S (Small), M (Medium) and L (Large), while the output variable is characterized by five singletons. Parameters used for determining membership functions $(V_{r+}, V_{r-}, V_{c+}, V_{c-})$ are calibrated by voltages that are set on designed IC pins. Since input membership functions have three parts, architecture is

a 3×3 fuzzy controller and nine rules are accessible. Control voltages set on IC pins, depending on various applications, can change these rules.

The architecture controller in Fig. 6 is constructed with CMOS components such as Membership function generator (MFG), MIN circuits and a defuzzifier (D blocks) circuit.

We especially used a novel ramp function generation circuit for MFG. In fuzzy interface, three basic circuits are used: a ramp generator (RG) circuit [5], minimum circuit, and fuzzy complementary circuit. We build membership functions for input and output variables of a controller by using two ramp functions. In Fig.5, a trapezoidal membership function with its parameters is depicted.

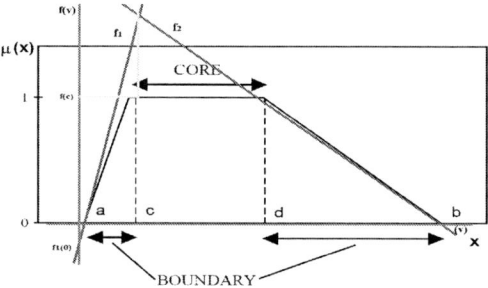

Fig.5. Trapezoidal membership function and its parameters

Considering these points it is necessary that we first select c and d equal to v_{r1} and v_{r2}, respectively, and then by changing v_{c1} and v_{c2}, a and b are determined. Diodes are used to cancel the extra parts. Since two ramp functions are needed, two RG circuits are used for each membership function. The typical circuit diagram complementary membership function generator (CMFG) circuit adopted from [5]. Considering, the output voltage will be $v_{out} = Ri_o$ where,

$$i_0 = \begin{cases} i_1 = k(v_{1+} - v_{in})v_{c1+}; \ v_{r1+}fv_{in} \\ i_2 = k(v_{in} - v_{r2-})v_{c2-}; \ v_{r2-}pv_{in} \end{cases}$$

Minimum circuits [5] can be used to clip extra curve and then reverse it by a fuzzy complementary circuit (FCC). In FCC, the two inputs are connected to the gates of the transistors. The output voltage v_{omin} always takes the smaller value of the two inputs v_{in1} and v_{in2} with a positive offset voltage v_{offset}

$$v_{min} = \min(v_{in1}, v_{in2}) + v_{offset}$$

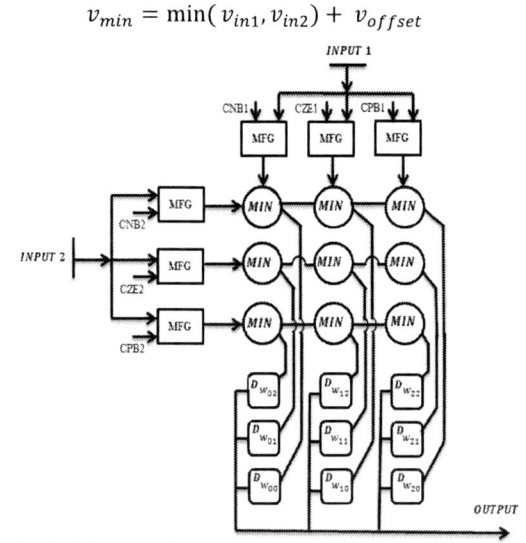

Fig.6. Block diagram of CMOS FLC

978-1-4673-9141-2/15 $31.00 © 2015 IEEE   156

Multi input minimum circuit using more than two transistors is needed for applications such as three-input fuzzy controllers. This result specifies that the offset voltage is about 1 Volt. In order to cancel the offset voltage, v_{off} a negative level shifter circuit is necessary. This shift will be provided by the next stage that is called fuzzy complement circuit with a reference voltage of E. The complementary membership function must be converted to an ordinary membership function achieved by a fuzzy complementary circuit of two g_m circuits.

In the circuit, assuming the reference voltage E = 0, we have

$$v_{comp} = R_i = R(i_1 + i_2) = R(-g_m v_{in} - g_m v_{comp}),$$

Therefore, this is opposite to the input voltage. The complement of the membership function is $E \neq 0$ in Figure, we can write

$$v_{comp} = E - v_{in} \ if \ Rg_m \gg 1$$

Note that in the final fuzzifier structure E must be set to cancel offset voltage associated with minimum circuit. The attenuation is due to limited gain of Rg_m that is equal to $-Rg_m = 1 + Rg_m$. This attenuation is the same in each one of the fuzzifiers used in the controller and the error due to attenuation does not affect fuzzy processing considerably. It is evident that if the input signal has a negative dc value, then E must be chosen to be greater than the Supremum value by $|V_{dc}|$.

B. Inference Engine

With Mamdani's inference technique, the inference is completed by a set of intersection and union operations. We used the Min- Product method for inferencing. Therefore, in this step we must specify minimum of two inputs permanently. Because the controller has two inputs with three membership functions for each one, nine two-input minimum circuits in inference engine are needed. These Min blocks and their synthesis in the complete controller structure will do the inferencing.

C. Defuzzification

We developed a novel defuzzifier in which the center of the area is calculated without employing a division circuit, therefore it occupies a small chip area. The main idea is based on parallel conductances g_n. This means that the output voltage of the circuit is the average value of the inputs. The contribution of each input to the output is weighted by the conductance of g_i that can be a controllable variable. The below equation effectively provides a non-fuzzy or defuzzifier output. We used a MOS transistor as a controllable g-element.

$$V_{defuzz} = \frac{g_1 V_1 + g_2 V_2 + \cdots + g_n V_n}{g_1 + g_2 \ldots \ldots + g_n}$$

V. IMPLEMENTATION AND RESULTS

The wearable ECG sensor node system was designed to fit perfectly into shirt. The main unit provides an extremely versatile framework for the incorporation of sensing, monitoring and information processing devices. Moreover, the designed node can be use in a variety of applications such as battlefield, public safety, health monitoring, sports and fitness, among others. The vital signal monitoring in wearing the designed smart shirt was tested to real time monitoring of the ECG. The inference performance test is done based on the physical activity under various conditions. The abnormal ECG signal is also measured and stored in inference engine. A wearable smart shirt with both physiological ECG transfers the signals without any troubles in wireless sensor network environment at the test. The rules were framed in the fuzzy decision making system such a way that an abnormal ECG signal is detected immediately and then an alert signal is sent to the gateway via microcontroller. The following sections depict the important results of the proposed system.

A. ECG Acquisition

The ECG measurements are obtained through wearable textile electrodes (Zero Resistance, 100% Silver fiber for conductive part). Fig. 7 shows the wearable electrodes which consist of a conductive fabric electrode pairs and the wearable sensor node system on the wearer's chest placement. The analog ECG output is driven directly by the preceding buffer stage without the need of additional hold amplifier, and sampled through a bootstrapped switch and held on the capacitive 12 bit DAC. The obtained digital codes are level-converted and passed to the SCU and QRS detection modules for further processing. QRS complexes and estimate R-R intervals are located by morphological filter which is used to remove noise artifacts in the ECG signal and to detect the QRS complex (As discussed in section 3.3). The R-R interval is measured by counting the number of clocks between R peaks using a binary counter. In addition, the heart rate is also calculated by counting the number R peaks in the last 60s. A parallel-to-serial converter is incorporated for sending out the heart rate through SPI interface. Interface control signals are generated by SCU for all blocks in the ECG on chip, based on the host CPU commands. The QRS complexes, estimate R-R intervals and other details are sent to FDM unit.

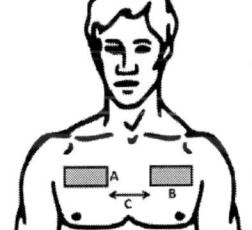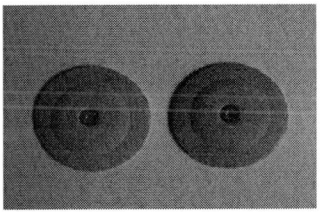

Fig.7. Placement of Wearable Electrodes

B. Fuzzy Decision Making

In FDM module, the Heart rate variability (HRV) is then computed from the time series R-R intervals (R-peak to R-peak) converted into a uniformly sampled time-spaced sequence. The power spectral density (PSD) of heart rate varies with the change from wakefulness to sleep. The low-to-high frequency ratio is a valid indicator of such change, because it reflects the balancing action of the ANS's sympathetic and parasympathetic nervous-system branches. PSD of HRV is then computed and the following three frequency bands have been carried out, are Very low frequencies (0-0.04 Hz), Low frequencies (0.04-0.15 Hz) and High frequencies (0.15-0.5 Hz). The features have been extracted from HRV and PSD used to feed the fuzzy logic engine that makes epoch-by-epoch (20 or 60 seconds per period) inferences. Fig.8. shows the major difference between the signals and it is noted that abnormal ECG data has elevated T wave. The fuzzy inference rules are based on the observed details of normal and abnormal ECG signals. The inputs of FDM chip are details of QRS complex and PSD results. The ranges of membership functions are tunable by changing the voltages of FDM IC pins, which is done by the user via microcontroller.

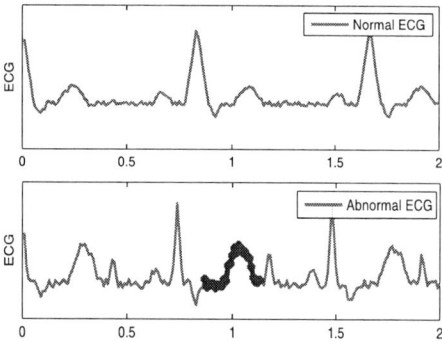

Fig.8. Measured ECG data for Inference Engines

The FDM chip provides the index values (i.e defuzzified output in the range of [0 - 4]), is sent to microcontroller. Hence we have five output states such as normal, sleep onset, fatigue, drowsiness and abnormal decoded by microcontroller unit. A set of meaningful rules has been hand tuned. The following are the strongest:

IF HRV (n) is Low **AND** LF/HF is Medium
THEN the Decision is SLEEP_ ONSET
****

IF HRV (n) is high **AND** LF/HF is high
THEN the Decision is NOR_WAKE
****

IF HRV (n) is Low **AND** LF/HF is Low
THEN the Decision is DROWSY

The status is continuously sent to remote unit every 30 minutes. Therefore the proposed system is capable of making the decisions based on the obtained details.

C. Results and Discussions

The results of the experiments confirmed our hypothesis that human health status can be predicted by the FDM module through features extracted. Detection may be successfully based on ECG signal captured from the wearable textile electrodes. When the sensed signal is captured in an uncontrolled environment (embedded) and with not invasive methods, the decision layer performs at least like the best subsystem. Measurement accuracy is almost equally to the clinical context, above all for ECG module, but decision system confirms to be robust enough to perform not under 90% successful early detections. The performance of each system has been validated by a human (expert). For clinical tests, a physician reads the bioelectric data and evaluates, for each module, if the identified transition from awake to asleep is acceptable or not.

For in the field tests this was done by visual observation of the awake/asleep status of the subject and the consequent qualification of the detection done by each system of the transition awake/asleep. Fig.9. depicts the designed graphical user interface for the proposed architecture. Timing of the early detection capability of each system was also evaluated during the tests.

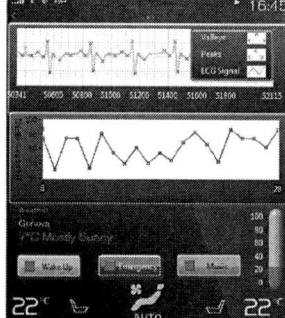

Fig.9. Designed GUI

VI. CONCLUSION

In this paper, we have proposed a wireless ECG on chip with integrated Fuzzy Decision making system for real-time ECG health monitoring. The proposed device is small in size wearable, battery life and also able to wirelessly transfer the patient's ECG signal to a remote station, where it can be analyzed in detail. The FDM chip is firstly integrated with ECG on Chip to take the decisions for altering the patients when necessary. The proposed device has already been tested with a reference ECG Holter for verification of accuracy and showed that the accuracy of proposed device is good enough, and the variation in key ECG parameters obtained from proposed device and the reference device is acceptable for clinical usage.

REFERENCES

[1] P.S. Pandian, K.P. Safeer, P. Gupta, D.T. Shakunthala, B.S. Sundersheshu, Wireless sensor network for wearable physiological monitoring, J. Networks 3 (5) (2008) 21–28.

[2] P.S. Pandian, K. Mohanavelu, K.P. Safeer, T.M. Kotresh, D.T. Shakunthala, P. Gopal, V.C. Padaki, Smart vest: wearable multi-parameter remote physiological monitorin system, Med. Eng. Phys. 30 (4) (2007) 466–477.

[3] U. Anliker, J.A. Ward, P. Lukowicz, G. Troster, F. Dolveck, M. Baer, F. Keita, E. Schenker, F. Catarsi, L. Coluccini, A. Belardinelli, M. Vuskovic, AMON: a wearable multi parameter medical monitoring and alert system, IEEE Trans. Inf. Technol. Biomed. 8 (4) (2004) 1–11.

[4] N. Halin, M. Junnila, P. Loula, P. Aarnio, The Life Shirt system for wireless patient monitoring in the operating room, J. Telemed. Telecare 11 (2005) 41–43.

[5] Hamed Peyravi, Abdollah Khoei, Khayrollah Hadidi, Design of an analog CMOS fuzzy logic controller chip, Fuzzy Sets and Systems 132 (2002) 245 – 260.

[6] Lobodzinski SS, Laks MM. New devices for very long-term ECG monitoring. Cardiol J (2012); 19(2): 210–4.

[7] Lobodzinski S S. ECG patch monitors for assessment of cardiac rhythm abnormalities. Prog Cardiovasc Dis 2013; 56(2): 224–9.

[8] Bass EB, Curtiss EI, Arena VC, Hanusa BH, Cecchetti A, Karpf M, et al. The duration of Holter monitoring in patients with syncope. Is 24 hours enough? Arch Intern Med 1990; 150(5): 1073–8.

[9] Manikandan, Pandiyan and Mani Geetha. "Takagi Sugeno fuzzy expert model based soft fault diagnosis for two tank interacting system."*Archives of Control Sciences* 24(3), pp.271–287 (2014).

Slack-aware Timing Margin Redistribution Technique Utilizing Error Avoidance Flip-Flops and Time Borrowing

Mini Jayakrishnan[1,2], Alan Chang[2], Jose Pineda De Gyvez[2], Kim Tae Hyoung[1]

[1]VIRTUS, School of Electrical and Electronic Engineering, Nanyang Technological University, Singapore
[2]NXP Semiconductors, Singapore
[1,2]mini001@e.ntu.edu.sg; [1]thkim@ntu.edu.sg

Abstract— There is much focus on timing error resilience for the speed critical paths of processors. In the context of growing parameter variations with technology scaling and voltage scaling, resilience helps to ensure functional correctness. Moreover it allows the chip to stretch its operating voltage and frequency beyond the conventional limits to meet the demand for high performance and low power. Conventionally, timing error resilience is achieved through variation tolerant circuitry at the cost of undesirable power, area and throughput overheads. Such overheads are aggravated by the presence of large number of critical timing paths in the design. In this paper, we propose a slack-aware timing margin redistribution technique for error resilience using time borrowing error avoidance flip-flops (EAFFs) while minimizing overheads. The proposed algorithm designs the processor critical paths ground up by inserting EAFFs at places where positive slack is available in the subsequent fan-out stage. Experiment results on an industrial processor design show that a timing margin improvement of 11% of the clock period can be achieved on 64% of the critical paths and a 55% timing margin on 45% of the critical paths without any throughput degradation. The area and power overheads of the additional flip-flops are 0.2% and 5.4%, respectively.

Keywords— *Timing error resilience; time borrowing; error avoidance flip-flop*

I. INTRODUCTION

With technology scaling, the nature of variations has changed from static to temporal and die-to-die to within-die. So we need additional timing margins for the speed critical paths in the design [1], [2]. Low power requirements tend to increase variations and affect the yield of chips [3]. The spread of variations has increased significantly, with leakage spread being larger than frequency spread [4]. Traditional chips have been designed to address worst case variations through conservative guard bands in operating voltage and frequency. However, providing enough guard bands becomes unacceptably expensive in deeply scaled process technology. *Better than Worst Case* (BTWC) design techniques together with resilient architecture try to recover the wasted design margins through typical case designs with less resources, lower voltage and higher frequency compared to their worst case counterparts [5], [6]. The effectiveness of *BTWC* designs is limited by the wall of slack which leads to massive timing errors on the near critical paths with voltage and frequency scaling. Variation Tolerant Design has become a one stop solution to achieve reliability as well as voltage and frequency

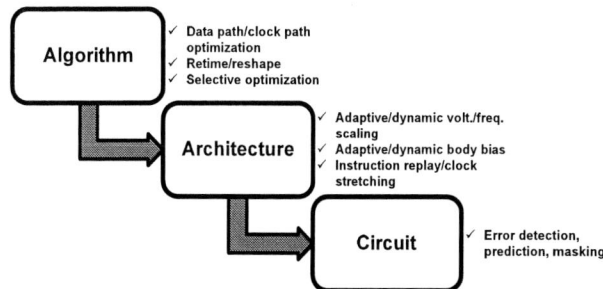

Fig. 1. Resilience, Power & Performance add-ons at different stages

scalability. We need to optimize performance, power, reliability and cost at all design levels to meet the design targets as summarized in Fig. 1. There has been considerable research going on for new variation aware sequential circuits. But we lack a proper design methodology which can maximize the gains offered by such circuits. This paper compares some of the existing resilience circuit and architecture, their pitfalls and how they can leverage on the huge slack margins offered by designs.

In this paper, we propose a technique to improve the timing margin of the critical paths in a processor pipeline proportional to the amount of slack available in the consecutive pipeline stages. We use a time borrowing flip-flop called error avoidance flip-flop (EAFF) based on *TIMBER* (Time Borrowing and Error Relay) [7] to replace the selected critical flip-flops. This method does not require complex design iterations and area overheads as in the cell resizing [8]. Compared to re-timing, our approach uses opportunistic time borrowing which can deal with dynamic process and environmental variations. In the proposed approach slack redistribution can be done by simply replacing the selected flip-flops with *EAFF* having a clock delay proportional to the amount of slack available. Compared to *TIMBER* design flow, we claim better timing margin improvement for the critical flip-flops as there is no need to divide the margin among multiple pipeline stages. *TIMBER* introduces additional overheads in terms of error propagation logic and error consolidation latency. On the contrary, the proposed method does not require error propagation to next stages for multi-stage time borrowing. The critical operating point of the endpoints is improved by the timing margin available to them which helps in aggressive voltage or frequency scaling. The key contributions of this paper are as follows:

978-1-4673-9141-2/15 $31.00 © 2015 IEEE

1) Our approach is based on available slack which can recover better timing margins and do not need an error relay logic for multi stage time borrowing as in *TIMBER*.

2) The proposed approach improves timing margin of the selected endpoints without any throughput degradation.

3) Opportunistic time borrowing is effectively done without costly area overheads and complex design iterations unlike resizing and/or retiming the combinational logic.

4) The critical operating point of the design is improved which has applications in aggressive voltage scaling.

The subsequent sections of this paper are organized as follows. Section II describes the related works in circuit, architecture and algorithmic level. Section III explains the motivation of the proposed approach. Section IV discusses the implementation details, the algorithm and the design methodology. Finally, section V presents simulation results and section VI draws the conclusions.

II. STATE OF THE ART METHODOLOGIES

Due to rise in process, voltage and temperature variations, chip reliability faces more challenges. Moreover, reliability is a prerequisite for *BTWC* design techniques. Resilience improvement techniques need to be extended to algorithm, architecture and circuit levels to reap maximum benefits with minimum overheads. In this section, we will briefly summarize the state-of-the-art error resilience techniques used in various design phases.

A. Circuit Level

We need timing error monitors at circuit level which can detect, predict or mask timing errors. Several papers propose error monitors in the form of flip-flops or latches. Most of them share a common architecture and can be categorized as Error Detection Sequential (EDS) circuits and Error Masking Sequential (EMS) circuits. EDS has a data path similar to a conventional flip-flop and a shadow path which captures input data using a delayed clock. Timing error detection is done by comparing a data path with a shadow path. *Razor* I [9], *Bubble Razor* [10], *Razor II* [11], *DSTB* and *TDTB* [12] belong to this category. EMS is similar to EDS except that the shadow path samples the data with a delayed clock and masks the timing error. EMS has an additional clock control block that provides the delayed clock for the data input. The width of the time borrowing window can be adjusted by the clock control circuitry. *TIMBER* flip-flops [7] and *Soft Edges* flip-flops [13] belong to this category. Using *TIMBER,* timing violation can be masked without the need for complex error recovery mechanisms. However, this design approach leads to shorter time borrowing intervals even at places where it can borrow a huge slack. With reliability, low power and performance demands on the rise, any circuit level error mitigation techniques should maximize the gains with minimal overheads as possible.

B. Architecture Level

For EDS circuits, once an error is detected at the circuit level, it needs to be corrected at the architecture level. Previous works used architectural replay and counter-flow pipelining to recover from the erroneous state, which incurs substantial latencies and lead to throughput degradation [9]. Architecture level solutions have to perform voltage and frequency scaling to mitigate high error rates [9], [12]. *TIMBER* based error masking avoids the need for complex architectural recovery and throughput loss. But it has additional overheads of error propagation logic and error consolidation latency for multistage time borrowing. *TIMBER* also suffer from meta-stability issues in the data path which affects the clock signal to the next pipeline stage.

C. Algorithmic Level

Several algorithms for error resilience and *BTWC* have been published so far. *EVAL* (Environment for Variation-Afflicted Logic) speeds up timing of critical paths through *Adaptive Body Bias* (*ABB*) as well as *Adaptive Supply Voltage* (*ASV*) [14]. It reshapes signal paths by speeding up slower paths and slowing down faster paths, which saves overall energy consumption but has significant area overheads. Another work *Blue Shift* uses *On-demand Selective Biasing* (*OSB*) and *Path Constrained Tuning* (*PCT*) to optimize selective critical paths [15]. The tuning knobs select frequently executed critical paths, apply forward body bias to some of the logic gates, and selectively tighten their timing constraints to achieve performance gains at the cost of significant power overheads. Power aware slack distribution (*SlackOptimizer)* [16] is another significant work which use cell sizing to distribute slack evenly in a power and cost efficient manner, but the benefits are small compared to the optimization effort. Another work in this area is *Selective End Point Optimization (SEOpt), Clock Skew Optimization (SkewOpt)* and *Combined Optimization (CombOpt)* [17]. *SEOpt* claims to reduce the cost of resilience by replacing error tolerant registers with conventional ones using additional margin insertion. *SkewOpt* migrates available timing slack from non-critical paths to critical paths. *CombOpt* is the combination of *SEOpt* and *SkewOpt,* and claims significant power savings. These techniques perform selective optimizations of the critical paths for performance and reliability enhancement while reduces cost of resilience in the form of area and power overheads. The efforts and benefits vary with the design and optimization approaches employed.

III. MOTIVATION

For our experiments we took an industrial processor design core in 40nm LP CMOS technology running on a clock period of 5.5ns. After synthesis and static timing analysis (STA), we filtered ~ 1000 most critical paths with worst slack of up to 0.1ns for slack improvement. Fig. 2 shows ~100 critical paths out of the 1000 paths in the processor pipeline with slack up to 6ps and Fig. 3 shows the corresponding consecutive stage path with enormous slack in the order of up to 4000ps. Our experiment results show that 85% of the critical paths has positive slack available in the consecutive stages to borrow from. Out of that, 64% paths can do coarse grain time borrowing with 11% to 77% timing margin improvement based on their available slack. We assign those critical paths into 7 different time borrow bands from *TB1* (11% timing margin) to

978-1-4673-9141-2/15 $31.00 © 2015 IEEE

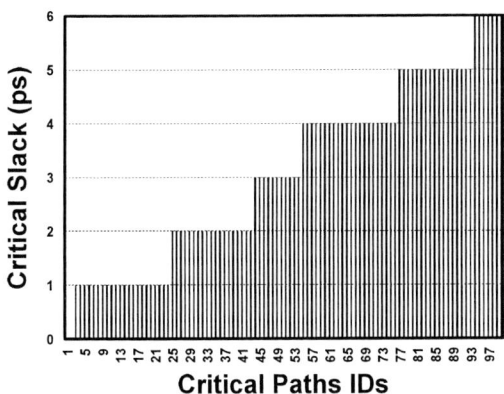

Fig. 2. Critical path slacks of the real time processor.

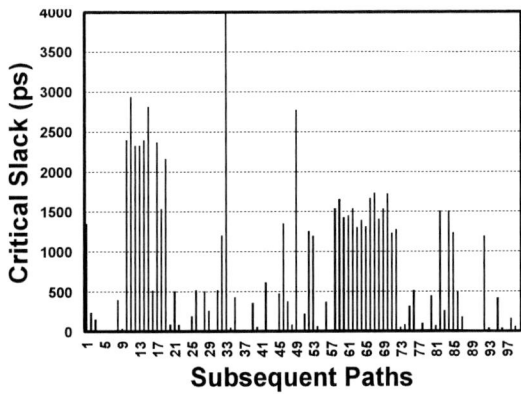

Fig. 3. Consecutive stage slacks of the real time processor.

TB7 (77% timing margin). 12% of the critical paths can do fine grain time borrowing with less than 11% timing margin improvement. Thus slack is redistributed effectively without any time consuming iterations as in [14], [15], [16] and [17].

Again, multistage time borrowing can be used for those critical paths which do not have enough slack to borrow from as shown in Fig. 4. Our approach can mask the variation induced timing errors opportunistically and will not degrade the throughput as in [9], [10], [11] and [12]. Since we place EAFF's based on slack analysis, we do not need error relay as in *TIMBER* [7]. The error signal can be used to monitor the error rate of the design. Compared to soft edge flip-flop [13], the delay clock buffer of *EAFF* is not embedded inside the standard library which leads to less clock buffer overheads with amount of time borrowed. Thus the proposed technique minimizes the throughput, power and area overhead.

IV. IMPLEMENTATION OF THE PROPOSED APPROACH

In this section, we will explain the implementation of the proposed technique. We aim to improve timing margin on selected critical paths in the design to make it more robust against process and environmental variations. We ran synthesis on an industrial processor design in *40nm LP CMOS* technology and did the timing analysis of the pipeline stages. This gave us potential locations which could use time borrowing to improve slack margins. We developed a standard cell library which has different flavours of time borrowing *EAFF*. Based on the results of static timing analysis, the

proposed algorithm replaces the critical flip-flops with *EAFF*. We used *RTL Compiler* for synthesis, timing analysis and to generate power and area reports. The library of EAFFs was generated using *Cadence Liberate*, a characterisation tool.

A. EAFF Library

Error Avoidance flip-flops (*EAFF*s) are based on the *TIMBER* flip-flop design [7] as shown in Fig. 5. Additional control logic is added to *TIMBER* flip-flop to generate the sampled error flag which is not shown here. The master latch (LATCH0) samples DATA while the shadow latch (LATCH1) captures DATA with the delayed clock DCK. The presence of shadow latch avoids data path meta-stability issues in the design. ERROR_FLAG indicates whether the data in LATCH0

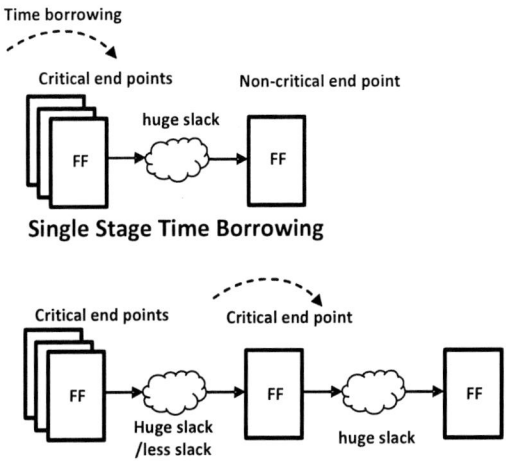

Fig. 4. Single stage and multi stage time borrowing.

is identical to that in LATCH1. ERROR_FLAG is sampled to avoid fake errors and glitches which give ERROR_SAMPLED. Fig. 6 shows simulation waveforms of *EAFF*. When timing error occur, LATCH0 and LATCH1 store different values. However the output is corrected by LATCH1, P0 and P1. The output signal Q is delayed by the delay value of DCK. Thus even if the output is corrected regardless of timing error, the design has to ensure enough slack in the consecutive stage of the *EAFF* so that the delayed output Q doesn't affect the chip functionality. The control signals, P0 and P1 are derived from the clock CK and the delayed clock DCK. To implement the proposed algorithm, a library of various *EAFF*s was generated using *Cadence Liberate* and the library validations were also performed. There can be hold time issues due to the delayed sampling of input data. This can be rectified by adding delay buffers in the corresponding short paths which is not part of this work.

B. Design Flow and Algorithm

The design flow used in our experiment is similar to the standard IC design flow. We use the post synthesis net list of an industrial processor designed for worst case for slack analysis. We also generate the standard cell library of a set of *EAFF*s. Our optimization algorithm finds the set of flip-flops

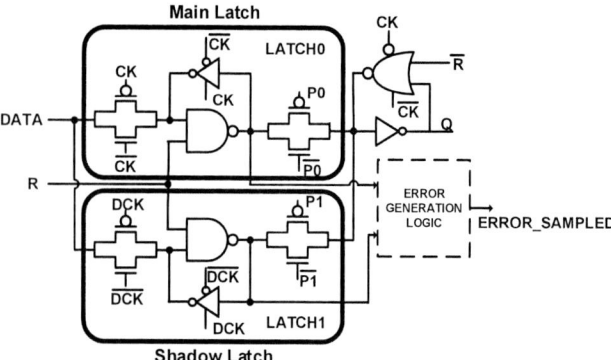

Fig. 5. Error Avoidance Flip-Flop (*EAFF*) based on *TIMBER* [7].

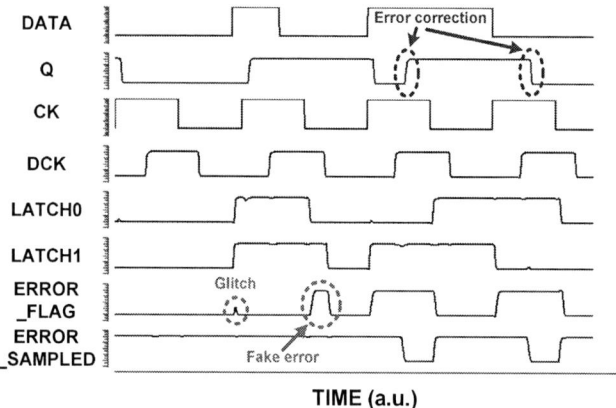

Fig. 6. Simulated waveforms of *EAFF*.

which are to be replaced with *EAFF*s along with the optimum time borrow value to be assigned to them. We confine our analysis to the most critical or near critical paths which are bound to fail when there is process or environmental variation.

Algorithm1 shows the pseudo code for the *EAFF* replacement based on the time borrow capability of the respective critical flip-flops. The algorithm is generic and can be used at any design stage from front end to back end. The critical paths analysed can be changed according to the analysis requirements. For the slack analysis, we consider the most critical paths with slack less than 2% of the clock period using *Find Analysis Paths* function. Now we report the subsequent stage timing for the critical paths using the function *Find Consecutive Slack* which show the time borrowing capability of each path. Based on this we group the critical flip-flops into different time borrowing bands from *TB1:TB7* (0.6ns ~ 4.2ns) using *Update TB Map*. *Replace Sequential* replaces the selected flip-flops in the critical paths with corresponding *EAFF*s. The proposed analysis starts with coarse grain time borrowing (*CoarseGrainTB*) which target a timing margin improvement in the range *TB1:TB7*. To further improve the critical path coverage, we use fine grain time borrowing (*FineGrainTB*) which targets a margin 0.1ns ~ 0.6ns. For paths which are not covered by *CoarseGrainTB* and *FineGrainTB*, we use multi-stage time borrowing (*MultistageTB*). The above three techniques ensure that the available slack is redistributed to the critical paths with minimum design iterations. The flip-

ALGORITHM1: PSEUDO CODE FOR COARSE GRAIN TB, FINE GRAIN TB AND MULTI-STAGE TB

1. **Procedure** *CoarseGrainTB (Initial Net list)*
2. # Find slack of consecutive stage for all critical paths
3. $P \leftarrow$ *Find Analysis Paths ( )*
4. **for** all $p \in P$ **do**
5. $S \leftarrow$ *Find Consecutive Slack (p)*
6. **end for**
7. # Replace all registers with consecutive slack by EAFF
8. **for all** *TB=TB1, TB<TB8, TB=TB+ TB1* **do**
9. **for** all $p \in P$ **do**
10. **if** *s >= TB* **then**
11. *Replace Sequential (p)*
12. *Update CoarseGrainTB Map (p)*
13. *Delete path (p)*
14. **end if**
15. **end for**
16. **return** (*net list2, P2*)
17. **Procedure** *FineGrainTB (net list2, P2)*
18. **for all** *TB=TB1/grain size, TB<TB1, TB=TB+ TB1/grain size* **do**
19. **for** all $p \in P2$ **do**
20. **if** *s >= TB* **then**
21. *Replace Sequential (p)*
22. *Update FineGrainTB Map (p)*
23. *Delete path (p)*
24. **end if**
25. **end for**
26. **return** (*net list3, P3*)
27. **Procedure** *MultistageTB (net list3, P3)*
28. $S2 \leftarrow$ *Find Next Consecutive Slack (p)*
29. **for** all $p \in P3$ **do**
30. **if** *s2 >= TB* **then**
31. *Replace Sequential (p)*
32. **end if**
33. **end for**
34. **return** (*net list final*)

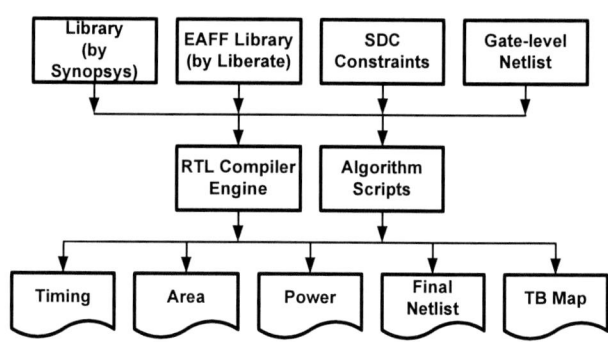

Fig. 7. Design Environment of the proposed optimization technique.

flop groups thus formed along with their time borrow values can be used to construct the delay clock tree in the later design phase, which is out of the scope of this work. After the flip-flop replacement, the area, the power and the timing are analysed in *RTL Compiler*.

Fig. 7 shows the design environment of the proposed optimization technique. The main inputs are the timing library generated by *Synopsys*, *EAFF* library generated by *Cadence Liberate*, *SDC* design constraints and initial gate level net list. The *RTL Compiler* engine is empowered with additional wrapper scripts for the replacement of flip-flops, optimum time borrow calculation for each flip-flop and analysis of results. We use the same engine for *STA and* power/area reports. We

978-1-4673-9141-2/15 $31.00 © 2015 IEEE

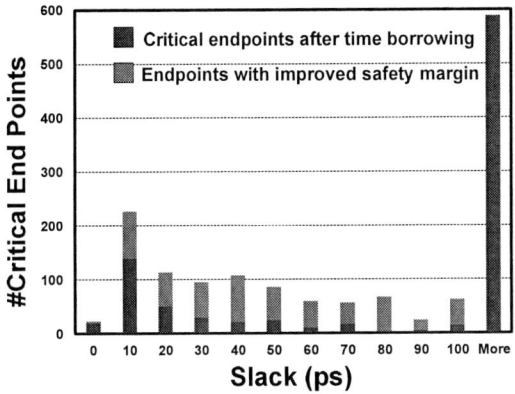

Fig. 8. Slack distribution after time borrowing

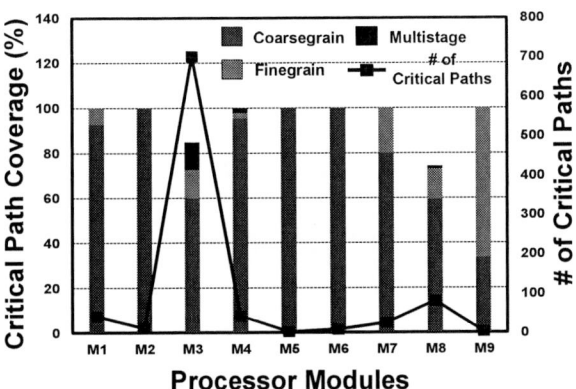

Fig. 9. % of critical paths improved using the proposed optimization scheme.

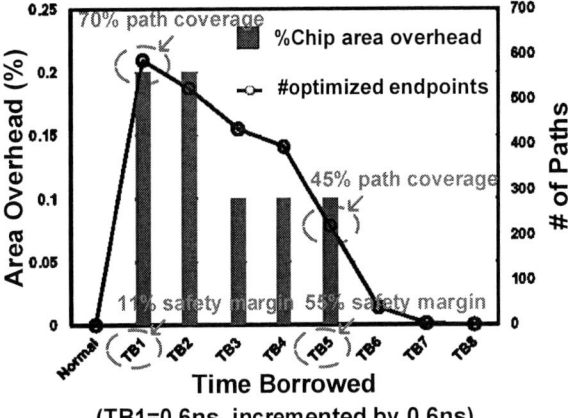

Fig. 10. Safety (timing) margin improvement and corresponding critical path coverage along with % chip area overhead caused by the proposed technique at different time borrowing values

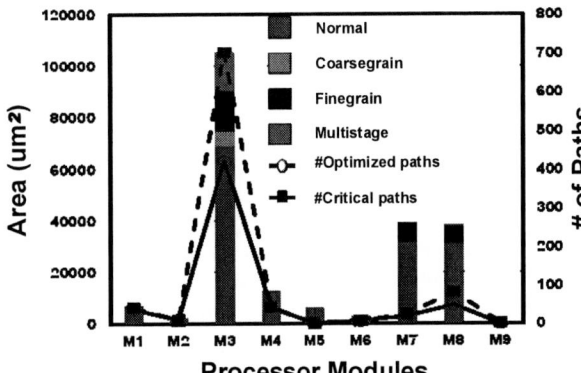

Fig. 11. Area overheads of various critical modules caused by the proposed technique.

do not use toggle rate or activity information of the critical flip-flops in our design where only those flip-flops with sufficient activity factor need to be replaced. This can reduce the power and area overheads further.

V. RESULTS AND ANALYSIS

Fig. 8 shows the slack re-distribution before and after *EAFF* replacement. The critical operating point shifts to right as we employ time borrowing in the design. This shows that a significant number of critical/near-critical paths can be protected from variations and voltage/frequency scaling by the proposed optimization scheme.

A. Timing Margin Improvement

The aim of the proposed algorithm is to improve the timing margin of the critical paths and to distribute the slack of the design more evenly with least overheads. Fig. 9 shows the percentage of paths out of the ~1000 paths which get a timing margin improvement using the proposed techniques. Results show that 7 out of 9 processor modules get timing margin improvement for all the critical paths. Modules M3 and M8 with most number of critical paths, get an added timing margin on 85% of the paths and 74% of the paths respectively as shown in Fig. 9. Overall, we get timing margin improvement on 85% of the critical paths using the three time borrowing

techniques with the coarse grain alone contributing 64%, fine grain contributing 12% and multi stage contributing 9%. The remaining 15% paths fan out mostly to memory write back stage where some slack can be created for timing margin improvement. Fig. 10 shows the trade-off between timing margin improvement and the number of critical paths that can be improved. As shown in Fig. 10, a safety (timing) margin improvement of 11% of clock period (at TB1=0.6ns) covers 70% of the critical paths. If we use higher time borrowing values we get a safety margin improvement of 55% of the clock period (at TB5=3.0ns) with 45% critical paths improved.

B. Overheads

The proposed technique incurs overheads because of the additional elements used in *EAFF*s for error resilience. The module wise area and power comparisons are shown in Fig. 11 and Fig. 12, respectively. The peak area and power overheads occur for the module M3, which has the maximum number of *EAFF* insertions. The chip wise area and power overheads of the *EAFF*'s for the timing margin improvement range 0.6ns ~ 4.8ns is depicted in Fig. 10 and 13 respectively. The maximum area and power overhead is around 0.2% and 6% respectively for this range. There is negligible architectural, error propagation or clock control overheads since we only need a

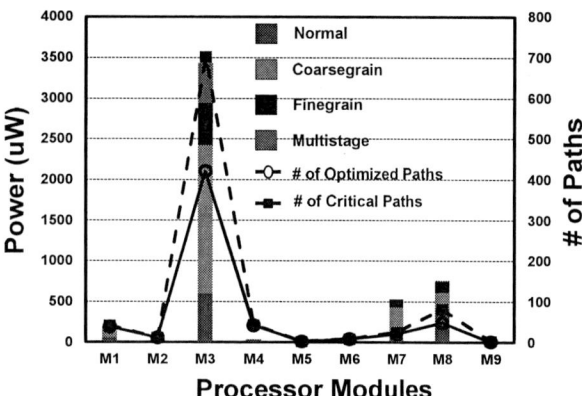

Fig. 12. Power overheads of various critical modules caused by the proposed technique.

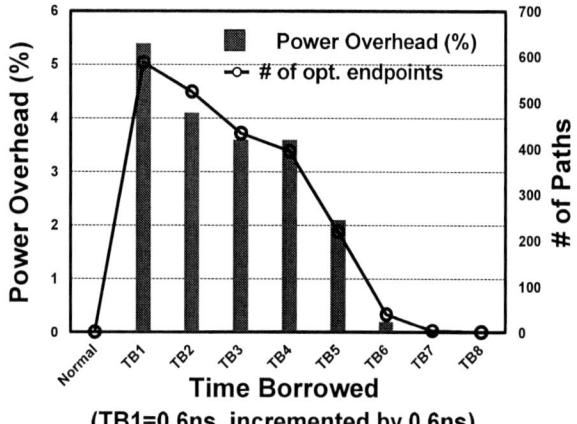

Fig. 13. Overall power overhead caused by the proposed technique at different time borrowing values.

clock delay network for *EAFFs* with the delay corresponding to the slack value calculated using the algorithm.

VI. CONCLUSIONS

The proposed technique is able to improve the timing margins of the critical paths in a fast and efficient way by using *EAFFs* with optimum time borrowing based on available slack in the subsequent stages. Simulation results show significant timing margin improvement using less design iterations. Architectural overhead is minimal and there is no throughput degradation because we use available slack and mask timing errors instead of clock stretching or instruction replay. There is minimal chip area overhead which makes it a viable choice compared to cell resizing. Finally the proposed technique helps in opportunistic time borrowing which can tolerate static and dynamic variations as and when required.

REFERENCES

[1] S. Ghosh and K. Roy, "Parameter variation tolerance and error resiliency: New design paradigm for the nanoscale era," *Proc. IEEE*, vol.98, no. 10, pp. 1718–1751, Oct. 2010.

[2] D. Bull, S. Das, K. Shivashankar, G. Dasika, K. Flautner, and D. Blaauw, "A power-efficient 32 bit ARM processor using timing-error detection and correction for transient-error tolerance and adaptation to PVT variation," *IEEE J. Solid-State Circuits*, vol. 46, no. 1, pp. 18–31, Jan. 2011.

[3] Ronald G. Dreslinski, Michael Wieckowski, David Blaauw, Dennis Sylvester, and Trevor Mudge, "Near-Threshold Computing: Reclaiming Moore's Law Through Energy Efficient Integrated Circuits," *Proc. IEEE*, Vol. 98, No. 2, Feb. 2010.

[4] Puneet Gupta *et al.*, "Underdesigned and Opportunistic Computing in Presence of Hardware Variability," *IEEE Transactions on Computer-Aided Design of integrated circuits and systems*, vol. 32, no. 1, january 2013.

[5] T. Austin, V. Bertacco, D. Blaauw and T. Mudge, "Oppotunities and Challenges for Better Than Worst-Case Design", *Proc. Asia and South Pacific Design Automation Conf.*, 2005, pp. 2–7.

[6] S. Moreno and J. Pineda de Gyvez, "A better than worst case circuit design using timing-error speculation and frequency adaptation," in *Proc. IEEE Int. SOC Conf.*, 2012, pp. 15–20.

[7] M. Choudhury et.al. "TIMBER: Time borrowing and error relaying for online timing error resilience," *Proc. of DATE*, Dresden, Germany, March 2010, pp. 1554-1559.

[8] A. B. Kahng, S. Kang, R. Kumar and J. Sartori, "Slack Redistribution for Graceful Degradation Under Voltage Overscaling", *Proc. ASP-DAC*, 2010, pp.825–831.

[9] D. Ernst, N. S. Kim, S. Das, S. Pant, R. Rao, T. Pham, C. Ziesler, D. Blaauw, T. Austin, T. Mudge, and K. Flautner. Razor: A Low-Power Pipeline Based on Circuit-Level Timing Speculation. *Proceedings of the 36th Symposium on Microarchitecture (MICRO-36)*, San Diego, CA, 2003.

[10] M. Fojtik *et al.*, "Bubble Razor: Eliminating timing margins in an ARM cortex-M3 processor in 45 nm CMOS using architecturally independent error detection and correction," *IEEE J. Solid-State Circuits*, vol. 48, no. 1, pp. 66–81, Jan. 2013.

[11] S. Das, C. Tokunaga, S. Pant, W. Ma, S. Kalaiselvan, K. Lai, D. Bull, and D. Blaauw, "Razor II: In Situ Error Detection and Correction for PVT and SER Tolerance," *IEEE J. Solid-State Circuits*, vol. 44, no. 1, pp. 32-48,2009.

[12] K. Bowman, J. Tschanz, N. Kim, J. Lee, C. Wilkerson, S. Lu, T. Karnik, and V. De, "Energy-Efficient and Metastability-Immune Resilient Circuits for Dynamic Variation Tolerance," *IEEE J. Solid-State Circuits*, vol. 44, no. 1, pp. 49-63, 2009.

[13] V. Joshi. David Blaauw. Deunis Sylvester, "Soft-edge flip-flops for improved timing yield: design and optimization," in *Proc. IEEFJACM Int. Conf. Computer-Aided Design*, Nov. 2007, pp. 667-673.

[14] S. R. Sarangi, B. Greskamp, A. Tiwari, and J. Torrellas. EVAL:Utilizing processors with variation-induced timing errors. In International Symposium on Microarchitecture, November 2008.

[15] B. Greskamp, L. Wan, W. R. Karpuzcu, J. J. Cook, J. Torrellas, D. Chen and C.Zilles, "BlueShift: Designing Processors for Timing Speculation from the Ground Up", *IEEE International Symposium on High Performance Computer Architecture*, 2009, pp. 213–224.

[16] A. B. Kahng, S. Kang, R. Kumar and J. Sartori, "Designing a Processor From the Ground Up to Allow Voltage/Reliability Tradeoffs", *IEEE International Symposimum on High-Performance Computer Architecture*, January 2010.

[17] A. B. Kahng, S. Kang, J. Li, "A New Methodology for Reduced Cost of Resilience", *Proc. GLSVLSI*, 2014, pp. 157-162.

978-1-4673-9141-2/15 $31.00 © 2015 IEEE

Circuit Performance Optimization for Local Intra-die Process Variations using a Gate Selection Metric

Victor Champac, Alejandra Nicte-ha Reyes, Andres F. Gomez
National Institute for Astrophysics, Optics and Electronics - INAOE,
Puebla, Mexico

Abstract—Process variations are imposing strong limits to performance of digital circuits at gigascale integration; they are classified in two types: inter-die and intra-die variations. Moreover, intra-die variations, which were ignored in the past, now have become significant. The present work proposes a statistical performance optimization methodology using a gate selection metric to enhance performance of digital integrated circuits in the presence of local intra-die process variations. The gate selection metric allows to select those gates to be re-sized for improving circuit performance at a lower area cost. This selection metric allows to optimize the circuit behavior using Lagrange method. The obtained results on ISCAS benchmark circuits show the benefits of the proposed methodology. The proposed optimization methodology allows to increases yield leading to better revenue.

I. INTRODUCTION

Process parameter variations have a significant impact on performance of digital integrated circuits. This issue has received considerable attention as technology scales down due to the increasing impact of process variability on propagation delays in critical paths [1]. Delay variability is commonly avoided adding time margins to the nominal clock period, penalizing the global performance. Therefore process variations impact on the maximum possible clock frequency to use inside the circuit [2], limiting performance, and therefore, the revenue achieved for each design.

Intra-die process parameter variations, which were ignored in the past, have become more significant, and they are now of great concern for the design of circuits with high yield. Random Dopant Fluctuations (RDF) is one of the main sources of intra-die variability. RDF is a statistical fluctuation in the number of dopants atoms of each transistor, which is translated into threshold-voltage variation. Its effect is presented locally at each transistor in the circuit. Therefore, it causes that digital gates, even placed very close to each other in the layout (the gates present very similar conditions during fabrication), can present very different timing responses.

There are several process variations-aware design techniques that aim to mitigate delay variability. Post-silicon tuning techniques [3] [4] adjust propagation delays according to a certain estimation and a control circuitry. However, they are

typically suitable for compensation of inter-die or global intra-die variations, but not for local intra-die variations. Process variations-aware gate library design techniques [5] [6] change the topology of standard gates to develop robust gates, in which single transistors are replaced with arrays of transistors in parallel, serial or mixed fashion. However, the design, characterization and insertion of those robust cells in large circuits can be complex. Gate sizing technique [7] [8] [9] [10] [11] is a widely used approach to cope with process variations. The aim is to find transistors channel width that makes the design to meet a required delay constraint with minimum area overhead. However, the formulation and solution of statistical optimization problems can become very complex for large circuits formed by millions of transistors.

This paper proposes a statistical performance optimization methodology using a gate selection metric for the design of digital integrated circuits robust to local intra-die process variations. The gate selection metric allows to focus optimization in those gates of the critical paths of a circuit that provide a higher improvement on delay variability at lower area cost. In such way, design complexity is reduced. Unlike conventional sizing approaches that aim to meet a specific delay constraint, the aim of the proposed optimization methodology is to minimize delay variability subject to a given extra area budget. This allows to increase yield, leading to better revenue, with low area cost.

The rest of this paper is organized as follows: Section II briefly introduces the effect of process variations on digital circuits. Section III, describes the selection metric formulation to focus critical paths optimization in a restricted set of gates. Section IV, presents the proposed methodology for statistical performance optimization using the proposed gate selection metric. Section V shows the application results of the methodology to some ISCAS benchmarks circuits, and Section VI presents the conclusions of this work.

II. EFFECT OF PROCESS VARIATIONS ON DIGITAL CIRCUITS

As technology scales down, process variations have become a challenge for circuit design. The effect of variations was typically handled by analyzing the circuits at different process corners of operation. However, for current state of art technologies, this is not a realistic approach, since variations are not fully correlated and independent delay variations (i.e. $V_t h$ variations due to RDF) are becoming greater. One manifestation of statistical process variations is variation of speed between different chips. Hence, if the circuits are designed using nominal parameters to run at a particular speed, some of them may fail to meet the desired frequency. Such variation in speed would lead to parametric yield loss. Thus, circuit performance needs to be analyzed taking into account statistical parameter variations.

In this work, Statistical Static Timing Analysis (SSTA) has been used to estimate the circuit performance response under process variations effects. In SSTA, the delay of each gate is a function of random process variables, whose behavior has to be described with a probability distribution. The focus of this work is on optimizing the circuit delay timing response under local intra-die process variations in transistors V_{th}. For each signal path in the circuit, the variance of the delay ($\sigma^2_{D,path}$) can be computed as the sum of the delay variance of each gate in the path:

$$\sigma^2_{D,path} = \sum_{i=1}^{N} \sigma^2_{Di} \qquad (1)$$

where σ^2_{Di} is the variance (due to RDF) of the delay of the i-th gate in the path and N is the number of gates in the path. Thus, locally improvement of delay variability on each gate improve overall signal path timing response. However, the contribution of each gate to the variance of the path delay can be very different depending on the load capacitance, input slope and gate size. Therefore, the use of gate selection metrics allows to identify the most influential gates with respect to delay variability, which allows to reduce design effort.

III. SELECTION METRIC FORMULATION

This section describes the selection metric formulation to focus critical paths optimization in a restricted set of gates.

A. Basic Metric Formulation

The derivative of the standard deviation of the delay with respect to the channel width sizing is analyzed as a first guess metric. This can be used to select the best gates to be upsized to reduce the standard deviation of the delay of logic paths with low area cost. Figure 1 plots the standard deviation of the delay of an inverter gate as function of the channel width for two load capacitance values. A large derivative value indicates that the standard deviation changes significantly for a small change in the channel width. In other words, the standard deviation can be reduced by sizing the gate with low area cost. The opposite is true for a low derivative value. This metric will be applied to two ad-hoc designed logic paths.

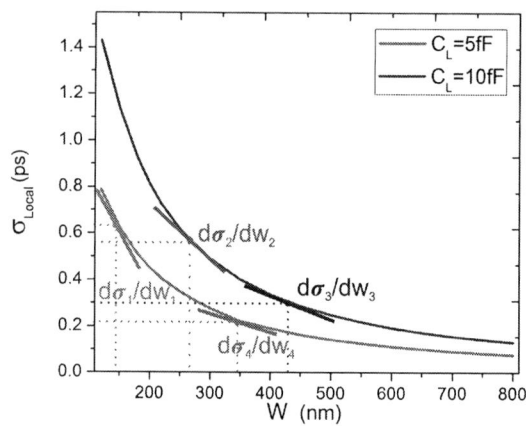

Figure 1. Standard deviation of the delay of an inverter gate as function of transistors channel width for two different capacitive load values.

1) Logic Path 1: Figure 2 shows the first analyzed logic path. This is composed of inverters, Nands and Nor gates. The transistor channel widths are also indicated. The derivative of the standard deviation of the delay with respect to the transistors channel width ($\frac{d\sigma_D}{dW}$) has been obtained for each gate using Hspice. To compute the derivatives, the channel width of the activated transistor in the first gate is increased by $\Delta W = 80nm$, while the other transistors remains with their original dimensions. Then, the standard deviation of the delay of the logic path is measured with Hspice by means of 500 Montecarlo runs. This process is repeated for the other gates. The derivatives can be approximated as the ratio of the change in the path standard deviation of the delay to the increase in the respective transistor size for each case. Table I shows $\frac{d\sigma_D}{dW}$ of each gate sorted from the higher to the lower value, and also the percentage of optimized path standard deviation of the delay when one transistor channel width of a gate is increased by $\Delta W = 80nm$. These results show that the metric works correctly to select the best gates to up-size. A higher degree of optimization corresponds to a higher value of $\frac{d\sigma_D}{dW}$.

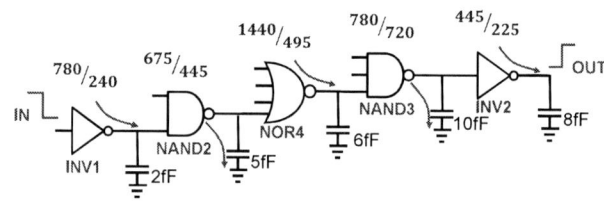

Figure 2. Logic Path 1.

2) Logic Path 2: Figure 3 shows the second analyzed logic path, which is composed of only inverters gates. The transistor channel widths are also indicated. Table II is obtained following a similar procedure to the Logic Path 1. However, a higher degree of optimization corresponding to a higher degree of $\frac{d\sigma_D}{dW}$ is not obtained for all the gates of this analyzed logic

Table I
DERIVATIVE OF THE STANDARD PATH DELAY DEVIATION WITH RESPECT
TO CHANNEL WIDTH SIZING AND DEGREE OF STANDARD DEVIATION OF
THE DELAY OPTIMIZATION ACHIEVED BY SIZING-UP EACH GATE OF THE
LOGIC PATH 1.

Sized gate	$\frac{d\sigma_D}{dW}$	% Optimization
$INV2$	9.71	6.39
$NAND3$	3.10	1.99
$NAND2$	2.35	0.62
$NOR4$	0.89	0.56
$INV1$	0.84	0.11

path. INV4 has a higher value of $\frac{d\sigma_D}{dW}$ than INV6, and however, sizing-up INV4 presents a lower value optimization of the standard deviation of the delay of the logic path than sizing-up INV6. Therefore, a refined metric will be proposed to deal with this issue.

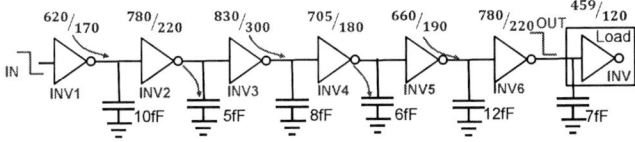

Figure 3. Logic Path 2.

Table II
DERIVATIVE OF THE STANDARD DEVIATION OF PATH DELAY WITH
RESPECT TO CHANNEL WIDTH SIZING AND DEGREE OF OPTIMIZATION OF
THE STANDARD DEVIATION OF THE DELAY, ACHIEVED BY SIZING-UP
EACH GATE OF THE LOGIC PATH 2.

Sized gate	$\frac{d\sigma_D}{dW}$	% Optimization
$INV4$	12.7	4.21
$INV6$	10.8	5.32
$INV2$	8.67	3.06
$INV5$	4.27	1.16
$INV1$	2.01	0.42
$INV3$	0.359	0.07

B. *Proposed Metric*

As was shown in previous experiments, the simple derivative of the standard deviation of the delay with respect to the channel width sizing should not be used as a selection metric, because it may fail to select the best gates to be optimized. This is due to different characteristics that the gates can present, such as load capacitance, topology and type of transistor involved in the logic path. For two gates having the same $\frac{d\sigma_D}{dW}$ and different σ, the gate with higher σ presents more reduction in the magnitude of the standard deviation of the delay than the gate with lower σ. Hence, in this work the product of the derivative of the standard deviation of the delay with respect to the channel width and the standard deviation of the gate delay is proposed. This allows to capture the impact of the magnitude of the standard deviation of the gate delay. The proposed metric will be called **Selection Factor (SF)**, and is given by,

$$SF = \frac{d\sigma_D}{dW}\sigma_D \qquad (2)$$

This metric will be applied to the Logic Path 2 to show that it provides a reliable classification of the gates to be optimized when the basic derivative metric failed. Table III shows $\frac{d\sigma_D}{dW}$ of each gate sorted from the higher to the lower value, and also the percentage of optimized path standard deviation of the delay when one transistor channel width of a gate is increased by $\Delta W = 80nm$. These results show that the proposed metric works correctly. A higher degree of optimization corresponds to a higher degree of SF for all the gates.

Table III
SELECTION FACTOR (SF) AND OPTIMIZATION DEGREE OF STANDARD
DEVIATION OF THE DELAY ACHIEVED BY SIZING-UP EACH GATE OF THE
LOGIC PATH 2.

Sized gate	SF	% Optimization
$INV6$	49.9	5.32
$INV4$	42.9	4.21
$INV2$	30.9	3.06
$INV5$	10.4	1.16
$INV1$	4.85	0.42
$INV3$	0.58	0.07

IV. CIRCUIT OPTIMIZATION METHODOLOGY

In this section, the methodology to obtain the statistical delay characteristics of the considered gates of the cell library technology is presented. This information is used to compute the Selection Factor of the gates. Then, our proposed optimization methodology using the the proposed Selection Factor is presented.

A. *Pre-characterization of the library cells*

All the library cells are pre-characterized through Hspice simulations using a 90nm CMOS commercial technology. Monte Carlo simulations were run using Hspice to obtain the delay probability density functions. Variations in transistors threshold voltage (V_{th}) due to RDF effects were assumed of $\pm20\%$.

The mean (μ_D) and standard deviation (σ_D) of the gate delay are modeled by an statistical approach based on *Design of Experiments* (DOE). DOE is a powerful statistical method that allows estimating, through an *Analysis of Variance* (ANOVA), the effect of each variable and their interactions on the gate delay which allows modeling accurately the behavior μ_D and σ_D. A full factorial DOE at 8 levels for each variable has been designed. The values μ_D and σ_D are extracted from HSPICE simulations for both transitions of each input of a gate. The result is two fourth degree polynomial functions, one for μ_D and other for σ_D. $\frac{d\sigma_D}{dW}$ is obtained by deriving the polynomial of σ_D with respect to the transistor channel width.

Although gate library characterization can be computational expensive it is required only once for each technology node. With this information, we can compute the selection factor of each gate in a path for any random logic design.

978-1-4673-9141-2/15 $31.00 © 2015 IEEE   167

B. Methodology Optimization

In this work, lagrangian multipliers are used to minimize standard deviation of circuit delay (σ_D) subject to a given extra area budget. Figure 4 shows the flow diagram of the proposed sizing algorithm for the optimization of standard deviation of the delay under some area constraint. The input is the set of critical paths, which are the paths likely to be critical for a given design. First, the overall circuit area budget is distributed within the critical paths of the circuit. Here, an heuristic partitioning scheme (see Equation 3), which is based on critical paths delay mean values is used. The heuristic assigns more area budget to the paths with the highest delays because their impact in the definition of circuit delay is more important.

$$A_{budget,i} = A_{budget,Circuit} \cdot \frac{\mu_i}{\sum_{j=0}^{N_{cp}} \mu_j} \qquad (3)$$

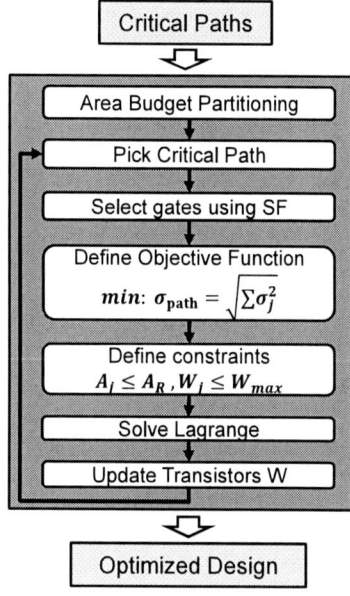

Figure 4.   Optimization Methodology.

The critical path set is optimized by one path at a time, starting from the most critical path of the circuit. The gates most favorable to be optimized are selected using the proposed *Selection Factor*. The optimization problem formulation is then developed using as design variables the channel width of transistors in the selected gates only. Due to the decreasing behavior of standard gate delay deviation as function of channel width (see Figure 1), a maximum channel width value, where the derivative is very low, is assigned for each gate. Thus, the restrictions of the optimization problem are given by the area budget assigned to the analized path and the maximum channel width allowed for each device in a gate. Once the problem formulation (design variales and restrictions definitions) is completed, the standard deviation of the path delay is minimized by lagrangian multipliers method. The optimum values of channel widths are updated and the process

continues with a new critical path. It should be highlighted that the optimized gates in one path are not selected again in other paths. In such way, previous optimization results remains nearly unchanged. The result of the methodology is the optimized design with minimum standard deviation of the delays for the given area budget.

V. Application to ISCAS85 Benchmark Circuits

The proposed optimization methodology with the developed metric has been applied to ISCAS85 benchmark digital circuits implemented in a CMOS 90nm commercial technology. Mentor Graphics suite of synthesis and layout tools are used to design the ISCAS85 benchmark circuits. First, detailed path results of the standard deviation delay are presented for ISCAS S510. Then, a summary results is presented for all the considered ISCAS.

A. Results for ISCAS510

Detailed results for the critical paths of ISCAS510 are presented. Table IV presents some important data for the optimization of ISCAS S510.

Table IV
ISCAS S510 Information

Number of gates	Number of critical paths	Nominal Area μm^2	Area Budget μm^2 1%	2%
186	20	326.4	329.6	331.2

Figure 5 presents the results for the optimization of the standard deviation of each path of circuit ISCAS510. Two area constraints cases (1% and 2%) have been considered. The nominal value for the standard deviation of each path is also given. For the area constraint of 1%, the standard deviation of all the logic paths is optimized significantly for most of the paths (See yellow bar in Figure 5). When the area constraint increases to 2%, the standard deviation is optimized even more. It should be noted that for the paths most affected by process variations (those with the highest standard deviation of nominal delay) a greater degree of optimization is achieved because the assigned area budget to those paths is greater than for the other paths.

B. ISCAS Summary Results

Some important data of the implemented ISCAS benchmark circuits are given in Table V. The critical paths have been obtained using statistical static timing analysis home-made tools. Figure 6 presents a summary of the obtained results for all the considered ISCAS benchmark circuits. For the entire circuit, its delay is computed as the statistical maximum of overall critical paths delays. The optimized standard deviation of the circuit and area cost are given. In these experiments, Case A has a lower area budget than Case B. The nominal case is also included in the plots. For constraint area Case A, it can be observed that the standard deviation of the circuit delay is optimized significantly for all the circuits (See Figure 6a), and it is further optimized for Case B, which has a greater area budget. Therefore, a trade-off can be established between

978-1-4673-9141-2/15 $31.00 © 2015 IEEE   168

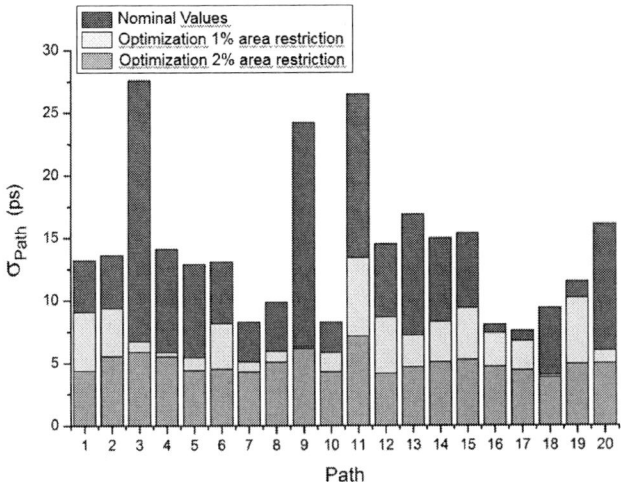

Figure 5. Comparison of the standard deviation of the delay for each path of ISCAS S510, for the nominal design and the optimized design. Results are shown for 1% and 2% of area budget.

maximum area budget and maximum yield improvement. The final area cost after optimization is given in Figure 6b. Most of circuits, except C1908, present an area increment lower than 2%. It should be noted that the area increment depends on the number of critical paths of the circuit. In our case, C1908 circuit has a high number of critical paths, therefore, most of the assigned area budget is required to improve each critical path standard deviation of the delay.

Table V
MAIN CHARACTERISTICS OF ISCAS CIRCUITS

ISCAS	Logic Depth	Number of gates	Number of critical paths	Area (pm^2)
C17	3	6	11	28.6
C1908	17	240	285	339.8
C880	27	274	113	395.4
S510	10	197	20	326.4
S820	13	280	10	278.9
S1196	25	481	124	764.1

Figure 7 shows the circuit delay variability ($3\sigma/\mu$) for the nominal design and for the two cases of area budget for optimization. Delay variability is a measure of the delay spread with respect to its nominal value. As can be seen for all the circuits its variability is improved. This means that possible delays after manufacturing are closer to the nominal design value, which can be translated in higher product yield and design profit.

VI. CONCLUSIONS

A statistical performance optimization methodology has been proposed. A gate selection metric was proposed to identify the most favorable gates where statistical optimization should be focused. Standard deviation of the circuit delay is then minimized using Lagrange multipliers method. An heuristic for area budget partitioning within the circuit critical paths was used to assign a less stringent area restriction for the

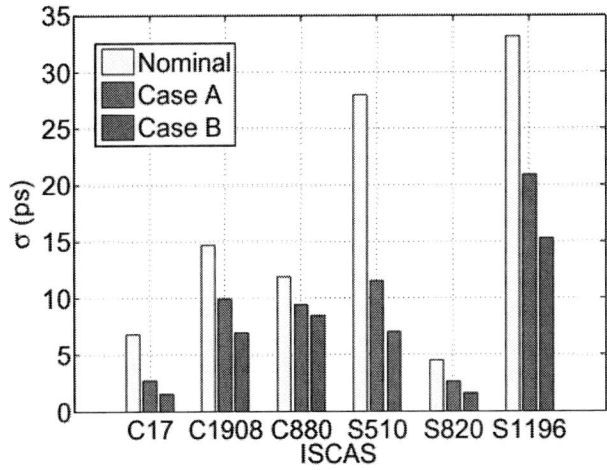

(a) Standard deviation of the delay

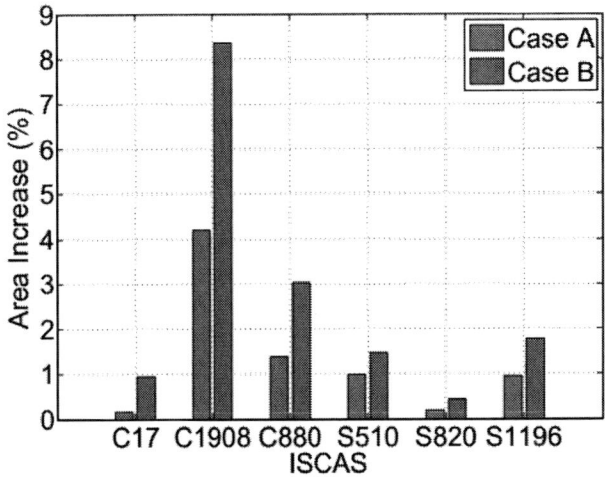

(b) Percentage of Area Increase

Figure 6. Optimization Results for ISCAS circuits.

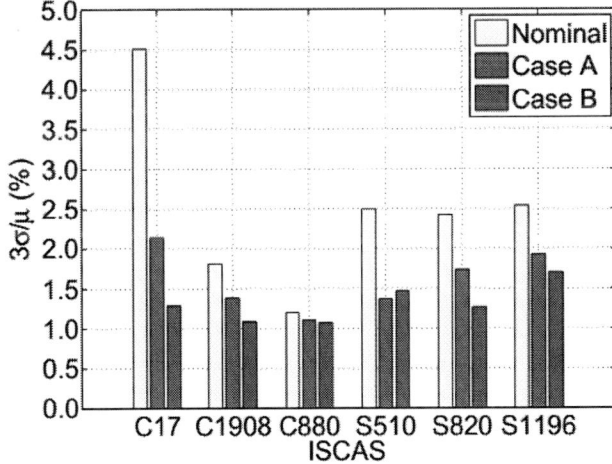

Figure 7. Delay variability of the optimized ISCAS.

most critical paths of the circuit, which have a greater impact on circuit delay. The results show that paths with a greater standard deviation of the delay are further optimized due to the area budget distribution. The effect in the overall circuit delay distribution is a significant reduction of standard circuit delay deviation (up to 75% for the greater area budget) for the allowed area overhead. The results also show that delay variability is reduced to less than 1.5% for all the circuits, which leads to higher product yield and design profit.

ACKNOWLEDGMENT – The work has been partially supported by CONACYT (Mexico) through the PhD scholarship number 420129/264560.

REFERENCES

[1] Blaauw, D.; Chopra, K.; Srivastava, A.; Scheffer, L., "Statistical Timing Analysis: From Basic Principles to State of the Art," Computer-Aided Design of Integrated Circuits and Systems, IEEE Transactions on , vol.27, no.4, pp.589,607, April 2008

[2] Bowman, K.A.; Duvall, S.G.; Meindl, J.D., "Impact of die-to-die and within-die parameter fluctuations on the maximum clock frequency distribution for gigascale integration," Solid-State Circuits, IEEE Journal of , vol.37, no.2, pp.183,190, Feb 2002

[3] Mishra, K.; Faraz, A.; Singh, A.D.; Chatterjee, A., "Path Delay Tuning for Performance Gain in the Face of Random Manufacturing Variations," VLSI Design (VLSI Design), 2011 24th International Conference on , vol., no., pp.382,388, 2-7 Jan. 2011

[4] Tschanz, J.W.; Kao, J.T.; Narendra, S.G.; Nair, R.; Antoniadis, D.A.; Chandrakasan, A.P.; De, V., "Adaptive body bias for reducing impacts of die-to-die and within-die parameter variations on microprocessor frequency and leakage," Solid-State Circuits, IEEE Journal of , vol.37, no.11, pp.1396,1402, Nov 2002

[5] Iparraguirre-Cardenas, D.; Garcia-Gervacio, J.L.; Champac, V., "A design methodology for logic paths tolerant to local intra-die variations," Circuits and Systems, 2008. ISCAS 2008. IEEE International Symposium on , vol., no., pp.596,599, 18-21 May 2008.

[6] Rajesh Garg, SunilP Khatri, "A Variation Tolerant Combinational Circuit Design Approach Using Parallel Gates", Book section, Analysis and Design of Resilient VLSI Circuits, pp 153-171 Springer US, 2010

[7] Visweswariah, C.; Ravindran, K.; Kalafala, K.; Walker, S.G.; Narayan, S.; Beece, D.K.; Piaget, J.; Venkateswaran, N.; Hemmett, J.G., "First-Order Incremental Block-Based Statistical Timing Analysis," Computer-Aided Design of Integrated Circuits and Systems, IEEE Transactions on , vol.25, no.10, pp.2170,2180, Oct. 2006

[8] Seung Hoon Choi; Paul, B.C.; Roy, K., "Novel sizing algorithm for yield improvement under process variation in nanometer technology," Design Automation Conference, 2004. Proceedings. 41st , vol., no., pp.454,459, 7-11 July 2004

[9] Datta, Animesh; Bhunia, S.; Jung Hwan Choi; Mukhopadhyay, S.; Roy, K., "Profit Aware Circuit Design Under Process Variations Considering Speed Binning," Very Large Scale Integration (VLSI) Systems, IEEE Transactions on , vol.16, no.7, pp.806,815, July 2008

[10] Aseem Agarwal, Kaviraj Chopra, David Blaauw, and Vladimir Zolotov. "Circuit optimization using statistical static timing analysis". In Proceedings of the 42nd annual Design Automation Conference (DAC '05). ACM, New York, NY, USA, 321-324. 2005.

[11] Davoodi, A.; Srivastava, A., "Variability Driven Gate Sizing for Binning Yield Optimization," Very Large Scale Integration (VLSI) Systems, IEEE Transactions on , vol.16, no.6, pp.683,692, June 2008

Scalable Algorithm for Structural Fault Collapsing in Digital Circuits

Raimund Ubar, Lembit Jürimägi, Elmet Orasson, Jaan Raik

Department of Computer Engineering, TTU, Ehitajate tee 5, 19086 Tallinn, Estonia
E-mails: raiub@pld.ttu.ee, lembit.jyrimagi@gmail.com,{elmet, jaan}@pld.ttu.ee

Abstract—The paper presents a new algorithm for structural fault collapsing to reduce search space for test generation, speed up fault simulation and make fault diagnosis easier in digital circuits. The proposed method is based on hierarchical topology analysis of the circuit description. First, the gate-level circuit will be converted into a macro-level network of fan-out-free regions each of them represented by a BDD. This conversion procedure represents the first step of fault collapsing, resulting in a compressed BDD model for representing the remaining set of fault sites. The paper presents an algorithm which implements a complementary step for further fault collapsing, and is carried out at the macro level by topological reasoning of equivalence and dominance relations between the nodes of BDDs. The algorithm has linear complexity and is implemented as a scalable fault collapsing procedure. We introduce higher and lower bounds for structural fault collapsing and provide statistics of distribution of fault collapsing results for a broad set of benchmark circuits. Experimental research has demonstrated better results for structural fault collapsing compared with state-of-the-art.

Keywords—combinational circuits, fault collapsing, fault equivalence and dominance, Binary Decision Diagrams, lower and higher bounds

I. INTRODUCTION

There are many reasons why fault collapsing is important for the VLSI community. A reduced set of only representative faults instead of a full set of faults allows pruning the search space during test generation, speeding-up fault simulation, and improving the efficiency of fault diagnosis. Fault collapsing methods are classified as structural and functional approaches. *Structural fault collapsing* uses only the topology of the circuit whereas *functional fault collapsing* is based on the functional equivalence [1].

Two classical ways are used for structural fault collapsing: *fault equivalence* based and *fault dominance* based collapsing. Two faults are *equivalent* if they are detected by the same tests. If two faults are equivalent, only one of them needs to be considered during test generation or fault diagnosis. A fault f_j is said to *dominate* a fault f_i if every test that detects f_i also detects f_j. If f_j dominates f_i, only f_i needs to be considered during test generation. When two faults dominate each other, they are called equivalent. *Structural fault collapsing* uses the topology of the circuit. For example, a *stuck-at* 0 fault (SAF y/0) on the output y of AND gate is equivalent to all of the SAF x/0 faults on its inputs x. In a similar way, SAF y/1 on the output of AND gate dominates all the input SAF x/1 faults. An approach based on *fault-folding* has been introduced in [2] for

structural collapsing faults, using the gate fault equivalence and dominance relations.

Two faults are functionally equivalent if they produce identical faulty functions [3] or we can say, two faults are functionally equivalent if we cannot distinguish them at primary output (PO) with any input test vector [4]. *Functional fault collapsing* uses the circuit functional information to establish equivalence and dominance relations. It is regarded as very difficult to compute because it deals with the whole function of the circuit under test, and its algorithmic complexity is similar to that of ATPG [5]. *Approximate* fault collapsing via simulation has been proposed in [6]. The potentials of hierarchical fault collapsing were shown in [7], and new algorithms based on *transitive closures* on *dominance graphs* were proposed in [8, 9], which enable more efficient hierarchical fault collapsing. In [10], functional dominance has been used to collapse the fault sets. However, this technique requires quadratic number of ATPG runs for fault collapsing. An improvement was proposed in [3], which has the linear complexity regarding the number of ATPG runs. Both these techniques are suitable only for small circuits.

A collapsed fault set contributes also to fault diagnosis. In [3], a novel *diagnostic fault equivalence and dominance* technique was proposed. A new method for fault collapsing for diagnosis called *dominance with sub-faults* was proposed in [11]. The method allows reducing the diagnosis search space. A framework where equivalence and dominance relations are defined for fault pairs is introduced in [12]. A technique to speed-up diagnosis via dominance relations between sets of faults using function-based techniques was proposed in [13]. Due to the high memory and time complexity this approach is applicable only for small circuits. All the listed techniques are fault oriented and use ATPG for identification of equivalence or dominance relations. One of the main limitations of the described methods is that there is no evidence that investing more effort in fault collapsing reduces the test total generation time [14]. The reason is that most of the methods are using for fault collapsing ATPG itself as a tool, or they are usable only for small circuits because of the high computing complexity.

In this paper we concentrate on the hierarchical structural fault collapsing based on the topology analysis of the circuit at two levels – gate and macro levels, where the Fan-out-Free Regions (FFR) are regarded as macros. Our ultimate target is to speed-up test generation and fault simulation by minimizing the fault set to be processed when solving these tasks.

978-1-4673-9141-2/15 $31.00 © 2015 IEEE

We propose a new method with linear complexity for fault collapsing, which provides a smaller collapsed fault set compared to the known structural methods, and which is usable for large circuits where the functional fault collapsing methods give up because of their complexity.

The approach we propose consists of two parts. During the first part, fault collapsing is carried out on the gate level by superposition of Binary Decision Diagrams (BDD) [15] of logic gates with the main goal of constructing a higher macro level model of the circuit in form of Structurally Synthesized BDDs (SSBDD) [16,17] where to each FFR an SSBDD corresponds. The fault collapsing can be regarded here as a side-effect (byproduct) of the SSBDD model synthesis. The contribution of this paper is the second, complementary part of the approach, and is carried out at the higher macro level by topological analysis of SSBDDs. Both parts of the fault collapsing procedure have linear complexity. It has been shown that SSBDDs can be efficiently used for fault simulation, outperforming in the speed state-of-the-art fault simulators [18, 19]. In this paper we show the possibility of additional fault collapsing using SSBDDs, which in its turn can lead to additional speed-up of fault simulation.

The paper is organized as follows. In Section 2 we give an overview of SSBDDs with explanations how fault collapsing is related to the synthesis procedure of SSBDDs. Sections 3 and 4 describe the theory and algorithm of fault collapsing using SSBDDs. In Section 5 we develop the lower and higher bounds for fault collapsing, Section 6 presents experimental data, and Section 7 concludes the paper.

II. STRUCTURALLY SYNTHESIZED BDDs

BDDs have become state-of-the-art data structure in VLSI CAD [15]. This model, however, suffers from the memory explosion which limits the usability of BDDs for large designs. Moreover, it is not well suited to represent directly in the model structural information about circuits like faults. In [16-17] SSBDDs were proposed for direct modeling of the structure of circuits in terms of signal paths and the faults causing errors on the paths. SSBDDs are generated by iterative superposition of library BDDs for simple or complex gates, guided by the structure of the given circuit. To avoid the explosion of the complexity of the SSBDD model, and to keep its size as minimum as possible, the superposition of BDDs is stopped at fan-out stems of the circuit. Using this restriction, to each FFR in the circuit an SSBDD will created where to each node in an SSBDD, a signal path in the FFR corresponds.

Example 1. An example of a combinational circuit with a single output and a single FFR is depicted in Fig.1. The FFR is described by the function

$$y = f(X) = (x_1 x_{21} \vee (x_{22} x_3 \vee x_4 (\overline{x_5} \vee \overline{x_{61}})) \overline{x_{71}}) x_{81} \vee$$
$$\vee \, x_{82} x_9 (x_{72} \vee \overline{x_{62}}) \overline{x_{10}}$$

which is represented by a single SSBDD in Fig.2. The nonterminal nodes of the SSBDD are labelled by the input variables of the FFR. To differentiate the fan-out branch variables from the fan-out stem variable we introduce for each of them a second subscript. When using SSBDDs for

calculating the output signals at given test patterns, we have to traverse the graph starting from the root node up to a terminal node guided by the input pattern. Let agree that we exit each node during simulation to the right if the node variable has value 1, and downwards if the value is 0. In this case we don't need to label the edges in the graph by the values of node variables on Figures. Entering the terminal node #1 as the outcome of graph traversing means the result of simulation $y = 1$, and entering the terminal node #0 means $y = 0$.

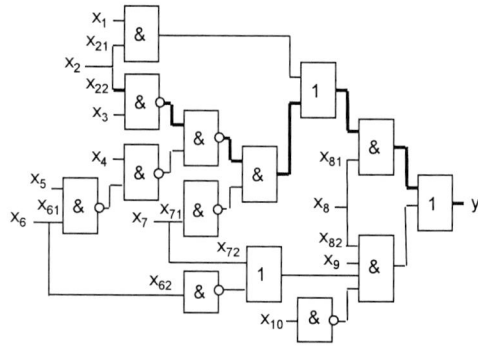

Fig.1. *Combinational circuit with a single output*

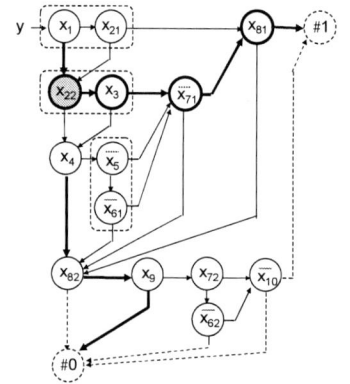

Fig. 2. *SSBDD for the circuit in Fig.1*

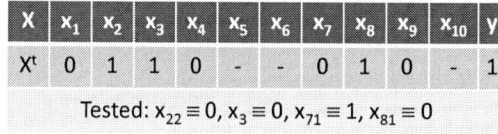

X	x_1	x_2	x_3	x_4	x_5	x_6	x_7	x_8	x_9	x_{10}	y
X^t	0	1	1	0	-	-	0	1	0	-	1

Tested: $x_{22} \equiv 0$, $x_3 \equiv 0$, $x_{71} \equiv 1$, $x_{81} \equiv 0$

Fig. 3. *Test pattern for detecting the faults SAF/0 or SAF/1*

Example 2. For the circuit in Fig.1 with function $y = f(X)$ the output signal y for the given input pattern X^t in Fig.3 will be $y = 1$. During simulation of this pattern on the SSBDD, the following nodes are traversed: x_1, x_{22}, x_3, $\neg x_{71}$, x_{81}, #1.

Definition 1. Let us call a path $L(a,b)$ in the SSBDD between two nodes a and b, *activated* by a given input pattern X^t, if by traversing the graph under guidance of X^t, the node b will be reached from a.

SSBDD model has several essential properties that makes the model very attractive compared to conventional BDDs or gate-level net-lists. The complexity of SSBDDs is linear in respect to the number of gates in the circuits, while it is exponential for traditional BDDs. Differently from traditional

BDDs [15], SSBDDs preserve structural information about the circuit: each node in the SSBDD represents a signal path in the related circuit. For example, the SSBDD in Fig.2 consists of 14 nodes (terminals are not counted) where each of them represents a related signal path of the total 14 paths in the circuit in Fig.1. The one-to-one mapping between the nodes of the SSBDD and the signal paths in the circuit is the result of SSBDD synthesis [17]. A node variable in the SSBDD is inverted if the related signal path has odd number of inverters. As shown in [20], the superposition procedure produces fault collapsing as a by-product of the SSBDD synthesis.

Example 3. The node x_{22} in the SSBDD represents the path from x_{22} to y in the circuit shown by bold lines in Fig.1. On the other hand, the stuck-at faults SAF $y/0$ and SAF $y/1$ dominate the faults $x_{22}/0$ and $x_{22}/1$, respectively. The same dominance relation stands for all the faults along the bold path from x_{22} to y, regarding to the faults at x_{22}.

From this dominance relation, it results that all the faults along the signal path from x_{22} to y, except $x_{22}/0$ and $x_{22}/1$, can be collapsed. The two faults at x_{22} will form the representative fault subset for the full signal path from x_{22} to y. But, exactly these faults are represented in the SSBDD as the faults of the node x_{22}. Hence, the SSBDD model can be regarded as the model where all the collapsed faults are removed and the fault sites are not visible as well. This fault collapsing result is similar to that of fault folding method presented in [2].

To summarize, the procedure of SSBDD synthesis can be regarded as the first part of fault collapsing for the given circuit. In the next section we will discuss the possibility of additional fault collapsing directly on the SSBDD model.

III. FAULT EQUIVALENCE AND FAULT DOMINANCE ON THE SSBDD MODEL

Consider a test pattern generation for the fault SAF $x_{22}/0$ with SSBDD in Fig.2.

Example 4. As explained in [17], to test a node in SSBDD we have to activate three paths in it: (1) $L(x_1, x_{22})$ from the root node x_1 to x_{22}, (2) $L(x_{22}, \#1)$ from x_{22} to the terminal node #1, and (3) $L(x_4, \#1)$ from x_4 to #0. The node x_4 is the neighbor of x_{22} not belonging to the path $L(x_{22}, \#1)$. The pattern X^t which activates these paths (bold lines in Fig.2) is depicted in Fig.3.

Definition 2. Let us call as *full path* in SSBDD which is activated from the root node up to one of the terminal nodes.

Definition 3. Let us call as 1-*path* (0-*path*) the full path activated by a given pattern and terminates in #1 (#0).

Definition 4. Let us call as 1-*nodes* (0-*nodes*) all the nodes traversed along the 1-path (0-path) in direction to 1 (0).

Example 5. The path $L(x_1, \#1) = (x_1, x_{22}, x_3, \neg x_7, x_{81}, \#1)$ in Fig.2, activated by the pattern in Fig.3, is 1-path, the node x_1 on this path is 0-node, and all other nodes are 1-nodes.

Property 1 [21]. If a test vector X^t activates in SSBDD a 0-path (1-path), then only 0-nodes (1-nodes) have to be considered as candidate fault sites.

The Property 1 can be taken into account to speed-up fault simulation. According to Property 1, the analysis of the 1-path in Fig.2 shows us that all the nodes, except x_1, may be

qualified as candidate fault sites. However, further analysis is needed to confirm which of the these candidate nodes are in fact detectable by the pattern. Since the faults at all 1-nodes for X^t (in Fig.3), will cause the direction change during graph traversing, then the faults at all 1-nodes are detectable by X^t.

Let us prove now two Theorems to determine the equivalence and dominance relations for the pairs of node faults on the SSBDD model.

Theorem 1. The faults at two connected nodes a and b are equivalent iff the following two conditions are satisfied: (1) the nodes have the same neighbor c, and (2) the node b has a single incoming edge from a.

Proof. The first condition refers to the fact that both nodes can be tested by the same test pattern which activates the paths $L(Root,a)$, $L(a,\#e)$ where $e \in \{0,1\}$, and the path $L(c,\#(\neg e))$. The second condition refers to that this test pattern is the only one which can test both of the node faults $a/\neg e$ and $b/\neg e$.

Example 6. For example the faults $x_{22}/0$ and $x_3/0$ are equivalent, and one of them can be collapsed.

Property 2 [21]. SSBDDs have always a single Hamiltonian path that visits all the nodes (except #0 and #1), and which determines a unique ranking of the nodes. The nodes a and b are in the relationship $a < b$ if the node a will be traversed before b along the Hamiltonian path.

Theorem 2. The fault $b/0$ dominates $a/0$ (or $b/1$ dominates $a/1$) iff the following conditions are satisfied: (1) there exists a single 1-path (or a single 0-path) through the nodes for detecting both of these faults, (2) $a < b$, and (3) the node b has more than 1 incoming edges.

Proof. The first conditions demands that these faults can be detected by a single test pattern (the condition of the equivalency). The second condition demands that there will be no other path for testing a and not testing b. The third condition is needed to give the possibility to test b and not to test a. From satisfying these conditions, it follows that any test for a must detect the related fault as well at b. Hence, the fault at a is dominated by b. If the third condition is not fulfilled, the related node faults at the nodes a and b are equivalent.

Example 7. The faults $\neg x_{71}/0$ and $x_{81}/0$ dominate $x_{22}/0$, and can be collapsed. Based on this result and taking into account Example 6, we can collapse 3 faults on the path $L(Root,\#1)$: $x_3/0$, $\neg x_{71}/0$ and $x_{81}/0$.

Corollary 1. The fault $a/0$ dominates $b/0$ ($a/1$ dominates $b/1$) iff the following conditions are satisfied: (1) there exists a single 1-path (0-path) through the nodes for detecting both of these faults, (2) $a < b$, and (3) the node a can be tested by activating another path where b is not tested.

For example, the fault $x_4/0$ dominates both of the faults $\neg x_5/0$, and $\neg x_{61}/0$, hence $x_4/0$ can be collapsed.

To use Theorems 1 and 2, and Corollary 1 directly may lead to an algorithm with quadratic complexity of pairwise analysis of all the nodes by a lot of traces in the graph. Taking into account the possibility of mapping sub-graphs in the SSBDD into the sub-circuits of the gate network, it would be possible to develop an algorithm of fault collapsing which will use a single trace through the SSBDD, which would provide linear complexity of the algorithm.

IV. FAULT EQUIVALENCE AND FAULT DOMINANCE FAST REASONING ON THE SSBDD MODEL

From the definition of the SSBDDs [21] we can derive the following rules for recognition of gates and sub-circuits in the SSBDD model, which will help us to develop a fault collapsing algorithm on SSBDDs with linear complexity.

Definition 5. Let us call the consecutive nodes on the Hamiltonian path of SSBDD as a *group* if they all have the same neighbor node, and all these nodes except the first one have a single incoming edge.

Example 8. Consider a circuit and its SSBDD model in Fig.4. The two consecutive nodes x_{22} and x_3, and the nodes $\neg x_5$ and $\neg x_{61}$ form two groups in the SSBDD in Fig.4. No more groups exist in this graph. The nodes $\neg x_{61}$ and $\neg x_{71}$ don't form a group.

Rule 1. A group of two nodes connected by horizontal edges (vertical edges) represents AND (OR) gate, and due to the fault equivalence, a fault at one of the inputs can be collapsed. The Rule 1 results directly from the method of synthesis SSBDDs by superposition of BDDs of gates [17].

Example 9. The nodes x_{22} and x_3 in the SSBDD in Fig.4 represent AND gate, and $\neg x_5$ with $\neg x_{61}$ represent OR gate. These gates can be recognized in the circuit. According to Rule 1, the faults $x_{22}/0$ (or $x_3/0$) and $\neg x_5/1$ (or $\neg x_{61}/1$) can be collapsed.

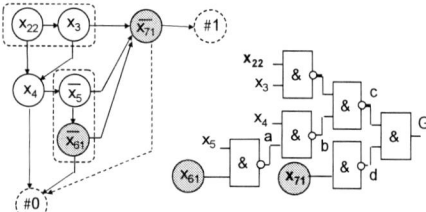

Fig. 4. Mapping SSBDD subgraphs into the circuit

Rule 2. If a node b in SSBDD has at least two or more incoming edges, it represents a path to a gate G where all the paths, represented by a subset of nodes $S(b) = \{a \mid a < b\}$, are joining. The fault of b dominates over the related faults of the nodes $a \in S(b)$ if the conditions of Theorem 2 are satisfied.

Example 10. the node $\neg x_{71}$ in SSBDD in Fig.4 has three incoming edges. It represents a path to the gate G joining with the paths represented by all other nodes a, $a < \neg x_{71}$, in this SSBDD. The nodes $\neg x_{61}$ and $\neg x_{71}$ don't form a group according to Definition 3 and don't represent AND.

The Rules 1 and 2 help to understand, how the fault equivalence and dominance relations in SSBDDs can be related to the equivalence relations in the circuit. If we have recognized a gate in SSBDD, the equivalence relations overlap for SSBDD and the circuit. The dominance relation in an SSBDD for a node with several incoming edges can be explained by transitive closure of dominance relations. For example, the dominance $\neg x_{71}/0 \rightarrow \neg x_{61}/0$ ($x_{71}/1 \rightarrow x_{61}/1$) in the SSBDD in Fig.4 can be explained by the following transitive

closures in the circuit: $x_{71}/0 \equiv d/0 \equiv c/0$, and $c/0 \rightarrow b/1 \rightarrow a/0 \rightarrow x_{61}/1$, from which $x_{71}/1 \rightarrow x_{61}/1$ results ($x_{71}/1$ dominates $x_{61}/1$).

In the following we present an algorithm with linear complexity for fault collapsing in a circuit represented by the SSBDD model. The algorithm is based on pairwise checking of Rule 1 (for equivalence) and Rule 2 (for dominance) by traversing along the Hamiltonian path in SSBDD.

ALGORITHM 1: Fault collapsing on SSBDDs

Input: SSBDD model for a given circuit
Output: Set of collapsed faults C
Notations:
    M – number of all nodes
    m – number of the first node of the current node pair
    n – number of the second node, $n = m + 1$, $n^* = n+1$
    $d(m)$ – direction from node m to n
    $n(d)$ – the neighbor of the node n in direction $d(m)$
    $C(m)$ – the type of the collapsed fault $C(m) \in \{0,1\}$
    $IN(m)$ / $OUT(m)$ – directions of incoming/outgoing edges
    $FI(m)$ – flag to remember the multiple fan-in for m
    $D(m)$ – flag to remember the direction of multiple fan-in
---

```
1:  for all SSBDDs in the model
2:    for all nodes m in the current SSBDD
3:      if m < M then go to 6
4:      if m = M then C(m) = ¬IN(m)
5:      go to 27
6:      if m(¬d) = n(¬d) then                    (checking of Rule 1)
7:        if FI(n) ≠ 1 then                       (checking of Rule 2)
8:          if [(IN(m) >1) & (IN(m) = OUT(m))] or D(m)=1
9:            then D(n) = 1
            end if
10:         if n = M then
11:           if D(n) = 1 then C(n) = ¬d(m), go to 13
12:           if OUT(m) ≠ OUT (n) then            (checking of Rule 1)
13:             C(m) = ¬d(m)
14:             if n(¬d) ≠ ∅ then FI(n(¬d)) = 1
15:           m = m + 2, go to 3
          else
16:           if FI(n*) = 1 then go to 13
17:           if n(¬d) = n*(¬d) then              (checking of Rule 1)
18:             C(m) = ¬d(m), m = m + 1, go to 3
19:           else C(n) = ¬d(m), go to 13
          end if
        else go to 22
20:     else if FI(m) = 1 then                    (checking of Rule 2)
21:       if FI(n) = 1 then                        (checking of Rule 2)
22:         C(m) = ¬IN(m)
23:         if m(¬d) ≠ ∅ then FI(m(¬d)) = 1
24:         m = m + 1, go to 3
25:       else  C(m) = ¬d(m), go to 23
26:     else go to 25
27:   end if
28:   end for
29: end for
```

Example 11. Consider the fault collapsing in the SSBDD in Fig.2 according to Algorithm 1. The initial number of gate level SAF faults in Fig.1 is 60 (2 faults per each of 30 lines). By synthesizing [17] the SSBDD model for the circuit we reduce the number of representative faults from 60 up to 36 (2 faults per each of 18 nodes in the SSBDD model (4 fan-out inputs where structural fault collapsing is not possible are represented by 4 single-node SSBDDs not shown in Fig.2). By using Algorithm 1, we further collapse 10 faults ($x_1/0$, $x_{22}/0$,

$x_4/0$, $x_5/0$, $x_{71}/1$, $x_{81}/0$, $x_{82}/0$, $x_9/0$, $x_{72}/1$, $x_{10}/1$) which results in the total number of remaining 26 representative faults.

V. LOWER AND HIGHER BOUNDS FOR FAULT COLLAPSING

In the following we will investigate the problem how to estimate the achievable size of the collapsed fault set.

Denote by N the number of all nodes in the SSBDD model of a circuit and by C the number of collapsed faults. The number of all SAF faults is $2N$, the number of representative faults after fault collapsing will be $R = 2N - C$, and the effect can be expressed by the ratio $R/2N$.

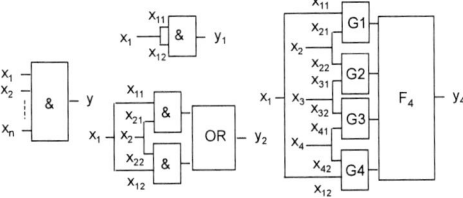

Fig. 5. Tree-like circuits with increasing complexity

Theorem 3. The effect of fault collapsing in the SSBDD model will be always in the boundary $1/2 < R/2N \leq 5/6$.

Proof. Any tree-like circuit with N inputs can be represented by SSBDD with N nodes. In the most simple tree, a single gate) with N inputs (gate y in Fig.5), we can collapse $N-1$ faults. Hence, $R = 2N-(N-1) = N+1$. Any partitioning of the set of inputs for more than one gate in this tree will reduce the total C by one fault per added gate and, hence, increase R. On the other hand, for increasing N, the lower bound for $R/2N$ is:

$$\lim_{n\to\infty}\frac{R+n}{2N+2n} = \lim_{n\to\infty}\frac{N+n+1}{2N+2n} = \lim_{n\to\infty}\frac{n}{2n} = \frac{1}{2}$$

Consider formally (neglecting the redundancy) a single-input logic gate y_1 in Fig. 5. The SSBDD model of the gate has $N=3$ nodes representing the fan-out stem with 2 branches. There are two equivalent faults on the gate inputs where one of them can be collapsed. Hence, the number of representative faults will be $R=2N-1=5$, and $R/2N = 5/6$.

Consider now the tree-like circuit y_2 in Fig.5 with two 2-input gates and two fan-out nodes. The SSBDD model of the circuit has $N=6$ nodes. There are again two equivalent faults on the gate inputs where one of them can be collapsed. Hence, the number of representative faults after fault collapsing will be $R = 2N-2 = 10$, and again we get $R/2N = 5/6$.

The circuit y_4 in Fig.5 illustrates how we can generalize the series of two circuits y_1 and y_2 into a series of expanding circuits y_n, $n = 1,2,3,4\ldots$, where each circuit will consist of an input sub-circuit IN_n as a chain of n 2-input gates, and a tree-like sub-circuit F_n. In each such a circuit, the ratio $R/2N = 5/6$ remains constant. In IN_n for each gate, only a single fault can be collapsed resulting in total in n collapsed faults.

It is easy to realize that any structural change inside the sub-circuit F_n will not change the ratio $R/2N = 5/6$. The reason is that all the faults in F_n will dominate the faults in IN_n. On the other hand, by adding $n = 1, 2,\ldots$ non-fan-out inputs to the sub-circuit IN_n we get $R/2N^* = (R+n)/(2N+2n)$, and by adding n fan-out inputs with 2 branches to IN_n we will get $R/2N^{**} = (R+2n)/(2N+6n)$. Each addition of a fan-out branch is equivalent to the case of adding a single input node where no faults can be collapsed. From above it follows:

$$R/2N^{**} < R/2N^* < R/2N \leq 5/6$$

Corollary 2. From Theorem 3, it directly follows that for any SSBDD with N nodes, the number of collapsed faults $C = 2N - R$ will belong to the interval $N/3 \leq C < N$. Hence, $N/3$ will serve as the lower bound for the number of collapsed faults.

VI. EXPERIMENTAL DATA

The fault collapsing experiments were carried out with Intel Core i5 3570 Quad Core 3.4 GHz, 8 GB RAM, using ISCAS'85, ISCAS'89 and ITC'99 benchmark circuits.

Table 1. Comparison with other methods

Circuit	# Faults	Fault set size				CPU time, s	
		[2]	[10]	[11]	New	[11]	New
c1355	2710	1234	1210	808	1210	46	0.003
c1908	3816	1568	1566	753	1243	14	0.008
c2670	5340	2324	2317	1853	1989	110	0.009
c3540	7080	2882	2786	2092	2340	831	0.010
c5315	10630	4530	4492	3443	3900	72	0.012
c6288	12576	5840	5824	5824	5824	4	0.019
c7552	15104	6163	6132	4707	5156	232	0.016

In Table 1, the sizes of fault sets after fault collapsing for the proposed method (New) with previous structural [2,10] and functional [11] methods are compared. The new proposed method has better results in fault collapsing than the previous structural methods. The functional method [11] is very slow and not scalable due to high computational cost of calculating transitive closures on dominance graphs whereas the proposed method has a very high speed due to the linear complexity and is well scalable. As an example, the difference in time costs for c3540 and c6288 in case of [11] is 200 times whereas the proposed method has nearly the same time cost.

Table 2. Fault collapsing for ISCAS'89 and ITC'99 circuits

Circ	GF	CF [23,24]	2N	R	R/2N	R/CF	Time s
s13207	24882	9815	10456	7933	75.9	1.24	0.04
s15850	29682	11727	12150	9178	75.5	1.28	0.04
s35932	65248	39094	39094	29797	76.2	1.31	0.26
s38417	69662	31180	32320	25162	77.9	1.24	0.20
s38584	72346	36305	38358	28016	73.0	1.30	0.18
b15	47414	21072	23498	17439	74.2	1.21	0.04
b17	154220	68037	81330	60684	74.6	1.12	0.12
b18	463570	206736	277978	205866	74.1	1.00	0.42
b18_1	453088	202812	264244	196179	74.2	1.03	0.40
b19	1345442	533142	560704	415251	74.1	1.28	0.84
b19_1	1275720	507476	534184	396151	74.2	1.28	0.80
b21	79556	35994	48182	35169	73.0	1.02	0.08
b21_1	63732	29091	34510	25359	73.5	1.15	0.06
b22	113308	51277	70464	51511	73.1	1.00	0.11
b22_1	98006	44771	52172	38359	73.5	1.17	0.08
Aver	290392	121902	138643	102804	74.1	1.2	0.24

The experimental results for larger ISCAS'89 and ITC'99 circuits (GF is the number of gates) are depiced in Table 2. The column R/CF shows the gain of achieved fault collapsing compared to the results in [23,24]. The last column shows that

Algorithm 1 has linear complexity, is scalable and can be efficiently used for large circuits. The linear complexity of the method is explained by the fact that the fault equivalence and dominance reasoning is reduced to pairwise node analysis in SSBDDs. The number of pairs to be analyzed, as it results from Algorithm 1, is in the interval $(N-1, N/2+1)$.

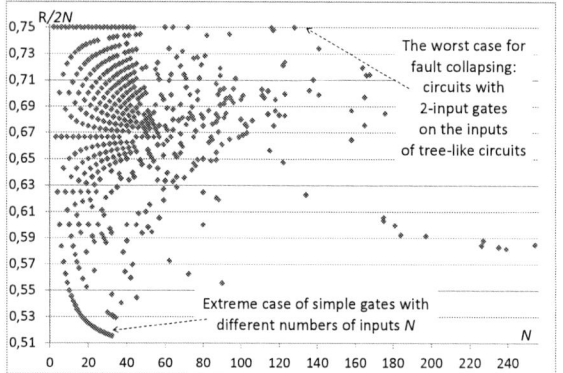

Fig. 6. Distribution of SSBDD cases with different characteristics (N, $R/2N$)

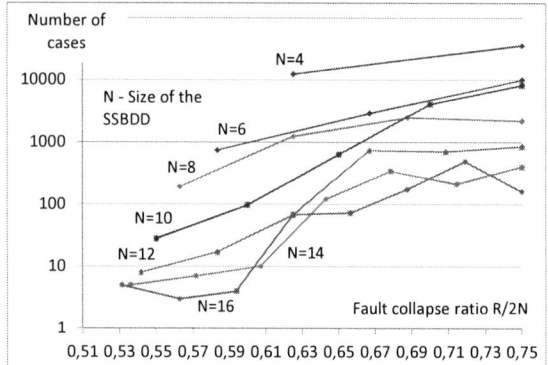

Fig. 7. Relations between the sizes of circuits, numbers of cases and fault collapsing rates

In Fig.6 and Fig. 7 we show statistical data collected from fault collapsing in 0.5 million of tree-like subcircuits in 111 circuits of ISCAS'85, ISCAS'89 and ITC'99 families. Fig.6 presents a plot of different subcircuit cases characterized by the number of nodes N in SSBDDs and the results of fault collapsing $R/2N$. Two extreme cases are highlighted: single-gate circuits (the best fault collapsing case) and the circuits with 2-input gates on the first level of tree-like circuits (the worst fault collapsing case). Fig.7 shows the distribution of the efficiency of fault collapsing in the interval between the lower (1/2) and higher (5/6) bounds, found in Section V. The number of cases and the efficiency of fault collapsing for the case are inversely proportional. When the size N of the circuit is increasing then the number of cases with high rate of fault collapsing is decreasing more rapidly.

V. CONCLUSIONS

In this paper we proposed a new structural fault collapsing method and algorithm with linear complexity. The method is based on using SSBDD model, and complements the fault collapsing achieved as a side-effect of SSBDD synthesis. We developed the lower and higher bounds of the effect of SSBDD based fault collapsing, and showed that the number of collapsed faults C belongs to the interval $N/3 \leq C < N$. where N is the number of nodes in the SSBDD model. Experiments showed that the proposed method is more efficient than the previous structural fault collapsing methods and due to high scalability makes it very promising for large circuits.

REFERENCES

[1] G.M.L.Bushnell, V.D.Agrawal. Essentials of Electronic Testing. Springer, Boston, 2000.

[2] K.To. Fault Folding for Irredundant and Redundant Combinational Circuits. IEEE Trans. on Comp, Vol. C-22, No. 11, 1973.

[3] Raja K. K. R. Sandireddy and Vishwani D. Agrawal. Using Hierarchy in Design Automation:The Fault Collapsing Problem. Proc. of the 11th VLSI Design and Test Symposium Kolkata, Aug. 8–11, 2007.

[4] A.Veneris, R.Chang, M.S.Abadir, S.Seyedi, Functional Fault Equivalence and Diagnostic Test Generation in Combinational Logic Circuits Using Conventional ATPG. JETTA. 21(5):495–502, 2005.

[5] A.Lioy. Advanced Fault Collapsing. IEEE Design and Test of Computers, Vol.9, No.1, pp. 64-71, Jan. 1992.

[6] H.Al-Assad, R.Lee. Simulation Based Approximate Global Fault Collapsing. Proc. of Int. Conf. on VLSI, pp.72-77, 2002.

[7] R.Hahn, R.Krieger, B.Becker. A Hierarchical Approach to Fault Collapsing. Proc. of EDTC, pp.171-176, 1994.

[8] A.V.S.S.Prasad, V.D.Agrawal, M.V.Atre. A New Algorithm for Global Fault Collapsing into Equivalence and Dominance Sets. Proc. of Int. Test Conference, pp.391-397, Oct.2002.

[9] R.Sethuram, M.L.Bushnell, V.D.Agrawal. Fault Nodes in Implication Graph for Equivalence Dominance Collapsing, and Identifying Untestable and Independent Faults. Proc. of VTS, pp.329-335, 2008.

[10] V.D.Agrawal, A.V.S.S.Prasad, M.V.Atre. Fault Collapsing via Functional Dominance. Int. Test Conf., pp.274-280, 2003.

[11] R.Adapa, S.Tragoudas, M.K.Michael. Sub-Faults Identification for Collapsing in Diagnosis. Int. Conf. ISCAS, pp.815-818, 2006.

[12] I.Pomeranz, S.Reddy. Equivalence and Dominance Relations Between Fault Pairs and Their Use in Fault Pair Collapsing for Fault Diagnosis. Int. Conf. on VLSI Design, pp. 1-6, 2007.

[13] R.Adapa, S.Tragoudas, M.K.Michael. Accelerating Diagnosis via Dominance Relations between Sets of Faults. Proc. of the VLSI Test Symposium, pp.219-224, 2007.

[14] I.Pomeranz, S.Reddy. Safe Fault Collapsing Based on Dominance Relations. Proc. of ETC, pp. 7-8, 2008.

[15] R.Bryant. Graph-based algorithms for Boolean function manipulation. IEEE Trans on Comp, 1986, vol. C-35, 677-691.

[16] R.Ubar. Test Generation for Digital Circuits Using Alternative Graphs. Proc. Tallinn Technical University, 1976, No.409, Tallinn, pp.75-81.

[17] R.Ubar. Test Synthesis with Alternative Graphs. IEEE Design & Test of Computers. Spring 1996, pp. 48-59.

[18] R.Ubar, S.Devadze, J.Raik, A.Jutman. Parallel X-Fault Simulation with Critical Path Tracing Technique. Proc. of DATE, 2010.

[19] M.Gorev, R.Ubar, S.Devadze. Fault Simulation with Parallel Exact Critical Path Tracing in Multiple Core Environment. Proc. of DATE, Grenoble, France, 2015, pp. 1-6.

[20] R.Ubar, D.Mironov, J.Raik, A.Jutman. Structural Fault Collapsing by Superposition of BDDs for Test Generation in Digital Circuits. IEEE ISQED, 2010, San Jose, CA - USA, pp. 250-257.

[21] R.Ubar. Overview about Low-Level and High-Level Decision Diagrams for Diagnostic Modeling of Digital Systems. Facta Universitatis (Nis) Ser.: Elec. Energ. vol.24, no.3, Dec. 2011, 303-324.

[22] D. Mironov, R. Ubar. Lower Bounds of the Size of S3BDDs. IEEE 17th Int.Symp. DDECS. Warsaw, April, 23-25, 2014, pp. 77-82.

[23] F.Brglez et al. Combinational Profiles of Sequential Benchmark Circuits. ISCAS'89, 1989.

[24] ITC'99. http://www.cad.polito.it/downloads/tools/itc99.html

Cost Reduction of System-level Tests with Stressed Structural Tests and SVM

Jing-Jia Liou*, Meng-Ta Hsieh[†], Jun-Fei Cherng[‡] and Harry H. Chen[§]

Email: *jjliou@ee.nthu.edu.tw [†]mthrandy@gmail.com [‡]jfcherng@gmail.com [§]harry-h.chen@mediatek.com

*[†][‡] Department of Electrical Engineering, National Tsing Hua University, Hsinchu 30013, Taiwan

[§] MediaTek Inc., Hsinchu Science Park, Hsinchu 30078, Taiwan

Abstract—System tests with boards are applied to capture defects in functional modes. Yet, these tests are usually costly with limitation on the production throughputs. Stressed structural tests (patterns produced by traditional ATPG) have been proposed to correlate with system tests and to replace them in production. However, due to low confidence level (small experimental samples and volatile chip variability conditions), we need a process to tune and apply stressed tests gradually. In this paper, we use SVM to classify stressed tests with the goal to select high-quality chips without the need of further system tests. The remaining (smaller batch of) chips will be processed by system tests for further defect screening. The proposed SVM method can be flexible in tuning the relative size of chip partitions.

I. INTRODUCTION

It is known that structural ATPG tests operate in a different condition than a chip's actual functional environments (including power voltage fluctuations, temperature distribution, clock trees, etc.). Therefore, to further reduce a chip's DPM and increase its reliability, it is desirable to apply system-level tests for packaged chips [1]. However, system tests are costly because of system board consumption (replacement necessary after a certain number of usage), and long software-based test sequences (affecting production throughputs).

In this paper, we would like to reduce the application of system tests with stressed ATPG tests. In the past, I_{DDQ}, minVDD, F_{MAX} and delay test have been used as stress mechanism to improve tests quality. For example, statistical clustering techniques, such as Principal Component Analysis (PCA) and Independent components analysis (ICA), have been applied on outlier screening with I_{DDQ} [2]–[4]. Authors in [5] introduced unsupervised Random Forests to identify outliers and supervised Random Forests to separate good/bad chips with the delay test signatures for the production test. Advanced analysis methods are also developed and applied to analog parameters successfully in [6]–[12]. To specifically address to issue of system-level tests, authors in [13] proposed to use stressed tests for predicting system-level fail or pass. The stressed tests are regular ATPG patterns applied with either lower voltage supply or faster sampling test clocks.

Even though stressed tests are effective in predicting failures in system tests, it has not been used in full-scale production. This is due to that (1) small chip sample size for experiments and (2) chip failures may variate from lot to lot. For the former

reason, the efforts are non-trivial to collect and analyze the stressed test for a large quantity of chips, so the size of the chip samples are usually too small with statistical significance. The major issue is actually with failed chips in system tests (called asymmetrical samples in training data). Considering that we are trying to control the defective parts in a few PPM, it may take months just to collect a few tens of bad chips. As for the second reason, we know that machine learning techniques cannot predict failure conditions not seen in previous lots. If we use solely the classifier built with stressed tests, we cannot guarantee that no extra escaped chip is leaked into customer's batch due to different variations.

To address the above issues, we proposed to use the stressed tests as a filter before the system tests, instead of replacing altogether the system tests. The responsibility of the filter is to select as many good chips as possible to avoid system tests. For example, if a chip has low mismatched scan outs even in deeper stressed level, it is highly unlikely that this chip needs another round of system tests. Initially, in actual production test, we can tune the filter with larger parts to be sent to system tests. And then gradually (with enough chips in statistically-significant number of samples), we tune the filter again to reduce the number for system tests. Note that we have removed the issue of asymmetrical samples, since we do not rely on the prediction of chip failures in system tests (which has usually low numbers).

The proposed method uses the support vector machine (SVM) to build the classifier (filter), but we tune the soft margin in SVM to produce a classifier with a high confidence level in selecting good chips that do not need system tests. In experiments, we apply the feature extraction method to two sets of selected production chips from MediaTek. The proposed classifiers demonstrate an effective and flexible way to reduce the need for a system tests.

II. TEST FEATURE EXTRACTION

To reduce the test costs, we would like to use a classifier to select a small subset of total chips in production and apply system-based tests only on these marginally-correct chips (or if the selected number is so small, we can completely ignore the system tests later). The generic procedure to create and use such a classifier is shown in Fig. 1. The training phase takes existing test data as inputs and produces a trained classifier

Published by the IEEE Computer Society

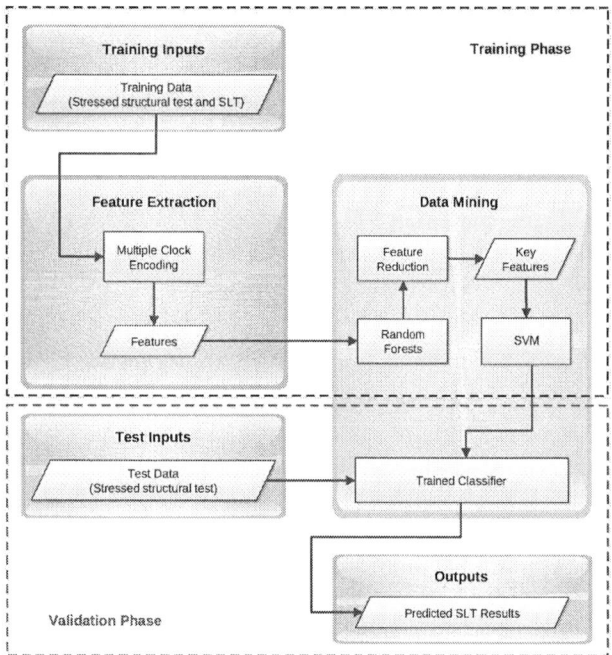

Fig. 1. Proposed Flow Chart for Test Classifiers

Label	Stressed corner	Effective frequency
f_1	(1.15V, 1482M Hz)	1574M Hz
f_2	(1.15V, 1508M Hz)	1601M Hz
f_3	(1.10V, 1482M Hz)	1677M Hz
f_4	(1.10V, 1508M Hz)	1707M Hz
f_5	(1.05V, 1482M Hz)	1795M Hz
f_6	(1.05V, 1508M Hz)	1826M Hz

TABLE I
AN EXAMPLE OF STRESSED CORNER ORDERING

A. Order Stress Corners

The input stressed test data are collected based on either lower supply voltages or higher sampling frequencies. To simplify the analysis, we map these test conditions to effective sampling frequencies. The higher effective frequency is, the deeper of the stressed tests are.

With a simplified propagation delay model for an inverter, we can derive the following mapping of the effective frequency:

$$f_{eff} = \frac{V_{DD}(V_{DD,nom} - V_{TH})^2}{V_{DD,nom}(V_{DD} - V_{TH})^2} \cdot f$$

As an example, we show the mapping for the case of six stressed corners $\{V = 1.15, 1.10, 1.05(V)\} \times \{F = 1482, 1508(MHz)\}$, in Table I. Here, the effective frequencies under nominal supply voltage are sorted. f_1 is the least stressed corner, while f_6 is the deepest stressed corner.

Note that our purpose is to sort the stressed corners with correct order as shown in the above example. As long as we retain the order properly, we can ignore the errors introduced in the process to derive the mapping with simplified switch model. We have also verified that the mapping keeps correct order with several circuit simulations.

B. Feature Encoding

After we convert the stressed conditions to effective sampling frequencies, we can compile all fail logs and find out all pass or fail results at these frequencies for each scan out. Then we records all toggling frequencies of a scan out. The toggle can be from F to P or P to F, since both conditions may be caused by failed paths merged at the scan out. As an example, when scanning the fail logs from f_1 to f_6 for a scan out, we found that that the results toggles at f_1, f_2, f_3 and f_5, so 1, 2, 3, 5 represents the test results of this scan out (under a test pattern). For easier processing in data mining tools, we further convert this set into a number with the function $\sum_{f_n \in f_{toggle}} 2^{n-1}$ (similar to a binary conversion). Thus, we have extracted as much failing information as possible from the fail log in a concise format.

C. Feature Reduction

After feature encoding, we use Random Forests (RF) algorithm to train the initial classifier model. Random Forests runs efficiently on large data sets with thousands of input variables without variable deletion. And it estimates feature ranking in the classification with "variable importance" [14].

through feature extraction and data mining stages. The feature extraction usually involves transformation or encoding of inputs for easier identifying structural information in the data. After proper encoding, data mining tools can be applied to create models for further evaluation. In this paper, random forest and SVM are applied to build classifiers. With the final trained model, we can apply inputs from evaluation sets to estimate the performance (through cross-validation setup).

In this paper, the input data are stressed structural ATPG patterns (i.e., stressed tests) with either lower voltage supply or faster sampling test clocks. And the output of the classifier is the prediction of pass/fail of system tests (labeled as SLTP or SLTF). Therefore, in the above training process (Fig. 1), a set of chips will be subjected to both stressed tests and system tests. We then use these data to create the classifier. The issues with the stressed test data are that the dimension of test data are usually large. We have to consider number of patterns, number of scan pins, number of scan cycles, and the number of stressed conditions (both voltage and sampling frequencies). To address these issues, we first illustrate an encoding scheme to assist the data mining process. And secondly, our classifiers are built in two iterations. In the first iteration, all test inputs (features) are used in the random forest. With the results, we can select fewer inputs as important features (feature reduction) and build another classifier with SVM, which would allow us to tune the classifier selecting SLTP chips with a high confidence level (details are in next section).

Our encoding method involves two steps: (1) stress corner ordering, and (2) feature encoding. There are discussed in details in the following sections.

978-1-4673-9141-2/15 $31.00 © 2015 IEEE

We can now only select a few of features to create a second classifier through SVM. Note that this step actually reduces dramatically the number of scan test patterns, which corresponds directly to test costs at this stage. Consider an empirical rule in supervised learning: it states that the ratio # features/# samples should be less than 1/10–1/15 to avoid data over-fitting [13]. We choose # features to build the classifier with lowest SLTF classification errors.

III. FLEXIBLE CLASSIFIER WITH SVM

A. Support Vector Machine

Support vector machine (SVM) [15] has been proved effective in classification work. SVM intends to find a separating hyper-plane with maximal margin between differently labeled data in a dimensional space. The hyper-plane or learned model $y = < \mathbf{x}, \mathbf{w} > +b$ is trained based on given set of data, and we use that learned model to classify unseen data. Each given training data is composed of instance-label pairs $\{\mathbf{x}_i, y_i\}$, $i = 1, ..., l$, where $x_i \in \mathcal{R}^n$, $y_i \in \{-1, 1\}$. Computationally, SVM solves the following quadratic optimization problem to obtain the learned model:

$$\min_{w,b,\xi} \frac{1}{2} w^T w + C^+ \sum_{k=1}^{m} \xi_k^+ + C^- \sum_{k=1}^{m} \xi_k^-.$$

$$\text{subject to } y_k(w^T \phi(x_k) + b) \geq 1 - (\xi_k^+ + \xi_k^-).$$

$$\xi_k^+, \xi_k^- \geq 0, k = 1, \ldots, m.$$

where data is mapped by the function ϕ, and C^+ and C^- are penalty parameters (weightings) on the training error ξ^+ and ξ^-, respectively. The kernel function $K(x_i, x_j)$ is defined as $K(x_i, x_j) \equiv \phi(x_i)^T \phi(x_j)$. If a kernel function is linear, then the method is Linear SVM. Otherwise, non-linear kernel functions can also be used to find a classifier. Yet, in order to solve the non-linear optimization problem, we resort to grid search algorithm to find parameters of kernel functions.

After building the primary classifier, we need to evaluate its performance for unseen data. For this purpose, we adopt classical K-fold cross validation to partition the training data into K sets. And we select one partition i sequentially among K sets as the test data We also build a classifier with the remaining $K - i$ sets, and evaluate the classifier with ith set. For all K classifiers, we calculate the average classification error as the evaluation criterion for current training data.

B. Classifier with Asymmetrical Data

In the above SVM formulation, training errors of ξ^+ and ξ^- are calculated based on mis-labeled data. In a regular training set, where we assume that sampled data are of equal size in both labels (+1 or -1), we often set $C^+ == C^-$ in the SVM process. However, in our test data, the percentage is heavily skewed towards the number of passing chips (in the order of 100000:1), where failed chips are scarce. For the unbalanced data, we can select C^-/C^+ to control the importance of a specific type of errors. For example, if ξ^- represents error

for mis-labeled failing chips, we may set $C^-/C^+ > 1$ such that more penalty are added to the constraint if ξ^- occurs. Therefore, we now can tune the SVM optimization process to avoid ξ^- errors as much as possible. Please note that even though there is less error by setting a higher value of C^-/C^+, we also put more chips into the bin for further system tests to validate the actual pass or fail. Obviously, there is a trade off between classification errors and percentage of chips sent for system tests.

C. Building SVM Classifiers

We use R package (libsvm [16] with R interface) in our implementation. To set different penalty parameters on the regularization terms in QP problem, in the svm function, we first set "cost" parameter and then change the class' weighting. By increasing the SLTF class weighting, svm will try to decrease the misclassification of SLTF chips. Intuitively, the decision hyper-plane will gradually move to the SLTP side. Therefore, our purpose is to select a high percentage of SLTP chips (**% bypass SLT**), while keeping the SLTF percentage in these chips low (**SLTF class.error**). As stated earlier, to validate the classifier, we use 10-fold cross-validation. So, we also report the cross validation error rate of predicted SLTP chips (**Pred.P CV accuracy**), which should be as high as possible.

To show the effectiveness of the tuning of penalty parameters in SVM, we first adapt and apply the method with a 2-class R example data, "cats". Here we label the two classes as SLTP and SLTF to match our discussion. Fig. 2 shows results of linear kernel, while Fig. 3 shows the results of RBF kernel. In both figure, we scan in X-axis the SLTF weightings from 0.01 to 0.99.

It can be seen from the figure that we can tune the weight of one label (SLTF) to trade off percentage of chips by-passing system tests and CV accuracy. With more chips sent to system tests (less **% bypass SLT**), we have higher CV accuracy. Also the RBF kernel performs better than Linear kernel, since in RBF less chips are sent to system tests while maintaining the same CV accuracy. Nevertheless, for less complexity of building ATE program, we prefer using linear kernel on training phase.

IV. EXPERIMENTAL RESULTS WITH SILICON DATA

In the following, several experiments are performed with test data from MediaTek. Two sets of test data are processed: ChipA and ChipB. Both chips are application processors with system tests. Note that chips are sampled from the production batches in order to perform stressed tests and to increase failed numbers for our experiments. For both chips, we compare several different setups:

A. Product Chip A

In the first part, the product (denoted as Chip A) has 292 packaged chips (200 SLTP + 92 SLTF). 60K on-chip-clock (OCC) scan patterns (60 pattern sets of 1K patterns per set) for the CPU block are applied to the DUT on the automated

978-1-4673-9141-2/15 $31.00 © 2015 IEEE

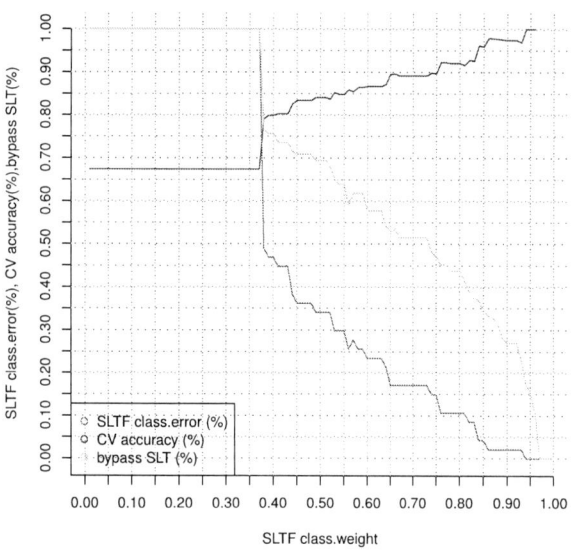

SLTF class.error, Pred.P CV accuracy, % bypass SLT

Fig. 2. Example Data: Trend of SLTF class.error; % bypass SLT; Pred.P CV accuracy v.s. SLTF class.weight (using linear kernel)

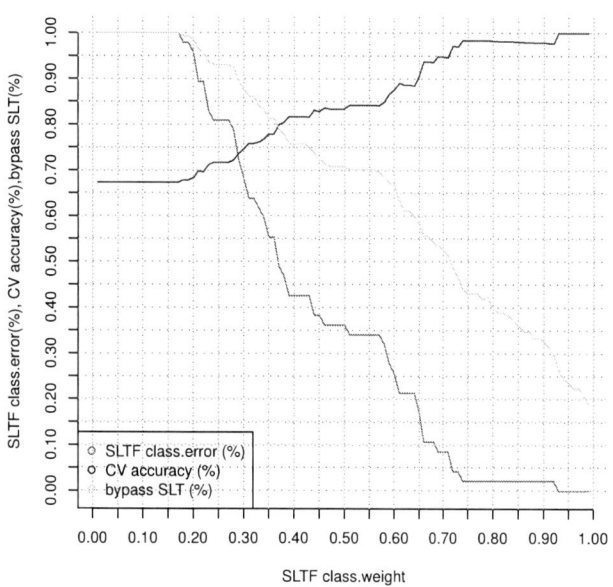

SLTF class.error, Pred.P CV accuracy, % bypass SLT

Fig. 3. Example Data: Trend of SLTF class.error; % bypass SLT; Pred.P CV accuracy v.s. SLTF class.weight (using RBF kernel)

test equipment (ATE) under stress conditions. Continue-on-fail test responses are shifted out of the DUT's scan output pins and all response mismatches are logged. The test data are in TetraMax cycle-based failure log format, which is processed from raw ATE fail log.

1) Order Stressed Corners: The nominal supply voltage is 1.2 V, and operating frequency is 1200 MHz. There are a total of 20 stressed conditions: (5 voltages × 4 frequencies).

$$V = \{1.15, 1.10, 1.05, 1.00, 0.95\}\text{Volt}$$

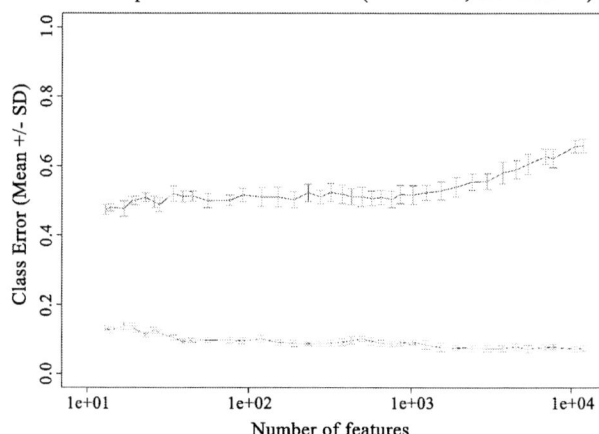

Fig. 4. Classification error of SLTF/SLTP devices v.s. number of selected features

$$F = \{1482, 1508, 1600, 1700\}\text{MHz}$$

By the mapping function, the calculated effective frequencies of 20 corners are obtained. We have also sorted them from the least stressed to the deepest stressed.

2) Feature Encoding and Extraction: To simplify analysis for feature extraction, each {Pattern Set, Cycle, SO} is encoded as with a path ID:

$$PathID = PatsetID \times 10000000 + Cycle \times 10 + SO$$

For all chips of chip A, there are 787616 unique path IDs. Therefore, the feature dimension for this device is also 787616. As stated above, we scan fail logs to obtain the multi-clock encoding of fail conditions for each path ID. At this stage, we have converted the test data into a table with each row composed of two entries: {path ID, feature encoding of fail logs}. Then, every chip has a feature vector in the above table and a tag label denoting SLTF or SLTP.

3) Feature Reduction: We use all 292 sample devices of chipA to train a Random Forest (RF) model with above encoded data set. We then apply feature selection before building a classifier, we plot classification errors of the both SLTF (red) and SLTP (green) v.s. number of selected features to build a Random Forest classifier in Fig. 4. In this initial model, Random Forest algorithm also estimate the variable importance. We also order the features according to the importance in X axis (from most to less important). For each number of features, multiple classifiers (40) are built, so we also plot the error bars to show estimated range. As it can be seen that errors of SLTP are stable beyond 20 features.

4) SVM Classifiers: Then we build a SVM model with linear kernel, as Fig. 5. Similar trends are observed that we can actually trade off chips bypassing system tests with higher CV accuracy. If we set the bypassing percentage to 50% (meaning a 2X improvements of test throughput compared with no filter with stressed tests), the linear kernel produce a classifier with 87% CV accuracy as compared with 78% of no tuning (50%

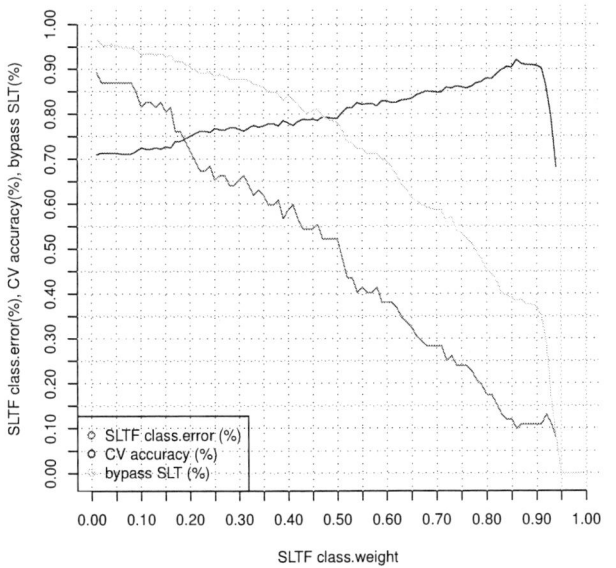

Fig. 5. Chip A: Trend of SLTF class.error; % bypass SLT; Pred.P CV accuracy v.s. SLTF class.weight (using linear kernel)

Module	Processor (ARM)		Multimedia (MM)	
Nominal Condition (Unit)	V_{DD} (Volt)	PLL (MHz)	V_{DD} (Volt)	PLL (MHz)
High Speed Bin (HS)	1.05	2000	0.90	700
Low Speed Bin (LS)	1.00	1700	0.90	700

TABLE II
THE NOMINAL CONDITIONS FOR CHIP B

Figure 6

LS-MM

MM PLL (MHz) nominal MHz	V1 0.850	V2 0.875	V3 0.900	V4 0.925
F1 700	✓			
F2 725	✓			
F3 750		✓		
F4 775		✓	✓	
F5 800			✓	
F6 825			✓	
F7 850				

MM-Vcc		MM-PLL		ominal Vcc	ffective PLL
V1	0.850	F1	700	0.90	766.73
V3	0.900	F4	775	0.90	775.00
V3	0.875	F3	750	0.90	784.18
V1	0.850	F2	725	0.90	794.12
V3	0.900	F5	800	0.90	800.00
V3	0.875	F4	775	0.90	810.32
V4	0.925	F7	850	0.90	814.40
V3	0.900	F6	825	0.90	825.00

LS-A M

A M PLL (MHz) nominal MHz	V1 1.000	V2 1.025	V3 1.050
F1 1900	✓		
F2 1950	✓		
F3 2000	✓		
F4 2050	✓	✓	
F5 2100		✓	✓
F6 2150		✓	✓
F7 2200			✓

A M-Vproc		A M-PLL		ominal Vproc	ffective PLL
V1	1.000	F1	1900	1.00	1900.00
V1	1.000	F2	1950	1.00	1950.00
V3	1.050	F5	2100	1.00	1953.22
V2	1.025	F4	2050	1.00	1975.83
V3	1.050	F6	2150	1.00	1999.72
V1	1.000	F3	2000	1.00	2000.00
V2	1.025	F5	2100	1.00	2024.02
V3	1.050	F7	2200	1.00	2046.23
V1	1.000	F4	2050	1.00	2050.00
V2	1.025	F6	2150	1.00	2072.21

HS-MM

MM PLL (MHz) nominal MHz	V1 0.825	V2 0.850	V3 0.875	V4 0.900
F1 700		✓		
F2 725		✓		
F3 750		✓		
F4 775		✓		
F5 800			✓	
F6 825			✓	✓
F7 850				✓

MM-Vcc		MM-PLL		ominal Vcc	ffective PLL
V2	0.850	F3	750	0.90	821.50
V4	0.900	F6	825	0.90	825.00
V3	0.825	F2	725	0.90	833.65
V3	0.875	F5	800	0.90	836.46
V2	0.850	F4	775	0.90	848.88
V4	0.900	F7	850	0.90	850.00
V3	0.875	F6	825	0.90	862.60

HS-A M

A M PLL (MHz) nominal MHz	V1 1.025	V2 1.050	V3 1.075	V4 1.100
F1 2150	✓	✓		
F2 2200	✓	✓		
F3 2250		✓		
F4 2300			✓	
F5 2350			✓	
F6 2400			✓	✓

A M-Vproc		A M-PLL		ominal Vproc	ffective PLL
V2	1.050	F1	2150	1.05	2150.00
V2	1.050	F2	2200	1.05	2200.00
V3	1.075	F4	2300	1.05	2222.13
V1	1.025	F1	2150	1.05	2227.94
V4	1.100	F6	2400	1.05	2242.68
V2	1.050	F3	2250	1.05	2250.00
V3	1.075	F5	2350	1.05	2270.43
V1	1.025	F2	2200	1.05	2279.75
V2	1.050	F4	2300	1.05	2300.00
V3	1.075	F6	2400	1.05	2318.74

Fig. 6. Mapped effective frequencies of all corners of chip B

Module	Processor (ARM)	Multimedia (MM)
High Speed Bin (HS)	60029	383521
Low Speed Bin (LS)	122788	347474

TABLE III
THE FEATURE DIMENSION OF MODELS FOR CHIP B

weighting for both classes). When we tune the weighting, we can achieve about 92% of CV accuracies and keep 38% of chips to bypass the system tests. In other words, we can achieve at least 1.6X improvements of SLT throughput.

B. Product Chip B

In the second part of this study, we sampled 329 packaged chips from Chip B. Before stressed tests, the chips are pre-sorted into two bins: high speed (HS) and low speed (LS). High speed bin has 150 SLTP + 22 SLTF, and low speed bin has 150 SLTP + 7 SLTF devices. Note that HS and LS are two different groups of packaged chips, so we apply our methods to both groups in the following.

Also for each chip, two distinct ATPG and system tests are generated for both ARM processors (ARM) and multimedia (MM) modules. 25K on-chip-clock (OCC) scan patterns (25 pattern sets of 1K patterns per set) are applied to the DUT on ATE under stress conditions. Similar to ChipA, continue-on-fail test responses are shifted out of the DUT's scan output pins and all response mismatches are logged. The given test data are in TetraMax cycle-based failure log file format pre-processed from raw ATE fail log.

1) Order Stressed Corners: Table. II are the nominal supply voltage and PLL settings for ARM and MM.

We also list the stressed conditions and their mapped effective frequencies under nominal supply voltages (sorted from least stressed to the deepest stressed) in Fig. 6.

2) Feature Encoding and Extraction: Similar to Chip A, each {Pattern Set, Cycle, SO} is encoded as with a path ID:

$$PathID = PatsetID \times 100000000 + Cycle \times 100 + SO$$

For all chips of chip B, we summarize the feature dimension of each model of both batches with Table. III.

3) Feature reduction: Similarly to Chip A, we apply the encoding technique to Chip B's test data and select features with Random Forest model. For high (or low) speed batch, we have two distinct ATPG and system tests for ARM and multimedia (MM) modules. Therefore, we can actually create two models and show results for both ARM and MM. Besides, since the two modules (ARM and MM) are on the same chip, we can actually cross-reference two models to predict full chip SLT pass or fail. If either models predict SLTF, we mark the chip as Fail. Or if both models predict SLTP, we then label the chip as Pass.

4) SVM filtering of high speed chip B: First we show the trend for the high speed batch after cross-referencing two models, as Fig. 7. If we set the bypassing percentage to 50% (meaning a 2X improvements of test throughput compared with no filter with stressed tests), the linear kernel produce a classifier with 95% CV accuracy as compared with 92% of no tuning (50% weighting for both classes). When we tune the weighting, we can achieve about 98.4% of CV accuracies

SLTF class.error, Pred.P CV accuracy, % bypass SLT

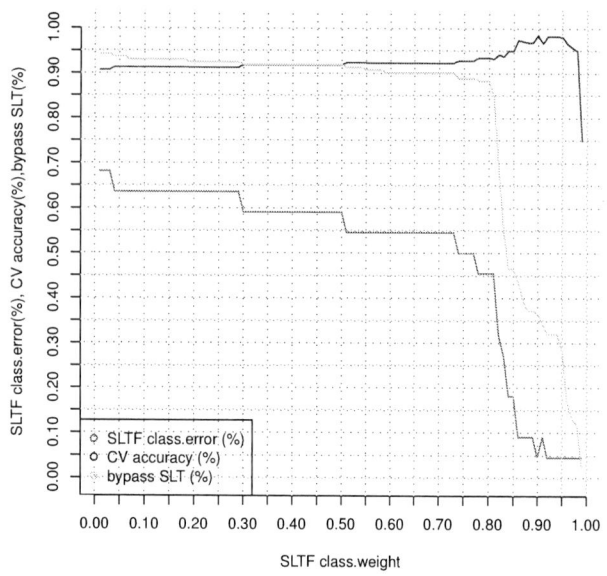

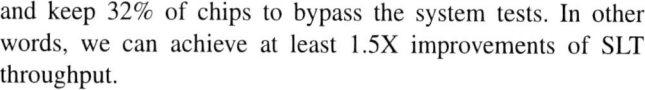

SLTF class.error, Pred.P CV accuracy, % bypass SLT

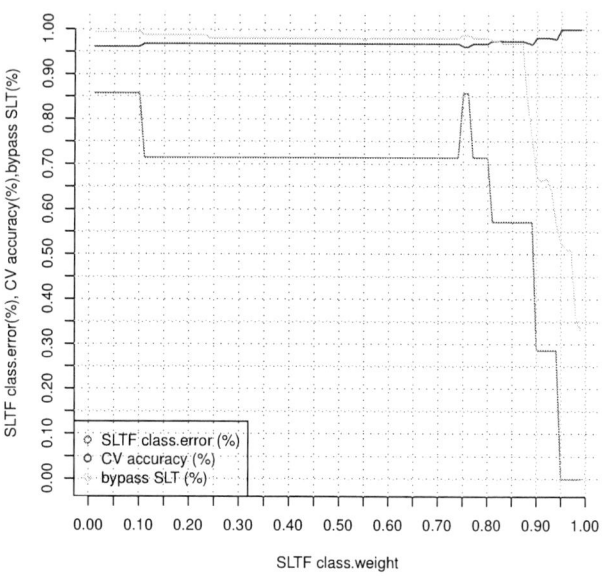

Fig. 7. Chip B-HS: Trend of SLTF class.error; % bypass SLT; Pred.P CV accuracy v.s. SLTF class.weight (using linear kernel)

Fig. 8. Chip B-LS: Trend of SLTF class.error; % bypass SLT; Pred.P CV accuracy v.s. SLTF class.weight (using linear kernel)

and keep 32% of chips to bypass the system tests. In other words, we can achieve at least 1.5X improvements of SLT throughput.

5) SVM filtering of low speed chip B: In the following, we show results for the low speed batch also after cross-referencing the two models, as Fig. 8. If we set the bypassing percentage to 50% (meaning a 2X improvements of test throughput compared with no filter with stressed tests), the linear kernel produce a classifier with 100% CV accuracy, representing no SLTF chips are leaked to the chips bypassing system tests.

V. CONCLUSION

In this paper, we propose a classifier to use stressed test data as a pre-filter to select partial chips for final system tests. As we tune the SVM process to produce a flexible classifier, we can trade off between test escape in chips without system tests and number of chips with system tests.

By the experiments of the three batches of chips, if we want to have 2X improvements of SLT throughput, the cross validation accuracy can achieve 87%, 92%, 100% for chip A, chip B's high-speed (HS) batch, chip B's low-speed (LS) batch, respectively. In addition, as tuning SVM class weighting, we can find that the percentage of chips bypassing system tests are kept at least 38%, 32%, 51% for chip A, chip B-HS, chip B-LS, respectively. Currently we are working with production test group in MediaTek to adopt and deploy the method and to further validate the results in production tests.

REFERENCES

[1] S. Biswas and B. Cory, "An industrial study of system-level test," *IEEE Design Test of Computers*, vol. 29, no. 1, pp. 19–27, Feb 2012.

[2] P. O'Neill, "Statistical test: A new paradigm to improve test effectiveness & efficiency," in *IEEE International Test Conference*, Oct 2007, pp. 1–10.

[3] A. Sharma, A. Jayasumana, and Y. Malaiya, "X-iddq: a novel defect detection technique using iddq data," in *IEEE VLSI Test Symposium*, April 2006.

[4] R. Turakhia, B. Benware, R. Madge, T. Shannon, and R. Daasch, "Defect screening using independent component analysis on iddq," in *IEEE VLSI Test Symposium*, May 2005, pp. 427–432.

[5] S. Wu, B. Lee, L.-C. Wang, and M. Abadir, "Statistical analysis and optimization of parametric delay test," in *IEEE International Test Conference*, Oct 2007, pp. 1–10.

[6] H.-G. Stratigopoulos, S. Mir, and Y. Makris, "Enrichment of limited training sets in machine-learning-based analog/rf test," in *IEEE Design, Automation Test in Europe*, April 2009, pp. 1668–1673.

[7] E. Yilmaz, S. Ozev, and K. Butler, "Adaptive multidimensional outlier analysis for analog and mixed signal circuits," in *IEEE International Test Conference*, Sept 2011, pp. 1–8.

[8] H. Chen, R. Hsu, P. Yang, and J. Shyr, "Predicting system-level test and in-field customer failures using data mining," in *IEEE International Test Conference*, Sept 2013, pp. 1–10.

[9] F. Lin, C.-K. Hsu, and K.-T. Cheng, "Learning from production test data: Correlation exploration and feature engineering," in *IEEE Asian Test Symposium*, Nov 2014, pp. 236–241.

[10] C.-K. Hsu *et al.*, "Test data analytics — exploring spatial and test-item correlations in production test data," in *IEEE International Test Conference*, Sept 2013, pp. 1–10.

[11] W. Zhang, X. Li, E. Acar, F. Liu, and R. Rutenbar, "Multi-wafer virtual probe: Minimum-cost variation characterization by exploring wafer-to-wafer correlation," in *IEEE/ACM International Conference on Computer-Aided Design*, Nov 2010, pp. 47–54.

[12] F. Lin, C.-K. Hsu, and K.-T. Cheng, "Feature engineering with canonical analysis for effective statistical tests screening test escapes," in *IEEE International Test Conference*, Oct 2014, pp. 1–10.

[13] H. Chen, "Perspectives on test data mining from industrial experience," in *IEEE Asian Test Symposium*, Nov 2014, pp. 242–247.

[14] L. Breiman and A. Cutler, "Random forests," http://www.stat.berkeley.edu/~breiman/RandomForests/cc_home.htm.

[15] C. Cortes and V. Vapnik, "Support-vector networks," in *Machine Learning*.

[16] C.-C. Chang and C.-J. Lin, "LIBSVM: A library for support vector machines," [Online]. Available: http://www.csie.ntu.edu.tw/~cjlin/libsvm.

Energy-Efficient Exclusive Last-Level Hybrid Caches Consisting of SRAM and STT-RAM

Namhyung Kim, Junwhan Ahn, Woong Seo* and Kiyoung Choi
Seoul National University   *Samsung Advanced Institute of Technology
nhkim@dal.snu.ac.kr, junwhan@snu.ac.kr, brad.seo@samsung.com, kchoi@snu.ac.kr

Abstract—This paper presents an energy-efficient exclusive last-level cache design based on STT-RAM, which is an emerging memory technology that has higher density and lower static power compared to SRAM. Exclusive caches are known to provide higher effective cache capacity than inclusive caches by removing duplicated copies of cache blocks across hierarchies. However, in exclusive cache hierarchies, every block evicted from the lower-level cache is written back to the last-level cache regardless of its dirtiness thereby incurring extra write overhead. This makes it challenging to use STT-RAM for exclusive last-level caches due to its high write energy and long write latency. To mitigate this problem, we design an SRAM/STT-RAM hybrid cache architecture based on reuse distance prediction. In our architecture, for the cache blocks evicted from the lower-level cache, blocks that are likely to be accessed again soon are inserted into the SRAM region and blocks that are unlikely to be reused are forced to bypass the last-level cache. Evaluation results show that the proposed architecture significantly reduces energy consumption of the last-level cache while slightly improving the system performance.

I. INTRODUCTION

In modern chip multiprocessors (CMPs), last-level caches (LLCs) have become increasingly important to mitigate the memory wall problem [1]. Over the last decade, the capacity of LLCs has continuously increased in order to filter out more main memory accesses, which leads to better performance and energy efficiency. However, conventional SRAM-based inclusive cache hierarchies may not be the optimal design in that (1) increasing SRAM size is detrimental in terms of area and energy efficiency and (2) a significant portion of on-chip cache capacity is wasted by storing multiple copies of a single cache block across multiple levels of caches.

Regarding the first inefficiency, the recent advancement of emerging memory technologies is opening up the possibility of enlarging LLCs in a more efficient manner. In particular, Spin-Transfer Torque RAM (STT-RAM) is gaining attention as a promising alternative to SRAM for LLCs. Compared to SRAM, STT-RAM shows much lower static power and smaller cell size due to its unique cell structures based on non-volatile, magnetic-based information storage. With these benefits, STT-RAM caches have been projected to realize very large on-chip LLCs while minimizing its impact on area and energy [2], [3].

Another dimension of increasing the on-chip cache efficiency is to employ better cache hierarchy management. In inclusive cache hierarchies, which are the most widely used, the inclusion property mandates each lower-level cache block (e.g., L1 cache block) to be duplicated across all upper-level caches (e.g., last-level cache), thereby degrading on-chip

storage efficiency. On the contrary, exclusive caches are free from such overhead as they store each block in only a single place. The performance gap between them is expected to be wider in modern CMPs due to deeper cache levels and larger lower-level caches. Because of these advantages, several commercial products already adopted exclusive cache hierarchies, including AMD's desktop and server processors [4].

As both STT-RAM and exclusive caches provide advantages over conventional SRAM-based inclusive caches, this paper for the first time explores STT-RAM-based exclusive LLC design. Our finding is that, while both STT-RAM and exclusive caches can improve the efficiency of LLCs, simply replacing SRAM with STT-RAM is very inefficient in exclusive LLCs. This is because enforcing the exclusion property greatly increases LLC writes (65% more writes compared to inclusive caches according to our experiments), which is harmful for STT-RAM caches whose drawback is high write energy.

To mitigate this, we propose a hybrid cache architecture composed of SRAM and STT-RAM based on reuse distance prediction. The key idea is to identify cache blocks with near/far reuse and then utilize the information to determine the block placement or bypassing. This greatly improves energy efficiency of STT-RAM exclusive LLCs, which has not been possible with existing approaches for STT-RAM inclusive/non-inclusive caches due to the fundamental differences in cache behavior on hits/misses between inclusive/non-inclusive and exclusive caches.

In summary, this paper makes the following contributions:

- For the first time, we evaluate STT-RAM in the context of exclusive cache hierarchies and identify the key challenges in designing an energy-efficient exclusive LLC based on STT-RAM.
- We propose a hybrid cache architecture for exclusive STT-RAM LLCs based on PC-directed reuse distance prediction. This is the first approach that utilizes the reuse distance prediction to improve energy efficiency of caches without any software modification.
- We evaluate our architecture based on a cycle-level simulator and show that it reduces energy consumption of the LLC by 55% compared to the STT-RAM baseline with slight performance improvement.

II. BACKGROUND AND MOTIVATION

A. STT-RAM

An STT-RAM cell consists of one magnetic tunnel junction (MTJ) and one NMOS transistor. An MTJ is composed of two

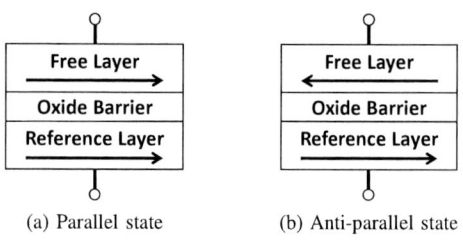

(a) Parallel state   (b) Anti-parallel state

Fig. 1. MTJ structure.

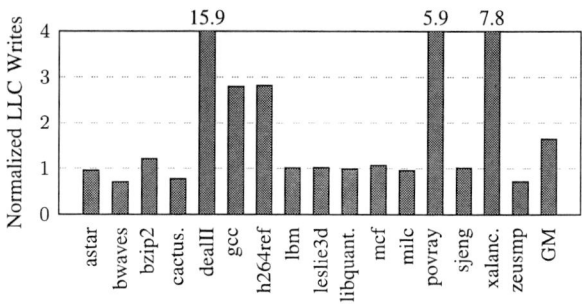

Fig. 2. The number of LLC writes in the exclusive cache hierarchy normalized to that of the inclusive cache hierarchy.

ferromagnetic layers and one oxide barrier layer between them. Magnetization direction of one ferromagnetic layer (reference layer) is fixed while that of the other ferromagnetic layer (free layer) can be changed by current flowing through the MTJ. The MTJ resistance is determined by magnetization direction of the free layer. Fig. 1 shows an example of (in-plane SLC) MTJ that can be in one of two states: parallel state and anti-parallel state. The MTJ resistance is low/high in parallel/anti-parallel state, respectively, which determines the bit data stored in it.

Compared to SRAM, STT-RAM consumes much lower static power due to its non-volatility, whereas it consumes higher write energy due to its high write current. In order to combine the benefits of these two memory technologies, researchers have proposed to construct a hybrid cache consisting of both SRAM and STT-RAM [5]–[9]. However, to the best of our knowledge, none of the previous proposals consider exclusive LLCs, which introduce several important challenges that do not exist in inclusive/non-inclusive cache hierarchies.

B. Exclusive Cache

Exclusive cache hierarchies enforce only one location per cache block (exclusion property) to maximize cache efficiency. To maintain such a property, some cache operations behave differently from those of inclusive/non-inclusive caches.

First, a cache block is inserted into the LLC *only after it is evicted from the lower-level caches*. In inclusive caches, a LLC miss loads the target block into both the LLC and the lower-level caches to maintain the inclusion property. On the contrary, exclusive caches bring cache blocks to the lower-level caches first, and move them to the LLC on eviction to prevent duplication.

Second, every LLC block is invalidated from the cache right after its first hit to preserve the exclusion property. This is different from inclusive caches where cache blocks are rather duplicated (instead of being moved to the lower-level caches) on hit for inclusion property.

Although these two differences improve cache efficiency by avoiding duplication of cache blocks, we observed that they incur significantly higher write overhead compared to inclusive caches, which is problematic in STT-RAM caches due to their inefficient write operations. As Fig. 2 shows, an exclusive LLC receives 65% higher average write traffic compared to its inclusive version (see Section IV-A for evaluation methodology).

Unfortunately, previous work on STT-RAM cache write reduction [5]–[11] is not compatible with exclusive caches since they are designed based on cache behavior under inclusive/non-inclusive caches. For example, most of the existing inclusive hybrid cache architectures keep track of access information (e.g., the number of writes per block) and use it as a metric to perform block placement or migration for write energy reduction [5]–[7], [9]. This, however, does not work for exclusive caches, since they invalidate corresponding cache blocks on LLC hit (i.e., every block receives only one hit before its eviction) and therefore it is impossible to keep track of their access information. This motivates the need for an architectural technique *specifically for exclusive cache hierarchies* to alleviate the write overhead of STT-RAM LLCs.

C. Reuse Distance

Our key observation for mitigating the write overhead of exclusive caches is that *a significant portion of blocks inserted into exclusive LLCs are either (1) accessed in a very near future or (2) not accessed at all until their eviction.* In the first case (i.e., near reuse), since such blocks reside in the cache in a very short amount of time, it can be stored in a small, auxiliary buffer that has low write overhead (e.g., SRAM cache) to reduce write energy. In the second case (i.e., far reuse), such blocks do not need to be stored into any cache as they will not be re-referenced at all until their eviction. Thus we can avoid writes to an STT-RAM cache in those two cases.

As empirical evidence of such behavior, Fig. 3 shows our experimental results on reuse distance profiling. *Reuse distance* of a target cache block is defined as the number of cache accesses to other blocks in the same set between two consecutive accesses to the target block. As can be seen in the figure, most of the cache blocks in the exclusive LLC exhibit either near reuse (reuse distance < 1, 19%) or far reuse (reuse distance > 14, 66%). This shows the potential of reuse distance as a metric for cache block classification for write overhead reduction in exclusive caches. In the

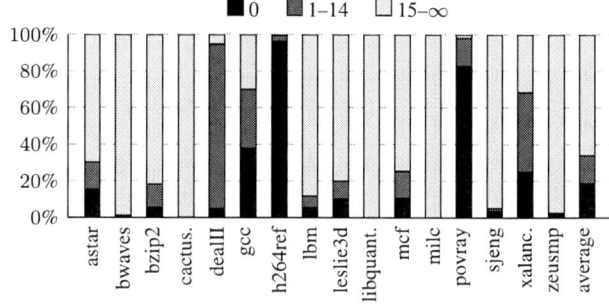

Fig. 3. Reuse distance distribution.

978-1-4673-9141-2/15 $31.00 © 2015 IEEE

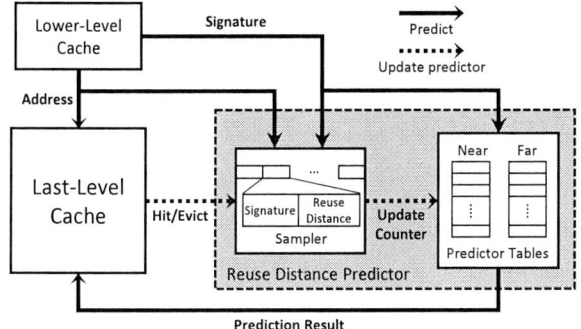

Fig. 4. Overview of the proposed architecture.

following section, we will describe our cache architecture that uses reuse distance prediction for energy-efficient exclusive hybrid cache management. Note that, unlike previous work on reuse distance prediction that focused only on improving performance of traditional SRAM caches [12], [13] and/or required software modifications [14], [15], our work borrows just the idea of reuse distance prediction and improves energy efficiency of exclusive hybrid caches with a hardware-based solution (i.e., no modifications to existing software).

III. ARCHITECTURE

Fig. 4 shows an overview of the proposed exclusive LLC architecture. The LLC is a hybrid cache combining a large STT-RAM region and a small SRAM region. This section explains the details of our hybrid cache management scheme.

A. Reuse Distance Predictor

Our reuse distance predictor shown in Fig. 4 is designed based on the previous work called Signature-based Hit Predictor using program counters (PCs) as signatures (SHiP-PC) [12]. To reduce the area overhead, it samples $\frac{1}{32}$ of the entire cache sets[1] [16] and maintains signatures and reuse distances only for the sampled sets in a separate hardware called sampler. For each access to a sampled set in the LLC, the corresponding reuse distance information maintained in the sampler is updated (i.e., reuse distances of all blocks in the set except the one for the accessed block are incremented by one). The reuse distance field is set to zero on block insertion.

As a signature of the predictor, we associate each block with the PC of the instruction that has issued the access to it in the lower-level cache (i.e., the last access to the block when the block was in the lower-level cache before it is moved to the LLC). For this purpose, each lower-level cache tag is equipped with an additional field that stores the last-access PC, which is transferred to the predictor on its eviction. We extract 13 least significant bits from the PCs as signatures instead of using full 32-bit PCs since the tables (predictor tables, explained below) addressed by the PC are composed of 8192 entries (which requires 13 bits for indexing).[2]

---

[1] According to our experiments, employing set sampling in our architecture increases LLC energy by a small amount (5.9% on average) compared to the one without sampling due to slightly more misprediction.

[2] We have also evaluated our architecture using 15 LSBs of PC (instead of 13 bits) and observed an ignorable improvement (only 0.3% more reduction of LLC energy consumption) at a cost of 4x larger tables.

For each signature value, the architecture tracks the reuse distance information records the trend in tables of 3-bit saturating counters (structurally similar to branch predictors). There are two predictor tables in our architecture, the near-reuse predictor table and the far-reuse predictor table. The former predicts if the next access with the given signature will have reuse distance shorter than a user-defined threshold Th_{near}, while the latter predicts if the next access will have reuse distance longer than a user-defined threshold Th_{far}.

The predictor tables are updated in two cases. First, when a cache block in the LLC is accessed, its reuse distance is compared with Th_{near} and Th_{far} to update the counters addressed by the signature of the current access. If the reuse distance of the block is shorter than Th_{near}, the corresponding counter in the near-reuse predictor table is incremented by one. Otherwise, the counter is decremented. Similarly, the counter in the far-reuse predictor table is incremented when the reuse distance is longer than Th_{far} and decremented otherwise. The second case of updating the counters is when a cache block is evicted from the LLC. In that case, the counters are updated as if the reuse distance of the evicted block were infinite and using the same mechanism described above.

Based on the counter values, the reuse distance prediction is made when the LLC receives a block insertion request. First, it accesses the two predictor tables and reads the counters corresponding to the PC from the request. According to the values of the near-reuse table counter (N) and the far-reuse table counter (F), we predict blocks with '$F \geq 4$' as far reuse distance and '$F < 4$ and $N \geq 4$' as near reuse distance (the value '4' is empirically determined). This information is used to determine block placement in our architecture, which will be discussed in the following section.

B. Hybrid Cache Architecture

Using predicted results from the reuse distance predictor, we develop a hybrid cache architecture for exclusive LLCs. Each cache set of our architecture is composed of many STT-RAM blocks and few SRAM blocks (e.g., 15 and 1 blocks per set, respectively, for a 16-way set-associative cache). This achieves better energy efficiency than homogeneous approaches since STT-RAM provides low static power, while SRAM shows much lower write energy and latency at the cost of increased static power.

When a cache block is evicted from the lower-level cache, our architecture needs to determine whether the block needs to bypass the LLC or to be inserted into the SRAM region or the STT-RAM region. For this purpose, we classify cache blocks into near, medium, and far reuse categories using reuse distance prediction and exploit such information to perform efficient block placement for our architecture as follows.

First, for cache blocks predicted as far reuse distance, we bypass the block insertions from the cache (and write them back to main memory if they are dirty). The key idea behind this is that inserting these blocks is wasteful since blocks with far reuse distance will most likely evicted before they receive hit (if any). Since the misprediction penalty of bypassing could be high due to extra cache misses, if there is a free space in the STT-RAM region (i.e., at least one invalid block exists in the set; note that unlike inclusive caches, exclusive caches can have invalid blocks even after warm-up since blocks are invalidated on hit), we insert blocks into the region even when they have far reuse distance. Those blocks are marked with one extra bit per block in the tag array and are evicted with higher priority on block replacement.

Second, cache blocks predicted as near reuse distance are inserted into the SRAM region. The rationale of this is that those blocks are accessed very soon after their insertion, which makes the small SRAM region sufficient to hold them until they are accessed. Inserting near reuse blocks into the SRAM region is particularly beneficial in exclusive caches since blocks are invalidated on cache hit, which makes room for another block to be cached in the SRAM region, thereby making the best use of the small SRAM region. Also, if a cache block in the SRAM region is not accessed at all before its eviction, we insert it into the STT-RAM region at its eviction instead of evicting them from the LLC to give a second chance since those blocks are likely to receive hits according to the prediction results.

Third, all other cache blocks (medium reuse distance) are inserted into the STT-RAM region as they are expected to be accessed before their eviction, but not in the near future. Through this, short-lived blocks exploit small write overhead of SRAM, while only long-lived ones pay high write cost of STT-RAM to benefit from its low static power.

IV. Evaluation

A. Methodology

Our architecture is evaluated by using MacSim [17], a trace-driven, cycle-level x86 simulator. The baseline system has a four-issue, out-of-order core with 256 reorder buffers operating at 4 GHz and 32 KB 8-way set-associative L1 instruction/data caches having 64-byte blocks, which use the LRU replacement policy. We use DDR3-1600 timing parameters with configuring two channels and eight banks per channel for the main memory of the system.

We use a 1 MB 16-way set-associative LLC with 64-byte blocks using the NRF (not recently filled) replacement policy [18] as the LLC of the baseline system. In the proposed hybrid LLC, each cache set is composed of one SRAM block and fifteen STT-RAM blocks (the numbers of ways are determined empirically) to minimize the high static energy consumption of the SRAM region.

For the reuse distance predictor, we use two 8192-entry predictor tables with 3-bit saturating counters. For reuse distance classification, we set Th_{near} to the number of SRAM ways in the hybrid cache (i.e., 1 in our current implementation) and Th_{far} to infinity to consider only the blocks that are evicted without reuse as having far reuse distance. Dynamic adjustment of the threshold according to application characteristics is part of our future work.

Table I shows characteristics of SRAM/STT-RAM used in our evaluation, which are modeled by CACTI [19] and NVSim [20], respectively, under the 45nm technology. We use LOP (Low Operating Power) cells for peripheral circuits

TABLE I
CHARACTERISTICS OF THE SRAM, STT-RAM, AND HYBRID CACHE

	SRAM	STT-RAM	Hybrid Cache (SRAM/STT-RAM)
Read Latency	12 cycles	12 cycles	12 / 12 cycles
Write Latency	12 cycles	44 cycles	12 / 44 cycles
Read Energy	0.04 nJ	0.10 nJ	0.04 / 0.10 nJ
Write Energy	0.04 nJ	0.63 nJ	0.04 / 0.63 nJ
Static Power	16.10 mW	5.02 mW	5.71 mW

of tag/data arrays. We refer to the previous work for the cell characteristics of STT-RAM [21], [22]. We also model energy overhead of the reuse distance predictor in our architecture, which is negligible compared to the LLC energy (5.0% on average as shown in Fig. 6). Since the reuse distance predictor is not on the critical path, they do not affect the performance.

As workloads for the evaluation, we choose 16 write-intensive benchmarks[3] from SPEC CPU2006 that have high L2 writes per kilo instruction (WPKI) in the baseline system. For benchmarks with multiple input sets, we choose the representative ones based on the previous work [23]. All benchmarks are run for 500 million instructions of the representative phases extracted by PinPoints [24].

B. LLC Energy Consumption

Fig. 5 compares the energy consumption of the LLC in the proposed architecture against the STT-RAM baseline. The baseline uses a monolithic STT-RAM cache without any improvement technique as a LLC (STT-RAM-Baseline). We show the effectiveness of our architecture by applying our bypassing scheme only (STT-RAM-Bypass) and then adopting our hybrid cache architecture on top of it (Hybrid-Bypass). We also include the results obtained by simply replacing STT-RAM with SRAM under our bypassing scheme (SRAM-Bypass) for comparison with the hybrid cache. Note that it is infeasible to compare with previously proposed hybrid caches since ours is the first hybrid cache architecture that can be applied to exclusive LLCs.

First, according to the evaluation results, our bypassing scheme alone reduces the LLC energy by 48%. Through our scheme, most of the block insertions bypass the LLC, which reduces the write energy consumption by 75% on average. This is particularly effective in bwaves, cactusADM, libquantum, milc, sjeng, and zeusmp where most of the accesses belong to the far reuse distance category as shown in Fig. 3.

Moreover, using SRAM/STT-RAM hybrid cache further reduces the LLC energy by 15% (55% from STT-RAM-Baseline). On average, the SRAM region absorbs 18% of cache writes to avoid costly STT-RAM writes for near reuse blocks. This allows us to further improve the energy efficiency of our scheme in benchmarks such as h264ref, povray, and xalancbmk where the bypassing scheme is ineffective. As can be seen in Fig. 3, such benchmarks have a considerable amount of blocks with near reuse distance, which should not bypass

---

[3]Although not shown in this paper, our architecture consumes 11% less LLC energy compared to the STT-RAM baseline in 13 other benchmarks.

978-1-4673-9141-2/15 $31.00 © 2015 IEEE

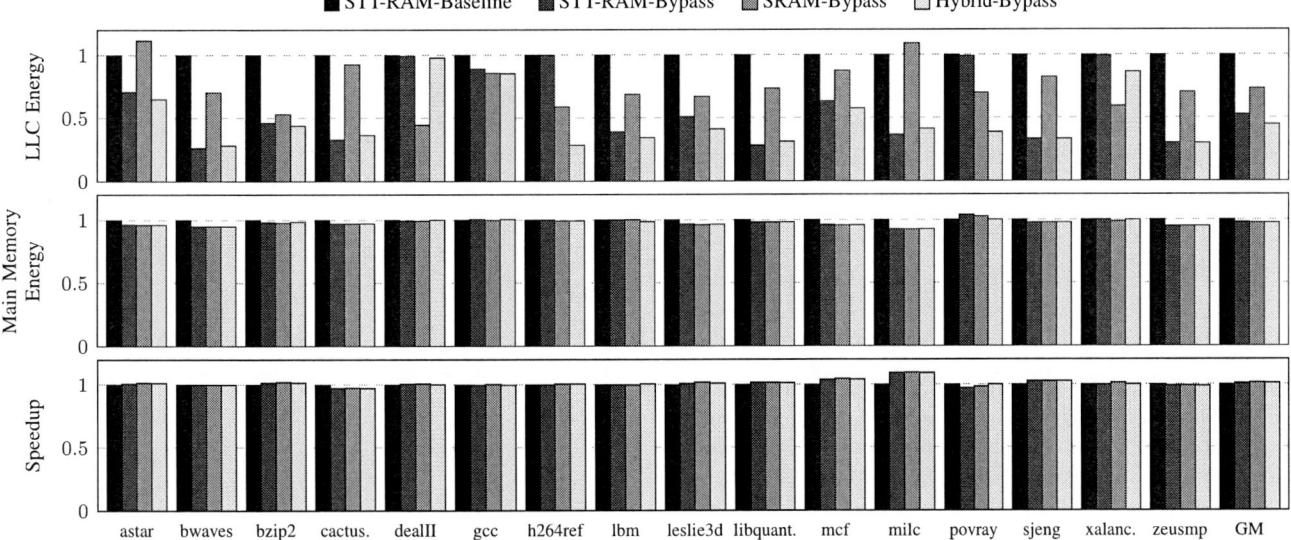

Fig. 5. Energy consumption and performance of our architecture normalized to the STT-RAM baseline.

the cache but can be well served by a small SRAM cache (e.g., 98% of cache writes are serviced by SRAM in h264ref).

In addition, our hybrid cache architecture also consumes less cache energy compared to the one that uses SRAM only (SRAM-Bypass). Simply replacing the entire STT-RAM with SRAM leads to a noticeable increase in LLC energy due to the higher static power consumption. Since such behavior highly depends on the write intensity of applications, there is no clear winner between STT-RAM-Bypass and SRAM-Bypass in terms of energy efficiency. On the contrary, our architecture substitutes only a small portion of the STT-RAM cache with SRAM and employs a mechanism that fully utilizes it thereby facilitating adaptation to application characteristics.

However, some benchmarks show a slight increase in LLC energy after applying the hybrid cache technique (e.g., bwaves, cactusADM, libquantum, and milc). This is because they do not have an enough amount of writes with near reuse distance to offset the increased static power of SRAM by exploiting its low write energy. This can be confirmed by the observation that far reuse distance dominates the cache accesses in those benchmarks as shown in Fig. 3. Also, dealII benefits from neither bypassing nor hybridizing since most of the cache accesses in dealII belong to medium reuse as shown in Fig. 3.

In summary, Fig. 6 compares the energy breakdown of our architecture against the STT-RAM baseline. Most noticeably, LLC energy consumption in STT-RAM-Baseline is dominated by the high write energy of STT-RAM. In contrast, our architecture greatly reduces the write energy consumption of exclusive LLCs at the cost of a modest increase in static power and negligible overhead from the reuse distance predictor.

C. Main Memory Energy Consumption

Since both bypassing blocks and using hybrid caches may affect the cache miss rate, we also measure the LLC miss rates across different configurations. On average, Hybrid-Bypass reduces the LLC miss rate by 0.53% compared to STT-RAM-Baseline because bypassing improves cache efficiency by not

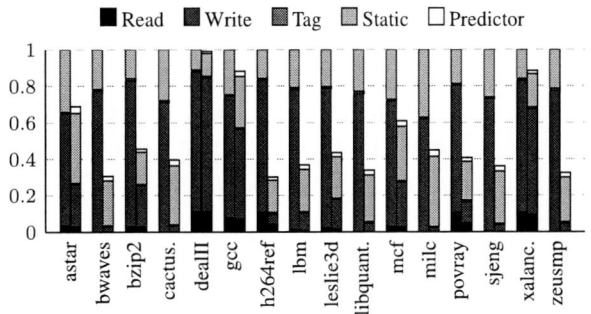

Fig. 6. Energy breakdown of STT-RAM-Baseline (left) and Hybrid-Bypass (right).

inserting cache blocks with no reuse. This saves main memory energy consumption by reducing main memory accesses. In addition, slight performance improvement (see Section IV-D) further reduces energy consumption of main memory. Accordingly, as shown in Fig. 5, our analysis on main memory energy consumption using Micron's Power Calculator [25] indicates that Hybrid-Bypass consumes less energy (2.4% on average) in the main memory compared to STT-RAM-Baseline.

D. Performance

Fig. 5 also shows the performance of our architecture in terms of instructions per cycle (IPC) normalized to that of the STT-RAM-Baseline. We observed that STT-RAM-Bypass and Hybrid-Bypass slightly improve the performance by 0.9% and 1.2%, respectively, compared to STT-RAM-Baseline. The performance improvement of our architecture is mainly due to the reduction in bank contention caused by long write operation and miss rate mentioned in the previous section.

E. Area Overhead

Our architecture incurs very small area overhead. In the LLC, we add one bit per block to record the replacement priority for blocks predicted as far reuse distance (see Section III-B). The reuse distance predictor is comprised of two predictor tables with 8192 3-bit counters and a sampler that stores a 13-bit signature and a 1-bit reuse distance[4] of each LLC block in the sampling sets (32 sets out of 1024 sets in our environment). Lastly, each L1 data cache tag is extended with a 13-bit last-access signature field for reuse distance prediction. In total, storage overhead of our technique is only 0.8KB/8.9KB (2.5%/0.9%) in the L1/L2 cache.

V. RELATED WORK

There have been many architecture-level researches on mitigating the high write overhead of STT-RAM caches. In this section, we will cover two representative categories of such techniques: cache bypass and hybrid caches.

STT-RAM Cache Bypass. Wang et al. [11] proposed an obstruction-aware cache management, which improves performance by bypassing cache blocks from cores that generate a large amount of writes and thus incur high bank contention. Although it reduces the LLC energy consumption, it greatly increases main memory accesses since it does not consider reuse information in bypass decision. Ahn et al. [10] proposed dead write prediction to bypass unnecessary writes from the LLC. However, such a technique is incompatible with exclusive caches since it relies on the concept of dead writes, which assumes inclusive/non-inclusive caches. The bypass scheme proposed by Wang et al. [8] also shares the similar limitation.

Hybrid Caches. Many researchers proposed to combine STT-RAM with other memory technologies (mostly SRAM) to overcome the poor write characteristics of STT-RAM [5]–[9]. In order to fully utilize the benefit of heterogeneous memory technologies, they mainly focus on (1) cache block placement algorithms that determine the blocks to be allocated to the SRAM region [5], [8] and (2) migration schemes to move cache blocks with high write intensity to the SRAM region [6], [7], [9]. To the best of our knowledge, these schemes are all based on inclusive/non-inclusive cache hierarchies, which makes their adoption to exclusive caches very challenging. For example, existing migration schemes first observe access patterns of each cache block and perform migration decision based on that information, which is infeasible in exclusive caches due to the lack of per-block access history. On the contrary, our architecture utilizes the distance from the block insertion to its access (or eviction) for hybrid cache management, which is well defined in exclusive caches.

VI. CONCLUSION

In this paper, we proposed an energy-efficient exclusive LLC architecture based on STT-RAM to take advantage of capacity benefit of exclusive caches and low static power of STT-RAM. The key challenge in designing such an architecture is the increased amount of LLC writes, which is detrimental for STT-RAM caches that show poor write characteristics. To address this issue, our architecture is composed of (1) a bypassing scheme to avoid write overhead for far reuse cache blocks, (2) a hybrid cache architecture that reduces write

energy consumption for near reuse blocks with a small SRAM cache, and (3) a reuse distance predictor that realizes such decision in a cost-effective manner. Our evaluations show that the proposed architecture reduces LLC energy by 55% with slight improvement in energy efficiency of main memory and system performance. We believe that our architecture can enlarge effective cache capacity in an energy-efficient way through appropriate combination of an emerging memory technology and better cache hierarchy design.

ACKNOWLEDGMENT

This work was supported by Samsung Advanced Institute of Technology, Samsung Electronics Co., Ltd.

REFERENCES

[1] W. A. Wulf and S. A. McKee, "Hitting the memory wall: Implications of the obvious," *ACM SIGARCH Comput. Archit. News*, vol. 23, pp. 20–24, Mar. 1995.

[2] X. Dong et al., "Circuit and microarchitecture evaluation of 3D stacking magnetic RAM (MRAM) as a universal memory replacement," in *Proc. DAC*, 2008.

[3] M.-T. Chang et al., "Technology comparison for large last-level caches (L^{3}Cs): Low-leakage SRAM, low write-energy STT-RAM, and refresh-optimized eDRAM," in *Proc. HPCA*, 2013.

[4] *Family 16h Models 00h - 0Fh AMD Opteron™ Processor Product Data Sheet*, Advanced Micro Devices, Feb. 2014.

[5] J. Ahn et al., "Write intensity prediction for energy-efficient non-volatile caches," in *Proc. ISLPED*, 2013.

[6] X. Wu et al., "Power and performance of read-write aware hybrid caches with non-volatile memories," in *Proc. DATE*, 2009.

[7] G. Sun et al., "A novel architecture of the 3D stacked MRAM L2 cache for CMPs," in *Proc. HPCA*, 2009.

[8] Z. Wang et al., "Adaptive placement and migration policy for an STT-RAM-based hybrid cache," in *Proc. HPCA*, 2014.

[9] J. Li et al., "STT-RAM based energy-efficiency hybrid cache for CMPs," in *Proc. VLSI-SoC*, 2011.

[10] J. Ahn et al., "DASCA: Dead write prediction assisted STT-RAM cache architecture," in *Proc. HPCA*, 2014.

[11] J. Wang et al., "OAP: An obstruction-aware cache management policy for STT-RAM last-level caches," in *Proc. DATE*, 2013.

[12] C.-J. Wu et al., "SHiP: Signature-based hit predictor for high performance caching," in *Proc. MICRO*, 2011.

[13] P. Petoumenos et al., "Instruction-based reuse-distance prediction for effective cache management," in *Proc. SAMOS*, 2009.

[14] Y.-T. Chen et al., "Static and dynamic co-optimizations for blocks mapping in hybrid caches," in *Proc. ISLPED*, 2012.

[15] C. Ding and Y. Zhong, "Predicting whole-program locality through reuse distance analysis," in *Proc. PLDI*, 2003.

[16] M. K. Qureshi et al., "A case for MLP-aware cache replacement," in *Proc. ISCA*, 2006.

[17] *MacSim Simulator.* http://code.google.com/p/macsim/

[18] J. Gaur et al., "Bypass and insertion algorithms for exclusive last-level caches," in *Proc. ISCA*, 2011.

[19] N. Muralimanohar et al., "CACTI 6.0: A tool to model large caches," HP Laboratories, Tech. Rep. HPL-2009-85, 2009.

[20] X. Dong et al., "NVSim: A circuit-level performance, energy, and area model for emerging nonvolatile memory," *IEEE Trans. Comput.-Aided Design Integr. Circuits Syst.*, vol. 31, pp. 994–1007, Jul. 2012.

[21] Y. Chen et al., "Design margin exploration of spin-transfer torque RAM (STT-RAM) in scaled technologies," *IEEE Trans. VLSI Syst.*, vol. 18, pp. 1724–1734, Dec. 2010.

[22] Y. Zhang et al., "The prospect of STT-RAM scaling from readability perspective," *IEEE Trans. Magn.*, vol. 48, pp. 3035–3038, Nov. 2012.

[23] A. Phansalkar et al., "Subsetting the SPEC CPU2006 benchmark suite," *ACM SIGARCH Comput. Archit. News*, vol. 35, pp. 69–76, Mar. 2007.

[24] H. Patil et al., "Pinpointing representative portions of large Intel® Itanium® programs with dynamic instrumentation," in *Proc. MICRO*, 2004.

[25] "Calculating memory system power for DDR3," Micron Technology, Tech. Rep. TN-41-01, 2007.

---

[4]We need to track reuse distances up to Th_{near} (i.e., 1), which requires one bit for each field, since Th_{far} is set to infinity.

JAIP-MP: A Four-Core Java Application Processor

Chun-Jen Tsai[†], Tsung-Han Wu, and Hung-Cheng Su
Dept. of Computer Science, National Chiao Tung University
†cjtsai@cs.nctu.edu.tw

Abstract—In this paper, we present the design of a four-core Java application processor, JAIP-MP. Each processor core in JAIP-MP is a hardwired Java core that supports dynamic class loading, two-fold bytecode execution, object-oriented dynamic resolution, method and object caching, Java exception handling, and temporal multithreading. For JAIP-MP, a global load-balancing task manager is used to evenly distribute Java threads among the local task queues of every processor cores. In addition, a data coherence controller is designed to enforce coherence across all data caches and to perform synchronization operations among Java threads of all processor cores. Since thread management and synchronization mechanisms are completely implemented in hardware, the single-core multi-tasking performance of JAIP-MP is much higher than that of a software-based VM running on a traditional OS kernel such as Linux. For execution of multithreading applications, the speedup of a four-core JAIP-MP system can be up to 3.69 times faster than a single-core JAIP system, tested using the JemBench parallel benchmark programs.

Keywords—Java processor; multi-core; embedded SoC; hardwired multi-threading; cache coherence

I. INTRODUCTION

Although there are many hardware Java core designs [1], very few of them have been synthesized in a real multi-core application processor [2]. One of the reasons is because previous work shows that a Just-in-time (JIT) based VM running on a high performance general-purpose processor can often outperform a hardwired Java processor. Therefore, it seems that there is no need to further pursue the development of hardwired Java processors. However, most of the JIT vs. hardwired VM comparisons are conducted using benchmarks where the application class files are not optimized for Java processor. For example, it has been shown that some popular benchmark class files can run much faster on a Java processor if bytecode optimizations in the class files are conducted [3]. Please note that an optimized Java class file still conforms to the Java specification and is portable across different Java platforms. In addition, many benchmarking processes discard the impact of the JIT compilation overhead [4]. Although ignoring the JIT overhead is reasonable for some applications, it is not valid for remote invocations that are common for object-oriented distributed computing.

In [4][5], we have presented a single-core hardwired Java processor, Java Application IP (JAIP), that contains a thread management circuit which is capable of single-cycle multi-thread context switching with a time quantum as small as 20 microseconds. For a traditional OS kernel such as Linux, the time quantum for a thread is usually around 10 milliseconds. As a result, for single-core multithread applications, hardware-based thread manager can achieve much smoother concurrent executions of all threads of equal priorities than a software-based thread manager. This is a very strong reason for the development of an efficient hardwired Java processor core. The Java language specification is one of the few programming languages that define standard programming interfaces for OS kernel services (e.g., process management and memory management). Other popular languages such as C and C++ do not standardize these functions. For example, thread creation API's are OS-dependent for C programs.

In short, a Java processor can be designed to implement the entire OS kernel system services in hardware while maintaining application portability. This is particularly important for embedded real-time applications where context-switching efficiency and dynamic memory management overhead are the key performance factors of a system. On the other hand, it is not so easy to "harden" the OS kernel for a traditional RISC processor due to lack of "standard" system calls for C/C++ applications.

Most Java processors support thread synchronization using software modules [2][6]. However, the execution time of a software-based synchronization operation, such as a mutex lock, can take more than a few hundred clock cycles since the lock objects are often accessed in conventional trap routines. PicoJava [7] uses a few special-purpose registers for the speedup of synchronization operations, but it still needs to use the main memory to maintain the information of all waiting threads and lock objects. Therefore, a high number of concurrent synchronized read/write operations can have significant synchronization overheads. JOP-CMP supports at most 8 processor cores with a software-based thread scheduler and a hardwired synchronization unit [2][8]. There is only one global lock register in the synchronization unit, which means that any threads trying to acquire the lock must wait until the lock is released. In addition, JOP-CMP does not have a coherent data cache. All data accesses will be directly issued to the external memory, which can hinder the multi-thread performance significantly.

In this paper, we have proposed a multi-core, multi-tasking architecture for JAIP. We have modified the local Thread Manager Unit for each JAIP core in [5] so that thread management can be handled across all local task queues by a centralized task manager. We have also proposed additional hardwired control modules, including the Data Coherence Controller (DCC) and the Inter-Core Communication Unit (ICCU) to allow seamless integration of four JAIP cores into a multi-core SoC called JAIP-MP. The DCC can handle read/write and synchronization requests from all threads across all JAIP cores. As described in [1], each JAIP core has its own

978-1-4673-9141-2/15 $31.00 © 2015 IEEE

local data cache. The DCC adopts a snooping scheme to ensure the coherence of data caches in JAIP-MP. The ICCU of each JAIP core is used to communicate with the Thread Manager Unit and the DCC.

The organization of this paper is as follows. Section II gives an introduction to the architecture of JAIP. Section III presents the design of the Inter-Core Communication Unit and the multi-core thread manager. In Section IV, we discuss the proposed Data Coherence Controller. The experimental results are presented in Section V, including the performance evaluation based on the JemBench benchmark suite [10] for embedded Java applications, and the analysis on the overhead of synchronization operations. Finally, some discussions are given in Section VI.

II. THE ARCHITECTURE OF THE JAIP CORE

A. System Overview

Fig. 1 shows the overall block diagram of the JAIP core. The complete SoC is composed of a RISC core and a JAIP core. For the execution of a Java program, the RISC core is only responsible for dynamic loading and parsing of the class files stored in a JAR file on the Compact Flash (CF) card. The RISC parser will convert the standard Java class files into runtime class images on-the-fly for direct execution by the JAIP core. In addition, the RISC parser will maintain the symbol cross reference table stored in the main memory for all loaded classes. The Java core is completely responsible for the two-fold execution of bytecodes, dynamic resolution, method and object caching, memory management, and multi-thread scheduling.

JAIP adopts a two-level method area design. All classes loaded at runtime will be stored in the DDR3 SDRAM (i.e., the second-level method area) using the late-resolution policy. However, a Java method (and its related symbol information) must be loaded into the on-chip method cache (the first-level method area) before it can be executed by the bytecode execution engine. In short, the complete class images of the Java applications are stored in DDR3 SDRAM while the most recently used methods and symbol information are stored in the on-chip method cache in a FIFO manner.

Since the Java VM is basically a stack machine, i.e., all the local variables and the intermediate values of operations are stored in the runtime stack, fast accesses to the most recent stack frames are essential to the performance of a Java processor. JAIP uses a special-purpose on-chip memory and three top-of-stack registers to form a two-level Java runtime stack. The design is a good tradeoff between performance and implementation cost as compared to the Java processor design with a large stack cache [7]. The special-purpose on-chip memory is a customized four-port memory device for the bytecode instruction set architecture. It is composed of two two-port memory blocks and four registers.

The two-level Java stack allows JAIP to perform two-fold instruction folding for frequent bytecode pairs such as load-load, ALU-store, etc. However, to simplify the microarchitecture, some folding patterns such as the ALU-ALU bytecode pairs are not allowed. According to our

empirical studies, the instruction folding rate of JAIP ranges from 10% to 40% for different applications.

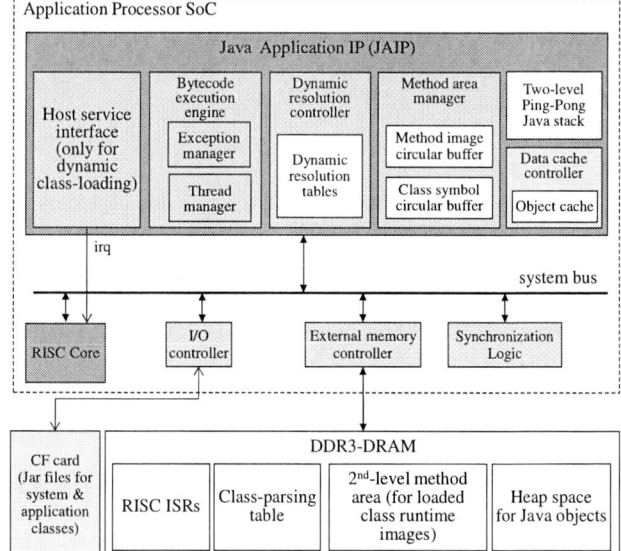

Fig. 1. The architecture of the JAIP core.

B. Single-Core Time Sharing Thread Management

For the execution of multi-thread Java programs, each thread must maintain its own registers and runtime stack. Typically, the register file of a Java processor is only composed of special-purpose registers (e.g., the program counter, the stack pointer, the local variable pointer, etc.) and can be swapped out to main memory quite efficiently. On the other hand, the Java runtime stack is much larger than the register file. If the runtime stack is stored in the main memory (e.g., DRAM), there is no need to save the runtime stack. However, most high-performance Java processors, including JAIP, use stack cache or on-chip memory to support instruction folding and to reduce the access delay of operands. In either case, the time it takes to swap out the on-chip stack would be non-negligible.

Saving/restoring the context of a JAIP thread involves transferring several stack frames (each ranging from a few bytes to a few hundreds bytes) to/from the DDR3 SDRAM memory. In order for JAIP to support hardware-based multi-threading, we have designed a low-cost hardwired thread manager unit with an on-chip ping-pong stack architecture to reduce the context-switching overhead. As a result, in most cases, switching from the current thread to the next active thread only takes a single cycle. This is much faster than any software-based preemptive multi-tasking operations where a context switching operation can take anywhere from a few hundreds to over a thousand cycles. The ping-pong stack architecture works as follows.

As soon as a thread is selected as the current thread and starts its execution, the multi-threading logic also picks the next thread to be executed and, while the first thread is running, swaps in the runtime stack of the second thread from the main memory. When the time slice of the first thread is up, JAIP can be switched to the second thread within a cycle since its stack has already been setup. In the rare case where the restoration of the runtime stack of the second thread takes longer than the predetermined time quantum of the first thread, the time quantum will be extended until the runtime stack of the second thread is in place. The average time it takes to backup or restore a runtime stack to/from the backing store (DDR3 memory) for the target system used in this paper (Xilinx ML 605) is less than 10 μs when the system clock is 83.3MHz.

When the execution is switched to the second thread, the runtime stack of the first thread will be saved to the main memory in parallel to the execution of the second thread. As soon as the stack of the first thread is saved, the multi-thread control logic will proceed to the setup of the third thread. With this design, the overhead of saving/restoring the runtime stack can be overlapped with the execution of the current thread. According to our experiments, the time slice of the proposed architecture can be as small as 20 μs and the only overhead in context switching is virtually the reset of the processor pipeline (similar to a branch instruction). Smaller time slice means the distribution of the CPU resources to each thread is more even. This level of multi-threading efficiency is very difficult to achieve with a software-based preemptive multitasking operating system.

III. THE HARDWARE MULTI-CORE THREAD MANAGER

In order to integrate multiple JAIP core into one application processor, we must modify the microarchitecture of JAIP. The multi-core capable JAIP core is shown in Fig. 2. The new addition to the original JAIP core is the Inter-Core Communication Unit (ICCU). The interactions between various components of the JAIP core and the ICCU are illustrated in Fig. 3.

In the Java programming language, an object belongs to the "Thread" class can register its own execution context by invocation of the Thread.start() method. At runtime, the Dynamic Symbol Resolution Unit (DSRU) of JAIP will resolve the method invocation of start() and trigger a hardwired signal to the Thread Manager Unit (TMU) of the local JAIP core that executes the start() method. Such direct invocation of a hardwired logic through the dynamic resolution unit is called the Hardware Native Interface (HNI). In the original single-core JAIP, the local TMU will handle the thread creation request by itself and register a new entry in its local task queue. However, for a multi-core capable JAIP, the thread creation request cannot be handled locally. Instead, the request will trigger the HNI invocation of the ICCU, and then the request signal will be passed to the DCC. The DCC then talks to a global thread manager to request for the creation of a new thread. The global thread manager will assign the new thread to a JAIP core based on the depth of its local task queue. The detail operations of the DCC will be discussed in section IV.

In addition to thread creations, the Java language also defines standard ways for synchronization. In short, each Java object contains a lock (similar to mutex in other programming language). Synchronization can be achieved explicitly through the acquisition of the lock in an object, or implicitly through invocation of a synchronized method. Similar to the thread creation problem, the acquisition of a lock cannot be handled locally since two threads requesting the same lock may be

running on different JAIP cores. Therefore, such locking requests will also be passed to the ICCU for multi-core mutex operations. However, this time, the ICCU is not activated by a HNI invocation from the DSRU because the lock request is triggered by the execution of a "monitor" bytecode. Therefore, the lock request is originated from the decode stage of the bytecode execution engine, as shown in Fig. 3.

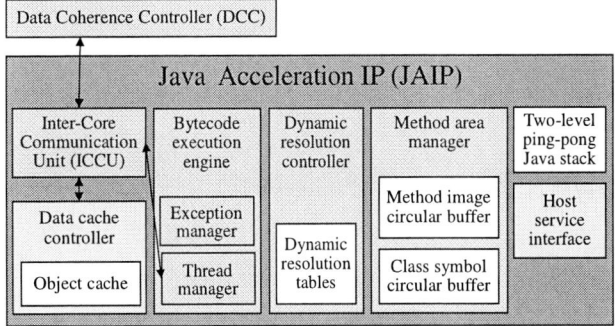

Fig. 2. Modification to the JAIP core for multi-core integration.

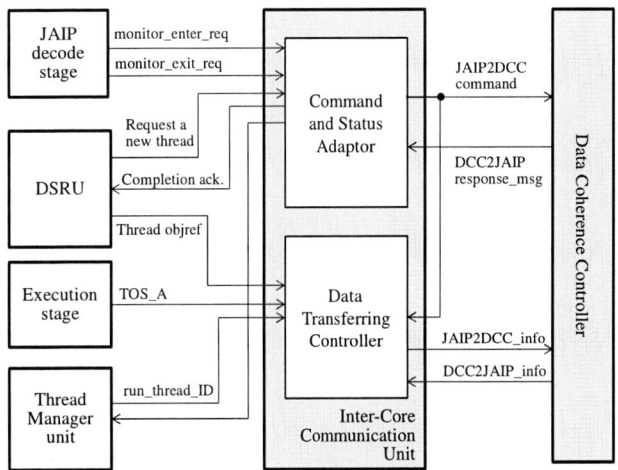

Fig. 3. The interactions between the Inter-Core Communication Unit and various components of the JAIP core.

The integration of four JAIP cores into a single application processor, JAIP-MP, is shown in Fig. 4. In the whole SoC, we only need one copy of DCC and global thread manager. The combination of these two hardware logic is referred to as the Multicore Coordinator of the JAIP-MP. Each JAIP core has its own ICCU. The local cache controller of each JAIP core will forward its cache block update status to the DCC so that the DCC can inform other cache controller to update their cache blocks if necessary. This is an efficient way to guarantee cache coherence when there are only few processor cores (such as four). However, to simplify the implementation of the coherent object cache, each cache controller adopts a write-through policy. This is different from the original single-core JAIP presented in [4], where a write-back policy is used. The write-through cache policy does hinder the single-core performance slightly. Nevertheless, the overall system performance still scales up fairly well.

978-1-4673-9141-2/15 $31.00 © 2015 IEEE

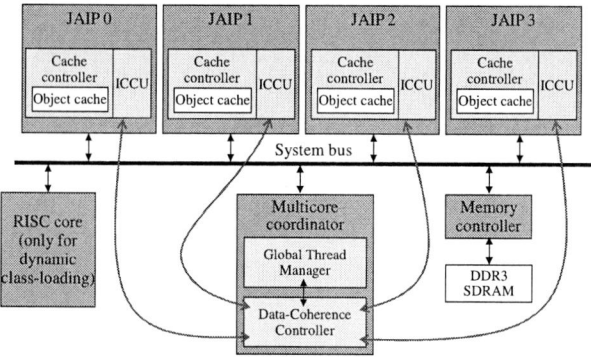

Fig. 4. System block diagram of the four-core JAIP-MP.

IV. THE DATA COHERENCE CONTROLLER ARCHITECTURE

The detail architecture of the DCC is shown in Fig. 5. It is composed of four sub-modules. The Cache Coherence Controller (CCC) maintains the data consistency across the Object Heap Controllers (OHC) of each core. The OHC adopts the LRU policy and write-through strategy for caching of Java heap objects. The Mutex Controller serially decodes requests sent by the JAIP cores and activates corresponding sub-module. The Thread Assignment Controller (TAC) is responsible for load balancing among all JAIP cores. When a JAIP cores invokes the Thread.start() method, the TAC will forward its special-purpose registers to the JAIP cores with the least number of ready threads. The Lock Object Accessing Controller (LOAC) maintains the information of waiting threads associated with each occupied lock object.

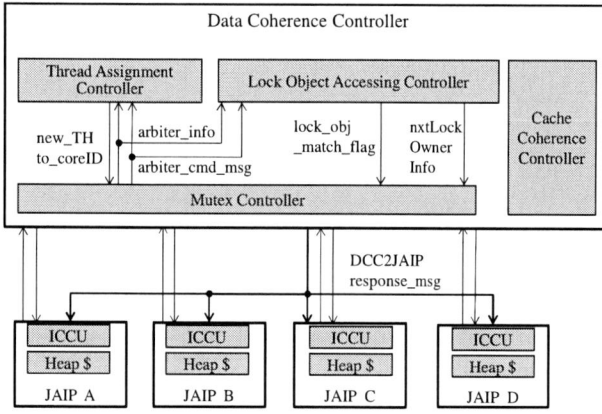

Fig. 5. The block diagram of the Data Coherence Controller.

When several JAIP cores try to request locks on the same mutex concurrently, the mutex controller uses a fixed-priority policy to determine which core can lock the mutex. Currently, the JAIP core with a smaller ID has a higher priority. The mutex controller supports three types of requests: dispatching a new thread, acquiring a lock object, and releasing a lock object. Either the TAC or the LOAC will be activated after the mutex controller determines the type of the request.

When any of the JAIP cores issues a request for the dispatching of a new thread, the TAC should determine a JAIP core to handle the new thread. In order to determine the current

number of active threads in each JAIP core, the TAC maintains a table. The table indexed is the ID of the JAIP core, and its entries store the current number of active threads of each core. The TAC will always assign the new thread to the lowest ID JAIP core that has the fewest number of ready threads. The TAC will inform the MHC to send a response signal to the chosen JAIP core with some essential information of the new thread. The ICCU of the JAIP core may process the information by decoding the response signal. Finally, the ICCU activates the Thread Manager Unit of the JAIP core for the addition of the new thread into its local thread queue. For the details of the local thread management within a JAIP core, please refer to [5].

Fig. 7 illustrates the design of the LOAC and Fig. 7 is an example of the link lists maintained by the LOAC, which consists of a Lock Object Table, a Waiting Thread Table, and a

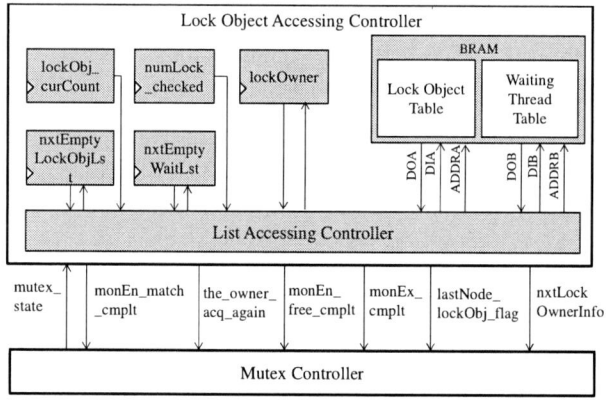

Fig. 6. The block diagram of the Lock Object Accessing Controller.

Lock Object Table

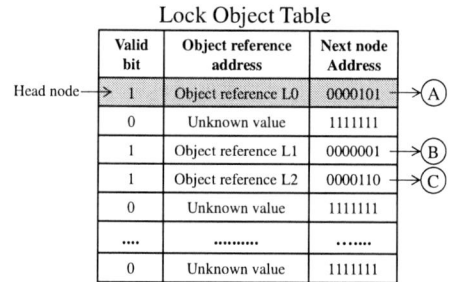

	Valid bit	Object reference address	Next node Address	
Head node →	1	Object reference L0	0000101	→(A)
	0	Unknown value	1111111	
	1	Object reference L1	0000001	→(B)
	1	Object reference L2	0000110	→(C)
	0	Unknown value	1111111	
	....	....	....	
	0	Unknown value	1111111	

Waiting Thread Table

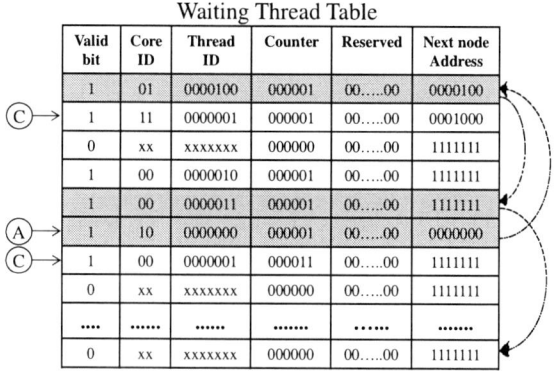

	Valid bit	Core ID	Thread ID	Counter	Reserved	Next node Address
	1	01	0000100	000001	00.....00	0000100
(C)→	1	11	0000001	000001	00.....00	0001000
	0	xx	xxxxxxx	000000	00.....00	1111111
	1	00	0000010	000001	00.....00	1111111
	1	00	0000011	000001	00.....00	1111111
(A)→	1	10	0000000	000001	00.....00	0000000
(C)→	1	00	0000001	000011	00.....00	1111111
	0	xx	xxxxxxx	000000	00.....00	1111111
	....	....	....	....	....	....
	0	xx	xxxxxxx	000000	00.....00	1111111

Fig. 7. Link lists maintained by the Lock Object Access Controller.

few internal registers. Each occupied lock object maintains a single-linked list in these two tables. The head node of the linked list of a lock object begins at an entry in the Lock Object Table, and the rest of the linked list nodes are entries of the Waiting Thread List. Each entry in the link list (except for the head node) represents a thread that is performing a lock operation on the object. The first thread in the linked list is the link list is the current owner of the lock. As soon as any thread in one of the JAIP cores tries to lock a Java object, the mutex controller will send a lock object L_n to the LOAC. The LOAC will look for the object address of an entry that matches L_n in the Lock Object Table. Once the matched entry is found, the information must be recorded in the Waiting Thread Table. Each entry contains the IDs of the JAIP core and the thread. The new entry is appended at the end of the link list. If the request from a thread is to release the lock object L_n, the LOAC will remove the thread from the link list. If any other thread is waiting for the same lock object L_n, the LOAC will make the second thread in the link list become the current owner of the lock object.

V. Experimental Results

The proposed architecture has been implemented on a Xilinx ML605 platform with a Xilinx Virtex6 XC6VLX240T FPGA. The RTL model of the JAIP core and the DCC logic are written in VHDL. Four JAIP cores and one DCC logic are integrated into the application processor using Xilinx XPS 13.4. The synthesis tool is Xilinx XST 13.4 and the target clock is 83.3MHz. According to the place-and-route timing report of the Xilinx tools, the critical path of the system is currently at the execution stage of JAIP, from the customized four-port stack memory to ALU and then back to the four-port memory. The target frequency is chosen at 83.3MHz due to some restrictions for DDR2/3 DRAM support on the development boards (ML605 and ML507). The FPGA resource usages of JAIP and DCC are shown in TABLE I. In TABLE I, LUT6 stands for a six-input lookup table in a Xilinx Virtex6 FPGA.

TABLE I.   Logic usage of JAIP-MP on a Virtex-6 FPGA

(a) JAIP (per core)

logic units	Number
LUT6	12,580
Flip-flop	5,912
2K BRAM	34
Max. Frequency	83.6 MHz

(b) DCC

logic units	Number
LUT6	663
Flip-flop	449
2K BRAM	1
Max. Frequency	83.6 MHz

In [4], we have presented the performance of the single-core JAIP against a JIT-based software VM and in [5], we have presented the multi-threading performance of JAIP comparing to a JIT-based software VM. In this paper, we do not use a software VM as a reference platform for performance evaluation because the software VM platform we used does not fully support multi-core multithreading execution of Java programs. We will leave this investigation for future work. Here, we focus on the evaluation of performance scalability of the proposed JAIP-MP.

A. Multithread Performance Evaluation

To evaluate the multi-threading performance of the proposed JAIP-MP, we used the multithreading benchmark programs from the JemBench Test Suites [10]. These test programs are explained as follows. The DUMMY test creates multiple threads to execute busy loops for 5000 iterations. For the MATRIX test, each thread computes the multiplication of two 20×20 matrices. The NQUEENS test solves the N-Queens puzzle for N = 13 in each thread, the time quantum used in the proposed architecture is set to 20 microseconds.

As TABLE II shows, when the total number of threads is less than or equal to four, the JemBench score scales up fairly well (up to 3.69× for the NQUEENS test). When the total number of threads is more than four, the score of each benchmark naturally drops as the temporal multi-threading mechanism of each JAIP core kicks in and there are time-sharing overheads due to the way the benchmarks are designed. This is especially true for the NQUEENS test. The overhead is explained in [5].

TABLE II.   The score of the multithreaded benchmark. Higher scores means better performance.

Benchmark #thread	Dummy Test	Matrix Test	N-Queens Test
1	151	167	116
2	298	240	225
3	374	395	330
4	491	498	428
5	410	425	311
6	373	412	251
8	362	399	262
12	359	366	212
16	340	327	195

B. Synchronization Overhead

In our current design, context-switching overhead is not the only reason to cause the performance drop. If several requests are sent concurrently to the DCC, it takes several cycles for the mutex controller to decode the requests sequentially. In addition, to maintain data cache coherency, as soon as each entry is updated in any of the Object Heap Controllers, the modified entry and its corresponding address are sent to the Cache Coherence Controller and the DDR memory for cache validation among JAIP cores.

TABLE III and TABLE IV show the synchronization overhead of the proposed architecture under the MATRIX test. The average overhead of a synchronization operation can be as small as tens of machine cycles.

978-1-4673-9141-2/15 $31.00 © 2015 IEEE

TABLE III. THE EXECUTION TIME (IN CLOCK CYCLES) OF ACQUIRING A LOCK OBJECT

Lock Object Overhead					
#thread	4	6	8	12	16
Average	23.1	25.4	25.4	28.9	28.9
Worst-case	43	89	108	108	110
Best-case	9	9	9	9	9

TABLE IV. THE EXECUTION TIME (IN CLOCK CYCLES) OF RELEASING A LOCK OBJECT

Unlock Object Overhead					
#thread	4	6	8	12	16
Average	20.2	21.6	21.8	28.4	28.7
Worst-case	43	76	85	92	98
Best-case	10	10	10	10	10

VI. CONCLUSIONS

In this paper, we have proposed an efficient hardware multithreading architecture for a multi-core Java processor. The architecture supports arbitrary number of threads (limited by on-chip TCB memory size), low context-switching overhead, small time quantum, and low synchronization overhead. The proposed architecture is implemented and verified on an FPGA platform. Experimental results show that the proposed design is suitable for embedded multithreaded applications.

For future work, we will look into the following directions. First of all, although the ping-pong buffer for context-switching is efficient performance-wise, it does impose heavy memory accesses. This may result in high power consumption. In the future, we will try to design a new architecture that can reduce the number of memory access per context-switch. Secondly, the coherent data cache in our current implementation only adopts one-level of cache hierarchy. Most general purpose processors nowadays adopt two to three levels of cache hierarchy. It would be interesting to study the effects of a multi-level cache on the object-oriented programming model of the Java language.

Finally, current thread management design only uses a round-robin policy to maintain load balance. We will look into the design of a new architecture that can customize the thread distribution policy at runtime and allow for thread migration across different JAIP cores so that better runtime load balance can be achieved.

ACKNOWLEDGMENT

This work was supported by Information and Communications Research Laboratories (ICL), Industrial Technology Research Institute (ITRI), Taiwan, under Grant 104-EC-17-A-24-0691, and by the Ministry of Science and Technology, Taiwan, under Grant MOST 103-2221-E-009-206-MY3.

REFERENCES

[1] M. Schoeberl, "A Java Processor Architecture for Embedded Real-Time Systems," *The EUROMICRO Journal of System Architecture*, **54**, 1-2, January–February 2008, pp. 265-286.

[2] F. Gruian and M. Schoeberl, "Hardware support for CSP on a Java chip multiprocessor," *Microprocessors and Microsystems*, 37.4, 2013.

[3] J. Tyystjaervi, T. Saentti, and J. Plosila, "Efficient Bytecode Optimizations for a Multicore Java Co-Processor System." Proc. of 12th Biennial Baltic Electronics Conf., Tallinn, Estonia, Oct. 4-6, 2010.

[4] C.-J. Tsai, H.-W. Kuo, Z. Lin, Z.-J. Guo, J.-F. Wang, "A Java Processor IP Design for Embedded SoC," *ACM Transactions on Embedded Computing Systems*, **14**, 2, Article 35, March 2015.

[5] H.-C. Su, T.-H. Wu, and C.-J. Tsai, "Temporal Multithreading Architecture Design for a Java Processor," *Proc. of IEEE Int. Symp. On Circuit and Systems (ISCAS 2014)*, Melbourne, Australia, June, 2014.

[6] J. Kreuzinger, U. Brinkschulte, M. Pfeffer, S. Uhrig, Th. Ungerer, "Real-time event-handling and scheduling on a multithreaded Java microcontroller," *Microprocessors and Microsystems*, 27.1, 2003.

[7] Sun Microsystems, picoJava-II Microarchitecture Guide, Sun Microsystems, Sun Microsystems, 1999.

[8] C. Pitter and M. Schoeberl, "Towards a Java multiprocessor," In Proc. of the 5th ACM int. workshop on Java technologies for real-time and embedded systems, 2007. pp. 144-151.

[9] Uhrig, Sascha, and Jörg Wiese, "jamuth: an IP processor core for embedded Java real-time systems," Proceedings of the 5th int. workshop on Java technologies for real-time and embedded systems. ACM, 2007.

[10] M. Schoeberl, T. B. Preusser, and S. Uhrig, "The Embedded Java Benchmark Suite JemBench," Proc. of JTRES'10, Prague, Czech Republic, Aug. 19-21, 2010.

Locality-Aware Vertex Scheduling for GPU-based Graph Computation

Hyunsun Park*, Junwhan Ahn[†], Eunhyeok Park[‡] and Sungjoo Yoo[‡]

*Department of Electrical
Engineering
Pohang University of Science and
Technology (POSTECH)

[†] Department of Electrical and
Computer Engineering
Seoul National University (SNU)

[‡]Department of Computer Science
and Engineering
Seoul National University (SNU)

Abstract—**Graph computation is becoming more and more popular in machine learning, big data analytics, etc. For such workloads, GPU is considered as an efficient execution platform since graph computation is characterized by massively parallel computation and high demand of memory bandwidth. In our investigation, existing GPU programming methods for graph computation do not fully exploit high memory bandwidth as well as high computing power in GPU. We propose a novel optimization called locality-aware vertex scheduling, which aims at minimizing memory requests by adjusting the order of vertex computations to improve temporal locality of vertex data stored in on-chip caches. Experiments with nine real-world graphs and three graph algorithms on the recent GPU platform show that the proposed method offers a significant speedup (average 46%) over the state-of-the-art graph algorithm implementation on GPUs.**

Keywords—*Graph computation; GPU; data locality; vertex scheduling*

I. INTRODUCTION

In the last decade, data analysis on computer systems has encountered a great challenge due to the explosive growth in the amount of data. With this 'big-data' trend, computer systems that can efficiently process and analyze huge data have been increasingly important to cope with data explosion.

In particular, large-scale graph processing is gaining attention as a key enabler to extract valuable information from relationships between data. For example, many web search engines represent web pages and their connections as vertices and edges of a web graph, respectively, to discover importance of pages based on their relationships (e.g., PageRank in Google). Other important examples that can benefit from graph processing include social networks, bioinformatics, citation relationship, machine learning, and so on.

However, developing a system that can efficiently handle large graphs is very challenging, despite the abundant computation parallelism in graph processing. This is because many graph algorithms are characterized as (1) poor data locality caused by random accesses to vertex data, (2) a small amount of computation per vertex, which is not sufficient to hide long memory access latency, and (3) severe workload imbalance across different vertices due to the nature of real-world graphs [1]. Due to these aspects, recent researches have actively developed software stacks for processing large graphs

[1-4], rather than relying on conventional big-data analysis frameworks such as MapReduce [6].

Another dimension of improving efficiency of large-scale graph processing is to utilize highly parallel hardware units. Specifically, GPUs have been considered as a promising platform for accelerating graph processing by leveraging massive parallelism inside them [7-11]. Regarding this, previous work has explored the way of structuring graph algorithms for GPUs to fully utilize their computation parallelism [7, 8]. However, little has been known about optimizing the memory access patterns of graph algorithms by considering characteristics of GPU memory hierarchies, which play an important role in processing large graphs due to the high memory bandwidth requirements.

In this paper, we propose locality-aware vertex scheduling (LAVER), a simple but versatile scheme to optimize the memory access behavior of graph processing on GPUs. The motivation of this work is that conventional GPU-based graph processing implementations determine the order of vertex processing without considering data locality. In order to tackle this inefficiency, we propose a methodology that schedules computations in a way that memory requests to the same vertex are issued closer in time to each other. Furthermore, our mechanism carefully considers the unique characteristics of large real-world graphs called power-law degree distribution in optimizing the vertex schedule. By doing this, our scheme facilitates much more efficient use of moderately sized on-chip caches in GPUs through higher temporal locality. Furthermore, our scheme is general enough to be applied to any GPU-based graph algorithm unlike many other optimizations. Evaluations using three different graph algorithms with nine real-world graphs show that our scheme improves the performance of GPU-based graph processing by 46% on average compared to the state-of-the-art GPU-based graph algorithm implementation.

II. RELATED WORK

There are several studies on large-scale graph processing focusing on how to exploit this parallelism to make it efficient and scalable. For this purpose, several software stacks have been proposed for automatically parallelizing graph processing workloads to reduce programming effort. Most notably, there have been several frameworks that scale user-defined vertex-centric computation across distributed servers [1-3] or GPUs [10, 11]. In addition, there have been a few researches to design domain-specific programming languages and compilers that generate efficient graph algorithm implementations from high-level descriptions [4, 12]. We expect that our scheme can be easily integrated into such frameworks without modifying the programming model.

Regarding graph processing on GPUs, Harish and Narayanan [8] proposed GPU implementation of several graph algorithms, in which computations for different vertices are parallelized across different threads of GPUs. Since this imposes workload imbalance between threads especially when the number of neighbors differs significantly across different vertices, Hong et al. [7] proposed to split computation for a single high-degree vertex (the one with many neighbors) into multiple threads, thereby balancing loads across threads to improve SIMD efficiency of graph processing on GPUs. Such a style of parallelization called vertex-centric computation is also used in developing efficient graph traversal algorithms [9] and even general frameworks for GPU-based graph processing [10]. Our work is also based on this concept, but focuses on optimizing memory access efficiency of graph algorithms on GPUs rather than improving computation efficiency in parallelization.

Apart from graph processing specific optimizations, researchers have proposed several ways to optimize memory access behavior of GPU applications [13-15]. Our work is orthogonal to those approaches and/or is distinguished from them in that it utilizes domain-specific knowledge (i.e., graph topology and memory access characteristics of graph algorithms) to optimize GPU-based graph algorithms.

III. PRELIMINARY

A. Graph Data Structure

A graph is composed of vertices representing data entities (e.g., users in social networks) and edges indicating relationships between two objects (e.g., friendships in social networks). In this paper, we use the term 'graph' as indicating a directed graph, in which edges have direction. An edge from u to v can be seen as an in-edge of v and an out-edge of u. Fig. 1(a) shows an example graph consisting of eight vertices (circles) and ten edges (arrows). The figure also shows the case where each vertex has its associated value called vertex property (e.g., 'a' for vertex 0), which can be utilized during graph algorithm execution.

One of the most popular ways to store graphs in memory is compressed sparse row (CSR) representation. In this scheme, a graph is stored with three arrays, which are row-offset array (R), column-index array (C), and vertex data array (V) as shown in Fig. 1(b). First, all vertex properties are stored in the V-array. Then, for each vertex, the list of its predecessors (or successors) is concatenated into the C-array. For example, as shown in Fig. 1(b), the C-array stores the predecessors of vertex 0 (i.e., vertex 4), vertex 1 (i.e., vertex 0), vertex 2 (i.e., vertex 4), vertex 3 (i.e., vertex 2 and 5), and so on in the C-array. Lastly, in order to distinguish the boundary of two predecessor (or successor) lists, the R-array stores the start index of the predecessor list in the C-array for each vertex. For example, the list of predecessors for vertex 3 is stored from fourth (i.e., R[3]=3) to fifth (i.e., R[4]-1=4) entries of the C-array.

B. Graph Computation on GPU

In order to accelerate graph processing for large graphs, graph processing usually exploits the parallelism among vertex computations. This is because computations for different vertices are mostly independent of each other. Based on this fact, many graph processing frameworks adopt the vertex-centric programming model [8], in which programmers describe computation for a single vertex with accesses to its neighbor vertices and then the framework automatically parallelizes the computation behind the scenes.

Since large real-world graphs have millions or even billions of vertices, it is beneficial to utilize hardware structures with massive parallelism such as GPUs. Graph processing on GPUs is similar to the one on multicore processors, except for the following two aspects. First, host processors are responsible for transferring graphs to GPUs and scheduling vertex computations across different threads. Second, graph processing on GPUs should be aware of GPU execution model where a number of threads (typically 32) share the same instruction stream called warp[1] and GPUs support the massive number of threads to be executed in parallel.

With that in mind, Fig. 2 shows the pseudocode of the PageRank algorithm on GPUs. Before executing the algorithm on GPUs, the host processor first constructs and loads the CSR data structure[2] to the GPUs (line 1). Then, the host processor maps vertex computations to threads by sending an array called scheduleQ (line 2), which stores the vertex id to be processed for each thread (i.e., i-th thread processes scheduleQ[i]-th vertex). Such information is accessed in each GPU thread (line 4) to determine which vertex to process. Threads with smaller indexes are invoked earlier by the runtime. Once each thread obtains the target vertex id, it traverses over its in-edges by accessing the C-array and the V-array to compute the PageRank value of the target vertex (line 7 and 8). Lastly, it updates the vertex property of the target vertex with the computed value (line 9). The bottleneck of this computation is in accessing the V-array in line 8, which generates random memory accesses over the entire graph through neighbor traversal (see indirect indexing with the C-array). Therefore,

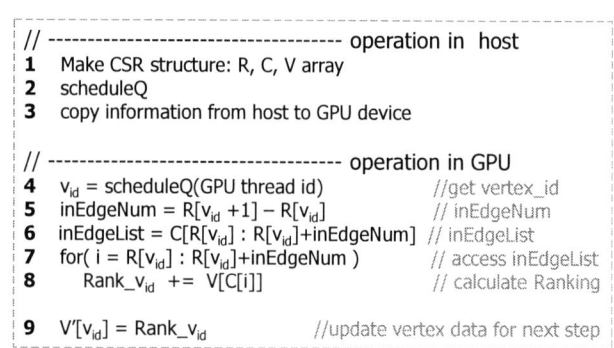

Fig. 2. PageRank algorithm on GPU.

---

[1] This indicates that, if a single thread in a warp stalls, all threads in the warp are stopped until it is resolved

[2] Since many graph processing applications use the same graph multiple times for different analyses, the overhead of loading graphs into the GPU is usually amortized over multiple runs of analysis kernels.

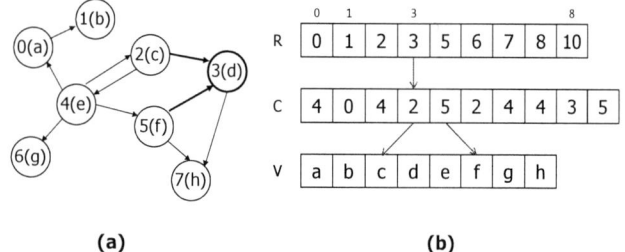

Fig. 1. (a) Example graph and (b) CSR data structure.

978-1-4673-9141-2/15 $31.00 © 2015 IEEE

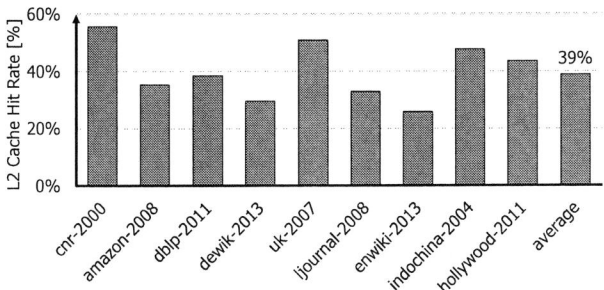

Fig. 3. L2 cache hit rate.

for efficient graph processing on GPUs, it is critical to optimize memory accesses for neighbor traversal.

In this paper, our focus is on determining the execution order of vertex computation through scheduleQ mechanism. In our baseline [7], scheduleQ is mainly used to enforce vertex-to-thread mapping and thus contains vertices to be processed in an arbitrary order. However, we observed that optimizing the order of vertex computation can improve the efficiency of GPU-based graph processing, especially in terms of data locality of neighbor traversal. In the following sections, we motivate and develop a simple method to schedule vertex computation in a way that maximizes data locality of neighbor traversal by considering topology of input graphs.

IV. MOTIVATION

A. L2 Cache Efficiency

As discussed in Section I, graph algorithms generate a huge amount of random memory accesses over the entire graph, which cannot be easily served by on-chip caches. To quantify this, Fig. 3 shows the L2 cache hit rate of the PageRank algorithm with nine different inputs measured based on hardware performance profiler (see Section VI-A for the detailed evaluation methodology). We use the state-of-the-art graph algorithm implementation [7], which considers workload imbalance by splitting computation for high-degree vertices into multiple threads.

As shown in the figure, the L2 cache covers only a small fraction of the memory requests generated by the PageRank algorithm (39% on average). The major contribution of this low hit rate comes from neighbor traversal (line 8 in Fig. 2), which involves random accesses to the V-array. More specifically, each iteration of the PageRank algorithm accesses the entire set of vertices in the graph, which often occupies very large memory footprint (29MB to 1.7GB in the case of real-word graphs used in our evaluations).

However, such low cache efficiency is not completely inevitable. Many researches state that real-world graphs typically show significant imbalance in the number of edges per vertex, which is called power-law degree distribution [16]. Considering a graph that represents Twitter users and their follower relationships as an example, celebrities would obviously have many more followers (i.e., predecessors) than other normal users. In graph algorithms that access out-edges to compute the vertex property of the current vertex, such high-in-degree vertices would be accessed much more frequently than other vertices. Therefore, we can conclude that graph algorithms do have some degree of data locality, but the current state-of-the-art implementations are not aware of it.

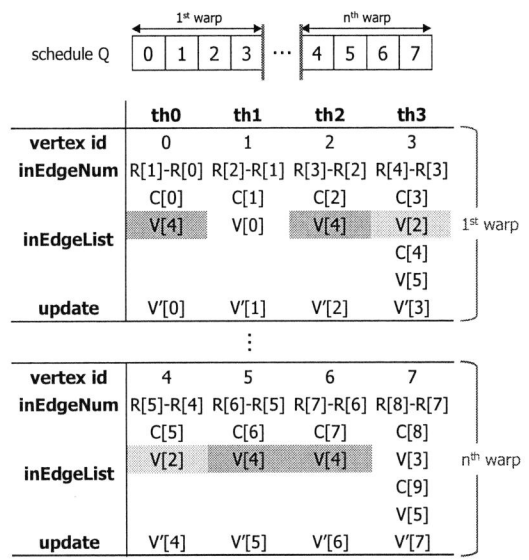

Fig. 4. Memory access sequence in a GPU with the default vertex schedule.

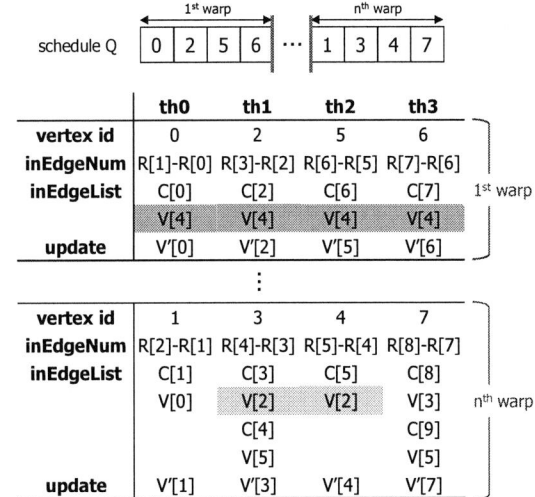

Fig. 5. Memory access sequence with the new vertex schedule that is aware of data locality of V-array accesses.

B. Vertex Scheduling for Better Locality

Our key observation for exploiting hidden data locality in graph algorithms is that changing the order of vertex computation (i.e., information stored in scheduleQ) can potentially improve data locality of neighbor traversal. We elaborate this observation with two motivational examples shown in Fig. 4 and 5. The examples depict memory access sequences of the PageRank algorithm (Fig. 2) when the input graph contains the graph shown in Fig. 1(a) as part of it.

Fig. 4 shows the execution schedule in a GPU having one streaming multiprocessor that can execute only one four-thread warp at a time. We assume that the default schedule processes vertex 0, 1, 2, and 3 in the first warp, while vertex 4, 5, 6, and 7 are handled in the n-th warp where $n \gg 1$ (we assume that other vertices in the input graph are processed between the first and the n-th warp).

Under this default schedule, it can be seen that accesses to V[4] are split across two different warps, which have many other warps in between (similar observation is possible for V[2]). This indicates that, although V[4] is accessed frequently due to its high out-degree (see Fig. 1(a)), the execution

978-1-4673-9141-2/15 $31.00 © 2015 IEEE   197

schedule does not fully utilize this fact due to the unawareness of such information in constructing vertex scheduling (i.e., building scheduleQ).

Such inefficiency can be alleviated by explicitly considering the locality of memory requests to the V-array. Fig. 5 shows an example schedule that allocates all successors of vertex 4 into the same warp (the first warp) to optimize the accesses to the V-array. As can be seen in the figure, all accesses to V[4] (and V[2]) are now close to each other in time, thereby improving temporal locality of memory accesses. What is more promising is that such optimization is completely orthogonal to other techniques since it does not modify the target graph algorithms at all. In the following section, we will describe the proposed algorithm developed for this purpose in detail.

V. LOCALITY-AWARE VERTEX SCHEDULING

In this section, we propose locality-aware vertex scheduling (LAVER). Since edges have directions in directed graphs, our scheduling mechanism distinguishes graph algorithms that access predecessors in their vertex computation (e.g., PageRank) and those accessing successors (e.g., single-source shortest path). The following describes our mechanism for the former type of graph algorithms (those accessing in-edges in their vertex computation). Nevertheless, our scheme can easily handle the algorithms accessing out-edges by considering the direction of edges as reversed.

The key idea of LAVER is to schedule vertices sharing the same predecessors as a group. This is because, if such vertices are processed in a short time window, multiple memory requests to the common predecessors have a great chance to be served by on-chip caches due to the shortened reuse distances. In other words, our method is to avoid the situation where V-array accesses to the same vertex to be spread over time, which in turn improves temporal locality of V-array accesses.

To accomplish this goal, our scheduling algorithm works as follows (also see Fig. 6). Our algorithm requires two data structures: scheduleQ to store the final schedule and a bit vector of length |V| to keep track of unscheduled vertices. First, it picks a new vertex v from unscheduled ones (line 5, see the next paragraph for the method to choose the start vertex). Then, it traverses all of its successors (line 6 to 8), finds the unscheduled ones among them by checking the bit vector (line 9), and adds them to scheduleQ with marking the corresponding bit vector entries to one (line 10 to 12). The rationale behind this is that vertex computations for those successors will access their predecessors including vertex v, and thus, scheduling those successors in a short time window will improve the data locality of the V-array element for vertex v. After the process for vertex v is done, the algorithm selects another start vertex among the unscheduled ones until there are no more unscheduled vertices.

The remaining task is to determine the way of choosing start vertices. Recalling that the objective of our algorithm is to place vertices that access the common predecessors to be close to each other in scheduleQ, the effectiveness of this mechanism highly depends on how many vertices share the common predecessor. This is because it implies the number of potential reuses in accessing the V-array element for that predecessor.

With that in mind, we propose to prioritize vertices with high out-degree over other vertices in start vertex selection. This is implemented by inserting all vertices into a temporary array, sorting it with decreasing order of out-degree, and taking

```
1    scheduleQ = []
2    bitvector = [0] * V.size()
3    scheduled = 0

4    while(scheduled < |V|)
5        v = selectStartVertex()
6        outEdgeNum = R_out[v+1] − R_out[v]
7        outEdgeList = C_out[R_out[v] : R_out[v]+outEdgeNum]
8        for( i = R_out[v] : R_out[v]+outEgeNum)
9            if(bitvector[i] == 0)
10               scheduleQ.append(i)
11               bitvector[i] = 1
12               scheduled += 1
```

Fig. 6. Pseudocode of LAVER.

Fig. 7. Example sequence of scheduling in LAVER.

one by one as a start vertex from the beginning. This directly relies on the aforementioned observation since, if a vertex has many successors, the corresponding V-array element will be accessed frequently in neighbor traversal (precisely, as many times as the number of successors). Therefore, it is beneficial to select such high-degree vertices first in our algorithm to maximize the opportunity to cluster many vertices that have a common predecessor. Furthermore, this policy is especially suitable for real-world graphs, which experience a large variation in degrees due to the power-law degree distribution [16].

Since our start vertex selection mechanism does not return the same start vertex more than once, our algorithm processes each edge only once during its execution. Therefore, the time complexity of our algorithm is $O(|V| \log |V| + |E|)$, which is arguably inexpensive, where the first term is the time complexity of the start vertex selection routine. Moreover, such processes need to be done only once for each graph, whereas an input graph is usually reused multiple times in practice to conduct different types of analyses.

Fig. 7 shows an example scenario in that our algorithm finds a schedule for the graph in Fig. 1(a). Initially, scheduleQ is empty and all entries in the bit vector are set to zero. The algorithm chooses vertex 4, 5, 2, 0, 3, 1, 6, and 7 (in order) as start vertices according to their out-degree. For the first start vertex, since all of its successors are not scheduled, the algorithm appends them to the scheduleQ. As can be seen in the figure, all vertices accessing vertex 4 are placed closely to each other in scheduleQ, thereby increasing the opportunity to exploit data locality in V-array accesses for vertex 4. The same process is done for vertex 5 as well. However, in the case of vertex 2, one of the successors (i.e., vertex 3) is already scheduled in scheduleQ (which can be checked by using the bit vector). Therefore, we add only the successors that are not scheduled yet (i.e., vertex 4) into scheduleQ. After applying the same method to vertex 0, the algorithm terminates with the final schedule (even if four more start vertices are left) since all vertices are now scheduled in scheduleQ.

VI. Experimental Evaluations

A. Methodology

We use a real GPU hardware (GTX 970) to measure performance of graph algorithms using the baseline and the proposed method. Execution time of graph algorithms is measured by real-time hardware clocks, while the L2 cache hit rate is obtained from the hardware profiler provided by NVIDIA. Table I shows the specification of the evaluation platform.

As target workloads, we implement three representative graph algorithms with different memory access behavior. The following briefly explains each algorithm. Although not shown in the paper, we also evaluated other algorithms (e.g., average teenage follower [5]) but omitted their results since they follow the same trends shown in the paper due to the similarity in memory access patterns.

- **PageRank** [4] calculates the importance of each vertex (e.g., website ranking) from the relationships between vertices.
- **Single-Source Shortest Path (SSSP)** [4] searches the shortest path from a given source to all other vertices in the graph.
- **Weakly Connected Components (WCC)** [17] finds groups of vertices that are reachable with each other when the direction of edges is ignored.

For input graphs, we choose nine real-world graphs [18] shown in Table 2 to evaluate a wide range of graph size and topology. These graphs are collected from social networks, citation networks, and web crawls. Memory footprint characteristics of input graphs are shown by calculating the total size of their CSR structures.

Our baseline is the state-of-the-art implementation for GPU-based graph processing [7], which splits high-degree vertices (degree ≥ 256 in our case) into multiple threads to mitigate the workload imbalance problem. Such a technique is applied to both baseline and ours since it is orthogonal to vertex scheduling.

TABLE I. GTX970 Configuration

Engine Specifications

CUDA Cores	1664
Base Clock	1050 MHz
L2 Cache Size	2 MB

Memory Specification

Memory Clock	7.0 Gbps
Memory Size	4 GB
Memory Interface	GDDR5
Memory Bandwidth	224 GB/sec

TABLE II. Characteristics of Input Graphs

Graph Name	Vertices	Edges	Footprint [MB]
cnr-2000	325,557	3,216,152	29
amazon-2008	735,323	5,158,388	46
dblp-2011	986,324	6,707,236	60
dewiki-2013	1,532,354	36,722,696	294
uk-2007	1,000,000	41,247,159	324
ljournal-2008	5,363,260	79,023,142	645
enwiki-2013	4,206,785	101,355,853	807
indochina -2004	7,414,866	194,109,311	1539
hollywood-2011	2,180,759	228,985,632	1765

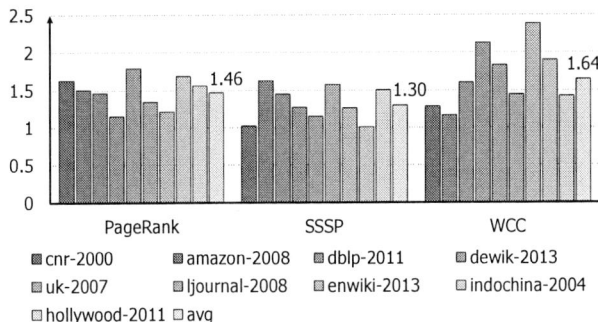

Fig. 8. Speedup achieved by the proposed method.

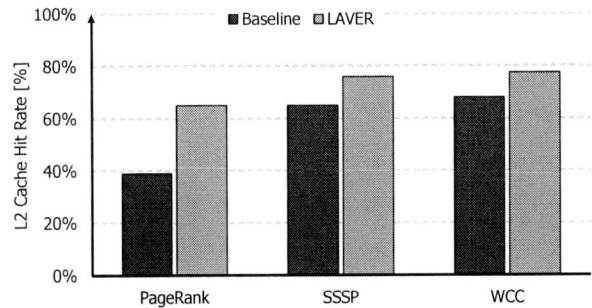

Fig. 9. Comparison in L2 cache hit rate.

B. Results

Fig. 8 shows speedup of our approach compared to the baseline. Our algorithm improves the performance of PageRank, SSSP, and WCC by 46%, 30%, and 64% on average compared to the state-of-the-art GPU-based graph algorithm implementations. More importantly, our scheme always achieves positive speedup across all input graphs and graph algorithms used in our evaluation. In the rest of this section, we analyze the source of this speedup in detail.

As the first step, Fig. 9 compares the L2 cache hit rate under the baseline and the proposed method (results are averaged over all input graphs). It is clearly seen that the L2 cache hit rate is significantly improved with the proposed mechanism (15% on average). This is especially noticeable in the PageRank algorithm since it generates the largest amount of V-array accesses in neighbor traversal among the three graph algorithms (the entire set of edges are traversed for each iteration), thereby having larger dynamic working sets.

Interestingly, we observed that our algorithm achieves the highest speedup in WCC despite its lowest improvement in L2 cache hit rate. This is because, in that case, our algorithm significantly reduces accesses to the L2 cache (23% on average), while other cases exhibit almost no difference in the number of L2 cache accesses. Such reduction is achieved by the memory request coalescing capability of GPUs, which merges the same memory requests in a warp into a single memory request. In terms of the number of L2 cache misses, WCC shows the highest reduction (41% on average) compared to PageRank (36%) or SSSP (26%), which is consistent with speedup results.

In order to further analyze the source of speedup, we choose the PageRank algorithm with two representative input graphs, *dewiki-2013* (lowest speedup) and *uk-2007* (highest speedup), for detailed evaluations. Since it is impossible to measure data locality of specific memory regions with the hardware profiler, we instead devise an analytical model of

978-1-4673-9141-2/15 $31.00 © 2015 IEEE

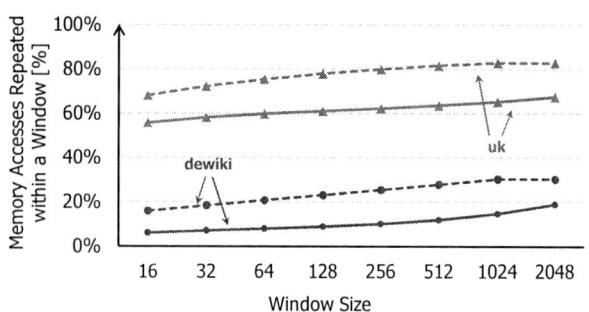

Fig. 10. Locality analysis for the baseline (solid lines) and the proposed method (dotted lines).

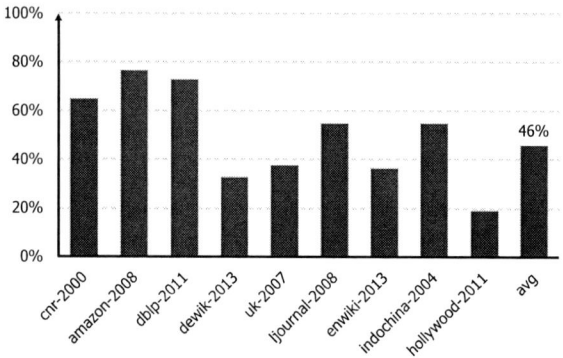

Fig. 11. Scheduling overhead normalized to the CSR construction time.

data locality of V-array accesses. For this purpose, we first chunk scheduleQ into equally sized windows (e.g., 16, 32, 64, …), which approximates the window of temporal locality that can be covered by the L2 cache. Then, for each window, we calculate the number of V-array accesses repeated within a single window as a proxy of data locality. This is because such memory accesses can be handled by the L2 cache as long as it does not overflow within the window.

Fig. 10 depicts the results for our locality analysis. Solid lines represent the baseline, while the dotted lines are for the proposed algorithm. From this figure, we draw three important conclusions. First, although V-array accesses in graph algorithms seem to be random at the first glance, they do have data locality. Even in the baseline with the smallest window size, 6% (*dewiki-2013*) to 56% (*uk-2007*) of V-array accesses (depending on the shape of graphs) are repeated within a single window. Second, we confirmed that the speedup of our mechanism indeed comes from the improved data locality of V-array accesses. In fact, this analysis explains why dewiki-2013 achieves lower speedup than uk-2007 (see PageRank in Fig. 8): in *dewiki-2013*, only a small fraction of V-array accesses are reused even after applying our technique (e.g., 16% under the smallest window size). Third, this evaluation result projects that the effectiveness of our technique will be consistent even if future GPUs incorporate larger L2 caches (i.e., larger window size). This is drawn from the observation that our algorithm maintains the benefit of higher data locality of V-array accesses even under very large window size.

Lastly, Fig. 11 shows the scheduling execution time, which is normalized to the time of constructing the CSR structure considering that scheduling needs to be done only once for each graph after CSR structure construction (note that the CSR structure is usually immutable due to its compact structure). We show only the results for the PageRank algorithm for brevity. As shown in the figure, our algorithm introduces

moderate execution time overhead (46% extra runtime to the CSR construction) in preprocessing steps for GPU-based graph processing.

VII. CONCLUSION AND FUTURE WORK

In this paper, we proposed locality-aware vertex scheduling to leverage hidden data locality in neighbor traversal of GPU-based graph processing workloads. Our key observation is that determining the order of processing vertices has a significant impact on data locality, and thus, on-chip cache efficiency. In order to schedule vertex computation in a way to maximize data locality of neighbor traversal, we devise a low-cost algorithm that places computation for vertices sharing the same predecessor (or successor) to be temporally proximate. Our algorithm also judiciously exploits the characteristics of large real-world graphs, which are skewed degree distribution, in obtaining the vertex schedule. Through this, our mechanism optimizes the data locality of neighbor traversals, which have been generally considered as showing random memory access patterns. Moreover, our algorithm can be applied to any graph algorithm executed on GPUs and is orthogonal to other GPU-graph-specific optimizations since it does not modify the target graph algorithms at all. Experimental results using nine large real-world graphs show that our algorithm improves the performance of three graph algorithms by 46% on average compared to the state-of-the-art GPU-based graph algorithm implementation. Our future work will explore the implication of our algorithm on multi-GPU systems and distributed servers with GPU-enabled nodes for scalability.

REFERENCES

[1] G. Malewicz et al., "Pregel: A system for large-scale graph processing," in *Proc. SIGMOD*, 2010.
[2] Y. Low et al., "Distributed GraphLab: A framework for machine learning and data mining in the cloud," *PVLDB*, vol. 5, no. 8, pp. 716-727, Apr. 2012.
[3] J. E. Gonzalez et al., "PowerGraph: Distributed graph-parallel computation on natural graphs," in *Proc. OSDI*, 2012.
[4] S. Hong et al., "Green-Marl: A DSL for easy and efficient graph analysis," in *Proc. ASPLOS*, 2012.
[5] S. Hong et al., "Simplifying scalable graph processing with a domain-specific language," in *Proc. CGO*, 2014.
[6] J. Dean and S. Ghemawat, "MapReduce: Simplified data processing on large clusters," in *Proc. OSDI*, 2004.
[7] S. Hong et al., "Accelerating CUDA graph algorithms at maximum warp," in *Proc. PPoPP*, 2011.
[8] P. Harish and P. J. Narayanan, "Accelerating large graph algorithms on the GPU using CUDA," in *Proc. HiPC*, 2007.
[9] S. Hong et al., "Efficient parallel graph exploration on multi-core CPU and GPU," in *Proc. PACT*, 2011.
[10] F. Khorasani et al., "CuSha: Vertex-centric graph processing on GPUs," in *Proc. HPDC*, 2014.
[11] J. Zhong and B. He, "Medusa: Simplified graph processing on GPUs," *IEEE TPDS*, vol. 25, no. 6, pp. 1543-1552, Jun. 2014.
[12] J. Seo et al., "SociaLite: Datalog extensions for efficient social network analysis," in *Proc. ICDE*, 2013.
[13] Y. Yang et al., "A GPGPU compiler for memory optimization and parallelism management," in *Proc. PLDI*, 2010.
[14] W. Jia et al., "Characterizing and improving the use of demand-fetched caches in GPUs," in *Proc. ICS*, 2012.
[15] J. Lee et al., "Many-thread aware prefetching mechanisms for GPGPU applications," in *Proc. MICRO*, 2010.
[16] A. Mislove et al., "Measurement and analysis of online social networks," in *Proc. IMC*, 2007.
[17] U. Kang et al., "PEGASUS: A peta-scale graph mining system implementation and observations," in *Proc. ICDM*, 2009.
[18] Laboratories for Web Algorithmics. http://law.di.unimi.it/datasets.php.

Traffic-Aware Buffer Reconfiguration in on-Chip Networks

Ramin Bashizade[*], Hamid Sarbazi-Azad[*†]
[*]Department of Computer Engineering, Sharif University of Technology, Tehran, Iran
[†]School of Computer Science, Institute for Research in Fundamental Sciences (IPM), Tehran Iran
rbashizade@ce.sharif.edu, azad@{ipm.ir, sharif.edu}

Abstract—Networks-on-Chip (NoCs) play a crucial role in the performance of Chip Multi-Processors (CMPs). Routers are one of the main components determining the efficiency of NoCs. As various applications have different communication characteristics and hence, buffering requirements, it is difficult to make proper decisions in this regard in the design time. In this paper, we propose a traffic-aware reconfigurable router which can adapt its buffers structure to the changes in the traffic of the network. Our proposed router manages to achieve up to 18.8% and 44.4% improvements in terms of postponing saturation rate under synthetic traffic patterns, and average packet latency for PARSEC applications, respectively, with respect to the conventional state-of-the-art router.

I. INTRODUCTION

Constantly shrinking feature sizes in addition to restrictions on clock rate, encourage the integration of many cores on a single chip. This in turn, highlights the requirement of a scalable interconnection structure among these cores. As a consequence, Network-on-Chip (NoC) [2] has become the most popular interconnection structure in many-core chips.

It is shown that traffic in NoCs is not uniformly distributed [17]. Some nodes play a larger role in generating and consuming traffic, while others have a smaller share. Therefore, buffering requirements for different routers may differ based on their location in the network and the tasks assigned to them. Additionally, such requirements are subject to change during different phases of a single application. This necessitates the existence of an efficient mechanism for adapting the buffers structure in input ports of NoC routers.

Bearing this fact in mind, there are works in the literature that focus on adapting the buffer structure of input ports with respect to the incoming traffic. For instance, in [7] a method is proposed in order to determine the proper size of buffers in each input port of the routers with respect to the traffic rate in application-specific NoCs. In [13], buffer slots are allocated to each input with regard to the requirements of the inputs extracted from applications communication task graph. There are other works such as [11], [14], in which the Virtual Channels' (VCs) structure in each port is subject to reconfiguration so that routers can adapt themselves with traffic rate and size of packets. In some other researches, a central buffer structure is used to overcome the issue of varying buffer requirements of input ports [5], [16], [18].

In this paper, in order to add the reconfiguration capability to the routers, we propose a new router architecture that allows for sharing buffers of input ports and dynamically reconfiguring their buffer structure based on a suitable yet simple monitoring scheme. Unlike works that implement a central buffer in the router, we try to impose minimal alterations to the microarchitecture of the router and the structure of input ports is similar to that of a conventional router. Instead, we add a relatively small crossbar right before input buffers to let all input channels access the buffers of all input ports (except Local port). By doing so, we managed to achieve up to 44.4% decrease in average packet latency for PARSEC applications. Additionally, our proposed router yields comparable performance to the conventional router having only half of its buffering resources.

The rest of this paper is organized as follows. Previous related work is presented in Section II. In Section III, the proposed architecture and its implementation are discussed in detail. Section IV is devoted to area and power analysis of our proposal. Evaluation of the proposed router is done in Section V. Ultimately, Section VI concludes the paper.

II. RELATED WORK

In [7], Hu *et al.* proposed a method to dedicate buffer to input ports by considering parameters such as packets sizes, time needed for a packet to pass a router without contention, routing algorithm, Communication Task Graph (CTG), and packet injection rate. They determine the probability of buffers being full based on the utilization of physical channels, and exploit a greedy algorithm to dedicate buffer to the port with the highest probability of its buffer being full. Taking advantage of CTG makes this work an application-specific method. Hence, unlike our proposal, it cannot be employed in general purpose designs.

Nicopoulos *et al.* presented ViChaR in [14]. There are no fixed-sized VCs in ViChaR. Instead, the size of each VC is variable between one flit and a maximum size determined by the size of the packet to which it is allocated. The idea is that upon receiving header flits, VCs are created, while upon arriving/departing body flits, the size of existing VC is regulated. Although ViChaR manages to adapt the buffer structure of each input to the needs of the traffic passing through it, it cannot overcome the problem of buffer underutilization in ports with light traffic loads. Our proposed router, however, targets this issue and successfully enhances the buffer utilization.

978-1-4673-9141-2/15 $31.00 © 2015 IEEE

A reconfigurable router is proposed in [13] by Matos *et al.* In this work, if an input port is identified as a hotspot using the Communication Task Graph (CTG) of the running application, it will borrow buffer slots (if the lender port is not a hotspot itself). In order to alleviate the overheads, each port can only borrow buffer slots from its neighboring ports (e.g., port North only can borrow from port East and port West), with the priority being with the port on its right side (e.g., for port North, the priority is to borrow buffer slots from port West). Reconfiguration is done only once in the beginning of the application in this design, and hence, it cannot efficiently handle the traffic in applications with varying requirements in different phases of execution. In addition, VCs are not exploited in this work. We address both issues in our proposal.

Authors in [5], [16], [18] implemented a central buffer structure to be shared among all the input ports. In [16], Ramanujam *et al.* proposed a router that rather than buffering data at the output ports, uses two crossbar stages with buffering sandwiched in between. Tran *et al.* use a similar buffer structure with two crossbars. However, they allow for bypassing the central buffers in low traffic loads. Both of these designs require quite large crossbars which incur considerable area and power overhead to the router. In [5], Hassan *et al.* deploy a central buffer alongside elastic buffered links. Central buffer in this design has one read and one write port, forcing it to serialize flits that need to travel in to or out of the CB in a single cycle.

III. RECONFIGURABLE ROUTER

A. Reconfiguration Method

A suitable yet simple mechanism for monitoring the incoming traffic is required to decide whether a reconfiguration should take place. To achieve this goal, we exploit two flags[1] in each port of the routers; *congestion flags* in output ports, which is responsible for making requests for more buffer slots when the downstream port is congested, and *underutilization flag* in input ports, which indicates the availability of free buffering space in its corresponding port.

A *congestion flag* is valid when there are requests for VCs in its corresponding output, but there are no free buffering space in the input port of the neighboring router. An *underutilization flag*, on the other hand, is set when at least two free buffer slots are available at a port that has a predefined minimum of buffering space. This serves three purposes, i) making sure that each input port has always buffering space available to assure connectivity, ii) eliminating the need for halting traffic flow during reconfiguration, and iii) avoiding aggressive reconfiguration. As it will be shown shortly, a reconfiguration takes place in two cycles and after that, one cycle is needed for the credit registers to get updated. If the lender port has two or more buffer slots available, then it is not necessary for it to stop receiving packets. This is due to the fact that one of the two free buffer slots will remain at this port and it can still

---

[1]These are not actual registers, and are computed combinationally. We refer to them as flags for the sake of simplicity.

---

store the flit that arrives in the next cycle. Additionally, setting a minimum for buffering space of input ports ensures that in less congested ports, there will be buffer slots available for incoming packets without needing to wait for reconfiguration. This eases the communication flows. It is obvious that the *congestion flag* of an output port and the *underutilization flag* of its adjacent input port will never be simultaneously set to "1".

When the *congestion flag* in an output port is set, that port sends a signal to its downstream router requesting buffer slots. In the downstream router, *underutilization flags* must be checked to see whether there are available free buffer slots in other ports. To save one cycle in the reconfiguration process, every cycle we determine the possible lender input port in each router. If such port exists, three operations are done: i) one buffer slot is subtracted from the credits of the output port of the neighboring router that is connected to the underutilized port, and it is added to the credits of the requesting port in the upstream router; ii) the *congestion flag* in the upstream router corresponding to the requesting port is reset; and iii) the *underutilization flag* of the port that lent the buffer slot is updated based on the number of remaining free slots in that port. If there are more than one port with available buffering space, the requesting port borrows buffer from the port that has more free buffer slots; if there are two or more ports that equally have the most free buffering space, priorities that are fixed at design time for each router are used (e.g., in a mesh topology, the distance of the port to the edge of the network can be exploited to assign priorities). On the other hand, if more than one port demands buffer, in order to avoid the complexities of lending to more than one port in a single cycle, an arbiter decides which port can borrow buffer at this cycle. Additionally, routers that have a reconfiguration process ongoing, do not accept any reconfiguration requests during the next one cycle required for completing the action. This way, the possibility of lending a buffer slot to an input port that were supposed to be lent to another input is eliminated.

B. Implementation

Figures 1-(a) and 1-(c) demonstrate the microarchitecture of the proposed router and its pipeline stages. Added or modified units and stages are shaded and their communication with other units of the router are also shown in the figure. We do not use a central buffer structure in this work. As a consequence, each input port needs to access the buffers of all other ports. This necessitates the existence of an extra crossbar before the input units. However, unlike designs proposed in [16], [18] that use large $P \times N$ crossbars (P is the number of router ports, and N is the number of central buffer queues, which is usually much greater that P) to write in and read from central buffers, we use an extra $(P-1) \times (P-1)$ crossbar which impose far less area and power overhead to the router (we exclude Local port from reconfiguration process). It must be noticed that the added crossbar does not add to the critical path of the router pipeline, as the critical path is determined by SW/VC Allocator and is long enough to let flits pass the

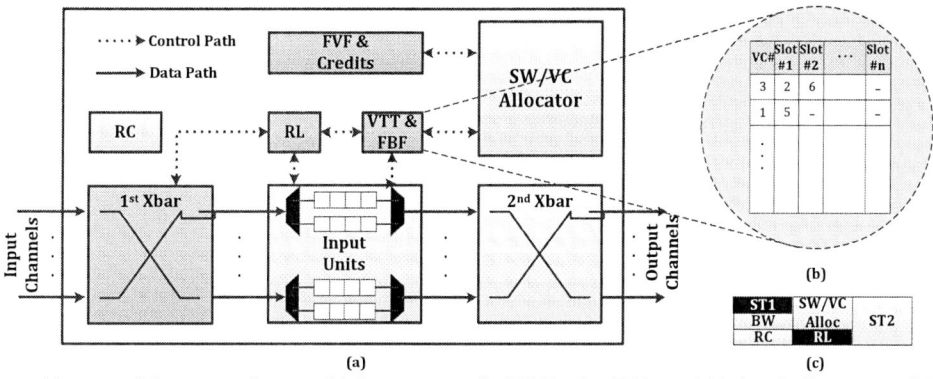

Fig. 1: (a) The microarchitecture of the proposed router, (b) the structure of a VC Tracker Table, and (c) the pipeline stages of the proposed router.

crossbar and be written in the buffers in a single cycle. In fact, the length of the critical path of this stage of the router pipeline is less than that of the last stage (Second Switch Traversal, ST2 in Figure 1-(c)).

A Reconfiguration Logic (RL) is required for the entire router. This unit is responsible for controlling the extra crossbar alongside handling the reconfiguration process. When a reconfiguration takes place, the lender and borrower ports might have flits ready to arrive at them. However, the buffers at each port can receive incoming flits only from one channel. Therefore, upon taking place of a reconfiguration, RL informs the router connected to the lender port that it cannot send flit in the next cycle, and sends proper signals to the first crossbar in order to configure it to connect the input channel corresponding to the borrower port, to the buffers of the lender port. The borrower port is prioritized for sending flit over the lender port due to the fact that it is congested at the time, and letting it send out a flit will probably help the flow of traffic in it and its neighboring routers.

In addition, RL handles the reconfiguration process, i.e., assigning proper values to *underutilization flags* and *congestion flags*, sending out and receiving reconfiguration requests and updating the VC Tracker Tables (VTTs). VTTs, depicted in Figure 1-(b), are needed at each input port to keep track of the buffer slots that compose each VC in the corresponding input. Also, a Free Buffer slot FIFO (FBF) is required in each input port to keep track of the buffer slots that belong to it. Thus, each buffer slot needs to have a unique ID in the router. Entries of VTT are used in Switch Allocation (SA). RL determines the need for sending out reconfiguration request in parallel with SW/VC allocation stage and hence, does not add to the length of the router pipeline. Additionally, upon receiving a request for more buffer from an input port, RL decides which input port lends buffer slot based on the constraints explained in section III-A. Afterwards, credit messages are sent to the adjacent routers.

To simplify reconfiguring the buffers structure, we adopt the design of [10], [14], in which VCs are created on-demand. We have also remove VC Allocation (VA) stage and instead use the winner of SA for the purpose of VC allocation [10]. In our design, Free VC FIFOs (FVF) are present at each output

port whose job is to track VCs in the neighboring input port. FVFs are exploited to dispense VCs when a header flit wins SA.

Other overheads associated with the proposed router are larger arbiters in the first stage of SA (SA1) arbiters. Due to the variable number of VCs in each port during execution time, it is necessary to take the worst case into account. Hence, we need to implement larger SA1 arbiters to support the case when a port have the maximum possible number of VCs. Although, larger SA1 arbiters means lower clock rate, which leads to performance degradation because arbitration logic has the longest critical path among all router modules [10], [14], [15]. Given that the maximum possible number of VCs in a port in our proposal may be up to multiples of that in a conventional router, clock rate may experience a dramatic decrease. To compensate the adverse effect of these large arbiters on performance, designs like [1] can be employed. This design reduces the critical path of large arbiters to match that of a small one. It cannot further improve the small arbiters, though. Hence, using it to enhance the clock rate of our design in order for it to match that of a conventional router is justified.

IV. AREA AND POWER ANALYSIS

We implemented the proposed and conventional routers in Verilog HDL and synthesized them using a 45nm library in CMOS technology. Power estimation was done at 1.0V and 110°C with 0.1 activity at each input of the modules [10]. We selected 2D mesh topology for networks. Thus, routers have five ports (North, East, South, West and Local). Each router has a 3-stage pipeline structure composed of First Switch Traversal (ST1)/Buffer Write (BW)/Route Computation (RC), SA, and Second Switch Traversal (ST2) stages. Both routers work at 2.8GHz clock rate. We set half of the buffering resources of a port as the lower bound of buffering that it may have. Flits and links between routers are considered to be 128-bit wide. The conventional router is equipped with four 4-flit VCs per port, and the proposed router have the same amount of buffering.

Table I shows the breakdown of area and power for the proposed and conventional routers. The synthesis results show that the proposed router has overheads of 38.3% and 19.4%

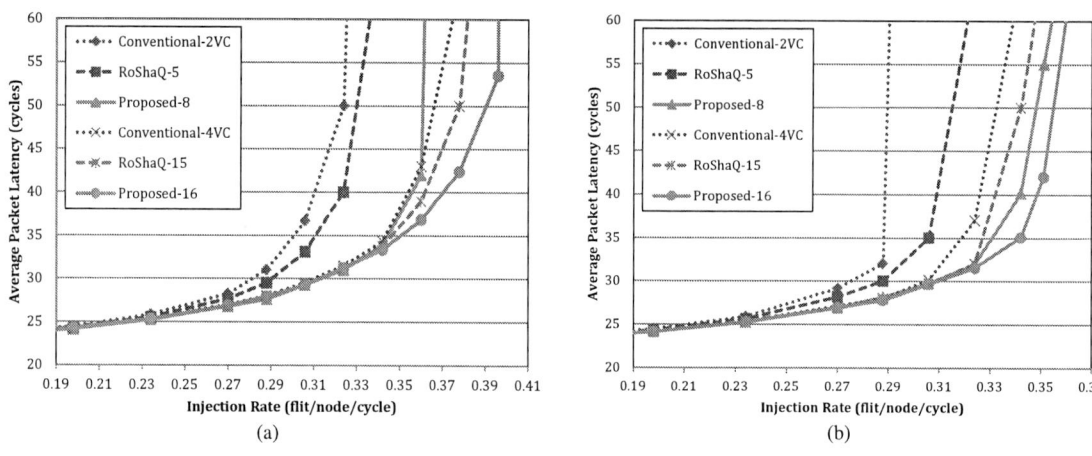

Fig. 2: Average packet latency under (a) uniform and (b) hotspot traffic patterns.

in area and power consumption, respectively, compared to the conventional router. Assuming that NoC consumes 10%-20% of the total power of the chip [16], the power overhead of our design varies from 1.9%-3.9% of the entire chip. These overheads are justifiable given the considerable improvements in performance that are shown in Section V. In fact, the significant performance improvements of our proposal result in decreasing the total energy consumption of the NoC for a given application (discussed in Section V). In addition, our proposal yields comparable performance to the conventional router, with half of its buffering resources. This translates into saving of power consumption, because on one hand, buffers consume a significant fraction of the static power of the router [4], and on the other hand, reducing the amount of buffers results in simpler control logic. In fact, our proposed router with 8-flit buffers at each input port has an area overhead of 11.6%, but it consumes 7.2% less power, with respect to a conventional router with four 4-flit VCs at each input port.

TABLE I: Area and power breakdown of the conventional and proposed routers with 16 flits of buffering in each port (4 VCs in the conventional router).

Module	Conventional		Proposed	
	Area $(10^3 \mu m^2)$	Power (mw)	Area $(10^3 \mu m^2)$	Power (mw)
Buffers	264	190	264	190
Crossbars	305	38	500	64
Allocator	2.4	4.1	7.9	15.9
Ctl. Logic	21.7	18	48.2	28.7
Total	593.1	250.1	820.1	298.6

V. EVALUATION

A. Simulation Environment

System level simulations were done using BookSim 2.0 [8] in order to extract latency results for the proposed and conventional routers. We also compared our proposal with RoShaQ [18], as the state-of-the-art design with a centralized buffer structure. The conventional router was simulated with two and four 4-flit VCs, and RoShaQ and the proposed router are equipped with the same amount of buffering. Results

were obtained for both synthetic traffic patterns and real applications. *Uniform* and *hotspot* (hotspot rate = 0.4 and 8 hotspot cores) traffic patterns were used for comparison of the performance under synthetic traffic patterns. Simulations were conducted for 500,000 cycles and the results for the first 20% of the execution time were discarded as warm-up period. The traffic load is composed of 80 percent 1-flit and 20 percent 5-flit packets [12]. Table II summarizes the simulation parameters.

TABLE II: Simulation parameters.

Parameter	Value
Network size and topology	8×8 mesh
Pipeline stages	3 (ST1/BW/RC, SA, and ST2)
VCs per port	2/4 4-flit VCs (equivalent buffering in RoShaQ and the proposed design)
Packet length	80% 1-flit and 20% 5-flit
Routing algorithm	XY

As for real applications, we used Netrace [6] traces for PARSEC applications [3]. These traces were collected from a 64-node cache-coherent CMP with in-order ALPHA cores which work at 2GHz. L1 instruction and data caches were 64KB each, 4-way associative with 64B lines and MESI coherence protocol. L2 cache was an 8-way associative 64-bank fully shared S-NUCA cache with a size of 32MB. Finally, main memory had 8 on-chip memory controllers. We conducted simulations for the whole Region Of Interest (ROI) of the applications (i.e., the parallel section of the applications).

In figures of this section, Conventional-2VC and -4VC represent the conventional router with two and four 4-flit VCs, respectively. RoShaQ-5 and RoShaQ-15 mean RoShaQ design with five and fifteen central 4-flit queues. Note that in RoShaQ design, total amount of buffering is $5 \times$ 4-flit queue (one per port) plus central buffer queues. Finally, Proposed-8 and Proposed-16 are respective representatives for the proposed router with eight and sixteen flits of buffering at each input port.

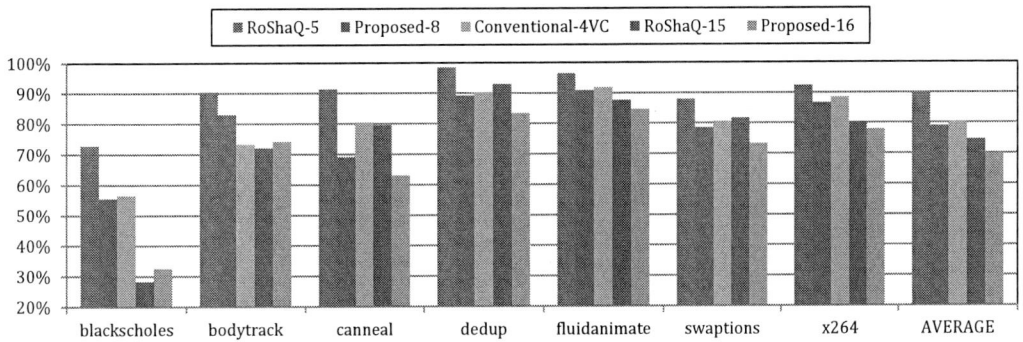

Fig. 3: Normalized average packet latency for PARSEC applications.

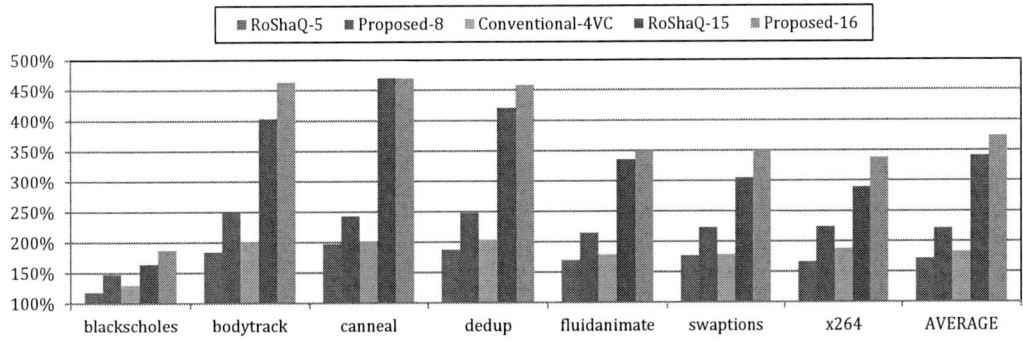

Fig. 4: Normalized average buffer occupancy for PARSEC applications.

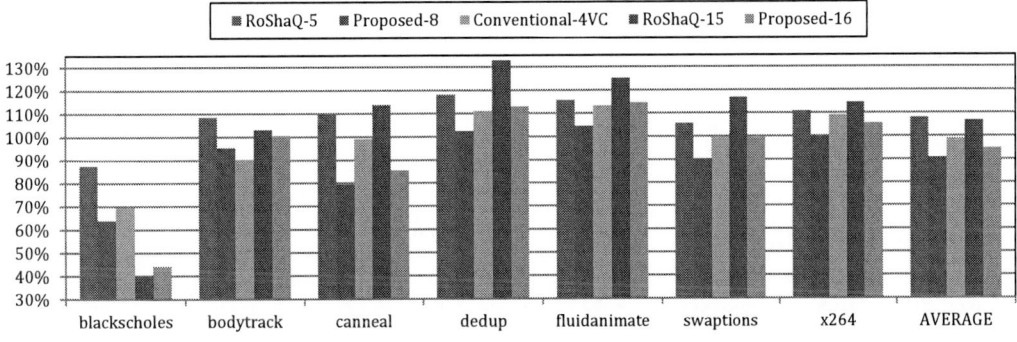

Fig. 5: Normalized power consumption for PARSEC applications.

B. Results for Synthetic Traffics

The average packet latency using the conventional router, RoShaQ, and the proposed router under uniform and hotspot traffic patterns are demonstrated in Figures 2-(a) and 2-(b), respectively. For designs with 8 flits of buffering at each input port, the results show improvements of 17.6% and 11.1% in terms of postponing the saturation rate, which is directly related to NoC throughput, with respect to the conventional router and RoShaQ, respectively, under uniform traffic pattern. However, with increasing the amount of buffers of each input to 16 flits, the improvements decrease to 5% and 2.7%.

In the case of hotspot traffic, our proposal achieves higher improvements due to the higher degree of diversity in traffic pattern and consequently, buffering requirements of input

ports. It manages to enhance the saturation rate by 18.8% and 11.8% in comparison with the conventional router and RoShaQ, respectively, for the designs with 8-flit buffers at their inputs. The decrease in improvement with extending buffering resources to 16 flits at each input port can be observed under hotspot traffic too. These improvements are 16.0% and 10.1% with respect to the conventional router and RoShaQ, respectively.

As it can be deducted from Figures 2-(a) and 2-(b), our proposed router manages to achieve the same level of performance as the conventional router with only half of its buffering resources. In fact, in case of hotspot traffic pattern, it even surpasses the conventional router with double buffering at its disposal.

978-1-4673-9141-2/15 $31.00 © 2015 IEEE   205

C. Results for PARSEC Applications

Figure 3 depicts the normalized average packet latency for PARSEC applications for the proposed router with 8 and 16 flits of buffering at each input, RoShaQ with 5 and 15 central queues, and conventional router with four 4-flit VCs with respect to the conventional router with two 4-flit VCs. As the figure demonstrates, the proposed router with 8-flit buffers decreases average packet latency by 21% and 12.2% on average, compared to the 2-VC conventional and RoShaQ-5 designs, respectively. Our proposal with 16 flits of buffering in its input ports, on the other hand, outperforms the conventional and RoShaQ designs with the same amount of buffering, by respective averages of 12.9% and 6.4%. It must be noticed that the proposed router with 8-flit buffers managed to obtain 1.5% better average packet latency in comparison with the Conventional-4VC design. Nevertheless, despite the overall improvements achieved by our proposal, in the cases of *blackscholes* and *bodytrack*, the one-cycle reconfiguration overhead imposed by our scheme results in slightly worse average packet latencies regarding RoShaQ-15 (and also Conventional-4VC for *bodytrack*).

Average buffer occupancy results normalized to the Conventional-2VC for the benchmarks are demonstrated in Figure 4. We used the summation of occupied buffer slots in all cycles of the execution of the application as the buffer occupancy metric. As the figure illustrates, our proposal significantly improves buffer occupancy in comparison with the conventional router. It enhances buffer occupancy by 2.21× and 29.2% with respect to Conventional-2VC router and RoShaQ-5, respectively. These results support the fact that using our proposal, half of the buffering resources of a conventional router suffices to achieve the same performance, in that the buffer occupancy of Proposed-8 is greater than that of Conventional-4VC.

Figure 5 illustrates the normalized power consumption values for PARSEC applications with respect to Conventional-2VC design. We used [9] in combination with BookSim 2.0 in order to extract power consumption of the NoC while running benchmark applications. The Proposed-8 design outperforms Conventional-2VC and RoShaQ-5 by 9.2% and 15.9%, respectively. These improvements are 5.3% and 11.2% for Proposed-16 with regard to Conventional-4VC and RoShaQ-15, respectively.

Proposed-8 consumes less power for four applications compared to Conventional-2VC. Nevertheless, for *dedup* and *fluidanimate*, it imposes negligible overheads of 2.5% and 4.5%, respectively. The results are pretty close in most cases for Proposed-16 and Conventional-4VC. Even in case of *bodytrack*, an 11.5% overhead is observed. On the other hand, both Proposed-8 and -16 outperform RoShaQ-5 and -15, except in *blackscholes* where RoShaQ-15 consumes 9.2% less power. These observations justify overheads of our design, given that for most of the applications it has a power consumption less than or equal to the conventional and state-of-the-art designs, with a better performance.

VI. Conclusion

In this paper, we proposed traffic-aware buffer reconfiguration in on-chip networks. A simple monitoring mechanism and reconfiguration method was suggested to enable the router to adapt its buffer structure to the changes of the traffic of the network. Our proposed router outperformed the conventional router and RoShaQ under both synthetic traffic patterns and PARSEC applications. Under synthetic traffic patterns, the proposed router postponed saturation rate by 18.8%, while it managed to improve average packet latency of PARSEC applications by an average of 21% with respect to the conventional router. Additionally, power consumption under PARSEC benchmarks was reduced by 9.2%, on average. These improvements came at the cost of 38.3% area compared to the conventional router. Additionally, our proposed router managed to yield the performance of the conventional router with double buffering resources. Buffer occupancy results supported this fact.

References

[1] R. Bashizade and H. Sarbazi-Azad, "P2R2: Parallel Pseudo-Round-Robin arbiter for high performance NoCs," *Integration, the VLSI Journal*, vol. 50, pp. 173 – 182, 2015.

[2] L. Benini and G. De Micheli, "Networks on chips: a new SoC paradigm," *Computer*, vol. 35, no. 1, pp. 70–78, Jan 2002.

[3] C. Bienia, S. Kumar, J. P. Singh, and K. Li, "The PARSEC Benchmark Suite: Characterization and Architectural Implications," in *PACT*. New York, NY, USA: ACM, 2008, pp. 72–81.

[4] L. Chen and T. Pinkston, "Worm-Bubble Flow Control," in *HPCA*, Feb 2013, pp. 366–377.

[5] S. Hassan and S. Yalamanchili, "Centralized buffer router: A low latency, low power router for high radix NoCs," in *NoCS*, April 2013, pp. 1–8.

[6] J. Hestness, B. Grot, and S. W. Keckler, "Netrace: Dependency-driven Trace-based Network-on-chip Simulation," in *NoCArc*. New York, NY, USA: ACM, 2010, pp. 31–36.

[7] J. Hu, U. Ogras, and R. Marculescu, "System-Level Buffer Allocation for Application-Specific Networks-on-Chip Router Design," *TCAD*, vol. 25, no. 12, pp. 2919–2933, Dec 2006.

[8] N. Jiang *et al.*, "A detailed and flexible cycle-accurate Network-on-Chip simulator," in *ISPASS*, April 2013, pp. 86–96.

[9] A. B. Kahng, B. Lin, and S. Nath, "Explicit Modeling of Control and Data for Improved NoC Router Estimation," in *DAC*, 2012, pp. 392–397.

[10] A. Kumar *et al.*, "A 4.6Tbits/s 3.6GHz single-cycle NoC router with a novel switch allocator in 65nm CMOS," in *ICCD*, Oct 2007, pp. 63–70.

[11] M. Lai *et al.*, "A Dynamically-allocated Virtual Channel Architecture with Congestion Awareness for On-chip Routers," in *DAC*. New York, NY, USA: ACM, 2008, pp. 630–633.

[12] S. Ma *et al.*, "Novel Flow Control for Fully Adaptive Routing in Cache-Coherent NoCs," *TPDS*, vol. 25, no. 9, pp. 2397–2407, Sept 2014.

[13] D. Matos *et al.*, "Reconfigurable Routers for Low Power and High Performance," *TVLSI*, vol. 19, no. 11, pp. 2045–2057, Nov 2011.

[14] C. Nicopoulos *et al.*, "ViChaR: A Dynamic Virtual Channel Regulator for Network-on-Chip Routers," in *MICRO*, Dec 2006, pp. 333–346.

[15] ——, "On the Effects of Process Variation in Network-on-Chip Architectures," *TDSC*, vol. 7, no. 3, pp. 240–254, Jul. 2010.

[16] R. Ramanujam, V. Soteriou, B. Lin, and L.-S. Peh, "Design of a High-Throughput Distributed Shared-Buffer NoC Router," in *NOCS*, May 2010, pp. 69–78.

[17] M. Rezazad and H. Sarbazi-Azad, "The effect of virtual channel organization on the performance of interconnection networks," in *IPDPS*, April 2005, p. 264a.

[18] A. Tran and B. Baas, "Achieving High-Performance On-Chip Networks With Shared-Buffer Routers," *TVLSI*, vol. 22, no. 6, pp. 1391–1403, June 2014.

978-1-4673-9141-2/15 $31.00 © 2015 IEEE

Design Optimization of Polyphase Digital Down Converters for Extremely High Frequency Wireless Communications

Gain Kim, *Student Member, IEEE*, Raffaele Capoccia and Yusuf Leblebici, *Fellow, IEEE*
Swiss Federal Institute of Technology
Lausanne, Switzerland
Email: gain.kim@epfl.ch, raffaele.capoccia@epfl.ch, yusuf.leblebici@epfl.ch

Abstract—In this paper, an area-optimized polyphase digital down converter (DDC) architecture is introduced, where the mixers can be completely merged into the polyphase decimation filter under certain conditions. We also introduce an interface architecture, called synchronizer, between the back-end of an extremely high-speed time interleaved ADC (TI-ADC) and the front-end of a polyphase DDC. The synchronizer enables safe downsampling for a polyphase DDC, when the TI-ADC's sampling rate is above tens of GS/s. We show that the proposed interface architecture prevents any potential timing constraint violations that might occur in the interface between a TI-ADC and a polyphase DDC for extremely high frequency (EHF) wireless communication applications.

I. INTRODUCTION

The increasing demand on high data transmission rates in telecommunication network is an important factor of rapid development of wireless communication technology. Various high frequency wireless local area network (WLAN) channels such as IEEE 802.11a/b/g/n/ac using carrier frequency range up to 5 GHz are standardized and widely used. Nowadays, extremely high frequency (EHF) band called millimetre wave (mmWave), such as IEEE 802.11ad standard, is an emerging frequency band for the future wireless communication, thanks to its wide bandwidth [1], [2].

The major building blocks of conventional radio frequency (RF) receivers (RX) include: a low noise amplifier (LNA), downconverting mixers, intermediate frequency (IF) filters/amplifiers, and analog-to-digital converters (ADCs). However, the analog components composing the RF receiver for GHz domain carriers should be very carefully designed to avoid some known problems such as DC offsets, AC coupling, I/Q mismatch, carrier leakage, mixer linearity, etc [3]. Moving towards mmWave frequency band, issues related to the high-frequency that analog circuits encounter, become difficult to be handled in standard CMOS or SOI process technologies [4].

One solution for the above-mentioned issues is to fabricate the EHF analog RX circuits in GaAs process, which is generally used for military and satellite communication applications. The drawbacks of this solution are more fabrication costs compared to the standard CMOS or SOI processes, the incompatibility with modern digital circuits, and lower yield. Another solution might be to fabricate the analog parts with SiGe process on a different substrate, that is separated from the chip where high density CMOS or SOI digital circuits are integrated. However, this also has a limitation because it requires two separate chips, making it necessary to have interconnections between the SiGe chip and the CMOS or SOI chip. A preferable solution is to build a digital direct conversion RX immediately following the RF front-end composed of the RF antenna and the LNA in standard CMOS or SOI process technology, which solves analog circuit issues related to EHF. Furthermore, the performance of a digital direct conversion RX can be enhanced without redesigning, directly following fabrication process node shrinkage.

There are three main challenges that should be addressed in the digital direct conversion RX approach. First, LNA should operate at tens of GHz frequency in CMOS or SOI. Second, a very fast ADC is required to properly sample RF signals coming from LNA. Third, the digital down converter (DDC) front-end must be able to handle such a high data rate. The first and second issues are no longer problematic as the advancement in CMOS, SOI process technologies and circuit design techniques has made very high-speed LNA and ADC feasible in nowadays technology [5], [6]. The third difficulty can be handled by implementing parallelised DDC architecture (polyphase DDC) with polyphase cascaded integrator-comb (CIC) decimation filters [7]–[10]. The research in [9] demonstrates that an 8.88 GS/s TI-ADC followed by a DDC can be implemented on a single standard CMOS chip. However, to directly sample mmWave and downconvert it with DDC, the interface between TI-ADC back-end and polyphase DDC front-end must be carefully considered.

In this paper, the design of an optimized polyphase DDC and the related timing constraints are presented. The interface between the TI-ADC and the polyphase DDC for EHF wireless communication applications is carefully analyzed. We show that the conventional TI-ADC and the polyphase DDC interface may suffer from the registers' timing constraint violations in EHF regime, and we propose an efficient solution to solve this issue. We also study an optimized polyphase DDC architecture of which mixers can be completely merged into the polyphase decimation filter, subject to certain restrictions, especially for $F_s = 4 \times F_c$, where F_s is the sampling frequency of the TI-ADC and F_c is the center frequency of the carrier wave.

The remaining parts of this paper are organised as follows:

978-1-4673-9141-2/15 $31.00 © 2015 IEEE

Section II reviews existing polyphase DDC architecture and highlights the parallel structure of the polyphase DDC. In Section III, an optimized polyphase DDC architecture, called mixer-less DDC, is proposed. Section IV then highlights a potential design issue for an integrated polyphase DDC with an extremely high-speed TI-ADC on a single chip, and proposes an efficient solution. Section V analyzes and validates the optimized polyphase DDC architecture and the proposed interface architecture, and the paper is concluded in Section VI.

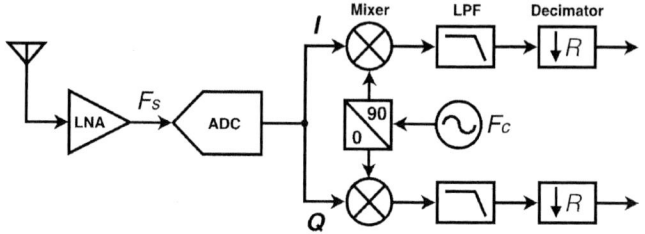

Fig. 1: Schematic diagram of a digital direct conversion RX

II. POLYPHASE DIGITAL DOWN CONVERTER

In this section, we will briefly review the general concept of direct digital downconversion, and highlight the parallel structure of the polyphase DDC.

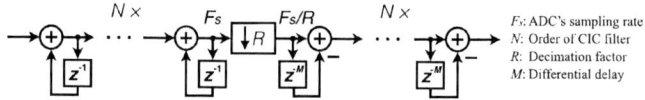

Fig. 2: Single-thread serial CIC decimation filter architecture

A. Direct Digital Downconversion

A simplified digital direct conversion RX architecture is illustrated in Fig. 1 [11]. The LNA amplifies RF signals received from RF antenna. Then the ADC samples the amplified RF signals. And a digital mixer using numerically controlled oscillator (NCO) mixes the digitized signals to move their frequency spectrum to baseband or near-baseband. The baseband signals are then low-pass filtered so that only the band-of-interest remains, and then decimated. The downsampled signals can be used for digital post-processing without loss of information.

For very high frequency wireless communication with a digital direct conversion RX, a very high-speed ADC is required and DDC must be able to successfully downconvert the massive data stream. The ADC should meet Nyquist criterion, i.e., the sampling frequency should be higher than at least twice the maximum input frequency, for safe sampling. For example, at least 12 GS/s ADC should be used for direct sampling of an RF signal which has maximum frequency components close to 6 GHz. Digital downconversion of such a high data rate using a single-thread architecture shown in Fig. 2 [12] is impossible, due to the technological limitation. Therefore, using a parallelised DDC structure is unavoidable for such a high-frequency application.

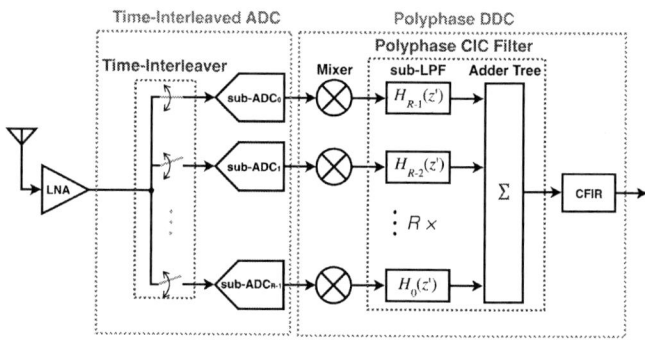

Fig. 3: Polyphase DDC architecture

where N is the order of the polyphase CIC filter and M is the differential delay. For the remaining parts of this paper, M is considered to be 1, which is the case most of the time. The number of sub-ADCs in the TI-ADC is considered to be equal to the decimation factor R, as well as the number of sub-transfer functions.

III. OPTIMIZED POLYPHASE DIGITAL DOWN CONVERTERS

In this section, an optimised polyphase DDC architecture is introduced. By choosing F_s four times higher than F_c, mixers are completely merged into the polyphase CIC decimation filter, resulting in global resource saving. The first subsection reviews existing polyphase DDC optimization techniques achieved with quadruple sampling. The mixer-less DDC architecture is shown in the second subsection.

A. Mixer Stage Simplification

The mixer stage usually consists of a NCO based on the look-up table (LUT), where sine or cosine values are stored, as illustrated in Fig. 4(a). Basically, an LUT-based NCO is controlled by a phase accumulator output word, which selects addresses of LUT to provide an amplitude output corresponding to the phase. The complexity of the LUT-based NCO architecture can be significantly reduced by making F_s four times higher than F_c. For $F_s = 4 \times F_c$, the design shown in Fig. 4(a) can be replaced by that in Fig. 4(b), thanks to the simplified LUT of the quadrature mixer as shown in Table I.

B. Polyphase DDC Architecture

The basic polyphase DDC based on a polyphase CIC decimation filter followed by a compensation FIR filter stage, is illustrated in Fig. 3. A CIC decimation filter followed by a compensation FIR (CFIR) filter is preferably used rather than only-FIR filter for DDC implementation when the decimation factor (R) is high, e.g., $R \geq 16$, thanks to its hardware-friendly structure [13], [14]. The transfer function of the polyphase CIC decimation filter, expressed in the sum of multiple sub-transfer functions, is equivalent to CIC recursive (1), and non-recursive (2) filter's transfer functions which are expressed as:

$$H(z) = \frac{(1 - z^{-RM})^N}{(1 - z^{-1})^N} \tag{1}$$

$$H(z) = \left(\sum_{k=0}^{RM-1} z^{-k} \right)^N \tag{2}$$

978-1-4673-9141-2/15 $31.00 © 2015 IEEE   208

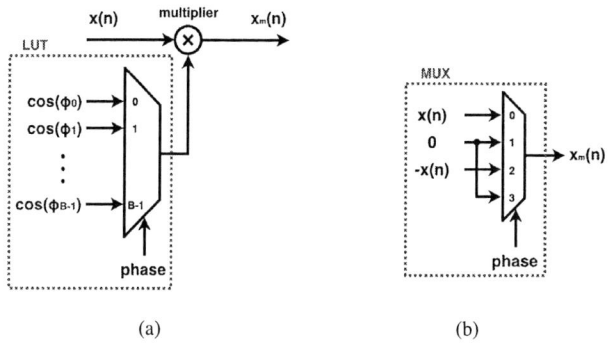

(a)   (b)

Fig. 4: (a) General LUT-based NCO. (b) Simplified LUT-based mixer architecture with quadruple sampling.

TABLE I: Output of a NCO with respect to phase.

ϕ (rad)	0	$\pi/2$	π	$3\pi/2$
$\cos(\phi)$	1	0	-1	0
$\sin(\phi)$	0	1	0	-1

This effectively removes the need of a multiplier for the mixer stage, as the outputs of the I/Q mixer are x, $-x$ or 0 depending on the phase, while x being the input of the I/Q mixer.

This architecture can be even more simplified when it is used for a polyphase DDC of which number of parallel lowpass sub-filters is positive integer multiples of 4, as implemented in [9]. We recall the non-recursive form of the CIC decimation filter's transfer function (2), where M is fixed to 1, and we develop that equation as:

$$H(z) = (1 + z^{-1} + z^{-2} + \cdots + z^{-(R-1)})^N. \quad (3)$$

Equation (3) can be rewritten as:

$$H(z) = c_0 + c_1 z^{-1} + c_2 z^{-2} + \cdots + c_{N(R-1)} z^{-N(R-1)} \quad (4)$$

where $c_0, c_1, \cdots, c_{N(R-1)}$ are the coefficients of the terms of the polynomial representing the transfer function of the CIC decimation filter.

Consider a numerical example of (4), and replace R by 4 and N by 3, as given in (5a). We can rearrange (5a) to form an equation expressing a transfer function composed of a sum of sub-transfer functions, as expressed in (5b).

$$H(z) = 1 + 3z^{-1} + 6z^{-2} + 10z^{-3} + 12z^{-4} \\ + 12z^{-5} + 10z^{-6} + 6z^{-7} + 3z^{-8} + z^{-9} \quad (5a)$$

$$= H_0(z') + z^{-1}H_1(z') + z^{-2}H_2(z') \\ + z^{-3}H_3(z') \quad (5b)$$

where

$$z' = z^4$$
$$H_0(z') = 1 + 12z'^{-1} + 3z'^{-2}$$
$$H_1(z') = 3 + 12z'^{-1} + z'^{-2}$$
$$H_2(z') = 6 + 10z'^{-1}$$
$$H_3(z') = 10 + 6z'^{-1}.$$

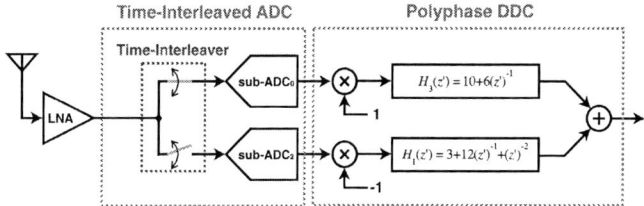

Fig. 5: I-path of the polyphase DDC architecture for $R = 4$ and $N = 3$ with $F_s = 4 \times F_c$. Same reduction is applicable to Q-path.

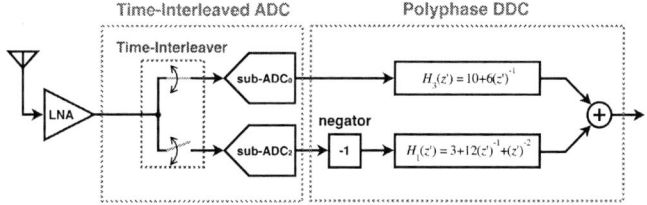

Fig. 6: Optimized architecture of Fig. 5

With quadruple sampling, for each of the in-phase (I-path) and the quadrature (Q-path), two sub-filters corresponding to the NCO output 0 can be deleted and hard-wired to the *gnd*. Therefore, the transfer function of I-path and Q-path can be represented as:

$$H_I(z) = -z^{-1}H_1(z') + z^{-3}H_3(z') \quad (6)$$

and

$$H_Q(z) = -H_0(z') + z^{-2}H_2(z') \quad (7)$$

, respectively. This halves the required number of sub-filters in each path as illustrated in Fig. 5, resulting in reductions in terms of power consumption and silicon area [9]. A major advantage of the polyphase DDC architecture is that, a multiplexer is no longer needed for the mixer stage. The paths corresponding to the NCO output 1 can be directly hard-wired to input x, and the paths corresponding to the NCO output -1 can introduce a binary negator, as shown in Fig. 6.

B. Mixer-Less DDC Design

So far, we have been considering a polyphase DDC with a mixer stage positioned before decimation filter stage. The positions of the mixers and the partial CIC filters can be swapped, as illustrated in Fig. 7(a), without any change in functionality. When the number of sub-ADCs is integer multiple of 4 and $F_s = 4 \times F_c$, the mixers are hard-wired negators and wires. Considering that each operator in Fig. 7(a) corresponds to the physical operator, such as adder and negator, it becomes more economical in terms of resource usage if the negator and the adder are merged into one single subtractor. The mixers can be completely merged into the decimation filter stage, as shown in Fig. 7(b), under three conditions:

- $F_s = 4 \times F_c$, i.e., the sampling rate of the TI-ADC must be four times higher than the center frequency of the carrier wave.

- $R = 4 \times k$ where $k \in \mathbb{Z}_+^*$, i.e., the decimation factor must be a non-zero positive integer multiple of 4.

- #(sub-ADCs) $= 4 \times l$ where $l \in \mathbb{Z}_+^*$, i.e., the number of sub-ADCs must be a non-zero positive integer multiple of 4.

978-1-4673-9141-2/15 $31.00 © 2015 IEEE

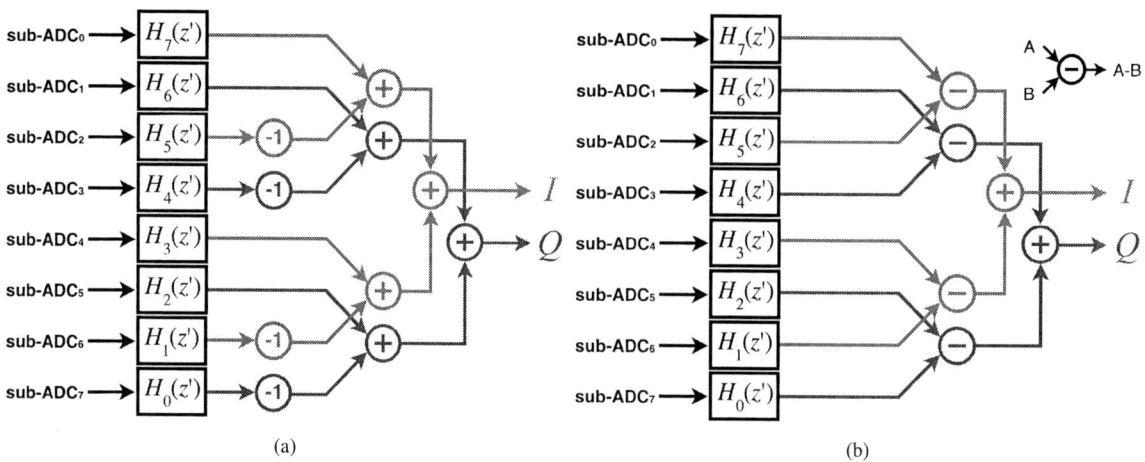

Fig. 7: Integrated TI-ADC and DDC architecture satisfying above three conditions with $R = $ #(sub-ADCs) $= 8$. (a) Switched. (b) Merged.

Applying these conditions, the resource usage can be reduced by removing negators that compose mixers, while a simple substitution of the adders in the first stage of the adder tree by subtractors is needed. The generalized transfer functions of the DDC's I-path satisfying these conditions, corresponding to architectures shown in Fig. 3, Fig. 7(a) and Fig. 7(b) are expressed in (8a), (8b) and (8c), respectively, while the same is applicable to Q-path.

$$
\begin{aligned}
Y_I(z) = \ & z^{-0} H_0(z') \cdot X(z) \cdot (0) + z^{-1} H_1(z') \cdot X(z) \cdot (-1) \\
& + z^{-2} H_2(z') \cdot X(z) \cdot (0) + z^{-3} H_3(z') \cdot X(z) \cdot (1) \\
& + z^{-4} H_4(z') \cdot X(z) \cdot (0) + z^{-5} H_5(z') \cdot X(z) \cdot (-1) \\
& + z^{-6} H_6(z') \cdot X(z) \cdot (0) + z^{-7} H_7(z') \cdot X(z) \cdot (1) \\
& + \cdots \\
& + z^{-R+2} H_{R-2}(z') \cdot X(z) \cdot (0) \\
& + z^{-R+1} H_{R-1}(z') \cdot X(z) \cdot (1)
\end{aligned}
\tag{8a}
$$

$$
\begin{aligned}
= & - z^{-1} H_1(z') \cdot X(z) + z^{-3} H_3(z') \cdot X(z) \\
& - z^{-5} H_5(z') \cdot X(z) + z^{-7} H_7(z') \cdot X(z) \\
& + \cdots \\
& - z^{-R+3} H_{R-3}(z') \cdot X(z) \\
& + z^{-R+1} H_{R-1}(z') \cdot X(z)
\end{aligned}
\tag{8b}
$$

$$
\begin{aligned}
= \ & (z^{-3} H_3(z') \cdot X(z) - z^{-1} H_1(z') \cdot X(z)) \\
& + (z^{-7} H_7(z') \cdot X(z) - z^{-5} H_5(z') \cdot X(z)) \\
& + \cdots \\
& + (z^{-R+1} H_{R-1}(z') \cdot X(z) \\
& - z^{-R+3} H_{R-3}(z') \cdot X(z))
\end{aligned}
\tag{8c}
$$

The fact that the mixer-less DDC requires no negator before the sub-LPF stage reduces the resource usage as well as a pipelining stage. However, the actual area saving is not as much as the reduced number of the negators and registers, because an M-bit subtractor requires M more inverters as compared to an adder. Note that we are considering a ripple carry adder/subtractor architecture for simplification of the analysis, while it can be built in faster but larger architectures

such as carry select adder/subtractor. The bit-width of subtractors may grow significantly depending on the decimation factor R and the order N of the DDC. This may introduce some undesired area overhead when the bit-width gain of subtractors caused by the high R and N becomes very large. This trade-off is analyzed in Section V.

IV. TI-ADC AND POLYPHASE DDC INTERFACE FOR EHF DOWNSAMPLING

This section provides a detailed analysis of the interface between a TI-ADC and a polyphase DDC. In the first subsection, we analyse a possible timing constraint violation on the interface when the TI-ADC's sampling rate exceeds several tens of GS/s. In the second subsection, a solution to solve this issue is proposed.

A. Design Consideration of TI-ADC–polyphase DDC Interface

When the sampling rate of ADC is limited by a few GHz, the timing constraints of DDC are not critical with modern CMOS/SOI process technologies. However, when ADC's sampling rate rises to 10 GS/s or even higher, timing constraints on the ADC-DDC interface become critical when polyphase decimation filter is used in the DDC. Consider a TI-ADC backend illustrated in Fig. 8(a), where registers are triggered at the rising edge of their clock [Fig. 8(b)]. The sub-ADCs and the *Sampling Registers* (SR)s are triggered by clocks shifted by T_{clk}/R one from each other. *Alignment Registers* (AR)s then collect all the sampled inputs coming from all sub-ADCs to send them to the mixer/decimation filter stage. The Δt of the sampling moments of the SR_i and AR_i can be expressed as:

$$
\Delta t_{i,SR-AR} = T_{clk} \times \frac{R-i}{R}
\tag{9}
$$

where $i \in \{0, 1, 2, \cdots, R-1\}$. The problem may occur when the SR_{R-1} samples the output of the sub-ADC_{R-1} and T_{clk}/R later, AR_{R-1} samples the output of the SR_{R-1}. In other words, once SR_{R-1} is triggered, its output must arrive and be stabilized before AR_{R-1} is triggered, for which only T_{clk}/R exists, i.e., $T_{clk}/R > T_{clk-q} + T_{su}$.

As a numerical example, consider a polyphase DDC with $F_s = 8$ GS/s and $R = 8$. Then the $T_{clk} = 1$ ns and $T_{clk}/R = $

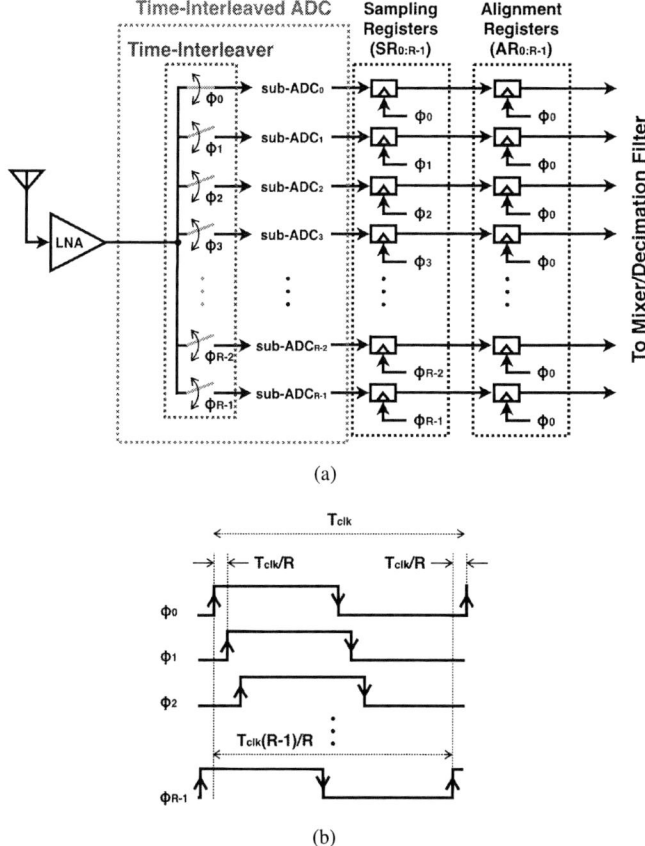

(a)

(b)

Fig. 8: (a) Close-up view of the TI-ADC's output registers and (b) their clocks.

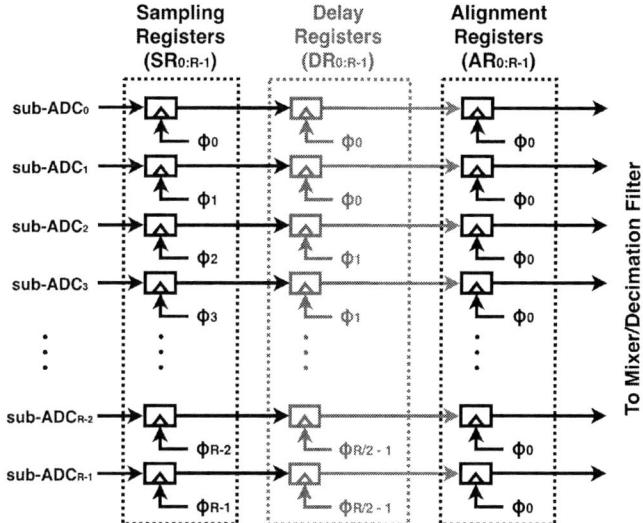

Fig. 9: The Latency Layer inserted between SR layer and AR layer

SR_i and DR_i can be expressed as:

$$\Delta t_{i,SR-DR} = \begin{cases} T_{clk} \times \frac{R-\frac{i}{2}}{R} & \text{if } i = 2k \\ T_{clk} \times \frac{R-\frac{i+1}{2}}{R} & \text{if } i = 2k+1 \end{cases} \quad (10)$$

and the Δt between DR_i and AR_i can be expressed as:

$$\Delta t_{i,DR-AR} = \begin{cases} T_{clk} \times \frac{R-\frac{i}{2}}{R} & \text{if } i = 2k \\ T_{clk} \times \frac{R-\frac{i-1}{2}}{R} & \text{if } i = 2k+1 \end{cases} \quad (11)$$

where $i \in \{0, 1, 2, \cdots, R-1\}$ and $k \in \mathbb{Z}_+$. We can understand that the introduced Latency Layer relaxes the worst-case timing constraint from T_{clk}/R to $T_{clk}/2$. Therefore, the synchronizer circuitry composed of SR, DR and AR layers can safely align and sample all the inputs that ADC converted, regardless of the F_s, as long as the sampling period (T_{clk}) of one sub-ADC is reasonably limited, e.g., $T_{clk} > 400$ ps.

V. ANALYSIS AND VALIDATION

In this section, the behaviour of the proposed synchronizer circuit is validated considering an example with $R = 4$. We then analyze the resource saving achieved by merging the mixer into the decimation filter.

The first digitized input $x(0)$ is sampled by SR_0 at $t = t_0$, with T_{clk-q} of all registers being one time step ($T_{clk}/4$), as shown in Fig. 10. Successive time-interleaved inputs shifted by $T_{clk}/4$ one from each other arrive corresponding 4 sub-ADCs. The input $x(3)$ is sampled by SR_3 at $t = t_3$ and stabilized at $t = t_4$. If there was no latency layer, AR_3 would have tried to sample $x(3)$ at $t = t_4$ to align all the input, but it would fail due to the violation of the setup time constraint. On the other hand, when the latency layer is inserted between the SR and AR layers, the worst-case timing constraint is stretched to $T_{clk}/2$, and as long as T_{su} is smaller than $T_{clk}/4$, the AR layer can safely align the data.

Fig. 11 shows the relative gain with respect to the maximum bit-width gain (BitGain) on the output of the CIC filter as compared to its input bit-width. The BitGain is caused by the multiplications of the inputs with the sub-filters' coefficients,

125 ps, meaning that $T_{clk-q} + T_{su}$ must be smaller than 125 ps in order to avoid any timing violation in the critical path. For this example, we would not expect any timing issue, as registers can be suitably designed for this particular timing specification in advanced CMOS/SOI process technologies. On the other hand, when a DDC is used for a mmWave receiver beyond 60 GS/s TI-ADC and a higher decimation factor, the above-mentioned timing issue becomes critical. Considering a polyphase DDC with $F_s = 64$ GS/s and $R = 64$, the T_{clk} is still 1 ns but T_{clk}/R is now 15.625 ps. However, $T_{clk-q} + T_{su}$ can not be made smaller than 15.625 ps in standard CMOS/SOI process technologies.

B. A Synchronizer Circuitry for Safe Signal Alignment

In this subsection, an efficient solution consisting of a column of registers that solves the introduced timing problem is proposed. The costs that we have to pay are slight area overhead, therefore slightly increased power dissipation, and latency of one clock cycle, which is not expensive considering the fact that the other parts are deeply pipelined. Basically, we add one T_{clk} between the sampling moment of SRs and the sampling moment of ARs, such that the critical path delay becomes longer, by introducing *Delay Registers* (DRs) called *Latency Layer*, as illustrated in Fig. 9. Note that the $R\times$ phase-shifted clocks exist intrinsically in the TI-ADC. Therefore, no additional phase-shifted clocks are needed. The Δt between

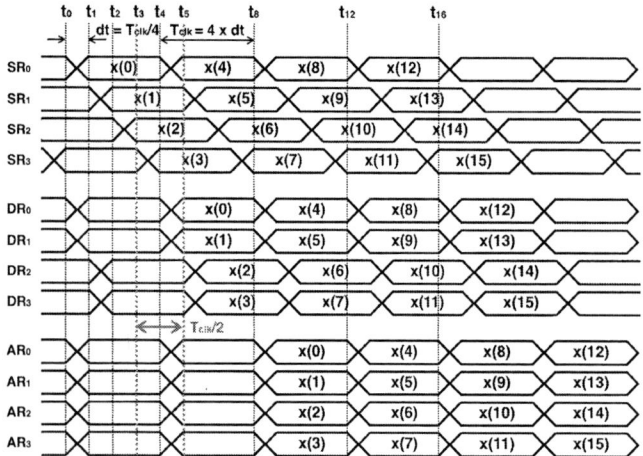

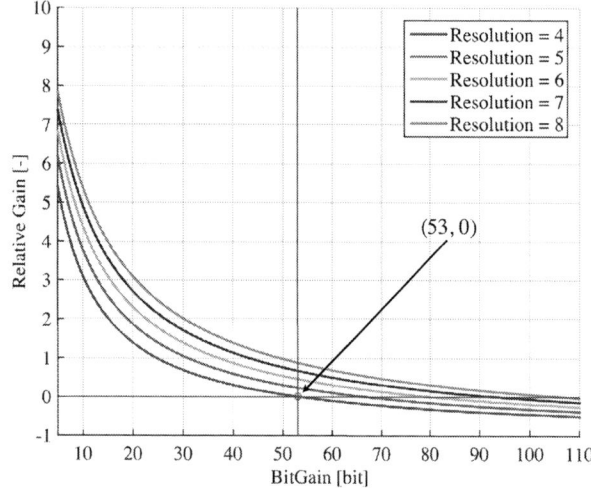

Fig. 10: Timing waveforms of the registers in the synchronizer, with $R = 4$

Fig. 11: Relative gain in terms of area with respect to the BitGain, for various ADC's resolutions

and their addition. The BitGain grows with the resolution of the ADC, the decimation factor and the order of the CIC filter. The relative gain is expressed as:

$$Relative\ Gain = \frac{Saved\ Resources}{Additional\ Resources} - 1 \quad (12)$$

where the *Saved Resources* are the eliminated negators and the *Additional Resources* are the additionally required inverters due to the replacement of some adders by subtractors. The resources are translated into the corresponding number of minimum width transistors. The mixer-less architecture is considered to be disadvantageous when the relative gain falls below 0, which means that the additional hardware overhead due to the inverters in subtractors is higher than the saved resources from mixers. Note that the counting of the number of transistors is performed in such a way that the mixer-less architecture has handicaps, e.g., XOR2 and OR2 gates to only four transistors for each and DFF to sixteen transistors while an INV is translated into three transistors, without considering driving strength.

From Fig. 11, we observe that the mixer-less architecture is advantageous as long as the BitGain does not exceed 53

bits, for ADC's resolution more than or equal to 4. It is worth noticing that the relative gain increases when the ADC's resolution increases, as the removed mixers should have been relatively large, while in the opposite case the mixers could be smaller. Considering that this result comes from a worst case estimation, i.e., giving handicaps to the mixer-less architecture, and considering that the bit-width growth more than 53 does not occur in practical DDC implementation, we conclude that the mixer-less implementation is advantageous.

VI. Conclusion

An interface architecture between an extremely high-speed TI-ADC and a polyphase DDC for mmWave receiver is presented, together with an optimized polyphase DDC of which mixer stage is completely merged into the decimation filter stage. The proposed interface architecture is promising to make sure that no timing violation appears in the ADC-DDC interface, even when the output data stream rate of the TI-ADC exceeds tens of Gb/s, especially for EHF wireless communication applications. And the silicon area of a polyphase DDC can be saved, by introducing a mixer-less polyphase DDC architecture. Worst-case simulation results show that the mixer-less architecture is advantageous in practice as compared to the conventional optimized form of the polyphase DDC.

References

[1] T. Rappaport et al., "Millimillimeter Wave Mobile Communications for 5G Cellular: It Will Work!," *IEEE Access*, vol. 1, pp. 335–349, 2013.

[2] H. Zhao et al., "28 GHz millimeter wave cellular communication measurements for reflection and penetration loss in and around buildings in New York City," in *IEEE Int. Commun. Conf.*, 2013, pp. 5163–5167.

[3] B. Razavi, *RF Microelectronics*, 2nd ed. Upper Saddle River, NJ, USA: Prentice-Hall, 2011.

[4] C. M. Ta et al., "Issues in the Implementation of a 60GHz Transceiver on CMOS," *The 2nd IEEE Int. Workshop on Radio-Frequency Integration Technology*, Singapore, 09-11 Dec., 2007.

[5] L. Kull et al., "A 90GS/s 8b 667mW 64 interleaved SAR ADC in 32nm digital SOI CMOS," *ISSCC Dig. Tech. Papers*, pp. 378–379, Feb. 2014.

[6] B. Razavi, "A 60GHz Direct-Conversion CMOS Receiver," *ISSCC Dig. Tech. Papers*, pp. 400-401, Feb. 2005.

[7] Y. Gao et al., "A Partial-Polyphase VLSI Architecture for Very High Speed CIC Decimation Filters," *Proc. 12th Annual IEEE Int. ASIC/SOC Conf.*, pp. 391–395, 1999.

[8] L. Ascari et al., "Low power implementation of a sigma delta decimation filter for cardiac applications," in *Proc. IEEE Instrum, Measur, Tech. Conf.*, Budapest, Hungary, pp. 750–755, May 2001.

[9] E. Martens et al., "RF-to-baseband digitization in 40 nm CMOS with RF bandpass modulator and polyphase decimation filter," *IEEE J. Solid-State Circuits*, vol. 47, no. 4, pp. 990–1002, Apr. 2012.

[10] X. Liu et al., "A 1.2 Gb/s Recursive Polyphase Cascaded Integrator Comb Prefilter for High Speed Digital Decimation Filters in 0.18-μm CMOS" *Proc. of the 2010 IEEE Int. Symp. on Circuits and Sys. (ISCAS)*, pp. 2115–2118, 2010.

[11] M. T. Hunter et al., "Wideband digital downconverters for synthetic instrumentation," *IEEE Trans. Instrum. Meas.*, vol. 58, no. 2, pp. 263–269, Feb. 2009.

[12] E.B.Hogenauer, "An economical class of digital filters for decimation and interpolation," *IEEE Trans. Acoust., Speech, Signal Processing*, vol. ASSP–29, pp. 155–162, Apr. 1981.

[13] A. Y. Kwentus et al., "Application of filter sharpening to cascaded integrator-comb decimation filters," *IEEE Trans. on Signal Process.*, vol. 45, no. 2, pp. 457–467, Feb. 1997.

[14] B. P. Brandt et al., "A low-power, area-efficient digital filter for decimation and interpolation," *IEEE J. Solid-State Circuits*, vol. 29, no. 6, pp. 679–687, June 1994.

978-1-4673-9141-2/15 $31.00 © 2015 IEEE

Dual-Mode Double Precision / Two-Parallel Single Precision Floating Point Multiplier Architecture

Manish Kumar Jaiswal and Hayden K.-H So
Department of EEE, The University of Hong Kong, Hong Kong
Email: {manishkj, hso}@eee.hku.hk

Abstract—Floating point multiplication is an integral part of any contemporary computing system. This paper presents a configurable dual-mode double precision floating point multiplier architecture, which can also process two-parallel single precision multiplication. This unified, double precision dual (two-parallel) single precision, architecture is named as DPdSP multiplier. The proposed architecture is based on the standard state-of-the-art flow of floating point multiplication, which can process normal and sub-normal operands along with exceptional case handling. The proposed architecture is aimed for a ASIC (*UMC 90nm*) implementation. The key single-mode design units in the computational flow (like mantissa multiplier, dynamic right/left shifters, leading one detector, etc) are re-designed for configurable dual-mode operation to enable efficient resource sharing. The proposed architecture is compared with the best available literature in terms of area, period and $\frac{area \times period}{throughput}$ complexity metric. The proposed dual mode architecture shows a significant improvement in design metrics and also provides more computation support.

Keywords-Floating Point Multiplier, Dual-mode Arithmetic, SIMD, ASIC, Digital Arithmetic.

I. INTRODUCTION

Floating point arithmetic is widely used in a large set of scientific, engineering and numerical processing computations. The hardware implementation of floating point arithmetic requires a large amount silicon area. The contemporary processing systems achieve high performance requirements for floating point computation by using vector-arrays of these arithmetic units. For eg., ARM VFU co-processor (VFU9-S) [1], Nvidia Kepler^{TM} GK110 [2], etc. These processing systems include separate vector-arrays for single precision (SP) and double precision (DP) arithmetic units. However, this paper is aimed towards the design of an unified dual-mode architecture which can be configured on-the-fly for either of double precision or dual (two-parallel SIMD) single precision computation. This approach can be used to save a large silicon area compared to the individual units of a double precision and two single precision arithmetic units.

Prior work presented in [3], [4], [5] includes dual-mode multiplier architectures, which are iterative in nature. An unfolded version of [5] architecture is available in [6], which requires much larger area. The prior literature mainly includes mantissa multiplication and rounding circuitry,

without any support for sub-normal operands, underflow and overflow processing. Further, the dual-mode mantissa multiplication is mainly achieved by array/rectangular multiplier by multiplexing all the partial products, which cost a large silicon area. [7], [8] are some recent literature on the dual-mode DPdSP architectures for adder and division arithmetic.

This paper proposes a dual-mode DPdSP (double precision with dual/two-parallel single precision) multiplier architecture. It follows the state-of-the-art computational flow for implementation, and its sub-components are designed with efficient resource sharing for configurable dual-mode support. The proposed architecture includes computational support for normal as well as sub-normal operands. It also incorporates underflow and overflow processing along with exceptional case handling, and with round-to-nearest rounding method. It is synthesized using *UMC 90nm* standard cell based library, and compared with the best available literature.

II. PROPOSED ARCHITECTURE OF DOUBLE PRECISION / DUAL (TWO-PARALLEL) SINGLE PRECISION (DPdSP) MULTIPLIER

A state-of-the-art computational flow of the single-mode floating point multiplier is shown in the Algorithm 1. It discusses the processing of normal as well as sub-normal operands. In this flow steps 6-7 and steps 14-23 are the specific components regarding the sub-normal processing. To support the dual-mode processing, each individual stage of the flow are re-constructed with efficient resource sharing and tuned data-path to minimize the multiplexing circuitry.

A floating point arithmetic computation involves computing separately the sign, exponent and mantissa part of the operands, and later combine them after normalization and rounding [9]. The standard format for single precision and double precision floating point numbers are as follows:

$$SP : \overbrace{1-bit}^{Sign} \overbrace{8-bit}^{Exponent} \overbrace{23-bit}^{Mantissa} \qquad DP : \overbrace{1-bit}^{Sign} \overbrace{11-bit}^{Exponent} \overbrace{52-bit}^{Mantissa}$$

The input/output register for the proposed dual-mode architecture is assumed as shown in Fig. 1. The two 64-bit input operands, contain either 1 set of double precision or 2 sets of single precision operands.

978-1-4673-9141-2/15 $31.00 © 2015 IEEE

Algorithm 1 Floating Point Multiplier Computational Flow [9]

```
1:  (IN1,IN2) Input Operands;
2:  Data Extraction and Exceptional Check-up:
3:      {S1(Sign1), E1(Exponent1), M1(Mantissa1)} ← IN1
4:      {S2, E2, M2} ← IN2
5:      Check for INFINITY, NAN, ZERO
6:      Check for SUB-NORMALs
7:      Update Exponents and Mantissa's MSB for SUB-NORMALs
8:  Core Computation:
9:      Sign Computation:
10:         S ← S1 xor S2
11:     Exponent Addition and Mantissa Multiplication
12:         E ← E1 + E2 - BIAS
13:         Mult_M ← M1 * M2
14:  SUB-NORMAL Processing
15:     Right-Shift Amount:
16:         Check if E is negative? (For Right Shifting)
17:         Right_Shift ← BIAS - (E1 + E2) + (Mult_M[MSB]&E̅)
18:     Left-Shift Amount:
19:         Left_Shift ← Leading-One-Detection (LOD) of M
20:         Left_Shift ← Check if less than or equal to E?
21:     Dynamic Right Shifting and Dynamic Left Shifting:
22:         Mult_M ← Mult_M >> Right_Shift
23:         Mult_M ← Mult_M << Left_Shift
24:  Normalization and Rounding:
25:     Mantissa Normalization and Compute Rounding ULP based on Guard,
        Round and Sticky Bit
26:         Mult_M ← Mult_M + ULP
27:         E ← E + Mult_M[MSB] - Left_Shift
28:  Finalizing Output:
29:     Update Exponent and Mantissa for Exceptional Cases
30:     Determine Final Output
```

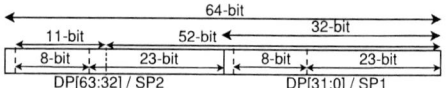

Figure 1: Input / Output Register Format.

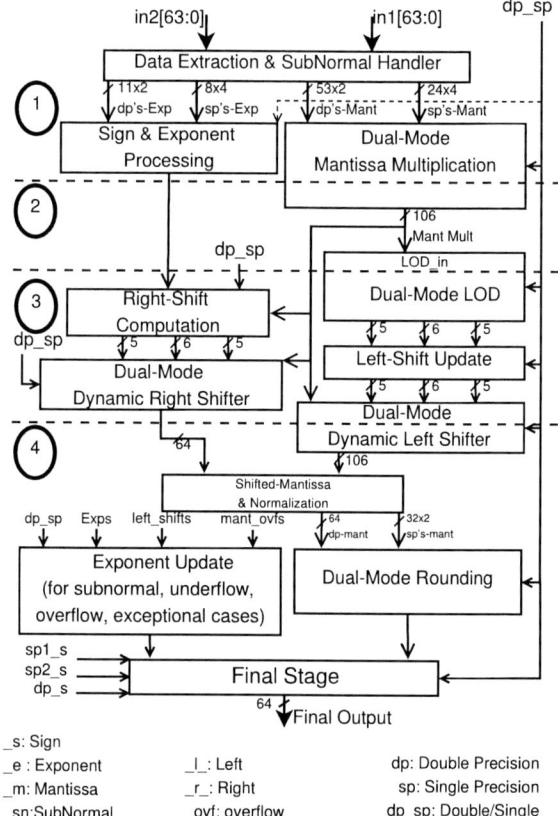

Figure 2: DPdSP Multiplier Architecture.

The proposed dual-mode DPdSP multiplier architecture is shown in Fig. 2. The signal in_1 and in_2 is fed as input operands. Based on the mode deciding control signal (dp_sp), the architecture process either a double precision or dual (two-parallel) single precision multiplication as follows:

- dp_sp: 1 → DP Mode
- dp_sp: 0 → Dual SP Mode

The architecture is designed with four pipeline stages as shown in Fig. 2. The detailed description of all the computational stages in architecture are discussed below.

The **first pipeline stage** includes the initial *Data Extraction and Subnormal Handler*, dual-mode *Exponent Processing* and part of dual-mode *Mantissa Multiplier* unit.

The *Data Extraction and Subnormal Handler* is shown in the Fig. 3. This unit first extracts the sign, exponent, and mantissa part for both single precision and the double precision, as per their format. It can be visualized from Fig. 1 that the exponent portion of DP and second SP (SP-2) operand are overlapped.

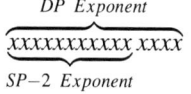

DP Exponent

$$\overbrace{\underline{xxxxxxxxxxx\ xxxx}}$$

SP−2 Exponent

Here, this opportunity is used to share the resources related to sub-normal, *infinity*, and *NaN* checks computations of DP and second SP operands. The checks of subnormal is shown in the Fig. 3, and similarly the checks for *infinity* and *NaN* are handled. Based on these subnormal and exceptional checks the respective exponents and mantissas are updated accordingly. In comparison to

a single mode DP computation, this unit requires extra related resources for first SP (SP-1) operand processing. Thus, this unit produce processed exponents and mantissas for double precision and both single precision operands set (DP Exponents: dp_e1, dp_e2; DP Mantissas: dp_m1, dp_m2; SP-1 Exponents: $sp1\_e1$, $sp1\_e2$; SP-1 Mantissas: $sp1\_m1$, $sp1\_m2$; SP-2 Exponents: $sp2\_e1$, $sp2\_e2$; and SP-2 Mantissas: $sp2\_m1$, $sp2\_m2$) along with their signs and exceptional signals (Signs: dp_s1, dp_s2, $sp1\_s1$, $sp1\_s2$, $sp2\_s1$, $sp2\_s2$; subnormal: $\_sn$, *infinity*, and *NaN*).

978-1-4673-9141-2/15 $31.00 © 2015 IEEE   214

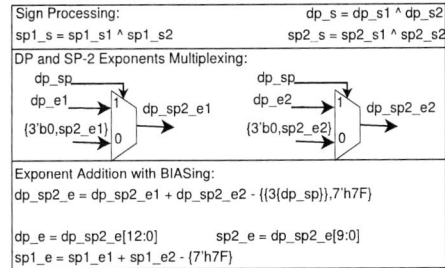

Sign Processing:	$dp\_s = dp\_s1 \wedge dp\_s2$
$sp1\_s = sp1\_s1 \wedge sp1\_s2$	$sp2\_s = sp2\_s1 \wedge sp2\_s2$

DP and SP-2 Exponents Multiplexing:

Exponent Addition with BIASing:
$dp\_sp2\_e = dp\_sp2\_e1 + dp\_sp2\_e2 - \{\{3\{dp\_sp\}\},7'h7F\}$

$dp\_e = dp\_sp2\_e[12:0]$ $sp2\_e = dp\_sp2\_e[9:0]$
$sp1\_e = sp1\_e1 + sp1\_e2 - \{7'h7F\}$

Figure 4: Sign and Exponent Processing.

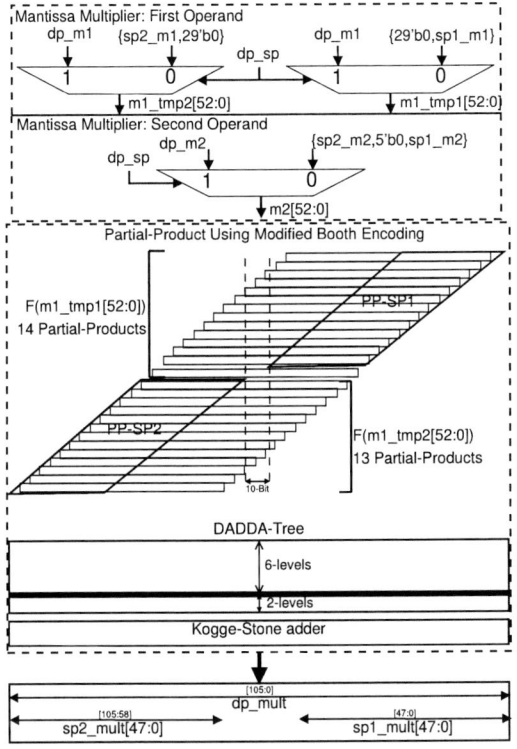

Figure 5: Dual-Mode Mantissa Multiplication.

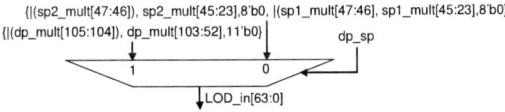

Figure 6: LOD_in.

In the *Sign and Exponent Processing* unit, the resultant sign is the *xor* operation among both operands sign-bits (as shown in Fig. 4). The dual-mode *Exponent Processing* unit first multiplex the exponents of DP and SP-2, which enables sharing of the resources required for DP and SP-2 exponents processing in the entire flow. With this multiplexing, the exponent addition, with proper BIASing, is shared among DP and SP-2 operands, which produce the resultant exponent (which still requires some processing in final stage). The computation of SP-1 exponent needs to be separately done;

and this SP-1 computation along with above two 11-bit 2:1 MUXs act as overhead over single-mode DP computation related to this unit.

The dual-mode *Mantissa Multiplier* is shown in the Fig. 5. The underlying mantissa multiplication is based on the modified Booth multiplication, which in present scenario generates 27 partial products. Initially, based on the mode-of-operation, effective operands for this multiplication are generated. Two sets of first operands, the multiplicand, are generated ($m1\_tmp1$ and $m1_tmp2$). The $m1\_tmp1$ contains either DP first mantissa or SP-1 first mantissa appended with 29-zeros at MSB side, and it acts as multiplicand for first 14 partial products. Whereas, $m1\_tmp2$, which acts as multiplicand for remaining 13 partial products, contains either of DP first mantissa or SP-2 first mantissa appended with 29-zeros at LSB side. The second operand $m2$, the multiplier operand, contains either DP second mantissa or combination of SP-2 and SP-1 second mantissas separated with in-between 5-zeros. Thus, with these sets of operands, the generated partial product contains data either for DP mantissa multiplication or two SPs mantissa multiplication. During dual single precision mode of operation, the partial product box $PP - SP1$ contains the data for SP-1, and the $PP - SP2$ contains the data for SP-2, with all other partial product data zero. Both the $PP - SP1$ and $PP - SP2$ box are separated enough to not corrupt each other in further computation. These partial products are then compress by 8-level Dadda-Tree [10]. The **first pipeline register** appears in Dadda-tree after 6-levels, and remaining 2-levels lies in **second pipeline** stage. The Kogge-Stone [11] adder is used as final adder, due to its smaller delay (however, it requires more area), which generates the dual mantissa multiplication result.

The mantissa multiplication result contains either DP-result or both SPs results (at both ends). Thus, compared to a single-mode modified Booth multiplication, the proposed dual-mode modified Booth multiplication just requires initial MUXs for operands generation. In the core multiplication flow, just the multiplicands are different for two sets of partial products, which does not affect any area/delay requirements. This approach for dual mode multiplication using modified Booth multiplication is one of the key contribution of this research work. All the prior literature have achieved dual-mode mantissa multiplication by multiplexing all the partial products (and have mainly used array multiplier method), which has a large area overhead as well as delay overhead in dual-mode operation.

After the dual-mode mantissa multiplication, input for the *leading-one-detector* (*LOD_in*) is generated as shown in Fig 6. It contains mantissa multiplication data either for DP or both SPs. In dual SP mode, each 32-bit portion of *LOD_in* contains data for each SPs, else contains data for DP only.

The **second pipeline stage** of the DPdSP multiplier ar-

978-1-4673-9141-2/15 $31.00 © 2015 IEEE   215

```
Right Shift Check:
r-shift_dp_sp2_chk = dp_sp2_e[12] | (~|dp_sp2_e)
r-shift_sp1_chk = sp1_e[9] | (~|sp1_e)

Right Shift Amount:
r-shift_dp_sp2 = {{3{dp_sp}},7'b7F} - (dp_sp2_e1 + dp_sp1_e2)
    + (dp_sp ? ~(mult_dp[MSB] & |dp_e) : ~(mult_sp2[MSB] & |sp2_e)

r-shift_dp = r-shift_dp_sp2[10:0]
r-shift_sp2 = r-shift_dp_sp2[7:0]
r-shift_sp1 = {7'h7F} - (sp1_e1 + sp1_e2) + ~(mult_sp1[MSB] & |sp1_e)
```

Figure 7: Right Shift Computation.

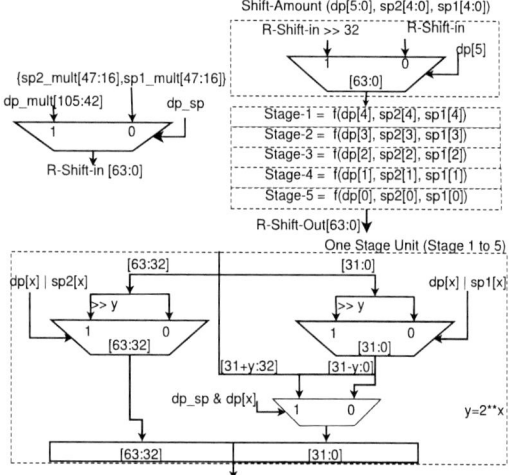

Figure 8: Dual-Mode Dynamic Right Shifting.

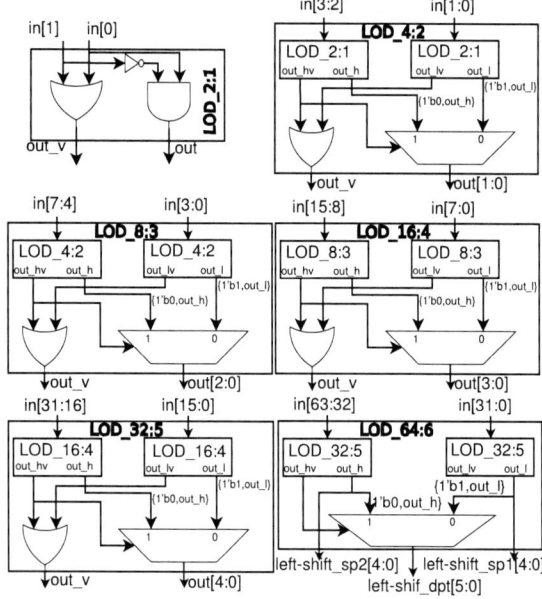

Figure 9: Dual-Mode Leading-One-Detector.

chitecture includes 2-levels of Dadda-tree, the Kogge-Stone adder, and the generation of *LOD_in*, which are discussed above. The **third pipeline stage** includes the processing of *leading-one-detection* (LOD) to compute left-shift-amount,

left-shift-update and part of *dynamic-left-shifting* (in case mantissa multiplication underflows, due to sub-normal input operand). It also contains the computation of *right-shift-amount* and *dynamic-right-shifting* (in case the resultant exponent in first pipeline stage falls below zero; step-12 in Algo 1).

The *right-shift-amount* computation is shown in Fig 7. It initially detects the exponents to check if it requires dynamic-right-shifting or not. Since the exponents of DP and SP-2 are shared, their checks are shared. Further, the amount of right shift computation is also shared for DP and SP-2. Similarly, the checks and shift-amount computation are done with SP-1 exponent. Further, based on the mode of operation this shifting amounts are updated. With true dp_sp both SPs' right shift are set to zero, and for false dp_dp the DP right shift is forced to zero.

The dual-mode *dynamic-right-shifter* is shown in Fig. 8. It first generates the input $(Right - Shift - in)$ for it from mantissa multiplication results, which either contains the multiplication results of DP or both SPs. The shifting amount is taken from previous computation. This *dynamic-right-shifter* is capable of right-shifting either a 64-bit data or two 32-bit data. The first stage of it works only for DP shifting, and remaining 5-stages work in dual-mode either for DP of both SPs, as shown in Fig. 8. The dual-mode stages are generic in nature and can be extended for any size dual-mode shifting.

On the other side of right shifting processing, further to the computation of *LOD_in* in second pipeline stage, *leading-one-detection* (LOD) is shown in Fig. 9. The LOD computes the amount of left-shift. A hierarchical architecture for it is presented, which processes either a DP or dual SPs data. It is similar to a DP only LOD. The left shift amount generated from LOD, is then updated for underflow case (if left shift amount exceeds or is equal to the exponent). For the underflow case, the corresponding left shift will be equal to corresponding exponent decremented by one. For the exponent decrements, the subtractor is shared for the DP and SP-2, as done in the case of computation of right shift amount. Further, with true dp_sp both SPs' left shift are set to zero, and for false dp_sp the DP left shift is forced to zero.

The left shift amounts are then fed in to the *dynamic-left-shifter*, which is shown in Fig. 10. It first generates the input for it from mantissa multiplication result as shown in Fig. 10. The input taken is 106-bit in order to prevent any data loss in final mantissa due to left shifting. The idea of dynamic-left-shifting is similar to the *dynamic-right-shifter*, except for the opposite shifting direction. Some part for left shifter circuitry occupies in **fourth pipeline stage** for balance pipelining.

The **fourth pipeline stage** includes half part of *dynamic-left-shifter*, finalizing *shifted-mantissa*, *normalization*, *rounding*, *exponent update* and *final processing*. The *shifted-mantissa and normalization* unit, shown in Fig. 11,

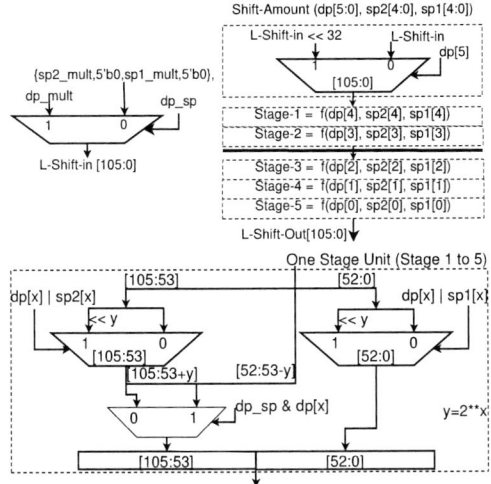

Figure 10: Dual-Mode Dynamic Left Shifter.

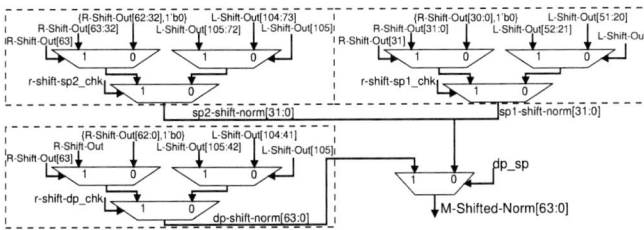

Figure 11: Shifted Mantissa and Normalization.

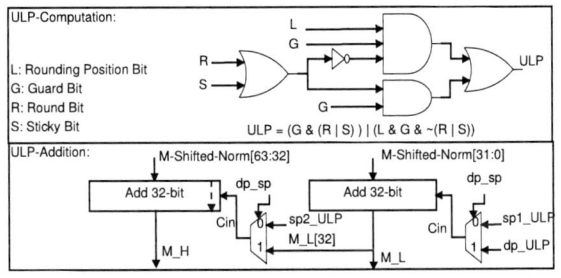

Figure 12: DPdSP Rounding.

Exponent Update
dp_sp2_eo = dp_sp2_e + dp_mult [105]
            - (dp_sp ? left-shift_dp : left-shift_sp2)
dp_eo = dp_sp2_eo[12:0]   sp2_eo = dp_sp2_eo[9:0]
sp1_eo = sp1_e + sp1_mult [47] - left-shift_sp1

Figure 13: Exponent Update.

takes the right and left shifting data, and decides which one to forward ahead with normalization (1-bit right shifting in case of mantissa overflow). It further generates the unified shifted and normalized mantissa ($M-Shifted-Norm$) which consists either of DP data or both SPs data (with required shifting and normalization).

The shifted and normalized mantissa further undergoes for rounding (ULP: unit-at-last-place computation and ULP-Mantissa addition). Rounding is shown in Fig. 12. In present

work, the round-to-nearest method is included, however, other method can be included easily. The rounding ULP computations are done based on LSB precision bit, Guard bit, Round bit and Sticky bit. Here, the ULP computation is done separately for each of DP and both SPs. However, the ULP-addition is shared among DP and both SPs, as shown in Fig. 11.

The *Exponent Update* unit, as shown in Fig. 13, update exponents for mantissa overflow or underflow. This is achieved by incrementing exponents by one or decrementing by left-shift-amount This computation is shared for DP and SP-2, and done separately for SP-1.

Finally, the exponents and mantissas are updated for underflow, overflow, sub-normal and exceptional cases to produce the final output, and each requires separate units for DP and both SPs. For overflow, the exponent will be set to infinity and mantissa will be set to zero, and for underflow case exponent will be set to zero and mantissa will take its related computed value. The computed signs, exponents and mantissas of DP and both SPs are finally multiplexed to produce the final 64-bit output, which either contains a DP output or two SPs outputs.

III. IMPLEMENTATION RESULTS AND COMPARISONS

The proposed architecture is synthesized using *UMC 90nm* standard-cell based ASIC library, and implementation details are presented in Table I. An architecture for DP only multiplier is also designed (using similar computational flow) and synthesized for area and period overhead measurements. Both of these architectures are constructed with four pipeline stages, and are synthesized for best possible period. The functionality is verified using 5-millions random test cases in each mode, with all possible pairs of operands combination (normal, sub-normal, exceptional cases). Compared to a single mode DP multiplier, the DPdSP multiplier needs approximately 11.6% more area and 6.74% more period.

A comparison with prior literature work on DPdSP architecture is shown in Table-II. Comparisons are carried out in terms of technological independent parameters (such as gate-equivalent, scaled-area equivalent, and "Fan-Out-of-4' (FO4) delay). An unified comparison of $\frac{area \times period}{throughput}$ metric is also performed (*throughput* is considered as the number of output results per clock cycle). Thus, the smaller value of $\frac{area \times period}{throughput}$ metric would be better for a design.

Akkas *et al.* [4] presented the DPdSP multiplier with two pipeline stages and throughput of 0.5 for DP. This architecture only consists of mantissa multiplier (2-cycle iterative) and rounding circuitry, without any support for sub-normal operands, underflow, overflow and exceptional cases. Compared to [4], the proposed architecture has 3-times smaller $\frac{area \times period}{throughput}$, with more computational support. Tan *et al.* [5] have also presented a similar architecture, however, it also consists of mantissa multiplier and rounding

978-1-4673-9141-2/15 $31.00 © 2015 IEEE   217

Table I: Implementation Details for Proposed Multiplier

	DP	DPdSP
Latency	4	4
Area(μm^2)	123330	137631
Area(gates)	41110	45877
Period(ns)	0.89	0.95
Period(FO4)	19.78	21.11
Power(mw)	58.56	62.72

Table II: Comparison of DPdSP Multiplier Architecture

	[4] 250nm	[5] 65nm	Proposed 90nm
Latency (DP/dSP)	3/2	4/2	4/4
Throughput[1] (DP/dSP)	0.5/1	0.5/1	1/1
Scaled-Area[2]	-	184198	137631
Gate-Count[3]	35103	-	45877
Period (FO4)[4]	40.56		21.11
$\frac{Scaled-Area \times Period(FO4)}{Throughput}$ (10^6)	-	-	2.9
$\frac{Gate-Count \times Period(FO4)}{Throughput}$ (10^6)	2.85	-	0.97

[1]Throughput (Output results per clock cycle)
[2]in μm^2 @ $90nm$ = (Area @ $65nm$) * $(90/65)^2$
[3]Based on minimum size inverter
[4]1 FO4 (ns) $\approx$ (Tech. in μm) / 2

circuitry only, with throughput of 0.5 for DP computation. Timing information is not provided in the Tan's paper, however, their scaled area requirement is much higher than the proposed architecture, also with poor throughput number. Huang *et al.* [12] have presented a dual-mode fused multiplier-adder architecture, however, they have also used array multiplier with the multiplexed partial products, which area requirement is very large, as shown in case of [4], [5]. The area requirement of only mantissa multiplier in Huang's architecture is similar to the total area requirement of proposed complete DPdSP multiplier architecture. Thus, the inclusion of proposed multiplication techniques will improve Huang's architecture design metrics.

In summary, the proposed DPdSP multiplier has shown better design metric, with more computational support than prior state of the art on it. The prior works basically used array/rectangular multipliers with multiplexed partial products which leads to the large area requirement along with more delay/period. Further, they also lack in the support for sub-normal processing, underflow and overflow processing and exceptional case handling.

IV. ACKNOWLEDGMENTS

This work is party supported by the "The University of Hong Kong" grant (Project Code. 201409176200), the "Research Grants Council" of Hong Kong (Project ECS 720012E), and the "Croucher Innovation Award" 2013.

V. CONCLUSIONS

This paper has presented a configurable dual mode floating point multiplier architecture, which is capable of processing a double precision or two-parallel (dual) single precision multiplication. Standard state-of-the-art floating point multiplication algorithm is the base of proposed architecture.

The computational support of normal as well as sub-normal processing is included, along with the exceptional case handling and underflow and overflow processing. The subcomponents of the flow are tuned for dual-mode operation with efficient resource sharing, which requires minimal multiplexing circuitry. The proposed DPdSP multiplier architecture approximately need 11.6% more area and 6.74% more period than its single mode DP only multiplication. Compared to prior literature, the proposed architecture outperform them in terms of area, period and $\frac{area \times period}{throughput}$ metric.

REFERENCES

[1] NXP Semiconductors, "AN10902 : Using the LPC32xx VFP," in *Application note*, Feb 2010. [Online]. Available: www.nxp.com/documents/application_note/AN10902.pdf

[2] Nvidia, "NVIDIA's Next Generation CUDA^{TM} Compute Architecture: Kepler^{TM} GK110," in *White Paper*, 2014. [Online]. Available: www.nvidia.com/content/PDF/kepler/ NVIDIA-Kepler-GK110-Architecture-Whitepaper.pdf

[3] A. Akkas and M. J. Schulte, "A quadruple precision and dual double precision floating-point multiplier," in *Proceedings of the Euromicro Symposium on Digital Systems Design*, ser. DSD '03, 2003, pp. 76–81.

[4] A. AkkaÅ§ and M. J. Schulte, "Dual-mode floating-point multiplier architectures with parallel operations," *Journal of Systems Architecture*, vol. 52, no. 10, pp. 549 – 562, 2006.

[5] D. Tan, C. E. Lemonds, and M. J. Schulte, "Low-power multiple-precision iterative floating-point multiplier with SIMD support," *IEEE Trans. Comput.*, vol. 58, no. 2, pp. 175–187, Feb. 2009.

[6] A. Baluni, F. Merchant, S. K. Nandy, and S. Balakrishnan, "A fully pipelined modular multiple precision floating point multiplier with vector support," in *Electronic System Design (ISED), 2011 International Symposium on*, 2011, pp. 45–50.

[7] M. Jaiswal, R. Cheung, M. Balakrishnan, and K. Paul, "Unified architecture for double/two-parallel single precision floating point adder," *Circuits and Systems II: Express Briefs, IEEE Transactions on*, vol. 61, no. 7, pp. 521–525, July 2014.

[8] ——, "Configurable architecture for double/two-parallel single precision floating point division," in *VLSI (ISVLSI), 2014 IEEE Computer Society Annual Symposium on*, July 2014, pp. 332–337.

[9] "IEEE standard for floating-point arithmetic," *IEEE Std 754-2008*, pp. 1–70, Aug 2008.

[10] L. Dadda, "Some schemes for parallel multipliers," *Alta Frequenza*, vol. 34, pp. 349–356, 1965.

[11] P. M. Kogge and H. S. Stone, "A parallel algorithm for the efficient solution of a general class of recurrence equations," *Computers, IEEE Transactions on*, vol. C-22, no. 8, pp. 786–793, Aug 1973.

[12] L. Huang, L. Shen, K. Dai, and Z. Wang, "A new architecture for multiple-precision floating-point multiply-add fused unit design," in *Computer Arithmetic, 2007. ARITH '07. 18th IEEE Symposium on*, 2007, pp. 69–76.

Hardware Implementation of Real-Time Multiple Frame Super-Resolution

Kerem Seyid, Sebastien Blanc, Yusuf Leblebici

Ecole Polytechnique Fédérale de Lausanne (EPFL) Lausanne, Switzerland

Email: {kerem.seyid, sebastien.blanc, yusuf.leblebici}@epfl.ch

Abstract—Super-resolution reconstruction is a method for reconstructing higher resolution images from a set of low resolution observations. The sub-pixel differences among different observations of the same scene allow to create higher resolution images with better quality. In the last thirty years, many methods for creating high resolution images have been proposed. However, hardware implementations of such methods are limited. In this work, highly parallel and pipelined implementation for iterative back projection super-resolution algorithm is presented. The proposed hardware implementation is capable of reconstructing 512x512 sized images from set of 20 lower resolution observations, with real-time capabilities up to 25 frame per second (fps). Explained system has been synthesized and verified via Xilinx VC707 FPGAs. To the best of our knowledge, the system is currently the fastest super-resolution implementation based on FPGA.

I. INTRODUCTION

From the early 70s on, usage of the charge-coupled devices (CCD) and CMOS sensors are widely increased as opposed to an exposure on photographic films. Back in 70s, these sensors were sufficient for most of the applications. However, today people want a digital handheld camera with high-resolution (HR) and affordable prices. Furthermore scientists often need a HR level close to an analog 35mm film that has no visible artifacts when an image is magnified [1]. This creates a need to find a way to increase the resolution level.

The most common method is to increase the pixel density, or in other words, reducing the pixel size (spatial resolution) by fabrication techniques. Spatial resolution refers to the pixel density in an image and measures in pixels per unit area [2]. It is common knowledge that the scaling effects in CMOS technology allow the semiconductor industry to make smaller devices [3]. This rule holds for CMOS imaging applications as well, in [3] it is indicated that CMOS image sensor technology is lagging behind the technology nodes in ITRS roadmap. The reason behind this lagging is very simple: current CMOS process is not imaging friendly. Reducing the pixel sizes mean less amount of light available per pixel. Furthermore, smaller size generates shot noise that reduces the image quality. There is a limit to reduce the pixel pitch without suffering the effects of shot noise. For $35\,\mu m$ CMOS process, estimated pixel area is around $40\,\mu m^2$ [1].

Another approach to enhance the resolution is to increase the chip size. However, that leads to an increase in capacitance, which makes it difficult to speed up a charge transfer rate [4]. Furthermore, cost of high precision optics and image

sensors are also playing an important role in commercial high resolution imaging.

As explained, consumer applications are in need of high resolution that can be obtained by smaller pixel sizes. Furthermore current CMOS technology has the means to satisfy the needs. However, smaller pixel size means degrading in the performance. In order to compete with the current needs of imaging technology, a new approach is required.

One approach is to use signal processing techniques to obtain an HR image where multiple low-resolution images of the same scene can be obtained. Recently, this approach has been one of the most attractive research areas. It is called Super-resolution (SR) image reconstruction or in another terms, image resolution enhancement. The term SR image reconstruction is referring to a signal processing approach that aims to overcome the limitations imposed by obtained low resolution images.

The major advantage of signal processing approach is that it may cost less and the existing image systems can be utilized [1]. After obtaining low resolution images with an inexpensive video recorders or handheld cameras, a higher resolution output image can be reconstructed in post processing.

Another approach related to resolution enhancement is the single-image interpolation approach. However, since there is no additional information and detail provided in this approach, the quality of the single-image interpolation is very much limited due to the ill-posed nature of the problem. The lost frequency components cannot be recovered, therefore, image interpolation methods are not considered as SR techniques.

In the example based SR methods, correspondences between low and high resolution image patches are learned from a database of low and high resolution image pairs. Corresponding patches are applied to a new low-resolution image to recover its most likely high-resolution version [5]. Furthermore, Glasner *et al.* [6] proposed a patch based SR resolution, which is database free single image super resolution. These methods are computationally expensive and real-time implementations are not feasible.

In the SR setting, multiple low-resolution observations are available for reconstruction, making the problem better constrained. The nonredundant information contained in the these LR images is typically introduced by sub-pixel shifts between them. These subpixel shifts may occur due to uncontrolled motions between the imaging system and scene, e.g., movements of objects, or due to controlled motions

978-1-4673-9141-2/15 $31.00 © 2015 IEEE

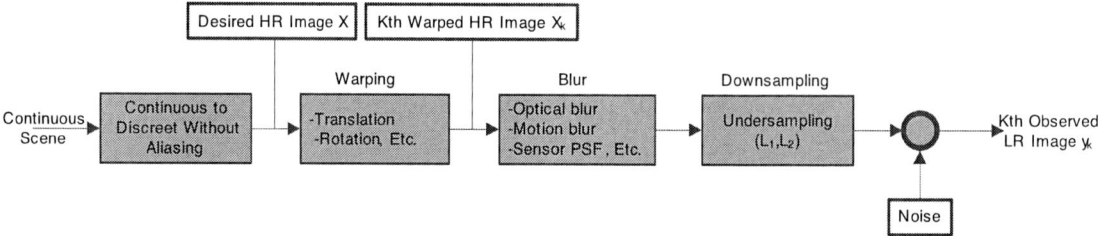

Figure 1. Observation model relating LR images to HR images

[2]. CCTV systems is now replacing DVRs and often they need object magnifications, such as extracting the car plate or focusing on the face of the suspect or region of interest (ROI) [1]. Also, the techniques can be utilized in every aspect of imaging technology where multiple images of same scene can be obtained, such as Magnetic Resonance Imaging (MRI), computed tomography (CT), satellite imaging, remote sensing and video standard conversion.

Multiple scenes can be obtained either from one camera with several captures that has relative motion in between frames or from multiple cameras located in different positions looking at same scene. After obtaining the motion estimation with sub pixel accuracy, SR image can be reconstructed.

In this work, a real-time super-resolution algorithm capable of creating 512×512 sized images from set of low resolution (LR) observations is presented. The paper is organized as follows, theoretical background for imaging process and previously implemented FPGA based super-resolution algorithms are given in Section II. Super-resolution algorithm implemented in this work is presented in Section III. Real-time hardware implementation is presented in Section IV. Implementation results is given in Section V and subsequently discussion on real-time implementation and future work is given in Section VI.

II. THEORETICAL BACKGROUND

SR reconstruction has been one of the most active research areas since its first mentioned by Tsai and Huang [7] in 1984. In the past 30 years, numerous approaches and techniques presented from frequency domain to spatial domain as well as from signal processing approach to machine learning techniques. In order to start talking about these techniques, one needs to firstly understand the image observation model, which represents the low resolution image capture process.

A. Observation Model

Image capturing process is not perfect due to the hardware limitations. During the process, there is a natural loss of spatial resolution caused by the optical distortions such as out of focus, diffraction limit etc., optical blur which can be modelled by Point Spread Function (PSF), motion blur due to limited shutter speed, noise that occurs within the sensor or during transmission [1]. Thus, recorded image usually suffers from blur, noise and aliasing effects. These degradations modelled

fully or partially in different SR techniques. The imaging process is presented in Fig. 1.

Assuming $\mathbf{X}$ is the desired $\mathbf{HR}$, sampled above the Nyquist rate and band limited. $\mathbf{y}_k$ is the $\mathbf{k^{th}}$ subsampled, warped and blurred LR observation of X. Furthermore, each image is corrupted by noise, capturing process can be expressed as

$$\mathbf{y}_k = \mathbf{SbB}_k\mathbf{W}_k\mathbf{X} + \mathbf{n}_k \qquad (1)$$

where $\mathbf{W_k}$ is the warping matrix, $\mathbf{B_k}$ is the blur matrix and $\mathbf{Sb}$ is the subsampling matrix, $\mathbf{n_k}$ represents the noise vector. A block diagram can be seen in Fig. 1.

The warp matrix, $\mathbf{W_k}$, represents all the motion among the captured images. It may contain global and local translation, rotation and etc. It varies among each scene and needs to be recalculated for each particular frame. Since the motion does not occur in integer pixel shifts, sub-pixel calculations and interpolations are necessary.

Bluring might be caused by the optical system, relative motion between the imaging system and the original scene, and Point Spread Function of the camera lens which is represented by the matrix $\mathbf{B_k}$. It is assumed to be known during the SR reconstruction, however, it is not easy to obtain this information. The subsampling matrix $\mathbf{SB}$ generates aliased images from warped and blurred HR image.

B. Related Work

Although the super-resolution reconstruction is an attractive field for over thirty years, there are not many research conducted for hardware implementation. Bowen and Bouganis [8] proposed a hardware architecture for FPGA implementation, however implementation needs 20 iteration stages for satisfactory results and device capabilities were not sufficient. Moreover, it was limited to the on-chip memory. Another implementation for iterative back projection algorithm was introduced in [9], which was designed for adaptive sensor and capable of outputting 25 frame per second (fps) in VGA resolution. The bottleneck of the system was explained as the triple buffering memory access scheme. One other implementation was presented by Szydik *et al.* [10]. The proposed algorithm was a non iterative approach. However, results was limited to QCIF to CIF super-resolution for 25 fps. Furthermore, authors

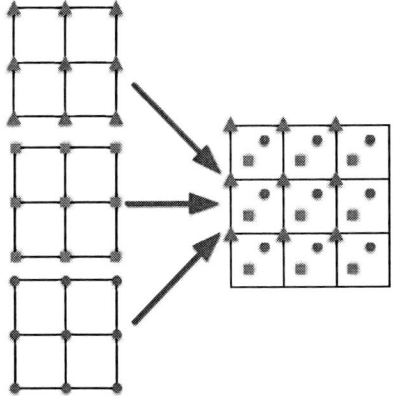

Figure 2. Sub-pixel shifts among the low resolution observations

stated that their implementation is not scalable and limited to work with only two reference frames.

III. SUPER-RESOLUTION

Super-resolution is computationally complex and ill-posed problem. The motion between the frames, the blur kernel(s), and the high-resolution image of interest are three interwoven unknowns that should ideally be estimated together rather than sequentially [11]. There are many methods proposed in literature to solve the ill-posed SR problem where multiple low resolution observations can be obtained. These methods are mainly divided into two domains, either in spatial domain or in frequency domain [1], [12], [13]. In this work, spatial domain approaches are investigated since they are more suitable for hardware implementation and can be easily parallized. Additionally, it will avoid the computational complexity of the frequency domain approaches.

Several algorithms have been investigated for real-time implementation. Farsiu *et al.* [14] proposed a fast and robust implementation where the aim is to reduce the modelling errors. Zomet *et al.* [15] proposed another method, utilizing pixel wise median. The iterative back-projection algorithm (IBP) proposed by Irani and Peleg [16] which aims to back project the error between observed and simulated images. Among these methods, IBP is suitable for hardware implementation, since it can be highly parallized and computational complexity is suitable for real-time implementation. In the iterative back projection algorithm, firstly an initial guess is created by using interpolation from one of the observed low resolution images. Afterwards the algorithm aims to find the high resolution image $\mathbf{X}$ by simulating the image capture process and creating the low resolution observations $\mathbf{y}_k$. The difference between low resolution observations $\mathbf{y}_k$ and created low resolution observations $\mathbf{y}'_k$ are iteratively back projected to the initial guess. The algorithm iteratively converges to an HR estimate until the error between simulated LR observations and obtained LR observations is negligable. A high quality image registration is needed in order to start the super-resolution process.

A. Image Registration

Image registration is the first and the crucial part of any super-resolution algorithm. The registration is transforming different sets of images to a single coordinate system. An example of three images registered to a single coordinate system can be seen in Fig. 2. Sub-pixel level registration allows super-resolution algorithms to create the HR image content. The registration methods can be divided into two, namely intensity based and feature based methods. Among the intensity based methods, majority of the algorithms use either block matching based methods or optical flow to calculate motion vectors in order to determine motion field. Optical flow based algorithms [17] provide superior motion vector quality over block matching based algorithms [18]. There are many well known works focusing on sub pixel image registration [19], [20]. Furthermore, there are works specifically focusing on motion estimation and image registration for super-resolution [21], [22].

In this work, images are assumed to be registered with respect to reference frame using the method explained in [20]. Horizontal shifts a, vertical shifts b and rotation angle θ are assumed to be known prior to starting the iterative image super-resolution process.

B. Iterative Reconstruction

Iterative algorithms start with initial high resolution approximation such as linear interpolation of the reference low resolution frame. For each iteration, the observation model is explained in Fig. 1 is applied to simulate the low resolution observation results $\mathbf{y}'_k$.

The aim of the IBP is to minimize the error $\mathbf{e}$ in every step (n) between simulated results $\mathbf{y}'_k$ and observed images set $\mathbf{y}_k$. If $\mathbf{X}_n$ is the correct high resolution image, then the simulated images $\mathbf{y}'_k$ and observed images $\mathbf{y}_k$ should be identical to one another.

$$\mathbf{e}^{(n)} = \sqrt{\sum_k \sum_{x,y} (\mathbf{y}_k(x,y) - \mathbf{y}''_k(x,y))^2} \qquad (2)$$

In each iteration, every pixel in high resolution image $\mathbf{X}$ is updated according to error of all low resolution pixels $\mathbf{y}'_k$ that it affects. The difference error $(\mathbf{y}_k - \mathbf{y}'_k)$ is multiplied by a factor λ and added to the initial high resolution estimate.

It is important to know that original high resolution frequencies may not be fully restored. For example, the blurring operation may filter out the high frequency components and make them impossible to restore. In such cases, there are more than one single high resolution images which result in same low resolution images after the imaging process. It has been stated in [16] initial guess does not influence the performance of the algorithm, but it will influence the HR estimate that the algorithm converges.

(a) Original Image   (b) First Shearing in the x axis

(c) Shearing in the y axis   (d) Second shearing in the x axis

Figure 3. Shearing based image rotation

For color images, the super-resolution algorithm is applied in $YCbCr$ domain. All the images in the RGB domain are initially converted to $YCbCr$ domain and the proposed iterative back projection algorithm is applied in Y domain, and Cb and Cr components are interpolated for HR image. Finally, the $YCbCr$ components are converted to RGB domain.

IV. PROPOSED HARDWARE IMPLEMENTATION

As stated previously, iterative back projection algorithm is suitable for hardware implementation. The iterative scheme suitable for pipelining and pipelined architectures can be implemented in parallel for each set of observed images. In this work, a pipelined hardware is designed in order to mimic the camera image acquisition process. The observation of low resolution images from high resolution estimate is designed in block based operations. Each block mimics the operation of capture process for 2×2 pixels. During the simulations, it has been realized that using block size bigger than 2×2 causes significant degradation in the output image quality.

Another important aspect is to choose which portion of estimated high resolution image should be applied to the imaging process. The best option is to choose the whole image and to create low resolution observation simulations. However, this method will drastically increase the memory bandwidth of the system. For every iteration, a new LR observation should be created and saved in the external memory and read back for the error calculations. Therefore, instead of saving the simulated results in the external memory, the calculation of the LR observations are created block by block in a pipelined architecture, and final results obtained for HR sequence is saved in external memory.

For calculating each $\mathbf{y}_k$ in pipelined stage, simulations have been conducted to find N×N block size such that it minimizes

the differences between operating in the whole image and the 2×2 sized pixel block. Throughout the simulations, the block size $N = 9$ gives the best results in the trade-off between block size and output image quality. For each 2×2 block, using a block size 9×9 for image acquisition process is sufficient for small horizontal shifts a, vertical shifts b and rotation angles θ.

The block diagram of imaging process designed for hardware implementation can be seen in Fig. 4. It can be seen in the first part of the diagram that the capture process is being mimicked. First of all, the warping operation between the HR image and the LR image is being conducted. After the HR image being shifted and rotated, it is blurred according to the model of the camera PSF. Once the blur operation is conducted, the final HR image is decimated by 2 in both directions to obtain the LR image. Once the process is finished, the warped and decimated image forms the $\mathbf{y}'_k$. Once the difference between the $\mathbf{y}'_k$ and $\mathbf{y}_k$ is calculated, the back projection of the error operation starts. The back projection is the inverse operation of the imaging process. The error block is back projected to HR image with inverse functions. Final result is saved to the external memory after adding the error to the current HR estimate of $\mathbf{X}(n)$.

A. Image Warping

In the image warping process, shifting and rotating operations are being conducted. Firstly, the sub-pixel shifts, that corresponds to integer pixel shifts in HR image, is applied. Sub-pixel shifts are calculated with half pixel precision, which are then applied to the HR image in integer pixel level.

After the images are shifted, the rotation operation with the angle θ is conducted. There are many methods proposed for image rotation operation among which Unser *et al.* [23] proposed a convolution based image rotation algorithm. The algorithm can be applied to operating blocks with angular motion. In [23] it has been stated that rotation can be defined as rotation matrix

$$\mathbf{R}(\theta) = \begin{bmatrix} cos(\theta) & -sin(\theta) \\ sin(\theta & cos(\theta) \end{bmatrix} \qquad (3)$$

Which is also equivalent to

$$\begin{bmatrix} 1 & -tan(\theta/2) \\ 0 & 1 \end{bmatrix} \times \begin{bmatrix} 1 & 0 \\ sin(\theta) & 1 \end{bmatrix} \times \begin{bmatrix} 1 & -tan(\theta/2) \\ 0 & 1 \end{bmatrix} \qquad (4)$$

The whole rotational translation can be decomposed to appropriate sequence of 1-D signal translations that can all be implemented via simple convolutions. It is a three pass implementation of rotation that can be seen in Fig. 3. The original image Fig. 3(a) is first sheared in x dimension Fig. 3(b) then in y dimension Fig. 3(c) and finally again in x dimension Fig. 3(d).

The algorithm can be implemented as first shearing among the x axis, using $-tan(\theta/2)$ and current macro block position.

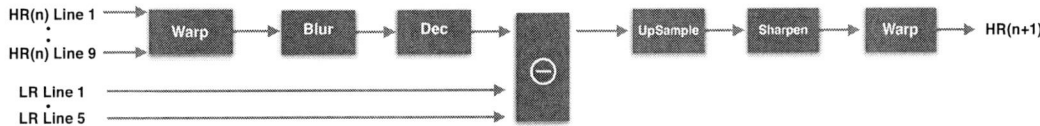

Figure 4. System Architecture Block Diagram of IBP Super-Resolution

The shearing values on the edges and the corners will not be equal to the shearing values in the center of the image. In order to solve this problem, appropriate values for $-tan(\theta/2)$ are stored in block rams. Depending on the values of $-tan(\theta/2)$ and current position of the macro block (X, Y), shearing operation can be calculated. Line values are delayed or forwarded depending on the calculated shearing operation.

Afterwards, the output image can be sheared in y axis using $sin(\theta)$. Similarly, $sin(\theta)$ values are stored in the block rams. Utilizing (X, Y) positions and $sin(\theta)$ values, the block is sheared among the Y axis. Output of the y axis shearing is extended in order for the image to be properly sheared. Finally the macro block sheared among the x axis with $-tan(\theta/2)$. Image dimensions are carefully calculated and increased with respect to shearing operations in the y dimensional shearing and second x dimensional shearing. Thanks to global motion, blocks do not coincide with each other. Final image is cropped to fit 9×9 macro block. The obtained macro block is the shifted and rotated version of the original HR macro block.

B. Blur and Decimate

After the image warping, the 9 line values coming from the image warping block are synchronized for the image blurring operation. The blur operation is applied to create the camera Point Spread Function. Camera PSF function is estimated as 3×3 matrix and blurring operation is applied accordingly to mimic both atmospheric and lens blur.

The output of the image filter block is subsampled in order to create the simulated observation of $\mathbf{y}'_k$. After the subsampling process, size of the macro block is reduced to 5×5. The obtained block $M_{obt}(x, y)$, corresponding to a portion of the simulated observation $\mathbf{y}_k$, is subtracted from the block of the observed LR $M_{obs}(x, y)$ of $\mathbf{y}'_k$. Results of the subtraction $M_{diff}(x, y)$ corresponds to error function that needs to be back projected into the initial HR image $X(n)$.

C. Back projection

Back projection is the inverse operation of the image observation process. The imaging process applied until subtraction block is now applied in reverse order. Firstly, $M_{diff}(x, y)$ block is interpolated in order to obtain $M'_{diff}(x, y)$, the 9×9 block. Simulations of the SR process showed that there were no significant difference between tested interpolation methods. Therefore, the averaging filter is applied to $M_{diff}(x, y)$ to obtain $M'_{diff}(x, y)$ block.

The upsampled block $M'_{diff}(x, y)$ is fed into the deblurring filter. The deblurring filter is different than the PSF function.

It can be chosen arbitrarily, where the camera PSF represents the camera blur parameters. 9 lines coming from the image interpolation blocks are filtered in parallel in order to reduce the noise to obtain $M'_{deblur}(x, y)$.

In the final stage of the pipeline, $M'_{deblur}(x, y)$ rotation with angle $R' = -\theta$ is conducted. The backwards warping block is the same as forward warping block, except the θ value which is the negative angle of the forward warping block. Same shearing process is applied on the deblurred block to obtain $M'_{warped}(x, y)$. From $M'_{warped}(x, y)$ block, the middle subset of 2×2 block is taken for the back projected error.

D. Parallelization

The system overview of the iterative back projection algorithm is shown in Fig. 4. The process can parallel for each image i and the same HR macro block can be used for each operation. Finally, the calculated errors are summed with an adder tree depending on the number of the LR observations as seen in Fig. 5. The final added error $\mathbf{E(n)}_{(X,Y)}$ is multiplied with a constant λ. The observed error is added to the current HR estimate $\mathbf{X(n)}$.

E. Simulation Results

The fixed point version of the proposed algorithm is implemented via MATLAB with different input image sets. Several shifted, rotated, blurred and subsampled sets of LR observations are created from a HR image and the HR image taken as ground truth. With different data sets, the output of the algorithm had an average PSNR value of 30.34 dB.

V. IMPLEMENTATION RESULTS

The proposed iterative back projection hardware is implemented using VHDL. The models are mapped to a Virtex-

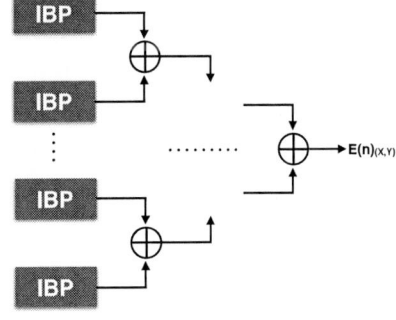

Figure 5. Parallelisation of iterative back projection blocks

978-1-4673-9141-2/15 $31.00 © 2015 IEEE

Table I
NUMBER OF ITERATIONS FOR DIFFERENT HR IMAGE RESOLUTIONS

HR	256×256	512×512	720×1280	1024×1024
Iterations	71	17	5	4

7 XC7VX485T FPGA. Single super-resolution module consumes %0.96 of LUT and %0.98 of DFF. The FPGA utilization scales proportionally depending on the number LR observations need to work in parallel. The proposed hardware operates at 265 MHz after place & route. The pipelined process flow can process 2×2 blocks in every $N_{cyc} = 9$ cycles. With the operating frequency, the system can construct 512×512 HR resolution image with 25fps, using up to $N_{it} = 17$ iterations and does not depend on the number of images, except the adder tree. The required number of cycles for SR implementation can be calculated as

$$\frac{M}{2} \times \frac{N}{2} \times N_{it} \times fps \times N_{cyc} + Lat_{tree} \qquad (5)$$

Where Lat_{tree} is corresponding to adder tree delay in the final stage.

Number of iterations that can be conducted for different HR image sizes can be seen in Table I. During the simulations, most of the LR sets are converged over the minimum error barrier in less than 10 iterations. The number of iterations as low as 5 can produce good results with sufficient number of LR observations.

VI. DISCUSSION AND FUTURE WORK

In this work, a novel hardware implementation for multiple frame super-resolution is presented. The presented hardware is easily scalable in terms of number of low resolution observations and final high resolution output size. The presented system is implemented in hardware description language VHDL; synthesized, placed and routed for Xilinx Virtex-7 FPGAs using Vivado Synthesis Tool. The implemented hardware operating real-time with 25 fps, and reconstructs a 512×512 sized HR image from 256×256 LR observations with 20 iterations. To the best of our knowledge, this is currently the fastest FPGA implementation for a real-time super resolution algorithm. Currently, an optical flow based real-time image registration algorithm is under development in order to create a real-time multiple image SR system.

REFERENCES

[1] S. C. Park, M. K. Park, and M. G. Kang, "Super-resolution image reconstruction: a technical overview," *Signal Processing Magazine, IEEE*, vol. 20, no. 3, pp. 21–36, 2003.

[2] P. Milanfar, *Super-resolution imaging*. CRC PressI Llc, 2010, vol. 1.

[3] A. J. Theuwissen, "Cmos image sensors: State-of-the-art," *Solid-State Electronics*, vol. 52, no. 9, pp. 1401 – 1406, 2008, papers Selected from the 37th European Solid-State Device Research Conference - ESSDERC. [Online]. Available: http://www.sciencedirect.com/science/article/pii/S0038110108001317

[4] T. Komatsu, K. Aizawa, T. Igarashi, and T. Saito, "Signal-processing based method for acquiring very high resolution images with multiple cameras and its theoretical analysis," *Communications, Speech and Vision, IEE Proceedings I*, vol. 140, no. 1, pp. 19–24, 1993.

[5] W. T. Freeman, T. R. Jones, and E. C. Pasztor, "Example-based super-resolution," *Computer Graphics and Applications, IEEE*, vol. 22, no. 2, pp. 56–65, 2002.

[6] D. Glasner, S. Bagon, and M. Irani, "Super-resolution from a single image," in *ICCV*, 2009. [Online]. Available: http://www.wisdom.weizmann.ac.il/ vision/SingleImageSR.html

[7] R. Tsai and T. S. Huang, "Multiframe image restoration and registration," *Advances in computer vision and Image Processing*, vol. 1, no. 2, pp. 317–339, 1984.

[8] O. Bowen and C.-S. Bouganis, "Real-time image super resolution using an fpga," in *Field Programmable Logic and Applications, 2008. FPL 2008. International Conference on*. IEEE, 2008, pp. 89–94.

[9] M. E. Angelopoulou, C.-S. Bouganis, P. Y. Cheung, and G. A. Constantinides, "Fpga-based real-time super-resolution on an adaptive image sensor," in *Reconfigurable Computing: Architectures, Tools and Applications*. Springer, 2008, pp. 125–136.

[10] T. Szydzik, G. Callico, and A. Nunez, "Efficient fpga implementation of a high-quality super-resolution algorithm with real-time performance," *Consumer Electronics, IEEE Transactions on*, vol. 57, no. 2, pp. 664–672, 2011.

[11] V. Bannore, "Iterative-interpolation super-resolution (iisr)," in *Iterative-Interpolation Super-Resolution Image Reconstruction*. Springer, 2009, pp. 19–50.

[12] K. Nasrollahi and T. B. Moeslund, "Super-resolution: a comprehensive survey," *Machine vision and applications*, vol. 25, no. 6, pp. 1423–1468, 2014.

[13] S. Farsiu, D. Robinson, M. Elad, and P. Milanfar, "Advances and challenges in super-resolution," *International Journal of Imaging Systems and Technology*, vol. 14, no. 2, pp. 47–57, 2004.

[14] S. Farsiu, M. Robinson, M. Elad, and P. Milanfar, "Fast and robust multiframe super resolution," *Image Processing, IEEE Transactions on*, vol. 13, no. 10, pp. 1327–1344, 2004.

[15] A. Zomet, A. Rav-Acha, and S. Peleg, "Robust super-resolution," in *Computer Vision and Pattern Recognition, 2001. CVPR 2001. Proceedings of the 2001 IEEE Computer Society Conference on*, vol. 1. IEEE, 2001, pp. I-645.

[16] M. Irani and S. Peleg, "Improving resolution by image registration," *CVGIP: Graphical Models and Image Processing*, vol. 53, no. 3, pp. 231 – 239, 1991. [Online]. Available: http://www.sciencedirect.com/science/article/pii/104996529190045L

[17] B. K. Horn and B. G. Schunck, "Determining optical flow," *Artificial Intelligence*, vol. 17, no. 13, pp. 185 – 203, 1981. [Online]. Available: http://www.sciencedirect.com/science/article/pii/0004370281900242

[18] S. Baker, D. Scharstein, J. P. Lewis, S. Roth, M. J. Black, and R. Szeliski, "A database and evaluation methodology for optical flow," *Int. J. Comput. Vision*, vol. 92, no. 1, pp. 1–31, Mar. 2011. [Online]. Available: http://dx.doi.org/10.1007/s11263-010-0390-2

[19] B. D. Lucas, T. Kanade *et al.*, "An iterative image registration technique with an application to stereo vision," in *Proceedings of the 7th international joint conference on Artificial intelligence*, 1981.

[20] D. Keren, S. Peleg, and R. Brada, "Image sequence enhancement using sub-pixel displacements," in *Computer Vision and Pattern Recognition, 1988. Proceedings CVPR'88., Computer Society Conference on*. IEEE, 1988, pp. 742–746.

[21] D. Barreto, L. Alvarez, and J. Abad, "Motion estimation techniques in super-resolution image reconstruction: a performance evaluation," in *Proceedings of the International Workshop Virtual Observatory: Plate Content Digitalization, Archive Mining and Image Sequence Processing*. Citeseer, 2005, pp. 254–268.

[22] G. Callico, S. Lopez, O. Sosa, J. Lopez, and R. Sarmiento, "Analysis of fast block matching motion estimation algorithms for video super-resolution systems," *Consumer Electronics, IEEE Transactions on*, vol. 54, no. 3, pp. 1430–1438, 2008.

[23] M. Unser, P. Thevenaz, and L. Yaroslavsky, "Convolution-based interpolation for fast, high-quality rotation of images," *Image Processing, IEEE Transactions on*, vol. 4, no. 10, pp. 1371–1381, Oct 1995.

978-1-4673-9141-2/15 $31.00 © 2015 IEEE

Timing and Robustness Analysis of Pulsed-Index Protocols for Single-Channel IoT Communications

Shahzad Muzaffar and Ibrahim (Abe) M. Elfadel

Institute Center for Microsystems (iMicro)
Masdar Institute of Science and Technology
Masdar City, Abu-Dhabi, UAE
{smuzaffar,ielfadel}@masdar.ac.ae

Abstract—**Pulsed-Index Communication (PIC) is a novel technique for single-channel, high-data-rate, low-power dynamic signaling that does not require any clock and data recovery. It is fully adapted to the simple yet robust communication needs of IoT devices and sensors. In this paper, we present a full quantitative analysis of the timing and robustness properties of PIC protocols, including the impact of important protocol parameters such as pulse width and inter-symbol delays on average data rate and protocol robustness with respect to clock variations. The main result of this paper is a theoretical upper bound on clock variability between transmitter and receiver below which the protocol operates with zero decoding error over an ideal channel. This bound is verified experimentally using a full FPGA implementation that includes point-to-point transmission between two TI MSP430 microcontrollers, acting as two IoT sensor nodes over a single-wire connection.**

I. INTRODUCTION

Serial communication is a popular, easy, and successful method of sending and receiving data from point to point or from point to multipoint over a variety of available communication media including wired, wireless, and human body channels [7]. Serial communication technology is used in modems, network cards, USB accessories (universal serial bus), and RS-232 serial ports. Most microcontroller chips are equipped with such serial communication ports.

Two types of serial communication are available, synchronous and asynchronous. In asynchronous serial transfer, a start and stop signals are sent before and after each code word so as to delimit the data stream and enable asynchronous decoding. The stop signal is also used to help the receiver recover and get ready for the next code word. In synchronous serial transfer, the data stream is sent on one wire called "DATA", and a periodic pulse stream is sent on a separate wire called the "CLOCK" or "STROBE". Machine-to-machine (M2M) communications are best satisfied with single-channel protocols that can be implemented either with single wires or wirelessly using asynchronous techniques.

Asynchronous techniques on single-channel ([1], [2], [3], [6]) are simple, offer faster setup, and are well suited for applications where messages are generated at irregular intervals. However, such techniques suffer from relatively large overhead (a high proportion of the transmitted bits carry no useful information), bit time dependency and low data rate.

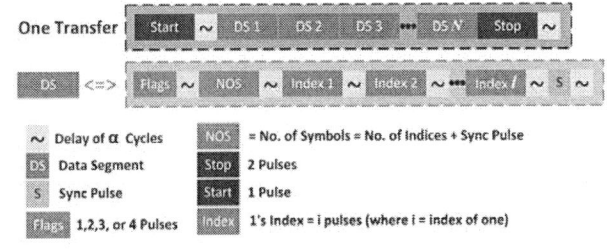

Fig. 1. PIC Packet Format

On the other hand, serial communication in the form of simple serial transfer provides a high data rate but requires a clock and data recovery (CDR) circuit to receive data successfully. The use of CDR resolves the issue of data rate and bit time dependency but at the cost of higher power consumption.

Recently, we have introduced a novel asynchronous single-channel communication protocol that we called Pulsed-Index Communication (PIC) [4] as a potential solution for the M2M IoT design space. The protocol's fundamental novelty is that it replaces the transmission of *data bits* with the transmission of *ON bit indices* in the packet stream. High-data rates are then achieved by enforcing data encoding rules in which only small indices are transmitted. Small foot print and ultra low-power are achieved by showing that such protocol does not require any circuitry for clock and data recovery. Furthermore, robustness with respect to device-to-device clocking uncertainties is achieved by using index pulse count (rather than data bit levels) for coding and decoding.

The objective of this paper is to provide a full theoretical framework that underlies the timing and robustness of PIC protocols. This framework is used to derive rigorously the performance parameters that we discovered empirically in [4], including those related PIC robustness with respect to clocking uncertainties. The most important result of the PIC theory presented in this paper is the crucial role played by the delay (expressed in transmitter clock cycles) between data segments in a given packet. For a given encoding scheme, this delay determines both the PIC average data rate and the maximum clock uncertainty tolerance. Rigorous timing and robustness analysis is provided to quantify the PIC robustness margin in the presence of transmitter-to-receiver clock variations as well

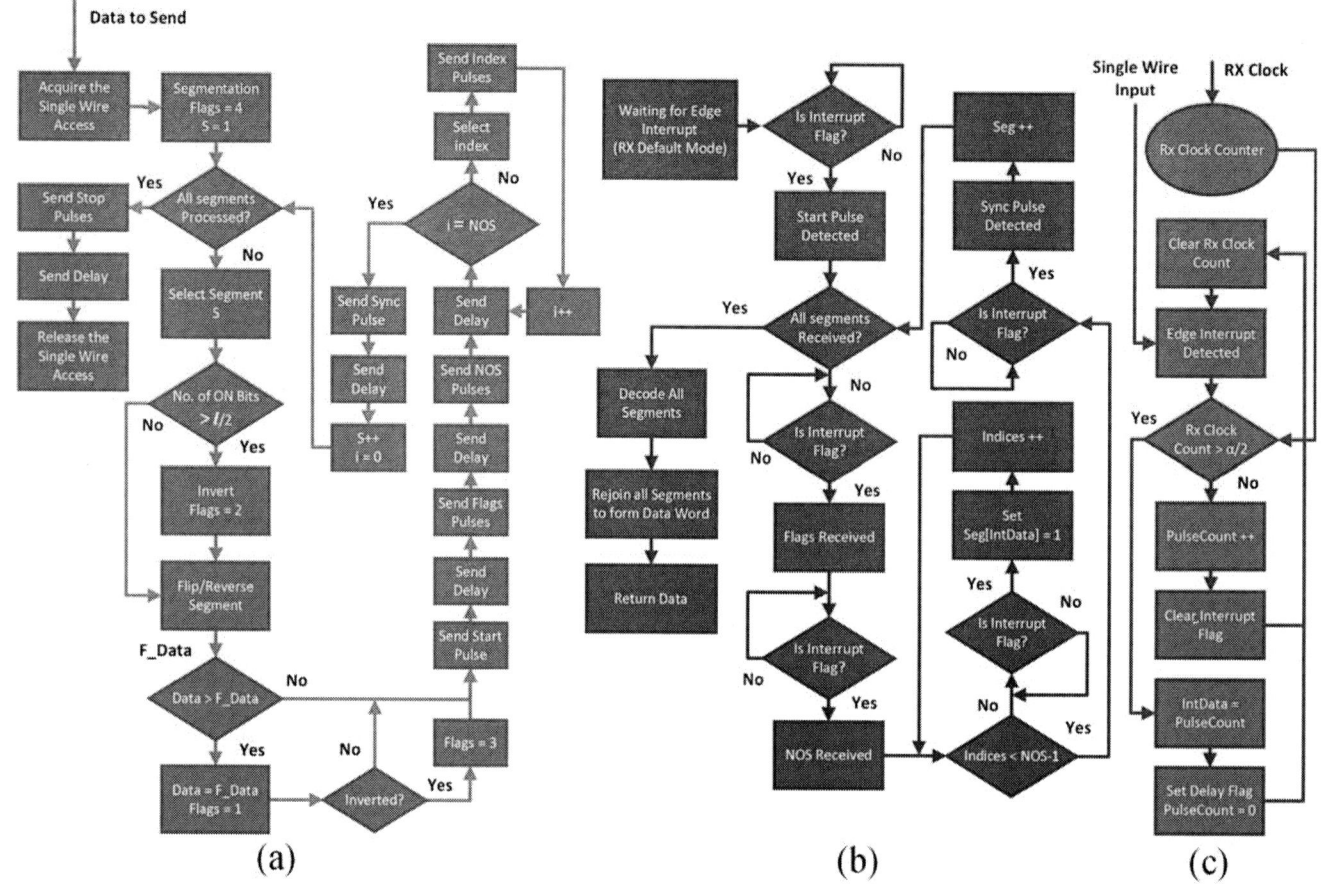

Fig. 2. PIC Transceiver Protocol Diagrams : (a) Transmitter (b) Receiver (c) Receiver Edge Detection

as clock skew and jitter within each clock. The PIC robustness margin is verified experimentally using a full FPGA implementation that includes point-to-point transmission between two TI MSP430 microcontrollers, acting as two IoT sensor nodes over a single-wire connection.

II. REVIEW OF PIC

Pulsed-Index Communication (PIC) [4] presents a novel single-channel protocol that does not require any CDR circuitry. The protocol is based on the novel concept of a pulsed index where data is encoded to minimize the number of ON bits, move them to the LSB end of the packet, and transmit the ON bit indices in the form of a pulse stream. The pulse count is equal to the index of the ON bit. Beside the elimination of CDR, the implementation of PIC is very area-efficient, low-power and highly tolerant of clocking differences between transmitter and receiver.

The equations presented in [4] will be used in section IV for clarifying the trade-off between robustness with respect to clock inaccuracies and data rate. As empirically found in [4], PIC is dynamic and provides data rates in the range of 3.1-8.5 Mb/s with an average of 4.1 Mb/s operating at 25 MHz of clock transmitting data word of 16-bits. For an ASIC implementation on 65nm technology, PIC can reduce area by more than 80% and power by more than 70% in comparison with a CDR-based serial bit transfer protocol.

III. PIC PROTOCOLS

PIC is carried out using transceivers which are comprised of a transmitter and a receiver. The transmitter and receiver are connected to the single-wire through tri-state buffers. The transmitter and receiver operate according to the protocols described below.

A. Transmitter Protocol

The PIC transmitter protocol is shown in Fig. 2(a). The receiver is the default mode when the transmitter is not active. To start transmission, it acquires access to the single-wire channel through tri-state buffers. Segmentation of data is performed, followed by the encoding of each segment. Inversion of segment bits is applied if the number of ON bits is larger than the number of OFF bits in the segment. The segment bits are reversed if the reversal results in a smaller binary number than the original segment. The encoded segments are then transmitted in the form of pulse streams separated by inter-symbol delays, according to the packet format shown in Fig. 1.

978-1-4673-9141-2/15 $31.00 © 2015 IEEE

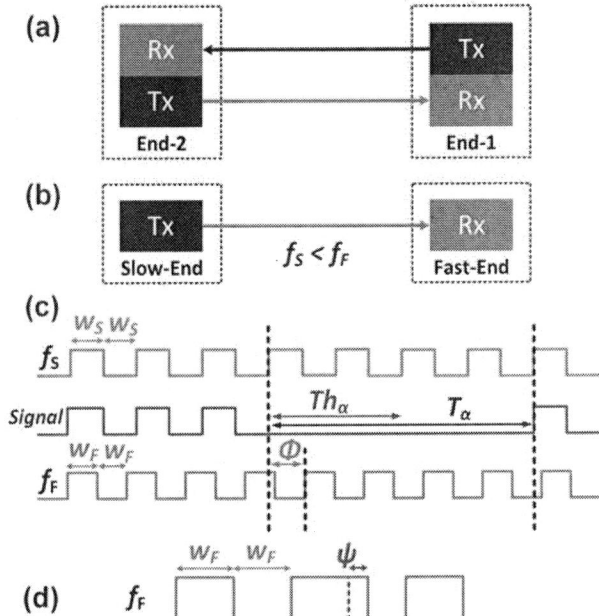

Fig. 3. (a) PIC communication link (b) PIC link with clock inaccuracy (c) PIC waveforms with clock inaccuracy (d) Receiver Clock Jitter

B. Receiver Protocol

The PIC receiver protocol is shown in Fig. 2(b) and (c). A counter is used to count the receiver clock cycles. It gets reset on each of the falling edges of the input data pulses. At each of the falling edges, the count is compared with the delay threshold coefficient, α_{Th}. If the receiver clock count is greater than α_{Th}, it is considered as an inter-symbol delay. Otherwise, it is a data pulse (Fig. 2(c)). The data pulses are counted using a separate counter and stored appropriately at the detection of inter-symbol delay. When all the pulses are received, the segment is decoded. At the end of a reception, all the decoded segments are rejoined to form the complete data word (Fig. 2(b)).

IV. TIMING AND ROBUSTNESS ANALYSIS

Robustness is directly related to the capability of behaving appropriately in the presence of different sources of errors. In this section, we first survey the sources of errors in PIC channels. Then we describe the intrinsic timing parameters of the PIC protocols and introduce some of the constraints they must satisfy for error-free operation. The last subsection is devoted to deriving a closed-form upper bound on clock discrepancy between transmitter and receiver for error-free operation in the presence of clock variations.

A. Sources of Errors

Clock inaccuracy between transmitter and receiver is one of the main sources of errors and creates significant trouble in digital communication systems. The problem becomes even more serious when single channel communication is used. The clocks of both ends need to be synchronized to limit the errors

for which a variety of techniques, including efficient clock-and-data recovery (CDR) circuits are available. PIC is an ultra-low power, single channel protocol, without CDR, but with unique robustness properties with respect to clock variations. This is a situation that is expected to be very wide-spread in the IoT environment where IoT devices with different clocking and performance requirements need to communicate. Each end of the PIC communication link is comprised of a transmitter and a receiver. All the subsequent formulation and calculations are carried out assuming the transmitter is running at a slow clock rate f_S while the receiver is running at the fast clock rate f_F. The link setup is shown in Fig. 3(a) and 3(b). In the remainder of this paper, we adopt the convention that all the parameters with S subscript are for the slow-end and with F subscript are for the fast-end.

The sources of timing errors in PIC are as follows:

1) Data pulse jitter is the main source of time difference between two adjacent data pulses. If such jitter is high, the pulses start overlapping which introduces a missing-pulse error at the receiver. The extent to which PIC can tolerate these jitters is explained using Fig. 4, where T_T is the transmitter clock time period. The faster the clock, the closer the data pulses are to each other, and therefore a small jitter may lead to successive pulse overlap. This explains the plot in Fig. 4 which shows that data pulse jitter tolerance is inversely proportional to transmitter clock rate. For a given clock rate, the data pulse jitter tolerance increases with the increase in pulse width coefficient (w).

2) Phase shift (Φ) is the time difference between the edge of the receiver clock (f_F) and the transmitter clock edge marking the start of the inter-symbol delay (T_α). The phase shift may affect the detection of inter-symbol delay especially in the presence of clock inaccuracies. Phase shift ranges from 0 to T_F, the period of the receiver clock, as shown in Fig. 3(c). In terms of shift percentage φ, we have $\Phi = \varphi T_F, 0 \leq \varphi \leq 1$.

3) Receiver clock jitter, Ψ, may also affect the detection of inter-symbol delay. The jitter value is related to the receiver clock period T_F by $\Psi = \psi T_F, 0 \leq \psi \leq 0.5$, where ψ represents jitter percentage. Receiver clock jitter is shown in Fig. 3(d).

4) Noise associated with the off-chip environment can have an effect on data pulses. The increase or decrease in pulse levels, due to the external noise, makes it difficult for the receiver to detect pulses correctly. Depending on the noise level, an extra pulse may be detected or a pulse can may be skipped. In both cases, one gets a decoding error. To analyze the performance of PIC in the presence of noise, the encoded pulse stream of data is exposed to white Gaussian noise. The noisy signal is filtered at the receiver end, then decoded according to PIC algorithm, and the number of errors encountered is counted. The results are plotted in Fig. 5 for different values of E_b/N_0 (the ration of energy-per-bit to noise

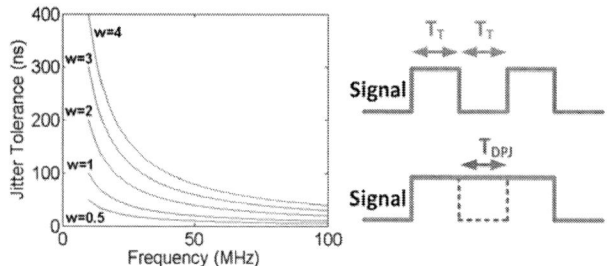

Fig. 4. Effect of frequency on Jitter Tolerance

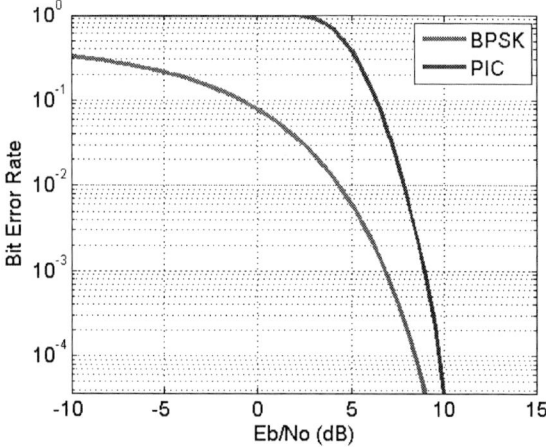

Fig. 5. BER Analysis

power spectral density) for both BPSK and PIC. PIC is less immune to noise as compared to BPSK, but its BER rapidly reduces to zero at E_b/N_0 of $\sim 10.5dB$. For the lower signal-to-noise ratios, a light-weight error correction scheme to improve BER albeit at the expense of a small increase in power consumption. A full system-level analysis of this tradeoff is the subject of on-going work. In this paper, we assume that the BER is zero.

B. Pulse Width Coefficient

The data pulse width coefficient, w, is the number of transmitter clock cycles during which the pulse remains high. An example is shown in Fig. 3(c). The inter-symbol delay increases linearly with the pulse width as will be explained in the next subsection. On the other hand, the data rate reduces with inter-symbol delay. For a 50% duty cycle, the total pulse duration $\Pi = 2wT$, where T is the clock period of the transmitter, see Fig. 3(c). A pulse width equal to half clock cycle (i.e. $w = 0.5$) results in the highest data rate and lowest requirement on the inter-symbol delay coefficient (see next paragraph). With the same pulse widths, for both the transmitter and receiver (i.e. $w_S = w_F$), the transmitted pulse time period would be equal to the time period at receiver (i.e. $T_S = T_F$). However, in the presence of clock inaccuracies, $w_S \neq w_F$, the transmitted data pulse with time period T_S will

not map exactly to the time period T_F at the receiver. Instead, we have

$$T_S = \frac{w_S}{w_F}T_F \implies w_S = \frac{f_F}{f_S}w_F. \tag{1}$$

This relationship expresses the invariance of the *number* of transmitted pulses. This number depends solely on the product of the pulse width and the corresponding clock rate.

C. Inter-symbol Delay Coefficient

The inter-symbol delay coefficient, α, is the number of transmitter clock cycles used to separate the pulse streams of two adjacent symbols. This is the most important parameter in PIC design as it affects both its data rate and reliability. A badly selected α will result in increased decoding errors and possibly complete failure. As will be seen in the following paragraphs, an efficient PIC protocol requires that α be an even number. The smallest possible α value is 4. T_{α_S}, the inter-symbol delay in terms of transmitter time period (T_S), is shown in Fig. 3(c) and is calculated as $T_{\alpha_S} = \alpha_S T_S$. To generate such delay, the transmitter pulls the line low and keeps it in that state for α number of clock cycles. For a successful reception, the inter-symbol delay at both ends must satisfy the *time* invariance principle:

$$\alpha_F T_F = \alpha_S T_S \tag{2}$$

In the presence of clock inaccuracies, phase shift, and clock jitter, the time invariance condition is expressed as

$$\alpha_F T_F = \alpha_S T_S + \varphi T_S + \psi T_S \implies \alpha_S = \frac{f_S}{f_F}\alpha_F - (\varphi + \psi) \tag{3}$$

As shown in Fig. 2, the correct PIC transmission and reception of data depends on several parameters that need to be selected judiciously. These parameters include the delay threshold and the inter-symbol delay.

1) Inter-symbol Delay Threshold: A portion of inter-symbol delay, the delay threshold (Th_α), shown in Fig. 3(c), is used at the receiver to discriminate between data pulses and inter-symbol delay. The optimal delay threshold is given by

$$Th_\alpha = \frac{T_\alpha}{2} = \frac{\alpha_F T_F}{2} = \alpha_{Th} T_F \tag{4}$$

where $\alpha_{Th} = \alpha_F/2$ is delay threshold coefficient and ensures that the receiver clock cycle count doesn't decrease during the reception of inter-symbol delay to the extent that the receiver stops detecting it as inter-symbol delay. Also, the cycle count should not increase to the extent that the receiver starts detecting even data pulses as inter-symbol delay. The absence of such optimal threshold will result in decoding errors due to either pulse undercounting or inter symbol interference.

2) Selection of Inter-Symbol Delay Coefficient: To distinguish the data pulses and the inter-symbol delay, the transmitter-generated delay should be longer than the duration of one data pulse. The inter-symbol delay coefficient should therefore satisfy

$$\alpha_S T_S > T_S + \varphi T_S + \psi T_F \implies \alpha_S > 1 + \varphi + \psi\frac{f_S}{f_F} \tag{5}$$

978-1-4673-9141-2/15 $31.00 © 2015 IEEE

where local clock inaccuracies such as receiver phase shift and clock jitter introduce are accounted for. Using the maximum possible values of 1 and 0.5 for φ and ψ, respectively, in (5) and assuming $f_F = f_S$, we get $\alpha_S > 2.5$.

In theory, the integer α_S can be chosen equal to 3, but from a hardware implementation view point we set it equal to 4 as it is easier to implement multiplication and division of power 2 numbers using left or right shift operations. In the next sub section, we study the interplay between inter-symbol delay and robustness with respect to clock variations.

D. Clock Inaccuracy Tolerance

Given the clock frequency of the slow-end (f_S) and the parameters $\alpha_S, \alpha_F, w_S, w_F, \psi,$ and φ of the PIC system, our goal now is to find the highest possible clock frequency for the fast-end (f_{Fmax}) above which decoding errors start to occur. To find f_{Fmax}, the condition for error-free operation should be fulfilled, namely, the pulse duration should be less than the inter-symbol delay threshold

$$2 w_S T_S \leq \frac{\alpha_S}{2} T_S \tag{6}$$

Using w_S from (1) and α_S from (3), we get

$$4 w_F f_F^2 + (\varphi + \psi) f_S f_F - f_S^2 \alpha_F \leq 0 \tag{7}$$

which is satisfied if and only if

$$\beta \equiv \frac{f_F}{f_S} \leq \left[\frac{\sqrt{(\varphi + \psi)^2 + 16 \, w_F \, \alpha_F} - (\varphi + \psi)}{8 \, w_F} \right] \equiv \beta_{max} \tag{8}$$

which is the main theoretical result of this paper. If $\beta \leq \beta_{max}$, PIC transmission will be error free. It is interesting to note that β_{max} is linear in the receiver clock jitter and phase shift but varies as the square root of receiver inter-symbol delay.

If we are to keep the inter-symbol delay coefficient (α) the same for both ends, the rate of the fast clock should not exceed the limit imposed by $f_{Fmax} = \beta_{max} f_S$. In Fig. 6(a), we plot f_{Fmax} for $f_S = 25 MHz$. Of course, for error-free transmission, $f_F \leq f_{Fmax}$. Fig. 6(b) identifies different regions of operation. The safe region of operation is marked up to the limit calculated using f_{Fmax}. Beyond this, there is a region of uncertainty in which errors start occurring randomly. At a certain level of clock discrepancy, total failure occurs due to the failure in detecting even a single inter-symbol delay. The recommend region of operation is of course the one delimited by f_{Fmax}.

E. Selection of Inter-Symbol Delay Coefficient

Another interpretation of the inequality in (7) is as a design formula for selecting an appropriate value of receiver inter symbol delay in the presence of clock discrepancies between transmitter and receiver and the presence of local clock inaccuracies at receiver. Solving for α_F, we get

$$\alpha_F \geq \lceil 4 w_F \beta^2 + \beta (\varphi + \psi) \rceil \equiv A_F \tag{9}$$

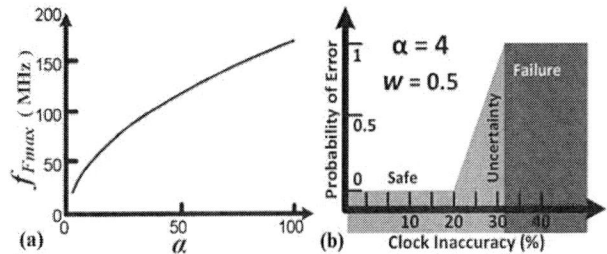

Fig. 6. (a) α vs Inaccuracy Limit (b) Regions of Operation ($f_S = 25MHz$)

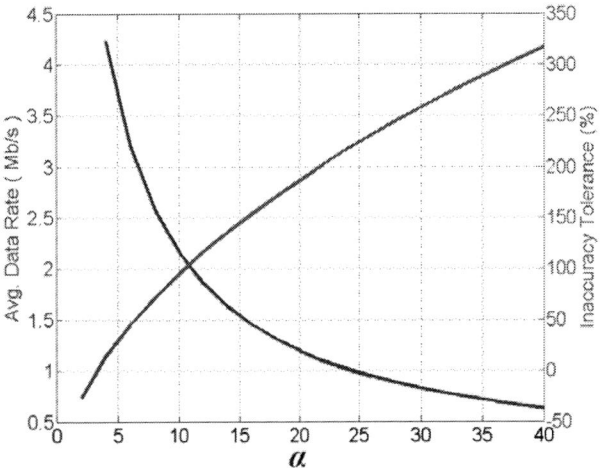

Fig. 7. Trade-off between Robustness and Data Rate

After selecting an inter-symbol delay coefficient for the slow end α_S, the inter-symbol delay coefficient for the fast end α_F is set equal to α_S if $f_F \leq f_{Fmax}$. Otherwise, α_F is a scaled version of α_S given by

$$\alpha_F = mod(A_F, 2) + A_F \tag{10}$$

This formula guarantees that α_F is the smallest even integer satisfying (7).

F. Summary on Inter-Symbol Delay

The larger α, the more resilient PIC is to timing variability. This however is achieved at the expense of significantly reduced data rates and increased power consumption. This tradeoff is illustrated in Fig. 7. A careful adjustment of α is necessary to meet the requirements of a particular application. The α value used in all our experiments is 4, with $w = 0.5$. Values of α less than 4 will result in transmission failure *even* if transmitter and receiver are running on ideal clocks with the same frequency. Therefore, the minimum allowable value of α is 4.

V. PROTOCOL FAILURE MODES AND ERROR CORRECTION

PIC transmits all the information in the form of a pulse train. A common case of erroneous transmission happens when there is a pulse miscount. In this case, PIC stops reception immediately and keeps it in a "halt" state until an explicit reset signal is sent, at which time transmission is resumed.

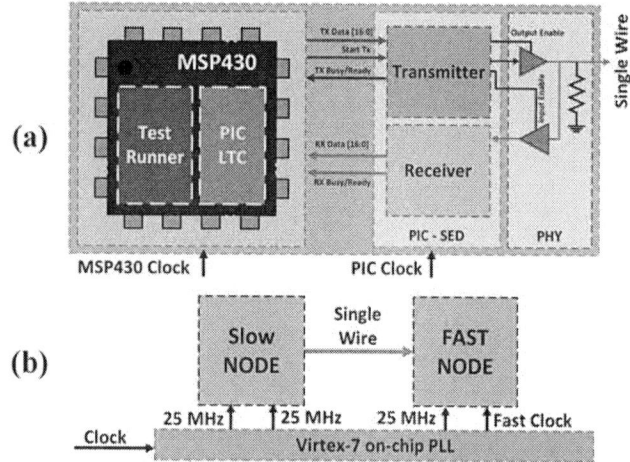

(a)

(b)

Fig. 8. Experimental Setup (a) Inside each Node (b) Clocks Generation

With the use of a simple counter, the halt state provides a useful error detection mechanism. The counter is activated via the receiver's busy signal which remains active when the error occurs. The counter gets reset at each falling edge and continues to count until the next falling edge. If the count reaches a threshold value, Th_{Error}, the state is considered an error and thus a reset signal is generated. An appropriate value of the error-detection threshold is

$$Th_{Error} = 2\left(\alpha + 2\,w\,l\right) \qquad (11)$$

which is twice the number clock cycles needed for the inter-symbol delay followed by pulse train representing a full segment of OFF bits.

A variety of available techniques [5] can be employed to handle an error condition. One simple recommendation is to request the transmitter to resend the data. Another recommendation is for the configurable layer of the protocol to send an acknowledgment for each of the successful transfers. The only error that cannot be detected using the count method is the distortion of middle pulses in the index pulse train. In such cases, there will be no halt state and the receiver will infer a wrong index number. This transmission error can be handled using a simple parity check or other similar methods. However, such pulse distortion error is very unlikely as it occurs only in the presence of excessive external noise.

VI. EXPERIMENTAL VERIFICATION

An experimental setup comprised of two IoT nodes communicating using PIC, as shown in Fig. 8, is used to verify the limitations imposed by (8) on the maximum tolerable clock frequency for the fast end. Each node is composed of a PIC protocol module (SED), a Logical Topology Control module (LTC), a PHY layer, and a Test Runner, as shown in Fig. 8(a). The SED and PHY are implemented in Verilog HDL. The LTC and test runner are implemented using Verilog IP of TI MSP430 microcontroller. The whole setup is implemented in Verilog on the Xilinx Virtex-7 FPGA platform. Two clocks,

one for each node, are generated with the help of a Virtex-7 on-chip PLL, as shown in Fig. 8(b). The slow-node clock is fixed at 25-MHz, but the rate of the fast-node clock is increased gradually from 25-MHz. Using $\alpha = 4$, $w = 0.5$, $\varphi = 1$, and $\psi = 0.01$ at both ends, we have $\beta_{max} = 1.2$. The LTC of the slow node directs the PIC transmitter to send the 16-bit data starting at 0 with an increment of 1 at each transmission. The fast end receives the data and echoes it back to the slow end. The returned and original data words are compared to verify the complete round-trip chain. The experiment confirms the results of (8) that the PIC transmission works flawlessly until the clock frequency of the fast node reaches $\approx$ 30-MHz (20% inaccuracy), which is in agreement with the theoretical bound of (7).

VII. CONCLUSIONS

In this paper, we have presented a detailed timing analysis for Pulsed-Index Communication protocols to meet robustness requirements. In the analysis, the inter-symbol delay parameter used to delimit the boundary between data segments in a packet plays a significant role. In addition to the gained insight into PIC, the analysis helps quantifying the trade-off between data rate and robustness. Based on our analysis, an inter-symbol delay coefficient α of 4 clock cycles and a pulse width w of half-clock cycle are recommended. These recommended design parameters have been experimentally verified using a Xilinx Virtex7 FPGA platform that illustrates both the simplicity, efficiency, and reliability of using PIC as a single-channel communication protocol for IoT devices. These additional PIC features augment the already proven ones of low-power, high data-rate, and small footprint [4].

VIII. ACKNOWLEDGMENT

This work has been supported by the Semiconductor Research Corporation (SRC) under the ATIC-SRC Center of Excellence on Energy-Efficient Electronic Systems (ACE⁴S), with funding from the Mubadala Development Company, Abu Dhabi, UAE.

REFERENCES

[1] C.A. dos Reis Filho, E.P. da Silva, E. de L. Azevedo, J.A.p. Seminario, and L. Dibb. Monolithic data circuit-terminating unit (dcu) for a one-wire vehicle network. *Solid-State Circuits Conference*, pages 228–231, September 1998.

[2] ETSI. *UICC - Contactless Front-end (CLF) Interface, Technical Specification, Version 7.3.0.* 2008-09.

[3] MAXIM. *OneWireViewer User's Guide, Version 1.4.* AN3358, 2009.

[4] Shahzad Muzaffar, Ayman Shabra, Jerald Yoo, and Ibrahim (Abe) M. Elfadel. A pulsed-index technique for single-channel, lowpower, dynamic signaling. *Design, Automation and Test In Europe(DATE)*, pages 1485–1490, March 2015.

[5] K.V.K.K. Prasad. *Principles Of Digital Communication System and Computer Network.* Dreamtech Press, New Delhi, 2003.

[6] Kee-Bum Shin, Ki-Hwan Seong, Dong-Hee Yeo, B. Kim, Jae-Yoon Sim, and H.J. Park. Verilog synthesis of usb 2.0 full-speed device phy ip. *International SoC Design Conference(ISOCC)*, pages 162–165, November 2013.

[7] S.-J. Song, N. Cho, and H.-J. Yoo. A 0.2-mw 2-mb/s digital transceiver based on wideband signaling for human body communications. *Journal of SolidState Circuits*, 42(9):2021–2033, September 2007.

978-1-4673-9141-2/15 $31.00 © 2015 IEEE

A Time Interleaved DAC Sharing SAR Pipeline ADC for Ultra-Low Power Camera Front Ends

Anvesha Amaravati, *Student Member, IEEE*, Manan Chugh, Arijit Raychowdhury, *Member, IEEE*,
School of Electrical and Computer Engineering, Georgia Institute of Technology.
email: {aamaravati3, manan.chugh}@gatech.edu, arijit.raychowdhury@ece.gatech.edu

Abstract—The growing need for ultra-low power cameras for sensors, surveillance and consumer applications has resulted in significant advances in compressed domain data acquisition from pixel arrays. In this paper we present a novel 64-input Successive Approximation (SAR) Pipeline analog-to-digital converter (ADC) suitable for compressed domain data acquisition in camera front-ends. The proposed architecture features a time interleaved capacitive digital-to-analog converter (DAC) shared between column parallel ADCs for area savings (2.28X); and a shared amplifier stage for power savings (60%). Simulations on a 130nm foundry process shows that the proposed SAR Pipeline ADC draws 31μW at 2MS/s having a target Figure-of-Merit (FOM) of 87fJ/conv. per step at Nyquist rate.

Keywords—*DAC sharing, SAR Pipeline, Compressive sensing.*

I. INTRODUCTION

Wearable devices for IOT (Internet of Things) require CMOS image sensor (CIS) with low power and area [1]. Traditional CIS for wearable devices consume power more than 50mW[2]. In a CMOS image sensor system column parallel ADCs along with digital processors consume most of the power[3][4]. In most of the reported image sensors, column parallel ADCs draw 50-65% of the power of the entire image sensor chip[5][1]. The power consumed by column parallel ADCs is proportional to the number of measurements to be performed by the ADC. It increases with the number of pixels. Cameras used in surveillance, continuous time monitoring, human machine interfaces etc., need to be "always on" and low-power is regarded as a key enabler for such next generation IoT devices.

Recently developed algorithms of compressive sensing (CS) promise to reduce the number of measurements with non-linear recovery at the back-end [6]. If the pixel values in a camera are represented as a discrete time signal $X = [x_1 x_2 x_3 x_4 \cdots x_n]^T$, the number of measurements needed in traditional column parallel ADCs will be equal to n. Instead of n samples, CS needs only m linear measurements ($m << n$). The CS measurement matrix is given by Eq. 1.

$$Y[m] = \phi[m,n] \times X[n] = \begin{pmatrix} 0 & 1 & \cdots & 1 \\ 1 & 1 & \cdots & 0 \\ \vdots & \vdots & \ddots & \vdots \\ 1 & 0 & & 0_{m,n} \end{pmatrix} \times \begin{pmatrix} x_1 \\ x_2 \\ \vdots \\ x_n \end{pmatrix} \quad (1)$$

Here $Y[m]$ is the m-dimensional measured array, ϕ is a random binary matrix of size $m*n$ and follows the "Independent and Identically Distributed (IID)" property. X is traditionally recovered at the back-end using an optmization algorithm, like determining the L_1 norm[6].

In this paper we present a novel pipeline-SAR ADC architecture with capacitive DAC sharing with the capability of acquiring linear combinations of 64 pixel data in a single conversion cycle. This is suitable for such compressed domain data acquisition.

II. ADC ARCHITECTURES FOR CS IMAGE ACQUISITION

In prior work for obtaining compressed domain data, both analog and digital techniques have been used to perform compressive measurements from the raw data. Typically, analog implementations of compressed sensing suffer from low Signal to Noise Ratio (SNR) and require an analog to digital converter to improve accuracy[7][8]. Resistor based compressed sensing multiplexor reported in [9], suffers from static power dissipation and the number of inputs (n) is limited, making it suitable for RF receiver applications only.

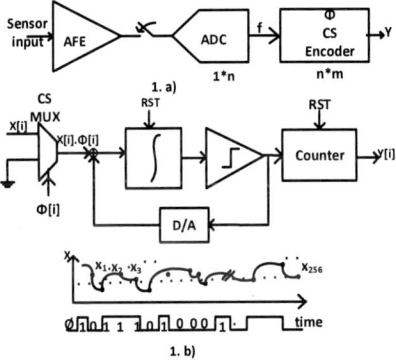

Fig. 1. a) CS encoder on acquired samples of ADC[10] b)Simultaneous compression and quantization within [5]

To overcome some of the disadvantages of analog CS circuits, [10] has proposed compression in the digital domain. Fig. 1. a) shows the technique proposed in [10]. The entire analog signal is converted into the digital domain by high-speed ADCs and the CS encoder does compression in the digital domain. This is mostly suited for bio-medical applications where the number of input channels are limited. However, for CIS of a typical 256*256 size, ADCs would need to acquire all the samples and then convert to the digital domain. The number of measurements by the ADC will not be reduced and it defeats the purpose of compressed domain data acquisition. Further, the size of digital CS encoder grows exponentially with the number of inputs. CS encoders will further add significant power along with the ADC making it infeasible for "always on" imaging applications.

To overcome the limitations of data acquisition followed by compressed domain measurements, Oike et.al, has proposed a CS camera through simultaneous averaging and quantization of pixels using a $\Sigma - \Delta$ ADC [5]. Fig. 1. b) shows the schematic of the resetting $\Sigma - \Delta$ ADC used for such linear measurements. Pixel values are multiplied with random numbers (from the ϕ matrix) sequentially and passed to the input of the $\Sigma - \Delta$ ADC. This approach requires m measurements; however it requires n conversion cycles for one measurement. This architecture requires a $16 * 16$ block for linear measurement. For each measurement of the block, the resetting $\Sigma - \Delta$ ADC needs 256 clock cycles. During this conversion period, all the high gain amplifiers will remain on and consume power. Hence, for lowering the total power dissipation, faster conversion with the opportunity for power gating once the conversion is complete, will be critical.

Once compressed domain data is acquired, the image is often used for online classification to detect potential trigger signals. For such in-situ classification [15] and trigger identification, 8bits of inputs are sufficient. Further, it has been shown that for most of the machine learning applications moderate resolution (6-8bits) is sufficient[1][11]. Fig. 2 shows the Energy per conversion with respect to the Signal to Noise and Distortion ratio (SNDR) for state of the art SAR, Pipeline and $\Sigma - \Delta$ ADCs. SNDR is related to effective number of bits ($ENOB = SNDR - 1.76/6$) [18]. From this plot we can observe that SAR ADC has best FOM (order of $10 - 1000 fJ/conv$) for moderate resolution (6-8 bits). Pipeline ADCs also have competitive FOM for moderate and high speed applications. Since most of the image sensors speed varies from 1MS/s to 10MS/s and we are interested in 8b of resolution for in-situ image processing applications, we propose a SAR-Pipeline ADC which achieves ultra-low power and high area efficiency.

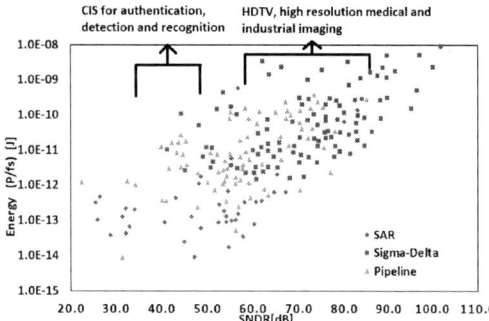

Fig. 2. Energy vs SNDR for state of the art reported SAR, Pipeline and $\Sigma - \Delta$ ADCs[14]

III. SAR-PIPELINE ADC ARCHITECTURE FOR CS MEASUREMENTS

For most of the low power applications SAR ADCs are used since they consume ultra-low energy per conversion(Fig. 2). But they are limited to low sampling rate & low resolution because of the DACs settling time for conversion (mainly set by MSB transition)[19] and large capacitance (for higher resolution). To alleviate this problem two-stage SAR Pipeline ADCs are proposed[19][20]. SAR-Pipeline uses two stage SAR-ADC and an amplifier which is used for amplifying residue generated by stage 1 SAR ADC (Fig. 3). Both the SAR ADC stages operates in parallel and each stage has to resolve lesser number of bits (lesser DAC settling time & capacitance as compared to traditional SAR). Therefore, SAR-Pipeline ADCs can operate at much higher speeds with high area efficiency [20]. One of the inherent advantage of SAR-Pipeline is residue voltage of Stag-1 SAR ADC is generated within its DAC after conversion phase. Hence this avoids extra DAC and clock phase to generate residue of Stage-1 unlike in traditional flash based Pipelined ADCs[20].

Fig. 3 illustrates the proposed ADC architecture. The design operates on a block size of $16 * 16$ (256 elements in pixel array). We propose SAR-Pipeline ADC consisting of 64-inputs Stage-1 SAR ADC resolving 4 bits (with 1bit redundancy) and Stage-2 SAR ADC resolving 5 bits. We propose time-interleaved DAC sharing for Stage-1 SAR ADC which provides a linear measurement of 64-inputs in a single cycle. We also share the amplifier (used for residue amplification) between 2 neighboring column parallel ADCs to save power. 64 inputs are simultaneously averaged and quantized using the SAR-Pipeline ADC. Post-conversion, 4 consecutive samples are averaged using a 10 bit accumulator and shift register. This allows us to average 256 samples in 4 sampling cycles.

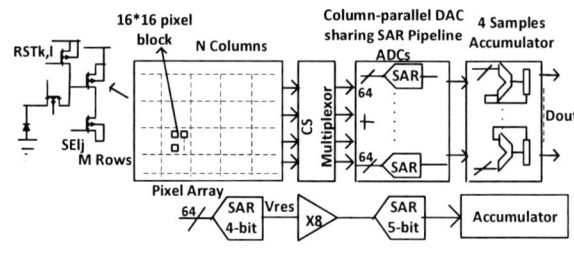

Fig. 3. Proposed CS front-end architecture for CIS

Fig. 4 shows a previously reported multi-input SAR ADC used for compressed sensing (with 8 bit resolution). It uses charge sharing. The MSB capacitor is equally divided among the inputs. Because of charge sharing the inputs will get averaged after the sampling cycle. [16] demonstrates a 4 input CS SAR ADC for wireless applications. This technique requires ($2^8 + 2^4 = 272C$) number of capacitors for an 8 bit ADC and measures 256 inputs (C is the unit capacitor). One of the main limitations of the proposed SAR ADC architecture for portable application is the area occupied by the sampling capacitors[17]. Dividing the MSB capacitors to accommodate 256 inputs requires 256 switches. For portable applications

978-1-4673-9141-2/15 $31.00 © 2015 IEEE

limited supply $\approx 1 - 1.3V$ provides high R_{ON}. This provides us the time constant (τ_{conv}) for conversion (min. sized capacitor of 50fF) of $\approx$ 220nsecs (DAC settling time). This allows a maximum sampling frequency of 730KHz. Hence, for high speed cameras (with 30frames/sec) the proposed ADC architecture will not be able to meet latency requirements.

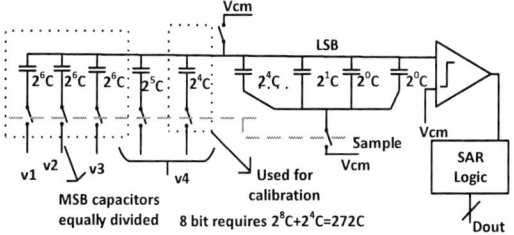

Fig. 4.   Reported multi-input SAR-ADC[16]

Fig. 5 is the proposed SAR-Pipeline with DAC sharing. We use 4 bit ADC as the first stage. Since 4-bit ADC has 16C capacitors, all the capacitors are divided into equal value of C and 16 inputs are applied. We have 3 instances of the same DAC which is used for accessing additional 48 inputs. Sampling is done in two phases. During sampling phase (S1) all 4 DAC's sample 16 inputs each. During second phase of sampling charge is redistributed between them. The averaged voltages across 4 DAC's during S1 phase given by Eq. 2.

$$V_{dac1} = \frac{(v_1 + v_2 + ... + v_{16})}{16}$$
$$\vdots \quad\quad\quad\quad\quad\quad (2)$$
$$V_{dac4} = \frac{(v_{48} + v_2 + ... + v_{64})}{16}$$

During the second sampling phase S2, averaging of V_{dac1} to V_{dac4} takes place. Therefore, the final voltage across DAC is given by Eq. 3.

$$V_{dacf} = \frac{(V_{dac1} + ... + V_{dac4})}{4}$$
$$V_{dacf} = \frac{(v_1 + ... + v_{64})}{64} \quad\quad (3)$$
$$V_{dacf} = \frac{(X[0].\phi[0] + ... + X[63].\phi[63])}{64}$$

We can observe form Eq. 3 that the final accumulated output represents the dot-product of the input pixel vector X with the sampling matrix, ϕ. ϕ can be random or programmed so that both random as well as structured compressed measurements can be obtained. As soon as S2 is done 3 DAC's are shared with neighboring column parallel ADC. Once the conversion in 4-bit SAR ADC is complete, we amplify the residue by 4x and pass it to a 5-bit fine ADC to resolve the LSBs. Ideally a gain of 16 is required for residue amplification. We use 1-bit digital redundancy in Stage 1 and half reference scaling for Stage 2 to reduce the gain requirement which helps to reduce the power in the high-gain op-amp [19].

Since all the capacitors we use are identical and of value C, calibration is not required (more details in section III). As 3 DAC's are shared with 4 ADC's, we need an additional capacitance of 12C. With 12C extra capacitance we can acquire

linear measurements of 64 inputs in each conversion cycle. This DAC shared method significantly improves area efficiency and enables simultaneous acquisition of multiple inputs. In this architecture, the conversion time-constant (τ_{conv}) is determined by the 4-bit ADC settling time even tough we are sampling 64-inputs. This makes the architecture suitable for high speed sensing with large number of inputs.

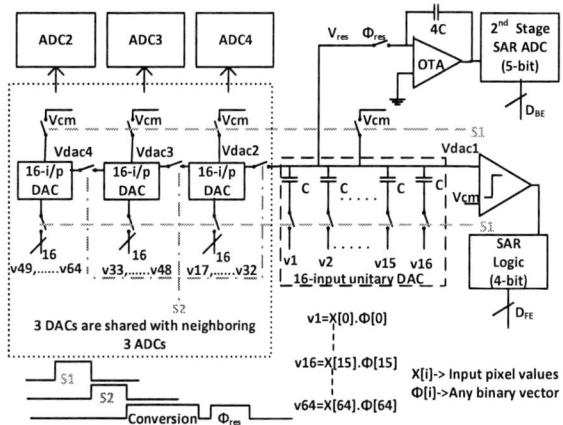

Fig. 5.   Proposed multi-input DAC sharing SAR ADC

Fig. 6 shows how the sampling schemes are time-interleaved for the entire column parallel ADC architecture. Conversion cycle for ADC is 8 clock cycles. During this period we share 3 DACs with 3 of the neighboring ADCs. S3 to S8 are sampling phases of ADC2 to ADC4. S3 to S8 phase operates during conversion period of ADC1. Pipelining facilitates overlapping of Stage-1 and Stage-2 conversion phases. 1-bit redundancy is added in the first stage to accommodate capacitor mismatch and offsets of the comparator, amplifiers[19]. We also share residue amplifier between two neighboring ADCs to reduce the total power[20]. Accumulator (10 bit) used to average 4 consecutive ADC output samples. The accumulator is reset after every 4 sampling cycles (F_s). The sampler operating at quarter sampling rate is used to capture the averaged output. Fig. 6 also shows the control logic used for proposed CS front-end ADC architecture. Global reset (RST) is used generate S1, S2 and conversion phase for ADC1. S2 phase of ADC1 is used to trigger sampling phase for neighboring column parallel ADC. This process is continued for all 4 ADCs.

IV. DESIGN COMPONENTS

In this section, the design details of the first and the second state of the ADC are discussed.

A. Stage 1 ADC and residue amplification

Fig. 7 shows the Stage 1 of the proposed SAR-Pipeline ADC. 64 inputs are acquired from S1 and S2. Residue is fed into an amplifier with gain of 4. Stage 1 of the ADC has 4 bit resolution with 1 bit digital redundancy. 1 bit redundancy is used to accommodate the residual offset of the comparator, op-amp and capacitor mismatch errors.

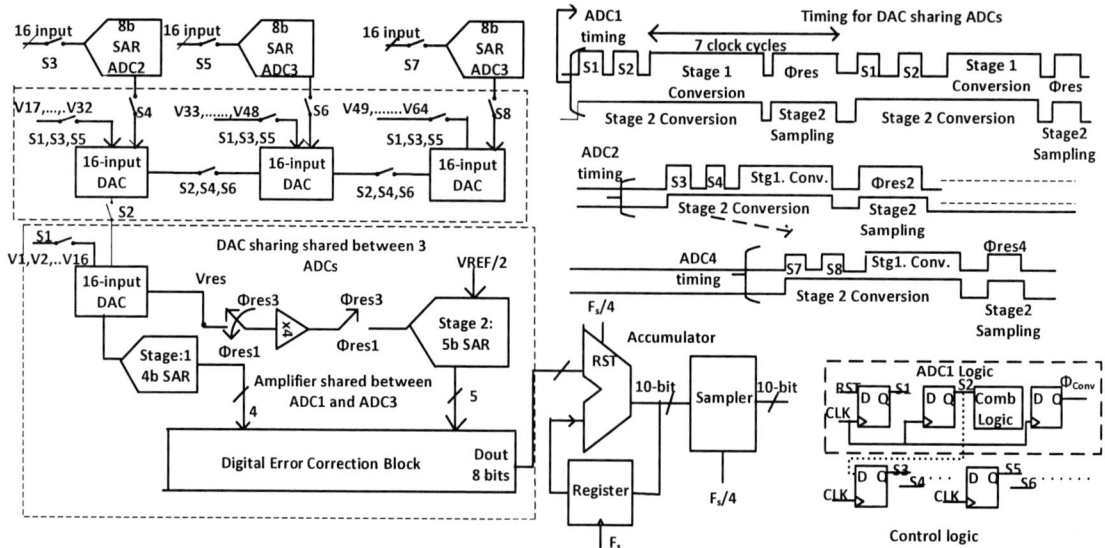

Fig. 6. Proposed SAR ADC with time-interleaved DAC sharing

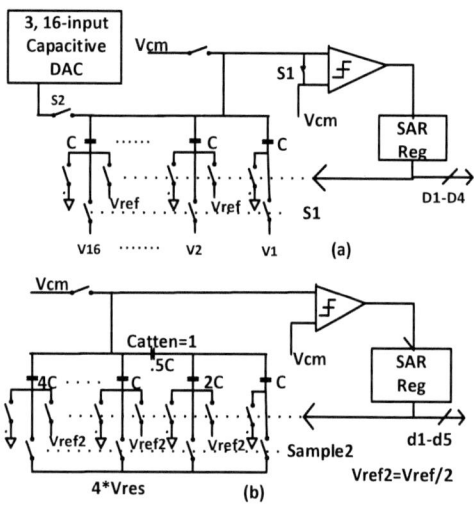

Fig. 7. Stage 1 and Stage 2 of the proposed ADC

The Op-amp open loop gain (A_{OL}), unity gain frequency (f_u) and swing ($Vp - p$) target based on the inter-stage gain is given in Table. I. The Target values are derived as per gain error, gain bandwidth (GBW) requirement of the OTA to be within 1/2LSB of the ADC error[20]. Simulated values across process corners are much above the Target values.

Inter-stage gain	Op-amp			2nd stage SAR
	A_{OL}	f_u	$Vp - p$	Offset
Target	42dB	42MHz	250mV	16.125mV
Simulated values	50dB	80MHz	300mV	8mV

TABLE I. DESIGN REQUIREMENT FOR AMPLIFIER AND 2ND STAGE OFFSET

We use pre-amplifier with output offset compensation to limit the offset of Stage 1 SAR ADC. The residual offset ($V_{os,res}$) is given by Eq. 4.

$$V_{os,res} = \frac{V_{os,pre-amp}}{A_p} + \frac{V_{os,latch}}{A_p} \quad (4)$$

where $V_{os,pre-amp}$ and $V_{os,latch}$ are the pre-amplifier offset and latch offset respectively. A_p is the pre-amplifier gain. The 3σ $V_{os,pre-amp}$ and $V_{os,latch}$ are 5mV and 30mV respectively. The gain amplifier features a cross coupled load which provides a high gain of 15. The residual offset is 2.33mV which is 0.25LSB of the sub-ADC.

We use telescopic cascoded OTA in the proposed design for residue amplification. Telescopic cascoded OTA has high power efficiency for a given gain bandwidth (GBW)[18]. Because of half gain and half reference implementation of the ADC, the open loop gain of the OTA is reduced. The OTA achieves a swing of 300mV_{p-p}.

B. Stage 2 ADC

Fig. 8 shows the Stage 2 of the proposed SAR-Pipeline ADC. We use a split capacitor architecture to reduce the area and power for the second ADC. Since the non-linearity of this ADC will get divided by the gain of the amplifier, it can be neglected. For the comparator in stage 2 of the proposed ADC, a pre-amplifier with gain of 3 is used since the offset requirement from it is 15mV. The total capacitance from the second ADC is 11C.

V. ANALYSIS OF CAPACITOR MISMATCH

Systematic variations has no effect of capacitor matching since all the capacitance in Stage 1 SAR ADC are equal to C. The capacitance mismatch standard deviation for metal-insulator-metal (MiM) is given by Eq. 5.

$$\sigma_{\Delta C/C} = \frac{A_{\Delta C/C}}{\sqrt{WL}} \quad (5)$$

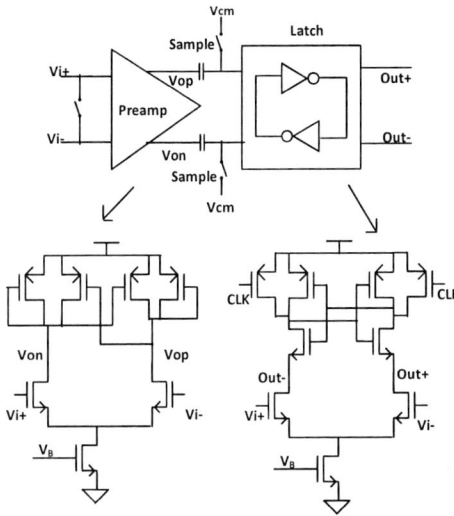

Fig. 8.   6- bit split cap SAR-ADC for Stage 2

where $A_{\Delta C/C}$ is process constant which is $1\%.\mu$ m for 0.13μm CMOS process[21]. W &/ L are width and length of the capacitor. The minimum size allowed in 0.13μm is $5\mu m *$ $5\mu m$. With minimum sized capacitor $\sigma_{\Delta C/C}$ obtained will be 0.002.

As per [22] maximum allowable capacitor mismatch for a resolution of n is given by Eq. 6.

$$\frac{\Delta C}{C}_{max} = \frac{2^n}{2^{2n} - 2^n + 1} \qquad (6)$$

For n=9, $\Delta C/C_{max}$ reaches close to 0.002. This shows the residue generated by first ADC will fall within the range of 1/8LSB of error. Hence the proposed architecture is robust towards capacitor mismatch.

VI.   SIMULATION RESULTS

Performance of the proposed SAR-Pipelined ADC is verified through design and simulations in the IBM 0.13μm CMOS.

Fig. 9 shows the normalized output frequency spectrum of the proposed ADC for input frequency (F_{in}) of 248.34kHz at sampling rate (Fs) of 1MSPS. A 1024-point FFT shows SNDR of 49.5dB which is equivalent to an ENOB of 7.9 for full scale input (dBFS). Fig. 10 shows the 64 inputs applied to ADC at each sampling cycle. Each 64 inputs corresponds to CS multiplexor output (Product of input vector with random number). Fig. 11 shows the ADC and accumulator outputs at each conversion cycles. For a particular case study, as shown in the figure, an ideal averaging without quantization results in a output of 270.11mV. The proposed ADC after accumulated 4 samples each provides an output of 269.53mV which is less than 1LSB of error. Fig. 12 shows the SNDR of the proposed ADC from input frequency range of 0.2MHz to to 0.98MHz. The ENOB at Nyquist frequency is 7.56. This ENOB achieves Walden FOM [18] of 85fJ/conv. step. Fig. 13 shows the DNL and INL of the proposed ADC across 256 digital codes. The

worst case DNL is within 0.4LSB. INL is within 1LSB across all digital codes.

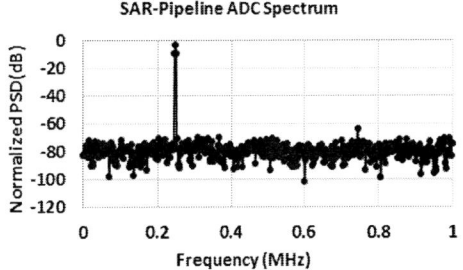

Fig. 9.   Frequency spectrum of the proposed ADC (Fin=248.34kHz & Fs=2MHz)

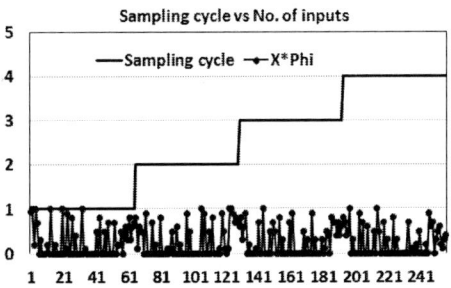

Fig. 10.   64 inputs for ADC for every 1 sampling cycle

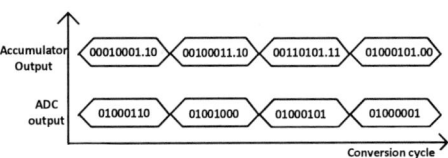

Fig. 11.   Output of ADC and accumulator for 4 conversion cycle

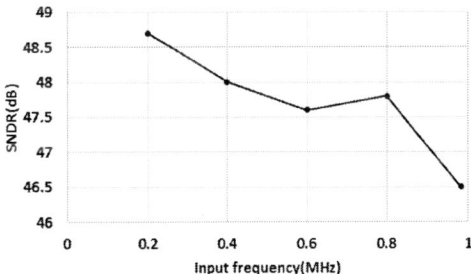

Fig. 12.   Simulation result of SNDR vs Input frequency at Fs=2MHz

The power budget for the proposed ADC is given in Table .II. Even though the total power consumed from the supply is 50μW, since the amplifier is shared between two ADC, the power for individual ADC's is 31μW.

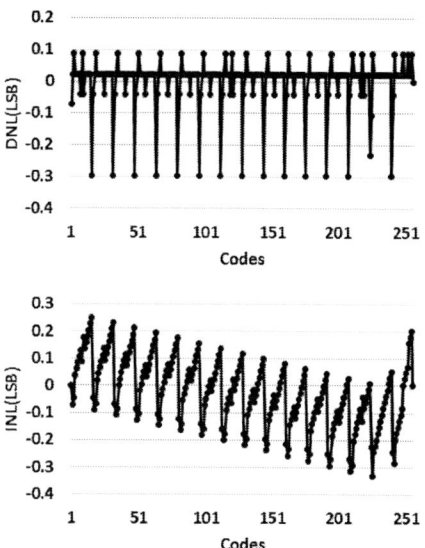

Fig. 13. DNL and INL of the proposed ADC across 256 codes

Block	Power/	Capacitance
SAR Stage1	7μW	28C
Amplifier	41μW	4C
SAR Stage1	2.5μW	10C
Accumulator	0.5μW	Nil

TABLE II.   POWER AND CAPACITANCE CONTRIBUTION FROM INDIVIDUAL BLOCKS

Table. III shows the comparison of the proposed design with state of the art CS architecture. Proposed design is scalable and can handle a large number of inputs at the same time.

	Oike [5]	Guo [16]	Chen [10]	This work
ADC type	$\Sigma - \Delta$	SAR	SAR	SAR-Pipeline
Technology	0.15μm	0.13μm	0.09μm	0.13μm
Design	Measured	Simulated	Measured	Simulated
No. of inputs	1	4	1	64
Sampling cycles	256	1	256	4
F_s	NA	1MHz	2kHz	2MHz
Capacitance	NA	272C	256C	40C
Power	NA	50μW	5μW	31μW

TABLE III.   COMPARISON WITH REPORTED WORKS

VII.   CONCLUSION

A novel time interleaved sharing DAC is integrated with multiple column parallel SAR Pipeline ADC. The proposed architecture can be used for wearable devices with ultra-low power requirements. Our design and simulation results show 87fJ/conv. step with an average power of 31μW.

REFERENCES

[1] J. Choi, J. Sin, D. Kang, Du-Sik Park , "A 45.5μW 15fps Always-On CMOS image sensor for Mobile and Wearable Devices," *IEEE Int. Solid-State Circuits Conf. (ISSCC) Dig. Tech. Papers*, pp. 114117, 2015.

[2] J. Deguchi et.al, "A 187.5Vrms-Read-Noise 51mW 1.4Mpixel CMOS Image Sensor with PMOSCAP Column CDS and 10b Self-Differential Offset-Cancelled Pipeline SAR-ADC," *IEEE ISSCC 2012*, pp. 494-496.

[3] J.-H. Park et.al, "A high-speed low-noise CMOS image sensor with 13-b column-parallel single-ended cyclic ADCs," *IEEE Trans. Electron Devices*, vol. 56, no. 11, pp. 24142422, Nov. 2009.

[4] T. Watabe et.al , "A 33 mpixel 120 fps cmos image sensor using 12 b column-parallel pipelined cyclic adcs," *IEEE ISSCC Dig. Tech. Papers, 2012*, pp. 388389.

[5] Yusuke Oike and Abbas El Gamal, "CMOS Image Sensor With Per-Column ADC and Programmable Compressed Sensing," *IEEE JSSC*, vol. 48, no. 1, Jan. 2013.

[6] D. L. Donoho, "Compressed sensing," *IEEE Trans. Inf. Theory*, vol. 52, no. 4, pp. 12891306, Apr. 2006.

[7] V. Gruev and R. E. Cummings, "Implementation of Steerable Spatiotemporal Image Filters on the Focal Plane," *IEEE TCAS-II*, vol. 49, no. 4, April 2002.

[8] R. Robucci et.al, "Compressive sensing on a CMOS separable-transform image sensor," *Proc. IEEE*, vol. 98, no. 6, pp. 10891101, Jun. 2010.

[9] J. P. Slavinsky et. al, "The Compressive Multiplexer for Multi-Channel Compressive Sensing ," *Proc. IEEE ICASSP*, pp.3980 -3983 2011.

[10] F. Chen, A. Chandrakasan and V. M. Stojanovic, "Design and Analysis of a Hardware-Efficient Compressed Sensing Architecture for Data Compression in Wireless Sensors," *IEEE JSSC*, vol. 47, no. 3, March. 2012.

[11] E. H. Lee et. al, "Factorization for Analog-to-Digital Matrix Multiplication," *Report, Standford University*, 2013.

[12] Y. Chae et.al , "A2.1Mpixel 120 frame/s CMOS image sensor with column-parallel ADC architecture," *IEEE JSSC*, vol. 46, no. 1, pp. 236247, Jan. 2011.

[13] T. Toyama et.al, "A 17.7 Mpixel 120 fps CMOS image sensor with 34.8 Gb/s readout," *IEEE Int. Solid-State Circuits Conf. (ISSCC)*, 2011, pp. 420422.

[14] B. Murmann, "ADC Performance Survey 1997-2015," *[Online]*, Available: http://web.stanford.edu/ murmann/adcsurvey.html.

[15] J. Zhang et. al, "A matrix-multiplying ADC implementing a machine-learning classifier directly with data conversion," *ISSCC*, 2015, pp. 1-3.

[16] W. Guo et. al, "A Single SAR ADC Converting Multi-Channel Sparse Signals," *Proc. IEEE ISCAS*, May 2013.

[17] D. G. Chen et.al, "A 64 fJ/step 9-bit SAR ADC Array With Forward Error Correction and Mixed-Signal CDS for CMOS Image Sensors," *IEEE TCASI*, vol. 61, no. 11, November 2014.

[18] M. Gustavsson et. al, "CMOS Data converters for communication," *Kluwer Academic Publishers*, 2000.

[19] C. C. Lee and M. Flynn, "A SAR-Assisted Two-Stage Pipeline ADC," *IEEE JSSC*, vol. 46, no. 4, April 2011.

[20] Y. Zhu et.al, "A 50-fJ 10-b 160-MS/s Pipelined-SAR ADC Decoupled Flip-Around MDAC and Self-Embedded Off. Cancel.," *IEEE JSSC*, vol. 47, no. 11, Nov. 2012.

[21] C. Diaz et. al, "CMOS tech. for MS/RF SoC," *IEEE Trans. Electron Dev.*, vol. 50, no. 3, pp. 8184, Mar. 2003.

[22] Z. Lin et.al, "Modeling of capacitor array mismatch effect in embedded CMOS CR SAR ADC," *Int. Conf. ASICs*, Oct. 2005.

High-Efficiency Voltage Regulation Stage in Energy Harvesting Systems

S. E. Kim, T. W. Kang, S. W. Kang, K. H. Park and M. A. Chung

Human Interface SoC Research Team
Electronics and Telecommunications Research Institute (ETRI)
Daejeon, Republic of Korea
sekim@etri.re.kr

Abstract— A high-efficiency voltage regulation stage in an energy harvesting system is presented. In an energy harvesting system, the availability of energy is uncertain. If energy is harvested from energy resources, the system operates; if not, the system does not work. Therefore, the energy harvesting system starts and stops repeatedly according to the presence or absence of energy. The repeated on and off state in the energy harvesting system decreases operating efficiency which is an important factor in such a system with the limited amount of harvested energy. To improve the efficiency of the system that repeatedly switch between on and off states, the energy dissipated in the transitional state should be minimized. In this work, a voltage regulation stage is implemented with two additional switches, two low power comparators and digital control logic in addition to the conventional regulation stage. The first switch determines the input voltage level of the regulation stage, and the second switch prevents undesired energy dissipation in the boosting state. When the regulation stage supplies power to a 100 ohm load resistor with a 100 uF load capacitor for 50 ms, the efficiency is improved up to 38%, and boosting time is reduced by 42% in comparison with the conventional regulation structure.

Keywords—energy harvester, power management, power switching, low power, regulation

I. INTRODUCTION

Energy harvesters generate electrical energy from natural resources, such as sunlight [1], vibration [2], heat [3], electromagnetic radiation [4], and so on, which are naturally replenished. An energy harvesting system collects and converts the energy generated from energy harvesters so that it can be utilized by electronic systems, such as wearable devices or WSN (Wireless Sensor Network). By using an energy harvesting system as a power source, the electrical energy required by powered systems can be obtained from nature, and the energy scarcity problem can be overcome.

However, energy harvesting systems have some features that distinguish them from conventional battery-powered systems. Among them, the uncertainty of energy availability is one of major differences. Their energy sources, such as sunlight, wind, and so on, are unstable, sporadic, and changing from time to time. According to the conditions of these natural resources, the frequency and the quantity of harvested energy vary, and it is hard to supply power continuously to a load in the energy harvesting system. Due to the uncertainty of energy

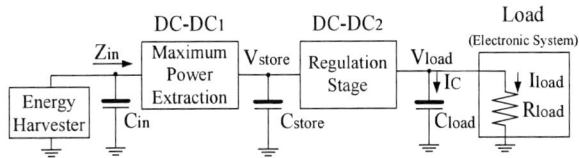

Fig. 1. Architecture of an energy harvesting system with an energy harvester

availability, the energy harvesting system turns on and off repeatedly. To reduce the energy dissipated in repeated switching between on and off states, the energy wasted in the transitional state should be minimized.

II. ANALYSIS OF TRADITIONAL ENERGY HARVESTING SYSTEM

The architecture of a traditional energy harvesting system consists of two stages of dc-dc converters [5], [6], as shown in Fig. 1. The first dc-dc converter extracts maximum power from an energy harvester by optimizing input impedance [6]. The energy extracted by the first converter is stored in a storage element, C_{store}. The second dc-dc converter regulates the output voltage to provide energy to electronic systems from the energy in C_{store}. The regulation stage adopts a switched boost topology, and the electronic system is simply modelled by a load resistor (R_{load}).

When the second dc-dc converter starts to operate, V_{load} rises from 0 V and reaches the regulated voltage after t_1 seconds. The regulated voltage is the voltage that can be applied to electronic devices. While V_{load} rises from 0 V to the regulated voltage, the energy continuously dissipated at the load (E_{load}) is defined as follows

$$E_{load}(joules) = P_{load} \cdot t_1 \qquad (1)$$

P_{load} is the wasted power at the load resistor, and t_1 is the time to make output voltage to the regulated voltage level. To save dissipated energy in the boosting state, P_{load} and t_1 have to be minimized in the boosting state. Here, I_{load} is the current through a load resistor, V_{load} is the output voltage, C_{load} is the load capacitor, and Q_c is the quantity of charge in C_{load}. The P_{load} is given as

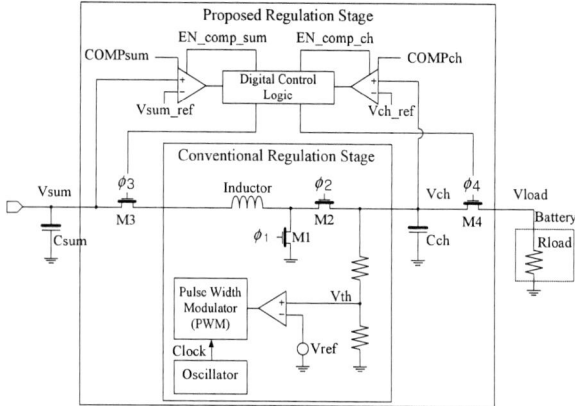

Fig. 2. Structure of the proposed regulation stage

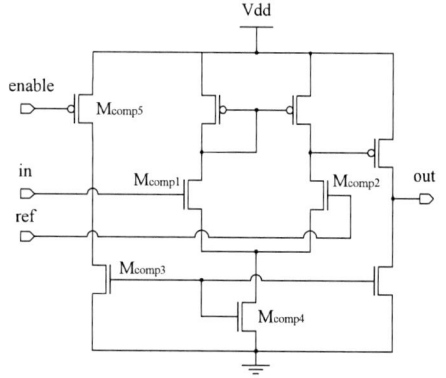

Fig. 3. The schematic of a comparator for the proposed regulation stage

$$P_{load} = I_{load} \cdot V_{load} = \frac{V_{load}^2}{R_{load}} \qquad (2)$$

By differentiating (3) with respect to t_1, the boosting time (t_1) can be derived as follows

$$Q_c = C_{load} \cdot V_{load} \qquad (3)$$

$$t_1 = \frac{C_{load} \cdot V_{load}}{I_c} \qquad (4)$$

In the boosting state, if R_{load} is small, lots of current flows into the R_{load}. As described in (2), small R_{load} increases P_{load}. Also small R_{load} causes small current to flow into the load capacitor. In (4), small I_c increases boosting time (t_1) in the boosting stage. Therefore, small R_{load} brings out more energy to be dissipated in boosting state. On the other hand, increased R_{load} minimizes the P_{load} during the boosting state and reduces boosting time, t_1. To prevent the energy to be wasted in the boosting state, the load resistance should remain high in the boosting state, and when the voltage reaches the regulated voltage level, a real load has to be connected to the energy harvesting system.

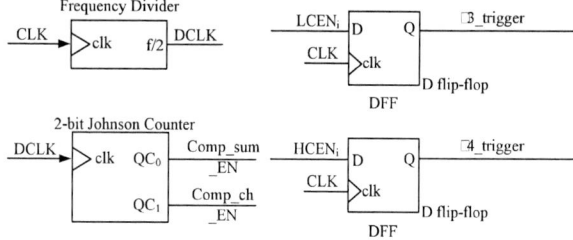

Comp_sum_EN : comparator_sum enable
Comp_ch_EN : comparator_ch enable

Fig. 4. The block diagram of trigger signal generator in digital control logic

III. THE STRUCTURE OF THE PROPOSED SYSTEM

In the proposed regulation stage, two switches (M3, M4), two low-power comparators ($Comp_{sum}$, $Comp_{ch}$) and digital control logic are added to the conventional regulation stage, as shown in Fig. 2. Low power comparators check periodically the magnitude of energy in the storage capacitors, C_{sum}, C_{ch}. The state of energy in C_{sum}, C_{ch} is informed to the digital control logic, and the digital control logic generates control signals, Φ_3, Φ_4, for analog path switches. Then the stored energy in C_{sum}, C_{ch}, is transferred through the analog path switches (M3, M4) according to Φ_3, Φ_4.

The analog path switches in the proposed system is designed with bilateral CMOS switch. Ideally, the path switches with no on-resistance, no distortion and zero time delay is desired. However, it is hard to implement real process technology. To minimize the limitations of real process technology, the path switches has to be designed with a bilateral CMOS device. The on-resistance of both NMOS and PMOS devices changes with channel voltage. This nonlinear resistance can causes errors in accuracy as well as distortion. The bilateral CMOS switch can solve this problem. On-resistance can be minimized, and linearity also can be improved.

The schematic of low power comparator is shown in Fig. 3. The comparators, $COMP_{sum}$, $COMP_{ch}$, check the magnitude of harvested energy which is stored in the storage capacitors, C_{sum}, C_{ch}. The role of the comparators is to inform the state of the stored energy in storage capacitors to digital control logic. When the stored energy is higher than threshold energy, the output of the comparator turn to high state, and digital control logic turns on the path switch. In this system, the path switch is controlled by the digital control logic, and there is no need for comparator's hysteresis. The comparators only inform the magnitude of the stored energy and digital logic determines the path to be connected. Without the need for hysteresis, the comparator becomes simple, and easy to implement.

The inputs of comparators are placed on gates of transistors, M_{comp1}, M_{comp2}, in order to show high impedance on the sensed nodes. High input impedance on the sensing node is important because it can minimize the effect of comparators on the harvester. Furthermore, to reduce power consumption in a

978-1-4673-9141-2/15 $31.00 © 2015 IEEE   238

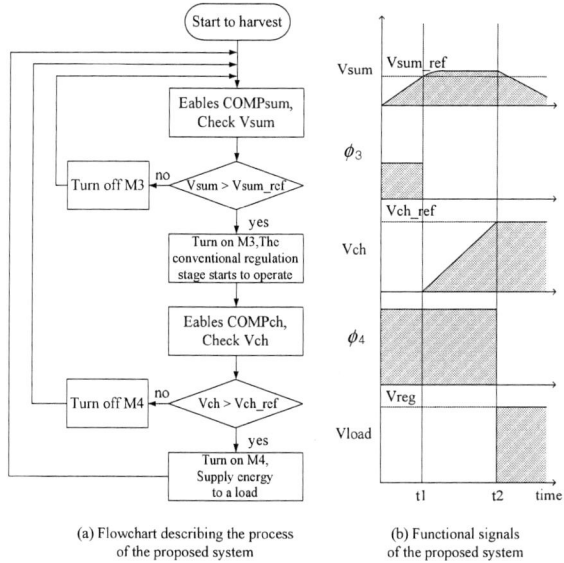

(a) Flowchart describing the process of the proposed system

(b) Functional signals of the proposed system

Fig. 5. Functional operation of the proposed system

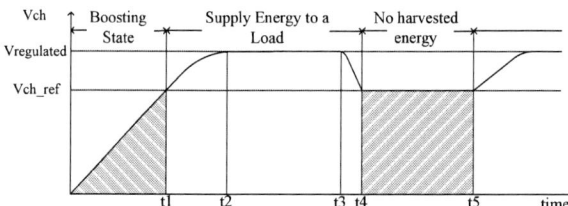

Fig. 6. Operation of the proposed regulation stage

comparator, the comparator only turns on, if operation is required. When operation is not requested by digital control logic, the bias current in comparators is blocked by M_{comp5}. By adapting enabling techniques in a comparator, the power consumption in a comparator can be minimized.

Digital control logic is composed of comparator enabling signal generator, and switch control signal generator. Of the two, the block diagram of comparator enabling signal generator is shown in Fig. 4. The clock signal generated from internal oscillator is divided by a frequency divider, and the johnson counter uses the divided clock as a digital clock. The comparator enabling signal is generated in 2-bit johnson counter. EN_comp_sum and EN_comp_ch signals enable the comparators to check the energy state of C_{sum}, C_{ch} periodically. After enabling the comparator, path switches, M3, M4, is triggered with D flip-flop.

IV. THE ALGORITHM OF THE PROPOSED SYSTEM

The flowchart in Fig. 5 (a) describes the process of the proposed system. When a harvester starts to generate electrical energy, electrical energy is stored in the storage capacitor, C_{sum}. The digital control logic periodically enables $COMP_{sum}$ to check the voltage in C_{sum}. By checking the voltage of the capacitor, the energy stored in capacitor can be estimated. As shown in Fig. 5 (b), at t_1, when V_{sum} is higher than the

threshold voltage, V_{sum_ref}, digital control logic turns on the analog path switch, M3. The M3 switch controls the input voltage level in order to improve operating efficiency. When V_{sum} is under the minimum input voltage level of the regulation stage, the energy in C_{sum} is exhausted without boosting output voltage. Low input voltage is also not good at the efficiency of the regulation stage because of the path resistance. To ensure that the energy is not exhausted, the M3 switch has to control the input voltage level of the regulation stage. When the input voltage is high enough to boost the output voltage, the M3 switch turns on.

To drive the conventional regulation stage, the power for starting up the pulse width modulator (PWM) is required. The value of V_{sum_ref} has to be determined based on the value of energy which is enough to drive the conventional regulation stage. After transferred the energy stored in C_{sum}, PWM starts to control M1 and M2 to boost the output voltage up to the regulated voltage level as in the conventional boost topology. During the boosting operation, the M4 switch remains in the off state and maintains a high load resistance. The high load resistance retrains the current into the load resistor until V_{load} reaches the regulated voltage. After the load capacitor is charged to the regulated voltage level, $COMP_{ch}$ check the energy state of C_{ch}, and the M4 switch turns on. The boosted energy is transferred to the load resistor through M4. The increased output load resistance during boosting state minimizes wasted power in transitional state. In Fig. 5 (b), at t2, when V_{ch} reaches the reference voltage, V_{ch_ref}, for M4, the M4 switch turns on, and the energy in C_{ch} starts to be transferred to the load resistor.

The waveform of V_{ch} in the proposed regulation stage is shown in Fig. 6. V_{ch} starts up to boost from 0 V, and at t_1, V_{ch} reaches the reference voltage for M4. The reference voltage for M4 is the minimum voltage available in electronic systems. From t_1, the M4 switch turns on, and the regulation stage starts to supply harvested energy to the load and keeps boosting V_{ch} up to the regulated voltage level. At t2, V_{ch} reaches the regulated voltage level. Then, the voltage regulation stage provides energy to the load with the regulated voltage until t3. At t3, the energy harvester stops providing energy. The energy stored in C_{store} starts to be reduced by supplying energy to the load, and V_{ch} starts to drop. When the output voltage drops under the reference voltage for M4, the regulation stage stops transferring energy to the load by turning off the M4 switch. Thus, the energy stored in C_{ch} can be maintained. The next time energy is harvested, the energy is transferred directly to the load without the boosting operation. Therefore, the boosting power wasted in the conventional regulation stage can be minimized.

V. SIMULATION AND MEASUREMENT RESULTS

The wasted energy in the boosting state of the conventional regulation stage is simulated and the results are shown in Fig. 7. Load resistance is increased from 100 ohm to 10 kohm. As mentioned in section II, the wasted energy in the boosting state increases, as the load resistance decreases. The required energy to boost output voltage is 8.325 mjoules with 100 ohm load resistance and 1 μjoule with 100 kohm. A small load resistance induces more current to flow to the load resistance before the

978-1-4673-9141-2/15 $31.00 © 2015 IEEE   239

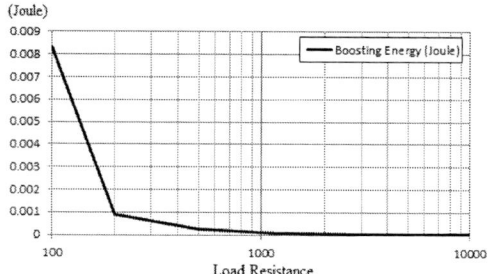

Fig. 7. Wasted energy in the boosting stage vs. load resistance

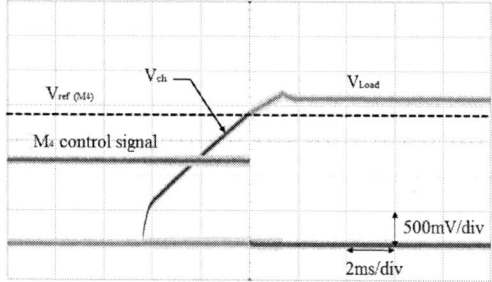

Fig. 8. Measured output waveforms of the proposed regulation stage

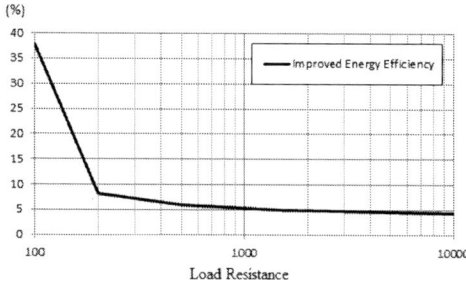

Fig. 9. Improved efficiency vs. load resistance

output voltage of the regulation stage reaches to the regulated voltage. Moreover, if the size of inductor and operating frequency of the conventional regulation stage are fixed, more time is needed to increase the output voltage.

Some parts of the proposed regulation stage were implemented in TSMC 0.13um CMOS process, and commercial dc-dc converter product is adopted. The measurement test was conducted with a 2.2 µH inductor and a 100 uF load capacitor. The measurement results were compared with the results of the conventional regulation scheme. The regulation stage has a regulated output voltage of 2.1 V. The reference voltages for M3 and M4, V_{sum_ref}, V_{ch_ref}, are 0.6 V and 1.9 V, respectively. When the input voltage is higher than 0.6 V, the M3 switch turns on and starts up the regulation stage to boost. When V_{ch} is higher than 1.9 V, the regulation stage starts to drive the load resistor.

The measurement result of output waveforms with a 200 ohm load resistor is shown in Fig. 8. Until V_{ch} boosts up to the reference voltage level for M4, the M4 switch keeps off-state maintaining high load resistance. The high load resistance in

the boosting state prevents undesired energy dissipation in the boosting state and reduces boosting time from 7 ms to 4 ms. When V_{ch} reaches the reference voltage level for M4, the M4 switch turns on and changes the load voltage (V_{load}) from 0 V to 1.9 V immediately. The energy stored in C_{ch} is transferred to the load, and V_{ch} keeps rising to 2.1 V, the regulated output voltage.

When the regulation stage supplies energy to the load for 50 ms, the energy efficiency is improved over that of the conventional structure as shown in Fig. 9. Depending on the size of the load resistance, the efficiency is improved by at least 4.4% up to as much as 38%. The smaller load resistance saves more energy in the boosting state because it dissipates more power in the boosting state. When the load resistance is 100 ohms, the regulation stage supplies 22 mJ to the load, and 8.3 mJ is saved in the boosting state. The efficiency can also vary according to the operating time. A shorter operating time of the regulation stage and a smaller load resistance make the efficiency better.

VI. CONCLUSIONS

A high efficiency voltage regulation stage for sporadic energy harvesters was presented. To prevent energy being wasted in the transitional state of the regulation stage, both input and output ports should be controlled. An improper input voltage level wastes energy without boosting the output voltage, and to minimize the energy dissipated during the boosting operation, the boosted output energy should be transferred to the load resistor after the load capacitor is charged up to the regulated voltage level. The proposed regulation stage improves efficiency up to 38% and reduces boosting time by 42% over the conventional regulation structure with a 100 ohm load resistor, a 100 µF load capacitor, and 50 ms operating time. The measurement results show that the efficiency of the regulation stage can be improved by saving boosting energy, and the energy harvesting system can supply more energy to load systems.

REFERENCES

[1] G. K. Ottman, H. F. Hofmann, and G. a. Lesieutre, "Optimized piezoelectric energy harvesting circuit using step-down converter in discontinuous conduction mode," IEEE Trans. Power Electron., vol. 18, no. 2, pp. 696–703, Mar. 2003.

[2] Y. Qiu, C. Van Liempd, P. G. Blanken, and C. Van Hoof, "5µW-to-10mW Input Power Range Inductive Boost Converter for Indoor Photovoltaic Energy Harvesting with Integrated Maximum Power Point Tracking," no. June 2007, pp. 300–301, 2011.

[3] Z. Wang, V. Leonov, P. Fiorini, and C. Van Hoof, "Realization of a wearable miniaturized thermoelectric generator for human body applications," Sensors Actuators A Phys., vol. 156, no. 1, pp. 95–102, Nov. 2009.

[4] M. Dini, M. Filippi, A. Costanzo, A. Romani, M. Tartagni, M. Del Prete, and D. Masotti, "A Fully-Autonomous Integrated RF Energy Harvesting System for Wearable Applications," pp. 987–990, 2013.

[5] N. J. Guilar, R. Amirtharajah, P. J.Hurst, and S.H. Lewis, "An energy aware multiple-input power supply with charge recovery for energy harvesting applications," in IEEE ISSCC Dig. Tech. Papers, Feb. 2009, pp. 298–299.

[6] Y. K. Ramadass and A. P. Chandrakasan, "A battery-less thermoelectric energy harvesting interface circuit with 35 mV startup voltage," IEEE J. Solid-State Circuits, vol. 46, no. 1, pp. 333–341, Jan. 2

A Fully On-Chip 25MHz PVT-Compensation CMOS Relaxation Oscillator

Hamed Abbasizadeh, Behnam Samadpoor Rikan, and Kang-Yoon Lee

College of Information and Communication Engineering

Sungkyunkwan University, Suwon, South Korea

Email: [hamed, behnam, klee]@skku.edu

Abstract—A fully on-chip, low-power and small area CMOS Relaxation Oscillator (ROSC) with voltage integral feedback structure and a new Bandgap Reference voltage (BGR) for accurate oscillation frequency independent of the PVT and comparator's delay variations is presented. The designed circuit uses a new bandgap reference to generate the reference voltage required by relaxation oscillator which allow variations due to voltage and temperature to be compensated. Another merit of this oscillator is that the phase noise at low-offset frequency is suppressed by the voltage integral feedback circuit. The frequency of the relaxation oscillator is determined by the RC response time. Thus, the current and capacitance are controlled by temperature and process compensation circuits to compensate for the frequency variation. The ROSC is implemented in a $0.18\mu m$ CMOS technology and its active area is $0.14mm^2$. The target frequency is 25MHz and current consumption is $22\mu A$, where V_{DD} is 1.8V. The oscillation frequency variation for V_{DD} ranges from 1.4 to 1.9V is 0.2% and for temperature ranges from -40 to 125°C is 0.18%.

I. INTRODUCTION

On-chip oscillators [1]−[7] are widely used in various low-power and low-cost applications such as biomedical devices, wireless sensor networks, micro-controllers,high speed interfaces and SoCs. A high reliable ROSC can replace a quartz crystal oscillator if it is designed as temperature and supply independent with digital trimming to counterbalance process variation. Since the quartz crystal oscillators cannot be integrated within a chip, A CMOS realization of a reliable oscillator reduces the power, the size and the cost for system-on-chip (SoC) implementations [10].

The objective of this paper is to design an oscillator with constant frequency over the variation of process, voltage and temperature. It is aimed for on-chip applications. The off-chip crystal oscillators are apparently not suitable. LC oscillators usually consume a large area, which is not a good option either for on-chip application. RC oscillators, including ring oscillators [8], Wien bridge oscillator [9] and relaxation oscillators, have been widely studied and developed for on-chip application due to their high compatibility with the standard CMOS process and their small area. In general situation, Wien bridge oscillator and ring oscillators show more frequency variation than relaxation oscillators. Many designs on Wien bridge oscillator, ring oscillators, relaxation oscillators or a hybrid of them have managed to maintain the frequency variation around 1% over process, voltage and temperature variation.

This paper presents a 25MHz relaxation oscillator, implemented in a standard $0.18\mu m$ CMOS process, as the on-chip

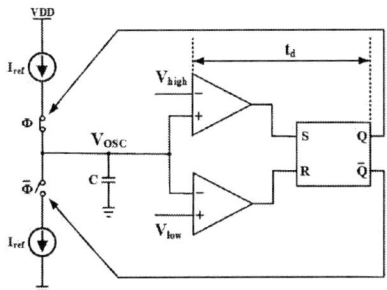

Fig. 1.   Conventional RC-oscillator

clock source for the ADC and Phase Locked Loop (PLL) in wireless power transfer system. The simulation results show that the oscillator with PVT compensation indeed performs better output frequency accuracy as well as jitter performance as compared to conventional oscillator. This paper is organized as follows; in Section II the principle of our ROSC is described, and then the proposed circuit design in $0.18\mu m$ CMOS is presented. Section III reports the performance results, followed by the conclusion.

II. ROSC CIRCUIT DESCRIPTION

A. Conventional Relaxation Oscillator

A conventional CMOS relaxation oscillator which shown in Fig. 1 operates by charging the capacitor C with a fixed current I_{ref} and comparing the ramp voltage ($V_C = V_{OSC}$) on the capacitor with a reference voltage ($V_{ref} = V_{high,low}$). When V_C reaches the V_{ref}, the ROSC resets the ramp voltage and generates it again. By repeating charge and discharge operations, ideally, the oscillator's periodic time is proportional to R * C. However, when V_C is higher than V_{ref}, the comparator will not flip its output immediately due to the comparator's delay time (t_d) and offset voltage (V_{off}). They degrade the frequency stability. The oscillator period T_{clk} can be written as

$$T_{\text{clk}} = 2(RC + t_{\text{d}} + V_{\text{off}}.C/I_{\text{ref}}), and (f_{\text{clk}} = \frac{1}{T_{\text{clk}}}) \quad (1)$$

where t_d and V_{off} are error sources varying with process, voltage, and temperature.

B. Proposed Relaxation Oscillator

1) Relaxation Oscillator Core: Fig. 2 shows a block diagram of the proposed relaxation oscillator. The equation

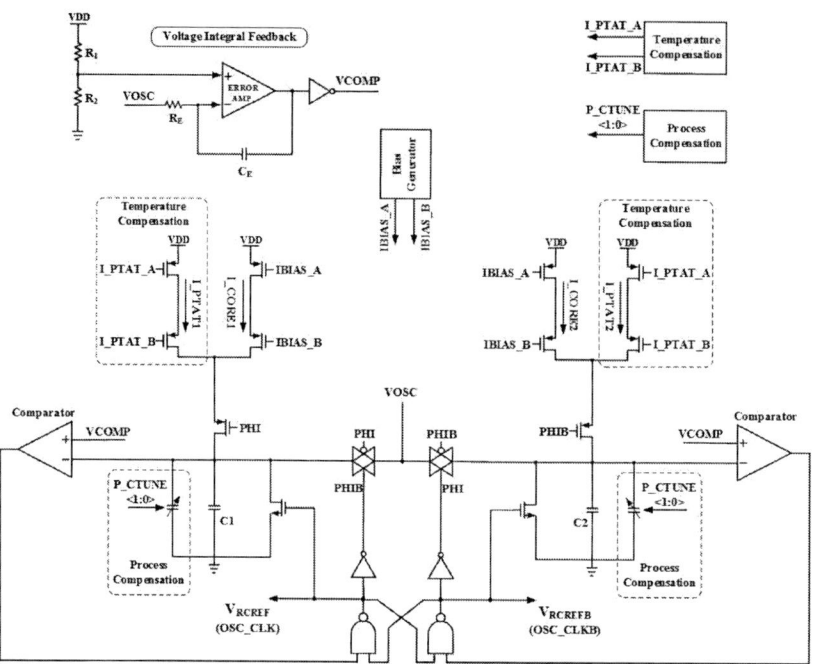

Fig. 2. A block diagram of the proposed Relaxation Oscillator

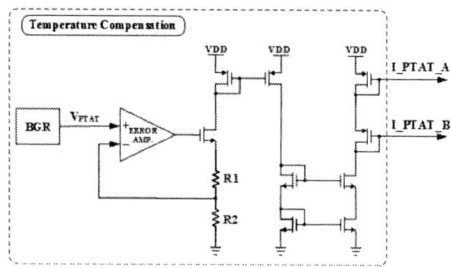

Fig. 3. The proposed temperature compensation circuit

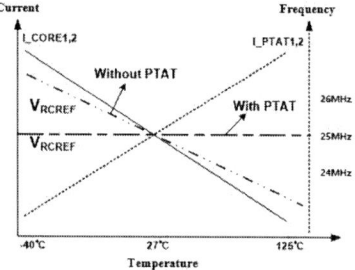

Fig. 4. The operation principle of the proposed temperature compensation using PTAT

of relaxation oscillator output frequency is as shown below. From Eq.(2), (3) and (4) relaxation oscillator is determined by current, capacitor, and supply voltage. By changing these 3 value we have different output frequency.

$$Q = CV = It \qquad (2)$$

$$t = \frac{CV}{I}(S) \qquad (3)$$

$$f = \frac{1}{t} = \frac{I}{CV}(Hz) \qquad (4)$$

where Q, C, I, V, f, t are quantity of electric charge, Capacitance, Current, Voltage, Output frequency and Output time of the ROSC respectively.

The external factors and temperature capacitor have a minimum effect on the supply voltage. The amount of current change is critical to the transistor characteristic. To compensate this effect, proportional to absolute temperature (PTAT) and Gain-Boosting circuit are adopted to BGR (Bandgap Reference) to control relaxation oscillator current, and P_CTUNE$< 1 : 0 >$ is added to control effect of process by BGR.

Adopting this two methods, the circuit generates constant frequency, ignoring the effect of temperature and process. The proposed Gain-Boosting circuit for temperature compensation is shown in Fig. 3. The internal current mirror of Gain-Boosting circuit is composed to cascade a current mirror, to minimize the effect of supply voltage variation.

Fig. 4 shows the operation principle of the proposed temperature compensation using PTAT. In Fig. 4, if PTAT is not applied, the frequency varies from 23.5MHz to 26.5MHz. To control the current of the relaxation oscillator, the PTAT of BGR senses temperature, and if it is relatively low, the current flow into the core increases, and V_{PTAT} lowers. The changed V_{PTAT} voltage compensates for the variation using the Gain-Boosting circuit in Fig. 3. On the other hand, if the temperature is high, the current flow into the core reduces, and the V_{PTAT} voltages increases. The changed V_{PTAT} voltage compensates for the reduced current by the Gain-Boosting circuit. Thus, these two methods keep the frequency constantly with high accuracy.

Fig. 5 shows the proposed process compensation circuit. To compensate for the process variation, voltage V_P is generated,

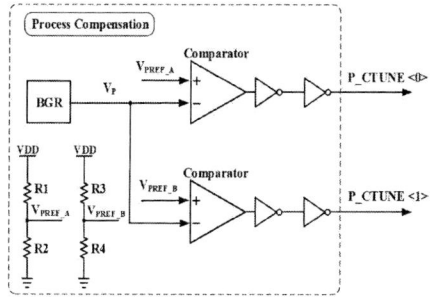

Fig. 5. The proposed process compensation circuit

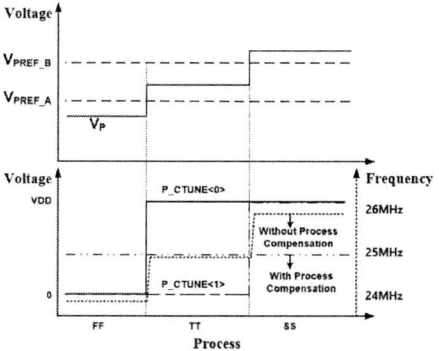

Fig. 6. The operation principle of the proposed process compensation circuit

which has the least dependency on temperature and is critical to the process variation by using the BGR. The role of voltage V_P is to compare V_{PREF_A} with V_{PREF_B}. The 2 bits signal (P_CTUNE$< 1 : 0 >$) generated from the comparators, controls the capacitor from the capacitor bank of the relaxation oscillator core.

Fig. 6 shows the operation principle of the proposed process compensation circuit. When the relaxation oscillator operates without the process compensation circuit, the frequency varies 24MHz to 26MHz with respect to the process variation. However, with the process compensation, the frequency is kept constant.

2) A New Bandgap Reference Voltage Design: In order to have a stable voltage reference with high PSRR and low temperature sensitivity with supply independency for temperature and process compensation circuits, a bandgap circuit is needed. Fig. 7 illustrates deployment of proposed low-voltage high PSRR bandgap reference circuit. This voltage reference must be able to completely remove supply voltage noise at lower frequencies and maintain this ability at higher frequencies. A current mode regulator for isolation of the area between supply source, and BG reference circuit is used. In fact a constant voltage, which is almost independent from supply source V_{reg}, is created used to supply BG reference circuit. Because of supplying operational amplifier with a voltage of V_{DD}, and inadequacy of op-amp PSRR, supply source ripple is initially transmitted to op-amp, then to V_{reg}. This results in incomplete independence of reference voltage generator circuit from supply voltage variations.

At proposed structure, the adjusted V_{reg} voltage, in addition to supplying BG reference circuit, also addresses sup-

plying error amplifier, which contributes to complete independence of circuit from supply source. In addition, modifications have been made to BG reference circuit configuration, which results in ease of reference voltage value determination, and easily achieving zero temperature coefficient, at desirable voltage, in comparison to previous configuration. Performing *KVL* at op-amp inlet circuit, will result in:

$$I_{d2} = \frac{V_{EBQ1} - V_{EBQ2}}{R_1} = \frac{V_T ln(N)}{R_1} \quad (5)$$

Considering $I_{d2} = I_{d3}$ we also will have:

$$V_{ref} = [R_2 \| (R_3 + R_4)]I_{d3} + (\frac{R_3 + R_4}{R_3 + R_4 + R_2})V_{EBQ3} \quad (6)$$

Additionally at bias node we will have:

$$V_{bias} = (\frac{R_4}{R_3 + R_4})V_{ref} = \alpha V_{ref} \quad (7)$$

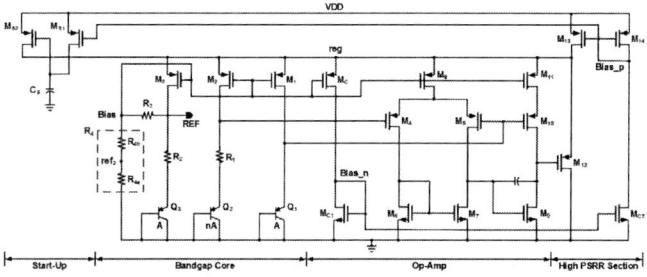

Fig. 7. The proposed low-voltage BG reference circuit

Which gives us the fact that variation of this voltage is alike to reference voltage. At Fig. 7 in order to guarantee activeness of M_8, and M_{11} transistors, V_{Bias} value is adjusted very closely to $V_{ref}(\alpha \approx 1)$. To deploy the op-amp, a two-stage amplifier is used which in addition to supplying sufficient loop gain, has acceptable phase margin, and bandwidth. At this design V_{ref}=800 mV, and at the event of requiring a less than 0.8V reference voltage, we can break R_4 resistance into two

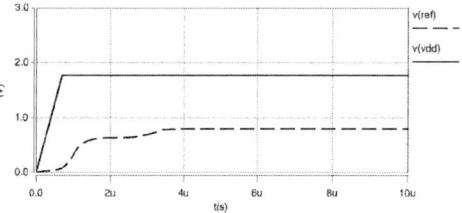

Fig. 8. V_{BG} during start-up

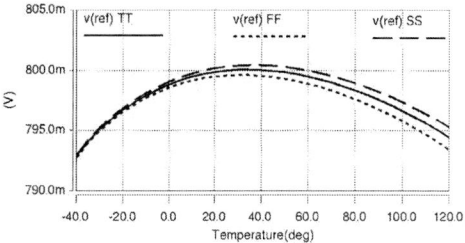

Fig. 9. Bandgap reference simulations given different process corners

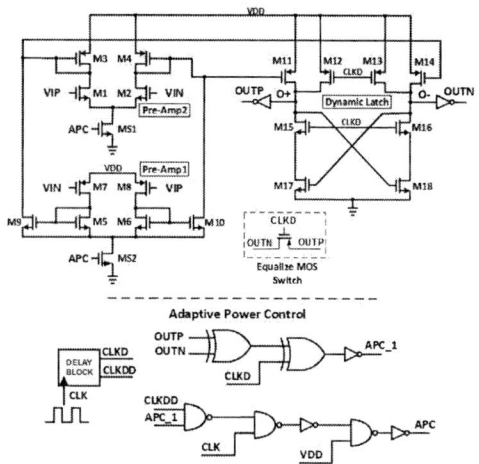

Fig. 10.   Schematic of the comparator

TABLE I.   SIGNALS STATE

CLK	O+	O-	APC	STATE
0	1	1	CLK	Reset
0→1	0	1	0→1→0	Evaluated
0→1	1	0	0→1→0	Evaluated

smaller ones, and without imposing any other modifications, we can create this voltage. For example V_{ref_2}=500 mV, at this design. M_{S1}, and M_{S2} transistors, and also C_S capacitor work as start-up circuit. Initially M_{S1} transistor is on, and reg node voltage is placed at an acceptable range, then generating adequate current at M_1, M_2 transistors, and transmitting it to M_{14}, M_{S2} capacitor is recharged, and M_{S1} transistor will go off.

In order to analyze start-up circuit, imposing the step input on supply voltage, the transient response, according to Fig. 8 is simulated, which shows in less than 4µsec, reference voltage would attain its infinite value. The output is a steady voltage of 800mV as input voltage varies between 1.2V to 1.8V, with a max variation of 8mV.

Besides, the temperature coefficient of the proposed circuit is only 60ppm/°C for temperature range from -40 to 125°C. In addition, the transient response of the BGR is 4µsec, and has PSRR of -120dB at DC frequencies. The topology of the circuit allows it to be portable to several different CMOS processes with minimal redesign effort. As shown in Fig. 9, V_{REF} has a variation of 8mV from the mean value 800mV given different process corners.

3) Comparator: Fig. 10 shows the schematic of the rail-to-rail comparator used in ROSC circuit which consists of a two pre-amplifier and a dynamic latch structure to achieve fast-

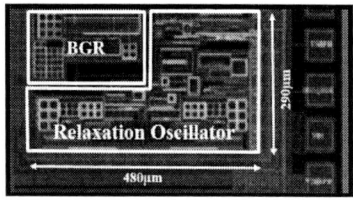

Fig. 11.   Chip Layout Pattern

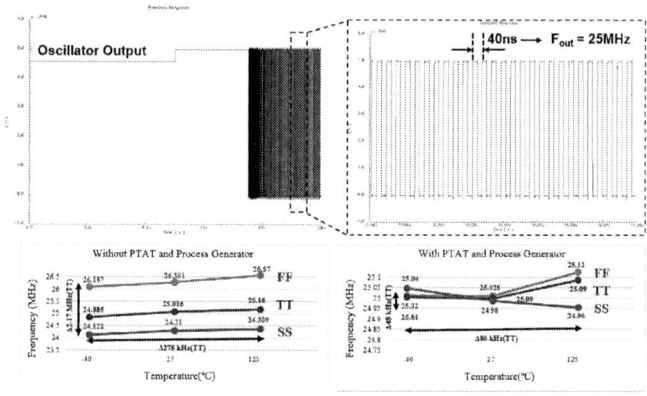

Fig. 12.   Transient and process corner simulations of relaxation oscillator With and W/O compensated circuits over the temperature range of -40 to 125°C

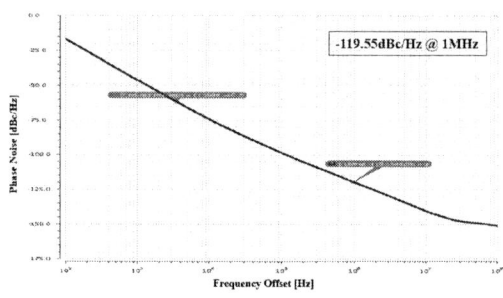

Fig. 13.   Phase noise performance of the ROSC @ 1.8V

decision and high-resolution [11]. An adaptive power control (APC) technique is used to reduce the kick-back noise and the power dissipation of the pre-amplifiers. Table I shows the signals state of the proposed comparator circuit. The simulation result shows that the proposed design exhibits a high resolution of 13bit and an ultra low power consumption with 0.2mV offset and 2.4nsec output propagation delay.

III.   POST-LAYOUT SIMULATION RESULTS

The proposed relaxation oscillator is simulated in 0.18µm CMOS technology and its active area is 0.14mm². The chip layout pattern of this oscillator is shown in Fig. 11. The target frequency is 25MHz and current consumption is 22µA, where V_{DD} is 1.8V and temperature varies from -40 to 125°C. Fig. 12 shows the transient and process corner post-layout simulations of relaxation oscillator with and without process voltage temperature compensation circuit over the temperature range of -40 to 125°C.

Fig. 13 illustrates the simulation phase noise performance of the relaxation oscillator at 1.8V. It can be clearly seen that, at a large offset frequency of 1MHz,the phase noise is -119.55dBc/Hz. Fig. 14 shows f_{osc} versus the supply voltage variation from 1.4V to 1.9V at 27°C. The variation in f_{osc} is ±0.2%. Fig. 15 shows f_{osc} versus temperature, f_{osc} varies by ±0.18% for a temperature variation from -40 to 125°C.

Advanced design techniques have been used to obtain a high performance oscillator, of which the specifications are listed in Table II. In Table II the FOM is defined as the ratio of the power consumption to the oscillation frequency (nW/kHz).

TABLE II.   PERFORMANCE SUMMARY AND COMPARISON

Parameter	[1]	[3]	[6]	[7]	This Work
Type	Relaxation	Relaxation	Relaxation	Relaxation	Relaxation
Technology	$0.18\mu m$	130nm	$0.25\mu m$	$0.18\mu m$	$0.18\mu m$
Area[mm^2]	0.04	0.016	1.6	0.09	0.14
Frequency [MHz]	14	1.2	7	0.00666	25
Current Consumption [μA]	24	5.8	625	0.63	22
Supply Voltage [V]	1.8	1	2.4	1.5	1.8
Freq. Variation[%] with Temp.	$\pm0.19@\text{-}40\sim125^{o}C$	$\pm1.8@\text{-}40\sim80^{o}C$	$\pm0.78@\text{-}40\sim125^{o}C$	$-0.62\sim+0.29@\text{-}40\sim120^{o}C$	$\pm0.18@\text{-}40\sim125^{o}C$
Freq. Variation[%] with V_{DD}	$\pm0.16@1.7\sim1.9V$	$\pm1.8@1.01\sim0.99V$	$\pm0.31@2.4\sim2.75V$	$-0.86\sim+0.12@0.8\sim1.8V$	$\pm0.2@1.4\sim1.9V$
FOM [nW/kHz]	3.085	4.833	214.285	141.141	1.584

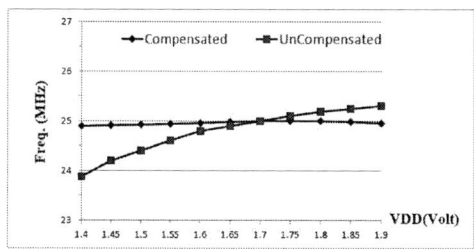

Fig. 14. Compensated and uncompensated oscillation frequency versus supply voltage

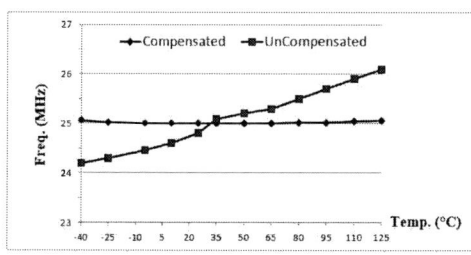

Fig. 15.   Compensated and uncompensated oscillation frequency versus temperature

The calculated FOM is 1.584nW/kHz for the proposed relaxation oscillator. It achieves the best FOM, which is at least three times better than other works. Its temperature coefficient is also good among the low-power relaxation oscillators.

IV. CONCLUSION

Table II summarize the performance of the proposed relaxation oscillator against prior arts. The voltage integral Feedback, Process, voltage and temperature compensation circuits optimize the performance of our ROSC. The circuit operates over wide temperature and supply voltage ranges. The oscillator offers a good balance between power, area, oscillation frequency. Our design achieves the best FOM for power, high output clock frequency and low process supply-voltage temperature sensitivity at the same time. The proposed circuit is possible to have equivalent compensation performances for shrinked technologies and for associated lower voltages. Finally, a 25MHz relaxation oscillator is realized with the FOM of 1.584nW/kHz.

ACKNOWLEDGMENT

This work was supported by the National Research Foundation of Korea (NRF) grant funded by the Korean government(MSIP)(2014R1A5A1011478).

REFERENCES

[1] Y. Tokunaga, S. Sakiyama, A. Matsumoto, and S. Dosho, *An on-chip CMOS relaxation oscillator with voltage averaging feedback*, IEEE J. Solid-State Circuits, vol. 45, no. 6, pp. 11501158, Jun. 2010.

[2] K. Ueno, T. Asai, and Y. Amemiya, *A 30-MHz, 90ppm/^{o}C fully-integrated clock reference generator with frequency-locked loop*, ESSCIRC, pp. 392-395, 2009.

[3] K.-K. Huang and D. D. Wentzloff, *A 1.2-MHz 5.8-μW Temperature-Compensated Relaxation Oscillator in 130-nm CMOS*, IEEE Trans. Circuits Syst. II, EXPRESS BRIEFS, vol. 61, no. 5, pp. 334338, May. 2014.

[4] K. Choe, O. D. Bernal, D. Nuttman, and M. Je, *A precision relaxation oscillator with a self-clocked offset-cancellation scheme for implantable biomedical SoCs*, in Proc. IEEE Int. Solid-State Circuits Conf. Dig. Tech. Papers, pp. 402403, Feb. 2009.

[5] U. Denier, *Analysis and design of an ultralow-power CMOS relaxation oscillator*, IEEE Trans. Circuits Syst. I, Reg. Papers, vol. 57, no. 8, pp. 19731982, Aug. 2010.

[6] K. Sundaresan, P. E. Allen, and F. Ayazi, *Process and temperature compensation in a 7-MHz CMOS clock oscillator*, IEEE J. Solid-State Circuits, vol. 41, no. 2, pp. 433442, Feb. 2006.

[7] K. Tsubaki, T. Hirose, Y. Osaki, S. Shiga, N. Kuroki, and M. Numa, *A 6.66-kHz, 940-nW, 56ppm/^{o}C, Fully On-chip PVT Variation Tolerant CMOS Relaxation Oscillator*, ICECS, pp. 97-100, 2012.

[8] J. Lee and S. Cho, *A 10 MHz 80 μW 67 ppm/^{o}C CMOS reference clock oscillator with a temperature compensated feedback loop in 0.18 μm CMOS*, in Proc. Dig. Symp. VLSI Circuits, pp. 226227, Jun. 2009.

[9] V. De Smedt, P. De Wit, W. Vereecken, and M. S. J. Steyaert, *A 66 μW 86 ppm/^{o}C fully-integrated 6 MHz Wien bridge oscillator with a 172 dB phase noise FOM*, IEEE J. Solid-State Circuits, vol. 44, no. 7, pp. 19902001, Jul. 2009.

[10] I. Taha, and M. Mirhassani, *A Temperature Compensated Relaxation Oscillator for SoC Implementations*, NEWCAS, pp. 373-376, 2014.

[11] S. Lan, C. Yuan, Y. Y. H. Lam, and L. Siek, *An Ultra Low-Power Rail-to-Rail Comparator for ADC Designs*, MWSCAS, pp. 1-4, 2011.

A time-window based approach for dynamic assertions mining on control signals

Alessandro Danese
Department of Computer Science
University of Verona, Italy
Email: alessandro.danese@univr.it

Francesca Filini
Department of Computer Science
University of Verona, Italy
Email: francesca.filini@studenti.univr.it

Graziano Pravadelli
Department of Computer Science
University of Verona, Italy
Email: graziano.pravadelli@univr.it

Abstract—Different mining approaches have been proposed in the past for automatic generation of assertions. However, in most cases, existing tools generate a set of over-constrained assertions. As a consequence, each assertion in the set is a long formula that describes a very specific behaviour of the design under verification (DUV). Thus, in the effort of covering as much DUV behaviours as possible, these approaches generate a huge amount of assertions with a negative impact on the total time required for their verification. To overcome this drawback, we introduce a dynamic approach that incrementally analyses control signals on DUV execution traces for mining more expressive temporal assertions that better capture the I/O communication protocol. Experimental results show that our approach allows generating a compact set of assertions without penalizing the coverage of DUV behaviours.

I. INTRODUCTION

Since the last decade, assertion-based verification (ABV) has emerged as one of the most promising solutions for electronic system level (ESL) and Register Transfer Level (RTL) verification [1]. ABV relies on the definition of assertions, i.e., logic formulas, generally written according to temporal logics, like LTL and CTL [2], and property specification languages, like PSL [3], that formalize the behaviours of the DUV by overcoming the ambiguity of natural languages and providing the engineers with precise and well-defined specifications. However, assertion definition is a time-consuming process that requires high expertise, and sometime terminates with an incomplete (some behaviours are not captured by assertions) but redundant (some assertions pass the verification vacuously, or they are logic consequence of the others) set of assertions, making the ABV process precarious (due to the incompleteness) and uselessly longer (due to the redundancy).

As a complementary approach to manual definition of assertions, several approaches and tools have been proposed for automatically extracting safety assertions in the form $always(antecedent \rightarrow consequent)$ from the implementation of the DUV [4], [5], [6], [7], [8]. These approaches either rely on static analysis of the DUV source code or they dynamically mine assertions from execution traces of the DUV. The first are accurate and provide assertions that are formally proved to be satisfied by the DUV, but they do not scale well for complex DUVs. The second provide only likely assertions, whose quality depends on the observed execution traces (i.e., likely assertions are guarantee to hold at least for the considered execution traces, however a counter example could be finally found), but they do not require the source code and guarantee a better scalability for complex DUVs. Independently from the adopted techniques, mined assertions can be compared with design intents to discover unexpected behaviours implemented in the DUV, to confirm that relevant behaviours are actually implemented, and for documentation purposes. However, existing approaches generally extract a high number of long assertions, but each of them covers a few specific behaviours of the DUV. This depends on the fact that mined assertions are over-specified, i.e., their antecedents and/or consequents include several atomic propositions that predicate almost on all primary inputs and outputs of the DUV. Moreover, mined assertions mix data and control signals making difficult the characterization of the I/O communication protocol of the DUV. Large sets of over-constrained assertions make impractical the analysis of mined assertions by verification engineers and greatly increase verification time.

To overcome drawbacks of existing works, this paper presents a dynamic approach that infers assertions by incrementally analysing time windows of the DUV's execution traces. The algorithm searches for recurrent temporal patterns among atomic propositions predicating on I/O control signals. Mined assertions are in average shorter, more expressive from the point of view of the I/O communication protocol, and then simpler to be understood by humans, compared, for example, to the approach proposed in [5]. This reduces verification time, and increases the effectiveness of verification engineers in discovering design errors by analysing mined assertions.

The rest of the paper is organized as follows. Section II presents some preliminary definitions and concepts relevant for the proposed approach, which is then detailed in Section III. Experimental results and concluding remarks are finally reported, respectively, in Section IV and Section V.

II. PRELIMINARIES

Before showing the proposed methodology, some preliminary definitions and concepts are reported to create the necessary background.

Definition 1. *(Execution trace) Given a finite sequence of simulation instants $\langle t_1,...,t_n \rangle$ and a model M working on a set of variables V, an execution trace of M is a finite sequence of pairs $T = \langle (V_1,t_1),...,(V_n,t_n) \rangle$ where $V_i = eval(V,t_i)$ is the evaluation of variables in V at simulation instant t_i.*

More informally, an execution trace describes for each simulation instant t_i the values assumed by each variable included in V during the evolution of the design under verification M. In this paper, variables in V are primary inputs and primary outputs representing control signals of the DUV. An example of an execution trace is reported in Figure 1 (left). By analysing an execution trace, we can extract a set of atomic propositions that predicate on variables included in V.

Definition 2. *(Atomic proposition) An atomic proposition is a formula that does not contain any logical connective.*

In our methodology, a set of atomic propositions is organized in an array $A = \{a_1, \ldots, a_m\}$ which is further divided in two sub-arrays: the array of *input atomic propositions* $A_I = \{a_0, \ldots, a_i\}$, whose elements are atomic propositions that predicate only on primary inputs of the DUV, and the array of *output atomic propositions* $A_O = \{a_{i+1}, \ldots, a_m\}$, whose elements predicate only on primary outputs of the DUV. We do not consider any atomic proposition that mixes primary inputs and primary outputs. Examples of atomic propositions are reported in Figure 1 (centre).

978-1-4673-9141-2/15 $31.00 © 2015 IEEE

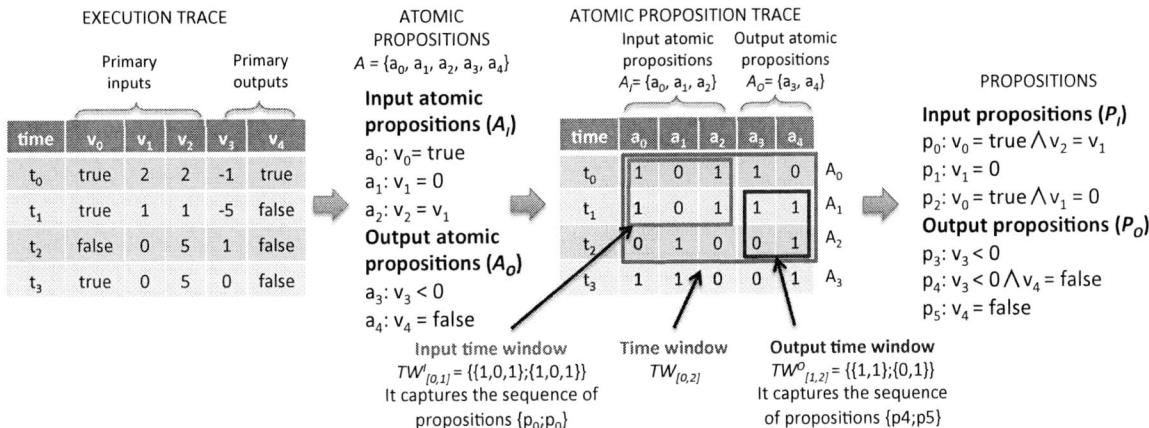

Fig. 1. Exemplification of execution trace, atomic proposition trace and time window.

From an execution trace and the corresponding set of atomic propositions, we can generate an atomic proposition trace.

Definition 3. (Atomic proposition trace) *Given an execution trace T and an array of atomic propositions A, an atomic proposition trace is a finite sequence of pairs $\omega = \langle (A_0, t_0), \ldots, (A_n, t_n) \rangle$ where $A_i = eval(A, t_i)$ is an array that represents the evaluation of atomic propositions in A at simulation instant t_i, i.e., $A_i[j] = 0$ if $A[j] = false$ at time t_i, $A_i[j] = 1$ otherwise.*

The atomic proposition trace is the data structure we use to mine propositions.

Definition 4. (Proposition) *A proposition is a composition of atomic propositions through logic connectives. An atomic proposition itself is a proposition.*

It this paper, we consider propositions involving only the logic *and* ($\land$) as connective, and we classify them in two different sets:

- The set of *input propositions* (P_I): a proposition p belongs to P_I if it is composed only of input atomic propositions and $\land$ connectives;
- The set of *output propositions* (P_O): a proposition p belongs to P_O if it is composed only of output atomic propositions and $\land$ connectives.

An example of an atomic proposition trace and a set of input/output propositions that can be extracted from it are shown in Figure 1 (right). To represent input and output propositions in a compact and efficient way, we use an array-based notation. Given an array of input (respectively, output) atomic propositions A_I (A_O), an input (output) proposition is represented by an array of Boolean values p such that $p[i] = 0$ if the input (output) atomic proposition $A_I[i]$ ($A_O[i]$) is not used in the proposition, $p[i] = 1$ otherwise. For example, the input proposition p_0 in Figure 1 can be represented by the array $\{1, 0, 1\}$.

Definition 5. (Time window) *Given an atomic proposition trace $\omega = \langle (A_0, t_0), \ldots, (A_n, t_n) \rangle$, and two simulation instants t_i and t_j such that $1 \leq t_i \leq t_j \leq n$, a time window $TW_{[i,j]} = \langle (A_i, t_i), \ldots, (A_j, t_j) \rangle$ is the subsequence of contiguous elements of ω included between instant t_i and instant t_j.*

Given a time window $TW_{[i,j]}$, and a simulation instant t_k such that $t_i \leq t_k \leq t_j$, we can separate $TW_{[i,j]}$ in two parts:

- The *input time window* $TW^I_{[i,k]}$, which is composed of elements of $TW_{[i,j]}$ included in the simulation instants between t_i and t_k, restricted to the input atomic propositions.

- The *ouput time window* $TW^O_{[k,j]}$, which is composed of elements of $TW_{[i,j]}$ included in the simulation instants between t_k and t_j, restricted to the output atomic propositions.

The input time window $TW^I_{[i,k]}$ and the corresponding output time window $TW^O_{[k,j]}$ overlap for exactly one simulation instant (t_k). Given a time window $TW_{[i,j]}$, we can generate $j - i + 1$ different couples of input/output time windows, one for each simulation instant $t_k \in [t_i, t_j]$. For example, in the atomic proposition trace of Figure 1, the green box highlights a time window composed of 3 simulation instants in the interval $[t_0, t_2]$, the red box corresponds to an input time window in the interval $[t_0, t_1]$, and finally the blue box shows the corresponding output time window in $[t_1, t_2]$. In the rest of the paper we will represent an input (output) time window by the sequence of arrays corresponding to the input (output) propositions it captures. For example, in Figure 1, the input time window corresponding to the interval $[t_0, t_1]$ is represented by the sequence of two arrays $\{\{1, 0, 1\}; \{1, 0, 1\}\}$.

By analysing input and output time windows in the atomic proposition trace we can mine temporal assertions.

Definition 6. (Temporal assertion) *A temporal assertion is a composition of propositions through temporal operators and logic connectives.*

In this paper, we consider Linear Temporal Logic (LTL) assertions in the form $G(antecedent \rightarrow consequent)$, where G is the LTL *always* operator[1], and *antecedent* and *consequent* may involve only X, i.e., the LTL *next* operator[2] and the $\land$ connective. Moreover, *antecedent* is composed only of an arbitrary number of propositions belonging to the set P_I extracted by analysing an input time window $TW^I_{[i,k]}$, while *consequent* includes a single proposition belonging to the set P_O extracted by analysing the corresponding output time window $TW^O_{[k,j]}$. We selected this specific form of assertion, since it is suited to describe the behaviour of the I/O communication protocol of a DUV, which, as reported in the introduction, represents the target for the current work. For example, from the atomic proposition trace of Figure 1 the following temporal assertion can be mined: $G((p_0 \land X^2(p_1)) \rightarrow X^3(p_5))$.

In according to the above definitions and notations, we define hereafter a set of operators working on propositions that will be used to illustrate the mining methodology proposed in Section III:

---

[1]Given a formula α, $G(\alpha)$ means that α is always true.
[2]Given a formula α, $X(\alpha)$ means that α is true at the next instant. As a short cut, we will write $X^n(\alpha)$ to represent the application of n consecutive next operators to the formula α.

978-1-4673-9141-2/15 $31.00 © 2015 IEEE

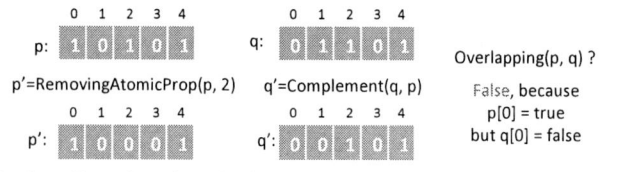

Fig. 2. Examples of application of *RemoveAtomicProp*, *Complement* and *Overlapping*.

RemoveAtomicProp: Given an array of Boolean elements p corresponding to a proposition, and an index i corresponding to the atomic proposition a_i, which represents the element at position i of p, $RemoveAtomicProp(p, i)$ returns a new array p' where $p'[j] = p[j]$ for all indexes $j \neq i$ and $p'[i] = 0$, i.e., the proposition corresponding to p' does not include a_i.

AddAtomicProp: Given an array of Boolean elements p corresponding to a proposition, and an index i corresponding to the atomic proposition a_i, which represents the element at position i of p, $AddAtomicProp(p, i)$ returns a new array p' where $p'[j] = p[j]$ for all indexes $j \neq i$ and $p'[i] = 1$, i.e., the proposition corresponding to p' includes a_i.

Overlapping: Given two arrays of Boolean elements p and q corresponding to two propositions predicating over the same array of atomic propositions, $Overlapping(p, q)$ returns $true$ if $q[i] = 1$ for at least all indexes i such that $p[i] = 1$, it returns $false$ otherwise.

Complement: Given two arrays of Boolean elements p and q corresponding to two propositions predicating over the same array of atomic propositions, $Complement(p, q)$ returns a new array p' where $p'[i] = false$ for all indexes i such that $q[i] = false$, and $p'[j] = p[j]$ for all indexes j such that $q[i] = true$.

For example, Figure 2 shows the application of operators *RemoveAtomicProp*, *Complement* and *Overlapping* to propositions. In a more general way, we apply *Complement* and *Overlapping* also to sequences of propositions of the same length. In this case, given two sequences of propositions s_1 and s_2, *Complement* and *Overlapping* operate iteratively on each couple of corresponding elements (p_i, q_i) such that $p_i \in s_1$ and $q_i \in s_2$.

III. TIME-WINDOW BASED ASSERTION MINING

Given a DUV and a related set of execution traces, the proposed methodology consists of two main phases: (i) mining of atomic propositions and generation of the corresponding atomic proposition traces, and (ii) extraction of input and output time windows and mining of temporal assertions. Since, the first phase has been already faced in [6], this paper focusses only on the second phase.

The main procedure of the proposed approach is showed in Algorithm 1. The function MAIN takes as parameters an atomic proposition trace ω and two thresholds: tw_len, and max_len, with $max\_len \leq tw\_len$. The first threshold (tw_len) advises how many clock cycles are required, at maximum, by the DUV to compute its functionality. This parameter, which can generally be obtained from the documentation or from simulation of the DUV, represents the length of the time windows that will be analysed by the miner inside ω. The second threshold (max_len) is used to bound the maximum number of instants that will be considered to mine the antecedent of a temporal assertion, i.e., it represents the maximum length for an input time window. Several DUVs need a high number of clock cycles for completely computing their functionality, but only few of them could be necessary to set control signals such that the computation phase starts. For instance, to mine interesting antecedents for an encryption/decryption component, it is relevant catching

```
1: function MAIN(ω, tw_len, max_len) return assertions_set
2:     len = 1
3:     assertions_set = ∅
4:     while len ≤ max_len do
5:         behaviours = getBehaviours(ω, tw_len, len)
6:         pruningBehaviours(assertions_set, behaviours)
7:         candidates = getAssertions(ω, behaviours)
8:         pruningAssertions(ω, assertions_set, candidates)
9:         len = len + 1
10:    end while
11: end function
```

Algorithm 1. Main procedure.

what happens to control signals during the initialization phase, before the computation proceeds without affecting control signals till outputs are ready. Similarly, in the case of a network component, it is relevant to capture in the antecedent of an assertion how control signals change during the phase that makes data transmission starting. The max_len threshold is then intended to guide the mining procedure such that, for mining antecedents, it focusses on the initial part of the considered time windows when the DUV is characterized, after an initialization phase, by a long elaboration phase that does not involve control signals any more till results are ready. In this case, generally max_len is small with respect to tw_len. On the contrary, max_len is set to be equal to tw_len for DUVs where control signals can change the functionality of the DUV at each simulation instant.

After initializing the *assertions_set* to the empty set, the main loop of Algorithm 1 cyclically calls the following four functions, by varying the length (len) of the considered input time window at each iteration, while $len \leq max\_len$.

$getBehaviours(\omega, tw\_len, len)$: it detects the behaviours exposed by the DUV in the atomic proposition trace. Each behaviour is represented by a sequence (with length between 1 and len) of input propositions associated to a unique output proposition. For example, in the atomic proposition trace of Figure 1, each time the input proposition $p_0 = \{1, 0, 1\}$ (corresponding to $a_0 \wedge a_2$) is true at a simulation instant t_i, the output proposition $p_5 = \{0, 1\}$ (corresponding to a_4) holds at simulation instant t_{i+1}. Thus, a behaviour can be identified that associates the input proposition p_0 to the output proposition p_5 with an *offset* of one simulation instant.

$pruningBehaviours(assertion\_set, behaviours)$: it removes all the behaviours returned by $getBehaviours$ that are already captured by assertions mined in the previous iterations of the main procedure. For example, if the assertion $\alpha = G(a_2 \to X(a_4))$ is already included in the *assertion_set*, a behaviour, which associates, with an offset of one simulation instant, the input proposition p_0 to the output proposition p_5, can be discarded, since it will create the assertion $G(a_0 \wedge a_2 \to X(a_4))$ that is a logical consequence of α.

$getAssertions(\omega, behaviours)$: it is in charge of creating a set of candidate temporal assertions, compliant with the form described in Section II, starting from the behaviours survived after $pruningBehaviour$ is called.

$pruningAssertions(\omega, assertion\_set, candidates)$: it analyses each candidate assertion returned by $getAssertions$ and preserve only those that allow improving the coverage with respect to the DUV behaviours.

The next sections describe in details how the previous functions work to mine temporal assertions.

A. Mining of interesting behaviors

The function $getBehaviours$ is implemented as shown in Algorithm 2. It takes as arguments an atomic proposition trace ω and the two thresholds tw_len and len. Its goal is to detect behaviours exposed in the execution traces of the DUV, in the form of associations between a sequence of input propositions

```
1:  function GETBEHAVIOURS(ω, tw_len, len) return behaviours
2:      behaviours = ∅
3:      dictionary = ∅
4:      t_i = 0
5:      while t_i ≤ (length(ω) − tw_len) do
6:          in = propSeq(TW^I_[t_i,t_i+len−1])
7:          out = propSeq(TW^O_[t_i+len−1,t_i+tw_len−len+1])
8:          ⟨in, old_out⟩ = findInDictionary(dictionary, in)
9:          if ⟨in, old_out⟩ = NIL then
10:             dictionary = dictionary ∪ ⟨in, out⟩
11:         else
12:             new_out = Complement(old_out, out)
13:             dictionary = dictionary \ ⟨in, old_out⟩
14:             dictionary = dictionary ∪ ⟨in, new_out⟩
15:         end if
16:         t_i = t_i + 1
17:     end while
18:     for all ⟨in, out⟩ ∈ dictionary do
19:         offset = 0
20:         for all p ∈ out do
21:             if p ≠ NIL then
22:                 behaviours = behaviours ∪ ⟨in, offset, p⟩
23:             end if
24:             offset = offset + 1
25:         end for
26:     end for
27: end function
```

Algorithm 2.   Extraction of DUV behaviours.

and a corresponding sequence of output propositions. In order to detect such associations, a time window $TW_{[t_i,t_i+tw\_len−1]}$ of length tw_len is analysed for each simulation instants t_i (lines 5-17). Such a time window is decomposed in an input time window of length len, and a corresponding output time window of length $tw\_len − len + 1$. Then, from the input time window, the function *propSeq* extracts the corresponding sequence of input propositions (in) that hold in the interval $[t_i, t_i + len − 1]$ (line 6). Similarly, *propSeq* extracts from the output time window the sequence of output propositions (out) that hold in the interval $[t_i+len−1, t_i+tw\_len−len+1]$ (line 7). For example, let us consider the atomic proposition trace of Figure 1 and let we fix $tw\_len = 2$ and $len = 1$. At time t_0, we have $in = \{1, 0, 1\}$ while out is represented by the sequence $\{\{1, 0\}; \{1, 1\}\}$. The function then searches if in is already present in the dictionary of the collected input/output propositions pairs (line 8-9). If this is not the case (line 10), a new association $⟨in, out⟩$ is added to the dictionary. On the contrary, when in is already present, the operator *Complement* is applied to return a new sequence of output propositions new_out, where atomic propositions excluded from elements of out are excluded also from the corresponding elements of new_out. Then, the couple $⟨in, old\_out⟩$ is replaced in the dictionary by $⟨in, new\_out⟩$ (lines 12-14). This replacement is necessary to refine the already collected behaviours by removing from their output propositions the atomic propositions that become false in the current output time window, such that only behaviours that are never violated throughout the trace are finally collected. The replacement happens, for example, at simulation instant t_1 of the atomic proposition trace of Figure 1. In fact, at t_1, $in = \{1, 0, 1\}$ is associated to $out = \{\{1, 1\}; \{0, 1\}\}$. However, in was already associated to $old\_out = \{\{1, 0\}; \{1, 1\}\}$ at t_0. Thus, in the dictionary $⟨\{1, 0, 1\}, \{\{1, 0\}; \{1, 1\}\}⟩$ is replaced by $⟨\{1, 0, 1\}, \{\{1, 0\}; \{0, 1\}\}⟩$ after the application of $Complement(old\_out, out)$. After the creation of the dictionary that collects associations between sequences of input and output propositions, the final loop (lines 18-26) creates a set of behaviours for each pair $⟨in, out⟩$. In particular, a behaviour, represented by a triplet $⟨in, offset, p⟩$, is extracted for each proposition p captured in the sequence of output propositions out, where $offset$ represents the distance, computed in simulation instants, between the last element of in and p. For example, for the pair $⟨\{1, 0, 1\}, \{\{1, 0\}; \{0, 1\}\}⟩$ the following two behaviours are extracted: $⟨\{1, 0, 1\}, 0, \{1, 0\}⟩$ and $⟨\{1, 0, 1\}, 1, \{0, 1\}⟩$ to represent respectively that the output proposition $\{1, 0\}$ holds exactly at the same time of input

proposition $⟨\{1, 0, 1\}$ (*offset* is 0), while the output proposition $\{0, 1\}$ holds one simulation instant later (*offset* is 1).

B. Pruning of behaviours

The function *pruningBehaviours* takes as arguments a set of assertions and a set of triplets representing behaviours collected by *getBehaviours* in the form $⟨in, offset, p⟩$. The goal of *pruningBehaviours* is to preserve only the triplets that are not already covered by assertions collected in previous iteration of Algorithm 1. A triplet is covered by an assertion $α$ when the following conditions are true concurrently:

1) All input atomic propositions included in the antecedent of $α$ are also present in in (i.e., values assigned to primary inputs of the DUV that satisfy the atomic propositions included in in satisfy also the antecedent of $α$).
2) All output atomic propositions included in the consequent of $α$ are also present in p (i.e. the consequent of $α$ is at least as detailed as p).
3) The distance, computed in simulation instants, between the last input atomic proposition of the antecedent of $α$ and the consequent of $α$ equals $offset$.

Triplets that falsify at least one of the previous conditions are preserve, the others are discarded.

C. Mining of assertions

Given an atomic proposition trace $ω$ and a set of triplets of the form $⟨in, offset, p⟩$ representing behaviours preserved by the *pruningBehaviours* function, the *getAssertions* function works as described in Algorithm 3. Its goal is to extract an assertion of the form $G(antecedent → consequent)$ from every triplet, such that the input propositions $\{p_0, \ldots, p_i\}$ captured inside the sequence in act as an antecedent of the form $(p_0 ∧ \cdots ∧ X^i(p_i))$, while $X^{i+offset}(p)$ represents the consequent. This is performed by the *makeAss* function at line 5. After an assertion is added to the set of candidates ass_set, the *pruningBehaviours* function is called to remove behaviours implicitly covered by the new assertion (line 6).

In order to increase the DUV behaviours covered by the mined assertions, before calling *makeAss*, the input proposition included in the sequence in is first simplified by removing atomic propositions from the antecedent such that the consequent can be verified by a higher number of simulation instants, thus enforcing the final assertion (line 4). For instant, let us consider the triplet $⟨in, offset, p⟩$ represented by $⟨\{\{1, 1, 0\}\}, 0, \{\{0, 1\}\}⟩$. Looking at Figure 1, we can see that the output proposition $\{0, 1\}$ is true at both simulation instants t_2 and t_3, but the input proposition $\{1, 1, 0\}$ is verified only at simulation instant t_3. However, if we set to $false$ the first input atomic proposition of $\{1, 1, 0\}$, we obtain the proposition $\{0, 1, 0\}$, which is true at both simulation instant t_2 and t_3. Thus, $in = \{\{1, 1, 0\}\}$ can be replaced by $s\_in = \{\{0, 1, 0\}\}$ in the triplet to cover a wider time window in the atomic proposition trace. In this way, the assertion $G(a_1 → a_4)$ can be extracted instead of $G(a_0 ∧ a_1 → a_4)$. The first is preferred because it implies the second.

The simplification of the sequence of input propositions in is performed by the function *SIMPLIFY* (lines 10-33). Given an input propositions q belonging to the sequence in, it makes a copy s_q of q (line 14), and then it performs the following steps for each atomic proposition a_i included in s_q (lines 16-31):

1) remove a_i from s_q (line 17);
2) create a new sequence of propositions new_in from in by replacing the proposition q with s_q (line-18);
3) check, for every simulation instant t_i, if the new sequence new_in is true on the atomic proposition

```
1:  function GETASSERTIONS(ω, behaviours, len) return ass_set
2:      ass_set = ∅
3:      for all ⟨in, offset, p⟩ ∈ behaviours do
4:          s_in = simplify(⟨in, offset, p⟩)
5:          ass_set = ass_set ∪ makeAss(s_in, offset, p)
6:          pruningBehaviours(ass_set, behaviours)
7:      end for
8:  end function
9:
10: function SIMPLIFY(ω, ⟨in, offset, p⟩)
11:     len = length(in)
12:     p_off = len − 1 + offset
13:     for all  q ∈ in do
14:         s_q = q
15:         i = 0
16:         while i < length(q) do
17:             s_q = RemoveAtomicProp(s_q, i)
18:             new_in = (in \ q) ∪ s_q
19:             t_i = 0
20:             while t_i ≤ (length(ω) − p_off set) do
21:                 temp = propSeq(TW^I_{[t_i,t_i+len−1]})
22:                 if Overlapping(new_in, temp) then
23:                     temp = propSeq(TW^O_{[p_off,p_off]})
24:                     if !Overlapping(p, temp) then
25:                         s_q = AddAtomicProp(s_q, i)
26:                         break
27:                     end if
28:                 end if
29:                 t_i = t_i + 1
30:             end while
31:         end while
32:         in = (in \ q) ∪ s_q
33:     end for
34:     return in
35: end function
```

Algorithm 3. Generation of temporal assertions.

trace (line-22), but the output proposition p is false (line 24). In this case, a counter example is found that shows we cannot remove the atomic proposition a_i from s_q, otherwise the association between s_q and p is not valid any more. Thus a_i is restored inside s_q (line 25). If a counter example is not found, a_i can be definitely removed.

D. Pruning of Assertions

In the final phase of the mining methodology, the procedure *pruningAssertions* is called to prune candidate assertions returned by the function *getAssertions*, such that at each iteration of the main procedure reported in Algorithm 1 only the "strongest" assertions are collected. The power of an assertion is measured according to the frequency of its activation on the considered execution traces (i.e., how many times the antecedent is fired) and its capability of formalizing new behaviours not yet covered by already collected assertions. Assertions are then ranked according to the following criteria. An assertion α_1 is ranked first with respect to an assertion α_2 when:

*   the number of atomic propositions in the antecedent of α_1 is lower than in α_2. In this way, we prefer assertions that describe the behaviours of the primary outputs of the DUV in relation with the minor number of constraints among its inputs.

*   the number of atomic propositions in the consequent of α_1 is higher than in α_2. In this way, we prefer assertions that describe in a more specific way how primary outputs of the DUV change during the execution according to variations on the primary inputs.

*   the *offset* between the antecedent and the consequent of α_1 is lower than the *offset* of α_2. In this way, we prefer assertions that, after the activation of the antecedent, capture the behaviour of primary outputs as soon as possible;

More sophisticated metrics for evaluating the power of candidate assertions will be part of future works.

IV. EXPERIMENTAL RESULTS

Experimental results have been carried out on an Intel Xeon E5649 @2.53Ghz equipped with 8 GB of RAM and running Linux OS. The benchmarks considered for evaluating the proposed mining strategy belong to the Open-Source-Test-Case (OSTC) platform developed as reference case study for the European project SMAC [9]. In particular, we considered the RTL implementation of the *UART* [10] and *BUS-APB* [11] components. These two benchmarks have been selected because they present different characteristics from the input/output latency point of view, i.e. the number of clock cycles required, at maximum, to compute the component's functionality. The I/O latency is an important parameter for mining approaches because longer is the I/O latency, higher is the time spent by the miner to create an assertion that puts in relation values provided to primary inputs with values obtained on primary outputs. *UART*, which is practically a parallel-to-serial/serial-to-parallel converter, requires 665 clock cycles before the output bit stream is produced, once data are provided in input for the conversion. On the contrary, the input/output latency of *BUS-APB* is 2 clock cycles.

Table I reports, for each component, the lines of code (*Lines*), the number of bits corresponding to control signals belonging to the primary inputs (*PIs*) and to the primary outputs (*POs*), and the input/output latency (*I/O latency*). Execution traces composed of 10,000 clock cycles have been generated for the two benchmarks by simulation.

The mining methodology proposed in this paper has been compared with a state-of-the-art approach presented in [5], which mines assertions from execution traces through an induction algorithm based on a decision tree [12]. The comparison between the two approaches is reported in Table II and Table III concerning, respectively the characteristics of the mined assertions and mining execution times, and the quality of the mined assertions measured in terms of mutation coverage.

Columns 2 and 3 of Table II report the configuration parameters, i.e., the length of considered time windows (*tw_len*) (which corresponds to the I/O latency of the DUV), the maximum number of propositions allowed in the antecedent of the mined assertions (*max_len*) for the time window approach (i.e., the maximum number of clock cycles that are observed in the antecedent), and the maximum depth of the analysed decision tree (*max_depth*) for the approach proposed in [5]. The parameters max_len has been selected according to the characteristics of the DUVs. For example, $max\_len = 1$ for *UART* because the values assigned to the input control signals to start the data elaboration are provided in a single clock cycle, while $max\_len = 2$ for *BUS-APB* since input control signals influence the bus functionality during the whole elaboration phase that always embraces 2 clock cycles. On the contrary, for the decision-tree based approach the maximum depth of analysed decision tree must be specified; we tested different values and we saw that for values higher than 10 and 12, respectively for *UART* and *BUS-APB*, the execution time of the algorithm increases without improving the quality (measured in terms of mutation coverage) of the mined assertions. Then, Columns 4-7 report the mining results, i.e, the number of extracted assertions (*# ass.*), the average number of input atomic propositions included in the antecedent of the extracted assertions (*# ant.*), the average number of output atomic propositions included in the consequent of the extracted assertions (*# cons.*), and the total time required for the mining procedure (*time*). Looking at the results, we see that the number of assertions generated by our approach is smaller than the number of assertions generated by [5]. However our assertions are composed of consequents with a higher number

DUV	Lines	PIs	POs	I/O latency
BUS-APB	390	6	12	2
UART	6819	10	9	665

TABLE I.   CHARACTERISTICS OF BENCHMARKS.

Time window-based approach

DUV	Configuration parameters		Results			
	tw_len	max_len	# ass.	# ant.	# cons.	Time
BUS-APB	2	2	24	3.3	11.1	1 s.
UART	655	1	21	2.94	6.47	720 s.

Decision tree-based approach [5]

DUV	Configuration parameters		Results			
	tw_len	max_depth	# ass.	# ant.	# cons.	Time
BUS-APB	2	12	86	2.82	1	1 s.
UART	665	10	39	5.6	1	5820 s.

TABLE II.   NUMBER OF ASSERTION EXTRACTED BY THE TIME-WINDOW APPROACH AND THE DECISION-TREE BASED APPROACH.

of atomic propositions, which reflects in a better description of the behaviours of the primary outputs of the DUV when an antecedent is fired. On the contrary, antecedents are generally compact (i.e., the number of involved atomic propositions is small), thus assertions cover a large number of behaviours from the perspective of the DUV's primary inputs. Finally, concerning execution time, our approach outperforms the decision tree-based algorithm when applied to mine assertions on DUVs, whose I/O latency (which impacts on the offset between antecedent and consequent) is very high, like in the case of *UART*.

Mined assertions have been then compared in terms of *mutant coverage*, i.e., the capability of discovering *mutants*, which represent small alterations of the DUV's source code that perturb its functionality. A mutant is observable if, in comparison with a mutant-free DUV, its effect is visible as an alteration in the DUV's primary outputs. A mutant is covered by an assertion if the assertion fails when the mutant is observed at primary outputs. The mutant coverage is then the ratio between covered mutants and observable mutants. Uncovered mutants highlight the incompleteness of the assertions set. Indeed, mutant analysis has been already adopted in the past to measure the quality of a set of assertions in terms of their ability to cover all the interesting functionalities of the DUV [13]. The well-known bit coverage fault model have been selected to inject mutants in the control signals of the DUVs [14]. Bit coverage alters, in single fault mode, each bit of the affected signal by fixing its value to 0 (stuck-at 0) or to 1 (stuck-at 1). Table III reports the results of the mutation analysis by showing the number of observable mutants (*# observ.*), the number of covered mutants (*# covered*), the average number of mutants covered by each assertion (*avg*), and finally the time required to simulate the OSTC platform connected to the set of checkers[3] corresponding to the assertions mined for *UART* and *BUS-APB* in presence of one mutant (*Time*).

The mutant coverage achieved for *BUS-APB* is 100% for both approaches, while the time window-based approach outperforms the decision-tree algorithm concerning mutant coverage of *UART*. Moreover, the number of mutants covered in average by each assertion mined with our approach is higher. Finally, concerning the simulation time, we observe that checkers corresponding to assertions mined by the decision tree algorithm require a longer simulation time, which greatly increases for assertions that predicate on DUV with a long I/O latency, as in the case of *UART*. We observed in particular, that antecedents of assertions generated according to [5] are composed of atomic propositions that could be removed, since they

Time window-based approach

DUV	# observ.	# covered	Avg	Time
BUS-APB	22	22	10.27	70 s.
UART	149	99	26.85	4208 s.

Decision tree-based approach [5]

DUV	# observ.	# covered	Avg	Time
BUS-APB	22	22	0.8	83 s.
UART	149	58	9.08	46853 s.

TABLE III.   COMPARISON BETWEEN THE PROPOSED APPROACH AND [5] BASED ON MUTANT COVERAGE.

do not affect the truth value of the assertions. This drawback is implicit in the use of a decision tree-based data structure, and it depends on the fact that an assertion generated at a leaf node necessarily includes atomic propositions predicating on variables involved in all previous levels of the tree. This leads to create assertions with longer antecedents, whose checkers require longer simulation times. Moreover, simulation times are affected by the total number of assertions which is higher in the case of the decision tree-based algorithm.

V.   CONCLUSIONS

The paper presents a mining algorithm for automatic generation of LTL temporal assertions. It relies on a time window-based analysis of execution traces that searches for behaviours that repeat periodically capturing the relation between primary inputs and primary outputs of the DUV. The approach is particularly suited for mining assertions that describe the behaviour of the control signals of the DUV, which are used to implement the I/O communication protocol surrounding the computation of the DUV core functionality. In comparison with a state-of-the-art assertion miner proposed in [5], experimental results show that our approach generates a more compact set of assertions, which achieves a higher mutant coverage and requires shorter times for the simulation of the corresponding checkers.

REFERENCES

[1] A. Gupta, "Assertion-based verification turns the corner." *IEEE Design & Test of Computers*, vol. 19, no. 4, pp. 131–132, 2002.

[2] A. Pnueli, "Linear and branching structures in the semantics and logics of reactive systems," in *Automata, Languages and Programming*, ser. Lecture Notes in Computer Science, W. Brauer, Ed.   Springer Berlin Heidelberg, 1985, vol. 194, pp. 15–32.

[3] "Standard for property specification language (PSL)," *IEC 62531:2012(E) (IEEE Std 1850-2010)*, pp. 1–184, 2012.

[4] G. Ammons, R. Bodík, and J. R. Larus, "Mining specifications," *ACM Sigplan Notices*, vol. 37, no. 1, pp. 4–16, 2002.

[5] S. Hertz, D. Sheridan, and S. Vasudevan, "Mining hardware assertions with guidance from static analysis," *IEEE Trans. Comp.-Aided Des. Integ. Cir. Sys.*, vol. 32, no. 6, pp. 952–965, 2013.

[6] A. Danese, T. Ghasempouri, and G. Pravadelli, "Automatic extraction of assertions from execution traces of behavioural models," in *In proc. of ACM/IEEE DATE*, 2015.

[7] "Jasper Activeprop," http://www.jasper-da.com.

[8] http://www.atrenta.com/solutions/bugscope.htm5.

[9] http://www.fp7-smac.org.

[10] http://opencores.org/project,a_vhd_16550_uart.

[11] http://www.arm.com/products/system-ip/amba/amba-open-specifications.php.

[12] J. R. Quinlan, "Induction of decision trees," *Machine learning*, vol. 1, no. 1, pp. 81–106, 1986.

[13] A. Fedeli, F. Fummi, and G. Pravadelli, "Properties incompleteness evaluation by functional verification," *IEEE Trans. Comput.*, vol. 56, no. 4, pp. 528–544, 2007.

[14] A. Fin, F. Fummi, and G. Pravadelli, "Amleto: A multi-language environment for functional test generation," in *Proc. of IEEE ITC*, 2001, pp. 821–829.

[15] https://www.research.ibm.com/haifa/projects/verification/focs/.

---

[3]A checker is an automaton that monitors the evolution of the DUV during simulation and raises a failure when the corresponding assertion is violated. We generated checkers for mined assertion by using IBM FoCs [15].

Physical-based Modeling and Fast Simulation of Wireline Links

Jun Guo, Peng Liu, and Weidong Wang
College of Information Science and Electronic Engineering, Zhejiang University
Hangzhou, 310027, China
Email: {guojun007, liupeng, wdwang}@zju.edu.cn

Abstract—To efficiently exploit the performance of wireline channels and alleviate the computation time due to the large amounts of simulation, this paper proposes a physical-based modeling mechanism for the performance studies of electrical links on multilayered printed circuit boards (PCBs). Our mechanism includes: 1) using a physical-based method to achieve high-speed and high precision simulation; and 2) combining configurable parameters and intermediate data of simulation to simplify evaluation process and support parameters sweeping. Experimental results show a good correlation with full-wave method up to 40 GHz and a significant acceleration in computation time of at least two orders of magnitude. Cooperating with system simulator, our approach also gives a good assistance in PCB channel design for high-speed interconnect systems.

Keywords—*Electrical wireline link, physical-based modeling, channel simulation*

I. INTRODUCTION

As the feature size of modern semiconductor technology scales down and date rates of high-speed systems continue to increase, higher densities and better performance are required for electrical interconnects on packages and PCBs [1]. However, thousands of off-chip interconnect elements, like vias, traces, and connectors, introduce serious effects of impedance mismatch and crosstalk. Degradation and distortion on signal integrity become the bottleneck for the maximum achievable data rate on signal path. Efficient modeling and simulation tools are highly demanded to assist the design, optimization, and validation of complex interconnect systems.

Different solutions range from full-wave 3D electromagnetic solvers to physical-based equivalent circuits [2]. Researchers focused on different algorithms to solve the Maxwell's equations and developed many full-wave commercial software based on momentum [3], finite integration (FI) [4], finite difference time domain (FDTD) [5], and finite element method (FEM) [6]. These solvers achieve high precision at the cost of computation speed. For complex systems which have a large number of via arrays or PCBs that have more than ten layers, full-wave simulation becomes impractical due to large computation resources required and unbearably long simulation time. Under these circumstances, physical-based method provides an attractive approach by means of constructing equivalent circuits to describe the component characteristics. This method also provides better flexibility and reconfigurability as the circuits can be parameterized for different dimensions and materials.

Vias and traces are two of the most important factors that determine the overall wireline link performance. In order to simulate the via performance, IBM proposed a multilayered via transition tool (MVTT) [7], the major engine behind its statistical and parametric link simulation environment (SPLSE) [8]. The main purpose of SPLSE is to analyze channels to achieve higher data bandwidth [9]–[11] and assist the design of serializer/deserializer (SerDes) circuits [8].

However, the trace model used in SPLSE is extracted from CZ2D, a 2D method of moment (MoM) tool [8]. The full-wave-alike tool CZ2D becomes new bottleneck of total calculation speed, especially in cases where simulation time of vias is shorter than traces with the help of MVTT. Besides, various of output files generated by different engines and the transformation between them [8] may confuse the users. Due to these considerations, we use the transmission line theory to solve the frequency responses of traces. Therefore, both traces and vias can be unified into physical-based analytical circuits, which simplify the subsequent process and help us to get rid of dependence on full-wave tools completely.

Based on the above considerations, a totally physical-based link simulation mechanism is proposed in this paper. The mathematical formula and ABCD matrices to calculate the equivalent circuits of channel components are introduced in Section II. Section III focuses on the other two main characteristics of the simulation tool, the parameter configuration script and intermediate results reuse, which simplify the experimental process and extend the simulation space. Some studied cases and correlation to full-wave simulations are presented in Section IV, which demonstrate the validity and efficiency of our approach. In addition, cooperating with system simulator, this tool can clearly show the influence of different channels and guide us in high-speed PCB designs.

II. PHYSICAL-BASED MODELS FOR WIRELINE LINK

The key steps to construct a physical-based link model include decomposing the channel into discrete elements, presenting them in closed-form equations, and combining them together. This section introduces the physical model of lines and vias, which are the two most important components that determine the overall link performance, and the ABCD matrices that are used to express and combine them.

978-1-4673-9141-2/15 $31.00 © 2015 IEEE

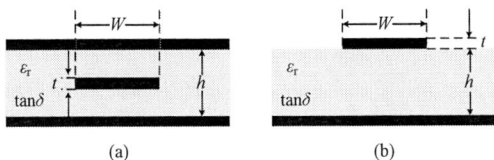

Fig. 1. Cross-sections of (a) stripline and (b) microstrip.

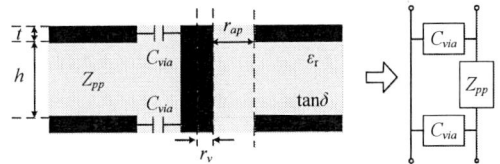

Fig. 2. Cross-section and equivalent circuit of via.

A. Trace Model

The transmission line theory is one of the most common methods to analyze traces. According to the transmission line theory, the fields of a transverse electromagnetic (TEM) transmission line are characterized by characteristic impedance Z_0, propagation constant γ, and line length l [12]. Typical cross-sections of stripline and microstrip are shown in Fig. 1.

As shown in Fig. 1, important physical and geometry parameters for traces include line width W, line thickness t, dielectric thickness h, relative dielectric constant ε_r, and dielectric dissipation factor $\tan\delta$. Take the centered stripline as an example, Z_0 and γ can be calculated as

$$Z_{0,stripline} = \frac{30\pi(1 - t/h)}{\sqrt{\varepsilon_r}(W/h + C_f/\pi)} \quad (1)$$

where

$$C_f = 2\ln(\frac{1}{1 - t/h} + 1) - \frac{t}{h}\ln(\frac{1}{(1 - t/h)^2} - 1) \quad (2)$$

and

$$\gamma = \alpha + j\beta = \frac{0.0231 R_s \varepsilon_r Z_0}{30\pi(h - t)} \cdot [1 + \frac{2W}{h - t}$$
$$+ \frac{h + t}{\pi(h - t)}\ln(\frac{2h - t}{t})] + \frac{27.3\sqrt{\varepsilon_r}\tan\delta}{\lambda_0} + j\beta \quad (3)$$

where R_s is the surface resistance, λ_0 is the wavelength in free space, and β is the phase constant.

Closed-form formulae to calculate Z_0 and γ of other traces like microstrip and coupled lines can be found in [12].

However, the traditional line models obtain non-causal time domain responses. Djordjević described ε_r as a function of frequency to fulfill the causality requirement [13]. We simplify his formula and express it as

$$\varepsilon(freq) = \varepsilon_\infty + \alpha \cdot \ln\frac{f_H + j \cdot freq}{f_L + j \cdot freq} \quad (4)$$

$\varepsilon(freq)$ represents the frequency-dependent permittivity; ε_∞ is the permittivity when frequency approaches infinity, α is a constant factor which can be measured for different substrates; f_H and f_L represent the high and low frequency for $\tan\delta$, which are usually set to 1THz and 1kHz, respectively.

B. Via Model

The physical-based via model within one cavity can be abstracted into a π-type RC circuit, which consists of two via-plate lateral capacitances C_{via} and parallel plate impedance Z_{pp} [14], as shown in Fig. 2.

The parallel plate impedance Z_{pp} describes the coupling between via and planes. The via-plate capacitance C_{via}

accounts for the displacement current between via barrels and planes. Important geometry parameters include plane distance h, plane thickness t, via radius r_v, antipad radius r_{ap}, vias locations, and plane dimensions. Formulae for Z_{pp} can be divided into different categories according to the boundary conditions, which are defined in [15]. C_{via} can be computed in analytical expression based on the Green's function [16]

$$C_{via} = \frac{2\pi\varepsilon t}{\ln(\frac{r_{ap}}{r_v})} + \frac{16\pi\varepsilon}{h\ln(\frac{r_{ap}}{r_v})}\sum_{n=1,3,5...}^{2N-1}\frac{1}{k_n^2 H_0^{(2)}(k_n r_v)} \quad (5)$$
$$\cdot [H_0^{(2)}(k_n r_{ap}) - H_0^{(2)}(k_n r_v)]$$

where $k_n^2 = \omega^2\mu\varepsilon - (n\pi/h)^2$ and $H_0^{(2)}$ is the Hankel function of second type and order 0.

C. Concatenation of Models

The discrete models for vias and lines can be applied to compute the response of each cavity between adjacent power/ground planes within a multilayer PCB. Next, these partial results need to be combined. Although this operation can be performed in terms of T-, Z-, or Y-parameters, ABCD matrix is more convenient to obtain the overall S-parameters.

Traces and vias can be presented in ABCD matrix forms as in equations (6) and (7), and the directions of voltage and current are illustrated in Fig. 3(a) and Fig. 3(b).

$$\begin{bmatrix} V_1 \\ I_1 \end{bmatrix} = \begin{bmatrix} \cosh(\gamma l) & Z_0\sinh(\gamma l) \\ \sinh(\gamma l)/Z_0 & \cosh(\gamma l) \end{bmatrix}\begin{bmatrix} V_2 \\ I_2 \end{bmatrix} \quad (6)$$

$$\begin{bmatrix} V_1 \\ I_1 \end{bmatrix} = \begin{bmatrix} 1 + j\omega Z_{pp}C_{via} & Z_{pp} \\ j\omega C_{via}(2 + j\omega Z_{pp}C_{via}) & 1 + j\omega Z_{pp}C_{via} \end{bmatrix}\begin{bmatrix} V_2 \\ I_2 \end{bmatrix} \quad (7)$$

Generally, the ABCD matrix representation is constrained to situations only when the input/output ports come in pairs. However, our model solves this issue by setting the directions of voltage and current vertical. If there are n traces and n vias, single elements V and I become vectors of size n, Z_{pp} and C_{via} become n-by-n matrices, as well as the trace matrix, and the ABCD matrix becomes a $2n$-by-$2n$ matrix. Thus, the matrix can be extended to include any numbers of vias and traces, regardless of the number of ports n is odd or even.

There are three kinds of via structure: through via (open ended), ground/power via (short ended), and vias connected by stripline. According to Fig. 3(c), the current directions of trace and via are orthogonal. Multi-conductor transmission line theory [17] can be used to address this issue. The transfer function can be calculated as

$$\begin{bmatrix} V_u \\ V_l \end{bmatrix} = \begin{bmatrix} k_u^2 Z_{pp} + Z_{sl} & -k_u k_l Z_{pp} + Z_{sl} \\ -k_u k_l Z_{pp} + Z_{sl} & k_l^2 Z_{pp} + Z_{sl} \end{bmatrix}\begin{bmatrix} I_u \\ I_l \end{bmatrix} \quad (8)$$

978-1-4673-9141-2/15 $31.00 © 2015 IEEE

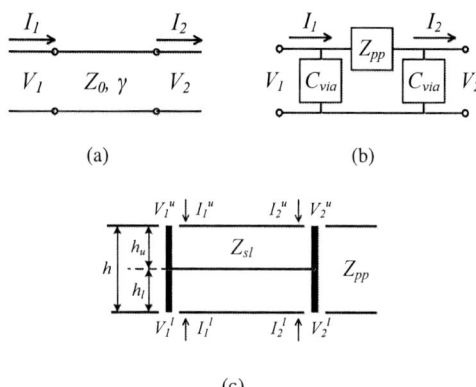

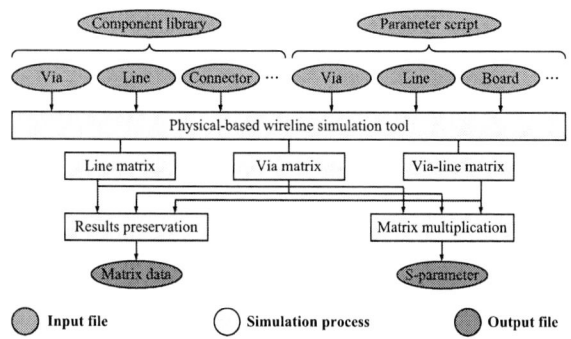

Fig. 3. ABCD matrix for (a) transmission line, (b) via, and (c) stripline-via model.

TABLE I.   AN EXAMPLE FORM OF PARAMETER SCRIPT

Line	Line Type	(stripline/microstrip/differential pair...)	
	Line Location		
	Line Size	Line Length	
		Line Width	
		Line Thick	
Via	Via Type	(through via/ground via/power via...)	
	Via Location		
	Via Size	Via Radius	
		Antipad Radius	
		Via Height	
		Stub Height	
Board	ε_r		
	$\tan\delta$		
	Plane Conductivity		
	Board Size	Board Length	
		Board Width	
		Layer Height	
Others	......	......	

where $k_{u(l)} = h_{u(l)}/h$, Z_{sl} and Z_{pp} represent the impedance matrices of stripline and parallel impedance.

Since each cavity has been formulated in terms of ABCD matrix, the final step to get the overall response of channel is to combine them by matrices multiplication. The advantage of using ABCD matrix is that the result can be computed very fast for an arbitrary number of layers and only a final transformation is required to convert to the S-parameters.

III. SIMULATION FLOW OF SIMULATOR

Fig. 4 shows the simulation flow of our tool. The first step to construct a physical-based model is to decompose the channel into discrete components. We build a component library, including models for various parts of PCBs, like vias and lines that have been introduced above, and complex structures like packages, connectors, and other elements. As all these components are expressed in the uniform ABCD matrix form, the matrices can be multiplied directly.

One important characteristic of this simulation engine is the use of a parameter configuration script to describe the PCB structure, rather than drawing the 3D model in full-wave software. The script file includes three main parts: parameters for transmission lines (line type, line location, and geometry size of lines); parameters for vias (via type, via location, and geometry size of vias and stubs), and parameters for board (ε_r and $\tan\delta$ of substrate dielectric, plane conductivity, and geometry size of board). Table I gives a basic form of the parameter script.

The simulation tool reads the input parameters, finds the equivalent circuits needed from the component library to constitute the channel, calculates the corresponding ABCD matrices and combines them with multiplication. The output file includes two parts: the reserved intermediate results and the whole link S-parameters. The reservation and reuse of intermediate results is another important characteristic of our engine. The intermediate result refers to the ABCD matrix for each element. In the physical-based model, structures with the same sizes produce the identical or similar ABCD matrices. For example, as mentioned in Section II, C_{via} and Z_{pp}, which make up the via model, is totally determined by the geometry size of via, which means that most of vias that have the same radius in the via cluster within one cavity actually share the same ABCD matrix. As the channel is described by the parameter script input, it is easy for the simulator to categorize them into one class, calculates their matrix only once and reuses it directly next time it meets the same structure, thus avoiding large amounts of repeated calculation and greatly accelerating the simulation speed. The final S-parameter can be used to analyze the links' attenuation and reflection; it can also be used as the input of the system simulator to evaluate the high-speed SerDes system's performance.

The use of the parameter script as the input file and the reservation and reuse of intermediate results are two important innovative points of this physical-based simulator, which make the engine have advantages of better flexibility and reconfigurability over the 3D full-wave tools. If we want to modify the previous design, we just need to rewrite the corresponding parameters that we want to change, rather than redrawing the whole 3D models every time for full-wave engines. The engine classifies the input parameters, only re-calculates the modified matrix, while the old data of the unchanged part can be reused directly, which can further improve the simulation efficiency. The two methods also make the parametric sweeping possible, which can be used to make further exploration on the maximum design space.

IV. EXPERIMENTAL RESULTS

Three sets of experiments were tested to evaluate the approach discussed above. First, we verified the rightness of the transmission line theory, especially at GHz frequency range. Second, in order to highlight the significant distortion caused by vias, we simulated a short channel with/without

Fig. 4. Simulation flow of the proposed engine.

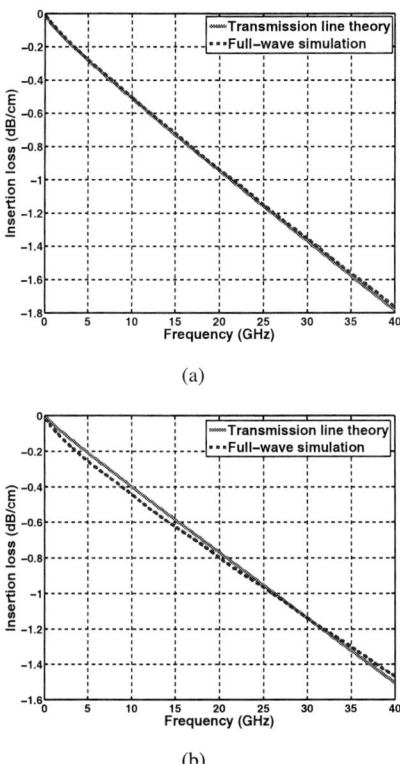

(a)

(b)

Fig. 5. Comparison of insertion loss between transmission line and full-wave simulation: (a) stripline and (b) microstrip.

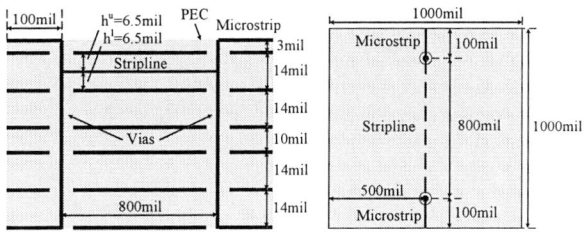

Via radius=5mil Antipad radis=15mil Line Width=5mil Line Thick=1mil ε_r =4.4 tanδ =0.02

Fig. 6. Cross-section and upper view of via-line evaluation structure.

long stubs. In these two experiments, we compared the results with a 3D full-wave electromagnetic software, HFSS [6], to demonstrate the validity and efficiency of our approach. In the last experiment, we swept the different lengths of channel, tried to find some guidelines for our PCB channel design to achieve higher data bandwidth. The physical-based simulator was implemented in MATLAB, and all experiments were tested on a 3.1 GHz Intel Xeon CPU, 4GB RAM machine.

A. Transmission Lines Test

The transmission line theory has been mature for decades. However, whether the theory is accurate enough at high-frequency over 1 GHz needs to be verified, especially when we take the frequency-dependent ε_r into consideration. Fig. 5 shows the comparison between the insertion losses of 1 cm long, 127 μm (5 mil) wide stripline and microstrip calculated from transmission line theory and HFSS. The characteristic impedances of lines are approximately 50 Ω. It proves that the line theory remains valid at high frequency. Considering the influence from the outer open space, the microstrip model is not as accurate as the stripline. However, the error of microstrip is still within 5% (about 0.06 dB/cm) even at 40 GHz. It indicates that we can use the transmission line theory to enhance the calculation speed.

B. Via-trace Links Test

The configuration of the second case is shown in Fig. 6, including two vias connected by a stripline on a six layers PCB. There are two kinds of transmission lines in this case: two microstrips on the upper surface of the first layer, acting

as the input/output ports, and one stripline centered in the second layer. All the lines are 127 μm (5 mil) wide, and the line lengths are 0.254 cm (100 mil) for microstrips and 2.032 cm (800 mil) for stripline, respectively. Two via constructions are considered here, including through holes (via-stub effect presented) and back-drilled holes (no stub effect). The via radius is 127 μm (5 mil), and the antipad radius is 381 μm (15 mil). We use our physical-based method and HFSS to calculate the final S-parameter separately.

The results obtained from the physical-based approach and comparisons against HFSS simulation are shown in Fig. 7 and Fig. 8. The results match quite well within 10 GHz, and good correlation can be stated up to 20 GHz within a difference margin of about 3 dB. The agreement for frequencies over 20 GHz is still fair, but the margin can be up to 5-8 dB in several frequency spots. The deviations may be owned to the fact that the physical model neglects inner inductance and resistance of vias as well as the impact of pads, which may cause impedance mismatch at high frequency.

C. Channel Characteristics for Sweeping Lengths

In typical link designs, driver and receiver technologies are determined in an early stage of integrated circuit design and cannot be modified later [11]. Under these constraints, whether the wireline links are suitable for applications under specific equalization schemes at a given data rate needs to be evaluated carefully before the on-board design. Among the controllable parameters, the line length is the most interest to designers because it has a significant impact on the channel performance and it can be adjusted within a large scale. Through the combination of physical-based method (accelerating the simulation speed), use of parameter configuration script (simplifying the modeling process), and reservation and reuse of intermediate results (avoiding repeated computation), the simulator can provide an accurate and fast sweeping simulation on different lengths of lines. Together with a system performance simulator, it can find the minimum and maximum channel lengths that meet the design requirements.

The trace lengths sweeping experiments were performed on a Megtron-6 (ε_r = 3.48, tan δ = 0.0068) board. The baud rate of data is fixed to 25.8 Gbps, and a three-tap feed-forward equalizer (FFE) and a ten-tap decision-feedback equalizer (DFE) are used for equalization with an automatic gain control. We simulated 100 different lengths of channel by sweeping the length from 1 cm to 100 cm. The sweeping simulations are evaluated for two different cases of layer stackups shown in Fig. 9. For both cases, the trace is routed on S3 layer.

978-1-4673-9141-2/15 $31.00 © 2015 IEEE

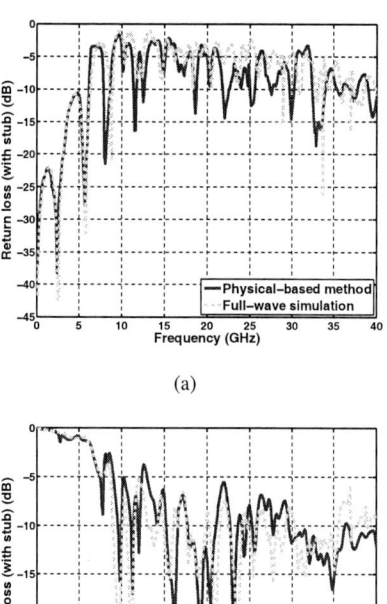

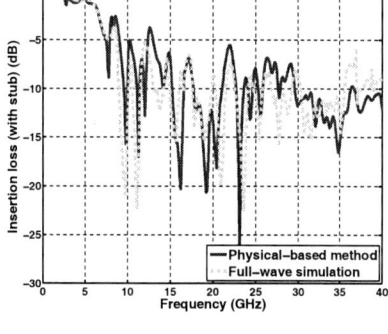

(a)

(b)

Fig. 7. (a) Return loss (S11) and (b) insertion loss (S21) for the evaluation structure with long via stubs, obtained by physical-based method and HFSS.

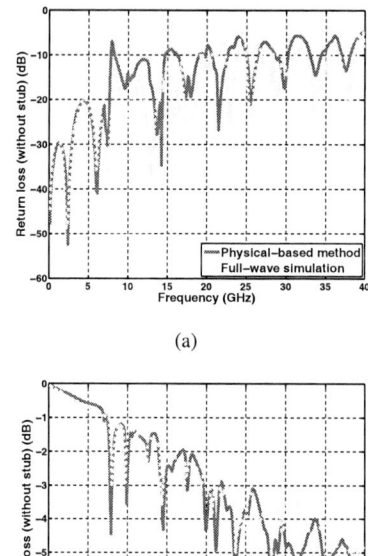

(a)

(b)

Fig. 8. (a) Return loss (S11) and (b) insertion loss (S21) for the evaluation structure without via stubs, obtained by physical-based method and HFSS.

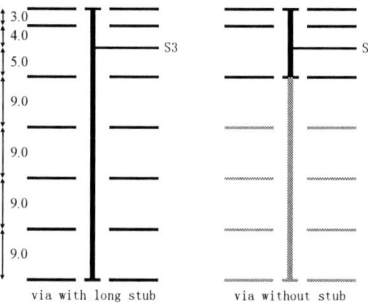

Fig. 9. Layer stack and dimensions used for sweeping simulation (in mils).

In one case, a long stub remains. While in the other cases, back-drilling eliminates most effect of stub. The simulations are also performed with and without equalizations.

Fig. 10 shows the results of the trace lengths sweeping simulations. We use the vertical and horizontal eye openings with the bit error rate of 1×10^{-12} as our evaluation criterion. The horizontal eye openings are normalized with the unit interval and the vertical eye openings are normalized with the mean eye opening which is the expected value of the eye openings at the eye center.

To recap,

- Via stubs cause severe reflections which strongly affect the performance of shorter channels, especially when the trace is shorter than 20 cm. Fig. 10(a) shows a relatively larger eye opening only the line is longer than 30 cm when there exist long stubs. The limitation of shorter channels prevents the increase of on board packaging density [11]. So the back-drilling process is necessary in high-speed applications.

- The maximum trace length is mainly determined by the attenuation of channels together with many other factors like dielectric materials, via stubs, and equalization schemes. Strong equalization scheme can compensate most of the signal distortion but needs complex circuits design and occupies more chip area. Multi-level signallings, such as four-level pulse amplitude modulation (PAM-4) and Duobinary, are also better choices, which reduce the Nyquist

frequency and alleviate the burden on circuit design.

D. Calculation Speed

The most important advantage of our physical-based approach is its computation efficiency. Compared to the full-wave method, the proposed solution can provide accurate results, meanwhile drastically reduce the simulation time by about two to three orders of magnitude in a conservative estimation. Table II summarizes the computation time needed using the physical-based method against the full-wave 3D solver, HFSS.

We can get the following insights from Table II.

- The use of transmission line theory indeed shortens the simulation time, especially when the calculation time of vias is shorter than lines.

- The speedup ratio improves with the increase of the PCB layers. This is attributed to the reservation

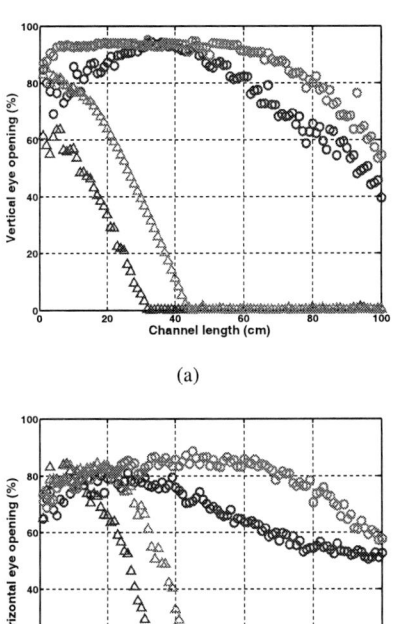

(a)

(b)

Fig. 10. Eye openings for 1 - 100 cm of channel lengths. Red: without stub. Blue: with stub. Circles: with equalization. Triangles: without equalization. (a) Vertical eye openings. (b) Horizontal eye openings.

TABLE II.   COMPUTATION TIME FOR PHYSICAL-BASED METHOD AND FULL-WAVE SIMULATION

Examples	Transmission lines (1 cm long)	Vias connecting lines (with stub)	Vias connecting lines (without stub)	Channel length sweeping
Physical-based method	< 1 s	23 s	8 s	6 min (without stub) 16 min (with stub)
Full-wave simulation	50s	7 hours	2 hours	> 1 week
Speedup	50x	1100x	900x	>2000x

and reuse of intermediate results. For the full-wave simulator, each circuit structure should be calculated from the start. While in our tool, components share the same geometry size need to be computed only once. This mechanism avoids great mass of repeated simulation, and the advantage on efficiency is specific obvious in the parameter sweep cases.

V.   CONCLUSION

In this paper, a physical-based simulation method for electrical wireline links is proposed. Experimental results validate our method good correlation to full-wave simulations up to 40 GHz with a significant improvement on the computation speed. Our tool is an efficient and reliable alternative to conventional 3D full-wave simulators. This method can also help us to improve PCB designs and layer planning through its fast sweeping on physical parameters for channels. With the aid of our tool, we can obtain a better understood for channel performance and design the appropriate PCBs rapidly, which offers a good assistance for further circuit design.

ACKNOWLEDGMENTS

This work was supported by the State Key Laboratory of Mathematical Engineering and Advanced Computing under grant 2013A04, and the State Key Laboratory of High-end and Storage Technology under grant 2014HSSA12. P. Liu is the corresponding author.

REFERENCES

[1] R. Rimolo-Donadio, H.-D. Brüns, and C. Schuster, "Including stripline connections into network parameter based via models for fast simulation of interconnects," in *Proc. 20th Int. Zurich Symp. Electromagn. Compat.*, Zurich, Switzerland, Jan. 2009, pp. 345–348.

[2] R. Rimolo-Donadio, A. J. Stepan, H.-D. Brüns, J. L. Drewniak, and C. Schuster, "Simulation of via interconnects using physics-based models and microwave network parameters," in *Proc. 12th IEEE Workshop Signal Propag. Interconnects*, Avignon, France, May 2008, pp. 1–4.

[3] ADS, ver. 2012, Agilent Corporation, Santa Clara, CA, 2012. [Online]. Available: http://www.agilent.com/

[4] Microwave Studio, ver. 2014, CST Corporation, Darmstadt, Germany, 2014. [Online]. Available: https://www.cst.com/

[5] AMDS, ver. 2007, Agilent Corporation, Santa Clara, CA, 2007. [Online]. Available: http://www.agilent.com/

[6] HFSS, ver. 13, Ansoft Corporation, Pittsburgh, PA, 2011. [Online]. Available: http://www.ansys.com/

[7] R. Rimolo-Donadio, G. Xiaoxiong, Y. H. Kwark, M. B. Ritter, B. Archambeault, F. De Paulis, Z. Yaojiang, F. Jun, H.-D. Brüns, and C. Schuster, "Physics-based via and trace models for efficient link simulation on multilayer structures up to 40 GHz," *IEEE Trans. Microw. Theory Tech.*, vol. 57, no. 8, pp. 2072–2083, Aug. 2009.

[8] K. J. Han, , M. B. Ritter, and X. Gu, "Electrical and physical parametric study of high-speed link performance," in *Proc. 19th IEEE Conf. Electr. Perform. Electron. Packag. Syst.*, Austin, TX, Oct. 2010, pp. 229–232.

[9] K. J. Han, X. Gu, Y. H. Kwark, Z. Yu, D. Liu, B. Archambeault, C. S. R., and J. Fan, "Parametric study on the effect of asymmetry in multi-channel differential signaling," in *Proc. IEEE Int. Symp. Electromagn. Compat.*, Long Beach, CA, Aug. 2011, pp. 131–136.

[10] S. Müller, X. Duan, M. Kotzev, Y. Zhang, J. Fan, X. Gu, Y. H. Kwark, R. Rimolo-Donadio, H.-D. Brüns, and C. Schuster, "Accuracy of physics-based via models for simulation of dense via arrays," *IEEE Trans. Electromagn. Compat.*, vol. 54, no. 5, pp. 1125–1136, Oct. 2012.

[11] K. J. Han, X. Gu, Y. H. Kwark, L. Shan, and M. B. Ritter, "Modeling on-board via stubs and traces in high-speed channels for achieving higher data bandwidth," *IEEE Trans. Compon., Packag. Manufact. Technol.*, vol. 4, no. 2, pp. 268–278, Feb. 2014.

[12] B. C. Wadell, *Transmission Line Design Handbook*. Norwood, MA: Artech House, 1991.

[13] A. R. Djordjević, R. M. Biljić, V. D. Likar-Smiljanić, and T. K. Sarkar, "Wideband frequency-domain characterization of FR-4 and time-domain causality," *IEEE Trans. Electromagn. Compat.*, vol. 43, no. 4, pp. 662–667, 2001.

[14] C. Schuster, Y. Kwark, G. Selli, and P. Muthana, "Developing a 'physical' model for vias," in *Proc. IEC DesignCon*, Santa Clara, CA., Feb. 2006, pp. 1–24.

[15] T. Okoshi, *Planar Circuits for Microwaves and Lightwaves*. Berlin, New York: Springer-Verlag, 1985.

[16] Y. Zhang, J. Fan, G. Selli, M. Cocchini, and F. De Paulis, "Analytical evaluation of via-plate capacitance for multilayer printed circuit boards and packages," *IEEE Trans. Microw. Theory Tech.*, vol. 56, no. 9, pp. 2118–2128, Sept. 2008.

[17] C. R. Paul, *Analysis of Multiconductor Transmission Lines*, 2nd ed. Hoboken, NJ: Wiley, Oct. 2007.

978-1-4673-9141-2/15 $31.00 © 2015 IEEE

Trace Signal Selection Methods for Post Silicon Debugging

Shridhar Choudhary*, Amir Masoud Gharehbaghi*, Takeshi Matsumoto[†] and Masahiro Fujita[‡]

*Dept. of Electrical Engineering and Information Systems, The University of Tokyo, Tokyo, JAPAN
Email: {shridhar, amir}@cad.t.u-tokyo.ac.jp
[†]Ishikawa National College of Technology, Ishikawa, JAPAN
Email: matsumoto@ishikawa-nct.ac.jp
[‡]VLSI Design and Education Center, The University of Tokyo, Tokyo, JAPAN
fujita@ee.t.u-tokyo.ac.jp

Abstract—In post-silicon debugging, only a limited number of states (flip-flops) can be traced, due to the area overhead that is introduced by trace buffers. Therefore, it is important to select the states which can restore most of the other states. There exist researches that try to heuristically select a set of flip-flops (FFs) which maximizes the number of restored FFs. We first show that those existing works are not so robust, as the cost functions used for selections do not work well in some cases. In this paper, we introduce a new signal selection that tries to improve the selection by swapping the FFs that are going to be traced. Furthermore, we introduce a hardware implementation of the method that is more than 3 orders of magnitude faster than software-based swapping. With the proposed methods, we can improve the signal selection and get consistent results even for large circuits.

I. INTRODUCTION

As integrated circuit technology has been closely following Moore's law, the complexity of silicon chips is increasing intensively, while the design cycle is decreasing. Because of shorter design cycle and increased complexity, it has become very hard to analyze the functional bugs during pre-silicon phase of verification. As a result, bugs are escaped from pre-silicon phase to the chip. The bugs which exist deep into design state space are very hard to detect during the manufacturing tests as well. There may be also some electrical bugs that are usually hard to detect. The traditional pre-silicon verification techniques such as formal verification and simulation based techniques are not efficient to identify these functional bugs as well as electrical bugs. As a result, post silicon validation methods have come into existence.

Post-silicon debugging tries to identify, locate and correct the bugs that exist in the chip after manufacturing to avoid shipping the buggy chips.

Mainly the post silicon debugging can be divided into two approaches.

- Bug localization techniques that try to automatically find cause of an observed erroneous state.

- Design for debug (DFD) techniques that improve observability and controllability of design.

One of the approaches to locate errors is to start from an erroneous state of the circuit and try to find the traces that cause the corresponding erroneous state by going back to the first corresponding error-free state. One such method is Backspace [3] which basically records a crashed state, then calculates its error free state, which is the target point. By repeating these steps on the new target point it reconstructs error trace backwardly. This method has shown promising results by reconstructing traces for hundreds of cycles, but it has high area overhead for the special on-chip hardware. Obtaining the previous state is represented as a SAT problem. SAT solvers are used to produce trace for such bugs that are detected after thousands of cycles of execution.

There are techniques that do not require any system level simulation and failure reproduction. For example, a technique called IFRA, Instruction Footprint Recording and Analysis [2] has been developed for localizing bugs and finding the instruction sequence that exposes the bug from a system failure, for instance a crash. A special hardware which records the instruction footprints, which contains special information about the flow of instructions, and what the instructions did as they passed through various micro-architectural blocks of the processor. During normal operation of processor, when a failure is detected, the recorded information is scanned out and analyzed for bug location. Trace based approaches require large amount of internal signal data to be analyzed. However, many of this data is irrelevant for reproducing the error and increases the processing time and on-chip memory. Therefore, DFD techniques are proposed to improve the controllability and observability of internal signals.

In this paper, we first introduce a method to select optimum number of traced states for a specific simulation vector using a Pseudo Boolean Optimization (PBO) based algorithm which can optimize the on-chip memory by restoring the unknown states from the fewer number of traced states stored in the on-chip buffer. Basically, we try to select some particular signals for tracing instead of tracing all the values, following some algorithms trying to reproduce other states of the design. Logical bugs can be detected using restorability based techniques. We can get optimum FF selections for small circuits with a single simulation vector by utilizing PBO techniques. Please note that the optimum solution basically changes if the simulation vector changes.

From the experiments of PBO we observed that there are cases where existing simulation based methods do not

978-1-4673-9141-2/15 $31.00 © 2015 IEEE

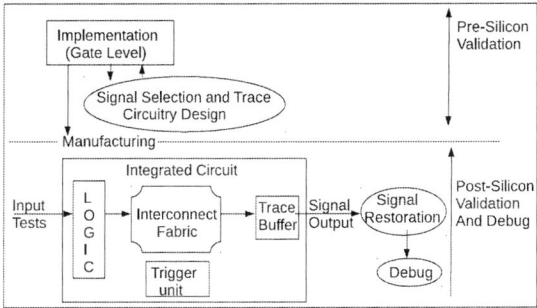

Fig. 1. Overview of system validation [5].

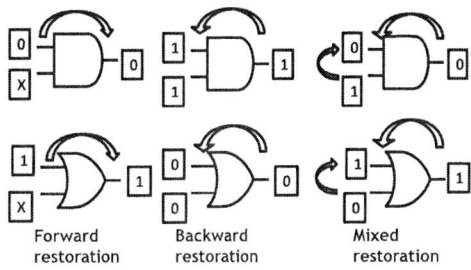

Fig. 2. Restoration examples.

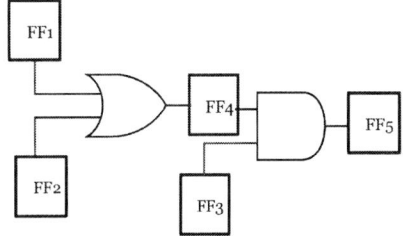

Fig. 3. Example circuit.

work well. That is because the cost functions of previous X-smiluations based methods are correct only if all of the FFs are fixed. The problem is that the cost function using incremental or decremental selection of FFs may not necessarily be a good criteria. In this paper, we show that the cost functions cannot be compensated to become correct neither by considering two FFs at the same time nor by considering internal signal values. Therefore, we need to fix all of the FF selection in order for the cost function to be correct or effective. We have introduced a swap based signal selection process. Furthermore, we have shown a FPGA based hardware implementation of swapping and compared it with its software implementation. We show that swapping can significantly improve the signal selection from the existing simulation based methods. However, many swaps are required for large circuits and software takes excessive time for those large circuits. Therefore, we propose to use FPGA that can achieve around 1000 times speed up over software.

The rest of the paper is organized as follows. Section II reviews the existing approaches for the post-silicon validation problems. Section III describes signal restorability and Section IV shows our PBO based implementation. Following it Section V describes various signal selection heuristics and Section VI explains our swapping based implementation. Finally Section VII shows experimental results and we conclude the paper in Section VIII by presenting future work.

II. RELATED WORK

To improve the internal signal observability, special DFD hardwares are proposed in the literature that improve data acquisition [8], [9], [10], [11]. They are either trace buffer based or scan based approaches. Fig. 1 shows the flow of design in case of using trace buffers inside the chip. Special hardware called ELA (Embedded Logic Analyzer) [7] is used on-chip which samples the data into trace buffers. An ELA has many trigger units that determine when the data should be sampled into trace buffers. This DFD hardware utilizes the sampled data by sending it through low bandwidth device pins and finally the errors can be identified off-chip using some special post processing algorithms.

Signal selection techniques that uses probabilistic metric [4], [5] exist but are not as effective as simulation based methods introduced in [14]. Simulation based technique introduced in [14] starts with all the FFs traced and reaches to the final selection by removing one FF at each step. Recently, a new ILP based signal selection approach, which uses incremental signal

selection rather than the decremental approach was introduced in [19]. In incremental approach they start with an empty set of traced FFs and incrementally selects one FF each time. Incremental approach needs less simulation cycles compared to the decremental approach to select trace FFs.

III. SIGNAL RESTORABILITY

In post-silicon debugging, ideally we want to observe every signal value in each cycle, while utilizing little chip area and consuming less time. However, it is unrealistic to observe each and every state of the signals at every cycle. While we use trace buffers to store these signals the amount of on-chip memory used by them should be kept as small as possible. Therefore, it is better to trace only limited number of signal values and try to restore other unknown values with off-line dependency analysis, which is called restoration. Restoration process utilizes the controlling value of a logic gate [14]. A controlling value at one of the inputs of a logic gate can be used to infer the value of some or all inputs of the corresponding gate. For example, value 1 on an input of OR gate determines the value of the output to be 1. Similarly, if we know one of the inputs and the output value of a 2-input gate, we can infer the other input value. Using similar technique for other logic gates we can restore unknown input states and output states for different cycles. The restoration can be divided into three types as shown in Fig. 2.

TABLE I. RESTORATION PERFORMED ON THE EXAMPLE CIRCUIT OF FIG. 3.

	1	2	3	4	5
FF1	0	x	0	x	x
FF2	0	x	0	x	x
FF3	x	x	x	x	x
FF4	0	0	1	0	x
FF5	x	0	0	x	0

978-1-4673-9141-2/15 $31.00 © 2015 IEEE

- *Forward restoration:* When one of the inputs of a gate has controlling value, the output can be inferred without knowing other input values.

- *Backward restoration:* When the output of a logic gate has the non-controlling value, we can infer that all the inputs have the non-controlling value.

- *Combined restoration:* When we know one input as well as the output of a 2 input gate, we can infer the other input from these values.

TABLE I shows the restoration process performed on the example circuit shown in Fig.3. Columns show the cycle values of FF1 to FF5 for 5 cycles. For evaluating the quality of restoration, State Restoration Ratio (SRR) is defined as:

$$\frac{sum\ of\ number\ of\ traced\ +\ number\ of\ restored\ states}{number\ of\ traced\ states}$$

In the example circuit Fig.3, we can see that FF2 is traced for 4 cycles and we were able to restore 4 states backwardly consisting of 1st and 3rd cycle values of FF1, FF2 from 2nd and 4th cycle values of FF4. Similarly 3 states consisting 2nd, 3rd and 5th cycle values of FF5 can be inferred from 1st, 2nd and 3rd cycle values of FF4 respectively. So SRR for this circuit equals to (7+4)/4 = 2.75. But if we trace a different signal the number of restored states would increase or decrease accordingly. It can be seen that restoration depends on the signals that are traced. Therefore, the trace signals should be carefully selected to get the optimum result. Which signals and how many signals to be traced have been an active research topic and a number of algorithms to select the best signals have been proposed for example in [5], [14] ,[19].

IV. PESUDO BOOLEAN OPTIMIZATION (PBO) FORMULATION

Based on the restoration technique mentioned above, in this section we present a PBO-based formulation for selecting the optimum number of traced signals for a given circuit. The PBO problem is the task of finding a satisfying assignment to a set of PB-constraints that minimizes a given objective function.

We introduce four Boolean variables

- $r_s(c), tr_s, p_{s \leftarrow t}(c)$ to have logical representation of restorability.

- $r_s(c)$ is 1 iff signal s is known (i.e. restored or traced) at c-th clock cycle.

- tr_s is defined only for state variables and indicates s is traced.

- $p_{s \leftarrow t}(c)$ is 1 iff signal s is restored using the values of signal in set $t = \{t1, ..., tn\}$ at c-th cycle. Variables $p_{s \leftarrow t}$ are defined to represent restoration for a logic gate, where each of s and $t1, ..., tn \in t$ are an input or output of a gate. An example of how p variables are defined is illustrated in Fig. 4. To avoid cyclic propagation, we add a constraint for p variables of each gate so that signal s is not used to restore other signals when s is restored by them, which can be seen in the Fig. 4.

- r_s is defined as disjunction of all p variables to restore s and tr_s (if s is a state variable). Note that p variables

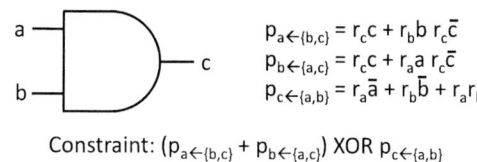

$$p_{a \leftarrow \{b,c\}} = r_c c + r_b b\ r_c \bar{c}$$
$$p_{b \leftarrow \{a,c\}} = r_c c + r_a a\ r_c \bar{c}$$
$$p_{c \leftarrow \{a,b\}} = r_a \bar{a} + r_b \bar{b} + r_a r_b$$

Constraint: $(p_{a \leftarrow \{b,c\}} + p_{b \leftarrow \{a,c\}})$ XOR $p_{c \leftarrow \{a,b\}}$

Fig. 4. Formulation example for AND gate.

to restore s may exist in different gates, since s can be restored both forwardly in a fanin gate of s and backwardly in fanout gates of s.

The logic formulas generated in the way described above are translated into PBO formulas. Since restorability is calculated for a given test pattern, we assign truth values of signals calculated by simulation, in order to simplify the formulas. Then, we convert the formulas into conjunctive normal form (CNF), and each clause of CNF formula is translated into an inequality which turns to be a constraint in PBO problem. For example, a CNF clause $(a, \neg b, c)$ is translated to an inequality $(a - b + c \geq 0)$. In addition, the number of traced signals and the objective function to maximize the number of restored state variables are added to PBO problem instances.

V. SIGNAL SELECTION HEURISTICS

We compared the results of the PBO-based method with those of X-simulation based methods [14] and found that cost function for selection in X-simulation based methods doesn't work well. Even the method which selects FFs incrementally [19] sometimes gives significantly inferior results than the PBO-based method. In order to overcome the problem of the existing methods which selects FFs heuristically, we have examined a couple of more complicated selection methods.

A. Selecting 2 FFs Incrementally At Once

Selecting FFs incrementally using X-simulations [19] is proved to be better than decremental selection [14]. We compared our implementation of the incremental approach where at every step, one FF is being traced, with the optimum selection from the PBO-based method. We observed that in some cases during the incremental selection of FFs at the second step or further steps, a wrong FF was selected in the sense that final pool of selection was not optimum. Therefore, as shown in the Fig.5, we changed the selection process by selecting 2 FFs at a step instead of one in the X-simulations based method presented in the previous works [19]. Experimental results showed that even selecting 2 FFs at a single step may finally select wrong FFs. That lead us to the conclusion that every position of FFs during the selection is needed to be fixed for the best selection of FFs which could significantly increase the processing time.

B. Selection Based On Cost Function With Intermediate Variables

Cost function in previous works just uses number of restored FFs for evaluation but we included restored intermediate signals as well, because they might affect the connected FFs. Unfortunately the results after the inclusion of intermediate variables in the cost function evaluation didn't change the final selection as expected.

978-1-4673-9141-2/15 $31.00 © 2015 IEEE

Algorithm 1 Simulation Based Selection Process

1: **procedure** SelectSignals(circuit,w,c)
2:     Create list of selected signals S => Initial selection
3:     **while** |S| < w **do**
4:         Gererate a random input vector I
5:         **for** every pair of FFs those are not in S **do**
6:             Calculate restoration difference
7:         **end for**
8:         Find FFs with maximum restoration difference
9:         Add those 2 FFs to the list S
10:    **end while**
11:    **return** S
12: **end procedure**

Fig. 5.   Simulation based selection process.

Algorithm 1 Swap Based Signal Selection Process

1: **procedure** SelectFFs(circuit,w(TraceBufferWidth,c(cycle))
2:     Create list of selected FFs S => Initial selection
3:     Create list of remaining FFs R
4:     Calculate # of restored FFs for S => N
5:         **for** (10000 times) **do**
6:             Swap 1 FF between S and R
7:             Calculate # of restored FFs for S => N_new
8:             if (N_new > N) accept the swap
9:             else re-swap the selection
10:        **end for**
11:    **return** S
12: **end procedure**

Fig. 7.   Swap based signal selection process.

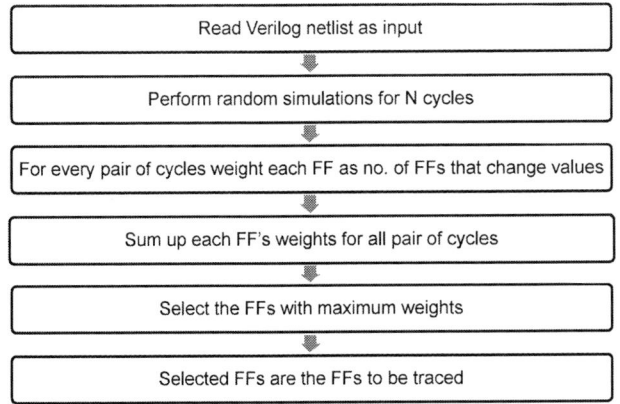

Fig. 6.   Signal selection process.

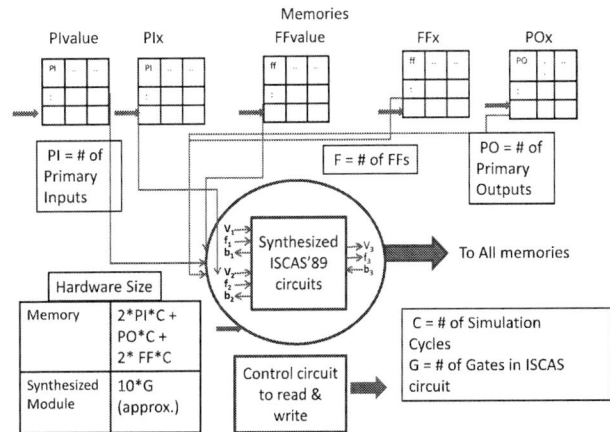

Fig. 8.   Swap based selection hardware.

C. Functional-Dependency Based Selection

As SRR based cost function might not be a good criteria, we came up with a new cost function that is completely different from the previous works. In our proposed heuristic, selection of FFs is based on functional dependency among the FF signal values. As the value of a FF changes in a given pair of cycles, it might affect other FFs by changing their values in the same pair because ultimately they are connected with each other through some logic gates in between. The signal selection process can be seen in the Fig.6.

VI.   SWAPPING BASED SELECTION IMPROVEMENT

As can be seen in the experimental results section, all the heuristics shown above could not improve the final FF selection much. However, based on the PBO formulation we know that there is a scope for improvement. Therefore, we proposed to start with an initial selection of best FFs as per existing simulation based methods and try to improve from there. For improving the FF selection, we introduce a swapping based heuristic. Basically we start with an initial list of selected FFs for restoration and a list of remaining FFs. Then we randomly swap FFs between the two lists and check whether the number of restored states is improved or not. We accept the swapping if the number of restored FFs is increased and continue swapping until we can not improve the restoration

amount further. Please note that the proposed swapping method is orthogonal to all the previous work on signal selection. In other words, we can accept the selected signals from any other method and improve the selection by our swapping method.

A. Software Implementation

We implemented the swapping based selection as a software program. As shown in Fig.7, algorithm for swapping accepts only improved selection. As, large numbers of swaps are required to reach to a much improved selection and the software takes some amount of time for each swapping, it is practically impossible to use the software for large circuits.

B. Hardware Implementation

As we need as many as possible swaps for larger circuits, we introduce a hardware implementation of swapping based signal selection. As shown in Fig.8, we introduce v, f, b variables for simulation values, forward restoration and backward restoration, respectively. Synthesized circuit modules contain all the formulas for v, f and b variables similar to our PBO formulas as shown in Fig.4. These formulas are used for X-simulation for computing restoration through the circuit. f and b variables are basically used to represent X values. We introduce 5 memories each for primary input value, X value as

TABLE II.   EXPERIMENT FOR INITIAL HEURISTICS.

circuit	# of FFs	Method1	Method2	Method3	Method4	PBO
s298	14	4.23	4.23	3.91	3.16	4.4
s344	15	3.24	3.31	3.13	2.73	4.14
s386	6	2.48	2.48	2.48	2.25	2.5
s444	21	4.81	5.78	3.76	3.81	6.05
s510	6	2.49	2.48	2.49	2.38	2.5
s526	21	5.95	5.96	3.74	3.71	6.04
s713	19	3.97	3.97	3.97	2.5	4.01
s832	5	2.15	2.15	2.06	2.14	2.17
s1423	74	7.8	T/O	6.8	5.56	T/O
s5378	179	15.3	T/O	15.1	12.1	T/O
s9234	211	10	T/O	9.59	5.18	T/O
s15850	638	27.8	T/O	T/O	23.8	T/O

TABLE III.   SOFTWARE SWAPPING IMPROVEMENT.

circuit	# at start	# of effective Swaps	# of restored
s298	802	3	850
s344	396	8	576
s349	396	7	590
s386	381	0	381
s444	929	6	1254
s510	384	0	384
s526	1269	1	1278
s713	754	2	821

PI and PIx, primary output X value as POx, FF value and FF X value as FFv and FFx respectively. Width of these memories are the number of variables they represent. For example, FFv memory width is the number of FFs in the circuit. Depth of these memories is the same as the number of cycles to be traced. In Fig.8 size of the hardware is shown. These memories are connected with the synthesized module through a control circuit. The total circuit size becomes around 10 times the original size. That is because, we have introduced three kinds of variables and their logics to the original circuit.

At the start, we give an initial selection vector to FFx memory while whole PI memory and FFv's 1st address is randomly initialized. Then X-simulation is performed in hardware for restoration by repeating forward and backward restoration until we reach to the maximum number of possible restored FFs. After restoration is finished, FFx memory is read to count the number of FFs to determine if the current swapping should be taken or not, which basically follows the procedure shown in Fig.7.

VII.   EXPERIMENTAL RESULTS

We applied our PBO formulation and all the three heuristics explained in section IV to ISCAS'89 circuits and performed trace signal selection. PBO based method for one random vector worked only for smaller circuits because of scalability issues. We could get optimum FF selections for small circuits with a single simulation vector by utilizing PBO techniques. For comparison of our results, we implemented one of the algorithms shown in existing simulation based method [19] with incremental selection.

TABLE II shows the comparison of various heuristics. Column 1 is name of the circuit. Column 2 is the number of FFs. Other columns show the SRR of different methods. In all the cases, we have used same 100 random input vectors for comparison of all the methods and reported the average SRR. We fixed the trace buffer width as 4 for circuits from s298 to s1423 and 8 for larger circuits with 64 simulation cycles. In TABLE II, Comparing Method1, which is our implementation of [19], with PBO shows that PBO gives better restoration in some of the cases excluding s510 and s832 where it was almost equal but not less. Simulation based approach provides better SRR compared to all the existing solutions. However, in some cases specially for small buffer width of 4, SRR was reduced because of fewer number of FFs after optimization using synthesis tools. When random inputs are used, according to [14], restoration capability can be obtained by actually simulating the restoration process on the circuit over a small number of cycles 64, and measuring the corresponding SRR. As trace signal selection is done only once during the design flow of circuit blocks of ELA, the run time of selection algorithm is less important than the quality of the selected signals [14].

TABLE II column Method2 shows the SRR for 2FFs at once heuristic. As the experiments could not finish for large circuits, comparison can be made only for smaller circuits. We can see there is not much improvement and even the restoration decreases in some cases. This shows that every position while selecting FFs should be fixed for optimum selection.

Similarly we compared the intermediate variable included heuristic's results, shown as Method3 in TABLE II, with Method1. Unfortunately, including intermediate variables in the cost function could not improve the final selection resulting in low or equal SRR compared to Method1.

For functional dependency based heuristic we performed different selections using 5 sets of random vectors and chose the best for comparison. Method4 shown in TABLE II represents the SRR for functional dependency based heuristic's result. Because of different cost function, SRR may not be a good criteria for the evaluation of Method 4 but comparing the final selection with PBO's selection, we can say that Method 4 can not achieve the best selection.

Table III compares the number of states restored at the start and at the end by performing 100000 swaps. The effective number of swaps in TABLE III shows the swaps which improve the selection. Comparison shows that, software based swapping could improve the initial selection of FFs, which is the final selection by Method1, in smaller ISCAS'89 circuits. The largest circuit for which we could achieve improvement with 2 hour time out was s713 circuit.

As software could take forever for larger circuits, we show improvement in the number of swaps using our FPGA based hardware in TABLE IV. The time hardware takes for one swap can be calculated for a particular circuit based on the following formulas. If C shows the number of simulation cycles and $Cres$ shows the number of clock cycles required for one complete process of restoration,

$$lock cycles for one swap(CC) = (FF number \times C) + Cres$$

$$Time for one swap = \frac{1}{FPGA\ clk\ freq.\ for\ the\ circuit} \times CC$$

Using above mentioned formulas TABLE IV shows the comparison between the number of swaps that can be performed in one hour using software and hardware implementations. The number of swaps calculated for hardware

TABLE IV.   EXPERIMENT COMPARING HARDWARE AND SOFTWARE SWAPS

circuit	# of Gates	# of FFs	# of Swaps Software (per hour)	# of Swaps Hardware (per hour)	Ratio
s5378	2779	179	10000	9309662	931
s9234	5597	211	6428	6209048	966
s15850	7951	638	3461	1821386	526

is shown based on wall clock time. Hardware, as expected, can perform more than 900 times swaps than software for some of the large ISCAS'89 circuits which in turn reach better selection as shown in TABLE V. TABLE V shows the SRR at the start, averaged over 100 different random vectors, which is calculated based on the best selection known in existing methods [19]. SRR at the end, again averaged over 100 different random vectors, the improvement by swapping from over the best selection known. Largest ISCAS'89 circuit for which we could perform swapping using hardware is s15850 as we are still improving the architecture to be able to embed very large circuits in Virtex 7 FPGA board available to us. So using hardware swapping, we expect to achieve improvement in signal selection even for industrial circuits using FPGA resources.

TABLE V.   EXPERIMENT SHOWING SRR FOR HARDWARE

Circuit	SRR at start	SRR at end	Improvement (in percent)
s5378	15.39	20.21	31.3
s9234	4.23	14.53	243.4
s15850	39.63	52.82	33.28

VIII.   CONCLUSIONS AND FUTURE WORK

Post silicon validation has become an essential step in the design flow of integrated circuits. In this paper, we presented different heuristics to select effective signals which can reconstruct other signals of the circuit. Through experiments, we came to know that signal selection can be improved by fixing FFs selected during each step of simulation based method. We did so by repeated local optimization on an early selection using swapping.

We have done experiments using swapping on software but it cost many simulation cycles and excessive runtime. Therefore we introduced a FPGA based swapping which can be thousand times faster than software and give us optimum results even for very large circuits because of high resources available in FPGA. We are planning to use sophisticated swapping algorithms like simulated annealing instead of random swapping with an improved FPGA architecture suitable for very large circuits so that more number of effective swaps can be done to get better signal selection.

REFERENCES

[1] N. Nataraj, T. Lundquist, and K. Shah, "Fault localization using time resolved photon emission and STIL waveforms," *Proc. International Test Conference, 2003*, pp. 254–263.

[2] S.-B. Park and S. Mitra, "IFRA: Instruction Footprint Recording and Analysis for Post-silicon Bug Localization in Processors," *Proc. of Design Automation Conference*, pp. 373–378, 2008.

[3] F. M. De Paula, M. Gort, A. J. Hu, S. Wilton, and Y. Jin, "BackSpace: Formal Analysis for Post-Silicon Debug," *Proc. of Formal Methods in Computer-Aided Design*, pp. 1–10, 2008.

[4] H. F. Ko and N. Nicolici, "Algorithms for state restoration and trace signal selection for data acquisition in silicon debug," *IEEE Trans. on Computer-Aided Design of Integrated Circuits and Systems*, vol. 28, no. 2, pp. 285–297, 2009.

[5] K. Basu and P. Mishra, "Efficient trace signal selection for post silicon validation and debug," *Proc. VLSI Design*, 2011, pp. 352–357.

[6] H. F. Ko and N. Nicolici, "Automated trace signals identification and state restoration for improving observability in post-silicon validation," *Proc. Design Automation and Test in Europe*, 2008, pp. 1298–1303.

[7] M. Abramovici, P. Bradley, K. Dwarakanath, P. Levin, G. Memmi, and D. Miller, "A reconfigurable design-for-debug infrastructure for SoCs," *Proc. Design Automation Conference (DAC)*, 2006, pp. 7–12.

[8] SignalTap II Embedded Logic Analyzer, Altera Verification Tool, 2006, http://www.altera.com/products/software/products/ quartus2/verification/signaltap2/sig-index.html.

[9] ChipScope Pro, Xilinx Verification Tool, 2006, http://www.xilinx.com/ise/optional prod/cspro.html.

[10] Sun Microsystems OpenSPARC, http://opensparc.net/.

[11] Embedded Trace Macrocells, ARM limited, 2007, http://www.arm.com/products/solutions/ETM.html.

[12] B. Vermeulen and S. K. Goel, "Design for Debug: Catching Design Errors in Digital Chips," *IEEE Design and Test of Computers*, vol. 19, no. 3, pp. 35–43, May 2002.

[13] Yeonbok Lee, Takeshi Matsumoto, Masahiro Fujita, "Generation of I/O Sequences for a High-level Design from Those in Post-silicon for Efficient Post-silicon Debugging," *Proc. of 28th IEEE International Conference on Computer Design*, pp. 402–408, October 2010.

[14] Debapriya Chatterjee, Calvin MacCarter, Valeria Bertacco, "Simulation-based signal selection for state restoration in silicon debug," *Proc. of the International Conference on Computer-Aided Design*, pp. 595–601, Nov. 2011.

[15] H. F. Ko and N. Nicolici, "Algorithms for state restoration and tracesignal selection for data acquisition in silicon debug," *IEEE Trans. on CAD*, vol. 28, no. 2, pp. 285?297, 2009.

[16] Xiao Liu and Qiang Xu,"On Signal Selection for Visibility Enhancement in Trace-Based Post-Silicon Validation," *IEEE Transactions on Computer-Aided Design of Integrated Circuits and Systems*, vol.31, no.8, pp.1263–1274, Aug. 2012.

[17] Kang ZHAO and Jinian BIAN, "Pruning-Based Trace Signal Selection Algorithm for Data Acquisition in Post-Silicon Validation," *IEICE TRANS. FUNDAMENTALS*, VOL.E95-A, NO.6 pp. 1030–1040, Jun. 2012.

[18] N. Eén and N Sörensson, "Translating Pseudo-Boolean Constraints into SAT," *Journal on Satisfiability, Boolean Modeling and Computation*, vol. 2, pp.1–26, 2006.

[19] Rahmani,K.Mishra, P.Ray, S., "Efficient trace signal selection using augmentation and ILP techniques," *Quality Electronic Design (ISQED), 2014 15th International Symposium on* , vol., no., pp.148,155, 3-5 March 2014.

978-1-4673-9141-2/15 $31.00 © 2015 IEEE

Timing Attack on NEMS Relay Based Design of AES

Samah Mohamed Saeed, Bodhisatwa Mazumdar, Sk Subidh Ali, Ozgur Sinanoglu
New York University Abu Dhabi (NYUAD)

Abstract—In deep submicron CMOS transistors, the static leakage current has become a significant contributor to power consumption with channel length and subthreshold voltage being continuously scaled down. Also, this increased leakage has recently led to the rise of side-channel attacks on CMOS based implementations. Nanoelectromechanical System (NEMS) relay technology is emerging as an alternative to CMOS with one of its most prominent advantages being the zero static leakage, providing an inherent defense against power side-channel attacks at the same time. On the other hand, this emerging technology introduces timing challenges in the design process; to minimize the timing delay of NEMS relays, binary decision diagram (BDD) based implementation is utilized to design combinational logic. What's important from a security perspective is that the timing delay of the BDD implementation of a NEMS relay based design is inherently input dependent. An adversary can therefore leverage the data dependency to identify secret information of the chip. We propose a timing delay based attack on NEMS relay based designs, use AES as a case study, and show that it can achieve a success rate of 1.0 for interconnect delay variations within a standard deviation of 0.0022. To the best of our knowledge, this paper is the first to expose an inherent security vulnerability of a NEMS relay based design.

Index Terms—Nanoelectromechanical System (NEMS) relays, Binary decision diagram (BDD), Timing attack, Security, AES, Side-channel attack

I. INTRODUCTION

CMOS technology has been scaled down to provide improvement in performance, area and power by scaling the channel length and the threshold voltage. This scaling is limited by the excessive levels of the off-state leakage for further threshold voltage reduction, which results in exponential growth of chip power density. Nanoelectromechanical System (NEMS) relay based designs can overcome such power dissipation issues. NEMS provides ideal switching characteristics, resulting in improvements in energy efficiency over CMOS technology [1]. Recent studies have shown that devices fabricated using NEMS relay remain functional for more than 60 billion cycles [2]. However, the switching speed of NEMS relay is much slower than CMOS transistor, which can degrade the performance of the design. Each NEMS relay takes a single long mechanical delay to move from the on-state to the off-state and a single short electrical delay to charge/discharge the current.

Technology scaling of CMOS also impacts the physical security of the chip. Security algorithms such as ciphers, implemented in CMOS technology can leak secret information through covert channels such as power [3], time [4], and electromagnetic emanation [5]. The attacks which exploit these

covert channels are referred to as *side-channel attacks*. Static power leakage of CMOS technology can also be exploited to reveal the secret key of a chip containing block ciphers such as AES [6]. Unlike CMOS, NEMS relay has zero static leakage, providing inherent resistance to static leakage based power analysis attacks.

To minimize the delay of NEMS relay based design, a circuit is implemented as a single complex logic gate with a minimum number of relays. As the mechanical actuation of NEMS relay results in a larger delay compared to CMOS technology, all the NEMS relays should actuate simultaneously to provide a single mechanical delay. Basic circuits, such as an adder, have been designed using NEMS relays and demonstrated to deliver energy efficiency at minimum delay [2]. Other NEMS relay based circuit design topologies have also been proposed to implement flip-flops, ADC/DAC, oscillators, RAM modules, and power gating circuits [1], [2], [7]. To exploit the trade-off between the power and performance, a hybrid CMOS-NEMS relay based circuit has been proposed [8], [9] and demonstrated [10].

Binary decision diagram (BDD) [11] based implementation provides a systematic approach to design any combinational logic using NEMS relays with a single mechanical delay [12]. BDDs were initially developed for switch-level simulation of MOS circuits [13]. Design verification also utilizes BDDs for model checking algorithms to verify the properties of the system [14]. Furthermore, BDDs can be used for testing sequential circuits [15].

While superior to CMOS technology in static leakage and the associated side-channel attacks, *in this work we expose, for the first time, an inherent security vulnerability of NEMS relay based designs implemented using BDDs.* Our work shows that such designs can be vulnerable to timing attacks, which form an important class of side-channel attacks. BDD based NEMS designs has input dependent delay across the combinational logic. In case of block ciphers, such as AES, the variation of the propagation delay is smaller across the output bits; yet our correlation based timing attack is strong enough to recover the last round key of AES by analyzing the critical path timing. We validated our attack in the presence of interconnect delay variation of Gaussian noise even with low standard deviation values.

The rest of this paper is organized as follows. Section II provides a background on NEMS relay structure. Section III presents the BDD implementation of NEMS relay based combinational design and the associated data dependency of

978-1-4673-9141-2/15 $31.00 © 2015 IEEE

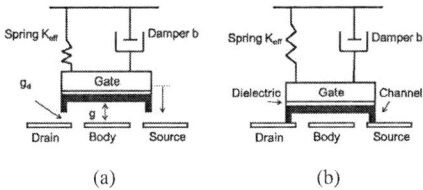

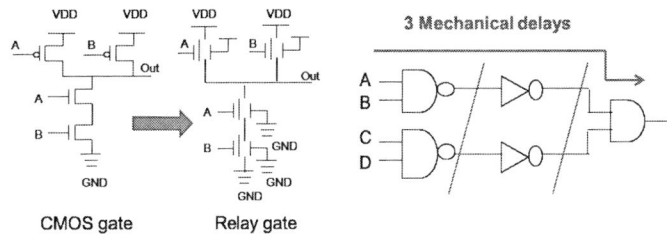

Fig. 1. Cross-sectional view of four-terminal NEMS relay design in (a) Off-state (b) On-state.

Fig. 2. CMOS style realization for NEMS technology.

the timing of such circuits. In Section IV, we perform a case study of timing attacks on an AES block cipher implemented using the NEMS relay design. Also, we propose a correlation based timing delay attack on the AES last round. We conclude the paper in Section V.

II. NEMS RELAY STRUCTURE

A typical folded-flexure style four-terminal NEMS relay is illustrated in Figure 1. Four-terminal relay consists of a movable gate (beam), body, source and drain. Unlike CMOS technology, the state of the relay is determined by the voltage difference between the gate and the body (V_{gb}). When sufficient voltage V_{gb} exceeds a threshold voltage, called pull-in voltage (V_{pi}), the electrostatic force pulls the mechanical beam toward the body. At certain point of displacement, the electrostatic force will increase faster than the spring force, resulting in an on-state ($|V_{gb}| > V_{pi}$). The channel creates a conductive path between the source and the drain. When no voltage is applied, an air gap separates the channel from the metallic source and drain electrodes, resulting in an off-state ($|V_{gb}| < V_{pi}$). The spring force keeps the gate and the channel suspended. The total delay of a single relay is the summation of the mechanical and the electrical delay, which are in nanoseconds and picoseconds, respectively [16]. Thus, NEMS relays suffer from a large delay compared to CMOS transistors.

There are several variations of the NEMS relays. A beam structure can be a cantilever or a fixed-fixed beam. In a cantilever structure, a projecting beam is fixed only at one end, while in a fixed-fixed beam structure the beam is supported from both ends. Fixed-fixed beam structure is more commonly used in a relay as this structure can cope with larger residual stress compared to the cantilever beam before it deforms catastrophically [16]. A four-terminal NEMS relay can be constructed using either one. A six-terminal cantilever based relay, which requires additional source and body compared to the four-terminal relay, has been proposed for combinational logic design [12].

III. NEMS RELAY BASED COMBINATIONAL CIRCUIT DESIGN

A. Challenges

While in CMOS technology every single gate is designed independently using transistors, such CMOS-style realization is unacceptable in NEMS technology. Mapping each transistor to a NEMS relay results in significant performance degradation. This is illustrated in Figure 2. The circuit input

can be propagated to the output through several logic levels, with each level causing an extra mechanical delay. In this example, three mechanical delays are required to propagate the inputs to the outputs. Thus, the implementation of the critical path using NEMS relay in CMOS-style realization results in poor performance. Functional logic circuits are carefully designed to implement NEMS relay based design with minimum mechanical delay [1].

It requires a significant effort to manually design a relay based circuit aiming at the minimum number of mechanical delays. Manual design also necessitates detailed design information of every single block of the design. These are challenging requirements for large and complex designs. Thus, there is an essential demand for a systematic design methodology that is capable of implementing a complex design using NEMS relays.

B. Combinational Logic Implementation Using BDDs

BDDs [17] can be used to synthesize combinational logic circuits to relay based design. BDD is a directed graph that can represent any combinational logic. Each node of the BDD represents a MUX that is implemented using two four-terminal relays or one six-terminal relay. A BDD example for the function $F = A + BC$ is given in Figure 3. In the BDD, terminal 0 is connected to the ground (GND), while terminal 1 is connected to the supply voltage (VDD). The gate terminal of each relay is connected to a primary input. Thus, V_{gb} for all the relays can be applied *simultaneously*, such that $|V_{gb}| > V_{pi}$. In other words, no source or drain terminal of a relay will drive the gate terminal of another relay. Therefore, all the gates will be actuated *simultaneously*, resulting in one mechanical delay. A conductive path from the source to the primary output will be created after one mechanical delay. The electrical delay of all the relays in the conductive path determines the time required to propagate the terminal value to the final output of the BDD. Therefore, the total delay is summation of one mechanical delay and n electrical delays, where n is the depth of the conductive path from the leaf to the root (i.e. final output) of the BDD.

C. Data Dependency of Timing in NEMS Relay Based Design Using BDD Implementation

The timing delay of the output in a BDD based implementation of a combinational logic is inherently input dependent. We use AND, OR, and XOR gates as examples. In Figure 4(a),

978-1-4673-9141-2/15 $31.00 © 2015 IEEE   265

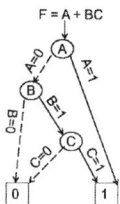

Fig. 3.   BDD implementation of function $F = A + BC$.

two-input AND gate $F = A.B$ is implemented using a BDD. If the signal A is 0, the signal F is directly determined after the arrival of the signal A. Thus, the timing delay of the AND gate when $A = 0$ is $T_{delay} = T_{elec} + T_{mech}$, where T_{elec} and T_{mech} are the electrical and mechanical delay of one relay, respectively. On the other hand, if the signal A is 1, the signal F is determined after the arrival of the signal B followed by the signal A. Thus, the timing delay of the AND gate when $A = 1$ is $T_{delay} = 2 * T_{elec} + T_{mech}$.

Similarly, for two-input OR gate $F = A + B$, if the signal A is 0 the timing delay of the OR gate is $T_{delay} = 2 * T_{elec} + T_{mech}$, while it is $T_{delay} = T_{elec} + T_{mech}$ when the signal A is 1 as shown in Figure 4(b).

Finally, in Figure 4(c) for two-input XOR gate $F = A \oplus B$, the timing delay is always $T_{delay} = 2*T_{elec} + T_{mech}$ regardless of the applied input signals. Thus, the timing delay of the output signals is data independent in the case of an XOR gate.

D. Timing Attack on Ciphers Implemented in NEMS

The data dependency of the timing of a NEMS based design using BDD implementation can be misused by an adversary to leak secret information of the chip. In Figure 4(a) and Figure 4(b), the delay across the NEMS relay based AND and OR gates inherently depends on the input data to the gate. So, the total timing delay across an entire combinational circuit is decided by the maximum timing delay of the signals across different paths in the circuit comprising such gates. The longest timing delay from the inputs to the outputs is referred to as the *critical timing delay* of the circuit. As the timing delay for each of the output bits of a combinational circuit comprising AND and OR gates is data dependent, so is the critical timing delay. Suppose a security-critical circuit such as AES is composed of such NEMS relay based devices. An adversary can then analyze the data dependency based on the structure of the circuit via simulation or implementation, and hence can extract the secret key information from the circuit.

IV. CASE STUDY: SECURITY OF AES IN NEMS RELAYS

To show the vulnerability of NEMS based design using BDD implementation, we propose a timing attack on the AES cipher, while we note that our attack is generic and can be adapted to any different cipher. We first provide a quick overview of the AES architecture, after which we illustrate the NEMS based AES S-box implementation. Then, we propose our timing attack with the guessing entropy plots for the last round key of AES.

A. AES Architecture

AES [18] is a well known block cipher that supports block lengths of 128-bits and key lengths of 128, 192 and 256 bits as shown in Figure 5. The AES algorithm consists of identical operations, i.e., rounds. The number of rounds depends on the key length; typical implementations consist of 10 rounds for 128-bit key, 12 rounds for 192-bit key and 14 rounds for 256-bit key. The AES encrypts the input, referred to as a plaintext, to the output, referred to as ciphertext after the desired number of rounds. Each round comprises the following four basic transformations, except for the last round, which omits MixColumns:

- SubBytes (SB) is a non-linear substitution operation. Each input byte to the SubBytes operation is replaced by another byte using one-byte substitution table, referred to as *S-box*.
- ShiftRows (SR) is the byte-wise permutation.
- MixColumns (MC) is the four-byte mixing operation.
- AddRoundKeys (ARK) is XOR operation of the state with the round key.

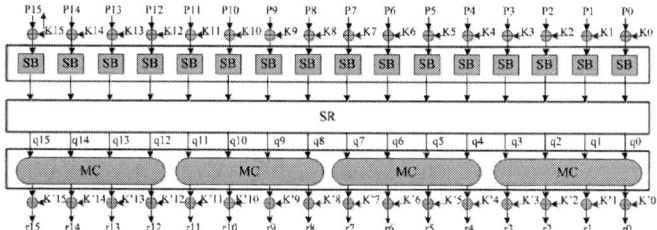

Fig. 5.   One round of AES: P_i is the plaintext byte, K_i is the initial key byte, q_i is the SR output byte, K'_i is the round key byte, and r_i is the round output byte.

B. AES S-Box Design Using NEMS

To perform security analysis on ciphers implemented using NEMS relays, we have designed the AES S-box using the

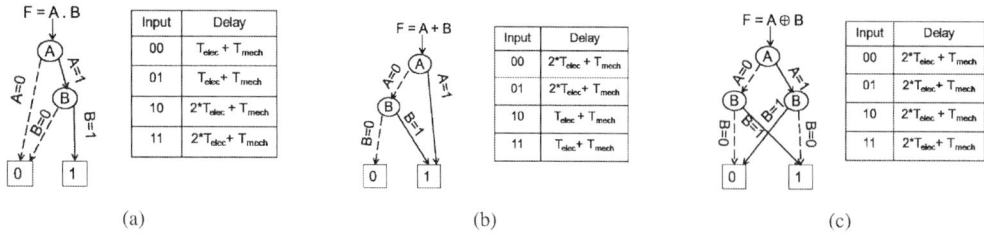

Fig. 4.   Delay of NEMS based (a) AND gate (b) OR gate (c) XOR gate using BDD implementation.

978-1-4673-9141-2/15 $31.00 © 2015 IEEE

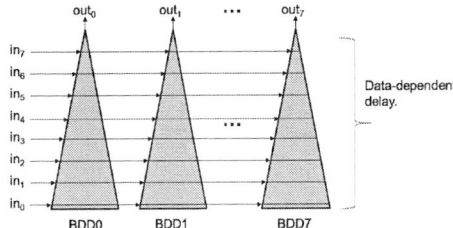

Fig. 6. S-box implementation using BDD.

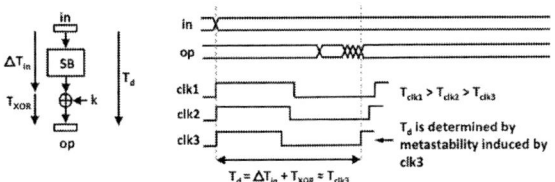

Fig. 7. Timing delay measurement across last round of AES.

BDD based implementation of NEMS relay. We analyze the data dependency of the critical timing delay for NEMS relay based AES S-box. Figure 6 shows the structure of the S-box implementation using BDDs. Each BDD corresponds to an output bit out_i of AES S-box, where i varies from 0 to 7. Each level of the BDD tree is controlled by an input bit in_i, where i varies from 0 to 7 as well. Thus, the depth of each BDD is eight. However, the length, and thus, the timing delay of the critical path can vary from one input byte to another. To build the BDD of the S-box, we converted the Verilog design of the S-box into Berkeley Logic Interchange Format (BLIF) by using Yosys tool [19]. CUDD [20] tool reads in the BLIF design and generates a BDD for each output bit. We developed a C++ script that parses the BDD and identifies the propagation delay (T_1) for each input byte across the AES S-box. The BDD of each output bit of the S-box can be implemented with a single mechanical delay by connecting all gates of the relays to the inputs of the S-box. The propagation delay is computed as $T = T_{mech} + n * T_{elec}$, where n denotes the number of nodes in the conductive path of the BDD. For each input byte (IN = $\{in_0, in_1, \ldots, in_7\}$), the propagation delay is determined by the maximum of (T_0, T_1, ... , T_7), where T_i is the propagation delay of the BDD of output bit i.

Each node is mapped to two four-terminal relays. A Verilog-A model has been utilized in this work to simulate the behavior of NEMS relay based on parameters published in [2], and shown in Table I. The mechanical and electrical delay can be obtained under the scaled device parameters through SPICE simulations.

TABLE I
SCALED NEMS RELAY DEVICE MODEL PARAMETERS [2]

Parameter	Scaled Model
$A_{ov}[\mu m^2]$	0.77
$g_0[nm]$	10
$g_d[nm]$	5
C_{gc}[fF]	0.9
C_{gb}[fF]	1.5
$C_{gd}(x=0) and C_{gs}(x=0)$[fF]	0.59
$T_{mech}[\mu s]$	0.02
T_{elec}[ps]	2.7
V_{pi}[V]	0.04

C. Timing Attack on the AES-128 Last Round

In this section, we first determine the timing characteristics of the last round of AES in which we show that the critical timing delay across the S-box depends on its input. The last round of AES omits the MixColumns operation. The ShiftRow operation is an arrangement of bytes without any byte operation. In Figure 7, we illustrate the timing characteristics in the last round of AES-128. In the figure, the register in contains an output byte of the penultimate round of AES-128, while the register op contains a ciphertext byte. The critical time delay from the register in to the register op is represented by T_d while the propagation delay across the XOR gate is T_{XOR}. We recall from the BDD diagram of Figure 4(c) that the propagation delay of an XOR gate is input independent.

As shown in the BDD of the AES S-box and from Table II, the propagation delays from the input to the output for all the output bits of the S-box are different. The number of input combinations across a row sum upto 256, which is the number of all possible input combinations to the S-box. Suppose for input byte IN, the propagation delay in the output bit j is denoted by $\Delta t_{IN,j}$. Now, the critical timing delay for the input IN can be determined as the maximum delay over all the output bits, i.e.,

$$\Delta T_{IN} = \max_{j=\{0,\ldots,7\}} \Delta t_{IN,j} \quad (1)$$

In Figure 7, the delay across the S-box for input in is shown as ΔT_{in}. This can be determined from the timing delay of the BDD for each of the output bits of the S-box. Now, we need to determine the value of T_d for the input in. In the timing diagram, the output op stops toggling after a certain duration of time in response to a change in in. This amount of time can be determined by gradually reducing the clock time period. At some point, the clock period gets smaller than the setup time, resulting in the capture of the metastable state of the register. Figure 7 illustrates this operation where T_{clk1} and T_{clk2} are greater but T_{clk3} is smaller than the setup time.

We also include the interconnect delay variations in the delays, modeled as a Gaussian random noise with zero mean

TABLE II
NUMBER OF INPUT COMBINATIONS WHICH LEAD TO 5, 6, 7, OR 8 T_{elec} ELECTRICAL DELAY VALUES FOR DIFFERENT OUTPUT BITS OF AES S-BOX

Output bit	$5T_{elec}$	$6T_{elec}$	$7T_{elec}$	$8T_{elec}$
out0	0	16	148	92
out1	0	40	124	92
out2	0	24	136	96
out3	8	28	128	92
out4	16	24	144	72
out5	16	60	104	76
out6	24	24	116	92
out7	8	48	124	76

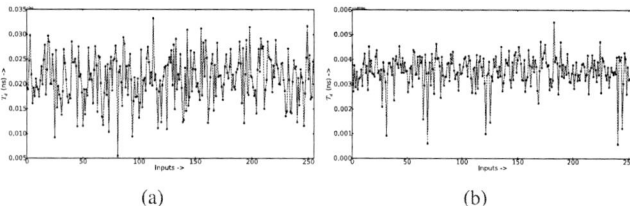

(a)   (b)

Fig. 8. T_d values across the AES last round for Gaussian noise with standard deviation (a) $\sigma = 0.005$, (b) $\sigma = 0.0005$.

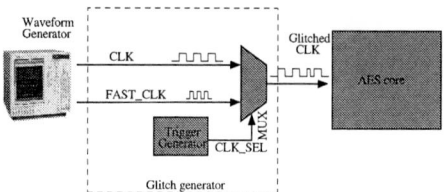

Fig. 9. Timing analysis setup for AES

and a low standard deviation, $\mathcal{N}(0, \sigma)$. In Figure 8, we show the timing delay values of T_d for all input combinations for the largest and smallest noise variations that we considered.

1) Threat model: The adversary performs a number of queries, referred to as nqueries, on AES and obtains the corresponding *op* bytes. The penultimate round output byte *in* is assumed to be unknown to the adversary. The adversary can, however, compute the propagation delay of the S-box for its different output values in terms of mechanical delay T_{mech} and electrical delay T_{elec}. The following are the assumptions for the proposed attack:

- The adversary knows the AES algorithm running inside the chip and has control over the clock.
- The adversary knows the BDD implementation of the S-box.

2) Experimental Setup: In Section IV-C, we showed the method of determining the delay of an AES round by varying the time period of the input clock. In this section, we show how an adversary can target the last round of AES. The following experimental setup model (Figure 9) is used by the adversary to precisely compute the propagation delay of the last round of AES [21]. The adversary has control over the input clock to the AES core. Two clocks are being used to inject a glitch at the last round input of AES encryption. The AES core is run by the normal clock (CLK), while at the tenth round the trigger module selects the fast clock (FAST_CLK) by activating the CLK_SEL line. The clock period of the fast clock defines the glitch width. The adversary keeps on increasing the frequency of the fast clock until he/she gets the faulty tenth round output due to the timing violation. The maximum frequency for which the adversary gets the fault-free output defines the critical timing delay across the AES S-box and the XOR gate (T_d). By varying input plaintext, the adversary can compute the corresponding last-round delay T_d corresponding to different S-box input bytes.

3) Timing attack: In this attack, the adversary first computes the critical timing delay $T_d[i]$ for each query $i, 0 \leq i \leq$ nqueries, from the timing characteristics of the design as shown in Figure 7. Thereafter, the critical timing delay across the AES S-box, $\Delta T_{SB}[i]$, is computed for each query as the difference of the critical timing delay $T_d[i]$ and the propagation delay across the XOR gate, T_{XOR}. In this attack, the component T_{XOR} is input-independent and is computed from the SPICE simulation.

Now, for each key guess k_g, we compute the corresponding S-box output $SB_g[i]$ for all queries i as $op[i] \oplus k_g$. For each of these S-box output values $SB_g[i]$, we can compute the corresponding critical timing delay of the S-box $\Delta T_{SBg}[i]$ from the timing delay computed from BDDs of all the eight S-box output bits in terms of mechanical delay T_{mech} and electrical delay T_{elec}. Suppose this delay computed from BDD for each output bit j, $0 \leq j \leq 7$ is denoted as $\Delta \tau_j$. The critical timing delay $\Delta T_{SBg}[i]$ can then be computed as,

$$\Delta T_{SBg}[i] = \max_{j = \{0, \ldots, 7\}} \Delta \tau_j \qquad (2)$$

Subsequently, we compute correlation ρ_{k_g} between the vectors $\Delta T_{SBg}[i]$ corresponding to the key guess k_g, and the the critical timing delay $\Delta T_{SB}[i]$ over all the queries $i, 1 \leq i \leq$ nqueries. The key guess k for which the correlation coefficient of the vectors is maximum is considered to be the actual key. The timing attack algorithm is shown in Algorithm 1.

Algorithm 1: Timing Attack on NEMS Relay Based MUX design of AES last round

---

Input: op[1,...,nqueries], T_d[1,...,nqueries], T_{XOR}
Output: AES last round key, k.

1 **for** *each query op[i] in op[1, ..., nqueries]* **do**
2   Compute critical timing delay $T_d[i]$ from timing chracteritics.
3   Compute corresponding critical timing delay across AES S-box $\Delta T_{SB}[i] = T_d[i] - T_{XOR}$.
4 **for** *each key guess* $k_g, 0 \leq k_g \leq 255$ **do**
5   **for** *each query op[i] in op[1, ..., nqueries]* **do**
6   Compute the AES S-box output $SB_g[i] \leftarrow op[i] \oplus k_g$.
7   Compute the $\Delta T_{SBg}[i]$ corresponding to $SB_g[i]$ from BDDs of the AES S-box output bits in terms of $T_{mech} + T_{elec}$ of NEMS relay device elements.
8   Compute correlation $\rho_{k_g}(\Delta T_{SBg}[i], \Delta T_{SB}[i])$ between the vectors, $\Delta T_{SBg}[i]$ and $\Delta T_{SB}[i]$.
9 Correct key k : $\rho_k \leftarrow \max_{0 \leq k_g \leq 255} (\rho_{k_g})$.

---

4) Results: In the attack results shown in Figure 10, it can be seen that the correct key k can be determined when the interconnect delay variation modeled as Gaussian noise distribution is low. From the results, we note that if the standard deviation $\sigma < 0.0022$, the correlation coefficient of the correct key is higher compared to the other incorrect key values. In addition, we also compute the guessing entropy of the last round AES key for different interconnect delay noise distributions as shown in Figure 11.

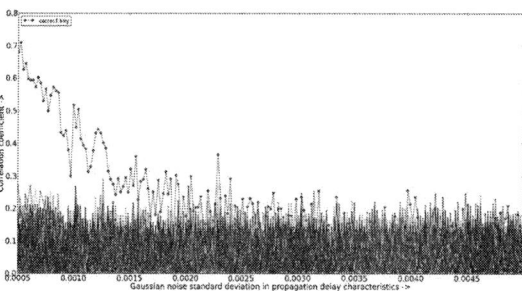

Fig. 10. Correlation coefficient plot of the vectors $\Delta T_{SBg}[i]$ and $\Delta T_{SB}[i]$ with respect to varying interconnect delay variations.

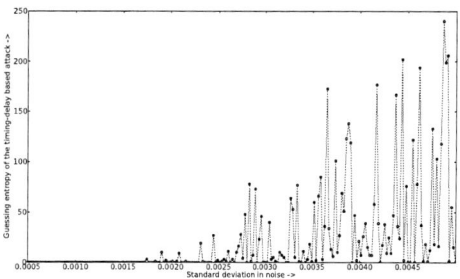

Fig. 11. Guessing entropy plot of the correct key over varying interconnect delay variations.

The guessing entropy gives the expected index or the ranking of the correct key after sorting the correlation coefficient values. As shown in the results for low noise with standard deviation $\sigma \leq 0.0018$, the rank of the correct key always comes at the top and so the residual uncertainty of the correct key falls to zero, indicating an attack success rate of 1.0.

V. CONCLUSION

Scaling down of CMOS technology has raised power consumption challenges and threats in the semiconductor industry. NEMS is a promising technology that can overcome these challenges at the cost of large mechanical delay. Thus, the implementation of NEMS relay based design targets minimizations of the mechanical delay in the circuit. It has been shown that BDD based implementations can mitigate this problem and deliver high-performance NEMS relay based designs. On the other hand, such implementations give rise to a security vulnerability. In this paper, we expose this vulnerability. We first show that the timing of such implementations are input dependent and then provide a correlation based timing attack by using AES as an example. Our results show that for low noise variations with standard deviation $\sigma < 0.0022$ in the interconnect delay, the AES last round key can be recovered by employing timing attacks. In the attack simulations, the guessing entropy of the AES last round key also falls to zero for low noise variations, indicating a successful attack. To conclude, this paper elicits the need to implement countermeasures to thwart timing based attacks on NEMS relay based designs, especially for security-critical applications such as block ciphers.

REFERENCES

[1] F. Chen, H. Kam, D. Markovic, T.-J. K. Liu, V. Stojanovic, and E. Alon, "Integrated circuit design with nem relays," in *Proceedings of IEEE/ACM International Conference on Computer-Aided Design*, 2008, pp. 750–757.

[2] M. Spencer, F. Chen, C. Wang, R. Nathanael, H. Fariborzi, A. Gupta, H. Kam, V. Pott, J. Jeon, T.-J. K. Liu, D. Markovic, E. Alon, and V. Stojanovic, "Demonstration of integrated micro-electro-mechanical relay circuits for vlsi applications," *IEEE Journal of Solid-State Circuits*, vol. 46, no. 1, pp. 308–320, 2011.

[3] K. et al, "Differential power analysis," in *CRYPTO'99,LNCS 1666*, 1999, p. 104.

[4] K. et al, "Timing attacks on implementations of diffie-hellman, rsa, dss, and other systems," in *CRYPTO'96,LNCS 1109*, 1996, p. 104.

[5] D. Agrawal, B. Archambeault, J. R. Rao, and P. Rohatgi, "The em side-channel(s)," in *Proceedings of Revised Papers from the 4th International Workshop on Cryptographic Hardware and Embedded Systems*, 2003, pp. 29–45.

[6] S. M. Del Pozo, F.-X. Standaert, D. Kamel, and A. Moradi, "Side-channel attacks from static power: When should we care?" in *Proceedings of the Design, Automation & Test in Europe Conference & Exhibition*, 2015, pp. 145–150.

[7] R. Venkatasubramanian, S. Manohar, and P. Balsara, "Nem relay-based sequential logic circuits for low-power design," *IEEE Transactions on Nanotechnology*, vol. 12, pp. 386–398, 2013.

[8] S. Chong, K. Akarvardar, R. Parsa, J.-B. Yoon, R. Howe, S. Mitra, and H.-S. Wong, "Nanoelectromechanical (nem) relays integrated with cmos sram for improved stability and low leakage," in *Proceedings of IEEE/ACM International Conference on Computer-Aided Design.*, 2009, pp. 478–484.

[9] H. Dadgour and K. Banerjee, "Design and analysis of hybrid nems-cmos circuits for ultra low-power applications," in *Proceedings of ACM/IEEE Design Automation Conference*, 2007, pp. 306–311.

[10] S. Chong, B. Lee, K. Parizi, J. Provine, S. Mitra, R. Howe, and H.-S. Wong, "Integration of nanoelectromechanical (nem) relays with silicon cmos with functional cmos-nem circuit," in *Proceedings of IEEE International Electron Devices Meeting*, 2011, pp. 30.5.1–30.5.4.

[11] S. B. Akers, "Binary decision diagrams," *IEEE Transactions on Computers*, vol. 27, pp. 509–516, 1978.

[12] D. Lee, W. Lee, C. Chen, F. Fallah, J. Provine, S. Chong, J. Watkins, R. Howe, H.-S. Wong, and S. Mitra, "Combinational logic design using six-terminal nem relays," *IEEE Transactions on Computer-Aided Design of Integrated Circuits and Systems*, vol. 32, pp. 653–666, 2013.

[13] R. Bryant, "Symbolic manipulation of boolean functions using a graphical representation," in *Proceedings of Conference on Design Automation*, 1985, pp. 688–694.

[14] J. Burch, E. Clarke, K. McMillan, D. Dill, and L. Hwang, "Symbolic model checking: 10^{20} states and beyond," in *Proceedings of IEEE Symposium on Logic in Computer Science*, 1990, pp. 428–439.

[15] H. Cho, G. Hachtel, and F. Somenzi, "Redundancy identification/removal and test generation for sequential circuits using implicit state enumeration," *IEEE Transactions on Computer-Aided Design of Integrated Circuits and Systems*, vol. 12, pp. 935–945, 1993.

[16] F. Chen and H. Kam, *Micro-Relay Technology for Energy-Efficient Integrated Circuits.* Springer, 2014.

[17] S. B. Akers, "Binary decision diagrams," *IEEE Transactions on Computers*, vol. 27, pp. 509–516, 1978.

[18] J. Daemen and V. Rijmen, *The Design of Rijndael: AES - The Advanced Encryption Standard.* Springer, 2002.

[19] N. C. Wolf. (2013, "Yosys application note 010: Converting verilog to blif [online]," in *Public Software [Online]. Available: http://www.clifford.at/yosys/download.html*.

[20] M. F. Somenzi. (2012, "Cudd: Cu decision diagram package," in *http://vlsi.colorado.edu/fabio/*.

[21] S. S. Ali, B. Mazumdar, and D. Mukhopadhyay, "A fault analysis perspective for testing of secured soc cores," *IEEE Design & Test*, vol. 30, no. 5, pp. 63–73, 2013.

A High Efficiency Rectifier for Inductively Power Transfer Application

Qiong Wei Low, Liter Siek and Mi Zhou
VIRTUS – Centre of Excellence in IC Design
School of Electrical and Electronic Engineering, Nanyang Technological University, Singapore
Email: qlow002@e.ntu.edu.sg

Abstract— **This paper presents a high efficiency rectifier for inductively power transfer application. The efficiency of the rectifier is optimized by utilizing unbalanced biasing technique in the proposed wide swing cascode comparator controlled switch to suppress the reverse leakage current. The design is implemented in standard CMOS 0.18um AMS process. The targeted application is for telemetry, for use in the ISM band between 125 kHz and 134 kHz. The proposed rectifier achieves a peak power conversion efficiency (PCE) of 94.9% in the frequency range of 125kHz – 2MHz with an AC amplitude ranges from 1.2 V to 2.5V under maximum load current condition. It can source a maximum load current of 55mA and operates well under all process corners conditions.**

Keywords—high efficiency rectifier, inductively power transfer, unbalanced biasing, wide swing cascode comparator, power conversion efficiency (PCE).

I. INTRODUCTION

In recent years, inductively coupled Wireless Power Transfer System (WPT) is gaining its popularity, especially in the research area. It is attractive as the lifetime of a system could be extended as it is no longer limited by the energy densities of the batteries. Usage of power wires could be avoided and the waste for the environment could be lessened as well. Wireless energy and data transmission is beneficial for applications such as RFID smart card, animal and goods tracking, access control, biomedical implantation, etc. [1]

Figure 1 shows a typical inductively coupled wireless power transfer system. It consists of two magnetically coupled inductors which are L1 and L2. Both of the resonant tanks in the transmitter and receiver resonate at the same frequency for maximum power transfer of the system. The AC signal in L1 generates a magnetic field which induces a voltage in the receiver coil [2]. The AC voltage at the receiver side needs to be converted into a DC voltage by a rectifier to supply for the next component blocks. Thus, a high efficiency rectifier plays a crucial role in the system as it dominates the overall power conversion efficiency (PCE) and voltage conversion efficiency (VCE) of the receiver system. Conventional CMOS rectifiers utilize the MOSFET diode-connected transistors to replace the diodes. The main factor that contributes to the degradation of the PCE and VCE of a rectifier is the due to the threshold voltage drop of the diode-connected transistor. Many innovative methods have been proposed to reduce this voltage

drop such as internal and external threshold voltage cancellation in [3] and [4] respectively, self-threshold voltage cancellation in [5], inverter-based rectifier in [6], etc. Even though the voltage drop has been minimized, but most of the designs have severe reverse leakage current which could in turn degrade the PCE and VCE as well.

In this paper, we propose a high efficiency rectifier which utilizes an unbalanced biasing wide swing cascode comparator controlled switch to solve the reverse leakage current issue.

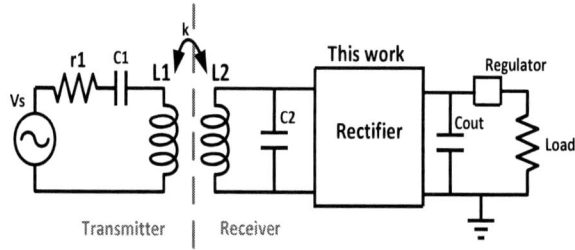

Fig. 1. An inductively coupled wireless power transfer system.

II. ACTIVE RECTIFIER

The implemented active rectifier uses the two-stage design concept [7]. It consists of a first passive stage and a second active diode stage. The two-stage design concept is adopted due to its simplicity. As compared to the CMOS rectifier in [8], only one comparator controlled switch is needed. As a result, it could reduce the total power consumed by the control circuits.

Figure 2 shows the first passive stage which is a negative voltage converter. It converts the negative half wave into positive half wave. By properly sizing the transistors in the negative voltage converter, the voltage drop could be designed to be less than a $V_{ds(sat)}$ drop to optimize the VCE of the rectifier at the expense of area consumption and dynamic power loss [9]. The second active diode stage consists of a comparator controlled switch. The delay in the comparator causes the reverse leakage current to flow for a short duration of time [7]. For large load current applications, the reverse leakage current may become a problem which could degrade both the PCE and VCE of the rectifier. Thus, an unbalanced biasing wide swing cascode comparator is proposed to suppress the reverse leakage current to enhance the rectifier efficiency.

978-1-4673-9141-2/15 $31.00 © 2015 IEEE

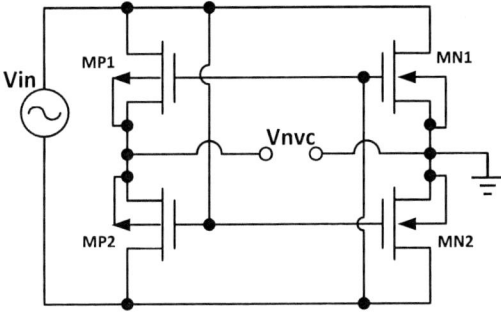

Fig. 2. The negative voltage converter in the first passive stage.

III. PROPOSED COMPARATOR

Figure 3 shows the proposed unbalanced biasing wide swing cascode comparator. The wide swing cascode structure is implemented to increase the accuracy of the current mirror comparator. The minimum voltage required for the comparator to operate is ($Vth_{n,p} + 4V_{ds(sat)}$). As the $Vth_n = 0.567V$ and $Vth_p = -0.527V$, the circuit is able to work at a supply as low as 1.2V. The resistors RB1, RB2 and RB3 are used to provide the required biasing currents. Node A is biased at $Vth_n + 2V_{ds(sat)}$ while node B is biased at VDD $- (|Vth_p| + 2V_{ds(sat)})$. By properly sizing the transistors, both the PMOS and NMOS are designed to have the same gate overdrive, $V_{ds(sat)}$.

The operation of the comparator can be explained by observing the changing input nodes Vin+ and Vin-. When input Vin- is greater than Vin+, more current flows to Mn3 which will be mirrored to Mn5 to pull the output node, VCMP low. At the same time, node C increases with the node Vin- which limits the current in Mp7 and Mp8 to speed up the pull down of the node VCMP. When node Vin- is lower than Vin+, node C decreases thus more current flows to MP7 and Mp8 to pull the output node VCMP high. At the same time, decreased node Vin- will limit the current flow to Mn3 and Mn5 to speed up the pull up of the node VCMP.

To further speed up node VCMP high in order to suppress the reverse leakage current, an offset can be inserted by having unbalanced biasing currents between transistors Mn3 and Mp9. Since both of the NMOS and PMOS have the same $V_{ds(sat)}$ as mentioned before, the offset can be created by having $I_3 > I_2$. The appropriate offset could be designed according to the Equation (1).

$$V_{offset} = V_{sg(Mp9)} - V_{gs(Mn3)} \qquad (1)$$

$$= \sqrt{\frac{2\,(I_3)}{\alpha}} - \sqrt{\frac{2\,(I_2)}{\beta}}$$

Where $\alpha = \mu_p C_{ox}(\frac{W}{L})_{Mp9}$, $\beta == \mu_n C_{ox}(\frac{W}{L})_{Mn3}$ and

$$I_2 = I_1 , \quad I_3 = \frac{(\frac{W}{L})_{Mn7}}{(\frac{W}{L})_{Mn1}} \times I_1$$

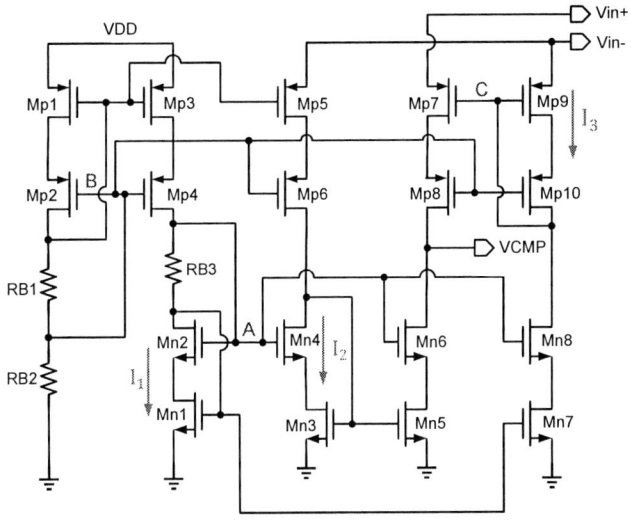

Fig. 3. The proposed unbalanced biasing wide swing cascode comparator

Current I_2 is designed to be the same as I_1 while current I_3 can be designed to be greater than I_1 by manipulating the ratio of the transistor sizes between Mn7 and Mn1. The VDD of the comparator is supplied from the output voltage of the rectifier, V_{rect} .The proposed comparator is supply dependent so that V_{offset} increases with the V_{rect} . The increase in the V_{offset} will increase the transition time and therefore eliminate any reverse leakage current that may happen when the output voltage increases since the delay time of the comparator increases with the input amplitude.

IV. PROPOSED RECTIFIER

Figure 4 shows the rectifier where the proposed comparator being used. It is imperative to use the dynamic bulk biasing (DBB) to keep the body of the power transistor at the highest potential as the two nodes of the active diode, V_{nvc} and V_{rect} are changing all the time. This is to ensure no latch up and to improve the power transmission efficiency as well [10].

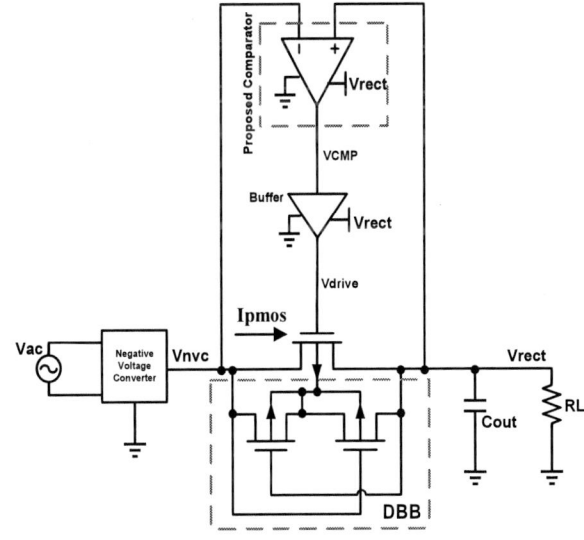

Fig. 4. The rectifier where the proposed unbalanced biasing wide swing cascode comparator being used.

978-1-4673-9141-2/15 $31.00 © 2015 IEEE

Figure 5 shows the leakage that happened when the conventional comparator is used. Due to the turn off delay of the comparator, the reverse leakage current flows for a short duration of time which incurs power losses. Figure 6 shows that with the proposed comparator, the turn off delay is minimized and the reverse leakage current has been suppressed. Both of the figures shown are simulated at the frequency of 125 kHz with input amplitude of 2.5V.

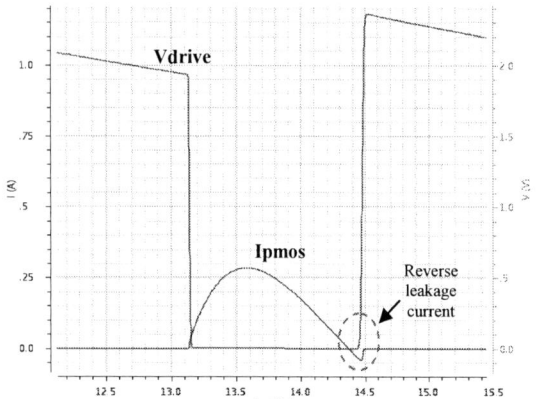

Fig. 5. The reverse leakage current (in dotted circle) happened when using conventional comparator.

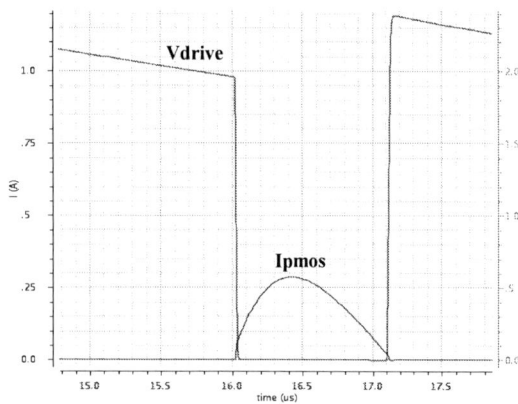

Fig. 6. The reverse leakage current has been suppressed when using the proposed unbalanced biasing wide swing cascode comparator.

V. SIMULATION RESULTS AND DISCUSSIONS

The power conversion efficiency (PCE) and voltage conversion efficiency (VCE) of the proposed rectifier has been simulated in 0.18um AMS CMOS process. The PCE and VCE can be defined as in the Equation (2) and (3) respectively.

$$PCE\ (\%) = \frac{P_{out}}{P_{in}} = \frac{V_{rect}^2/RL}{\frac{1}{n.T}\int_{to}^{to+n.T} V_{ac}(t).\ I_{ac}(t)dt} \times 100\% \qquad (2)$$

$$VCE\ (\%) = \frac{V_{out}}{|V_{in}|} \times 100\% \qquad (3)$$

The proposed rectifier is simulated with $R_L = 39\Omega$ at a maximum load current of 55mA. The output storage capacitor of 330nF with ESR of 50mΩ is used to store the charge and to smoothen the output voltage ripple. Figure 7 shows the simulated PCE and VCE versus the input amplitude which ranges from 1.2V to 2.5V at the frequency of 125 kHz. A peak PCE is achieved at 94.9% with an input amplitude of 2.5V. The peak VCE is at 87.24%.

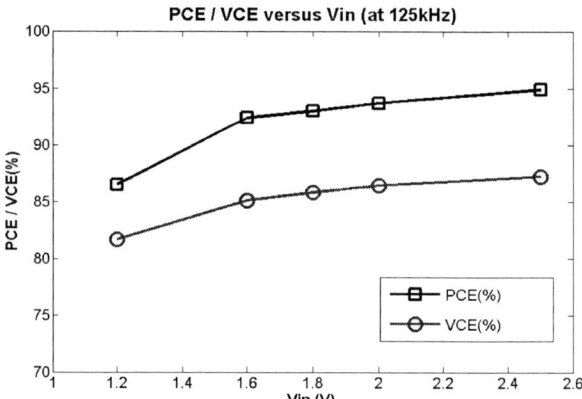

Fig. 7. The graph of PCE and VCE versus varying input amplitudes at 125kHz.

Figure 8 shows the simulated PCE and VCE at the frequency of 134 kHz with the varying input amplitudes from 1.2V to 2.5V. A peak PCE of 91.7% is observed at 2.5V input amplitude. Both the PCE and VCE are of at least 81.7%.

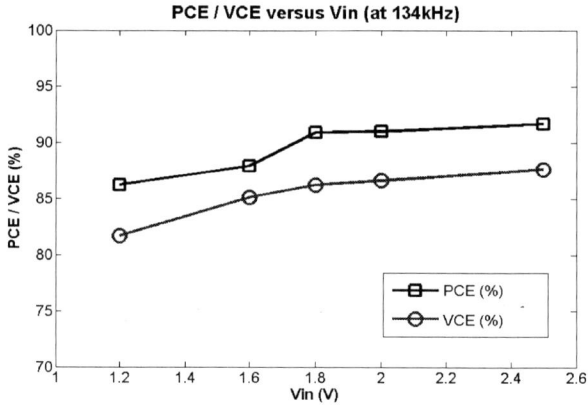

Fig. 8. The graph of PCE and VCE versus varying input amplitudes at 134kHz.

The targeted application of the rectifier is for telemetry, for use in the ISM band between 125 kHz and 134 kHz. However, the proposed rectifier is able to work up to the frequency of 2 MHz with at least 82.5% of PCE. Figure 9 shows the performance of the rectifier with the frequency ranges from 125 kHz to 2 MHz. As the frequency increases, the PCE decreases as the offset voltage value designed was intended to optimize the performance at the frequency of the targeted application.

978-1-4673-9141-2/15 $31.00 © 2015 IEEE

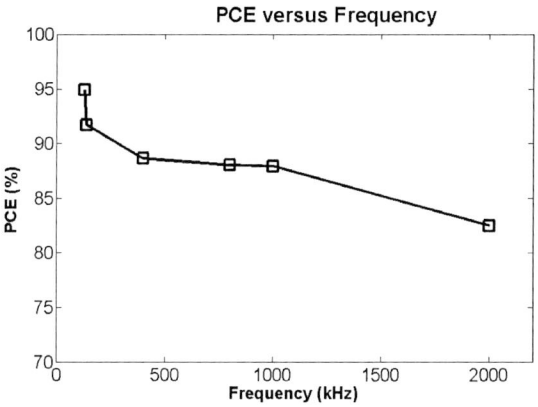

Fig. 9. The PCE versus frequency with input amplitude of 2.5V.

Table I shows the performance of the proposed rectifier under different process corner conditions. It is simulated with input amplitude of 2.5V and frequency of 125 kHz. The PCE achieved is of at least 89.9% while the VCE is of at least 86.6%. Table II shows the summary and the comparison results of the proposed rectifier with the start-of-the-art designs.

Process Corners	PCE (%)	VCE (%)
TT	94.9	87.24
SF	90.7	87.20
FS	93.5	86.90
FF	94.6	87.60
SS	89.9	86.60

Table I: PCE and VCE under different process corner conditions.

Table II
Comparison to State-Of-The-Art Rectifiers

	[11] ISCAS 2005	[8] JSSC 2009	[12] TBCS 2008	This Work (Simulated)
Technology	0.35μm CMOS	0.35μm CMOS	0.5μm CMOS	**0.18μm CMOS**
Input Amplitude	3.2V	1.2V - 2.4V	5V	**1.2V – 2.5V**
Output Voltage	2.88V (simulation)	0.98V - 2.08V	4.36V (max)	**0.98V – 2.18V**
R_L	2kΩ	100Ω	1kΩ	**39Ω**
$I_{RL(MAX)}$	1.44mA (simulation)	20mA	4.36mA	**55mA**
Frequency	5MHz	200kHz-1.5MHz	125kHz - 1MHz	**125kHz – 2MHz**
PCE (%)	80 (simulation)	87 (max)	84.8 (max)	**94.9 (max)**
VCE (%)	90 (simulation)	84	87.2 (max)	**87.24 (max)**

VI. CONCLUSIONS

The rectifier for inductively coupled wireless power transfer application is designed and simulated in 0.18μm AMS CMOS process. By utilizing the proposed unbalanced biasing wide swing cascode comparator, the reverse leakage current has been successfully suppressed. Comparing to the state-of-the-art designs, the power conversion efficiency of the proposed rectifier has been enhanced. The simulation results show that it could achieve a power conversion efficiency of as high as 94.9% at the maximum load current of 55mA.

ACKNOWLEDGEMENT

The authors would like to thank VIRTUS and Maxim Integrated for supporting this work.

REFERENCES

[1] N. Karmakar, "RFID Readers – Review and Design," Wiley-IEEE Press, 1st Edition, ISBN: 9780470872178, 2010.

[2] S. Ullerich, W. Mokwa, U. Schnakenberg, "Micro coils for improved power transfer in telemetry system," Microtechnologies in Medicine & Biology 2nd Annual International IEEE-EMB Special Topic Conference on, 2002, pp:420-423.

[3] H. Nakamoto, et al. "A passive UHF RF Identification CMOS Tag IC using Ferroelectric RAM in 0.35-μm Technology," Solid-State Circuits, IEEE Journal of, Volume 42, Issue 1, Jan. 2007, pp: 101-110.

[4] T. Umeda, H. Yoshida, S. Sekine, Y. Fujita, T. Suzuki, and S. Otaka, "A 950-MHz rectifier circuit for sensor network tags with 10-m distance," IEEE J. Solid-StateCircuits, Volume 41, Issue 1, Jan. 2006, pp: 35-41.

[5] K. Kotani, et al. "High effieicny CMOS rectifier circuit with self-Vth-cancellation and power regulation functions for UHF RFIDs," Proc. IEEE Asian Solid-State Circuits Conf., Nov 2007, pp:119-122.

[6] H. C. Lin, D. S. Wu, C. M. Kung, Y. S. Hwang, and J.J Chen, "New CMOS inverted-based voltage multipliers," Proc. IEEE Int. Conf. Electron Devices Solid-State Circuits, Dec. 2010, pp:1-4.

[7] C. Peters, M. Ortmanns and Y. Manoli, " Fully CMOS Integrated Active Rectifier without Voltage Drop", Proc. MWSCAS, 2008, pp:185-188.

[8] Song Guo, Hoi Lee, "An Effieicny Enhanced CMOS Rectifer with unbalanced-biased comparators for Transcutaneous-Powered High-Current Implants", IEEE J. Solid-State Circuits, Volume 44, Issue 6, June. 2009, pp: 1796-1804.

[9] C. Peters,J. Handwerker, D. Maurath, Y. Manoli, "A Sub-500mV Highly Efficient Active Rectifier for Energy Harvesting Applications," IEEE Transactions on Circuits and Systems-I , Volume 58, Issue 7, June. 2011, pp: 1542-1550.

[10] M. Ghovanloo, K. Najafi "Fully Integrated Wideband High-Curent Rectifiers for Inductively Powered Devices," IEEE J. Solid-State Circuits, Volume 39, Issue 11, Nov. 2004, pp: 1976-1984.

[11] T. Lehmann, Y. Moghe, "On-chip Active Power Rectifier for biomedical applications," IEEE Int. Symposium on Circuits and Systems (ISCAS) 2005, pp:732-735.

[12] G. Bawa, M. Ghovanloo, " Active High Power Conversion Effieicny Rectifier with Built-In Dual Mode Telemetry in standard CMOS Technology", IEEE Transactions on Biomedical Circuits and Systems, Sept. 2008, pp:184-192.

978-1-4673-9141-2/15 $31.00 © 2015 IEEE

Design and Analysis of Search Algorithms for Lower Power Consumption and Faster Convergence of DAC Input of SAR-ADC in 65nm CMOS

Ananthanarayanan Parthasarathy, *Student Member, IEEE*

Department of EE, Stanford University, CA - USA

Abstract- **We propose a new approach in reducing the power consumption of the Successive approximation register Analog to Digital Converter (SAR-ADC) by changing the convergence algorithm of the Digital to Analog converter (DAC) input of the SAR-ADC. Different search algorithms such as binary search tree, moving binary search tree (BST), least significant bit shifter (LSB), adaptive algorithm and split-register moving BST algorithm are designed and analyzed for faster convergence of the DAC input. In this paper, we design a 0.8 GS/s, 8 bit (Effective number of bits (ENOB) – 7.42), 8.352 mW SAR ADC with a proposed moving BST algorithm in 65nm CMOS which ranks amongst the current state of the art ADCs with a FOM 65.25fJ/step.**

Index Terms- **Moving binary search tree, SAR-ADC, Low Power**

I. INTRODUCTION

Analog to Digital converters (ADC) are among the most important electronic structures due to the technological shift and advances in digital electronics. There are many tradeoffs in designing ADCs including power, cost, sampling speed, resolution; the most challenges are in designing a high frequency and low power ADC. In some applications, the size of device becomes the main concern; an example would be for imaging sensors that require an ADC for each pixel of input. Successive approximation register (SAR) analog-to digital converters (ADCs) require several comparison cycles to complete one conversion, and therefore have limited operational speed [1] [2]. SAR architectures are extensively used in low-power and low-speed (below several MS/s) applications. In recent years, with the feature sizes of CMOS devices scaled down, SAR ADCs have achieved several tens of MS/s to low GS/s sampling rates with 5-bit to 10-bit resolutions. To judge the performance of an ADC this parameter has been defined to consider everything in one unique parameter. This parameter is the main characteristic that monitors the quality of ADC.

The good thing about this parameter is that it is not dependent on the structure [3]. This metric calculates the raw energy efficiency of ADC using equation [1]. The benefits of this equation are that it can be quickly calculated from a few ADC parameters and applied to assess the relative performance across different design approaches. The pioneers in designing ADC have considered remodeling the different state of the art components like the comparator, S/H circuit and DAC (figure 1) to reduce the power consumption.

$$\text{Figure of Merit (FoM)} = \frac{P}{2^{ENOB} \times f_s} \qquad [1]$$

Here in this paper, we looked at a different idea of reducing the power consumption by decreasing the iterations and the time it takes for the DAC to stabilize to its input value. We propose different search algorithms for SAR-ADC and analyze the design and performance tradeoffs. We feel that this could potentially alter be used to design low power consumption ADCs.

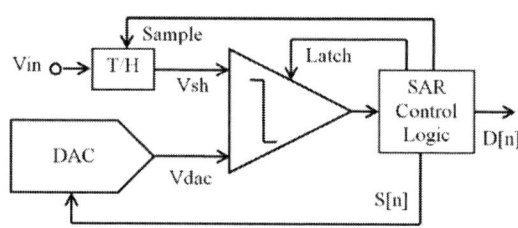

Fig. 1. Typical block diagram of SAR-ADC

Most commonly [4] [9], binary search algorithm is used for DAC input convergence since it is fastest search algorithm existing for finding a value in a random group of arranged numbers with O (log n) as its Big O notation. Though this might not be suitable if the input signal is slowly varying or if the values are within a range (like audio signals) then we can come up with algorithms that can converge faster to the required value. In all our cases in this paper, we are assuming the number of bits to be 10 (0 to 1023) for explaining the proof of concept.

978-1-4673-9141-2/15 $31.00 © 2015 IEEE

II. BINARY SEARCH TREE ALGORITHM

The regular binary search tree (figure 2) always starts with the middle value 1000000000 (i.e. 512) and compares with the actual input. If the output of the comparator (Cp) happens to be 1, then it means that the value lies between 512 and 1053; and if it is 0, then the value lies between 0 and 512. If Cp equals 1, then the value is readjusted to 768 and the comparison is done again. This is repeated until the value equals the input value to the comparator. One of the biggest disadvantages is that, the BST always starts comparing from the center value even if the signal is close to the boundary conditions; also if the current value is very close to the previous value, then the number of comparisons done to reach the actual target is higher than the difference between the two numbers.

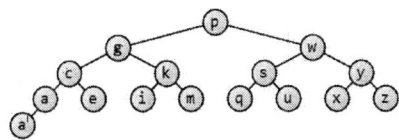

Fig. 2. A general binary search tree

The following example illustrates the mechanism of the circuits. For instance, the two consecutive digital signals, D, are:
D [10] = 1001100001 (current signal-609)
D [10] = 1001100010 (previous signal-610)

The number of conversions for the binary search to reach to the value of 609 is 9

(10000000000-> 1100000000->1010000000->1001000000- >1001100000->1001110000->1001101000-> 1001100100- >1001100010->1001100001)

Irrespective of what the previous signal is, the binary search always starts from the middle value 1000000000 (512). This happens to be a concern because if we come up with algorithms that take this into consideration, then much power can be saved.

III. MOVING BINARY SEARCH TREE ALGORITHM

In order to overcome the limitation of BST, we have proposed moving BST. The moving binary search tree starts with the previous value and compares with the actual input. If the output of the comparator (Cp) happens to be 1, then it means that the value lies between previous value and 1053; and if it is 0, then the value lies between 0 and previous value. If Cp equals 1, then the value is readjusted to (previous value+1053/2) and the comparison is done again. This is repeated until the value equals the input value to the comparator. How can this method be better than the regular binary search tree? The number of iterations it takes for the regular binary search tree to converge to the current value is

higher than the moving binary search tree because the span or the range in which the moving binary search tree searches is lower. A moving BST can be implemented using the control structure in figure 3.

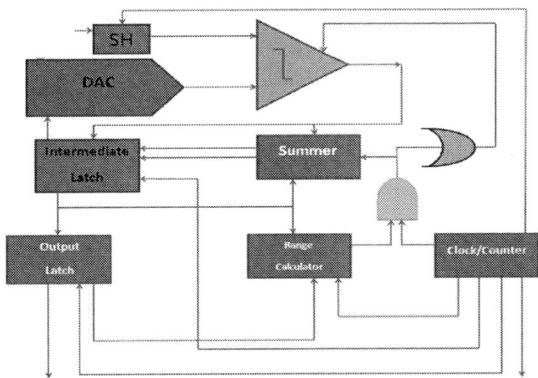

Fig. 3. SAR-ADC with moving BST algorithm

The following example illustrates the mechanism of the circuits. For instance, the two consecutive signals are:

D [10] = 1001100001 (current signal-609)
D [10] = 1001011111 (previous signal-607)

The number of conversions for the moving binary search to reach to the value of 609 is 7.

(1001011111-> 1101011111->1011011111->1001111111- >1001101111->1001100111->1001100011->1001100001)

The number of conversions has reduced from 9 to 7 which reduce the power consumption. Using MATLAB it is shown that the number of iterations Moving binary search tree takes to converge to the current value from previous value is on an average 44.12% of that taken by Binary search tree algorithm (as shown in figure 14)

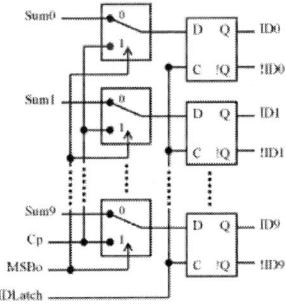

Fig. 4. Intermediate latch

Intermediate latch (figure 4) drives the DAC during the conversion. In addition, it retains its result after each conversion for range calculator and provides this result as the starting point of moving binary search tree in subsequent

978-1-4673-9141-2/15 $31.00 © 2015 IEEE   275

conversion. The multiplexers serve to limit the range of moving binary search tree to avoid the search tree from going beyond upper or below lower limits of the ADC full scale range. During each conversion process, the output latch (figure 5) first provides the preceding result for range calculator and then latches the current result from intermediate latch.

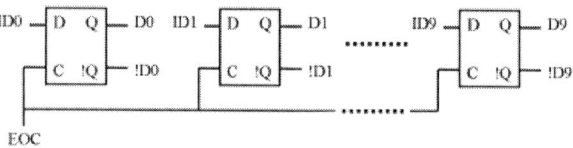

Fig. 5. Output latch

Fig. 6. Layout of Difference calculating unit and DAC Module

Range calculator consists of a difference-calculating circuit (figure 6) and a decision-making circuit. The difference-calculating circuit, implemented by full adders, finds the absolute difference between two consecutive signal levels using the complement notation and then passes the result to decision- making circuit. The decision-making circuit decides the size of binary search tree required to cover the changes in signal level for the following conversion. But the moving BST has its own limitations because if the current value is very close to the previous value, then the number of comparisons done to reach the actual target is higher than the difference between the two numbers.

IV. ADAPTIVE ALGORITHM

A general linear search method takes a longer time to converge than a BST because the big O notation is of the order of n, which is the worst of all the search algorithms. But why don't we use the advantage of linear search? When the difference between two consecutive signals is very close, then linear search seems to be a good method to resort to. To implement the moving binary search tree which uses previous result as starting point for conversion, the proposed search tree requires to move "up or down" at any level of the tree. Therefore, a summer (fig. 8) which performs both addition and subtraction is needed. During the successive approximation process, the summer performs addition at the selected bit if the comparator result is logical "1" (i.e. sampled analog input > DAC input) or subtraction at the selected bit if the comparator result is logical "0" (i.e. sampled analog input < DAC input). So now why don't we combine both, i.e. the advantages of binary and linear

search methods? The following example illustrates the mechanism of the circuits. For instance, the two consecutive signals are:

D [10] = 1001100001 (current signal-609)
D [10] = 1001100010 (previous signal-607)

The number of conversions for the adaptive algorithm (figure 7) to reach to the value of 609 is 2.

1001011111->1001100000->1001100001

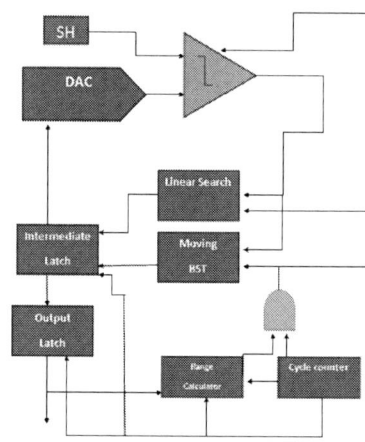

Fig. 7. SAR-ADC with adaptive algorithm

This happens because the consecutive signals are so close to each other that the difference between them is less than the number of bits. Hence the linear search is chosen and the value converges in 2 iterations. Figure 2 depicts clearly that for the value to move from m->q, BST takes 6 iterations i.e. m->k->g->p->w->s->q. But linear search can finish it in 2 iterations, i.e. m->p->q. Hence we use a range calculating circuit that finds the difference between current and previous signal, if the difference is less than n-1 [n-number of bits] then linear search will be used, otherwise binary search is used. Since this algorithm has two search algorithms that are used based on the speed at which the input signal changes, it is called as adaptive algorithm.

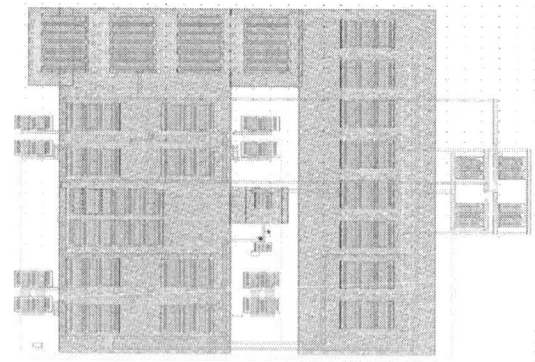

Fig. 8. Layout of Summer/subtractor unit

978-1-4673-9141-2/15 $31.00 © 2015 IEEE

V. LSB SHIFTER ALGORITHM

This is another interesting algorithm we have proposed (fig. 9) which predicts the value based on changing lease significant 0 (fig. 10) or least significant 1 (fig. 11). Just like a moving BST, this algorithm also stores the previous value but instead of doing a regular binary search with the previous value, this algorithm alters individual bits until the required value is reached. When Cp (comparator value) is 1, the least significant 0 is changed to 1 and if the Cp is 0, the least significant 1 is changed to 0. The following example illustrates the mechanism of the circuits. For instance, the two consecutive signals are:

D [10] = 1000011111 (current signal-767)
D [10] = 1000001100 (previous signal-524)

The number of conversions for the LSB shifter to reach to the value of 767 is 3

(1000001100-> 1000001101->1000001111->1000011111)

The Moving BST takes 6 conversions to do the same.

(1000001100->1100001100->1010001100->1001001100->1000101100->1000011110->1000011111). While the regular BST takes 9 conversions to do the same (1000000000->1100000000->1010000000->1001000000->1000100000->1000010000->1000011000->1000011100->1000011110->1000011111)

But this is not true for all cases. There is a big limitation: Moving BST and BST will be faster if the difference between current and previous signal is not (2^n-1) or if the current signal has a difference in two consecutive bits of altering values (eg- previous signal-0000001100 and current signal-0000010110).

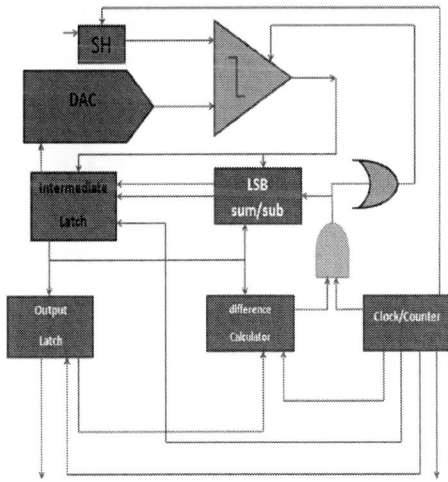

Fig. 9. SAR-ADC with LSB shifter algorithm

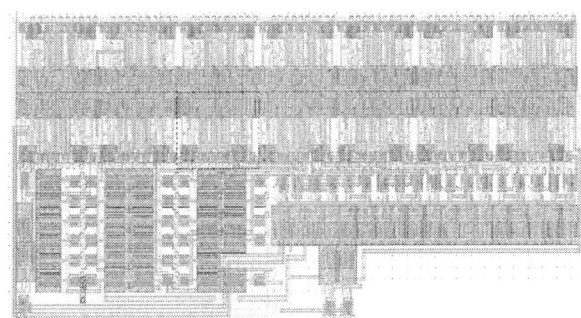

Fig. 10. Layout of least significant zero modifier

The input from comparator acts as the select signal to the multiplexer shown in the circuits which changes either least significant zero/one based on the analog input to the comparator. This circuit when implemented gave a better FOM than a BST algorithm with a limitation that it gave better performance only for above mentioned input condition and it increased the area.

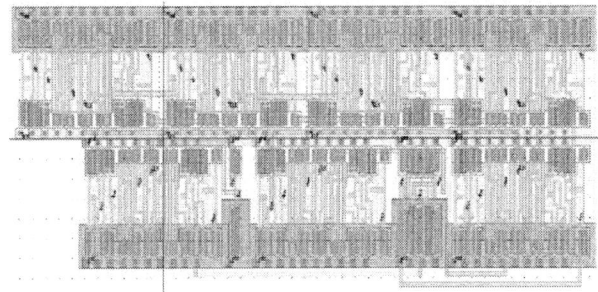

Fig. 11. Layout of least significant one modifier

VI. SPLIT REGISTER ALGORITHM

When the input signal is slowly varying or if the sampling rate is very high, then the current value may be close to the previous value, then why do we need to do the regular BST or moving BST for the entire range of 0 to 1023? Instead, we can do the same moving BST for a smaller range which could eventually reduce the iterations and hence power consumption. So, in this method we are planning to use 8 registers of size 128 each which give 8 boundary values (0, 128, 256 ... 896). For example, if the current value is 169 then moving BST is done between 128 and 255. Since the signal is slowly varying, the next value will either be in the same register or in the adjacent registers (i.e. first register (0-127) or second register (128-255) or third register (256-383)).

Now, how does the algorithm decide which register the value is in? We store the 8 boundary conditions in a memory cell and they are selected using a multiplexer whose control is based on the comparator value. The previous value is in a register and that particular register is known to us. Now three boundary conditions are compared with the analog input, i.e. present, previous and next registers. When the comparator value is 0, it

978-1-4673-9141-2/15 $31.00 © 2015 IEEE

means that the value is in previous register and if it's one, then it can be in the same or next register. So based on these comparisons, the register of interest is selected and the moving BST is performed. Will these 8 registers increase the area to a large extent? So these 8 registers are not physical registers, they are just an imaginary assumption hence the initial thought that blocks our mind regarding increase in area is negated. Only the memory cell which stores those 8 boundary conditions is an additional component but the advantage is that, it reduces the number of iterations by 8. Now, why is that we choose 8 imaginary registers and not any other number? Again that depends on how slow the input signal is changing or how fast the sampling rate is. This algorithm is comparable to another similar approach which assumes a short span around the previous value in which it searches and expands its search area if the value is not within that span. But the only difference in this split-register algorithm is that, it finds the search area and then searches thereby saving a lot of iterations (i.e. if the current value is not very close to the previous value then there can be unwanted search iterations used).

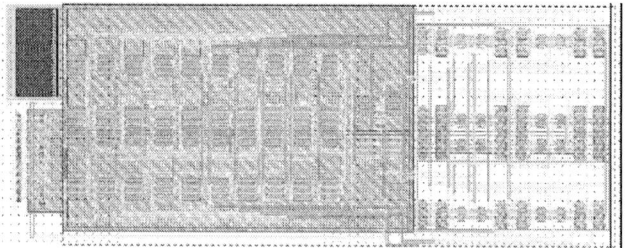

Fig. 12. Layout of Split Register Logic Module

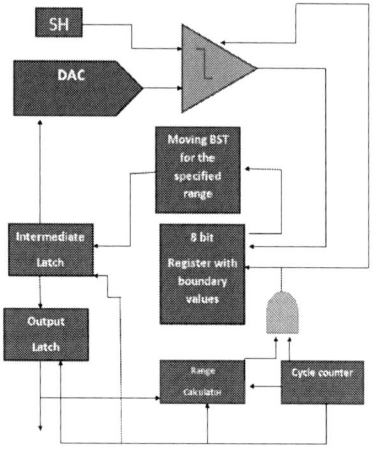

Fig. 13. SAR-ADC with split register algorithm

VII. RESULTS

The proposed algorithms are better in terms of power consumption but they would increase the area because of the complex logic structures used and the overall speed is a question which could limit the sampling rate. Hence there is a tradeoff between power, sampling rate, area and cost.

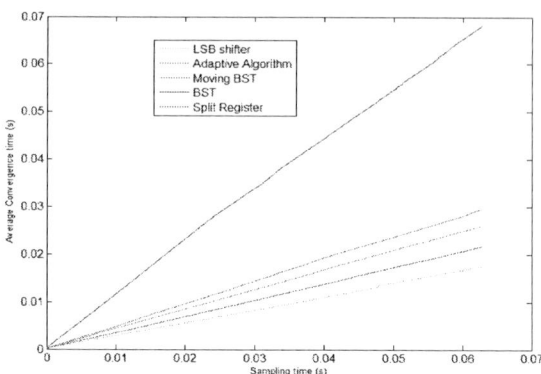

Fig. 14. Convergence time of proposed algorithms

If power consumption was the only concern, then a possible approach of combining all the algorithms would be feasible. Using a range calculator and a difference calculator unit in identifying the type of input signals and the difference in consecutive signals could help in choosing which algorithm to be used. Hence a select signal can be extracted which chooses which algorithm to use for which input signal.

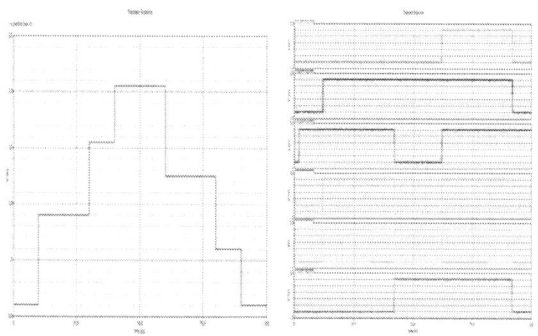

Fig. 15. Post Layout Input/Output results at 0.5GS/s

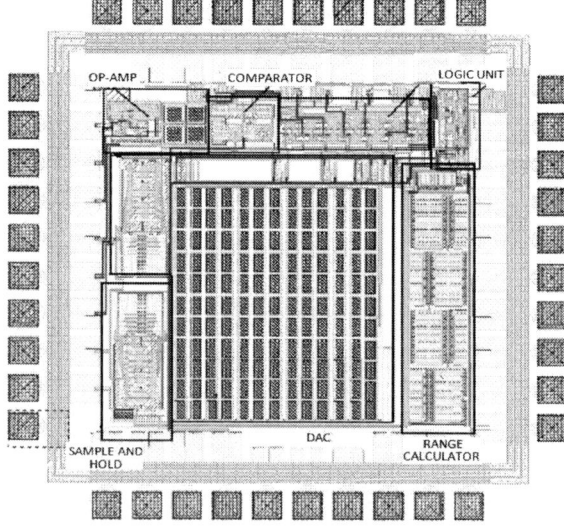

Fig. 16. SAR-ADC chip with MBST algorithm Logic Unit in 65nm CMOS

978-1-4673-9141-2/15 $31.00 © 2015 IEEE   278

REFERENCES	ARCHITECTURE	PROCESS	BITS	Fs (Gs/sec)	POWER (mW)	FOM (fJ/Step)
Shan J [10]	Pipelined	180nm	8	0.2	22	430
Flynn [8]	Folding-Flash	65nm	6	0.4	200	7813
Yahya [5]	Delay-line	65nm	4	1	2	196
Chun [11]	SAR	130nm	10	0.05	0.826	29
Chandrakasan [7]	SAR	65nm	5	0.25	1.2	240
J. Yang [6]	SAR	65nm	6	1	6.7	210
Our Work	**SAR**	**65nm**	**8**	**0.8**	**8.352**	**65.25**

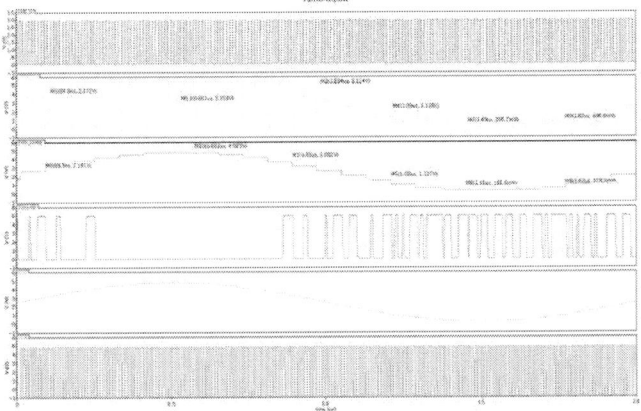

Fig. 17. Measurement results of the ADC chip

VIII.   CONCLUSION

A new approach to low power consumption with different search algorithms and associated control logic has been proposed for the SAR ADC. The proposed modified BST SAR-ADC design chip with 8 bit at 0.8GS/s consumes 8.532mW of power in 65nm CMOS which achieves a power saving of about 56% compared to that of a regular binary search tree SAR-ADC.

IX.   REFERENCES

[1] P. W. Li, M. J. Chin, P. R. Gray, and R. Castello, "A ratio independent algorithmic analog-to-digital conversion technique." *IEEE J. Solid-State Circuits,* vol. SC-19, no. 6. pp. 1138-1143. Dec. 1984.

[2] H. Ohara *et al.,* "A CMOS programmable self-calibrating 13-b eightchannel data acquisition peripheral," *IEEE J . Solid-Stare Cif-cuits,* vol. SC-22, pp. 93C938, Dec. 1987.

[3] C. C. Shih *er al.,* "Reference refreshing cyclic analog-to-digital and digital-to-analog converters," *IEEE J. Solid-Stare Circuits,* vol. SC-21, pp. 544-554, Aug. 1986.

[4] Wen-Sin Liew, Libin Yao and Yong Lian, "A Moving Binary Search SAR-ADC for Low Power Biomedical Data Acquisition System" 978-1-4244-2342-2/08

[5] Y.M.Tousi, G. Li, A. Hassibi and E. Afshari "A 1 mW 4 bit 1Gs/s Delay-line Based Analog to digital converter" *Proc. of IEEE International Symposium on Circuits and Systems(ISCAS),* pp.1121-1124, 24-27 May 2009.

[6] J. Yang, T. L. Niang, and R. W. Broderson, "A 1 GS/s 6 bit 6.7 mW successive approximation ADC using asynchronous processing," *IEEE J. Solid-State Circuits,* vol. 45, no. 8, pp. 1469–1478, Aug. 2010.

[7] B. P. Ginsburg and A. P. Chandrakasan, "Highly interleaved 5-bit, 250- MSample/s, 1.2-mWADC with redundant channels in 65-nm CMOS," *IEEE J. Solid-State Circuits,* vol. 43, no. 12, pp. 2641–2650, Dec. 2008.

[8] S. Park, Y. Palaskas, and M. P. Flynn, "A 4-GS/s 4-bit flash ADC in 0.18um CMOS," *IEEE J. Solid-State Circuits,* vol. 42, no. 9, pp. 1865–1872, Sep. 2007.

[9] Wen-Sin Liew, Libin Yao and Yong Lian, "A Moving Binary Search SAR-ADC for Low Power Biomedical Data Acquisition System" 978-1-4244-2342-2/08

[10] Shan J., M.A. Do, K. S. Yeo, and W. M. Lim, "An 8-bit 200-MSample/s Pipelined ADC With Mixed-Mode Front-End S/H Circuit" *IEEE TRANSACTIONS ON CIRCUITS AND SYSTEMS—I: REGULAR PAPERS, VOL. 55, NO. 6, JULY 2008.*

[11] Chun-Cheng Liu, Soon-Jyh Chang, Guan-Ying Huang. A 10-bit 50MS/s SAR ADC with a Monotonic Capacitor Switching Procedure. *IEEE J. Solid-State Circuits,* 2010, 45(4): 731

Efficient Signature-Based Sub-Circuit Matching

Amir Masoud Gharehbaghi
Dept. of Electrical Engineering and Information Systems
The University of Tokyo
Tokyo, JAPAN 113–0032
Email: amir@cad.t.u-tokyo.ac.jp

Masahiro Fujita
VLSI Design and Education Center
The University of Tokyo
Tokyo, JAPAN 113–0032
Email: fujita@ee.t.u-tokyo.ac.jp

Abstract—**We introduce a new approach for circuit matching using signatures. We have introduced a signature based on the topology of the fanin cone of each circuit element. First, we generate the signature for each circuit element. Then, we find all the circuit elements with unique signature among the given circuits. Finally, we expand the matching area by our expansion rules. Our experiments on IWLS2005 benchmark suite show that our method is able to find the perfect matching between two 160,000-gate IP in 43 minutes. In addition, our method is more than one order of magnitude faster than a state of the art graph matching based approach, while the size of the matched area is more than 30% larger.**

I. INTRODUCTION

Finding common sub-circuits among multiple circuits has many applications in the design phase as well as maintenance of designs. It can help the EDA CAD tools and the designers to reuse the existing results of synthesis, verification, test generation, etc., from one circuit to another; hence, it avoids redoing the same tasks. In addition, finding similarities and differences among multiple circuits has other applications in design maintenance, fraud detection, reverse engineering, redundancy detection, and so forth.

In this paper, we have considered the problem of finding common sub-circuits, having similar structure, among multiple circuits. To solve the problem, we have introduced a new efficient approach based on signatures. The basic idea is to generate signature for each circuit element based on the input cone of the elements connected to it. In the first step, we assign a basic signature for each circuit element. Then, we expand the signature by considering the input cone of each element towards inputs until we reach to the primary inputs or flip-flops, which are considered as pseudo primary inputs. After that, we begin matching circuit elements by finding all the elements that their largest cone's signature uniquely matches with each other. This way, we find all the elements with largest possible matched cones. Finally, we expand the matching area by applying our expansion rules.

During the process of matching, it is possible that we cannot find a unique match for some elements. In this case, we randomly match an element with one of the candidate matching elements, and continue to find unique matches based on the current match, and expand the matching area as much as possible. Therefore, it is possible that we may not be able to find the largest possible matched area in a circuit. However, based on our experiments on ISCAS and IWLS2005 circuits

[5], in all the cases we could find perfect or near perfect matching with less than 1% unmatched gates.

The problem of circuit matching, in general, can be solved with two different approaches. The first approach is modeling the circuit as a graph and solve the problem of graph matching. However, graph matching problem belongs to NP class of algorithms [1], which is not scalable for large graphs, or large circuits in our case. Therefore, heuristic methods are proposed for graph matching problem. However, the quality of the results are very low. The most famous work in this category is SubGemini [2] that works at transistor-level. It tries to find the isomorphic graphs between two input graphs. However, in many cases it fails to decide about the isomorphism. In the general case of graph matching, one of the state-of-art graph manipulation programs, NetworkX [6], could generate matching for the largest ISCAS85 circuit in around 2 minutes, with matched area more than 30% smaller than the actual one. However, our method finishes in around 1 second and the matching area is greater than 99% of the perfect match.

Another approach for circuit matching is signature-based. In [3], a signature-based algorithm for finding isomorphic outputs of circuits in And-Invert Graph (AIG) is proposed. It begins from primary inputs of a given circuit and tries to assign unique labels to the AIG nodes by traversing the circuit multiple times from inputs to outputs and then from outputs to inputs. This way, the isomorphic outputs can be identified. Our method, however, generates signature for any internal node with multiple depth of the input cones. Therefore, we can find sub-circuits in a circuit that are not directly connected to primary inputs or primary outputs. In other words, we are somehow solving the problem of homomorphism that is a superset of isomorphism problem.

Another category of related work is Boolean matching that tries to find the equivalency between two Boolean functions under permutation of inputs [4]. Our circuit matching can be seen as a special case of Boolean matching when structure of the circuits are the same (isomorphic). However, our method can be applied to find the subcircuits that are isomorphic, not just the whole circuit. Moreover, we can find the subcircuits that are structurally isomorphic but functionally different by an appropriate signature, for example by ignoring the gate types, as will be explained later. Hence, our method can be used for other purposes other than Boolean matching.

The main contributions of the paper are: 1) Proposing an efficient signature for sub-circuit matching. and 2) Proposing an algorithm based on our proposed signature for effective sub-circuit matching. Finally, our experimental results on ISCAS

978-1-4673-9141-2/15 $31.00 © 2015 IEEE

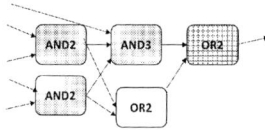

Fig. 1. An Example Circuit Representation

circuits as well as IP cores from IWLS2005 benchmarks [5] shows efficiency and effectiveness of our method. For example, our method could find the perfect match between two IPs of 160,000 gates in around 43 minutes.

The rest of the paper is organized as follows. Section II introduces our signature and how to generate it. Section III presents our signature-based matching method. Section IV shows our experimental results. Section V concludes the paper.

II. SIGNATURE GENERATION

Signature and the way it is generated is very important for our method in two ways. First, the signature should uniquely represent the circuit elements so that the matching can be done perfectly. Second, it should be easy to generate and manipulate so that the runtime of the matching method is acceptable, even for large circuits. However, usually the above two factors are in trade-off and it is very hard to achieve both. In this section we introduce our signature and the way we generate it. We have shown in our experimental results that our method results in near perfect match with an acceptable runtime even for large circuits of 100,000 gates or more.

A. Circuit Representation

In our method, we represent the circuit as a graph. The circuit elements are considered to have one or more inputs and one output. Note that the output may be connected to several circuit elements. Each circuit element is a node of the graph. The functionality of the nodes are labels of the nodes. Therefore, circuit elements with the same functionality have the same label. For example, all the nodes of the graph representing 3-input AND gates have the same label. Each node is connected to other nodes with a directed link. Direction is from output of a node to an input of the other node. Fig. 1 shows part of a circuit and how it is represented.

For primary inputs and primary outputs we have special nodes that does not have any input or output, correspondingly. Flip-flops, or memory elements in general, are also special nodes and are considered as a pseudo primary input and a pseudo primary output. Note that we do not generate signature for (pseudo) primary inputs/outputs. However, they are used in the matching phase to find the mapping of (pseudo) primary inputs/outputs at the end of the matching process, if possible. The main reason that we may not be able to match (pseudo) primary inputs/outputs of two circuits is that the matched sub-circuits may have no direct link to (pseudo) primary inputs/outputs, and this situation may only happen if the two circuits are essentially different.

B. Node Signature

As mentioned above, each circuit element becomes a node with a label representing its functionality. In our method, we

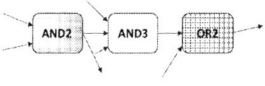

Fig. 2. An Example Path

have considered labels, and consequently signatures as integer numbers so that we can process them more efficiently than strings. Given a circuit, we enumerate all the different elements and assign a unique integer number to each element type as its basic signature. Therefore, two nodes can be potentially matched if their basic signatures are the same.

Note that we can change the signatures to do more than just simple matching. For example, if we assign the same signature to all N-input elements, it means that we allow matching N-input OR gate with N-input AND gate, or any N-input element. Also, we may ignore the number of inputs to match the elements that have similar functionality with each other. For example, we may match a 2-input AND gate with any AND gate with any number of inputs. This kind of matching is specifically useful when doing correction or performing a slight change to the original circuit, like ECO.

C. Path Signature

A path is defined as a finite sequence of nodes such that: all nodes are connected, and each node is connected to exactly one output node and one input node, except for the first node in the sequence that is only connected to an output node, and the last node in the sequence that is only connected to an input node. We call the first node in the sequence *Head*, and last node in the sequence *Tail*. Path length is defined as the number of nodes in the path. Path signature is generated by concatenating all the nodes' signatures from head to tail, in order. Fig. 2 shows a sample path in a circuit. The head node is shown with dotted background and the tail node with grid background. The path length is 3, and the corresponding signature is $(AND2, AND3, OR2)$.

D. Cone Signature

We have defined cone as the set of all the paths with the same tail. This means that we have considered paths beginning from a circuit element towards inputs. Cone length is defined as the maximum length of its paths. For example in Fig. 1 we have shown a cone with the tail OR2 gate with grid background. The cone length is 3. Note that there is a path of length 2, assuming the other input of AND3 is a primary input.

Here we define cone signature as the set of all the signatures of its paths with maximum length plus the set of all the head nodes. Note that the set of paths and the set of nodes does not contain redundant elements. Following we will explain how to generate the cones and also expand them. For example, in Fig. 1 we have shown cone of an element, which has 1 tail, 3 heads, and 5 paths. The signature of the cone is $\{(AND2, AND3, OR2), (AND3, OR2), (AND2, OR2, OR2)\}$ and the set of heads are the nodes shown with dotted background. Note that although there are 5 paths in Fig. 1, only 3 path signatures exists. This way of representing the paths helps us to compact the number of signatures that are kept and compared with others, especially for the cones of longer length.

E. Cone Generation & Expansion

Now, we explain how to generate a cone. Given a node T and the desired cone length L, we do the following:

- If $L = 1$, the cone consists of only node T (T is both head and tail)

- If $L > 1$, generate cones of length L-1 for all the parent nodes (inputs) of node T and combine them with T.
  Also, if at least one of the inputs is (pseudo) primary input, add itself to the list of heads (a path with length 1)

Similarly, we can generate signature for a cone of length L and tail T as follows:

- If $L = 1$, the signature is signature of the node T and the head list contains only T.

- If $L > 1$, generate signatures for the cones of length L-1 for all the parent nodes (inputs) of the node T and concatenate them with its node signature. List of heads also becomes union of all the heads of the cones of its parent nodes (inputs).
  Also, if at least one of the inputs is (pseudo) primary input, add itself to the list of heads. Moreover, add its node signature to the list of signatures.

The above process can be considered as expansion of length 1. Similarly, we can define cone expansion of length Lx. Given a cone C of length L and the list of heads H, we can expand the cone by length Lx as follows. Note that the expanded cone become of length $L + Lx$.

- For each node Ti in the head list H generate cones of length $Lx + 1$ and combine them with the paths ending in Ti.
  List of heads in the expanded cone is union of all the heads of the cones with tail Ti and length $Lx + 1$

The signatures for the expanded cone can be also generated easily, similar to the above procedure. As an example consider the circuit shown in Fig. 3.A. The cone of length 3 for the tail shown with grid background is $\{(AND2, AND3, OR2), (AND3, OR2), (AND2, OR2, OR2)\}$ and the set of heads are the nodes shown with dotted background. Assuming that we want to expand the cone by length 2, first we generate the cone of length 3 and tail with grid background shown in Fig. 3.B. Similarly we generate the cone for the other node in the head list as shown in Fig. 3.C. The final cone signature after combining them becomes Fig. 3.D. The signature is $\{(AND2, AND2, AND2, AND3, OR2), (AND2, OR2, AND2, AND3, OR2), (AND3, OR2), (AND2, AND2, AND2, OR2, OR2), (AND2, OR2, AND2, OR2, OR2)\}$ and the list of heads is shown with dotted background.

III. SIGNATURE-BASED MATCHING

In the previous section, we explained how to generate signatures for cones of different length for each node. Based on the signatures that are generated for each node, we present how to match different nodes from two circuits, or find common sub-circuits.

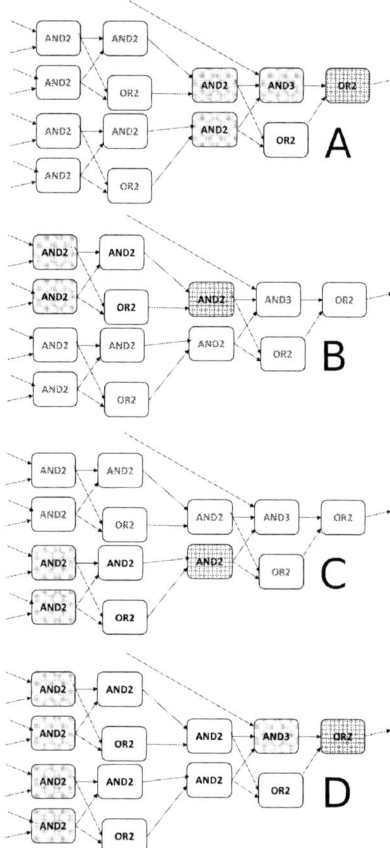

Fig. 3. A Cone Expansion Example

A. Matching Overview

Assuming that we have two circuits, that are represented as graphs, as explained before, the basic idea for matching between the two circuits is to generate the cone signature for every node (circuit element). Then, based on the cone signatures, we find the nodes that can be uniquely matched between the two circuits. Finally, starting from the matched points, we expand the matched area by considering the neighboring nodes, in an iterative process. If at some point it is not possible to find the unique matches, we randomly select one of the possible choices and continue until all the nodes are matched, or no further matching is possible.

Note that because of selecting one of the possible choices randomly as well as not performing backtracking to try the other matching choices, it is possible to reach to a final matching that is not perfect. It means that the final matching compared to the perfect matching may have a larger number of unmatched nodes, or a smaller number of connected matched nodes. However, as we will show in our experimental results, practically, we can reach to perfect or near perfect matching, even for large circuits of more than 100,000 gates.

The step by step algorithm for matching is as follows. In the subsequent subsections we explain each step in more details.

1) Read circuit1 and circuit2 and generate the corresponding graphs, G1 and G2, respectively

2) For every node of G1 generate the cone signatures for cone length from 1 to the maximum possible length
3) For every node of G2 generate the cone signatures for cone length from 1 to the maximum possible length
4) Mark all the nodes of G1 and G2 as unmatched
5) While there is an unmatched node do the following
   a) For every pair of unmatched nodes from G1 and G2 find the largest cone depth with the same signature
   b) For each unmatched node of G1 (G2), list all the candidate matched nodes with the maximum cone length
   c) Select all the unmatched nodes of G1 (G2) from the above list with exactly one candidate matched node and marked them as matched
   d) While there are new matched nodes do the following
      i) For each matched node of G1 (G2) that has any unmatched neighbor, try to expand the matching area to its neighbors, uniquely
   e) Do the following, while there are new matched nodes
      i) For each matched node of G1 (G2) that has any unmatched neighbor, try to expand the matching area to its neighbors, if possible uniquely, otherwise randomly beginning from the nodes with less number of candidates for selection
   f) If no further matching is possible, exit
6) Report the matched nodes

B. Finding the Largest Matched Cone Depth

Given two nodes, we want to find the length of the largest cone with the same structure with the two nodes as the tail nodes. To find the largest cone, we start with the cone of length one and gradually increase the cone length until we find the length of the largest cone. Now the problem is to find out that given two nodes and a length L, the cones of Length L for the two nodes have the same structure or not.

- For $L = 1$ the problem is that node signatures are the same or not.

- For $L > 1$ the problem is that node signatures are the same or not AND there is a one-to-one mapping between inputs of the two nodes such that the inputs of the nodes have the same structure (signature) for the cones of length L-1.

C. Expanding the Matched Area

As mentioned above, after finding some nodes that are matched, we want to expand the matching area to the neighboring nodes. We do expansion in two phases. At first, we try to expand the matching areas uniquely. It means that we match only if there exists a unique matching. However, we may not be able to match uniquely, since there may be more than one node with exactly the same signature to be matched.

When there is no more node that its neighboring nodes can be matched uniquely, we enter the second phase of the matching. In this phase, we just choose the first node from the list of the matching candidate nodes that we have found. Note that in the second phase, it is still possible to find unique matches as new nodes are matched.

Following, we define a number of rules for unique expansion, followed by the rules for non-unique expansion. Given two nodes V1 and V2 that are matched, we can uniquely expand the matching as follows:

1) If all the parents of V1 and V2 are matched except one of them, then we can match those unmatched parents, if they have the same node signature. Note that a special case is when V1 and V2 have only one parent.
2) If all the children of V1 and V2 are matched except one of them, then we can match those unmatched children, if they have the same node signature. Note that in this case V1 and V2 have the same number of children.
3) If all the parents of V1 and V2 are matched except k number of them, and we can uniquely find a one-to-one match between those unmatched parents, then we can match those unmatched parents. Note that this is a generalized case of 1 (k = 1).
4) If all the children of V1 and V2 are matched except k number of them, and we can uniquely find a one-to-one match between those unmatched children, then we can match those unmatched children. Note that V1 and V2 have the same number of children. Also, note that this is a generalized case of 2 (k = 1).

Given two nodes V1 and V2 that are matched, we can non-uniquely expand the matching as follows:

1) If there are k1 unmatched and k2 unmatched parents of V1 and V2, respectively, and we can find a match between m1 unmatched parents of V1 with m2 unmatched parents of V2 (k1 ≥ m1 ≥ 1, k2 ≥ m2 ≥ 1), we match the first node from m1 to the first node from m2. Note that all the m1 and m2 nodes have the same cone signature with the maximum possible depth.
2) If there are k1 unmatched and k2 unmatched children of V1 and V2, respectively, and we can find a match between m1 unmatched children of V1 with m2 unmatched children of V2 (k1 ≥ m1 ≥ 1, k2 ≥ m2 ≥ 1), we match the first node from m1 to the first node from m2. Note that all the m1 and m2 nodes have the same cone signature with the maximum possible depth.

IV. EXPERIMENTAL RESULTS

We have conducted experiments on ISCAS benchmarks as well as IWLS2005 IPs [5]. We have implemented our method in C++, around 1800 lines of code. The experiments are performed on a PC with Intel core i5-3.1GHz processor and 8 GB memory running Linux kernel 2.6.32.

Before going into details of the results, we want to highlight some of the implementation details of the method, as

TABLE I. MATCHING IWLS2005 IPS

Name	OG	RG	NI	LIS	MNI	MLIS	U	Time
vga_lcd	158708	154140	9	154122	9	154122	0	2577
des_perf	102918	97826	257	97570	257	97570	0	877
ethernet	88571	87519	2	87518	2	87518	0	1798
wb_conmax	46534	46224	1	46224	5	46220	0	311
pci_bridge	26339	26003	159	25800	159	25800	0	54
aes_core	24371	23772	1	23772	1	23772	0	43
usb_funct	15640	15532	50	15135	50	15135	0	7.72
ac97_ctrl	15675	15376	29	14835	29	14835	0	9.23
mem_ctrl	15152	14788	7	14638	47	14558	0	12.02
systemcaes	12842	12641	1	12641	1	12641	0	10.22
tv80	9926	9725	2	9724	2	9724	0	6.07
des_area	6001	5573	1	5573	1	5573	0	7.7
wb_dma	4477	4389	37	4329	37	4329	0	1.19
spi	4340	4220	1	4220	1	4220	0	1.67

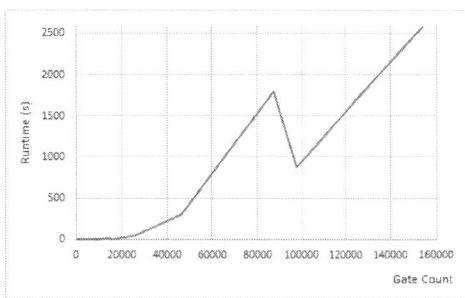

Fig. 4. Matching Time vs. Gate Count for IWLS2005 circuits

without proper considerations the runtime of the method can increase significantly.

- Signatures Format: Signatures are the key part of the method. Efficient generation and comparison of signatures are crucial for efficiency of the method. We generate the node signatures and path signatures as integer numbers to make the comparison very fast. Depending on how many kinds of nodes (or in our case, gate types) exists, we can select the minimum required number of bits and keep the path signatures up to some length in one integer by concatenating the node signatures. For example, in our case we needed maximum 5 bits (maximum 32 gate types) for node signature. Therefore, with integer width of 64, we could hold paths of maximum 12 levels in one integer word. Those numbers can be adjusted based on the problem and the system that is used for the experiment. For the paths of more than 12 levels, we just kept multiple integers. For example, for a path of length 30, we used 3 integer words.

- Signature Comparison: As stated above, longer paths need more number of integers to be kept, and also more number of integer comparison operations. Therefore, in our implementation we first tried to find the unique matches only for paths up to 12 levels, which require only one integer comparison operation. After that, we considered longer paths that require more number of integer comparison operations. According to our experiment, in more than 85% of the cases we could find the unique matches by considering only paths up to 12 levels (one integer word). For the rest of the cases, we could continue with non-unique matching, but obviously it would lead to non-perfect matching. We intend to study in more depth the

relationship between the path length and its effect on the runtime as well as matching quality in our future work.

- Efficient Unique Matching: As we explained in the flow of our method, we have to initially find some unique matches so that we can expand them. As the initial matching is done globally among all the nodes in the input circuits, it is necessary to be implemented very efficiently. Therefore, during our search for unique matches, whenever we find that there is more than one possible match for a node, we give up on that node and continue with the next one. Therefore, the total number of comparisons is reduced drastically.

In addition, we have added a preprocessing phase to our program. The task of the preprocessor is to remove the redundancies that are found in the original netlist. This helps to improve the quality of matching. That is because when we remove the redundant nodes, there is more chance of finding unique matches. The functionality of the circuit after the preprocessing phase is exactly the same as the original circuit. Preprocessor does the following tasks:

1) removing the redundant inputs of a gate and keeping only one of them (inputs that are the same).
2) removing the gates which have exactly the same inputs and merging their outputs.
3) removing the buffers.

In our experiments, we have reported a number of factors about the quality of the matching. One factor is the number of unmatched gates, shown with U symbol in the tables.

Another important factor is about the connectivity of the matched area. We define a term "island" to refer to a set of gates that are connected to each other. Each gate in an island is connected to an input or an output of at least one gate in the island. Basically we are not just interested to match all the gates. But, we are interested to match in a way that more gates are connected to each other. In other words, less number of islands and larger islands. For example, consider two circuits that are implemented with only 2 input NAND gates, and with all the gates connected, forming one large island. Obviously, we can easily match every gate of circuit 1 with every gate of circuit 2, and the result will be 100% matched gates. But, this kind of matching can be far from the perfect matching if the matched gates generate lots of disjoint islands.

Therefore, we have reported number of islands (NI) in the original circuit, largest island size (LIS) in the original circuit, number of islands in the matched circuit (MNI), and largest island size in the matched circuit (MLIS)

Moreover, we have reported the number of gates in the original circuit (OG) and the number of gates after removing redundancies (RG) by the preprocessor.

Another thing is that in our experiments, we reversed the order of inputs of all the gates in one of the input circuits. This way, we generate a new circuit that is functionally the same and structurally isomorphic to the original circuit. For example, in a 2-input gate, we change the input 1 to input 2, and input 2 to input 1.

TABLE II. MATCHING ISCAS85/89 IN AIG FORMAT

Name	OG	RG	NI	LIS	MNI	MLIS	U	Time
s38584.1	23443	23014	10	22984	10	22984	0	118.59
s38584	23446	23014	10	22984	10	22984	0	178.73
s35932	22467	22467	1441	18210	1441	18210	0	487.15
s38417	17456	16900	53	16709	53	16709	0	47.43
s15850.1	6620	6454	71	4847	71	4847	0	4.11
s15850	6636	6451	71	4847	75	4847	0	5.07
s13207.1	5402	4930	81	1944	83	1944	0	2.37
s13207	5464	4928	79	1944	79	1944	0	2.3
c6288	4628	4628	2	4627	2	4627	0	1.24
c6288nr	4620	4620	2	4619	2	4619	0	1.31
c7552	3867	3808	4	3797	12	3777	0	2.56
c7552nr	3614	3555	4	3544	14	3518	0	1.76
s9234.1	3599	3504	9	3224	9	3224	0	1.17
s9234	3602	3503	8	3224	8	3224	0	1.2
c5315	3008	2981	19	2883	19	2883	0	9.26

TABLE III. MATCHING ISCAS85/89 IN ORIGINAL FORMAT

Name	OG	RG	NI	LIS	MNI	MLIS	U	Time
s38417	22179	19749	192	17870	202	17848	0	41.89
s38584	19253	16139	146	15472	147	15472	0	9.75
s35932	16065	14895	1162	11880	1162	11880	0	143.24
s15850.1	9772	9630	88	8375	98	8361	0	9.25
s15850	9772	8259	86	6992	86	6992	0	4.94
s13207.1	7951	7832	136	5166	140	5162	0	5.58
s13207	7951	6890	125	4467	129	4455	0	3.15
s9234.1	5597	5502	20	5025	20	5025	0	2.64
s9234	5597	4268	18	3833	18	3833	0	1.17
c7552	3517	2489	4	2479	8	2471	0	0.99
s5378	2794	2425	13	2381	13	2381	0	0.38
c6288	2416	2416	2	2415	2	2415	0	3.04
c6288nr	6186	2399	2	2398	2	2398	0	3.06
c7552nr	6976	2357	4	2347	8	2339	0	0.83
c5315	2313	1729	20	1666	20	1666	0	16.17

TABLE IV. MATCHING ISCAS85 REDUNDANT/NON-REDUNDANT IN AIG FORMAT

C1/C2	G1	G2	LIS1	LIS2	MLIS	U1	U2	Time
c6288/c6288nr	4628	4620	4627	4619	4566	8	0	1.21
c7552/c7552nr	3808	3555	3797	3544	3502	253	0	2.44
c5315/c5315nr	2981	2924	2883	2849	2842	57	0	2.19
c3540/c3540nr	1764	1638	1764	1638	1324	126	0	2.04
c2670/c2670nr	1231	1129	1208	1107	1084	102	0	0.18
c1355/c1355nr	930	930	930	930	834	2	2	0.91
c499/c499nr	794	794	794	794	786	2	2	0.55
c1908/c1908nr	740	718	740	718	715	22	0	0.14
c432/c432nr	364	270	364	270	256	94	0	0.05

TABLE V. MATCHING ISCAS85 REDUNDANT/NON-REDUNDANT IN ORIGINAL FORMAT

C1/C2	G1	G2	LIS1	LIS2	MLIS	U1	U2	Time
c7552/c7552nr	2489	2357	2479	2347	2281	140	8	0.74
c6288/c6288nr	2416	2399	2415	2398	2318	35	18	3.99
c5315/c5315nr	1729	1707	1666	1650	1641	32	10	2.86
c3540/c3540nr	1038	1042	1038	1042	909	34	38	3.99
c2670/c2670nr	769	715	751	697	677	57	3	0.08
c1908/c1908nr	513	512	513	512	506	5	4	0.11
c1355/c1355nr	482	482	482	482	346	0	0	1.88
c499/c499nr	170	170	170	170	162	8	8	0.02
c432/c432nr	155	152	155	151	150	5	2	0.01

In the first experiment, we have compared each IP in IWLS2005 to its structurally isomorphic IP. Table I shows the results for the 15 largest circuits. As it can be seen in all the circuits, there is no unmatched gate. Moreover, in all the cases the number of islands and the size of the largest islands are the same, except for two cases (wb_conmax, mem_ctrl) that the size of the largest island is decreased less than 1 percent. The increase in the number of islands in those two cases is for the very small islands, and it is due to the fact that basically we have many choices and unique matches may not exist.

Also, Fig. 4 shows the runtime of our matching tool vs. number of gates. As it is seen, the runtime increases near linearly. However, there are cases that matching is very easy despite large number of gates (like des_perf). Also note that it is possible that in some cases the matching may be more difficult. However, the general trend is near linear.

In the next experiment, we repeated experiment one on ISCAS85/89 circuits. Table II and Table III show the results for the original circuit and the same one converted to AIG (And-Invert-Graph), which has only 2-input AND and inverter. What is seen is that in general the quality of results is better for the AIG circuits. Although a circuit in AIG format has fewer gate types and consequently more possibilities for initial matching, finally the matching is closer to perfect in more cases. Note that because of lack of space, we have only reported the results for the largest 15 circuits.

In the next experiment, we compared each ISCAS85 circuit versus its non-redundant version. The results are shown in Table IV and Table V. As it can be seen, AIG format leads to matching that is almost perfect and better quality than the original format of the circuits.

In the next experiment, we generated carry-lookahead (CLA) circuits by cascading 1-bit CLAs. We generated CLAs of size 4, 8, 16, 32, 64, and 128 bits and tried to match all the possible pairs (for example: 4-bit CLA with 64-bit CLA, 8-bit CLA with 16-bit CLA, ...). In all the cases we could match perfectly the smaller size CLA in the larger or equal size CLA. This shows that although CLA circuit has lots of similar structures, our algorithm can find the perfect matches.

In the last experiment, we used NetworkX [6], a state-of-the-art graph manipulation toolkit for comparison. To have fair comparison, we reimplemented the C++ version of the algorithm that was written in Python from the package by a straight forward translation from Python to C++. For ISCAS85 circuits, the runtime was at least 1 order of magnitude slower than our method, while the matched area was at least 30% smaller than our method.

V. CONCLUSIONS & FUTURE WORK

We have presented a signature-based method for sub-circuit matching. Our experimental results on ISCAS and IWLS benchmarks shows efficiency and effectiveness of our method. Our future work includes improving the runtime of our method as well as incorporating different kinds of signatures for different applications of the method.

REFERENCES

[1] D. A. Basin, *A Term Equality Problem Equivalent to Graph Isomorphism,* Elsevier Information Processing Letters, Vol. 51, pp. 6166, 1994.

[2] M. Ohlrich, C. Ebeling, E. Ginting, and L. Sather, *SubGemini: Identifying SubCircuits using a Fast Subgraph Isomorphism Algorithm,* Design Automation Conference, pp. 31-37, 1993.

[3] A. Mishchenko, N. Een, R. Brayton, M. Case, P. Chauhan, and N. Sharma, *Semi-Canonical Form for Sequential AIGs,* Conference on Design, Automation and Test in Europe, pp. 797-802, 2013.

[4] H. Katebi, and I.L. Markov, *Large-scale Boolean Matching,* Conference on Design, Automation and Test in Europe, pp. 771-776, 2010

[5] *http://iwls.org/iwls2005/benchmarks.html*

[6] *http://networkx.github.io*

Reversible Circuit Rewriting with Simulated Annealing

Nabila Abdessaied[*]　　Mathias Soeken[*†]　　Gerhard W. Dueck [‡]　　Rolf Drechsler[*†]

[*]Cyber-Physical Systems, DFKI GmbH, Bremen, Germany
[†]Department of Mathematics and Computer Science, University of Bremen, Germany
[‡]Faculty of Computer Science, University of New Brunswick, Canada
{nabila,msoeken,drechsle}@informatik.uni-bremen.de, gdueck@unb.ca

Abstract—This paper presents a rule based approach to optimize the quantum cost of reversible circuits using circuit rewriting rules that handle positive and negative controls. Since incremental optimization cannot guarantee optimality, we consider the application of simulated annealing to find further sub-circuits that could be replaced with smaller ones.

Experimental evaluations show that simulated annealing not only can significantly improve the quality of reversible circuits but also is more efficient than a comparable greedy approach. Using the rewriting rules combined with the proposed method quantum cost reductions by up to 80% can be achieved.

I. Introduction

Synthesis describes techniques of finding a reversible circuit that realizes a given reversible function. Typically, synthesis approaches are not aware of the technology in which the circuit is applied. In order to optimize synthesis results with respect to some target technology, post-synthesis optimization approaches are used to reduce the circuits with respect to a given cost, e.g., transistor costs in CMOS architectures or quantum costs in quantum computing architectures.

Due to the inherent complexity of reversible circuits, usually local optimization strategies are implemented, i.e., sub-circuits are analyzed for possible reductions. The most employed optimization approaches can be categorized into *rule-based optimization* and *template matching*.

Rule-based optimization (see, e.g., [1]–[3]) suggests a specific set of sub-circuits together with cheaper replacements. Rule-based approaches are typically motivated based on some synthesis approaches and exploit often reoccurring circuit structures. The approaches are not complete, i.e., they cannot guarantee an optimal circuit after optimization. However, the approaches are greedy meaning that optimization is only applied if a reduction can be achieved. Consequently, they cannot escape local optima.

Template matching approaches (see, e.g., [4])are more powerful than rule-based approaches. A template is a generic circuit that realizes the identity function. Generic means that *one* template *represents* a (theoretically infinite) *set* of identity circuits. If a sub-circuit matches one part of an instance of a template it can always be replaced with the inverse of the remaining part, since they represent the same function.

Optimization with templates is also greedy, however, it is still unknown whether template matching is complete.

In order to find suitable sub-circuits to apply a rule or a template, *moving rules* [5] have been proposed that allow gate movement without changing the function. These moving rules change the order of gates but not the gates themselves. Moving rules are not complete, i.e., one cannot necessary obtain all function-preserving permutation of gates in the circuit by repeatedly applying the moving rules. Since the rule-based and template matching approaches are incremental greedy algorithms, they can achieve further improvements on synthesized circuits but cannot guarantee optimality. In fact, these approaches can guarantee only local optimization but not global optimization.

Recently, *circuit rewriting* [6] has been proposed that exploits negative control to rewriting sub-circuits while preserving the function. Circuit rewriting is based on a small set of elementary rules which can be extended to more complicated rules. It is conjectured that circuit rewriting is complete based on a very small set of rules, i.e., one can rewrite a circuit into all other circuits that represent the same functionality. Circuit rewriting is not greedy, i.e., intermediate steps may increase the cost. In fact, no guided approach for circuit optimization using the rewrite rules has been proposed so far.

In this paper, we present rule based optimization approaches using the rewrite rules from [6] instead of the moving rules [5]. Two quantum cost optimization approaches are presented. The first method aims to reduce the cost by applying a greedy approach, whereas the second method is based on simulated annealing. The application of simulated annealing can attain not only local optimum but also global optimum [7], hence it can find further sub-circuits to be replaced with smaller ones. The rewrite rules support this freedom by allowing a higher flexibility in gate movement. As confirmed by an experimental evaluation, improvements on quantum cost of up to 17% in the first case and up to 30% in the latter case can be observed. This clearly demonstrates that the simulated annealing approach outperforms the greedy approach on optimizing reversible circuits.

The remainder of the paper is structured as follows: first the basics on reversible circuits are recaptured in Sect. II. The next

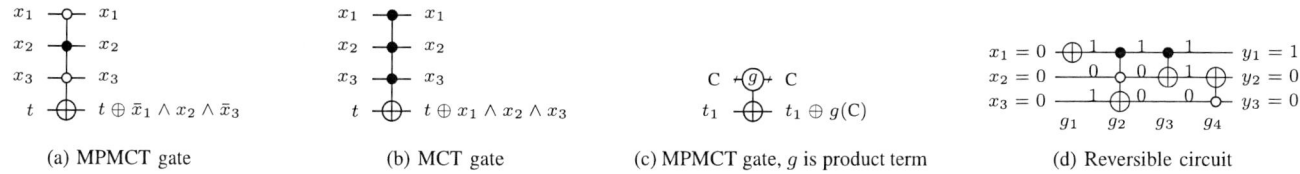

Fig. 1: Reversible circuitry

(a) MPMCT gate   (b) MCT gate   (c) MPMCT gate, g is product term   (d) Reversible circuit

section outlines the simulated annealing approach. Section IV introduces the rewriting rules and Sect. V gives a detailed description of the implementation of the presented approach, and experimental results are evaluated and interpreted in Sect. VI. The paper is concluded in Sect. VII.

II. BACKGROUND

To keep this paper self-contained, this section briefly introduces the basics on reversible logic, quantum circuits, and their cost metrics.

A. Reversible Logic

Let $\mathbb{B} = \{0, 1\}$ denote the *Boolean values*. Then we refer to $\mathcal{B}_{n,m} = \{f \mid f \colon \mathbb{B}^n \to \mathbb{B}^m\}$ as the set of all *Boolean multiple-output functions* with n inputs and m outputs.

Definition 1 (Reversible function): A function $f \in \mathcal{B}_{n,m}$ is called *reversible* if f is bijective, i.e., if each input pattern is uniquely mapped to an output pattern, and vice versa. Otherwise, it is called *irreversible*. Clearly, if f is reversible, then $n = m$.

Reversible functions are realized by reversible circuits that consist of at least n lines and are constructed as cascades of reversible gates from some gate library. The most common gate library consists of multiple control Toffoli gates [8].

Definition 2 (Toffoli gate): Given a set of variables $X = \{x_1, \ldots, x_n\}$, a *mixed-polarity multiple-control Toffoli (MPMCT)* gate $\mathrm{T}(C, t)$ has *control lines* $C = \{x_{j_1}, x_{j_2}, \ldots, x_{j_l}\} \subset X$ and a *target line* $t \in X \setminus C$. The gate maps $t \mapsto t \oplus g(x_{j_1}, x_{j_2}, \ldots, x_{j_l})$ where g is defined as:

$$g : (x_{j_1}, x_{j_2}, \cdots, x_{j_l}) \mapsto (x_{j_1}^{p_1} \wedge x_{j_2}^{p_2} \wedge \cdots \wedge x_{j_l}^{p_l})$$

with each *literal* $x_{j_i}^{p_i}$ is either a propositional variable $x_{j_i}^1 = x$ or its negation $x_{j_i}^0 = \bar{x}$. All remaining other lines are passed through unaltered. *Multiple-control Toffoli gates (MCT)* are a subset from MPMCT gates in which the product terms in h can only consist of positive literals.

Example 1: Figure 1(a) shows a Toffoli gate with mixed polarity control lines, Fig. 1(b) shows a Toffoli gate with n positive controls. The control lines are either denoted by solid black circles to indicate positive controls, white circles to indicate negated controls or represented by a one product terms Boolean function g as depicted in Fig. 1(c). The target line is denoted by $\oplus$. Figure 1(d) shows different Toffoli gates in a cascade forming a reversible circuit. The annotated values demonstrate the computation of the gate for a given input assignment.

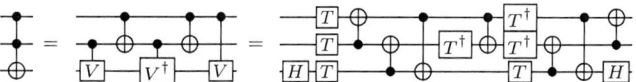

Fig. 2: Mapping a 2-controlled Toffoli gate to NCV and Clifford+T quantum gates

B. Cost Metrics

The quantum cost of a reversible or a quantum circuit is varying with respect to the quantum library used in the technology mapping. When using the NCV library, then the quantum cost is called the *NCV-cost* while it is called *T-depth* when the Clifford+T library is used. The motivation for that cost measure originates from the fact that the T gate is significantly larger compared to the other gates in the circuit.

The common gate library NCV and the Clifford+T gate libraries are composed of the elementary gate set {NOT, CNOT, V, $\mathrm{V}^\dagger$} and {NOT, CNOT, H, Z, S, $\mathrm{S}^\dagger$, T, $\mathrm{T}^\dagger$}, respectively. The two libraries are universal for quantum computation, however only the gates of the Clifford+T library can be implemented in a fault-tolerant way.

Definition 3 (NCV-Cost): The *NCV-cost* is the total number of elementary gates used in a quantum circuit.

Definition 4 (T-depth): The *T-depth* is the number of T-stages in a quantum circuit where each stage consists of one or more T or $\mathrm{T}^\dagger$ gates that can be performed concurrently on separate qubits. The total number of incorporated T or $\mathrm{T}^\dagger$ gates in the whole circuit is denoted by *T-count*.

Example 2: Consider a Toffoli gate with two control lines as shown in Fig. 2. A functionally equivalent realization in terms of quantum gates using the NCV library is depicted in the second network. The NCV-cost of this circuit is 5 since there are 5 elementary gates in the circuit. The third cascade represent the quantum realization of a Toffoli gate using the Clifford+T library. It has a *T-count* of 7 and *T-depth* of 3. The reversible circuit depicted in Fig. 1(d) has an NCV-cost of 9 and a *T-depth* of 3.

Based on [9], the following table summarizes the quantum cost for MCT gates where c denotes the number of controls:

Number of controls	< 2	2	3	4	≥ 5
NCV-cost	1	5	20	50	$40(c-3)$
T-depth	0	3	12	30	$24(c-3)$

978-1-4673-9141-2/15 $31.00 © 2015 IEEE

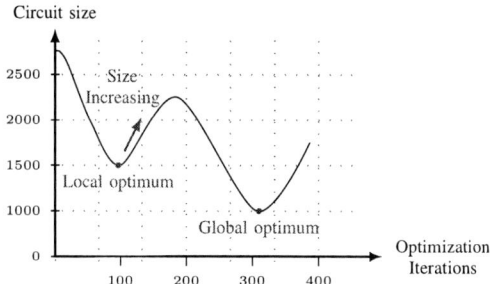

Fig. 3: Greedy heuristics via simulated annealing algorithm

III. SIMULATED ANNEALING

Assume that S is a solution to a given problem. Let M be a "move" that can be performed on S. M will have an effect on the "cost" of S. The objective of optimization is to to repeatedly perform "moves" to S such that the cost of S is reduced. The greedy algorithm is a simple heuristic optimization technique. With the problem at hand, it would only perform moves that lower the cost. The obvious problem with such an approach is that it can get stuck in a local optimum (see Fig. 3). Kirkpatrick et. al. [7] suggest an optimization technique called *simulated annealing* that is based on statistical mechanics.

The idea of simulated annealing is simple. Randomly select moves (or transformations) on a given solution. Each move will have an effect on the cost of the solution. The moves can be cost-increasing, cost-decreasing, or cost-neutral. Cost-decreasing moves are always accepted. Simulated annealing uses the concept of *temperature* (it plays a significant role in statistical mechanics) to deal with cost-increasing moves. Initially the temperature is high. Cost increasing moves are accepted with a probability that depends on the temperature. Initially, many cost-increasing moves are accepted. As the temperature decreases, fewer and fewer moves are accepted. Finally, the procedure stops when the system reaches a stable state, that is, no moves are accepted.

The process of reversible circuit optimization is well suited for simulated annealing. Gates can be moved within the circuit. The move may be associated with an increased cost of the circuit. On the other hand, a gate may be moved adjacent to an other gate that is identical. In this case both gates can be removed since Toffoli gates are self inverse, hence the cost will be reduced. In dealing with reversible circuits, the NCV-cost or the T-depth may be considered as cost metric during the simulated annealing process.

Simulated annealing has been used in the synthesis of reversible circuits, e.g, in [10], the authors have presented a simulated Annealing based Quine-McCluskey approach to synthesize a reversible circuit. Also, the synthesis algorithm presented in [11] have used simulated annealing to transform ESOP cubes.

IV. REWRITING RULES

Most of the existing synthesis approaches for reversible functions do not generate optimal circuit realizations with

Fig. 4: Rewriting rules

respect to quantum cost. Therefore optimization approaches are applied to reduce the quantum cost. These post synthesis techniques attempt to apply reduction rules by deleting identical gates or replacing cascades of gates with smaller ones [4]. To do so, the gates are rearranged to match the reduction rules. The moving rules were originally defined in [5] as the following property and illustrated in Fig. 4(a).

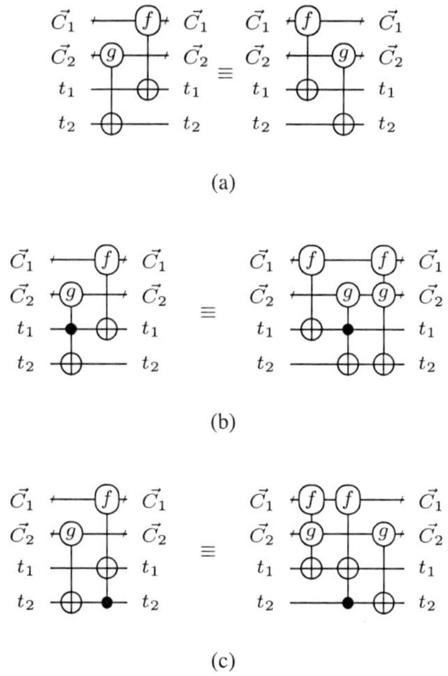

Moving rule. Two adjacent gates can be interchanged if and only if the target for each gate is not a control for the other gate, i.e., in a reversible circuit, gate $T_1(C_1, t_1)$ can be interchanged with gate $T_2(C_2, t_2)$ if and only if $t_2 \notin C_1$ and $t_1 \notin C_2$.

This moving rule is very restrictive, therefore, the conventional moving rule was extended as described in [12] allowing further moving abilities for each gate in the circuit. The extended moving rule is defined in the following property.

Extended moving rule. A gate can be moved from one end of a cascade of gates to the other end if its controls are on lines that are invariant with respect to the cascade and its target is on a non-controlling line.

To identify whether a line is invariant, an algorithm called *line labelling procedure* (LLP) labels the line segments in a circuit. A label is an assignment of numbers to each line after a gate. If the label is the same at the beginning and the end of the segment, then the line is invariant.

Rewriting rules. However, the above moving rules are restraining the movement of gates into a circuit. On the other hand, the moving rules introduced in [6] are general for both MCT and MPMCT cascades and have more freedom for gate rearrangement. Based on the rules, we extract three scenarios for moving a gate from one position to another as shown in Fig. 4.

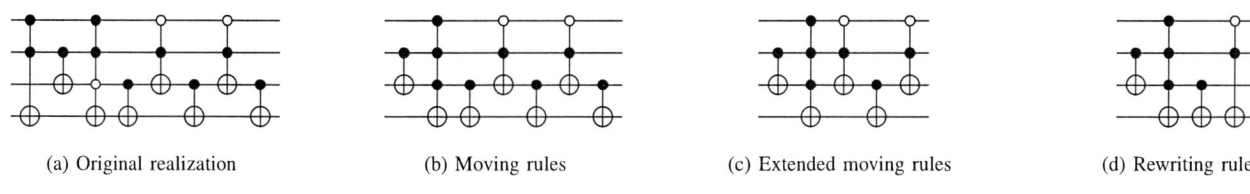

| (a) Original realization | (b) Moving rules | (c) Extended moving rules | (d) Rewriting rules |

Fig. 5: Circuit optimization wrt. different moving rules

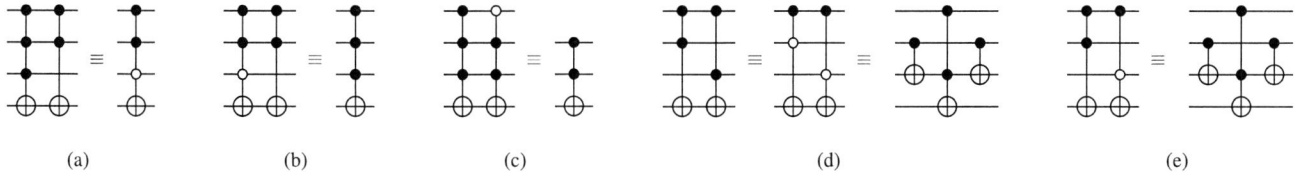

| (a) | (b) | (c) | (d) | (e) |

Fig. 6: Applied reductions

Example 3: Consider the reversible circuit depicted in Fig. 5(a). Its NCV-cost is 39 while its T-depth is 21. When this circuit is optimized using the conventional moving rules, then the gate in position 1 could be moved to position 2. Hence the gates in position 2 and 3 could be merged into one gate as shown in the circuit depicted in Fig. 5(b). No more reductions are possible. The obtained circuit has an NCV-cost of 34 and a T-depth of 18. But if we consider the extended moving rules, one can move from the circuit in Fig. 5(b) also the gate in position 3 to the position 7 since its control line is invariant and its target line does not contain any controls between position 3 and 7. Thus the moved gate would be removed with its neighbour because they form an identity circuit. The obtained circuit is shown in Fig. 5(c) and has NCV-cost of 32 and a T-depth of 18. Now if we use the rewriting rules, one can move the gate in position 3 from the circuit in Fig. 5(c) to position 4 by applying the rule shown in Fig. 4(c). As a result the moved gate form an identity with gate in position 5 and get both removed. The resulting circuit is presented in Fig. 5(d). It has an NCV-cost of 27 and T-depth of 15.

In the following, we will consider the rewriting rules for optimizing reversible circuits and show through the experimental results its efficiency in reducing the quantum cost of reversible circuits.

V. ALGORITHMS

To show the advantages of applying simulated annealing to reduce the quantum cost for reversible circuit, we compare it with the well known approach for optimizing reversible circuits based on exhaustive search. To do so, we introduce in this section as a first step a greedy approach combined with the rewriting rules. Then, as a second step, we define the simulated annealing approach. Both approaches are applied to gate cascade with a common target. Note that gates can be rearranged to create such a cascade using rewriting rules. Each MPMCT optimization procedure finds possible reductions in the circuit by moving gates across the circuit and making them

adjacent. The gates may either be cancelled when they are identical or may be reduced to a less expensive cascade using the five different rules sketched in Fig. 6.

A. Greedy Approach

Given is a reversible circuit $G = T_1(C_1, t_1) \dots T_k(C_k, t_k)$ with k gates over variables $x_1, \dots, x_n$. This algorithm optimizes the circuit by applying a greedy approach.

For each gate, this technique searches over the circuit for a gate that has the same target. A found gate can be merged with the requested gate only when the rewriting process reduces the quantum cost of the targeted reversible circuit. After a reduction is applied, the optimization restarts the same process from the first gate of the circuit.

B. Simulated Annealing Approach

Given is a reversible circuit $G = T_1(c_1, t_1) \dots T_k(c_k, t_k)$ with size k over variables $x_1, \dots, x_n$. This algorithm optimizes the circuit by applying simulated annealing. For the computation, we are making use of the variables k, T, frozen, l, and Δ_{cost} to denote the size of the circuit, the used temperature, the stopping criterion, the number of generated perturbation, and the cost variable, respectively. The remaining variables (i, j, and optimized) are used to control the algorithm loops.

The algorithm is listed in Algorithm 1. We have chosen the initial temperature and the number of perturbation as factors of the number of gates in the initial circuit while the stopping criterion is set to 0 regardless of the size of the circuit and should not exceed the value 5 (see lines 2, 3, and 4). The algorithm generates, for a predetermined number of times l (line 9), two different positions of gates (line 10 and 11) denoted by loc and pos. Then, it calculates the rewriting cost for rearranging them together (line 12). If the cost is decreased then the solution is accepted, i.e, the gates are merged together (line 14). Otherwise the solution is accepted with a certain probability (line 18). After each loop the temperature T is

978-1-4673-9141-2/15 $31.00 © 2015 IEEE

Algorithm 1: Simulated annealing

Input: Reversible circuit G
Output: Optimized circuit G'

1   $k \leftarrow \texttt{Size}(G)$
2   $T \leftarrow 10k$
3   frozen $\leftarrow 0$
4   $l \leftarrow 100k$
5   **while** frozen > 5 **do**
6      $j \leftarrow i + 1$
7   optimized $\leftarrow$ False
8      $i \leftarrow 0$
9   **for** $i = 0$ *to* l **do**
10   pos $\leftarrow \texttt{Random}(0,k)$
11   loc $\leftarrow \texttt{Random}(0,k)$
12        $\Delta_{\text{cost}} \leftarrow \texttt{RewriteCost}(G, \text{pos,loc})$
13   **if** $\Delta_{\text{cost}} < 0$ **then**
14          $\texttt{RewriteMerge}(G, \text{pos}, \text{loc})$
15   optimized $\leftarrow$ True
16   **else**
17          $q \leftarrow \texttt{Random}(0,1)$
18   **if** $q < e^{-\frac{\Delta_{\text{cost}}}{T}}$ **then**
19            $\texttt{RewriteMerge}(G, \text{pos}, \text{loc})$
20   optimized $\leftarrow$ True
21   **end**
22   **end**
23   **end**
24      $T \leftarrow 0.8T$
25   **if** optimized **then**
26   frozen $\leftarrow 0$
27   **else**
28   frozen $\leftarrow$ frozen $+ 1$
29   **end**
30 **end**

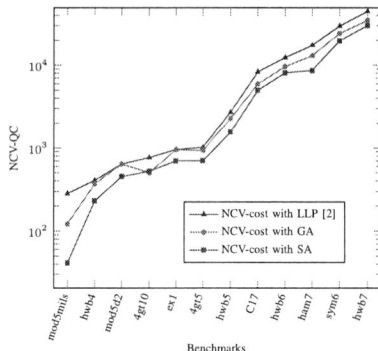

Fig. 7: Evaluation of Optimization approaches wrt NCV-cost

decreased (line 24) and the stopping criterion frozen is reset to 0 when the circuit has changed (line 26) otherwise the variable is incremented (line 28). This process is repeated until the stopping criterion reaches the value 5.

VI. EXPERIMENTAL RESULTS

In this work, we proposed rule based approaches for optimizing MPMCT circuits using rewriting rules. The first is using a greedy search algorithm while the second is using a simulated annealing algorithm. The proposed ideas described above in Sect. V have been implemented in the open source toolkit *RevKit* [13]. The experimental evaluation has been carried out on an Intel Core i5 Processor with 4 GB of main memory using the benchmarks taken from [14], [15]. We have observed that our approaches lead to reversible circuits with smaller NCV-cost and T-depth.

The experimental results presented graphically in the plot depicted in Fig. 7 show the NCV-cost of optimized reversible circuits with respect to different optimization algorithms. The values of x-axis and the y-axis denote the name of

Benchmarks and the NCV-cost, respectively. The plot contains three different scenarios: the NCV-cost of optimized reversible circuits based on LLP approach [2] (in blue), the NCV-cost of circuits optimized with greedy approach (in red), and the NCV-cost of optimized circuits based on simulated annealing (in green). One can clearly see that the greedy approach outperforms the LLP approach [2] based on extended moving rules. While the simulated annealing approach produces circuits with the smallest NCV-cost in comparison with the other approaches. In the rest of the paper we consider only the results of the greedy and simulated annealing approaches.

Our results capture the following values: (1) the results of the greedy optimization approach (GA), (2) the results of optimized reversible circuits using simulated annealing approach (SA), (3) improved quality with respect to NCV-costs and T-depths of resulting circuits from the greedy approached compared to the original benchmarks (GA/OB), (4) the quality of optimized circuits using simulated annealing compared to the original circuits (SA/OB), and (5) the improvement of the simulated annealing approach over the greedy approach (SA/GA).

In Table I, the results are summarized as follow: for each benchmark we show the name (ID), the number of lines (L), the NCV-cost (NCV), and the T-depth (TD). In addition to these metrics, for each approach, we add the required runtime in seconds (Time). The NCV-cost and the T-depth reduction and improvement are provided in the columns denoted by ΔNCV, I_{NCV}, ΔTD, and I_{TD}, respectively.

The results from the simulated annealing approach are given in third, fifth, and sixth columns of Table I. Our proposed second approach leads to significant *T-depth* and NCV-cost reductions. Over all circuits, reductions up to 184589 NCV gates can be obtained. Also, it enables further improvements of the overall *T-depth*. The *T-depth* is reduced by 30% on average and in the best case (*mod5mils*) by 86%.

As can be clearly seen, the effect of incorporating the simulated annealing algorithm for reversible circuit optimization is significant. By rearranging the gates in a random way, further reductions are applied which confirms the idea outlined in Sect. III. Therefore this approach outperforms the greedy approach for most of the functions. Consider as an

TABLE I: Experimental Results For Greedy and Simulated Annealing Approaches

Original Benchmark (OB)				Greedy Ap. (GA)			Simulated An. (SA)			GA/OB				SA/OB				SA/GA			
ID	L	NCV	TD	NCV	TD	Time	NCV	TD	Time	Δ_{NCV}	I_{NCV}	Δ_{TD}	I_{TD}	Δ_{NCV}	I_{NCV}	Δ_{TD}	I_{TD}	Δ_{NCV}	I_{NCV}	Δ_{TD}	I_{TD}
Aj-e11	4	141	84	131	78	0.0	91	54	0.9	10	7%	6	7%	50	35%	30	36%	40	31%	24	31%
mod5mils	5	281	168	121	72	0.0	41	24	0.2	160	57%	96	57%	240	85%	144	86%	80	66%	48	67%
hwb4	4	403	240	368	219	0.0	229	135	0.8	35	9%	21	9%	174	43%	105	44%	139	38%	84	38%
mod5d2	5	641	384	636	381	0.0	451	270	1.3	5	1%	3	1%	190	30%	114	30%	185	29%	111	29%
4gt10	5	760	456	500	300	0.0	520	312	0.8	260	34%	156	34%	240	32%	144	32%	-20	-4%	-12	-4%
ex1	5	955	573	945	567	0.0	695	417	3.0	10	1%	6	1%	260	27%	156	27%	250	26%	150	26%
4gt5	5	1010	606	930	558	0.0	700	420	1.1	80	8%	48	8%	310	31%	186	31%	230	25%	138	25%
hwb5	5	2700	1617	2265	1356	0.1	1544	924	4.2	435	16%	261	16%	1156	43%	693	43%	721	32%	432	32%
C17	6	8332	4998	5937	3561	0.3	4977	2985	12.7	2395	29%	1437	29%	3355	40%	2013	40%	960	16%	576	16%
hwb6	6	12340	7404	9620	5772	0.7	8090	4854	23.9	2720	22%	1632	22%	4250	34%	2550	34%	1530	16%	918	16%
ham7	7	17340	10404	13020	7812	2.0	8580	5148	30.9	4320	25%	2592	25%	8760	51%	5256	51%	4440	34%	2664	34%
sym6	7	29492	17694	24022	14412	4.1	19567	11739	79.2	5470	19%	3282	19%	9925	34%	5955	34%	4455	19%	2673	19%
hwb7	7	44435	26661	34645	20787	13.3	29995	17997	154.3	9790	22%	5874	22%	14440	32%	8664	32%	4650	13%	2790	13%
con1	8	99306	59580	80356	48210	45.4	68956	41370	508.4	18950	19%	11370	19%	30350	31%	18210	31%	11400	14%	6840	14%
z4	8	105380	63225	86990	52191	86.3	74260	44553	555.7	18390	17%	11034	17%	31120	30%	18672	30%	12730	15%	7638	15%
hwb8	8	150315	90189	118795	71277	289.3	103445	62067	1121.3	31520	21%	18912	21%	46870	31%	28122	31%	15350	13%	9210	13%
sqrt8	9	282759	169650	232954	139767	927.2	205669	123396	351.7	49805	18%	29883	18%	77090	27%	46254	27%	27285	12%	16371	12%
radd	9	292495	175494	241895	145134	1166.7	211416	126846	383.0	50600	17%	30360	17%	81079	28%	48648	28%	30479	13%	18288	13%
plus63	12	317821	190692	313871	188322	10183.9	311026	186615	342.4	3950	1%	2370	1%	6795	2%	4077	2%	2845	1%	1707	1%
urf1	9	381172	228702	321182	192708	2386.8	285702	171420	632.6	59990	16%	35994	16%	95470	25%	57282	25%	35480	11%	21288	11%
hwb9	9	449579	269745	377349	226407	4451.1	316135	189678	889.5	72230	16%	43338	16%	133444	30%	80067	30%	61214	16%	36729	16%
x2	15	511366	306816	499346	299604	3007.6	474546	284724	297.3	12020	2%	7212	2%	36820	7%	22092	7%	24800	5%	14880	5%
5xp1	10	517028	310212	433568	260136	6341.1	378968	227376	1544.3	83460	16%	50076	16%	138060	27%	82836	27%	54600	13%	32760	13%
root	10	536719	322029	447239	268341	4182.2	394789	236871	1498.0	89480	17%	53688	17%	141930	26%	85158	26%	52450	12%	31470	12%
max46	10	652290	391374	539680	323808	7096.7	484300	290580	2219.2	112610	17%	67566	17%	167990	26%	100794	26%	55380	10%	33228	10%
dist	10	677078	406242	560483	336285	10876.7	492489	295488	2360.5	116595	17%	69957	17%	184589	27%	110754	27%	67994	12%	40797	12%
9symml	10	902254	541347	759794	455871	3459.4	665089	399048	4862.9	142460	16%	85476	16%	237165	26%	142299	26%	94705	12%	56823	12%
sym9	10	912693	547611	781628	468972	3491.1	673958	404370	4548.1	131065	14%	78639	14%	238735	26%	143241	26%	107670	14%	64602	14%
urf3	10	962832	577698	815702	489420	4560.5	724112	434466	4708.2	147130	15%	88278	15%	238720	25%	143232	25%	91590	11%	54954	11%
sqr6	12	1157296	694374	979286	587568	4185.7	898006	538800	5980.3	178010	15%	106806	15%	259290	22%	155574	22%	81280	8%	48768	8%
rd84	11	1747516	1048503	1493116	895863	19181.5	1322236	793335	16395.8	254400	15%	152640	15%	425280	24%	255168	24%	170880	11%	102528	11%
Average										40274	17%	24165	17%	65872	30%	39523	30%	25598	16%	15359	16%

example the function *hwb4*, its realization is reduced by 8% when the greedy approach is applied. Then additional 38% of improvement is achieved by applying simulated annealing. In general, the latter approach leads to additional quantum cost reductions of 16% in average compared to realizations optimized via the greedy approach.

VII. CONCLUSION

In this paper we introduced optimization approaches for reversible circuits based on rewriting rules. We presented two different strategies; a greedy approach and a simulated annealing approach. On our set of functions we showed that significant reductions (with respect to the NCV-cost and T-depth) can be achieved, specially when simulated annealing is considered.

REFERENCES

[1] M. Arabzadeh, M. Saeedi, and M. S. Zamani, "Rule-based optimization of reversible circuits," in *Asia and South Pacific Design Automation Conference*, 2010, pp. 849–854.

[2] M. Soeken, Z. Sasanian, R. Wille, D. M. Miller, and R. Drechsler, "Optimizing the mapping of reversible circuits to four-valued quantum gate circuits," in *International Symposium on Multiple-Valued Logic*, 2012, pp. 173–178.

[3] Z. Sasanian and D. M. Miller, "Reversible and quantum circuit optimization: A functional approach," in *Reversible Computation*. Springer, 2013, pp. 112–124.

[4] D. Maslov, G. Dueck, and D. Miller, "Simplification of Toffoli networks via templates," in *Symposium on Integrated Circuits and Systems Design*, 2003, pp. 53–58.

[5] D. M. Miller, D. Maslov, and G. W. Dueck, "A transformation based algorithm for reversible logic synthesis," in *Design Automation Conference*, 2003, pp. 318–323.

[6] M. Soeken and M. K. Thomsen, "White dots do matter: rewriting reversible logic circuits," in *Reversible Computation*. Springer, 2013, pp. 196–208.

[7] S. Kirkpatrick, C. D. Gelatt, and M. P. Vecchi, "Optimization by simulated annealing," *SCIENCE*, vol. 220, no. 4598, pp. 671–680, 1983.

[8] T. Toffoli, "Reversible computing." Springer, 1980.

[9] A. Barenco, C. H. Bennett, R. Cleve, D. DiVinchenzo, N. Margolus, P. Shor, T. Sleator, J. Smolin, and H. Weinfurter, "Elementary gates for quantum computation," *The American Physical Society*, vol. 52, pp. 3457–3467, 1995.

[10] M. Sarkar, P. Ghosal, and S. P. Mohanty, "Reversible circuit synthesis using aco and sa based quine-mccluskey method," in *International Midwest Symposium on Circuits and Systems*, 2013, pp. 416–419.

[11] K. Datta, A. Gokhale, I. Sengupta, and H. Rahaman, "An esop-based reversible circuit synthesis flow using simulated annealing," in *Applied Computation and Security Systems*. Springer, 2015, pp. 131–144.

[12] D. M. Miller and Z. Sasanian, "Lowering the quantum gate cost of reversible circuits," in *International Midwest Symposium on Circuits and Systems*, 2010, pp. 260–263.

[13] M. Soeken, S. Frehse, R. Wille, and R. Drechsler, "Revkit: A toolkit for reversible circuit design." *Journal of Multiple-Valued Logic & Soft Computing*, vol. 18, no. 1, 2012, RevKit is available at http://www.revkit.org.

[14] R. Wille, D. Große, L. Teuber, G. W. Dueck, and R. Drechsler, "RevLib: an online resource for reversible functions and reversible circuits," in *International Symposium on Multiple-Valued Logic*, 2008, pp. 220–225, RevLib is available at http://www.revlib.org.

[15] D. Maslov. Reversible logic synthesis benchmarks page. Available at http://webhome.cs.uvic.ca dmaslov/, last accessed January 2011.

[16] M. Soeken, R. Wille, C. Hilken, N. Przigoda, and R. Drechsler, "Synthesis of reversible circuits with minimal lines for large functions," in *Asia and South Pacific Design Automation Conference*, 2012, pp. 85–92.

A Hybrid Embedded Compression Codec Engine for Ultra HD Video Application

Seongmo Park, Kyungjin Byun, and Nak-woong Eum

SoC Research Department, ETRI
Multimedia Processor Research Section
Yuseong-gu, Daejeon, KOREA
{smpark, kjbyun, nweum}@etri.re.kr

Abstract— **We proposed an efficient VLSI hardware architecture of the High Efficiency Video Coding (HEVC) using a hybrid embedded compression algorithm for reducing the frame memory bandwidth. This architecture was designed to reduce the memory bandwidth using an adaptive prediction lossy/lossless algorithm. We saved about 50% of the memory access cycles for the reference data compared to a previous algorithm. The PSNR degradation of 0.12 dB on average was proposed algorithm at the compression ratio of 50%. The architecture was implemented in Verilog HDL and synthesized using a Synopsys Design Compiler with a 65nm cell library; the gate count was about 25,000 gates.**

Keywords—lossy compression, lossless compression, embedded compression, video coding

I. Introduction

The HEVC, H.264/AVC, and MPEG-2 of video codec is used for huge data from external DRAM memory to internal SRAM memory. High resolution video system is the core technique for UHDTV (Ultra High Definition TV: 4K (3840x2160 resolutions), 8K (7680x4320 resolutions) services, in which a large amount of data is supported. To solve the memory bandwidth problem, lossy and lossless algorithms in high level design are used [1-2]. Embedded compression and Ultra High-Definition (UHD) videos transmitted over a wireless network to a remote display device for the required communication bandwidth can have tremendous sizes when raw pixel data are used [2]. Many studies have been done on reusing the overcalled reference frame data on various levels and improving the DRAM access efficiency by optimized memory controller architectures [1], etc. External Memory access to DRAM dominates the overall performance and power consumption in a modern complex video codec chip, which becomes worse with increasing demands of higher resolution videos and an increasing number of complex video compression algorithms [3]. For lossless compression, simple adaptive differential data and modified Golomb-Rice (GR) Coding for compression with decoding error resilient scheme for uncompressible case, can achieve HD size video compression with much lower hardware cost [3]. An algorithm

and hardware architecture of high performance, lossy embedded compression is designed as a high throughput, lossy embedded compression with relatively low hardware area costs. The random access unit is a 16-by-16 macro-block (MB), and it is transformed, quantized, and then encoded by parallel variable length coding (VLC) with area-optimized Huffman table [4]. Lossless Embedded Compression (EC) algorithm for HD video sequences and related hardware architecture is proposed. The first steps is, a hierarchical prediction method based on pixel averaging and copying, and the second step involves significant bit truncation (SBT) which encodes prediction errors in a group with the same number of bits so that the multiple prediction errors are decoded in a clock cycle [5]. The memory bandwidth power consumption is reduced by the size of the reference frames using hybrid frame buffer compression that is low resolution and high resolution components [6]. The memory-efficient EC algorithm with two level rate control scheme is designed and belongs to spatial based EC [7]. To precisely control the reduction ratio of memory bandwidth, the proposed EC algorithm not only guarantees the target compression ratio (TCR) but also maintains the visual quality [7]. Video Codec system has a large display resolution and frame rate which is a significant overhead on memory bandwidth. To reduce the memory bandwidth, the embedded compression (EC) compresses the display data and handles frame memory. The EC algorithm is designed for an effective coding flow, parallel encoding/decoding processing, and a lossless/lossy coding method with to solve huge throughput problems.

To solve the problems described above, various embedded compression algorithm and architecture have been proposed such as lossless and lossy compression. The lossless compression produces high image quality, but it is hard to design the algorithm in hardware. In contrast, the compression ratio for lossy/lossless compression algorithm is low, but it is easy to design and implement in hardware. The effective coding scheme on the EC consists of processing speed and the parallel encoding/decoding with a real-time architecture. Besides, the EC should offer the lossless/near-lossless coding to maintain higher visual-quality and rate control to save memory bandwidth and capacity [8]. A lossy EC algorithm uses hybrid coding scheme of feed forward DPCM and modified four-level BTC, which aims at the target specification of the 4 x 2 block-wise random access at 50 % compression ratio [9]. The lossless embedded compression engine used compacted fast, efficient, lossless image

978-1-4673-9141-2/15 $31.00 © 2015 IEEE

compression system (FELICS) algorithm, which primarily consists of adjusted binary code and Golomb-Rice code with real-time VLSI architecture. The lossless embedded compression engine can achieve Full-HD 1080p@60Hz [10]. Examples of the lossless frame memory recompression scheme is a lossless pixel compression algorithm, an efficient address table organization method for random accessibility, and a frame memory placement scheme for compressed data to reduce the effective access time of SDRAM by suppressing row switching in compressing pixel data before storing in off-chip frame memory [11]. The proposed algorithm decomposes a frame into 4x4 blocks which are then compressed into 64-bit segments. The recompression algorithm is implemented in hardware and integrated with an H.264 encoder using select the scan order of the 4x4 blocks and DPCM (Differential Pulse Code Modulation) algorithm [12]. The implementation consists of an efficient pipelined JPEG-LS encoder, which operates at a significantly higher encoding rate than any other JPEG-LS hardware or software implementation while keeping area small with high-performance JPEG-LS encoder [13]. Lossless mode, and lossy modes with rate control modes and quality control modes are all supported by single algorithm with an algorithm and hardware architecture of a new type EC codec engine with multiple modes [14]. A lossless EC algorithm for HD video sequences and related hardware architecture is a hierarchical prediction method based on pixel averaging and copying and the second step involves significant bit truncation (SBT) which encodes the theoretical lower bound of the compression ratio of the SBT coding was also derived [15].

This paper proposes a low-power lossy embedded compression scheme as shown in Fig. 1. The proposed method is as follows: 1) a simple lossy compression method with adaptive prediction algorithm, 2) an efficient variable length coding method, and 3) reduction of the memory access throughput and power consumption. The proposed architecture is scalable with compression unit size, and it is adopted a 8x8 access unit, 128-bit bus width, and Ultra HD application with a video codec system. We propose a simple lossy compression algorithm, adaptive variable length coding, and high throughput hardware architecture for compression and decompression to solve memory bandwidth.

The implementation shows it can achieve Ultra HD size video compression easily with much lower hardware costs. Section II describes the proposed algorithm. Section III shows the corresponding hardware architecture. Section IV shows the simulation and implementation results. Conclusion is made in Section V.

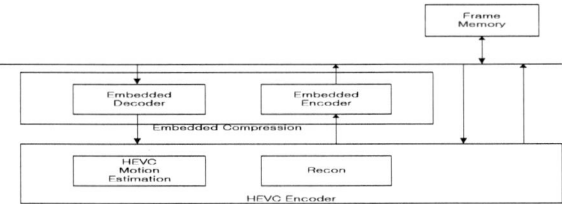

Fig.1. Architecture block diagram for embedded compression

II. PROPOSED ALGORITHMS

Fig. 2 shows the CU, PU, and TU depth information of HEVC encoder using Dual HEVC Bit-stream Analyzer. It is used sequence of BQ_Terrace (full-HD, 32x32 CTU). In general image pixel value is correlated with neighborhood pixel according to coding depth information. The same depth information is correlated to similar pixel information, and thus it can reduce the compression ratio. Fig. 3 shows the 3-step adaptive algorithm. CUx is defined as a current coding unit, CUda is a left coding depth, CUdb is an upper coding unit depth, and CUdc is an upper right coding unit depth. First, if CUda, CUdb, and CUdc are all zero, and then step 1 prediction processing is activated. Second, if either CUda, CUdb, or CUdc are not zero, then step 2 prediction processing is activated. Final, if the first and second steps are not activated, the step 3 processing is activated. This algorithm adopts 3-step prediction using the depth of coding depth to improve compression ration with lossy algorithm. There is an access data unit (8x8, 8 bit pixel block)

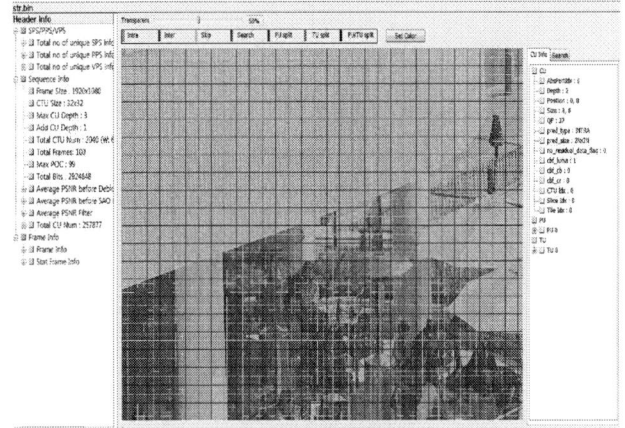

Fig.2. Bit stream Analysis for HEVC encoder

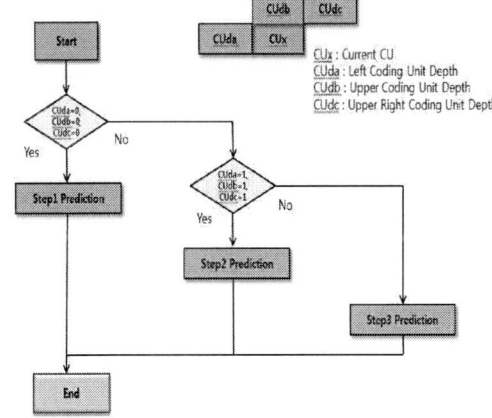

Fig.3. 3-Step Adaptive Lossy algorithm

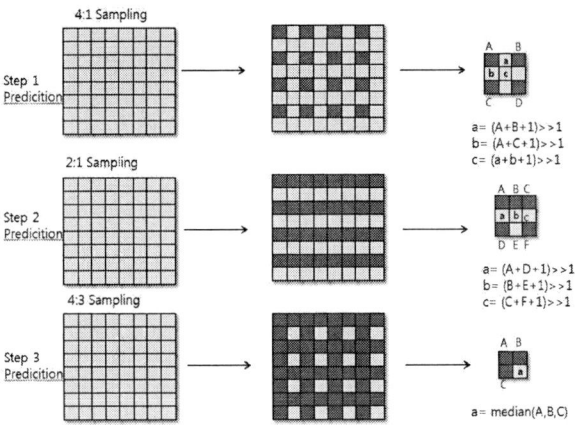

Fig.4. 3-Step prediction Compression Sampling

Fig. 4 shows 3-step prediction compression sampling as follows. The result of the step 1 prediction algorithm is 4:1 sampling at compression flow. The result of the step 2 prediction algorithm is 2:1 sampling at compression flow. The result of the step 3 prediction algorithm is 4:3 sampling at compression flow. At decoding processing, it is determined decoding values of interpolation as follows.

For Step 1 Prediction

$$a = (A+B+1) >> 1 \qquad (1)$$
$$b = (A+C+1) >> 1 \qquad (2)$$
$$c = (a+b+1) >> 1 \qquad (3)$$

For Step 2 Prediction

$$a = (A+D+1) >> 1 \qquad (4)$$
$$b = (B+E+1) >> 1 \qquad (5)$$
$$c = (C+F+1) >> 1 \qquad (6)$$

For Step 3 Prediction

$$a = \text{median} \{A, B, C\} \qquad (7)$$

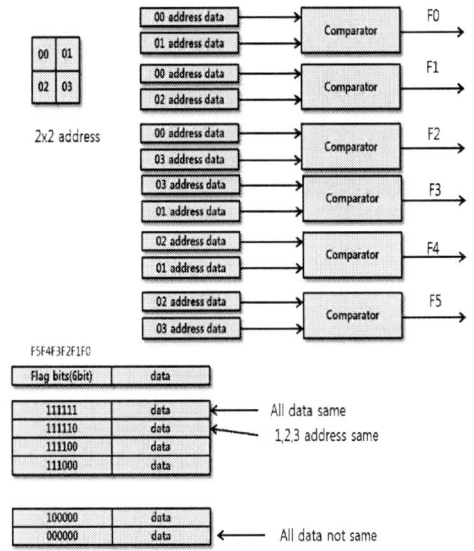

Fig. 5. Same Data Encoding

Fig. 5 shows the same data reducing scheme. After the

prediction of encoding compression ends, and the coding unit is divided into 2x2 unit. If four block data have the same value, that value is assigned to all blocks. If all data do not have the same value, zeros are assigned to all blocks. If the three data have the same value (0, 1, and 2), the last bit will have zero, and other bits will have one.

III. HARDWARE ARCHITECTURE DESIGN

Fig. 6 shows the hardware architecture for compression/decompression block in encoder. It consists of compression module and decompression module. The Compression module has internal memory, compression algorithm processing unit, and output of AXI bus interface unit. Then, the bit streams of final coding are packed based on the same data encoding unit. The decompresssion module has internal memory, interpolation of adaptive method, and output of internal reference data of the motion estimation module.

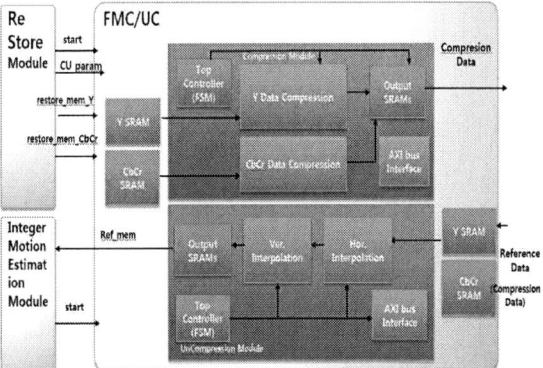

Fig. 6. Hardware architecture

IV. EXPERIMENTAL RESULTS

In order to evaluate the performance of the proposed adaptive prediction lossy/lossless compression scheme, and the methods are implemented on the recent HEVC reference software with HPNC (ETRI Reference Test Model). The proposed algorithm is evaluated with QPs 22, 27, 32 and 37 using test sequences recommended by JCT-VC (Joint Collaborative Team on Video Coding) in four resolutions (formats). The high-level design was implemented in the HPNC test model as an anchor. Experiments were conducted on the encoder configuration using low delay P. An increase in the encoding and decoding times compared to the HM was reported. 64-bit Cent OS Linux PCs with CPUs similar to Intel(R) Quad-Core i7 (3.3 GHz) were used in the of test platform of experiments. Fig. 7 shows PSNR vs. Bitrate calculation using Low Delay P. The proposed scheme has 0.01 to 0.15 dB degradation in video quality. Fig. 8 shows result of video for Kimono test sequence. The average BDBR increment and BD-PSNR drop are 2% (Y).

Table I shows analysis of PSNR for proposed algorithm compare to reference (HPNC) using test sequences at 300 fps. Table II shows the performance comparison of the proposed scheme and the reference design. We developed a design motion estimation module using Verilog Hardware Description

Language (HDL) and verification. The synthesis was designed using 0.65μm design library (TSMC).

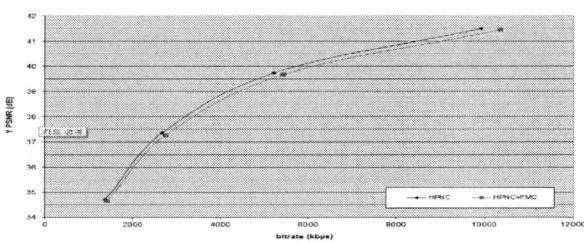

(a) Test sequence : Kimono

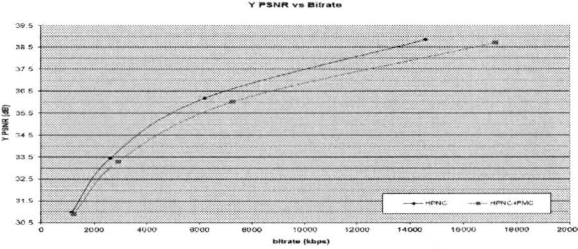

(b) Test sequence : ParkScene

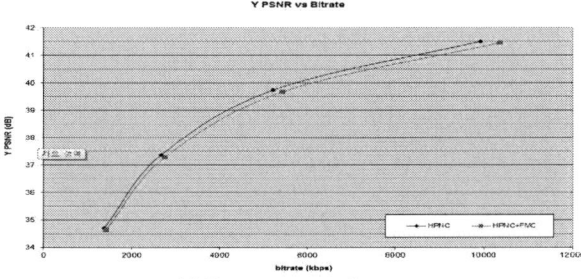

(c) Test sequence : Cactus

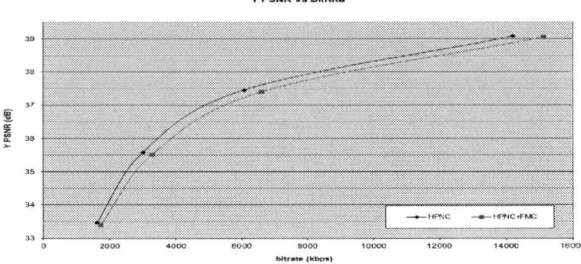

(d) Test sequence : BasketballDrive

(e) Test sequence : BQTerrace

Fig.7. Comparisons of coding efficiency

TABLE I.   ANALYSIS OF PSNR

Sequences	Parameters				
	Qp	HPNC (dB)	Proposed (dB)	Degradation (dB)	Comp.
Kimono	22	41.50	41.46	0.04	
	27	39.74	39.67	0.07	
	32	37.35	37.27	0.04	
	37	34.71	34.63	0.08	
ParkScene	22	38.84	38.71	0.13	
	27	36.18	36.02	0.17	
	32	33.46	33.32	0.15	
	37	30.99	30.90	0.09	
Cactus	22	37.98	37.93	0.01	300 fps 1080 p
	27	36.03	35.91	0.13	
	32	33.82	33.64	0.18	
	37	31.47	31.29	0.18	
BasketballDrive	22	39.08	39.05	0.02	
	27	37.46	37.41	0.05	
	32	35.59	35.51	0.07	
	37	33.47	33.39	0.08	
BQTerrace	22	37.53	37.50	0.03	
	27	34.86	34.69	0.18	
	32	32.55	32.37	0.28	
	37	30.29	29.97	0.32	
Sum		35.64	35.53	0.12	

Original   Frame Memory
                        Compression

Fig.8. Original vs Frame Memory Compression

TABLE II.   COMPARSION OF PreVious WORKS

Items	References				
	[12]	[13]	[14]	[15]	Proposed
Data reduction ration-YUV	50	71.4	49.7	60.4	60.4
Throughput	2.6	1	0.45	14.2	26.2
CMOS tech.(um)	0.18	0.18	0.18	0.18	0.65
Gate Count (K)	28.0	27.7	26.9	36.1	25.3

Items	References				
	[12]	*[13]*	*[14]*	*[15]*	*Proposed*
Max Freq.(MHz)	14	183	10	180	300
PSNR (dB)	-0.12	0.0	0.0	0.0	-0.12
Random Access Unit	4x4	N/A	16x16	16x8	8x8
Throughput (pixels/cycle)	2.6	1	0.45	14.2	20.1
Modeling Method	H.264 Intra	MED (Median edge detector)	2-L-2-D DWT	HCAP	Adaptive 3-Step
Coding Method	Golumb -Rice	JPEG -LS	SPIHT	SBT	SDR
Compression Type	Lossy	Lossl ess	Lossy /Lossless	Lossless	Lossy /Lossless

V. CONCLUSION

We proposed an efficient VLSI hardware architecture of High Efficiency Video Coding (HEVC) using a hybrid embedded compression algorithm to reduce the frame memory bandwidth. This architecture was designed to reduce the memory bandwidth using an adaptive prediction lossy/lossless algorithm. We saved about 50% of the memory access cycles for the reference data compared to previous algorithm. The PSNR degradation of 0.12 dB on average is proposed algorithm at the compression ratio of 50%. The architecture was implemented in Verilog HDL and synthesized using a Synopsys Design Compiler with a 65nm cell library, the gate count is about 25,000 gates.

ACKNOWLEDGMENT

This study was supported by IT R&D program of MOTIE/KEIT [10039214, Video Codec SoC for Ultra High Definition].

REFERENCES

[1] Li Guo, Dajiang Zhou, and Satoshi Goto, "Lossless Embedded Compression Using Multi-mode DPCM & Averaging Prediction," Signal Processing Conference (EUSIPCO), Proceeding of 21st European, pp. 1-5, 2013.

[2] Yin-Tsung Hwang, Ming-Wei Lyu, and Cheng-Chen Lin, "A Low-Complexity Embedded Compression Codec Design With Rate Control for High-Definition Video," IEEE Trans. Circuits Syst. Video Technol., vol. 25, no. 4, pp. 674-687, Apr. 2015.

[3] Liang-Chi Chiu and Tian-Sheuan Chang, "A Lossless Embedded Compression Codec Engine for HD Video Decoding," VLSI Design, Automation, and Test (VLSI-DAT), 2012 International Symposium on , pp. 1-4. 2012.

[4] Wei-Yin Chen, Li-Fu Ding, Pei-Kuei Tsung, and Liang-Gee Chen, "Architecture Design of High Performance Embedded Compression for High Definition Video Coding" Multimedia and Expo, IEEE International Conference on ICME, vol.1, pp. 825-828, 2008.

[5] Jaemoon Kim, and Chong-Min Kyung, R. Nicole, "A Lossless Embedded Compression Using Significant Bit Truncation for HD Video Coding," IEEE Trans. Circuits Syst. Video Technol., vol. 20, no. 6, pp. 848-860, Jun. 2010.

[6] Zhan Ma and Andrew Seall, "Frame Buffer Compression for Low-Power Video Coding," IEEE International Conference on Image Processing, pp. 757-760, 2011.

[7] Yu-Hsuan Lee, Yi-Cheng Chen, and Tsung-Han Tsai, "A Bandwidth-Efficient Embedded Compression Algorithm Using Two-Level Rate Control Scheme for Video Coding System," Proceedings of IEEE International Symposium on Circuits and Systems (ISCAS), pp. 1149-1152, 2010

[8] Tsung-Han Tsai and Yu-Hsuan Lee, "A 6.4 Gbit/s Embedded Compression Codec for Memory-Efficient Applications on Advanced-HD Specification," IEEE Trans. Circuits Syst. Video Technol., vol. 20, no. 10, pp. 1277-1291, Oct. 2010.

[9] Kiwon Yool, Changsu Hanl, Useok Kangl, and Kwanghoon Sohn, "Embedded Compression Algorithm using Error-Aware Quantization and Hybird DPCM/BTC coding," IEEE International Conference on ICME, vol.1, pp. 1-6, 2011.

[10] Yu-Yu Lee, Yu-Hsuan Lee, and Tsung-Han Tsai, "An Efficient Lossless Embedded Compression Engine Using Compacted-FELICS Algorithm," SOC Conference, IEEE International Conference on SOCC, vol.1, pp. 233-236, 2008.

[11] Sang-Heon Lee, Moo-Kyoung Chung, Sung-Mo Park, and Chong-Min Kyung, "Lossless Frame Memory Recompression for Video Codec Preserving Random Accessibility of Coding Unit", IEEE Trans. On Consumer Electronics, vol.55, no. 4, pp. 2105-2113, Jun. 2009.

[12] Y.Lee, "A New Frame Recompression Algorithm Integrated with H.264 Video Compression", in Proceeding IEEE Int. Symp. Circuits Syst. (ISCAS), pp.1621-1624, May 2007.

[13] Papadonikolakis, "Efficient High-Performance ASIC Implementation of JPEG-LS Encoder", in Proceeding IEEE Conf. on Design, Automation & Test in Europe Conference & Exhibition(DATE), , pp. 1-6, 2007.

[14] Chih-Chi Cheng, "Multimode Embedded Compression Codec Engine for Power-Aware Video Coding System", IEEE Transaction on Circuits and systems for video Technology, vol.19, no. 2, Feb. 2009.

[15] Kim, " A lossless Embedded Compression using Significant Bit Truncation for HD Video Coding", IEEE Transaction on Circuits and systems for video Technology, vol.20, no. 6, Feb. 2010.

A New Sizing Approach for Lifetime Improvement of Nanoscale Digital Circuits due to BTI Aging

Andres Gomez, Victor Champac

Dept. of Electronic Engineering, National Institute for Astrophysics,
Optics and Electronics - INAOE, Mexico.

Abstract—**Bias Temperature Instability (BTI) has become a major aging issue for circuit lifetime reliability in deeply scaled CMOS technologies. Due to BTI, circuit delay increases as time progress, which may lead to a timing constraint violation before the end of the expected lifetime. This paper proposes a new sizing approach to mitigate BTI induced delay degradation of digital circuits. The approach is based on the observation that the delay sensitivity to transistor sizing of a digital gate is composed by a nominal delay sensitivity and a delay degradation sensitivity components. By exploiting the differences between these two components, one can size some gates in the critical paths of a circuit, in such way that the delay degradation due to BTI is reduced while the nominal delay remains nearly unchanged. By using our sizing approach, the reduction of delay degradation allows to further extend the lifetime of a circuit with negligible area and power overhead.**

I. INTRODUCTION

Delay degradation due to Bias Temperature Instability (BTI) has been shown to have a significant impact in lifetime reliability of digital circuits [1]. Design of reliable circuits that need to operate correctly for a long period of time (i.e 10 years or even more) under harsh conditions is a challenge in deeply scaled CMOS technologies [2]. The conventional approach to cope with aging effects (i.e. BTI) is to add timing guardbands to delay specifications. However, greater guardbands are required as technology scales down because aging effects are aggravated due to higher electric fields and elevated temperatures [3]. Hence, new design strategies to overcome aging degradation, and consequently, to improve circuits lifetime reliability with low performance penalization are required.

Gate sizing is a widely used approach to cope with aging effects. In order to improve circuit lifetime under BTI aging, it is possible to decrease either the nominal delay (to leave a larger time margin for aging) or the delay degradation of the circuit (to reduce the aging rate of the circuit). The former can be seen as a compensation method while the latter is an actual BTI mitigation method [4]. Aging compensation by gate sizing has been explored in the past. In [5] the authors computed the expected percentage of delay degradation due to BTI of a circuit and formulate a gate sizing optimization problem to reduce the nominal delay in the required amount to tolerate aging. In [6] a finer-grained transistor-level sizing method was proposed, in which the pull-up and pull-down networks in the same gate could be sized to different ratios. The authors aim was to make the delay of the circuit to be under a specific

delay constraint. In [7] a gate selection metric was proposed to select the most favorable gates in a circuit to be resized to improve circuit lifetime reliability. These approaches ([5] [6] [7]) compensate for the increased delay due to BTI, but they do not focus on limiting the amount of delay degradation the circuit may suffer. Therefore, nominal delay specifications are changed but delay degradation due to BTI during operational lifetime could still be large. On the other hand, the aim of mitigation approaches is to reduce the expected amount of delay degradation due to BTI. In [9] and [10] a method for identifying the most influential gates with respect to BTI-induced aging is proposed. The selected gates are replaced by BTI-robust counterparts, i.e. gates from an aging-aware gate library [8]. In [11] an optimization problem is formulated to find the minimum area of the gates to ensure that delay degradation of all the gates in a circuit are below some threshold value. These approaches ([9], [10], [8], [11]) are attractive, but they could result in large area overhead. Moreover, they also could modify the nominal (initial) specifications of the circuit, which may result in over-design.

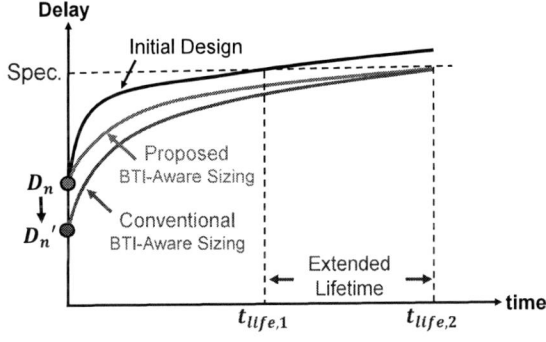

Figure 1. Delay behavior of a circuit using conventional and the proposed BTI-aware sizing approaches.

In this work, a new sizing approach for BTI aging mitigation is proposed. Figure 1 shows a comparison between the delay degradation due to BTI of a circuit as function of operational lifetime using a conventional BTI-aware sizing approach and using our proposed BTI-Aware sizing approach. As mentioned previously, conventional approaches can modify both nominal delay and delay degradation of a circuit. Although timing response and lifetime of the circuit is improved, the nominal delay specification is changed and delay degradation may still be large. The aim of the proposed technique in this work is

978-1-4673-9141-2/15 $31.00 © 2015 IEEE

to mitigate the expected amount of delay degradation due to BTI while the initial delay remains unchanged. In such way, the lifetime of the circuit can be extended for the same delay constraint. Moreover, monitoring effort for adaptive techniques [12] may become less complex, as time between sensing intervals can be larger, because the circuit degrades at a slower rate than its original design. The proposed approach take advantage of the fact that gate delay sensitivity with respect to transistor sizing is composed by a nominal delay sensitivity component and a delay degradation sensitivity component [7]. By exploiting the differences between these two components, one can size-up and size-down some gates in the critical paths of the circuit in such way that the delay degradation is reduced while its nominal delay remains unchanged. By using our sizing approach, the reduction of delay degradation allows to further improve the circuit lifetime reliability.

The rest of the paper is organized as follows: Section II describes the impact of BTI mechanism on circuit delay and its respective modeling. Section III presents the proposed sizing approach to mitigate delay degradation of digital circuits. Section IV presents the application results to some ISCAS benchmark circuits. Finally, the conclusions of this works are exposed in Section V.

II. DELAY DEGRADATION DUE TO BTI AGING

This section briefly reviewes the BTI mechanism modeling for aging-aware timing analysis of digital circuits.

A. Transistor Threshold Voltage Shift Due to BTI

BTI phenomena occur during normal transistor operation at specific stress conditions: positive gate bias for NMOS (PBTI) and negative gate bias for PMOS (NBTI) transistors. In previous technology nodes, PBTI effect was negligible in comparison to NBTI. However, since the introduction of high-k metal-gate technologies, PBTI has also become a reliability concern [13]. BTI is usually described as the generation of interface-traps in the S_i-O_2 interface due to reaction-diffusion (R-D) processes [14], which take place at stress conditions. The generation of interface traps increase devices threshold voltage (V_{th}). When stress is removed a partial recovery of V_{th} degradation is presented.

A long-term model to predict the threshold voltage shift due to BTI ($\Delta V th_{BTI}$) of a device after alternate stress and recovery phases was proposed in [15],

$$\Delta V th_{BTI} = A \cdot \alpha^n \cdot t^n \qquad (1)$$

where A is a fitted constant, which mainly depends on technology parameters, supply voltage, and temperature; α is the stress duty cycle of the device, which can be understood as the ratio of the time the device is under the stress condition to the total operational time; t is the operational time; and n is the time exponent related to reaction and diffusion process ($n = 1/6$ is used in this work [15]).

In this work, it is assumed, without loss of generality, that both NBTI and PBTI mechanisms induce the same

$\Delta V th_{BTI} = 80mv$ at worst case stress conditions ($\alpha = 1$). The constant A is fitted according to this assumption.

B. Duty Cycle Computation

The stress duty cycle that each device in a circuit experiences has a significant impact on the timing response of the circuit under BTI aging [15]. The stress duty cycle of each device depends on the workload pattern applied at circuit main inputs during lifetime operation. Duty cycles can be estimated at design phase by *Logic Simulation*. A large number of test vectors are applied to the circuit inputs and it is counted how many times each device is settled at stress condition. Stress duty cycle of each device can be approximated as the ratio of the number of times the device is found at stress condition to the total number of generated vectors. The input vectors used for logic simulation are randomly generated. However, a typical workload input pattern can be used if it is available.

C. Aging-Aware Delay Model

Once the threshold voltage shift of each device is known, the timing response under aging effects of each gate in the circuit can be estimated. Assuming that the amount of ΔV_{th} is small enough for each transistor in the gate, a first order Taylor approximation at the nominal transistor parameters values can be used to obtain gate delay as a linear function of $\Delta V th_{BTI}$ [16]. Then, the aged delay of a gate (D_{age}) can be expressed as:

$$D_{age} = D_n + \Delta D_{BTI}. \qquad (2)$$

where D_n is the nominal (initial) gate delay, and ΔD_{BTI} is the BTI induced delay degradation, which is function of the operational time.

Depending on the amount of $\Delta V th_{BTI}$ of each device in a gate and how much sensitive is the gate delay to V_{th} instability, the delay degradation over time (ΔD_{BTI}) of each gate in a circuit can be very different. As shown on Figure 2, the delay degradation due to BTI of a signal path can be obtained by adding the contributions of each of its gates.

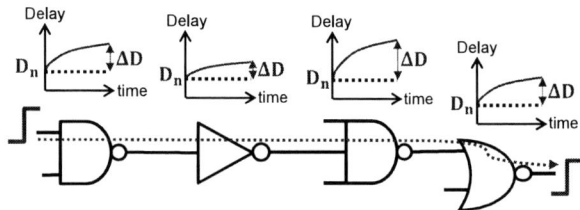

Figure 2. Gate contributions to the delay degradation due to BTI of a logic path.

III. PROPOSED SIZING APPROACH FOR BTI AGING MITIGATION

Due to aging effects, the timing response of the signal paths in a circuit increase over time. If the delay of a path becomes larger than the clock period (delay constraint), the signal propagated through the path would reach memory elements after the clock edge, and consequently, an incorrect logic value would be stored. Therefore, sizing techniques for the design of reliable circuits with extended lifetime are required.

As mentioned before, in conventional aging-aware sizing approaches, when one of nominal delay and delay degradation is reduced, the other may be affected [4]. Therefore, those approaches are not completely for aging compensation neither completely for aging mitigation. The aim of the proposed sizing approach is to mitigate BTI induced delay degradation while nominal delay remains unchanged (the approach is completely for aging mitigation). In such way, the expected lifetime of a circuit is extended for the same delay constraint (See Figure 1). Moreover, delay monitoring effort for adaptive techniques [12] may become less complex because time between sensing intervals can be larger, because the circuit degrades at a slower rate than the original design.

In the rest of this section, first the sizing effect on gate delay is analyzed. Then, the proposed approach for delay degradation mitigation at path level and circuit level are presented.

A. Sizing Effect On Gate Delay

When channel width of transistors that belong to a gate are resized, gate delay behavior over time is modified. Figure 3 shows the initial (nominal) delay and the gate delay degradation due to BTI after 10 years of aging, for the rising output transition of a $NOR2$ gate as function of their PMOS transistors channel width. As can be observed both nominal delay and delay degradation of the gate decrease as gate size is increased. Nominal delay reduces because transistors current capability to charge or discharge capacitive loads increases, while gate delay degradation reduces because gate delay becomes less sensitive to V_{th} instability. From Figure 3 we can conclude that overall aged gate delay sensitivity to channel width sizing is composed by a nominal delay sensitivity component ($S_w^{D_n}$) and a delay degradation sensitivity component ($S_w^{\Delta D}$) [7], as expressed in Equation 3.

$$S_w^{Dage} = S_w^{D_n} + S_w^{\Delta D} \qquad (3)$$

The two delay sensitivities can be obtained by the derivative of Equation 2 with respect to channel width. Gate delay degradation sensitivity depends on time and aging conditions of devices in the gate. Therefore, it is possible that sizing-up gates with similar nominal sensitivities results in different aged delay behaviors.

B. Proposed Sizing Approach

The key idea of the proposed sizing approach is to exploit the differences between the sensitivity components of the gates that belong to the critical paths of a circuit to reduce the amount of delay degradation due to BTI while nominal delay remains almost unchanged. Figure 4 illustrates the key idea of the proposed sizing approach in a single critical path. In order to reduce its delay degradation, it is possible to size-up the gate with the highest delay degradation sensitivity to gate sizing, because this would result in the highest delay

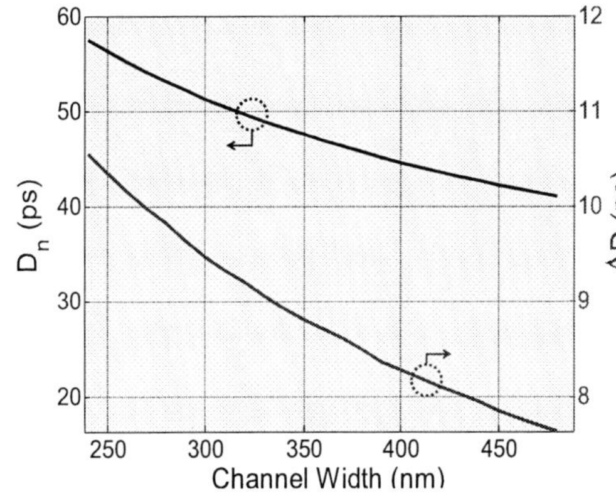

Figure 3. Nominal delay and Delay degradation for the rising transition of a $NOR2$ gate as function of PMOS channel width ($TSP = 1$, $time = 10$ years).

degradation reduction. As mentioned before, this also reduce the nominal delay of the path, which may be unnecessary to improve circuit lifetime. Therefore, some of the area overhead due to sizing-up a gate can be compensated by sizing-down one or more gates to adjust the nominal delay to its original value. Thus, some area is borrowed to the gates that are sized-up, from the gates that are sized-down, which allows to reduce overall area overhead. As shown on Figure 4, the gates that are sized-down should have a low impact on delay degradation (low delay degradation sensitivity to gate sizing) to keep some of the delay degradation reduction achieved by previous sizing-up. Therefore, the overall effect is that delay degradation due to BTI is mitigated, with negligible area overhead.

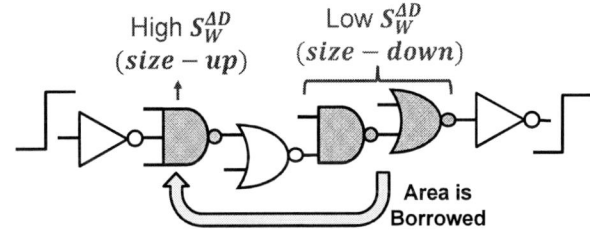

Figure 4. Signal path to illustrate the proposed sizing approach.

C. Methodology for Aging Mitigation on a Single Path

Figure 5 shows the proposed methodology to mitigate delay degradation of a single critical path based on the idea explained above. The input is the critical path information such as topological gates, original delay and delay degradation, etc. As explained before, the overall sizing approach is composed by two main steps. First, the sensitivity components of each gate in the path are computed. Then, the gate with the highest delay degradation sensitivity is identified and sized-up in a small step size. Second, sensitivities are updated because previous sizing-up modifies the timing response of the gates in the path. The gate with the lowest delay degradation sensitivity is identified and sized-down in a small step size. The sizing-down process is repeated until the nominal delay of the path match well with its original nominal delay. The sizing-up and sizing-down steps are iteratively repeated until a stop criteria is met. Here, a maximum number of iterations and a minimum delay change tolerance are used as stop criteria. The result of the methodology is a path with an improved design, which has an extended lifetime.

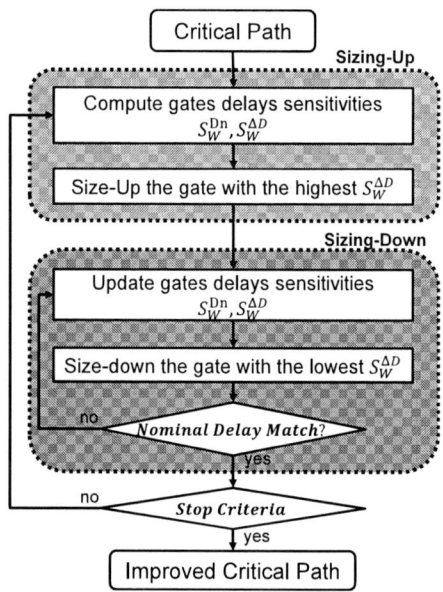

Figure 5. Methodology for Aging Mitigation by Gate Sizing in a Single Path.

Figure 6 shows an example of the two delay sensitivity components of the gates in the original design of the Longest Critical Path (LCP) of ISCAS C1908 circuit, which is composed by 19 gates. Figure 6(a) shows the delay degradation sensitiviy components and Figure 6(b) shows the nominal delay sensitivity components of each gate, respectively. As can be seen, Gate 11 is the one with the highest delay degradation sensitivity. Therefore, Gate 11 should be sized-up to mitigate delay degradation. However, Gate 11 also has a significant component of nominal delay sensitivity. Then, most of its impact on nominal delay, which may be unnecessary to improve circuit lifetime, can be compensated by sizing down other gates. For example, by sizing down Gate 12 four times approximately ($S_{W,11}^{Dn} \approx 4S_{W,12}^{Dn}$). Since Gate 12 has a low delay degradation sensitivity, sizing down this gate would not impact significantly in the delay degradation. Therefore, most of the delay degradation reduction achieved by sizing-up Gate 11 is kept ($S_{W,11}^{\Delta D} >> S_{W,12}^{\Delta D}$). The overall effect is that delay degradation is reduced while nominal delay remains almost unchanged. Note that not only a single gate could be sized-down, a sizing-down combination of gates numbers 3, 7, 9, 12 and 16, which have the lowest delay degradation sensitivities, could be performed too (See Figure 5).

Figure 7 shows both delay degradation due to BTI and nominal delay of the LCP of ISCAS C1908 circuit, after each iteration of the proposed methodology is accomplished. An iteration is composed by the sizing-up and its corresponding

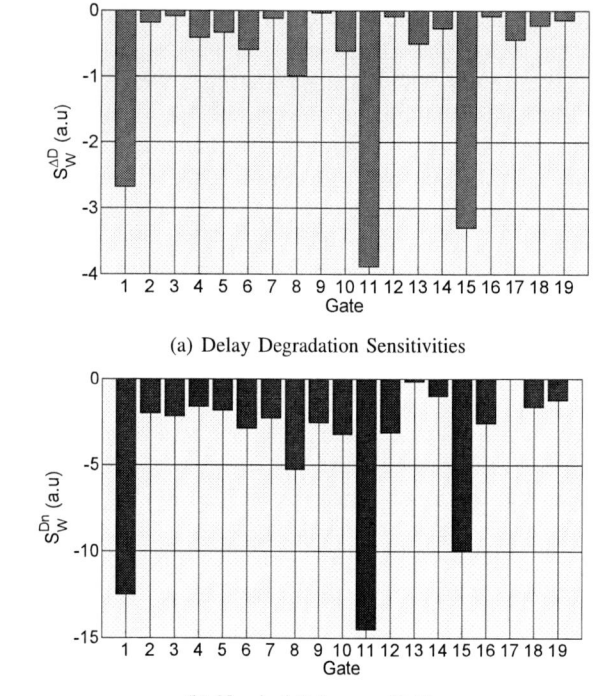

(a) Delay Degradation Sensitivities

(b) Nominal Delay sensitivities

Figure 6. Delay Sensitivities to channel width sizing of the gates in the LCP of circuit C1908.

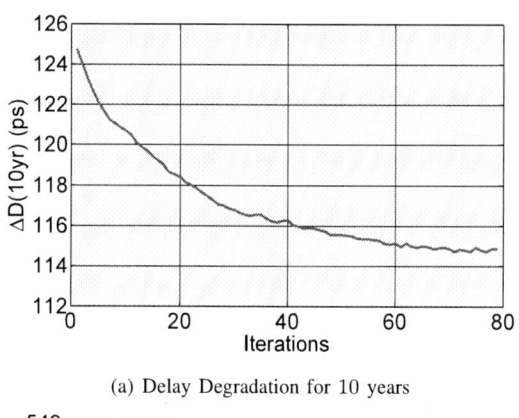

(a) Delay Degradation for 10 years

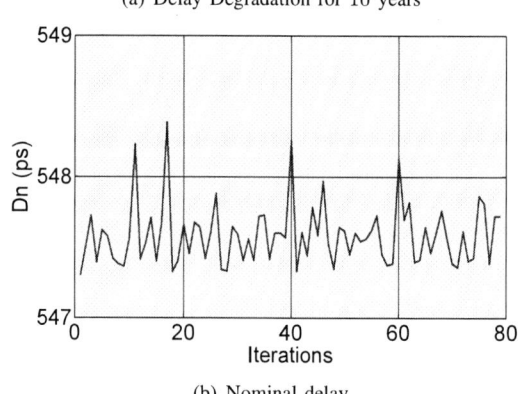

(b) Nominal delay

Figure 7. Delay degradation reduction and nominal delay (nearly unchanged) of the LCP of ISCAS circuit C1908 after each iteration.

sizing-down steps. As can be seen, delay degradation is reduced as each iteration is performed, while nominal delay fluctuates around its original value ($\approx 547.5ps$). The total delay degradation reduction is of $10ps$ ($\approx 8\%$). Although this reduction may seem small, it is enough for further lifetime improvement.

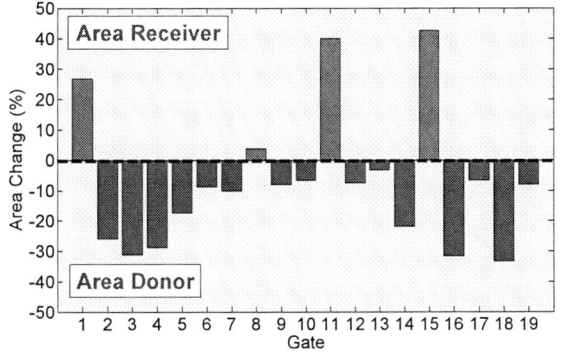

Figure 8. Area change in percentage of the gates in the LCP of circuit C1908 using the proposed sizing methodology of Figure 5.

The gate area change in percentage after sizing the LCP of circuit C1908 with the proposed methodology is shown on Figure 8. As can be seen, some gates are *Area Receivers* (the ones which were sized-up), those gates were the ones with the highest delay degradation sensitivity (See Figure 6(a)). Furthermore, the other gates in the LCP become *Area Donors*, the ones which are sized-down to borrow area to the Area Receivers. Those gates has a low delay degradation sensitivity (See Figure 6(a)). In this example, the number of gates that are sized-down is greater than the number of gates that are sized-up to mitigate aging, therefore, the proposed methodology provides some area savings while delay degradation is mitigated.

D. Aging Mitigation on Entire Circuits

Previous analysis has been focused on improving lifetime of a single critical path (i.e. the LCP of the circuit). However, to effectively mitigate delay degradation in an entire circuit, the sizing approach need to be applied to the set of paths that may define the amount of circuit delay degradation. This set is composed by those paths that can be either the paths with highest nominal delay or the paths with highest aged delay.

Figure 9 shows the possible behaviors that define which others signals paths should be improved after sizing the LCP of a circuit. Initially, the LCP is the path with the longest nominal delay, which defines the nominal delay of the entire circuit. The time guardband (GB) is settled to tolerate a given time of aging (i.e. $10yr$). The delay constraint (D_{cons}) is the maximum delay allowed for correct operation. After improving the LCP with the afforementioned methodology (See Figure 5), its nominal delay remains nearly the same while its delay degradation is reduced. Therefore, the new aged delay of the LCP after $10yr$ of aging is smaller than the delay constraint. In other words, the LCP can age for extra time since some extra ΔD is allowed. Once the LCP has been improved, other critical paths are identified. There are two cases of critical paths that should be improved: Case

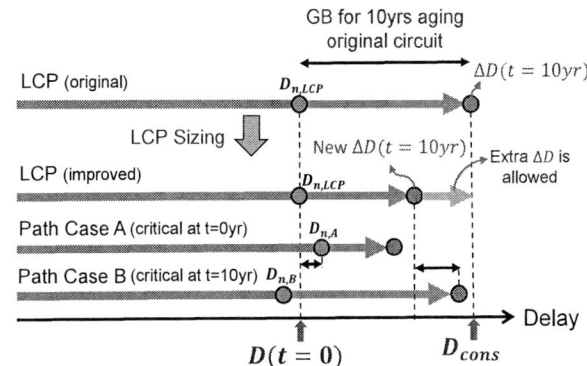

Figure 9. Possible paths cases that should be improved after sizing the LCP of a circuit

A and Case B. The paths in Case A are those whose nominal delay become greater than the nominal delay of the LCP because of loading effects introduced by sizing the LCP. Those paths are critical at design time (t=0yr). Therefore, they are also improved with the proposed methodology to move their nominal delay towards their original value and to reduce their amount of delay degradation. The paths in Case B are those whose aged delay become greater than the new aged delay of the LCP. Those paths could become critical and eventually limit the lifetime of the circuit. Therefore, they are also improved with the proposed sizing methodology to reduce their delay degradation. By sizing the paths in either Case A or Case B of Figure 9, we assure that the initial circuit delay is still defined by the LCP of the circuit while its aged delay is defined by an improved path with extended lifetime.

IV. APPLICATION TO ISCAS CIRCUITS

The proposed sizing approach has been applied to some ISCAS benchmark circuits to validate its effectiveness. Original circuits have been implemented with a standard gate library using only NANDs, NORs and NOTs gates. The delay constraint has been taken as the delay of the original design for 10 years of operational lifetime.

Figure 10 shows the delay behavior of the C1908 ISCAS circuit as function of aging time for both the original design and the improved design obtained with the proposed approach. Note that the improved design starts from the same initial delay as the original design while its delay degradation rate is lower than the original. Therefore, the proposed approach has mitigated aging-induced delay degradation. Figure 10 also shows that due to the exponential behavior of BTI as function of time, the small reduction in delay degradation allows to further increase the lifetime of the circuit from 10 years to around 16 years ($\approx 60\%$).

Table I summarizes the obtained results. The change in percentage for nominal delay, delay degradation, lifetime, area and power are given. As can be seen, for all the circuits the nominal delay remains nearly unchanged. It increases only up to 0.15%, while delay degradation is reduced up to 8% for almost all the circuits. For the same delay constraint of the

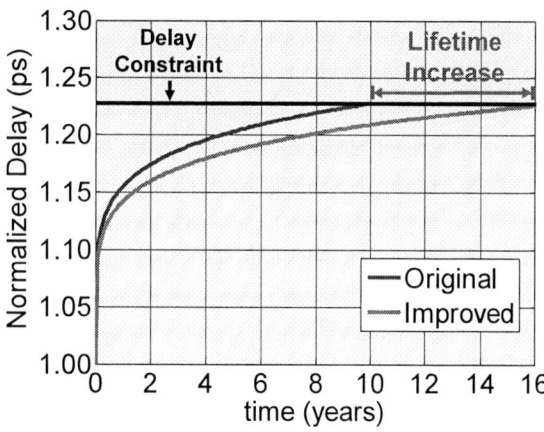

Figure 10. Normalized Delay as function of aging time.

original design, this reduction in delay degradation represents a significant increase in the expected lifetime (more than 40%) for almost all the circuits. Only for circuit $C2670$, its lifetime is not further extended, because the original design was close to the improved design. Table I also shows that the proposed sizing approach provides some small area and power reduction due to the effect of sizing-down some gates. The obtained results show that the proposed approach is efficient to improve circuits lifetime under BTI aging effects.

Table I
RESULTS: CHANGE IN PERCENTAGE OF CIRCUIT PERFORMANCE METRICS

Circuit	D_n Inc. (%)	ΔD Red. (%)	Lifetime Inc. (%)	Area Red. (%)	Power Red. %
C1908	0.07	8.15	64.16	0.64	0.81
C432	0.15	6.24	41.66	0.41	0.51
C499	0.11	6.97	49.16	1.34	1.20
C880	0.01	6.94	54.16	0.24	0.25
C5315	0.06	6.81	50.81	0.25	0.17
C2670	0.15	1.49	4.16	0.42	0.34

V. CONCLUSIONS

A new sizing approach for BTI aging mitigation has been proposed. The aim of the proposed approach is to reduce delay degradation of circuit critical paths while their nominal delay remains nearly unchanged. Gate delay sensitivity to gate sizing is descomposed in a nominal delay sensitivity and a delay degradation sensitivity components. By exploiting the differences between these two components, the proposed approach borrows some area from the gates with low delay degradation sensitivity to the gates with high delay degradation sensitivity. In such way that the overall effect is to mitigate delay degradation. Note that unlike conventional sizing approaches

where gates are sized-down only in the non-critical paths, here, gates are either sized-up or sized-down in the critical paths of the circuit. The results shows that our approach can improve circuit lifetime reliability up to 64% with negligible area and power overhead. If further reduction of aged delay responses is required, conventional sizing approaches can be performed first, and next our proposed sizing approach can take place to further improve circuit reliability.

ACKNOWLEDGMENT – This work was supported by CONACYT (Mexico) through the PhD scholarship number 420129/264560.

REFERENCES

[1] J.H. Stathis, M. Wang, K. Zhao, "Reliability of advanced high-k/metal-gate n-FET devices", Microelectronics Reliability, Volume 50, Issues 911, SeptemberNovember 2010.

[2] M.S. Khan, S. Hamdioui, N.Z.B. Haron, CMOS scaling impacts on Reliability, What do we understand?, 19th Annual Workshop on Circuits, Systems and Signal Processing (ProRISC 2008), 27-28 November 2008.

[3] M.S. Khan, S. Hamdioui, "Temperature Impact on NBTI Modeling in the Framework of Technology Scaling" 2nd HiPEAC Workshop on Design for Reliability, 24 January 2010.

[4] Xiaoming Chen; Yu Wang; Huazhong Yang; Yuan Xie; Yu Cao, "Assessment of Circuit Optimization Techniques Under NBTI," Design & Test, IEEE , vol.30, no.6, pp.40,49, Dec. 2013.

[5] Xiangning Yang; Saluja, K., "Combating NBTI Degradation via Gate Sizing," Quality Electronic Design, 2007. ISQED '07. 8th International Symposium on , vol., no., pp.47,52, 26-28 March 2007.

[6] Kunhyuk Kang; Kufluoglu, H.; Alain, M.A.; Roy, K., "Efficient Transistor-Level Sizing Technique under Temporal Performance Degradation due to NBTI," Computer Design, 2006. ICCD 2006. International Conference on , vol., no., pp.216,221, 1-4 Oct. 2007.

[7] Gomez, A.; Champac, V., "Effective selection of favorable gates in BTI-critical paths to enhance circuit reliability", 16th Latin-American Test Symposium (LATS), 2015, vol., no., pp.1,6, 25-27 March 2015.

[8] Kiamehr, S.; Firouzi, F.; Ebrahimi, M.; Tahoori, M.B., "Aging-aware standard cell library design," Design, Automation and Test in Europe Conference and Exhibition (DATE), 2014 , vol., no., pp.1,4, 24-28 March 2014.

[9] Wenping Wang; Shengqi Yang; Yu Cao, "Node Criticality Computation for Circuit Timing Analysis and Optimization under NBTI Effect," Quality Electronic Design, 2008. ISQED 2008. 9th International Symposium on , vol., no., pp.763,768, 17-19 March 2008

[10] Kostin, S.; Raik, J.; Ubar, R.; Jenihhin, M.; Vargas, F.; Bolzani Poehls, L.M.; Copetti, T.S., "Hierarchical identification of NBTI-critical gates in nanoscale logic," Test Workshop - LATW, 2014 15th Latin American , vol., no., pp.1,6, 12-15 March 2014.

[11] Khan, S.; Hamdioui, S., "Modeling and mitigating NBTI in nanoscale circuits," On-Line Testing Symposium (IOLTS), 2011 IEEE 17th International , vol., no., pp.1,6, 13-15 July 2011.

[12] Agarwal, M.; Paul, B.C.; Ming Zhang; Mitra, S., "Circuit Failure Prediction and Its Application to Transistor Aging," VLSI Test Symposium, 2007. 25th IEEE , vol., no., pp.277,286, 6-10 May 2007.

[13] Zafar, S.; Kim, Y.H.; Narayanan, V.; Cabral, C.; Paruchuri, V.; Doris, B.; Stathis, J.; Callegari, A.; Chudzik, M., "A Comparative Study of NBTI and PBTI (Charge Trapping) in SiO2/HfO2 Stacks with FUSI, TiN, Re Gates," VLSI Technology, 2006.

[14] M.A. Alam, H. Kufluoglu, D. Varghese, S. Mahapatra, "A comprehensive model for PMOS NBTI degradation: Recent progress", Microelectronics Reliability, Volume 47, Issue 6, June 2007.

[15] Wenping Wang; Shengqi Yang; Bhardwaj, S.; Vrudhula, S.; Liu, F.; Yu Cao, "The Impact of NBTI Effect on Combinational Circuit: Modeling, Simulation, and Analysis," Very Large Scale Integration (VLSI) Systems, IEEE Transactions on , vol.18, no.2, pp.173,183, Feb. 2010

[16] Dominik Lorenz, Georg Georgakos, Ulf Schlichtmann, "Aging-aware Timing Analysis of Combinatorial Circuits on Gate Level", IT-Information Technology, Volume 52, Issue 4, Pages 181187, July 2010.

Virtual Prototype Based on Aldebarn CPU Core

Jae-Jin Lee, Chan Kim, KyungJin Byun, and NakWoong Eum
Multimedia Processor Research Team, ETRI
{ceicarus, ckim, kjbyun, nweum}@etri.re.kr

Abstract—This paper proposes a virtual prototype based on the Aldebaran CPU core developed independently by ETRI. The virtual prototype provides instruction and function profiling functionality for software optimization as well as standard integration emulation interface (SystemC, Verilog, Netlist, etc.) compatibility, and architecture performance analysis for efficient adoption into a system-level design environment.

Keywords—Virtual Prototype, Aldebaran, Emulation

I. INTRODUCTION

In SoC design, RTL is the final hardware design output and RTL simulation is the only way to precisely simulate the design. But to simulate a complex SoC with all the more complex software takes too much time consuming so it is not appropriate to use RTL code for hardware and software co-simulation especially from the point of software development.

Furthermore, hardware oriented design approach has a critical problem that software development cannot start until after most of the hardware design is ready and thus hardware and software co-design environment is mandatory in today's complex SoC design flow. Virtual prototype based design approach increases the design productivity tremendously and enables SoC performance analysis and software development through high speed hardware emulation ultimately providing higher level of design completeness in much shorter design time.

Embedded software is getting more complex due to the parallel processing from multi-core processors and ever increasing complexity of the applications running on them. Therefore virtual prototype based design methodology using software modeling of the CPU core and the hardware IPs has been receiving much attention as an effective solution for designing optimal SoC and software. Virtual prototype based design utilizes reusable IPs and through over 90% IP reuses reduces design time, thus increasing product competitiveness through system level optimization.

This paper proposes a virtual prototype based on the Aldebaran CPU core developed independently by ETRI. Linux Kernel, RTOS (RTEMS, FreeRTOS), AUTOSAR OS, etc. have been ported onto Aldebaran virtual prototype providing integrated emulation interface with hardware models written in various languages such as C/C++, SystemC[1], RTL and Netlist. The multi-core Aldebaran virtual prototype also enables SoC performance optimization, design time and cost reduction through HW/SW co-development and increases the reusability of silicon IPs and software drivers.

II. VIRTUAL PROTOTYPE TECHNOLOGY

Many major companies around the world are now adopting virtual prototype to their top-down design flow for developing optimal SW-SoC system architecture and reusing IPs used in their products. But many small sized companies are still using traditional design flow due to high cost and time budget incurred for setting up the virtual prototype.

In 2014, ETRI has developed a multi-core virtual prototype composed of many multimedia processors and applied it to MPEG-4, H.264 and HEVC decoding and is currently performing research on developing and applying the virtual model of the bus and On-chip-network to various applications.

OVP(Open Virtual Platforms)[2] forum is developing many processor models including ARM, MIPS, PowerPC, SPARC, ARC, Open Cores and peripherals. OCPIP(Open Core Protocol International Partnership) defined a high performance interface between IP cores, and SPRINT(Open SoC Design Platform for Reuse and Integration of IPs) consortium is conducting a research on various standardized platforms but does not provide tools for analyzing the system power and area.

Google provides a virtual prototype called "Android Emulator" which is a virtual hardware emulation model on which one can run the software. AVD(Android Virtual Device) permits one to change the hardware component for feature exploration but it is hard to add a virtual model of an new hardware IP thus provides low expandability.

Synopsys provides a commercial virtual prototype development tool "Virtualizer". It enables fast time to virtual prototype availability with the largest portfolio of transaction-level models and reference designs, intuitive graphical prototype assembly and debug, and standard-based (SystemC/ TLM) support and publishing capabilities. But it requires high license fee and you should put in much efforts to construct a target virtual prototype.

III. MULTI-CORE ALDEBARAN VIRTUAL PROTOTYPE

A. Aldebaran : 32-Bit RISC CPU Core

Aldebaran is the CPU core developed independently by ETRI and provides a micro-architecture minimizing the power consumption and open source based software tool chain (C/C++ compiler, assembler, linker, debugger, boot loader, libraries and applications). The Aldebaran CPU core design consists of a 13 stage dual-issue superscalar architecture that includes performance enhancing mechanisms such as branch prediction and an instruction queue, as well as in-order

978-1-4673-9141-2/15 $31.00 © 2015 IEEE

execution, cache access optimization, dynamic voltage-frequency scaling mechanisms for optimized power efficiency. With its high performance and minimal power consumption architecture, Aldebaran is optimal for wearable smart device applications.

Fig 1. shows Aldebaran SoC fabricated using TSMC 65nm design technology. It provides maximum operating clock frequency of 800MHz with energy efficiency of 0.24 mW/MHz. The user can use OCD(On-Chip Debugger) to load and debug the final linked execution program using USB2JTAG interface at source and assembly level.

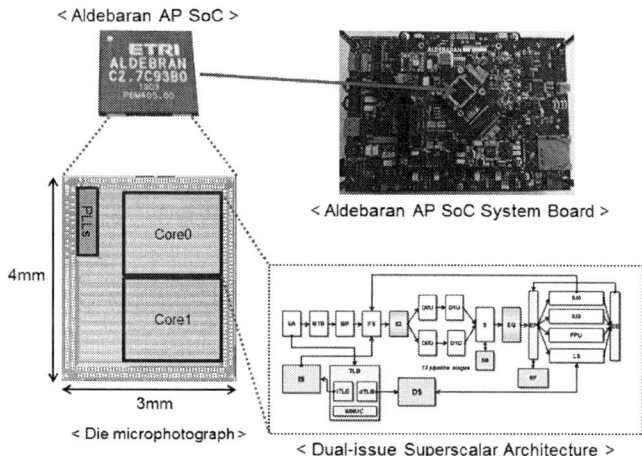

Fig. 1. Dual-Core Aldebaran SoC

The Aldebaran CPU core is fully compatible with AXI protocol based On-Chip Bus and constructing a SoC is greatly simplified. IPs listed in Table I are furnished for companies that wish to collaborate with ETRI to develop a product based on the Aldebaran CPU core.

TABLE I.   ALDEBARAN PLATFORM IPs

IP	Description	Specification
VC	Video Display Unit	Internal DMA, HDMI Support
iROM/iRAM	Internal ROM/RAM	Bootloader in ROM/RAM
ARESET	System init. Controller	System initializer
PMU	Power Management Unit	CPU power-down mode control
NFC	NAND Flash Controller	128M-32Gbytes, 400Mbps
SDC	SD Controller	SD card/SDIO/SPI
SMC	Host Interface Controller	Ethernet(LAN9220)
DMA	DMA Controller	Multi-channel/dimension DMA
USBHS	USB Host Controller	USB 1.1
Timer	Timer	Periodic/Ont-Shot, 4sets
WDT	Watch-dog Timer	Watchdog interrupt
RTC	Real-Time Clock	BCD, 32.768khz XTAL
UART	Serial 8-bit Transceiver	UART 16550
AC97	AC97 codec	Audio output
I2C	Inter-IC Control	7-bit/10-bit, Master/Slave
PWM	Pulse-Width odulation	PWM signal generation
GPIO	General-Purpose I/O	Bidirectional, upto 64 GPIOs

B. Multi-Core Aldebaran Emulator

There are three types of processor core emulation techniques. Interpretive instruction set simulation technique simulates the pipeline model of the processor core and Compiled ISS emulates the instruction processing without pipelining. The third method, DCT(Dynamic Code Translation)[3][4] technique is to convert the machine code to host machine code providing the highest speed. For development of drivers, firmware and application programs on virtual prototype, 10~100MIPS core emulation speed is required and nowadays emulation based on DCT is the most widely used method.

As shown in Fig. 2, the multi-core Aldebaran emulator imports the binary execution image and looks up the TB(Translation Block) which is a continuous sequence of instructions from the cache, and if not found, it translates the code block into a code block written in host instruction set through reverse assembly processing. The translated TB is stored in the TB cache and used when the same code block is to be executed. This way it can achieve more than 10 times the speed of emulation through ISS method.

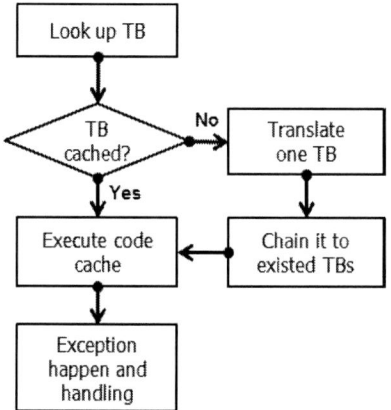

Fig. 2. Aldebaran CPU Core Emulation Based on DCT

Aldebaran multi-core emulator translates the target instruction code to the host instruction code using the Dynamic Code Translation technique and incorporates the "multi-core emulation manager" which handles the context switching between cores and it runs the emulation of a core after context switching. Our work expands the multi-core emulation scheme further and emulates dynamic clock frequency[5] setting by allocating different workloads in time sharing manner and this can be regarded as a high level multi-core emulation control method. Fig. 3 shows the results of running a random line drawing graphic library testbench on a quad-core Aldebaran system emulation with cores running at 100MHz, 200MHz, 300MHz, and 400MHz respectively (clockwise).

The virtual prototype proposed in this paper provides high speed emulation of multi-core Aldebaran. Fig. 4 shows the result of running JPEG image decoding application on a octa-core Aldebaran system emulation.

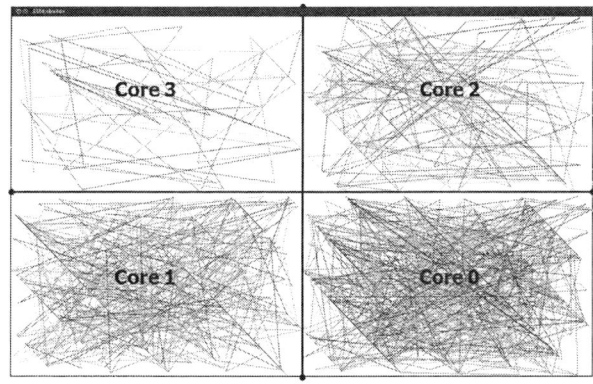

Fig. 3.   Screenshot of Quad-Core Emulation

Fig. 4.   Screenshot of Octa-Core Emulation

C. Virtual prototype

The Virtual Prototype from this work consists of a function accurate emulator and a cycle accurate simulator. The function-accurate emulator of the hardware platform was developed to enable simultaneous hardware and application software development. The cycle-accurate simulator provides a developer-friendly platform for rapid IP integration, verification, and performance profiling.

The function-accurate virtual models in the emulator is written in C/C++ and the emulator also provides instruction and function profiling. It also provides a standard interface for integrated emulation with the existing models written in SystemC or Verilog.

The virtual models used in the simulator were implemented using SystemC transaction level modeling, and supports platform performance analysis and integrated simulation with other IPs modeled in various levels. Fig. 5 shows the integrated simulation environment with virtual prototype and external FPGA board. The simulator runs on the workstation and communicates to the Dynalith iNCITE[6] FPGA board through USB 2.0 interface. Inside the FPGA board are implemented video controller, touch sensor, frame buffer and the simulator receives the detected touch information from the FPGA board, generates a fractal image[7] and displays it on the LCD of the board.

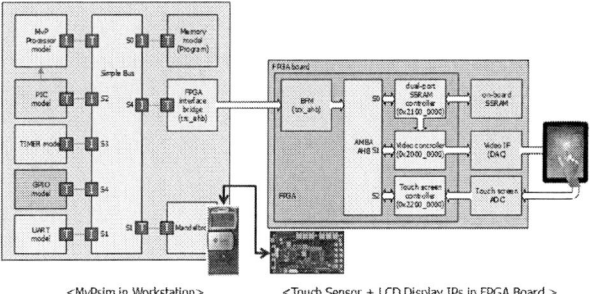

Fig. 5.   Co-Simulation with FPGA Board

Profiling is a form of dynamic program analysis measuring the time complexity, usage of specific instructions or functions, frequency or period of function calls, etc. The profiler of virtual prototype produces detailed performance analysis reports that include instruction usage statistics, function call frequency and time complexity for each function type. Fig. 6 shows the instruction profiling report of the "Sieve" benchmark, and function profiling report of the "Rasterization" benchmark respectively.

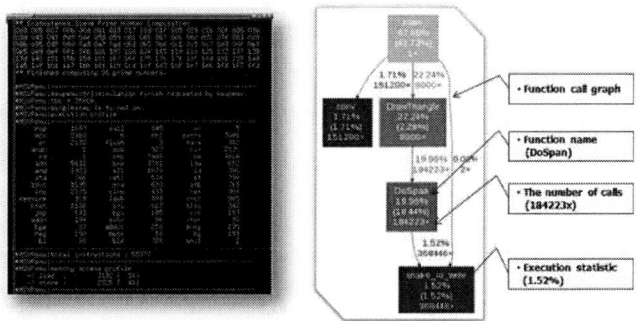

Fig. 6.   Instruction and Function Profiling

A GUI user-interface provides a developer-friendly medium for virtual prototype design and integration. Fig. 7 shows a screenshot of the GUI environment with a sample design that consists of a variety of IP's. The modeled IPs are categorized into different graphic-based libraries to be maneuvered in drag & drop fashion.

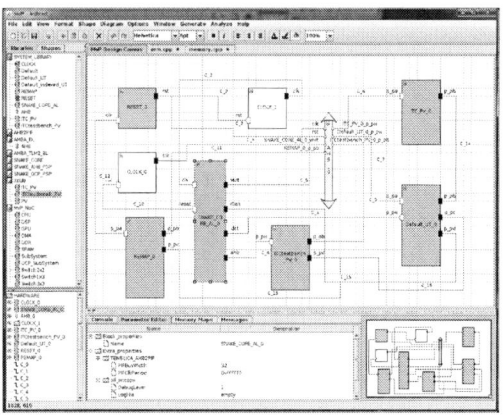

Fig. 7.   Screenshot of Virtual Prototpye GUI environment

978-1-4673-9141-2/15 $31.00 © 2015 IEEE

Virtual model can be attached to GDB, The GNU debugger, to allow user to debug code running on the Aldebaran core and monitor console for interacting with virtual prototype.

IV. IMPLEMENTATION AND EXPERIMENTAL RESULTS

The function of the virtual prototype from this work has been verified using MPEG2, MPEG4, H.264, AVS, VP9, HEVC video decoders, various multimedia applications including motion and lane detection, Linux and RTEMS operating systems. Table II shows that the emulation provide 4 times faster speed than FPGA board running at 80MHz.

TABLE II.   SPEED : VIRTUAL PROTOTYPE VS. FPGA

Benchmark Program	FPGA	Virtual Prototype
Linux Kernel 3.3 booting speed	6.3 sec.	1.7 sec.
RTEMS Kernel boogint speed	2.9 FPS	0.8 FPS
MPEG2 decoding (VGA)	4.3 FPS	19.8 FPS
H.264 decoding (VGA)	2.1 FPS	8.3 FPS
HEVC decoding (VGA)	0.8 FPS	3.1 FPS
Lane detection (CIF)	5.2 FPS	22.1 FPS
Motion detection (CIF)	8.3 FPS	28.8 FPS

Virtual prototype emulation was run on Linux Ubuntu 12.04, Intel® Core™ i7-4770K CPU/3.50GHz environment and the performance was measured for OS booting speed, the number of processed frames per time(FPS : Frames Per Second) in multimedia application.

RTEMS, an open source real-time OS, and Qt, an open source GUI framework, were implemented on both Aldebaran FPGA and virtual prototype. Fig. 8 shows the screenshot of RTEMS multi-task emulation with diagram showing how multi-tasking is performed.

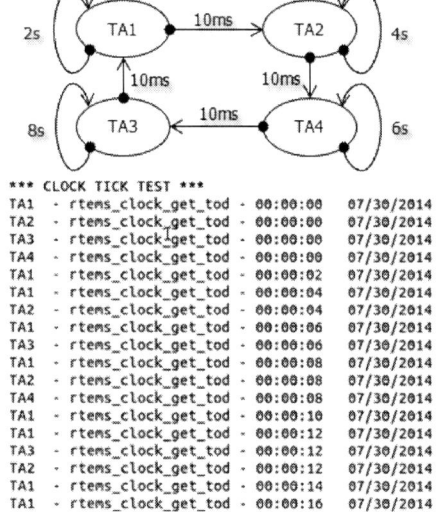

Fig. 8.   Screenshot of RTEMS Multi-Tasking

RTEMS provides multi-tasking environment with file system and networking support. Basic drivers including timer, UART, interrupt controller, NAND flash controller (YAFFS file system), RAM file system, I2C and touch screen drivers were developed for Aldebaran virtual prototype. Qt is a C++ cross-platform development framework for application, UI and device creation. Fig. 9 shows snapshot of the ETRI thin client virtual prototype running on RTEMS and Qt.

Fig. 9.   Screenshot of Thin Clinet Virtual Prototype

V. CONCLUSIONS

In this paper, we propose Aldebarn virtual prototype whose emulation speed is about 4 time faster than that of real hardware FPGA. The proposed virtual prototype supports quad-core dynamic context switching and octa-core emulation, instruction and function profiling, and multi-level co-emulation with IPs modeled by various abstraction levels.

ACKNOWLEDGMENT

This work was supported by the IT R&D program of MOTIE/KEIT [10048843, Automotive ECU SoC and Embedded SW for Multi-domain Integration]

REFERENCES

[1]   O.S.Initiative., IEEE 1666-2005 Standard SystemC Language Reference Manual [Online]. Available: http://www.systemc.org

[2]   http://www.ovpworld.org

[3]   F. Bellard, "QEMU, A Fast and Portable Dynamic Translator," Proc. USENIX Ann. Technical Conf., pp.41-46, 2005

[4]   D.-Y.Hong,C.-C.Hsu, P.-C.Yew,J.-J. Wu, W.-C. Hsu, P.Liu, C.-M. Wang, and Y.-C. Chung., "HQEMU: A Multi-Threaded and Retargetable Dynamic Binary Translator on Multicore," Proc. 10th Int'l Sympo. Cod Generation and Optimization (CGO), pp.104-113, 2012

[5]   Larsson, E.G., Gustafsson, O.: 'The Impact of Dynamic Voltage and Frequency Scaling on Multicore DSP Algorithm Design', IEEE. Signal Processing Magazine, 2011, 28, (3), pp. 127-144 doi 10.1109/MSP.2011.940410

[6]   http://www.dynalith.com/

[7]   L.Lazareck, G.Verch, and J.F.Peters, "fractals In Circuits," canadian Conference on Electrical and Computer Engineering, v1, pp.589-594, 2001

A Generic Clock Controller for Low Power Systems: Experimentation on an AXI Bus

Chadi Al khatib[1,2], Claire Aupetit[1,2], Cyril Chevalier[3], Chouki Aktouf[4], Gilles Sicard[1,2], Laurent Fesquet[1,2]

[1]Univ. Grenoble Alpes, TIMA, F-38000 Grenoble, France
[2]CNRS, TIMA, F-38000 Grenoble, France
Email: firstname.name@imag.fr
[3]STMicroelectronics, F-38000 Grenoble, France
Email: firstname.name@st.com
[4]DeFacto Technologies, F-38430 Moirans
Email: firstname@defactotech.com

Abstract—Today, high performance and low power consumption are important requirements for the embedded SoCs. The variation in transistors characteristics is increasing as CMOS transistors are scaled to nanometer sizes. Indeed, the MIPS per Watt ratio are more and more an important requirement for digital systems. This makes the power consumption constraint a relevant design criterion. This paper illustrates a new architecture based on an asynchronous approach able to easily reduce the power consumption without performance degradation on an existing design. The evaluation results demonstrate the effectiveness of the proposed technique. This new technique can be considered as generic for systems based on busses or NoCs. Experimentation has been done on the industrial AXI bus.

Keywords— power consumption; asynchronous logic; clock gating; asynchronous clock controller; bus AXI.

I. INTRODUCTION

The today's requirements for low power consumption must be met with several requirements such as high chip density and high speed. Therefore low-power digital design is becoming a very active field of research. In this paper, we target a novel approach for reducing the dynamic power consumption. Several techniques exist to reduce this kind of consumption such as DVFS (Dynamic Voltage and Frequency Scaling) [1][2] and clock gating techniques [11][12][13]. In synchronous systems, the clock tree is always active. Indeed, it synchronizes the whole chip without discarding the inactive circuit parts. For this reason, the implementation of gated clock structures is a good way to limit the dynamic power consumption. Nevertheless, this needs extra computation to determine which parts of the circuit have to be switched off. In order to dynamically determine these blocks, local synchronization by the data between the blocks is a viable solution which is currently implemented in asynchronous circuits [3]. For implementing such an approach, it is possible to design asynchronous distributed controllers that carry out data signaling. The local synchronization signals can advantageously be exploited for controlling the clock or the power.

The clock gating approach is widely applied to limit the impact of the clock tree consumption. Indeed, it reduces the switching activity in latches and flip-flops by stopping the clock in the blocks which do not process data. This induces a power reduction compared to their equivalent fully synchronous version. In this paper, we adopted a clock gating technique controlled by asynchronous distributed extra blocks ensuring communications between the blocks. This approach is complementary to the usual gated clock techniques because it is helpful for improving the power consumption of an existing design or to insert hard IP blocks: no need to redesign and resynthesize the system. This insertion of asynchronous interfaces instead of synthesizing gated clock in an existing system is one way to quickly provide a robust framework for power reduction. In the sequel, section 2 gives a short state of the art of asynchronous systems. Section 3, the distributed control structure is detailed with a general view of AXI bus, its features and its operations. Section 4 illustrates the generic structure used and its behavior. Section 5 presents the simulation results of simple systems.

II. THE ASYNCHRONOUS APPROACH

The digital systems are divided into two main groups: synchronous and asynchronous. Synchronous systems have been adopted by industry to solve the problem of the data synchronization because it was very simple and convenient to implement even if timing assumptions are made. In synchronous systems, a global clock synchronizes the data on each clock positive (or negative) edge. The communication mechanisms are rather trivial but the high speed, the system complexity and the technology shrink make the timing assumptions more and more difficult to verify. This is the reason of a renew of interest for asynchronous systems. Nevertheless, the adoption of asynchronous logic faces a lack of tools and competencies and this approach cannot be instantly and largely adopted. A first approach is to use small asynchronous blocks helping in supporting large and complex systems.

Indeed, the addition of asynchronous interfaces to each synchronous block strengthens their integration and can convey the data independently. The main difference with

978-1-4673-9141-2/15 $31.00 © 2015 IEEE

synchronous systems is the locality of the control. The communications in asynchronous designs are data-driven and each block is controlled locally by the data on its inputs; each block is active if and only if the data are present at its inputs [4] [5]. Thus, the blocks are interconnected by a communication channels and each operator has its own communication protocol (4-phase protocol in our case) [6]. Fig.1. illustrates the basic structure of an asynchronous circuit.

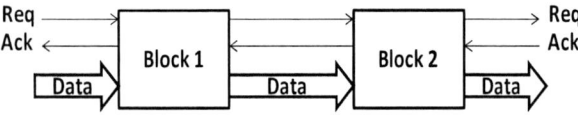

Fig. 1. Basic structure of asynchronous circuit

In Fig. 1, we see a bidirectional signaling, which indicates the presence of data at the inputs of the processing blocks. Indeed, asynchronous logic locally synchronizes data transfers thanks to the request (*Req*) and acknowledgement (*Ack*) signals. Each request event must be acknowledged by the receiver, after that, the transmitter can send data again. This type of communication is based on "Handshakes". Fig. 2 shows a typical controller WCHB (Weak Condition Half Buffer) which is capable to implement a 4-phase communication protocol.

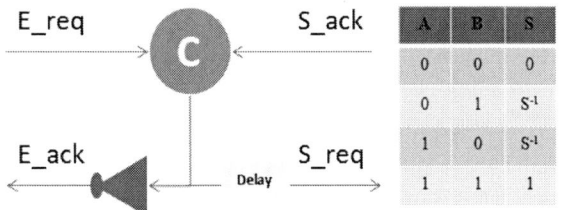

Fig. 2. WCHB Controller with the Muller gate truth table

In order to implement such a controller, two specific gates are required: a delay and a C-element.
- The delay can be implemented by an analog block or several inverters implemented in series.
- The C-element (or Muller gate) is obtained with a 3-input majority gate where the output is fed back to one of the input. The Reset and Set functionalities are easily obtained by respectively inserting an AND gate and an OR gate in the feedback.

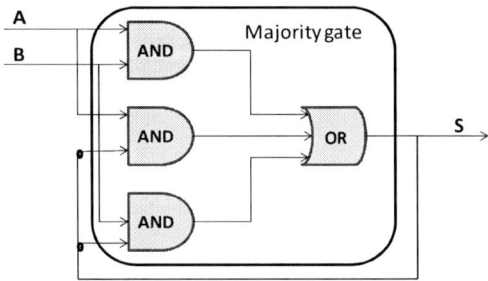

Fig. 3. The C-element implemented with a majority standard cell.

These controllers are very simple and delay insensitive. This means that they are able to guaranty the functional correctness of a 4-phase WCHB protocol whatever the delays. This is particularly interesting for implementing micropipeline circuits [7][10]. Indeed, micropipeline circuits are quite similar to synchronous circuits, but their synchronization mechanism is based on asynchronous controllers which replace the clock tree.

III. PRINCIPLES OF THE ASYNCHRONOUS CLOCK CONTROLLERS

A. principle of the clock controllers

In synchronous systems, the clock tree is always active. At each clock edge, the data are conveyed from one stage of registers to the next stage of registers through a combinational part. This is exactly what is described at the RTL level. This also can be interpreted as a global synchronization with no specific knowledge where the data are really processed. Indeed, the global clock synchronizes the whole chip without discarding the inactive circuit parts. For this reason, the implementation of a structure based on gated clocks is a good way to limit the dynamic power consumption. Nevertheless, this needs extra computation to determine which part of the circuit has to be switched off.

Another approach is to locally synchronize the data between synchronous blocks. For implementing such an approach, it is required to design asynchronous controllers, similar to the WCHB controllers, which carry out data signaling. This gives the opportunity to exploit the local synchronization signals for locally controlling the clock. The architecture principles of such a system are given in Fig. 4

Request signals are active high and acknowledgment signals are active low. The occurrence of a rising edge on the request indicates that data are ready to be processed and, thus the clock signal "gclk" has to be provided to the synchronous receiving sub-block. The acknowledgment signal sent by the receiver indicates that he has completed the processing and that he is ready again to receive new data.

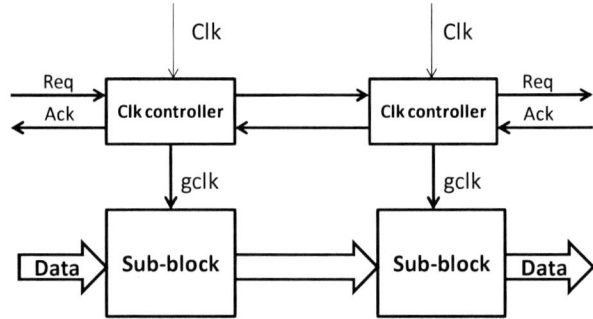

Fig. 4. General view of distributed clock in synchronous blocks

B. Clock partitioning and criteria (IPs/Busses)

The idea is to exploit the existing communication elements (Busses, NoCs) that already own signalization to implement their communication protocols. Indeed, the protocols are carried out by specific signals such as *request*, *wait request, acknowledge, etc.* These signals are perfect to provide the right information to our event-driven distributed control blocks which will be in charge of managing the gated clocks of the system sub-blocks.

C. Application to the AXI bus

AXI Busses are frequently used in the industry. They can support high-performance, high frequency system designs and include optional extensions that cover signaling for low power operations. Among the most important characteristics of the AXI busses, they can support for issuing multiple outstanding addresses and for out-of-order transaction completion [8]. They consist of 3 main buses: a data bus, an address bus and a control bus. Figure below shows a general view of an AXI bus.

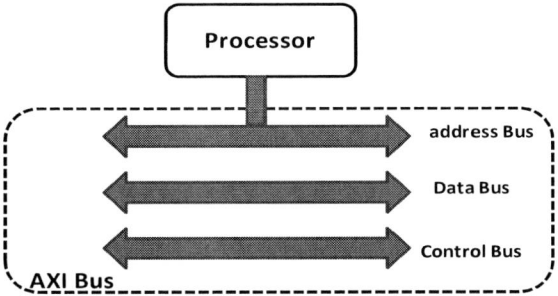

Fig. 5.   General view of AXI bus

In our study, we only focus on the control bus signals to insert the low power structure and to generate its inputs. Noting that, a simple algorithm that automatically inserts the low power structure has been developed. It is presented in the sequel.

IV.   DESIGNING AN ASYNCHRONOUS CLOCK CONTROLLER

A.   The Clock controller structure

One way to limit the dynamic power is to locally synchronize the data between blocks. Asynchronous controllers like WCHB can allow this protocol. Fig.6 shows the clock controller structure.

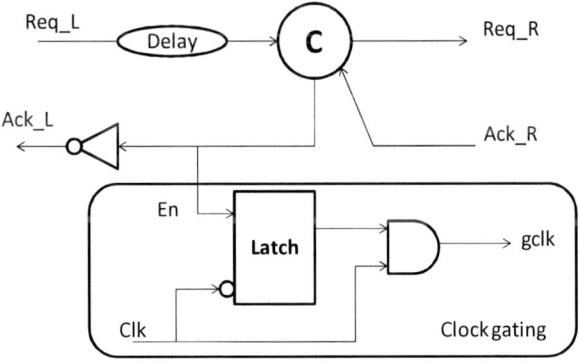

Fig. 6.   Clock controller structure

As already mentioned, the main difference between synchronous and asynchronous systems is the local control.

Request signals are active high and acknowledgement signals are active low. The idea to apply our asynchronous principle at a coarser granularity is:

- Distribute clock controllers in the system to manage the gated clocks.
- Locally activate synchronous blocks / IPs thanks to our asynchronous clock controller.

This structure of Fig.6 is very simple and is able to control pipelined synchronous sub-blocks. It consists of two main blocks: a WCHB controller and a clock gating block. The operating principle is very simple. As an acknowledge signal, Ack_R is initialized at one. When Req_L goes to one to indicate valid data, the output of the Muller gate goes to one. Thus the "En" input of the latch goes to one too. The sampling is done on the falling clock edges and the output clock is generated through an AND gate. In this case, data are ready to be processed and the clock signal has to be provided to the sub-block. The delay (timing assumptions) determines the end of the processing and can cover the timing constraints imposed by the synchronous block. Alternatively, an acknowledgement of the processing block indicating the end of the computation is more appropriated. Nevertheless, the end of the computation has to be provided by a delay or the synchronous bloc itself. This is required to adjust the synchronization instants between the beginning and the end of the block computation, i.e. between Req and Ack signals. When En goes back to low, the latch samples the new value of "En" and the clock will no longer propagate.

B.   Inserting the clock controller in an existing system

Once the clock controller has been designed, the structure has to be inserted in the system. The insertion method consists of detecting the specific control signals carrying the bus protocol, cutting them and inserting the controller. Our case study is based on the AXI bus. As already indicated, only the control part of the AXI bus is of interests for our technique. Indeed, some of the control bus signals are used to generate the synchronization signals of the clock controller (request and acknowledgement). In order to be compliant with the bus protocol, a wrapper is designed to adapt the bus (or the NoC) to the clock controllers.

More specifically, in a system using an AXI bus, two types of communications occur: read communications and write communications [9]. The first one is to transmit data from a slave peripheral to a master and the second is to do the opposite. For the read communications, the following signals are required: Rready (active low), Rvalid and Rlast (active high). The slave generates Rvalid to indicate that there is a valid data ready to process. Rready is generated by the master goes to '0' after a short time meaning that the master is ready to receive it. Then the transfer begins when these two signals (Rvalid and Rready) are active. When burst mode transactions are used, an additional signal, Rlast (in read mode), is required to manage the protocol. It goes to '1' to indicate the last transferred data. For the write transactions, the signals used for the write communication are: Wvalid and wlast (active high), Wready (active low). The data are sent from the master to the slave. Wvalid goes to '1' to indicate that the data are ready to be transmitted.

The slave peripheral generates the Wready signal after a while to indicate that it is ready to receive the data. Similarly to the read mode, the master generates a Wlast signal when using a burst mode for writing. Notice that the transfer only takes place when the AXI "valid" and "ready" signals are active. Each signal returns to its initial state after the end of the transaction. The idea is here to map these signals on a wrapper able to adapt the AXI bus protocol and the asynchronous controller protocol. The asynchronous controller can be considered as generic. Only the wrapper has to be rethought when another bus or NoC is used. We can notice that both protocols rely on handshaking and more especially for the case study on a four-phase protocol. Therefore we just have to identify the signals that have to be exploiting with the clock controller structure. Fig. 7 illustrates the wrapper used for the AXI bus that has to be added to the generic asynchronous controller. As shown on Fig. 7, the wrapper is rather simple in this case, but this is not really a surprised considering the similarity of the protocols.

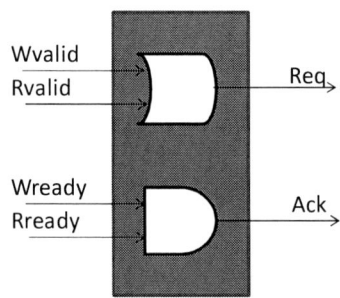

Fig. 7. : The wrapper used with the AXI bus

To generate the input signals of the generic clock controller, we just add one OR and one AND gates. This generates the request and acknowledgment signals as shown in the fig. 7.

C. Insertion algorithm for the clock controllers

The insertion of clock controller in a synchronous system makes the circuit clock gated. To insert the structure in the system, we only have to identify the signals Wvalid, Rvalid, Wready and Rready for the AXI bus and the slave peripheral. Fig. 8 illustrates the general use case for the distributed clock controllers.

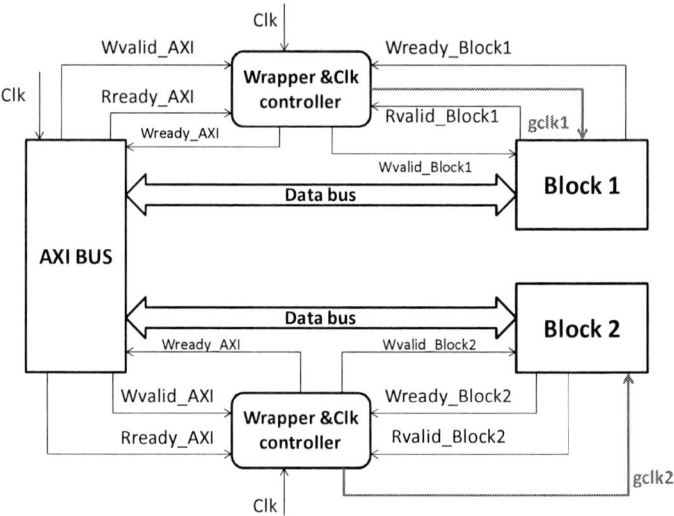

Fig. 8. Distributed clock controllers in a system using an AXI bus

In order to define a generic insertion algorithm, it is important to describe in a database a specific knowledge related to the system and to clock controllers. The database should contain the hardware interfaces between the interconnect system and the peripherals (the name of AXI slave interfaces for instance), the control signals that have to be cut or use, the asynchronous generic controllers and the wrappers for making compliant the interconnect system and the clock controllers. With such knowledge, the algorithm behaves as follow:

(a) Explore the hierarchical design
(b) Detect the interconnect interfaces and the control signals
(c) Cut the control signals and insert the wrappers
(d) Insert the generic clock controllers
(e) Add the connections between the communication system, the controller (wrapper + generic controller) and the peripheral
(f) Generate a new block netlist

This algorithm has been implemented as an extension of the STAR tool from DeFacto Technologies. Fig. 9 shows in the STAR graphical user interface the result of an automatic insertion example (the industrial test case given in the sequel).

978-1-4673-9141-2/15 $31.00 © 2015 IEEE   310

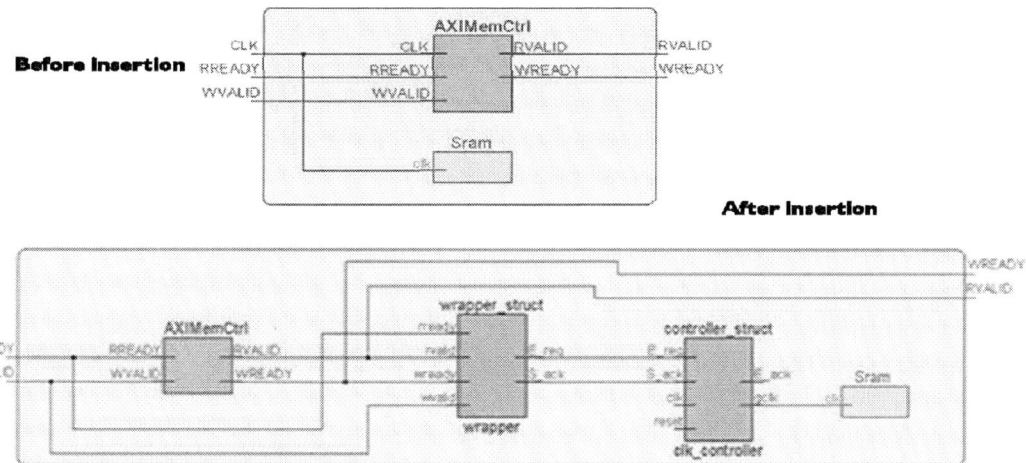

Fig. 9. Automatic insertion of the clock controller in an AXI bus system using AXI bus

V. SIMULATION RESULTS

To demonstrate the validity of our approach, this section presents two circuit examples. In these examples, the bus supports different modes and provides high-performances for the integrated communication system. The first one uses a CMOS 0.35 µm from AMS technology and the second one the 28 nm FDSOI from STMicroelectronics. The first step after the insertion is to validate the behavior correctness of our design. Then the dynamic power and the area are estimated.

A. First use case in AMS technology

This first study experiments a small circuit composed by two memories Spram0 and Spram1 which are interconnected through an AXI bus. Fig. 10 shows the circuit after the insertion of the clock controller.

We can notice the presence of a bus interface for each memory block. Indeed, in systems using the AXI bus, each block uses a bus interface (also called AXI slave or AXIMemCtrl in Fig. 9) to allow the communication between the AXI bus and peripherals as shown in fig. 10.

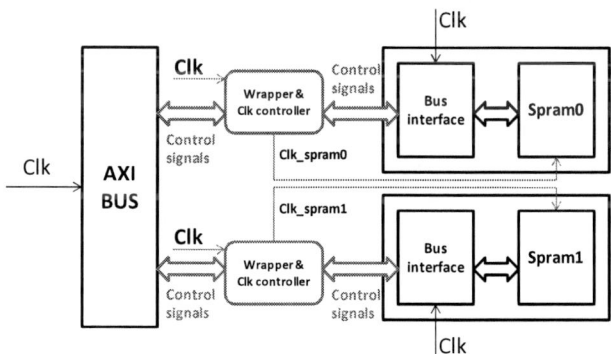

Fig. 10. Test memories circuit

The Wrapper defined in Section IV is used to connect the generic clock controller to the system. The AXI control bus provides all the required synchronization signals to the asynchronous control structures (signals generated by the AXI control bus from the AXI bus and the bus interface) are exploited to enter in the wrapper which generates the input signals of the clock controller (Req and Ack). Then the clock controller generates the clock for each peripheral: clk_spram0 or clk_spram1. Firstly, the circuit has been simulated with Modelsim to check its behavior. Thanks to Design Vision from Synopsys, we estimated the power and area of the two circuits (with and without clock controllers). The fully synchronous circuit consumes 11 mW for. With the inserted clock controllers, the power consumption falls to 7 mW with our test case scenario. Indeed, the activity of the Spram blocks influences the simulation results. In term of area, the clock controller represents less than 3% of the overall circuit area.

B. Industrial test case in 28 nm FDSOI from STMicroelectronics

The industrial test case used to validate the efficiency of our distributed asynchronous clock management system is also based on the AXI bus. Three memory banks (1024 words stack and 32 bit width) are connected to the AXI bus.

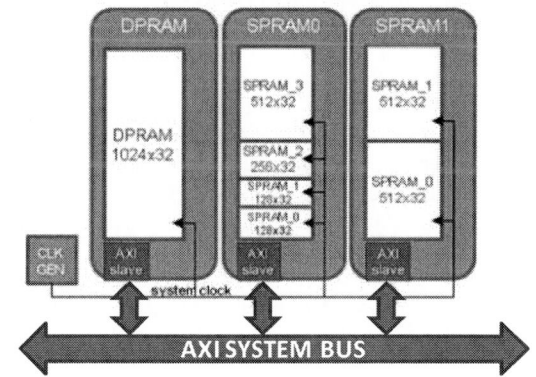

Fig. 11. Testcase validation

For each memory bank, the interface with the bus is done through an AXI memory controller able to manage any Read/Write access in respect with the AXI protocol. All is sequenced by a common system clock. Nothing is initially gated.

The test bench emulates a master that successively accesses for Write and Read operations to the blocks DPRAM, SPRAM0 and SPRAM1. For any W/R operation, the bus provides the control signals to the required slave block and waits for the acknowledgement signal from the slave block which informs that the operation has been achieved. This only requires a few clock cycles per operation; the system clock is applied to each memory block all along the test bench.

The design implemented in 28 nm FDSOI technology, a power consumption analysis has been performed. Table 1 shows the simulation results before and after inserting the clock control structures. All the results have been obtained thanks to Primetime PX.

TABLE I. SIMULATION RESULTS

	DPRAM	*SPRAM0*	*SPRAM1*
Synchronous (µW)	13	24	60
Asynchronous (µW)	0.6	2	12.8
Area (%)	0.05	0.0008	0.28

The results presented in this table reflect the efficiency of such clock controllers. Notice that, the supply voltage was 0.9V. After inserting the clock controllers, the decay of consumption of each block is remarkable. This power reduction is correlated to the circuit activity. Indeed, the bus activity is split between the different memory blocks. In term of area, the clock controllers are negligible compared to the memory blocks where they are applied.

VI. CONCLUSION

This paper illustrates the advantages of using small asynchronous circuits to help controlling classical synchronous circuits, especially their capabilities to reduce the dynamic power by gating a clock tree in an existing synchronous design.

The presented approach describes a new strategy for the gated clock technique. This can be considered as complementary strategy which can be applied to existing designs (but not only). Moreover, the speed performances are kept while the dynamic power can be drastically reduced. The presented structure can be inserted in any system using a bus or a NoC. This technique really limits the redesign effort because there is no needs to resynthesize the system and the clock controllers can automatically be inserted in synchronous systems. This provides a robust framework for dynamic power reduction at a negligible cost area. The approach was successfully tested on two circuits using an AXI bus. The obtained results indicate that the insertion of such generic structures has a significant impact

on the dynamic power consumption at a very low redesign cost. Future works target a more general strategy for also reducing the static power leakage in the advanced technologies.

ACKNOWLEDGMENT

This work is supported by BPI France in the Framework of the HiCool project (F1211048V). The authors thank Alexandre Chagoya from Grenoble INP - CIME Nanotech for his constant support for especially validating our design in the FDSOI technology.

REFERENCES

[1] H. Zakaria, L. Fesquet, "Designing Process Variability Robust Energy-Efficient Control for Complex SoCs" IEEE Trans. On Emerging. On Emerging and Selected Topics in Circuits and Systems (JETCAS), Vol. 1, No. 2, June 2011.

[2] K. Flautner, D. Flynn, D. Roberts and D.Patel: 'Iem926: An energy efficient SoC with dynamic voltage scaling', Design, Automation and Test in Europe Conference and Exhibition, 3, pp. 324-327, 2004.

[3] Muttersbach, J. Villiger and Fichtner, W.: 'Practical design of globally-asynchronous locally-synchronous systems' In Proceedings of the International Symposium on Advanced Research in Asynchronous Circuits and Systems ASYNC'00, pp. 52-59, 2000.

[4] M. Renaudin and J. Fragoso: 'Asynchronous Circuits Design: AnArchitectural Approach'. In: Guntzel J. and Reis R. V Escola deMicroeletrônica da SBC-Sul, Rio Grande, Brazil, pp. 1-43, 2003.

[5] J. Sparsø and S. Furber (eds.), Principles of asynchronous circuit design - A systems perspective. Kluwer Academic Publishers, 2001

[6] M. Renaudin, P. Vivet, F. Robin, "ASPRO-216: a standard-cell Q.D.I. 16-bits RISC asynchronous microprocessor", in Proc. IEEE of the Fourth International Symposium on Advanced Research in Asynchronous Circuits and Systems, 1998, p.22-31.

[7] A. M. Lines. Pipelined Asynchronous circuits. M. Sc. Thesis, California Institute of technology, June 1995, revised 1998.

[8] Y. Liao, "system design and implementation of AXI bus," National Chiao Tung University in partial Fullfilment of the Requirements for the Degree of Master.

[9] F. Xiao, D. Li, G. Du, Y. Song, D. Zhang, M. Gao, "Design of AXI bus based MPSoC on FPGA," *Anti-counterfeiting, Security, and Identification in Communication, 2009.ASID 2009. 3rd International Conference on*, vol, no., pp.560, 564, 20-22 Aug. 2009.

[10] Ivan Suherland, "micropipelines", *Communications of the ACM June 1989 Volume 32 Number 6, pp 720-738*

[11] Robert Mullins and Simon Moore, "Demystifying Data-Driven and Pausible Clocking Schemes", ASYNC 2007, 13th IEEE International Symposium on Asynchronous Circuits and Systems, 12-14 March 2007, pp 175-185.

[12] Carlsson, J.; Palmkvist, K.; Wanhammar, L., "A Clock Gating Circuit for Globally Asynchronous Locally Synchronous Systems," Norchip Conference, 2006. 24th , vol., no., pp.15,18, Nov. 2006

[13] Amini, E.; Najibi, M.; Pedram, H., "Globally asynchronous locally synchronous wrapper circuit based on clock gating," Emerging VLSI Technologies and Architectures, 2006. IEEE Computer Society Annual Symposium on , vol., no., pp.6 pp.,, 2-3 March 2006

978-1-4673-9141-2/15 $31.00 © 2015 IEEE

Fast Global Interconnnect Driven 3D Floorplanning

Artur Quiring, Markus Olbrich, and Erich Barke

Institute of Microelectronic Systems
Leibniz Universität Hannover
Appelstraße 4, 30167 Hannover, Germany
Email: {quiring, olbrich, eb}@ims.uni-hannover.de

Abstract—**Reduction of interconnect length and thus reduction of power dissipation, and improvement of chip-performance is a key benefit of three-dimensional integrated circuits (3D ICs). However, to profit from 3D, carefully planning global interconnects (e.g. buses or wide IO) already at the floorplanning stage is very important. To this end, we present a new global interconnect driven 3D floorplanner which, different from previous work, simultaneously optimizes global interconnect routes, performs TSV placement, and accounts for fixed-outline floorplanning. For our 3D floorplanner, which is based on Simulated Annealing, we introduce an approach that efficiently guides the optimization process towards a valid and low-cost 3D floorplan solution. Experimental results on GSRC benchmarks demonstrate the efficiency of our tool.**

I. INTRODUCTION

In future integrated circuits (ICs), interconnects will more and more dominate overall power consumption and performance. Three-dimensional integrated circuits (3D ICs) offer the opportunity to realize designs with smaller interconnect length compared to 2D ICs and thus allow for more efficient ICs. However, to fully benefit from 3D, carefully planning interconnects is indispensable.

The experiments in [11] reveal that the main advantage of 3D ICs would be the reduction of very long global interconnects, which heavily determine the performance. Furthermore, it would be beneficial to reuse 2D IP modules to reduce design cost for near-term 3D ICs [7]. Both studies indicate that 3D interconnect planning should be already performed at block-level (floorplanning stage), where several 2D modules have to be arranged on various layers, and to be connected by global interconnects (inter-module connections).

Accounting for growing communication demands, interconnect structures like buses, NoCs and wide IO are frequently used to realize global interconnects. They have to be very wide to satisfy communication demands and hence occupy a non-neglectable routing area. Therefore, several research studies [15], [3], [10], [5] have been published, addressing the bus-driven floorplanning problem. All of them, however, consider only 2D designs and thus lack 3D specific design objectives like TSV placement.

Several research studies [6], [7], [9], [14] have been recently published, addressing TSV placement at block-level. Since TSVs occupy non-neglectable device area, they have to be considered as blockage area. In [6] and [7] two methods

have been proposed where TSV placement is performed after floorplanning, using the whitespace of an initial floorplan (post-floorplan TSV placement). Even though the methods are capable of whitespace redistribution and thus allow minor changes to the initial floorplan, they still heavily depend on the quality of the initial floorplan. In [14] and [9] TSV placement is performed directly during floorplanning rather than in a post-floorplan step. The modules can be placed more appropriately so that sufficient whitespace is available to position the TSVs optimally in terms of wirelength and other design objectives. However, the drawback of all mentioned TSV placement aware optimization methods is, that they do not consider routing of global interconnects.

Even though there exist 3D floorplanners [8], [9], [14] which account for interconnect planning, to the best of our knowledge, no 3D floorplanner performs concrete global interconnect routing with TSV placement, and simultaneously considers further important design objectives such as fixed-outline. A simple reuse of previously mentioned approaches is not easily possible since no strategy exists to combine the different approaches with each other.

Our major contributions are as follows:

- We solve the mentioned problem by providing a new global interconnect driven 3D floorplanner, introducing a methodology that allows to simultaneously consider routing of global interconnects, placement of TSVs, and fixed-outline.
- To this end, we add problem specific adaptions to classical Simulated Annealing (SA). We propose for each design objective an analysis function, to guide the SA towards a routable and low-cost 3D floorplan solution.
- We propose a very efficient global routing algorithm.
- We provide our 3D floorplan benchmarks with global interconnect information to allow for comparison with our results.

Experimental results show the efficiency of our 3D floorplanner.

II. PROBLEM FORMULATION

In this paper, the commonly used F2B multi-chip stacking process is assumed [14]. This implicates that a TSV connecting two layers only occupies device area on the upper layer. Furthermore, to simplify routing, it is expected that single global interconnects are hierarchically arranged, e.g., in buses [15]. They are linked across layers using TSV islands (throughout this paper, we use the term TSV representatively for TSV island). The size of a TSV depends on the width of the respective global interconnect. An illustration of a 3D floorplan with one TSV and two global interconnect routes is shown in Fig. 1.

Based on the preliminary assumptions we define the 3D floorplanning problem as follows. Given:

- A fixed-outline.
- One horizontal and one vertical routing layer for each device layer.
- A set of modules B, where each $b_i \in B$ has a fixed width and a fixed height.
- A set of global interconnects G, where each global interconnect $g_i \in G$ connects a set of modules $g_i = \{b_k, ..., b_z\}$, has a geometrical width w_{g_i}, and should not exceed a maximal length $l_{i_{max\_c}}$.

Find a 3D floorplan, minimizing total global interconnect area, and satisfying the following constraints:

- If a g_i traverses several layers, it is extend by a set of TSVs T_i. A TSV $t_j \in T_i$, connecting two layers, consists of two vertically aligned modules $t_j = \{b_{t_j}, b_{v_j}\}$, where b_{v_j} is the module on the lower layer and b_{t_j} is the module on the upper layer. Only b_{t_j} occupies device area. A global interconnect with TSVs $g_i^* \in G^*$ is defined as $g_i^* = \{g_i, T_i\}$.
- All modules B and all TSVs $T = \bigcup_i T_i$ have to be placed overlap-free within the fixed-outline.
- Two sets, one for horizontal segments S_h and one for vertical segments S_v, have to be determined so that no overlap exists on a device layer between any two segments in S_v and any two segments in S_h. Furthermore, for each $g_i^* \in G^*$ all modules $b_j \in g_i^*$ have to be connected by segments $S_{h_i} \subseteq S_h$ and $S_{v_i} \subseteq S_v$.
- For two modules $b_j, b_k \in g_i^*, b_j \neq b_k$ the accumulated length of all segments connecting b_j and b_k has to be less than the allowed maximal length $l_{i_{max\_c}}$.

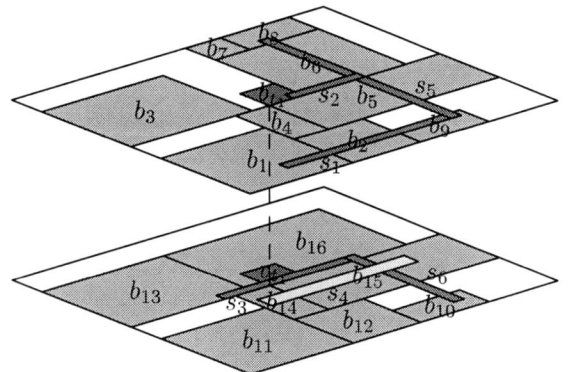

Fig. 1: 3D floorplan with TSVs and global interconnects.

III. GLOBAL INTERCONNECT ROUTING METHOD

Our overall 3D floorplanning optimization algorithm is based on Simulated Annealing (SA). In each SA step, place-ment of modules and TSVs is altered (using a *move*, *swap*, or *rotate* operation; commonly used in 3D floorplanning) and global interconnect routing is performed. The vertically aligned modules $\{b_{t_i}, b_{v_i}\}$ of a TSV are used to connect the respective global interconnect between two layers. Thus, routing (computation of S_v and S_h) can be performed on each layer separately.

For routing, we use our method *Fast-ST-Routing*. It is based on the approach proposed by Kim and Lim [5], however, we made considerable modifications to significantly improve runtime and solution quality. In Section V experimental results show that *Fast-ST-Routing* performs significantly better than the approach in [5]. The major modifications of *Fast-ST-Routing* are as follows:

- Instead of a congestion based topology generation, we use Steiner trees which can be determined very fast using *FLUTE* [4].
- Before solving the time-consuming Linear Program (LP) in each SA step, our method *Fast-Check* examines in advance if a solution to the LP exists. *Fast-Check* notably improves runtime of the overall optimization algorithm.
- In [5] the routability—which is considered as a term in the cost function—is represented by 1 or 0 (routable or not routable). This oversimplification leads to unsatisfying results when used in an SA-based optimization flow. Therefore, we use a more appropriate routing overflow estimation, computed with our *Fast-Check* method. It allows to better approximate the routability of a floorplan.

Fast-ST-Routing can be divided into four steps. They are described in more detail in the following subsections.

A. Selection of Terminals

To simplify alignment of global interconnects, Hanan grids are used as routing grids [5]. For each global interconnect the respective Hanan grid (only modules of the global interconnect are considered) is determined. An example of a floorplan and a Hanan grid for the global interconnect g_1^* is depicted in Fig. 2a and 2b, respectively. Each grid tile gt_i of the Hanan grid contains an intersection value $gt_{i_{hv}}$, representing the total number of modules which can be connected by a horizontal or a vertical segment with gt_i. To avoid too many bends, it is preferable to connect as many modules and TSVs as possible with one segment. Therefore, the grid tile with the highest $gt_{i_{hv}}$ is selected for each module. For g_1^* in Fig. 2b the selected grid tiles are shaded dark.

Given the selected grid tiles, a topology connecting them has to be found. For this step, the selected tiles are represented by their center points (terminals) k_i as shown in Fig. 2c.

B. Topology Generation

Kim and Lim [5] propose a method for topology generation based on a congestion map to increase the probability of finding a legal routing. This method is very time-consuming, especially when used in the inner loop of an SA. Furthermore, we have observed that routable solutions are not necessarily

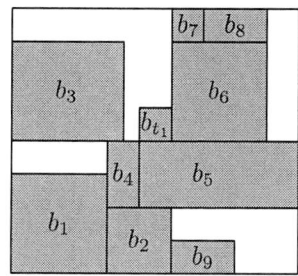

(a) Unrouted floorplan.

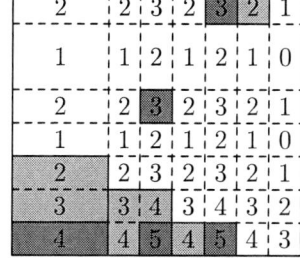

(b) Hanan grid for g_1^* with intersection values $gt_{i_{hv}}$ and selected grid tiles.

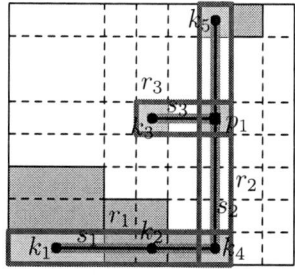

(c) Selected terminals with Steiner tree.

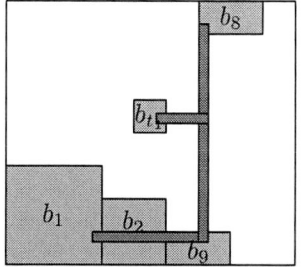

(d) Final routing.

Fig. 2: Global interconnect routing. The modules b_1, b_2, b_8, b_9 and b_{t_1} belong to the global interconnect g_1^* and thus have to be connected. The Hanan grid of g_1^* contains only the modules of g_1^*.

needed in the beginning of the optimization when the floorplan is way off a final solution.

It has been shown that it is sufficient to estimate the interconnect area, in case no legal routing exists, and to consider routability (congestion information) only in the cost function of our SA-based optimization. Therefore, our routing method uses Steiner trees for topology generation, without directly considering congestion information. A Steiner tree can be determined very fast with *FLUTE* [4] and ensures minimal interconnect length for a given set of terminals. Hence, it is a good candidate to reduce interconnect area and maximal interconnect length of a global interconnect. A Steiner tree is illustrated by blue shaded rectangles and an additional Steiner point p_1 in Fig. 2c.

The segments can be obtained from the Steiner tree, where continuous connections like $k_1 - k_2$ and $k_2 - k_4$ are merged together to form one segment s_1. The starting point and the endpoint of a segment have to be placed within the respective tiles, highlighted as light red shaded rectangles in Fig. 2c. The bounding box of the tiles represents the allowable rectangle r_i (allowable area) for the placement of a segment s_i.

C. Routability Check

Now, the segments determined in the topology generation step have to be placed according to the allowable rectangles. This can be formulated as a Linear Program (LP), with total interconnect area being optimized. To prevent segments of different global interconnects from overlapping, the segments have to be ordered based on the center coordinates of their allowable rectangles. The ordering is represented in the LP formulation by additional inequalities. For the rectangles r_1 and r_2 of the horizontal segments s_1 and s_2 in Fig. 3, the center y-coordinate of r_1 is smaller than the one of r_2, hence, s_1 has to be placed below s_2.

Before solving the LP and thus dissipating runtime, it is preferable to perform a fast routability check in advance. Our method *Fast-Check* is demonstrated for five horizontal segments of different global interconnects in Fig. 3. It can be shortly described as follows:

1) Sort all allowable rectangles r_i according to their center x-coordinate (y-coordinate).
2) Iterate over all r_i starting with the rectangle which has the lowest x-coordinate (y-coordinate).
   a) For each r_i, place the respective segment s_i at the lowermost possible position. This is illustrated by the gray shaded rectangles in Fig. 3b.
   b) If s_i would not fit into r_i, place s_i at the uppermost possible position within r_i, and determine the occurring overlap $\hat{o}_{s_i}$. In Fig. 3b it can be seen that s_6 can not be placed within r_6, leading to the overlap $\hat{o}_{s_6}$, represented by a dark gray shaded rectangle.
3) Return the total overlap $\hat{o} = \sum \hat{o}_{s_i}$ after all r_i have been processed.

If $\hat{o}$ is larger than zero, it is very likely that no solution can be found for the LP, whereas if it is zero, a solution for the LP always exists. Thus, in case of a non-zero $\hat{o}$, the placement of the segments (determined in *Fast-Check*) is used to estimate global interconnect area; no LP is solved. The estimated global interconnect area and $\hat{o}$ are used in the cost function to evaluate the current 3D floorplan solution. If, instead, $\hat{o}$ is zero, an LP formulation is set up and solved like described in the next subsection.

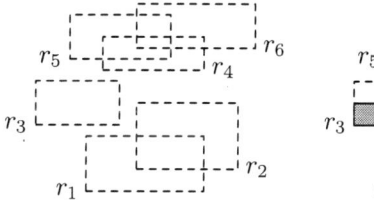

(a) Allowable rectangles of segments.

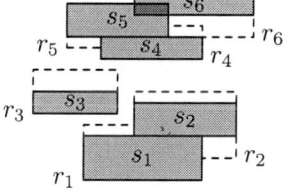

(b) Placed segments with overlap occurring.

Fig. 3: Routability check.

D. Placement of Routing Segments

To determine the exact segment positions an LP formulation similar to [5] is set up. The formulation consists of the following three constraint groups.

1) For each segment s_i, variables x_i^l, y_i^l, x_i^h and y_i^h are created, representing the lower left and upper right coordinates. The ranges for the variables are deduced from the respective tiles (light red shaded rectangles in Fig. 2c).

2) Constraints are added, ensuring that two connected segments belonging to the same global interconnect overlap with each other.
3) Constraints are added, preventing overlaps between segments belonging to different global interconnects. The constraints are based on the order of the segments (see Section III-C).

Finally, the LP is solved with the state-of-the-art LP solver *SCIP* [1]. The obtained legal routing is used to evaluate the current 3D floorplan solution.

IV. IMPROVEMENT OF OPTIMIZATION ALGORITHM (GUIDED SA)

In [2] it has been shown for the fixed-outline floorplanning problem that it is beneficial to guide the SA, because it is hard for the classical SA approach to find valid solutions. Our 3D floorplanning problem is even harder, since it has to consider the constraints for routability, max length and fixed-outline simultaneously.

Therefore, we have modified the SA perturbation operations to improve the local search characteristics of our SA-based optimization algorithm. Our algorithm uses analysis values, obtained for the design objectives fixed-outline, routability, interconnect area and max length, to favor the selection of certain modules for the next SA perturbation operation, which are good candidates to improve the current 3D floorplan solution. The computation methods for the analysis values are described in the subsequent subsections. Since we use the well-known B^*-Tree [16] data structure for our 3D floorplanner, all computation methods have been adapted to the B^*-Tree.

A. Fixed-outline

For fixed-outline floorplanning, the approach described in [13] is used. It computes longest path values lp_i for each module and for each TSV, using a depth-first search in the B^*-Tree. Modules and TSVs with a large lp_i are suitable to reduce fixed-outline overflow and, therefore, should be selected for the next SA perturbation operation with a higher probability than modules with a low lp_i. The longest path values lp_i are used as analysis values for fixed-outline (FO A.).

B. Routability

To guide SA towards a routable solution, the routing overflow values $\hat{o}_{s_i}$ determined in *Fast-Check* (see Section III-C) are used. Each $\hat{o}_{s_i}$ corresponds to a segment s_i, which in turn corresponds to multiple modules b_j or TSVs $\{b_{t_j}, b_{v_j}\}$. Hence, for each b_j and $\{b_{t_j}, b_{v_j}\}$ all respective $\hat{o}_{s_i}$ can be summed up, leading to the overflow value $\hat{o}_{b_j} = \sum_{s_i \in S_{b_j}} \hat{o}_{s_i}$, where S_{b_j} is the set of segments which connect b_j to the respective global interconnects. Moving a b_j which is responsible for a high routing overflow (i.e. $\hat{o}_{b_j}$ is high) can significantly improve routability. Therefore, the overflow values $\hat{o}_{b_j}$ are used as analysis values for routability (Rout A.).

C. Interconnect Area

Fig. 4 depicts a 3D floorplan with two global interconnects g_1^* and g_2^*. Modules not connected with g_1^* and g_2^* are omitted for clarity. To reduce interconnect area, it is useful to place modules and TSVs next to the centroid of the respective global interconnects, especially if they have a high distance to the centroid. Hence, for each module b_j (and TSV) the centroid c_{b_j} is computed based on the centroids of the respective global interconnects g_i^* of b_j. The centroid c_{g_i} for each global interconnect g_i^* is determined based on the center points of the respective modules and TSVs.

In Fig. 4 the centroids c_{g_1} and c_{g_2} are depicted for g_1^* and g_2^*. Thus, since the module b_{10} is connected with g_1^* and g_2^* the centroid $c_{b_{10}}$ is the centroid of c_{g_1} and c_{g_2}. Moving b_{10} next to $c_{b_{10}}$ would reduce the interconnect area of g_1^* and g_2^*.

Given the current position of a module b_j (center point of b_j) and the centroid c_{b_j}, the distance between these two points is defined as d_j. A module is a good candidate to reduce interconnect area, if it is connected with many global interconnects (number of connections: n_{b_j}), and if it has a high d_j. The values $ar_j = d_j \cdot n_{b_j}$ are used as analysis values for interconnect area (Interc. Area A.).

D. Maximal Length

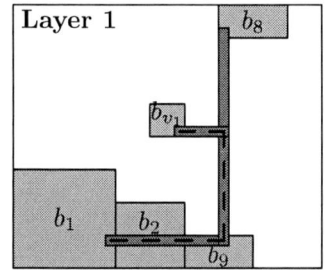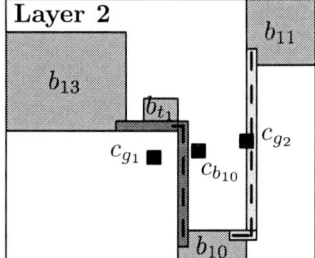

Fig. 4: Routing example for a 3D floorplan with two layers and two global interconnects $g_1^* = \{b_1, b_2, b_8, b_9, b_{10}, b_{13}, b_{t_1}, b_{v_1}\}$ and $g_2^* = \{b_{10}, b_{11}\}$. The centroids for g_1^* and g_2^* are c_{g_1} and c_{g_2}, respectively, and for module b_{10} the centroid is $c_{b_{10}}$.

For a global interconnect g_i^* we define the length $l_{i_{j,k}}$ between any two modules b_j and b_k based on the segments connecting b_j and b_k; it is $l_{i_{j,k}} = \sum len(s_l), s_l \in S_{h_i} \cup S_{v_i}$, where $len(s_l)$ is the length of segment s_l. Furthermore, for a global interconnect g_i^*, the max length $l_{i_{max}}$, the modules responsible for the max length $L_{i_{max}}$, and the overflow l_{i_o} are defined as follows:

$$l_{i_{max}} = \max\{l_{i_{j,k}} \mid b_j, b_k \in g_i^*\}$$
$$L_{i_{max}} = \{b_j \mid b_j, b_k \in g_i^*, \ l_{i_{j,k}} = l_{i_{max}}\}$$
$$l_{i_o} = \max(0, l_{i_{max}} - l_{i_{max\_c}}),$$

The overflow o_{b_j} for a module b_j is then defined as follows:

$$o_{b_j} = \sum_{g_i^* \in G^* \mid b_j \in L_{i_{max}}} l_{i_o}$$

Fig. 4 illustrates a 3D floorplan with two routed global interconnects g_1^* and g_2^*. The dashed line for g_1^* depicts the path between b_1 and b_{10}, representing the maximal length $l_{1_{max}}$, and the dashed line for g_2^* depicts the path between b_{10} and b_{11}, representing $l_{2_{max}}$.

For a legal 3D floorplan the maximal length of each global interconnect g_i^* has to be smaller than $l_{i_{max\_c}}$ (i.e., all l_{i_o} have to be zero), where $l_{i_{max\_c}}$ is the individual maximal length constraint for g_i^*. In case of a non-zero l_{i_o} it is preferable to move the respective modules b_j with high o_{b_j} close to its centroid c_{b_j}—especially, if the distance d_j between b_j and c_{b_j} is high—in order to reduce the maximal lengths of the respective global interconnects. The values $ml_j = d_j \cdot o_{b_j}$ are used as analysis values for maximal length (Max. Length A.).

E. Combination of Analysis Values

For the combination of the analysis values the approach presented in [13] is used. The analysis values lg_i (FO A.), o_{b_i} (Rout A.), ar_i (Interc. Area A.), and ml_i (Max. Length A.)—having different dimensions—need to be normalized. For instance, each lg_i is divided by the maximum of all lg_i; that way all lg_i are normalized to the range $[0, 1]$ (normalized values are $\bar{lg}_i$), while the relations are still preserved. The other analysis values are normalized the same way.

Subsequently, the normalized values are summed up $av_i = \bar{lg}_i + \bar{o}_{b_i} + \bar{ar}_i + \bar{ml}_i$, and we obtain a discrete distribution like depicted in Fig. 5. Modules or TSVs with a high av_i have a higher probability to be selected for the next SA perturbation operation than modules with a low av_i.

Using the discrete distribution based on analysis values, our guided SA is more adapted to the 3D floorplanning problem than classical SA which simply uses a uniform distribution. Therefore, it should find a legal 3D floorplan with a good solution quality faster than classical SA.

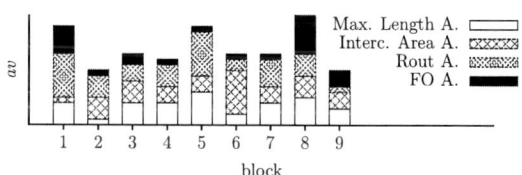

Fig. 5: Discrete distribution used to guide SA.

V. Experimental Results

Our global interconnect driven 3D floorplanner has been implemented in C++. All experiments have been performed on a Linux workstation with an Intel 3.4 GHz CPU and 16 GB of memory. The resulting runtime values are given in seconds. We use the well-known GSRC benchmarks to evaluate our algorithm. However, since they do not contain global interconnect information, we add random global interconnects to the benchmarks (marked with gi). Previous studies have not published their benchmarks, making a comparison difficult. Therefore, we publish all our benchmarks in [12], to allow for comparison with our results.

A. Routing

We have performed experiments to compare the performance of the following routing methods:

- *H-LP* is a reimplementation of the floorplanner described in [5]. It uses a congestion-based topology generation.
- *H-LP-Fast-Check* is an extension of *H-LP*. It uses *Fast-Check* rather than only 1 or 0 (routable or not routable) to estimate routability.
- *Fast-ST-Routing* uses Steiner trees for topology generation and *Fast-Check* to estimate routability.
- *Guided Fast-ST-Routing* is an extension of *Fast-ST-Routing*. It uses guided SA based on the analysis values described in Section IV.

Since the approach in [5] only considers 2D floorplanning, we restrict the number of layers to one for these experiments. We set whitespace to 15 % and an aspect ratio to one. The cost function consists of interconnect area and routability. The pitch for the congestion map in *H-LP* and *H-LP-Fast-Check* is set to twice the minimal bus width. The weights in the cost function and all other SA specific parameters are set to equal values to allow for a fair comparison.

The routing results are shown in Table I. It can be seen that *H-LP-Fast-Check* produces on average equal results to *H-LP* in terms of interconnect area (Interc. Area is averaged over all legal 3D floorplans) with 10 % less runtime, and with a three times higher probability of finding a legal 3D floorplan. The runtime overhead of *H-LP* is basically due to the LP, which has to be solved in each SA step, and the lower probability of finding a legal 3D floorplan is due to the poor consideration of routability in the cost function, misguiding the SA optimization process.

Through the usage of *Fast-ST-Routing* instead of *H-LP-Fast-Check* runtime and interconnect area can be further improved, while probability of finding a legal 3D floorplan is about equal. The runtime overhead of *H-LP-Fast-Check* is basically due to the computation of a congestion map in each SA step. Even though *H-LP* and *H-LP-Fast-Check* have usually a higher probability of finding a routing for a particular 3D floorplan, they perform worse than the Steiner tree based approach, when embedded in an SA based optimization flow.

Comparing *Guided Fast-ST-Routing* with *Fast-ST-Routing*, it can be seen that solution quality can be further improved. *Guided Fast-ST-Routing* finds legal 3D floorplans with a 34 % higher probability, and with 9 % less interconnect area, with only little runtime overhead.

Overall, the results demonstrate that our routing method *Guided Fast-ST-Routing* is superior to the method *H-LP* presented in [5].

B. Classical SA compared to Guided SA

To the best of our knowledge, there are no other fixed-outline driven 3D floorplanner considering TSV placement and routing. Therefore, we solve the 3D floorplanning problem described in Section II using *Guided Fast-ST-Routing* and *Fast-ST-Routing* to demonstrate the efficiency of our guided

TABLE I: Results for routing-driven floorplanner

Bench-mark	H-LP			H-LP-Fast-Check			Fast-ST-Routing			Guided Fast-ST-Routing		
	Run-time	Interc. Area	Valid (%)	Run-time	Interc. Area	Valid (%)	Run-time	Interc. Area	Valid (%)	Run-time	Interc. Area	Valid (%)
n100-gi	2104	93636	26	1854	92304	72	564	90340	71	611	87103	97
n200-gi	4402	141658	21	3945	132473	60	711	119797	64	759	101537	94
n300-gi	5924	165375	14	5499	162572	63	1050	152504	62	1167	131267	93
Avg. ($\Delta_\%$)	0	0	0	-10	-3	2.3×	-79	-9	240	-78	-18	3.9×

TABLE II: Results for classical SA and guided SA.

Bench-mark	Fast-ST-Routing			Guided Fast-ST-Routing		
	Run-time	Interc. Area	Valid (%)	Run-time	Interc. Area	Valid (%)
n100-gi	735	79981	65	786	75275	89
n200-gi	813	95438	60	842	89269	87
n300-gi	1245	117385	51	1311	101275	83
Avg. ($\Delta_\%$)		0	0	-9	48

SA compared to classical SA. For each benchmark 100 independent runs have been performed with 15 % whitespace, an aspect ratio of one, and three layers. Fixed-outline, maximal length of global interconnects, and routability have been considered as constraints. Global interconnect area has been considered as optimization goal. The maximal number of SA steps has been the same for *Guided Fast-ST-Routing* and *Fast-ST-Routing*, and it has been set high enough so that both methods show an asymptotic behavior. The results are shown in Tab. II. It can be seen that *Guided Fast-ST-Routing* has a 48 % higher probability of finding a legal 3D floorplan. Furthermore, interconnect area of legal 3D floorplans is on average 9 % lower compared to *Fast-ST-Routing*. Hence, our guided SA approach notably improves solution quality with only little runtime overhead compared to classical SA. An exemplary 3D floorplan with legal routing is shown in Fig. 6.

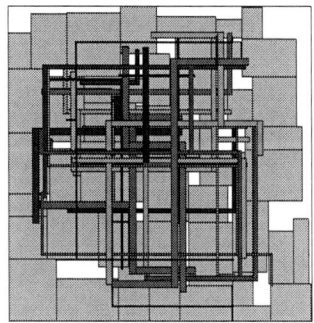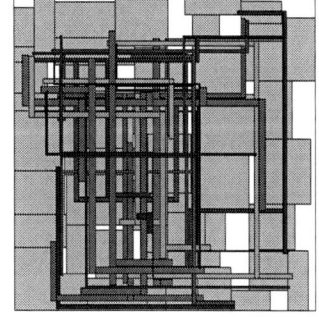

Fig. 6: 3D floorplan (left: layer 1; right: layer 2) with legal routing (n100-gi).

VI. CONCLUSION

In this paper, we have addressed the routing of global interconnects in 3D ICs, which is critical for the IC's power consumption and performance. We formulate a 3D floorplanning problem, provide appropriate benchmarks, and present a global interconnect driven 3D floorplanner. Experimental results demonstrate that our 3D floorplanner solves the problem efficiently, using a fast Steiner tree based routing method, and a guided SA-based optimization flow.

REFERENCES

[1] T. Achterberg. SCIP: Solving Constraint Integer Programs. *Mathematical Programming Computation*, 1(1), 2009.
[2] S. N. Adya and I. L. Markov. Fixed-Outline Floorplanning: Enabling Hierarchical Design. *IEEE Transactions on Very Large Scale Integration Systems*, 11(6), 2003.
[3] T.-C. Chen and Y.-W. Chang. Modern Floorplanning Based on B*-Tree and Fast Simulated Annealing. *IEEE Transactions on Computer-Aided Design of Integrated Circuits and Systems*, 25(4), 2006.
[4] C. Chu and Y.-C. Wong. FLUTE: Fast Lookup Table Based Rectilinear Steiner Minimal Tree Algorithm for VLSI Design. *IEEE Transactions on Computer-Aided Design of Integrated Circuits and Systems*, 27(1), 2008.
[5] D. H. Kim and S.-K. Lim. Global Bus Route Optimization with Application to Microarchitectural Design Exploration. *Proceedings of the IEEE International Conference on Computer Design*, 2008.
[6] D. H. Kim, R. O. Topaloglu, and S.-K. Lim. Block-level 3D IC Design with Through-Silicon-Via Planning. *Proceedings of the Asia and South Pacific Design Automation Conference*, 2012.
[7] J. Knechtel, I. L. Markov, J. Lienig, and M. Thiele. Multiobjective Optimization of Deadspace, a Critical Resource for 3D-IC Integration. *Proceedings of the IEEE/ACM International Conference on Computer-Aided Design*, 2012.
[8] J. Knechtel, E. Young, and J. Lienig. Planning Massive Interconnects in 3D Chips. *IEEE Transactions on Computer-Aided Design of Integrated Circuits and Systems*, PP(99), 2015.
[9] C.-R. Li, W.-K. Mak, and T.-C. Wang. Fast Fixed-Outline 3-D IC Floorplanning With TSV Co-Placement. *IEEE Transactions on Very Large Scale Integration Systems*, 21(3), 2012.
[10] T. Ma and E. F. Y. Young. TCG-Based Multi-Bend Bus Driven Floorplanning. *Proceedings of the Asia and South Pacific Design Automation Conference*, 2008.
[11] M. Pathak, Y.-J. Lee, T. Moon, and S.-K. Lim. Through-Silicon-Via Management during 3D Physical Design: When to Add and How Many? *Proceedings of the IEEE/ACM International Conference on Computer-Aided Design*, 2010.
[12] A. Quiring. Global Interconnect Benchmarks. http://users-ea.ims.uni-hannover.de/quiring/benchmarks, 2015.
[13] A. Quiring, M. Olbrich, and E. Barke. Improving 3D-Floorplanning Using Smart Selection Operations in Meta-heuristic Optimization. *Proceedings of the IEEE International 3D Systems Integration Conference*, 2013.
[14] M.-C. Tsai, T.-C. Wang, and T. Hwang. Through-Silicon Via Planning in 3-D Floorplanning. *IEEE Transactions on Very Large Scale Integration Systems*, 19(8), 2011.
[15] H. Xiang, X. Tang, and M. D. F. Wong. Bus-Driven Floorplanning. *IEEE Transactions on Computer-Aided Design of Integrated Circuits and Systems*, 23(11), 2004.
[16] Y.-C. C. Yao-Wen, Y.-C. Chang, Y.-W. Chang, G.-M. Wu, and S.-W. Wu. B*-Trees: A New Representation for Non-Slicing Floorplans. *Proceedings of the ACM/IEEE Design Automation Conference*, 2000.

Filtering Dirty Data in DRAM to Reduce PRAM Writes

Hyunsun Park*, Chanha Kim*, Sungjoo Yoo[†] and Chanik Park[‡]

| *Department of Electrical Engineering
Pohang University of Science and Technology (POSTECH) | [†] Department of Computer Science and Engineering
Seoul National University (SNU) | [‡]S/W Development Team,
Memory Business
Samsung Electronics |

Abstract— **Phase-change RAM (PRAM) is a promising candidate of emerging memory technologies which provides large capacity and low leakage power to compensate for the limitations of DRAM in the hybrid DRAM/PRAM memory subsystem. However, for practical applications of PRAM in the hybrid main memory, we need to reduce write traffics to PRAM in order to overcome the write-related limitations in PRAM such as write endurance. In our work, we propose a concept called in-DRAM write buffer for the hybrid main memory to reduce PRAM write traffics. The cache line-level dirty data are filtered out from the evicted DRAM row and stored in the write buffer which occupies a portion of DRAM by sacrificing the capacity of DRAM cache. In order to reduce PRAM writes, the write buffer tries to maximize write coalescing by avoiding the PRAM write-back of soon-to-be-accessed dirty data. In order to adapt to the dynamically changing program behavior in PRAM writes, we also propose a method to adjust the write buffer size dynamically during runtime. Experimental results show that the proposed dynamic method offers up to 91.92% reduction in PRAM writes and gives results (average 14.81% and 9.47% reduction in PRAM writes and program runtime, respectively) comparable to the best of static write buffer size cases.**

I. INTRODUCTION

As DRAM is faced with its scaling limit, new memory technologies are gaining more and more attention [1]. Among them, phase-change RAM (PRAM) is a promising candidate to complement DRAM in the main memory subsystem. PRAM is non-volatile and can contribute to reducing the leakage power of main memory. It is expected to give lower cost than DRAM since it can be fabricated with a smaller area (4F2 in [2] compared with 6F2 in DRAM) and simpler fabrication (no need to fabricate expensive capacitors required in DRAM). Large capacity prototypes are already available (8Gb in [2], and 1Gb in [3, 4]).

For real applications especially in main memory subsystem, however, PRAM needs to overcome its disadvantages of high write overhead (endurance, latency and power) and slow read latency. In order to make best use of the advantages of both DRAM and PRAM, a hybrid DRAM/PRAM memory subsystem is proposed [5, 6, 8]. In this case, DRAM plays the role of working memory while PRAM is a large background memory to reduce expensive accesses to the storage and leakage power in the DRAM. The hybrid memory subsystem helps mitigate the limitations of PRAM. For instance, most traffics are served by DRAM thereby significantly reducing average memory access latency. However, in the hybrid memory subsystem, DRAM to PRAM write traffics can be still significant, especially, when small DRAM is used (for cost and

power reasons) or the memory footprint of application is larger than DRAM capacity. Significant evictions from DRAM, i.e., PRAM writes can degrade PRAM lifetime as well as performance due to long PRAM write latency. Considering the trend of increasing performance and capacity requirements in the main memory [7], such a problem will become more exacerbated in the hybrid DRAM/PRAM memory subsystem. A simple solution to address this problem is to increase DRAM size. However, it increases cost and leakage power consumption [8].

In this paper, we propose a concept called in-DRAM write buffer to reduce PRAM write traffics in the hybrid DRAM/PRAM memory subsystem. The write buffer tries to maximize write coalescing by avoiding the PRAM write-back of soon-to-be-accessed dirty data. The basic idea is that, first, we reserve an area in DRAM for write buffering purpose. When evicting a dirty DRAM page[1] (in 2KB), we filter out dirty cache lines (in 64B) from the victim page and store them in the reserved area in DRAM without writing them back to PRAM. When fetching a page from PRAM to DRAM, if it has associated dirty cache line(s) in the write buffer, the page is updated with the dirty cache lines. In this way, dirty data with short reuse distance continue to reside in DRAM, which ultimately enables write coalescing, and, finally, reduction in PRAM write-backs.

This paper is organized as follows. Section II reviews related work. Section III shows motivation of our works. Section IV gives our basic idea. Section V describes the details of in-DRAM write buffer. Section VI reports experimental results. Section VII concludes the paper.

II. RELATED WORK

In this section, we review recent studies on PRAM write reduction. They are categorized into two groups based on write reduction methods, encoding and write coalescing. In the group of encoding methods [9-12], we try to reduce the number of bit updates per request. In the write coalescing methods, we typically assume a hybrid memory subsystem, e.g., DRAM/PRAM hybrid main memory and try to delay PRAM writes, as much as possible, expecting future over-writes.

In [9], Yang et al. present a differential write method called data comparison and write (DCW) which compares existing data in PRAM and new write data and updates only different bits in PRAM thereby reducing bit updates. In [10], Cho et al. apply invert coding to achieve further reduction in PRAM bit updates. In [11], Sun et al. present a method of bit update reduction utilizing a frequent value-based data encoding. In [12], Mirhoseini et al. propose applying data encoding to reduce PRAM writes.

---

[1] We use two terms, row and page, interchangeably.

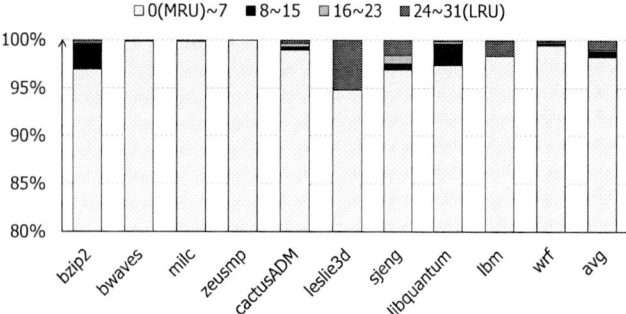

Fig. 1. Hit ratio at priority stack.

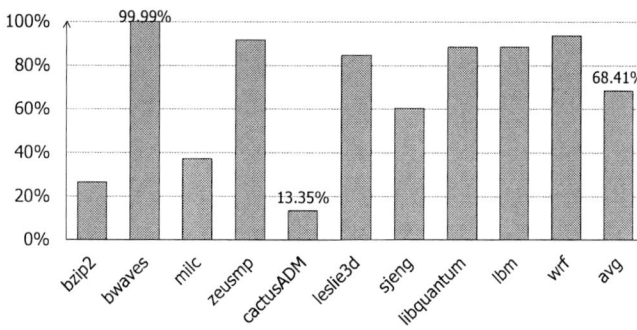

Fig. 2. Dirty data rate in DRAM cache row evictions.

In the write coalescing methods, PRAM writes can be reduced by adopting a hybrid structure where DRAM or SRAM is located in front of PRAM in order to absorb most of writes as well as to provide fast accesses. In [5], Qureshi et al. present a hierarchical hybrid DRAM/PRAM main memory subsystem where DRAM is used as a working memory and PRAM is a large background main memory. They show that the DRAM can mitigate the limitations of PRAM (long latency and write endurance) and the PRAM main memory provides the advantage of large capacity thereby reducing expensive accesses to the disk. In [6], Dhiman et al. present a method of dynamic data migration which identifies and moves frequently written data from PRAM to DRAM in the flat hybrid memory subsystem where DRAM and PRAM are allocated for different address ranges in the memory space (not used as a cache and a main memory). In [5, 6], the authors propose dirty data managements where dirty data are identified with dirty bits and keep dirty data in DRAM as much as possible, expecting the benefit of write coalescing. In [8], Park et al. present a power management method for the hierarchical hybrid DRAM/PRAM memory which decays data in DRAM, i.e., evicts old data from DRAM in order to avoid DRAM refreshes thereby reducing refresh energy consumption. Each DRAM row has a decay counter which is set when accessed. The initial value of decay counter is dynamically determined to minimize total energy consumption in the hybrid DRAM/PRAM main memory. In [13], Wu et al. present a hybrid SRAM/PRAM cache where write-oriented data are identified and allocated on SRAM in order to reduce PRAM writes. In [24], when choosing the victim data from the DRAM page cache, dirty data has higher priority than clean data, which delays PRAM writes. Recently, there have been many studies on the other issues of PRAM (wear leveling, error correction, write performance improvement, R drift in MLC PRAM, etc.). For details, refer to related works, e.g., in [14-19].

Our idea, in-DRAM write buffer is similar to the ideas of word-oriented cache [20] and micro-pages [21]. In [20], Qureshi et al. propose filtering accessed words from evicted cache lines and storing them in the cache. The cache consists of two types of caches, word-oriented and block-oriented ones. In [21], frequently accessed DRAM data are identified and stored in a dedicated area of DRAM expecting future reuses. The ideas in [20, 21] utilize a fixed resource to accommodate the (frequently) accessed data. Compared with [20] and [21], our difference is two-fold. First, our proposed method targets only dirty data in a victim row, not (frequently) accessed data in order to reduce writes to PRAM. Second, we propose a runtime method which dynamically adjusts the write buffer size, which enables us to adapt to dynamically changing program behavior thereby improving the effectiveness of our proposed method.

III. MOTIVATION

In the hybrid DRAM/PRAM main memory, DRAM needs to be judiciously utilized for performance. We assume a hierarchical structure where DRAM plays a role of cache in the hybrid memory subsystem. Fig. 1 illustrates the efficiency of DRAM resource usage in the hybrid main memory. We use a 32-way set-associative 16MB DRAM cache and large-footprint write-intensive SPEC2006 CPU benchmarks (for more details, see Section VI). The figure shows a decomposition of cache hits on the priority stack where the LRU policy is used. In the future, '0' represents highest priority (MRU) in the priority stack. The figure shows that average 98.25% (99.46% in case of wrf) of DRAM cache hits occur in the MRU priority position. The low priority positions (16~31) give only 1.13% of cache hits. The figure shows that DRAM resource, especially, that occupied by lower priority data is not efficiently used. Therefore, we proposed this space to use effectively in DRAM/PRAM-based main memory.

Fig. 2 shows a breakdown of dirty data ratio in the evicted DRAM rows in the same memory subsystem as in Fig. 1. The figure shows that average 68.41% of evicted data, i.e., PRAM write data is dirty, i.e., different from that in PRAM. The figure also shows that the variation of dirty ratio is large. *CactusADM* gives only 13.35% while *bwaves* shows 99.99%. The size of DRAM resource dedicated to filtered dirty data, i.e., in-DRAM write buffer depends on the amount of write coalescing on the filtered dirty data which might correlate with the amount of dirty data in the evicted DRAM rows. Our method needs to be both selective to identify hot write data from evicted ones, and adaptive to resize the in-DRAM write buffer in order to minimize PRAM write traffics.

IV. BASIC IDEA

Fig. 3 illustrates overall operations of our ideas. DRAM consists of two parts: DRAM cache and in-DRAM write buffer (the thick rectangle in the figure). In the figure, we assume that the DRAM cache is managed at the granularity of DRAM row (2KB in our work). The in-DRAM write buffer stores dirty cache lines extracted from the DRAM row which is evicted from the DRAM cache. It has three key operations: copying dirty cache lines to one of in-DRAM buffer rows (arrow 1 in the figure), evicting a set of cache lines from the write buffer (arrow 2) and providing a new row in the DRAM cache (arrow 3) with previously stored dirty cache lines (arrow 3').

Assume that row a is evicted from the DRAM cache in Fig. 3. A DRAM cache row (row a in the figure) has only one associated row (row b in the figure) in the write buffers (details on calculating the associative row address will be given in Section V). As arrow 1 in the figure shows, three dirty cache

978-1-4673-9141-2/15 $31.00 © 2015 IEEE

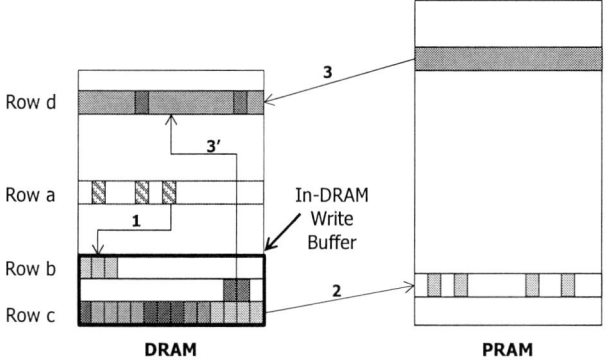

Fig. 3. In-DRAM write buffer.

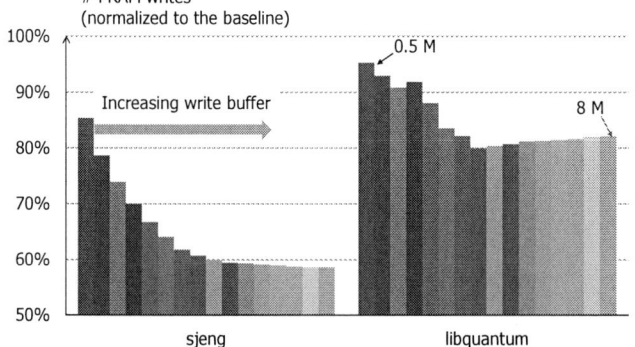

Fig. 4. PRAM writes vs. write buffer size (increasing unit is 0.5MB).

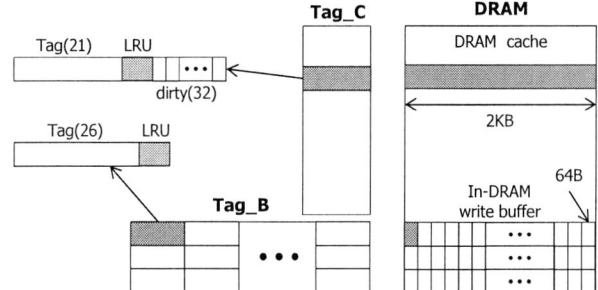

Fig. 5. Tag structure.

to adapt to the dynamically changing behavior across different programs and different phases of single program run.

V. IN-DRAM WRITE BUFFER

A. Overall Structure

Fig. 5 shows the tag structure of hybrid DRAM/PRAM memory system with the in-DRAM write buffer. DRAM is divided into DRAM cache and in-DRAM write buffer. The DRAM cache is managed at the granularity of DRAM row (in 2KB) while the in-DRAM write buffer at the granularity of cache line (in 64B). We assume 40b address and 32 way set-associative DRAM cache. As the figure shows, tags for DRAM cache (Tag_C) consist of address tag, LRU bits, and dirty bits. A dirty bit represents the dirtiness of a cache line. Thus, as the figure shows, a tag entry has 32 dirty bits for 32 cache lines (in 64B) on a row (in 2KB).

Given the data address (Addr_evicted) of a row evicted from the DRAM cache, in order to filter and store its dirty cache lines, we select the target row (index) in the write buffer as follows.

$$\text{Target row index} = \text{Addr\_evicted} \% \text{N\_rows\_WB} \quad (1)$$

where N_rows_WB represents the number of DRAM rows in the write buffer. Thus, a DRAM row has only one associated row in the in-DRAM write buffer. This direct mapping guarantees that only one row in the write buffer is accessed when retrieving dirty cache lines, from the write buffer, which belong to the same row in the DRAM cache. This also simplifies the operation of searching the in-DRAM write buffer for the existence of dirty data associated with a DRAM row newly fetched from the PRAM.

Fig. 5 shows the tags (Tag_B) for the in-DRAM write buffer. The in-DRAM write buffer is managed at the granularity of cache line. Thus, a tag entry has a longer address tag. The tag entry also has an LRU field for the priority management of cache lines on a DRAM row in the write buffer as will be explained in more details in Section V-B. The area overhead of tag structure for the entire DRAM is small, i.e., 483KB in the case of 16MB DRAM used in our experiments (the details will be given in Section VI).

B. In-DRAM Write Buffer Operations

In this subsection, we explain how read/write operations and replacement are performed in the hybrid DRAM/PRAM memory with the in-DRAM write buffer. Fig. 6 illustrates the flow chart of finding the request data in DRAM. When a read request arrives at the main memory, first the DRAM cache tag structure (Tag_C in Fig. 5) is looked up for a tag match. In case of mismatch, we look up the tag structure of in-DRAM write buffer (Tag_B) for a tag match. In case of tag match in either

lines (small rectangles in row a) are copied to the empty space of row b of in-DRAM write buffer. When there is no space to accommodate newly arrived dirty cache lines, we perform a replacement by selecting a victim row in the write buffer and evicting the associated dirty cache lines to PRAM as arrow 2 shows in the figure.

When allocating a new row in the DRAM cache, we copy the entire data from PRAM to the DRAM cache (arrow 3). If there are any associative dirty cache lines in the write buffer which belong to the newly allocated DRAM row, then the dirty cache lines move to the DRAM row (arrow 3'). In this case, the cache lines are invalidate in-DRAM buffer. Note that, in such a case, additional DRAM accesses are required to read the associated dirty data stored in the in-DRAM write buffer. However, the overhead of additional DRAM accesses is small because all the dirty cache lines belonging to the same DRAM row are stored in the same row of in-DRAM write buffer as will be explained in Section V. Thus, only one DRAM row in the write buffer is accessed to retrieve the associated dirty cache lines. In addition, the retrieval of dirty data from the write buffer can be performed in parallel with the data fetch from PRAM to DRAM. Thus, the performance impact due to additional DRAM accesses for the retrieval of dirty data is small (as will be experimentally shown in Section VI).

The in-DRAM write buffer reduces PRAM writes by write coalescing. Fig. 4 illustrates the effects of in-DRAM write buffer obtained by running two SPEC2006 benchmark programs on the hybrid DRAM/PRAM memory subsystem as in Fig. 1 and 2. The figure shows that the in-DRAM write buffer significantly reduces PRAM writes. We can also observe that optimal size of the write buffer is program-specific. *Libquantum* has minimum PRAM writes with a medium-sized (4MB) write buffer while sjeng requires the largest write buffer (8MB) for PRAM write reduction. Fig. 4 shows the necessity of dynamic adjustment of write buffer size

978-1-4673-9141-2/15 $31.00 © 2015 IEEE   321

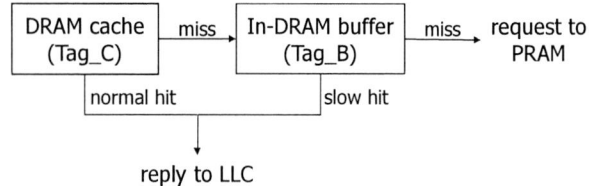

Fig. 6. Request flow chart in hybrid main memory.

Tag_C or Tag_B, the required data are read and transferred to the last level cache (LLC).

When the required data are not found in DRAM (neither DRAM cache nor write buffer), we fetch the required data, at the granularity of DRAM row, from PRAM. In order to allocate a new DRAM row, we perform replacement in the DRAM cache. In our experiments, we use LRU policy. When fetching a new row from PRAM to DRAM, we check to see if the in-DRAM write buffer contains any dirty data associated with the newly fetched row. If there is such dirty data in the write buffer, we update the row, fetched from PRAM, with the dirty cache lines in the write buffer. Note that, in this case, the operation of checking and reading the write buffer tag and data can be performed in parallel with PRAM access. Note that the dirty data in the write buffer are invalidated to maintain only one data copy in DRAM.

When a new row (which is evicted from the DRAM cache, arrow 1 in Fig. 3) arrives at the in-DRAM write buffer, a target row is selected according to Eq. (1). If the target row has available space to accommodate all the dirty cache lines in the evicted row, then we store those dirty data without replacement in the target row. If there is no enough space to accommodate all those dirty cache lines, then we evict cache lines (arrow 2 in Fig. 3), with LRU policy, from the target row until enough space becomes available to accommodate all the new dirty cache lines. Note that serving required data could be performed in parallel with eviction operation, which means evict operations, generally, do not stall other read requests.

C. 5.3 Dynamic Adjustment of Write Buffer Size

The optimal size of in-DRAM write buffer varies across programs and program phases as mentioned in Section IV. Design-time characterization of write buffer size for each program will be expensive. We propose a runtime solution which adjusts the write buffer size to the currently running programs. It can also adapt to phase behavior since the adjustment is performed periodically during runtime.

The basic idea of proposed method is that depending on the relative amounts of hits on young (recently accessed) and old data in the write buffer, we adjust the write buffer size to maximize the utility of write buffer. To be specific, if write buffer hits go to old data more frequently than young data, then we increase the write buffer size since the current write buffer size is not large enough. If the numbers of write buffer hits on young and old data are comparable, then we decrease write buffer size since the current write buffer does not capture hot dirty data.

Fig. 7 shows the pseudo code of the proposed method. First, we count the numbers of write buffer hits on young and old data in the write buffer (lines 1 – 6 in Fig. 7). We use two counters, old_hit and young_hit. The data in the write buffer are classified to old and young data based on their priority in the cache replacement policy (LRU). We divide priority levels into two halves (#valid_lines/2 in line 3) and the data in the higher (lower) priority half are considered to be young (old).

```
1    // In case of buffer hit
2    If (write buffer hit)
3        If (hit_priority > #valid_lines/2)
4            old_hit++;
5        Else
6            young_hit++;
7    ...
8    // Every 40,000 accesses (lines 9 – 21)
9    // No buffer access check
10   If (no_buf_hits == 0) buf_no_access++;
11   Else buf_no_access=0;
12   If (buf_no_access == 8)
13       DecreaseBufSize();
14
15   // Long and short RRI counters check
16   If (0.9 < old_hit/young_hit < 1.1)
17       DecreaseBufSize();
18   Else if (1.1 <= old_hit/young_hit)
19       IncreaseBufSize();
20   Else
21       // Buffer size does not change
```

Fig. 7. Dynamic write buffer size adjustment.

Periodically (on every 40,000 accesses to the main memory in our experiments), we determine the write buffer size as shown in lines 9 – 21 in Fig. 7. First, we check to see if there are any hits on the write buffer during the period. If there is no hit during the past 8 consecutive periods (320,000 accesses in our experiments), then we decrease the write buffer size (lines 12 – 13) since the write buffer is not useful. If the number of hits on young and old data are comparable (line 16), then we decrease the write buffer size. If old data have much more hits than young data (line 18), then we increase the write buffer size. For the remaining cases where young data have much more hits than old data (line 20), we keep the current buffer size since the write buffer does well in capturing hot write data. We determined the parameters (the period and thresholds) in Fig. 7 experimentally after sensitivity analyses in our experiments.

VI. EXPERIMENTS.

A. Experimental Setup

Table I shows the architectural parameters used in our experiments. We use a Pin-based simulation environment called McSim [22]. We run SPEC2006 benchmarks on the in-order x86 core model with the hybrid DRAM/PRAM main memory subsystem.

We used CACTI v5.3 [23] and 32nm low power

TABLE I. ARCHITECTURAL PARAMETERS

Component	Details
CPU core	In-order x86 core, 1.2GHz
L1 cache	4-way 16KB/32KB I/D cache, 64B cache line
L2 cache	16-way, 1MB cache, 64B cache line, LRU policy
Memory Controller	FR-FCFS, closed page scheme
DRAM (cache)	16MB^2, LPDDR2-800[26], $t_{CL}/t_{RP}/t_{RCD}$ = 12.5ns, 32-way, LRU policy, array read = 0.26 pJ/bit, array write = 0.09 pJ/bit
PRAM	1GB, LPDDR2-800, $t_{CL}/t_{RP}/t_{RCD}$ = 12.5ns/150ns/55ns, array read = 0.55 pJ/bit, array write = 3.74 pJ/bit differential write

^{2} We use 16MB of total DRAM capacity, similar ratio between DRAM and PRAM in [5].

TABLE II. FOOTPRINT OF SPEC2006 BENCHMARK

Benchmark	size [MB]	WPKI	Benchmark	size [MB]	WPKI	Benchmark	size [MB]	WPKI
perlbench	25.778	0.005	leslie3d	126.77	0.258	libquantum	34.916	0.299
bzip2	864.514	0.548	namd	7.53	-	h264ref	4.248	-
gcc	46.538	0.047	gobmk	18.818	0.003	tonto	4.622	-
bwaves	773.402	0.379	dealII	4.496	-	lbm	419.176	0.378
gamess	3.332	-	soplex	25.518	0.004	omnetpp	11.05	-
mcf	12.912	-	Povray	3.594	-	astar	33.186	0.035
milc	347.356	0.654	calculix	9.164	-	wrf	323.558	0.192
zeusmp	222.782	0.232	hmmer	13.01	-	sphinx3	4.402	-
gromacs	11.956	-	sjeng	178.862	0.442	xalancbmk	9.37	-
cactusADM	559.036	3.491	GemsFDTD	0.626	-			

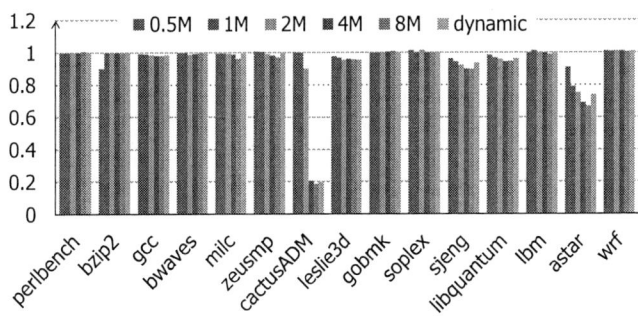

Fig. 9. Total execution cycles (normalized to the baseline).

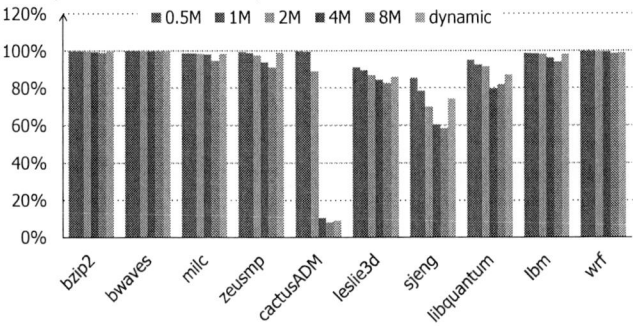

Fig. 8. PRAM writes comparison.

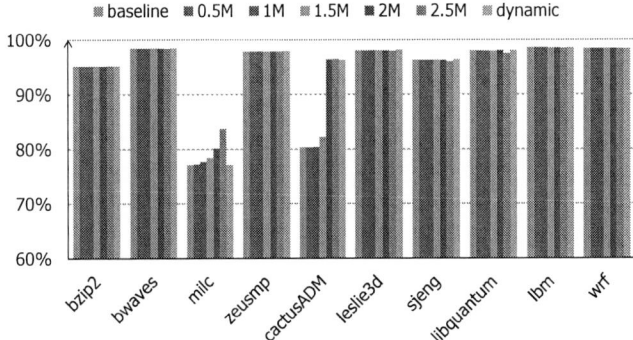

Fig. 10. DRAM hit rate.

technology to estimate the energy consumption of SRAM tag (Tag_C in Fig. 5) accesses. In case of DRAM and PRAM access energy, we used the parameters from [25]. They assume 90nm PRAM technology, but currently PRAM is scaled down under 20nm technology [2]. Thus, we linearly scale down the energy parameters as shown in Table 1.

Table II shows the footprint size and PRAM writes per kilo instructions (WPKI) of SPEC2006 benchmarks. First, because the target DRAM size of our model is 16MB, we chose the benchmarks which have footprint larger than 16MB. Since, the size of DRAM is 16MB, WPKI of small footprint size benchmarks is zero in the below table. In addition, since we target reducing PRAM writes, we used 10 write-intensive benchmarks with high WPKI (shaded in blue) in Table II to evaluate our proposed method.

B. Results

We use as the baseline the hybrid DRAM/PRAM-based main memory subsystem without the in-DRAM write buffer. Fig. 8 compares the amount of PRAM writes in the baseline and proposed methods. In the figure, we use both static and dynamic write buffer cases. In the case of static write buffer, we do not change the write buffer size during runtime. We use the static cases to show the effectiveness of dynamic adjustment method. For the dynamic write buffer cases, we set the maximum size to 8MB, 50% of total DRAM capacity.

Fig. 8 shows that, compared with the baseline, the proposed method offers average 19.20% (up to 91.92% in case of *cactusADM*) reduction in PRAM writes with the best cases of static write buffer size. *Zeusmp, cactusADM, leslie3d, sjeng* and *libquantum* show a similar behavior that as the static write buffer size increases, we obtain more reduction in PRAM writes. However, some benchmarks, *bzip2, bwaves, lbm* and *wrf*, have no noticeable change in PRAM writes as the write buffer size varies. According to Table II, these benchmarks have larger footprint, more than 300MB, than the size of DRAM (16MB). Thus, the in-DRAM write buffer size (maximum 8MB) may not be large enough to hold the hot write data. We expect they could benefit from the in-DRAM

write buffer in case of larger DRAM.

Fig. 8 shows that the dynamic adjustment method gives average 14.81% reduction in PRAM writes compared with the baseline. In most of programs, it gives results comparable to the best of static write buffer cases. Note that the static method has a significant limitation that a pre-characterization (to determine the optimal write buffer size) is required for each program or program mixes. The dynamic method does not need such a pre-characterization.

Fig. 9 compares total execution cycles (normalized to the baseline). In the hybrid DRAM/PRAM memory subsystem, the latency of PRAM write is critical factor of total cycle, since it is (about 6 times in [5]) longer than the DRAM write latency. Thus, especially for write-intensive programs, total execution cycles tend to follow the amount of PRAM writes.

Fig. 9 shows that the proposed dynamic method gives performance improvement by 9.47% on average. However, in some programs such as *wrf*, total execution cycle increases slightly, by 0.41%~1.04% for *wrf*. It is due to small PRAM write reduction (1%~2% in Fig. 8) and additional DRAM accesses for write buffer accesses.

Fig. 10 shows DRAM hit ratio including both cache and write buffer hit ratios. For most of programs, the hit ratio does not change significantly even though the effective capacity of DRAM cache is reduced due to the in-DRAM write buffer. In the case of *cactusADM*, the trend of DRAM hit ratio matches that of PRAM writes and total execution cycles, which shows that the program has significant write traffics with high temporal locality (considering its large footprint in Table II).

Fig. 11 shows the total energy decomposition of baseline and dynamic adjustment methods. The DRAM energy of dynamic method is larger than that of baseline, due to additional in-DRAM write buffer accesses. The proposed method reduces PRAM energy by up to 84.9%. However, *bwaves, libquantum* and *wrf* have slightly larger PRAM energy than in the baseline. In those benchmarks, the energy gain from reduction in PRAM writes (0.85~13.5%) is smaller than the

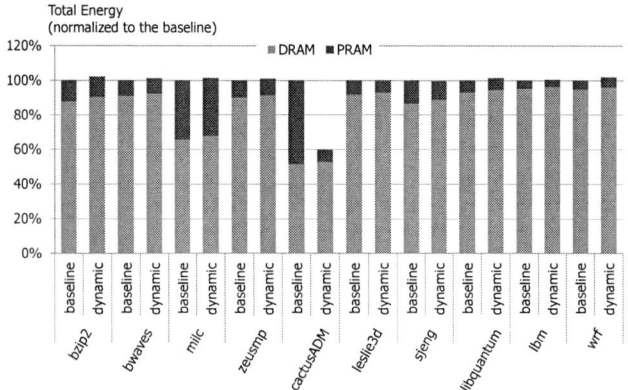

Fig. 11. Total energy decomposition.

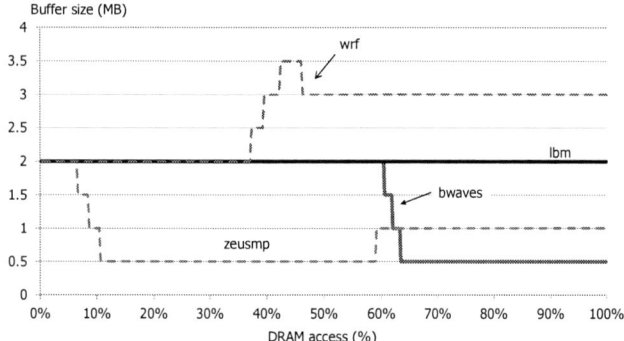

Fig. 12. Write buffer size tracking.

additional energy consumption (for additional row activation and precharge operations in PRAM) due to reduced row buffer hit ratio (3.66%~9.26% reduction) in PRAM.

Fig. 12 illustrates how the write buffer size changes during runtime in the dynamic adjustment method. We set the initial size of in-DRAM write buffer to 2MB. The figure shows representative examples, *bwaves*, *zeusmp*, *lbm* and *wrf*, from the 10 benchmarks. We observe that the programs tend to settle down on steady state write buffer sizes after the initial transition periods.

The proposed in-DRAM write buffer method requires additional tag resource as described in Fig. 4. Compared with the conventional DRAM cache (where the tag and dirtiness are managed at the granularity of DRAM row), the DRAM cache needs a dirty bit per 64B cache line, in our case, total 32b per 2KB DRAM row. The in-DRAM write buffer requires cache line (64B)-level tags, i.e., 32 tags for a DRAM row. In the case of 16MB DRAM having a maximum 8MB in-DRAM write buffer, the tag structure of proposed method requires 483.3KB. The area overhead can be reduced by limiting the maximum size of write buffer.

VII. CONCLUSION

In this paper, we presented an in-DRAM write buffer in the hybrid DRAM/PRAM main memory subsystem. It filters out dirty data from DRAM data evicted to PRAM and stores them in the write buffer area in DRAM thereby enabling write coalescing in future DRAM accesses. We also proposed a dynamic method to adjust write buffer size in order to adapt to dynamically changing behavior in PRAM writes. Our experiments show that the dynamic method gives average 14.81% and 9.47% reduction in PRAM writes and runtime for large-footprint memory-intensive programs.

REFERENCES

[1] International Technology Roadmap for Semiconductors (ITRS), available at www.itrs.net.

[2] Y. Choi, et al., "A 20nm 1.8V 8Gb PRAM with 40MB/s program bandwidth," Proc. ISSCC, 2012.

[3] C. Villa, et al., "A 45nm 1Gb 1.8V Phase-Change Memory," Proc. ISSCC, 2010.

[4] H. Chung, et al., "A 58nm 1.8V 1Gb PRAM with 6.4MB/s Program BW," Proc. ISSCC, 2011.

[5] M. K. Qureshi, V. Srinivasan, and J. A. Rivers, "Scalable High Performance Main Memory System Using Phase-Change Memory Technology," Proc. ISCA, 2009.

[6] G. Dhiman, R. Ayoub, and T. Rosing, "PDRAM: A Hybrid PRAM and DRAM Main Memory System," Proc. DAC, 2009.

[7] S. Dumas, "Mobile Memory Forum: LPDDR3 and WideIO," JEDEC Mobile Forum, June 2011.

[8] H. Park, S. Yoo, and S. Lee, "Power Management of Hybrid DRAM/PRAM-based Main Memory," Proc. DAC, 2011.

[9] B. D. Yang, et al., "A Low Power Phase-Change Random Access Memory Using a Data-Comparison Write Scheme," Proc. ISCAS, 2007.

[10] S. Cho and H. Lee, "Flit-N-Write: A Simple Deterministic Technique to Improve PRAM Write Performance, Energy and Endurance," Proc. MICRO, 2009.

[11] G. Sun, et al., "A frequent-value based PRAM memory architecture," Proc. ASPDAC, 2011.

[12] A. Mirhoseini, M. Potkonjak, and F. Koushanfar, "Coding-Based Energy Minimization for Phase Change Memory," Proc. DAC, 2012.

[13] X. Wu, et al., "Power and performance of read-write aware Hybrid Caches with non-volatile memories," Proc. DATE, 2009.

[14] M. K. Qureshi, et al., "Enhancing Lifetime and Security of PCM-Based Main Memory with Start-Gap Wear Leveling," Proc. MICRO, 2009.

[15] N. H. Seong, et al., "Security Refresh: Prevent Malicious Wear-out and Increase Durability for Phase-Change Memory with Dynamically Randomized Address Mapping," Proc. ISCA, 2010.

[16] S. Schechter, et al., "Use ECP, not ECC, for Hard Failures in Resistive Memories," Proc. ISCA, 2010.

[17] N. Seong, et al., "SAFER: Stuck-At-Fault Error Recovery for Memories," Proc. MICRO, 2010.

[18] [18] M. K. Qureshi, et al., "Improving Read Performance of Phase Change Memories via Write Cancellation and Write Pausing," Proc. HPCA, 2010.

[19] M. Awasthi, et al., "Efficient Scrub Mechanisms for Error-Prone Emerging Memories," Proc. HPCA, 2012.

[20] M. Qureshi, M. A. Suleman, Y. N. Patt, "Line Distillation: Increasing Cache Capacity by Filtering Unused Words in Cache Lines," Proc. HPCA, 2007.

[21] K. Sudan, et al., "Micro-Pages: Increasing DRAM Efficiency with Locality-Aware Data Placement," Proc. ASPLOS, 2010.

[22] Manycore simulation infrastructure, available at http://cal.snu.ac.kr/mediawiki/index.php/McSim.

[23] HP Labs., CACTI 5.3, available at http://www.hpl.hp.com/research/cacti/.

[24] A. P. Ferreira, M. Zhou, S. Bock, B. Childers, R. Melhem, and D.Mosse, "Increasing PCM main memory lifetime," in Proc. Design Autom. Test Euro. (DATE), 2010, pp. 2–4.

[25] B. C. Lee, et al., "Architecting Phase Change Memory as a Scalable DRAM Alternative," Proc. ISCA, 2009.

[26] JEDEC Standard, Low Power Double Data Rate 2 (LPDDR2), JESD209-2E, April 2011.

On the estimation of assertion interestingness

Tara Ghasempouri
Department of Computer Science
University of Verona, Italy
Email: tara.ghasempouri@univr.it

Graziano Pravadelli
Department of Computer Science
University of Verona, Italy
Email: graziano.pravadelli@univr.it

Abstract—The definition of assertions is a fundamental phase for formal and semi-formal verification strategies as well as for documenting purposes. Assertions are generally manually defined, but several (semi-) automatic approaches have been also proposed that mine assertions directly from execution traces of the design under verification (DUV). In both cases, assertion qualification is necessary to evaluate the quality of the defined assertions. Current approaches evaluate the interestingness of a set of assertions by measuring the percentage of DUV's behaviours covered by the assertions, mainly by adopting techniques based on mutation analysis, which require long simulation time. On the contrary, this work proposes an automatic technique to estimate the interestingness of assertions by ranking them according to metrics typically adopted in the context of data mining, which reveals to be a faster approach. Experimental results that compare the proposed assertion ranking strategy with assertion qualification based on mutation analysis are reported.

I. INTRODUCTION

Assertion based verification (ABV) arises as one of the most popular candidate solution for verifying the consistency between design intents and design implementation [1]. It relies on the definition and checking of assertions, i.e., logic formulas, generally defined by adopting temporal logics like LTL and CTL, which formally express the expected behaviours of the DUV. While several static and dynamic approaches exist for checking assertions, their definition still represents a crucial phase in the ABV flow.

Manual definition of assertions requires high expertise and it is an error-prone and time-consuming activity. Main problems are related to the risk of defining assertion sets that are incomplete (i.e., unable to cover all expected behaviours of the DUV), inconsistent (i.e., with contradicting assertions), redundant (i.e., with assertions that are logical consequence of others), and including vacuous assertions (i.e., assertions that are true independently from the DUV, and thus irrelevant). As a result, a false sense of security is induced by an ABV campaign conducted with a low-quality set of assertions.

As opposite to manual definition, some works have been done recently to automatically generate assertions from the DUV implementation [2], [3], [4], [5]. In these approaches, execution traces obtained by simulating the DUV are dynamically analysed to mine significant assertions. Independently from the abstraction level of the DUV (e.g.. TLM, RTL, gate level) execution traces pass through an assertion miner, whose output is a set of candidate assertions capturing the behaviours exposed by the DUV during simulation, according to a set of predefined temporal patterns. Extracted assertions may highlight shortcomings of the original specifications, which may lead to distinguish design's errors and unpredictable behaviours implemented in the DUV. While vacuity and inconsistency in the set of generated assertions are generally avoided by the mining approach itself, assertion incompleteness and redundancy may still affect the outcome of assertion mining. Thus, a qualification phase for evaluating the degree of interestingness of extracted assertions is still necessary. As the number of mined assertions can be very high, their manual qualification is almost impractical. For this reason a strategy to automatically evaluate the interestingness of extracted assertions and rank them accordingly is necessary. Unfortunately, current approaches for assertion mining are still unsatisfactory from this point of view. In [2], a stressing phase is proposed only to verify the likelihood that mined assertions are globally satisfied (and not only for the execution traces analysed by the miner), but no strategy is proposed to measure their interestingness in covering DUV behaviours. In [3], interestingness estimation is based on the number of propositions included in the antecedent of the assertion, according to the fact that an assertion with a lower number of propositions in its antecedent has an higher input space coverage than one with many propositions in its antecedent. However, the correlation between the antecedent and the consequent of an assertion is not considered. To solve this drawback, in [4] a ranking function is proposed that evaluates the quality of the mined assertions in terms of cause-effect relationship between antecedent and consequent of an assertion. Finally, in [5], mined assertions are said to be generally ranked according to their frequency of occurrences and time of first occurrence but no specific approach is presented.

As an opposite class of approaches, coverage metrics have been widely studied for qualification of assertions [6], [7], [8], [9]. Most of these works relies on mutation analysis, which requires perturbing the DUV implementation by injecting mutations (faults) to check, either statically [7], [8] or dynamically [9], whether they change the truth values of the assertions; mutations that do not cause a change are said to be not detected. Assertions that detect a few mutations are less interesting than assertions detecting an higher number of mutations. Not detected mutants generally highlight area/behaviours of the DUV that are not covered by any of the defined assertions showing a hole on the coverage. Dynamic approaches like [9] scale better with respect to static techniques, however, they still require long simulation runs for checking each assertion for each mutation with a significant set of testbenches. When the number of assertions is very high, as in the case of assertions extracted automatically, evaluating their interestingness through mutation analysis becomes a very time-consuming activity.

To fill in the gap and overcome the limitations of current approaches, this work proposes an automatic technique to estimate the interestingness of assertions by ranking them according to probabilistic metrics typically adopted in the context of data mining (i.e., support and correlation coefficient) [10], [11], which we adapt here for the specific case of assertion mining. From the point of view of the general concept, data mining and assertion mining share the same idea (extracting rules from data), but they have several differences that make practically different how these metrics are computed and interpreted for evaluating the interestingness of assertions.

As shown by experimental results, the proposed solution reveals to be a faster approach without loss in the accuracy with respect to mutation-based techniques.

The rest of this paper is organized as follows. Section II introduces some preliminary definitions and concepts necessary to understand the methodology. Section III and Section IV describe, respectively, the adopted metrics and how they are computed and used to rank assertions. Section V reports experimental results. Finally, Section VI is related to conclusions and future works.

II. PRELIMINARIES

Before describing the proposed methodology for the evaluation of assertion interestingness, some definitions and concepts concerning data mining and assertion mining are reported to create the necessary background.

A. Definitions

Data mining deals with itemsets, transactions and association rules, whose definitions are as follows.

Definition 1: Let $I = \{i_1, i_2, \ldots, i_n\}$ be a set of items. Let $D = \{d_1, d_2, \ldots, d_m\}$ be a data set, i.e., a set of observations, called transactions, with respect the set of items I. Each element in D contains a subset of the items in I. An association rule is defined as an implication of the form $X \to Y$ where $X, Y \subseteq I$ and $X \cap Y = \emptyset$. X and Y are called itemsets.

Figure 1a shows an example of a data set which describes the behaviours of customers in a supermarket with respect to a set of items (i.e., milk, bread, ..., coffee). Data mining approaches are generally intended to extract association rules from data sets, which are then used to predict non trivial, implicit, previously unknown and potential useful information, like, for example,"when milk is bought bread and coffee are generally bought too", which is expressed by the association rule $Milk \to Bread \wedge Coffee$.

Assertion mining deals instead with execution traces and temporal assertions which are formalized as follows.

Definition 2: Given a finite sequence of simulation instants $T = \langle t_1, \ldots, t_n \rangle$ and a model M working on a set of variables V, an execution trace of M is a finite sequence of pair $E = \langle (V_1, t_1), \ldots, (V_n, t_n) \rangle$, where $V_i = eval(V, t_i)$ is the evaluation of variables in V at simulation instant t_i.

Definition 3: An atomic proposition is a logic formula that does not contain logic connectives.

Example of atomic propositions are for example $v_1 = true$, $v_2 > v_3$, $v_4 = false$, etc.

Definition 4: A proposition is a composition of atomic propositions through logic connectives. An atomic proposition itself is a proposition.

Examples of a propositions are $p_1 : (v_1 = true) \wedge (v_2 > v_3)$, $p_2 : v_4 = false$, $p_3 : (v_4 = false) \vee (v_1 = true)$, etc.

Definition 5: A temporal assertion is a composition of propositions through temporal operators according to some temporal logics.

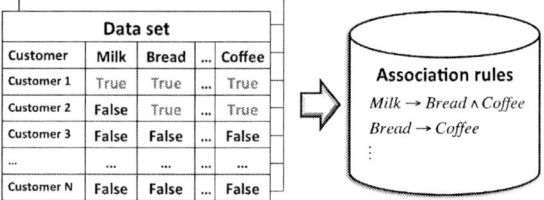

(a) Data mining concept.

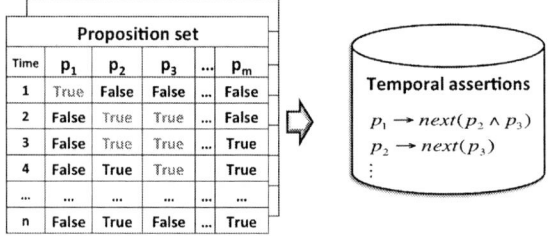

(b) Assertion mining concept.

Fig. 1: Similarities between data mining and assertion mining

An example of a temporal assertion in Linear Time Logic is $always(p_1 \to next(p_2 \wedge p_3))$ which states it always happens that p_2 and p_3 are satisfied one simulation instant later than p_1 becomes true (Figure 1b).

In this paper, without lack of generality, and to preserve the independence of the proposed methodology from specific temporal patterns instantiated in the analysed assertions, we generically consider formulas in the form $A \to C$, where the antecedent A and the consequent C are composed by propositions, logic connectives, and temporal operators according to the selected temporal logic. The initial hypothesis is that the analysed assertions are true on the DUV. The way they are generated (manually, or automatically) is irrelevant.

B. Data mining and assertion mining

The overall goal of data mining is to extract information from a data set and transform it into an understandable and useful structure. This structure allows user analysing data from many different dimensions, categorizing them and summarizing correlations between items in a database. For example, analysing data from behaviours of different customers as reported in Figure 1a leads to obtain useful information and helps analysers to decide which trend is more interesting for marketing. Association rules can also be extracted when data are referred to time sequences. In this case, temporal data mining strategies are adopted, whose goal is to discover hidden relations between sequences and sub sequences of events [12]. In any case, the mined (temporal) association rules are a prediction for future behaviours, which may be true or not. Metrics are thus used to estimate the probability that rules extracted from past observations can be valid also in the future.

On the contrary, the main goal of assertion mining consists of extracting formulas that exactly describe the functionality implemented in the DUV, which is not ambiguous and does not not vary in the future, except in the case the implementation is changed. Assertion mining is thus not intended to predict the future, but to formalize the actual set of DUV behaviours.

Summarizing, main similarities among data mining and assertion mining are the presence of a set of data that represents observations with respect to past behaviours exposed by the observed target (customers, DUV, ...), and the need of extracting association rules that formalize such observations. As shown in Figure 1, items, data sets, and association rules in data mining correspond, respectively, to propositions, execution traces, and temporal assertions in assertion mining. Meanwhile, the main difference between data mining and assertion mining is represented by the concept of transaction (i.e., a row in a data set), which does not have a direct correspondence with a row of an execution trace, because a temporal assertion is composed by one antecedent and one consequent that are true in different instants inside the execution trace. This difference impacts on the way metrics typically adopted for evaluating association rules in data mining can be reused for measuring the interestingness of assertions. Finally, another difference is related to the final goal of the mining: in one case the prediction of future behaviours, in the other the formalisation of actual (unmodifiable, except in the case the DUV functionality is changed) behaviours.

III. METRICS

Several metrics have been proposed in data mining for evaluating the interestingness of association rules. The use of metrics allows analysers evaluating the rules from different points of view [10], [13]. For instance, *odds ratio* and *entropy* are appropriate for estimating the probability of distribution of items, *support* and *confidence* are able to calculate the interestingness of an association rule based on the number of item's occurrences; while the *correlation coefficient* is suited to determine the dependency between set of items.

In the context of assertion qualification, metrics that provide information about the degree of accuracy of a rules with respect to the probability it will hold in the future (like for example, confidence, which estimates the joint probability between occurrences of the antecedent and the consequent in the data set) are not relevant, because we know that assertions under analysis are always true on the DUV. We are instead interested in metrics that measure the interestingness of an assertion with respect to covered behaviours, number of activations, and correlation between antecedents and consequents. For this reason, we identified *support* and *correlation coefficient* as the most interesting metrics for assertion evaluation. Their definition in the context of data mining are hereafter reported together with considerations related to how they can be adapted to be suited for assertion evaluation.

Definition 6: Given a set of items I, and the corresponding set of transactions D, a rule $X \rightarrow Y$ has support S if X and Y occur concurrently in S percent of transactions in D.

In practice, to compute the support of an association rule, it is necessary to count how many rows in the transaction set table contain both X and Y. In case of temporal assertions, the support corresponds instead to the number of times a temporal assertion occurs (i.e., its antecedent is fired and then its consequent is satisfied) in the execution traces with respect to the total number of occurrences corresponding to the other temporal assertions under analysis. This requires a different computation approach with respect to data mining as described in Section IV. For example, let us consider a temporal assertion $A \rightarrow C$ that occurs 10 times in a set of execution traces. If it belongs to a set of temporal assertions that globally occur 1000 times in the same execution traces, the support of $A \rightarrow C$ is $10/1000 = 0.01$.

Definition 7: Given a set of items I, and the corresponding set of transactions D, the correlation coefficient of the rule $X \rightarrow Y$ is the covariance of X and Y divided by the product of their individual standard deviations.

More informally, the correlation coefficient can determine if antecedent and consequent are related or not by observing whether occurrences of the antecedent depend on occurrences of the consequent and vice versa. For example, Figure 2 graphically shows the meaning of the correlation coefficient with respect to the association rule $X \rightarrow Y$. On the left, X and Y has a positive correlation, i.e., an increment in occurrences of X corresponds to an increment in occurrences of Y. In the middle, a negative correlation is shown. Finally, on the right, no dependence between X and Y exists. Higher is the correlation coefficient higher is the interestingness of the analysed rule.

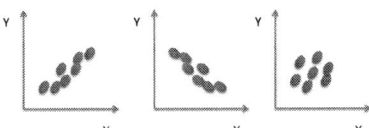

Fig. 2: The correlation coefficient: positive correlation (on the left), negative correlation (in the middle), no correlation (on the right).

IV. ASSERTION RANKING

For estimating the interestingness of assertions, we implemented an assertion ranker based on support and correlation coefficient. It works independently from the way assertions are defined/generated, from their meaning, and from the adopted temporal logic. The hypothesis is that they are in the form $A \rightarrow C$ where A and C are compositions of propositions through temporal operators and logic connectives.

Inputs of the ranker are the set of assertions to be evaluated and a set of execution traces on which assertions hold. Higher is the quality of the execution traces in exposing the complete set of DUV behaviours, higher is the quality of the interestingness estimation. In case assertions have been automatically generated through assertion mining, execution traces could be the same used by the miner. The work flow of the proposed methodology is then divided in 3 main steps (Figure 3):

1) *Counting of occurrences*: In this phase, the number of times an assertion is verified in the execution traces is computed. Then, each assertion is decomposed in antecedent and consequent and their respective frequencies in the execution traces are computed too.

2) *Computation of contingency tables*: the information collected in step 1 is then organized in contingency tables (one per each assertion) that represent the ingredients for the computation of the evaluation metrics in the final step. Contingency tables make simpler the extraction of information like how many times an antecedent and the corresponding consequent occur in the execution trace, how many times an antecedent occurs but the corresponding consequent does not, and how many times a consequent occurs but the corresponding antecedent does not.

3) *Evaluation of interestingness*: The final step, starting from the contingency tables, computes support, correlation coefficient, and their linear combination to obtain a final metrics that considers both of them. Their combination is necessary because support and

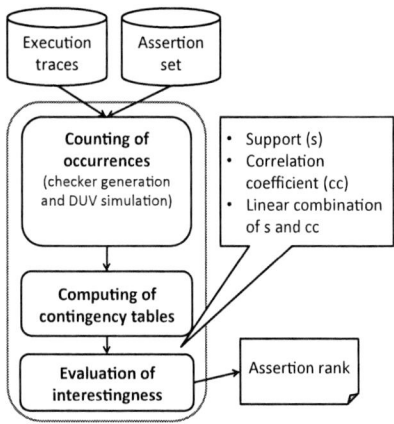

Fig. 3: Overview of methodology.

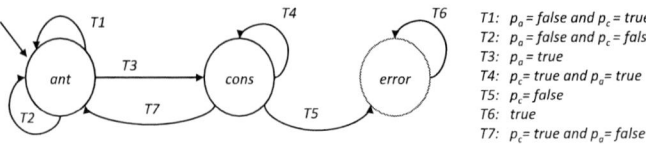

T1: p_a = false and p_c = true
T2: p_a = false and p_c = false
T3: p_a = true
T4: p_c = true and p_a = true
T5: p_c = false
T6: true
T7: p_c = true and p_a = false

Fig. 4: Example of the checker for assertion $always(p_a \rightarrow next(p_c))$.

correlation coefficient separately may provide very different estimations, which only partially characterise the quality of each assertion, as clarified in Section IV-C.

In the following of this section the three steps of the proposed methodology are described.

A. Counting of occurrences

To count occurrences of assertions, antecedents and consequents, we generate a checker for each assertion. A checker can be considered an automaton that monitors the evolution of the DUV during simulation and raises a failure when the corresponding assertion is violated [14]. To perform such a verification, the checker exactly knows when the antecedent is fired and when the consequent is then satisfied. Thus, it can be used for counting of occurrences as required for our estimation.

For example, the automaton generated for counting occurrences for an assertion like $always(p_a \rightarrow next(p_c))$ is reported in Figure 4. The automata starts in the initial state *ant*. It remains in this state (corresponding to a vacuous satisfaction of the assertion) till the antecedent p_a is finally fired (transition $T3$). Then, it moves to the state *cons*, where it stays by continuously traversing $T4$ at each simulation instant while p_a remains true and p_c is also satisfied. This represents the case in which the assertion is activated and satisfied (non vacuously) for several consecutive simulation instants. The assertion is non vacuously satisfied also when the automaton exits *cons* by traversing $T7$, which corresponds to the case p_c still holds but p_a stops to be fired. Alternatively, the automaton exits *cons* to reach the *error* state through $T5$ in case p_c stops holding. In this case the assertion is falsified, but according to our assumption (assertions are true in the DUV) this never happens in our methodology. The number of occurrences of the assertion corresponds to the number of traversals of transitions $T4$ and $T7$. The number of times the antecedent is fired corresponds to the number of traversals of $T3$ and $T4$. Finally, the number of times the consequent is fired corresponds to the number of traversals of $T1$ and $T4$.

B. Computation of contingency tables

Support and correlation coefficient can be effectively computed by relying on a 2×2 frequency count matrix called contingency table [15], whose computation derived from the counting of occurrences performed in the previous step. Given

an assertion $A \rightarrow C$, its contingency table represents the relation between A and C. The cells of the table contain the following information (Table I):

- Cell f_{11} is the number of times where A is true and consequently C is true in the execution traces;

- Cell f_{10} is the number of times where A is true but consequently C is false and other consequents than C are true in the execution traces, i.e., it is the sum of occurrences of assertions $A \rightarrow C'$ included in the considered assertion set with $C \neq C'$. It is worth noting that $A \rightarrow C$ and $A \rightarrow C'$ are not inconsistent, because C and C' refer to different temporal instants. For example, $always(p_1 \rightarrow next(p_2))$ and $always(p_1 \rightarrow next(next(p_3)))$ can be both true for the same DUV.

- Cell f_{01} is the dual of f_{10}, i.e., it is the number of times where A is false but A' different from A is true and consequently C is true in the execution traces, i.e., it is the sum of occurrences of assertions $A' \rightarrow C$ included in the considered assertion set with $A \neq A'$. In this case, A and A' can also be conflicting because this doest not represent an inconsistency for the assertion set. For example, $always(p_1 \rightarrow next(p_2))$ and $always(p_3 \ until \ p_4 \rightarrow next(p_2))$ can be both true for the same DUV.

- Cell f_{00} is the number of times an assertion is true, whose antecedent and consequent are both different, respectively, from A and C, i.e., it is the sum of occurrences of the other assertions included in the analysed set.

- Cell f_{1X} is the sum of cells f_{11} and f_{10}.
- Cell f_{0X} is the sum of cells f_{01} and f_{00}.
- Cell f_{X1} is the sum of cells f_{11} and f_{01}.
- Cell f_{X0} is the sum of cells f_{10} and f_{00}.
- Cell f_{XX} is the grand total.

As an illustrative example, let us consider assertions reported in Table II. For sake of clearness, and without loss of generality, the table does not show the atomic propositions composing antecedents and consequents of assertions, but only the temporal relations between them in PSL syntax [16]. The corresponding contingency tables are reported in Table III. For example, for assertion $A1$, f_{11} correspond to the total number of occurrences of $A1$ in the analysed execution traces; f_{10} is equal to 0, since antecedent A does not appear in none of the other assertions; f_{01} is 0 since consequent $A \ until \ F$ does not appear in none of the other assertions; and finally, f_{00} is obtained by summing the occurrences of all the other assertions except $A1$. Cells f_{10} for assertions $A5$, $A6$ and $A7$ are not zero since they share the same antecedent E. Thus, f_{10} for $A5$, $A6$ and $A7$ are, respectively, the sum of occurrences of $A6$ and $A7$, $A5$ and $A7$, and $A5$ and $A6$. Similar considerations allow computing values for all the other cells of Table III.

	C	$\bar{C}$	
A	f_{11}	f_{10}	f_{1X}
$\bar{A}$	f_{01}	f_{00}	f_{0X}
	f_{X1}	f_{X0}	f_{XX}

TABLE I: Contingency table for $A \rightarrow C$.

AssertionID	Assertion	Occurrences
A1	$always(A \rightarrow A\ until\ F)$	468
A2	$always(B \rightarrow B\ until\ G)$	436
A3	$always(C \rightarrow C\ until\ H)$	481
A4	$always(D \rightarrow D\ until\ I)$	361
A5	$always(E \rightarrow next(J))$	524
A6	$always(E \rightarrow next[2](J))$	516
A7	$always(E \rightarrow next[3](J))$	509

TABLE II: An assertion set with the corresponding number of occurrences in the execution traces.

AssertionID	f_{11}	f_{10}	f_{01}	f_{00}
A1	468	0	0	2827
A2	436	0	0	2859
A3	481	0	0	2814
A4	361	0	0	2934
A5	524	1025	0	1746
A6	516	1033	0	1746
A7	509	1040	0	1746

TABLE III: Contingency tables of assertions reported in Table II.

C. Evaluation of interestingness

Contingency tables provide basic ingredients for the computation of support and correlation coefficient of a temporal assertion. Concerning support, according to Definition 6, it is simply computed with the following formula:

$$s = \frac{f_{11}}{f_{XX}}. \qquad (1)$$

The computation of the correlation coefficient for an assertion $A \rightarrow C$, according with Definition 7, is obtained instead by means of the following formula:

$$\rho = \frac{cov(A,C)}{\sigma A \cdot \sigma C} \qquad (2)$$

where $cov(A,C)$ is the covariance of A and C, while σA and σC are the standard deviation, respectively, of A and C. Disregarding mathematical steps, the correlation coefficient can be computed in terms of the cells of a contingency table as follows:

$$\rho = \frac{f_{11} \cdot f_{00} - f_{10} \cdot f_{01}}{\sqrt{f_{1X} \cdot f_{0X} \cdot f_{X1} \cdot f_{X0}}} \qquad (3)$$

According to equation (1) the support ranks in the highest positions assertions that occur frequently in the execution traces. However, we can have very interesting assertions that occur a few times because they refer to corner cases. On the other hand, the correlation coefficient privileges assertions where the number of occurrences of the antecedent better matches the number of occurrences of the consequent, but assertions where these numbers are low could be extracted by chance without representing a real behaviour of the DUV. For this reason a combination of support and correlation coefficient provides a more accurate estimation of assertion interestingness. Thus, we propose the measure the interestingness of an assertion A through the following formula:

$$I(A) = \alpha * s_n(A) + (1 - \alpha) * \rho_n(A) \qquad (4)$$

where, $\alpha \in [0, 1]$, and $s_n(A)$ and $\rho_n(A)$ are the value obtained by normalizing , respectively, the support s and the correlation coefficient ρ of A with respect to the whole set of analysed assertions. At varying of α the role of support becomes more or less important with respect to the role of the correlation coefficient in determining the final estimation of assertion interestingness. In our experiments best results have been obtained with $\alpha = 0.4$.

V. EXPERIMENTAL RESULTS

Experimental results have been carried out on an Intel (R) Xeon E5649 @2.53Ghz process equipped with 8 GB of RAM and running Linux OS. Efficiency and effectiveness of the proposed ranking methodology have been evaluated by considering assertions generated for an UART component and an APB bus through an assertion mining tool [17]. The accuracy of the interestingness estimation measured according to the metrics I, defined in Section IV-C, has been compared with a mutant analysis-based approach that ranks assertions according with their achieved *mutant coverage*, i.e., the capability of discovering *mutants*. A mutant is a small alteration of the DUV's source code that perturbs its functionality. A mutant is *observable* if, in comparison with a mutant-free DUV, its effect is visible as an alteration in the DUV's primary outputs. A mutant is *covered* by an assertion if the assertion fails when the mutant is observed at primary outputs. Then, the mutant coverage C is the ratio between covered mutants and observable mutants. Uncovered mutants highlight the incompleteness of the assertions set [9]. The hypothesis is that assertions with the highest mutant coverage are ranked in the highest positions also according to the proposed metrics I.

To experimentally prove the previous hypothesis, after the computation of the metrics I (with $\alpha = 0.4$) and the mutant coverage C, we divided assertions in 4 groups, respectively, $Q1_I, \ldots, Q4_I$ for I, and $Q1_C, \ldots, Q4_C$ for C. The division in groups has been done according with *quartiles* computed on I and C. In this way, the top 25%-ranked assertions with respect to I and C are included, respectively, in $Q4_I$ and $Q4_C$, while the worst 25%-ranked assertions are included in $Q1_I$ and $Q1_C$. Similarly, $Q3_I$ and $Q3_C$ include assertions between the first and the second quartile, while $Q2_I$ and $Q2_C$ include assertions between the second and the third quartile. Then, we analysed the impact of assertions belonging to the different groups in covering mutants. Results are reported in Table IV. After the DUV name, the second and the third Columns report, respectively, the number of analysed assertions ($\#assertions$) and the number of mutants totally covered by assertions ($\#mutants$). Then, Columns under *Preserved mutants* show how many mutants are still covered by preserving assertions belonging to only $Q4_I$ and only $Q4_C$, and to only $Q4_I \cup Q3_I$ and only $Q4_C \cup Q3_C$. Finally, Columns under *Loss mutants* show how many mutants remain uncovered by removing assertions belonging to $Q2_I \cup Q1_I$ and $Q2_C \cup Q1_C$, and to only $Q1_I$ and only $Q1_C$. It is evident from the results reported in Table IV that measuring the interestingness of assertions according to the metrics I proposed in this paper ranks in the highest positions assertions that cover the most of mutants, while in the lowest positions remain assertions that very rarely cover mutants not yet covered by better ranked assertions. In this context, the ranking provided by I is even better than

DUV	#assertions	#mutants	Preserved mutants				Loss mutants			
			$Q4_I$	$Q4_C$	$Q4_I \cup Q3_I$	$Q4_C \cup Q3_C$	$Q2_I \cup Q1_I$	$Q2_C \cup Q1_C$	$Q1_I$	$Q1_C$
UART	21	99	76	73	97	97	2	2	1	2
BUS-APB	24	22	18	NA	22	21	0	1	0	0

TABLE IV: Comparison between assertion ranking based on metrics I and mutation coverage C.

DUV	I time	I + sim time	C Time
UART	2 s.	4208 s.	26400 s.
BUS-APB	2 s.	70 s.	940 s.

TABLE V: Execution time for computing I and C.

the ranking provided by C, since, for example, in the case of UART, 76 mutants are covered by assertions included in $Q4_I$, while only 73 mutants are covered by assertions included in $Q4_C$; on the opposite, only one mutant remains uncovered by discarding assertions in $Q1_I$, while 2 mutants remain uncovered by discarding assertions in $Q1_C$.

It is worth noting also that in the case of *BUS-APB*, the number of mutants covered only by assertions belonging to group $Q4_C$ cannot be computed, because due to a particular distribution of covered mutants among assertions, the third quartile correspond exactly to the fourth (i.e., to the maximum number of mutants covered by the assertions with the highest mutant coverage). In particular, this happens because, by chance, 8 assertions on 24 cover the same (highest) number of mutants. In this situation, due to the low variability of mutant coverage among assertions there is no distinction between $Q3_C$ and $Q4_C$. This represents a drawback of the mutant-based analysis, which is instead outcome by the approach proposed in this paper that can effectively distinguish between $Q3_I$ and $Q4_I$. A further analysis has been conducted by measuring the time required for the computation of I and C.

Results are reported in Table V. It is evident that measuring I (I time) requires a few seconds, independently from the complexity of the DUV. On the contrary, mutation analysis requires a longer verification time I (C time) to simulate DUV and checkers for each mutant. This is particularly evident for complex designs like *UART*, where assertions predicate on large time windows (up to 665 clock cycles). For sake of clarity, the time reported for I does not include the time spent for counting assertion occurrences in the execution traces, since the result of such a counting is already available when assertions are automatically generated through assertions mining. If this information was not available, or assertions were manually defined, the time for computing I would include the time spent for one simulation run to compute assertion occurrences on the execution traces (I + *sim time*), while computation of C always requires a number of simulation runs equal to the number of mutants.

VI. CONCLUSIONS

In this paper we proposed a metrics to evaluate the interestingness of assertions. The approach re-adapts metrics typically adopted in data mining, i.e., support and correlation coefficient, to measure the importance of an assertion on the basis of both its activation frequency during simulation runs and the correlation between its antecedent and consequent. Experimental results showed that, compared to traditional mutant coverage-based techniques, our metrics provides a better estimation of assertion interestingness by ranking in the top positions assertions that cover the major number of mutants and in the lowest positions assertions that cover mutants detected also by better ranked assertions. Finally, concerning estimation time, we outperform the mutant coverage-based approach of one order of magnitude, by considering also the time required for computing the frequency of assertions by simulation. When such frequencies are already available (e.g., when provided by an assertion mining tool) the computation of the proposed metrics is almost negligible (a few seconds).

REFERENCES

[1] H. D. Foster, A. C. Krolnik, and D. J. Lacey, *Assertion-based design*. Springer, 2004.

[2] A. Danese, T. Ghasempouri, and G. Pravadelli, "Automatic extraction of assertions from execution traces of behavioural models," in *Proc. of ACM/IEEE DATE*, 2015.

[3] S. Hertz, D. Sheridan, and S. Vasudevan, "Mining hardware assertion with guidance from static analysis," *IEEE Trans. on CAD*, vol. 32, no. 6, pp. 952–965, 2013.

[4] M. Bertasi, G. Di Guglielmo, and G. Pravadelli, "Automatic generation of compact formal properties for effective error detection," in *Proc. of ACM/IEEE CODES+ISSS*, 2013, pp. 1–10.

[5] W. Li, A. Forin, and S. A. Seshia, "Scalable specification mining for verification and diagnosis," in *Proc. of ACM/IEEE DAC*, 2010.

[6] S. Katz, O. Grumberg, and D. Geist, "Have I written enough properties? — A method of comparison between specification and implementation," in *Proc. of ACM CHARME*, 1999, pp. 280–297.

[7] H. Hoskote, T. Kam, P. H. Ho, and X. Zao, "Coverage estimation for symbolic model checking," in *Proc. of ACM/IEEE DAC*, 1999, pp. 300–305.

[8] N. Jayakumar, M. Purandare, and F. Somenzi, "Dos and don'ts of CTL state coverage estimation," in *Proc. of ACM/IEEE DAC*, 2003, pp. 292–295.

[9] A. Fedeli, F. Fummi, and G. Pravadelli, "Properties incompleteness evaluation by functional verification," *IEEE Trans. on Computers*, vol. 56, no. 4, pp. 528–544, 2007.

[10] P.-N. Tan, V. Kumar, and J. Srivastava, "Selecting the right interestingness measure for association patterns," in *Proc. of ACM/SIGKDD KDD*, 2002, pp. 32–41.

[11] P.-N. Tan and V. Kumar, "Interestingness measures for association patterns: A perspective," in *Proc. of Workshop on Postprocessing in Machine Learning and Data Mining*, 2000.

[12] C. M. Antunes and A. L. Oliveira, "Temporal data mining: An overview," in *Proc. of Workshop on Temporal Data Mining*, 2001.

[13] R. J. Bayardo Jr and R. Agrawal, "Mining the most interesting rules," in *Proceedings of the fifth ACM SIGKDD international conference on Knowledge discovery and data mining*. ACM, 1999, pp. 145–154.

[14] M. Boulé and Z. Zilic, *Generating Hardware Assertion Checkers: For Hardware Verification, Emulation, Post-Fabrication Debugging and On-Line Monitoring*. Springer Publishing Company, Incorporated, 2008.

[15] K. Pearson and L. N. G. Filon, "Mathematical contributions to the theory of evolution. IV. on the probable errors of frequency constants and on the influence of random selection on variation and correlation," *Philosophical Transactions*, vol. 191, pp. 229–311, 1898.

[16] "Standard for property specification language (PSL)," *IEC 62531:2012(E) (IEEE Std 1850-2010)*, pp. 1–184, 2012.

[17] A. Danese, F. Filini, and G. Pravadelli, "A time-window based approach for dynamic assertions mining on control signals," in *Proc. of IFIP/IEEE VLSI-SOC*, 2015.

Hardware/Software Partitioning of Embedded System-on-Chip Applications

Jia Wei Tang
Faculty of Electrical Engineering
Universiti Teknologi Malaysia
Email: jwtang2@live.utm.my

Yuan Wen Hau
Faculty of Biosciences and Medical Engineering
Universiti Teknologi Malaysia
Email: hauyuanwen@biomedical.utm.my

MN Marsono
Faculty of Electrical Engineering
Universiti Teknologi Malaysia
Email: nadzir@fke.utm.my

Abstract—HW/SW partitioning is an important development step during HW/SW co-design to ensure application performance in embedded System-on-Chip (SoC). This paper formulates the optimization of HW/SW partitioning aiming at maximizing streaming throughput with predefined area constraint, targeted for multi-processor system with hardware accelerator sharing capability. Two software-oriented and the second hardware-oriented greedy heuristic algorithms for HW/SW partitioning are proposed and tested on several random graphs and one multimedia application (MP3 decoder). Results show that the best result from both proposed greedy algorithms produce 93.6% near-optimal solution compared to brute force ground truth with faster HW/SW partitioning time.

Keywords—Embedded system, hardware/software partitioning, streaming applications.

I. INTRODUCTION

Hardware/software (HW/SW) co-design has been widely used in numerous embedded system-on-chips (SoCs). Moreover, modern embedded SoCs allow software parallelism as well as hardware accelerations. Thus, efficient HW/SW partitioning is required to ensure high throughput and cost-effective embedded SoCs while satisfying the shorter time-to-market.

An embedded SoC application can be represented by pipelined execution of multiple tasks. HW/SW partitioning plays an important role for selecting software execution or hardware acceleration for each task. Hardware-executed tasks usually perform faster at a cost of increased hardware area and higher power consumption. As tasks with similar functionality often exist in streaming applications such as MP3 decoder [1] and JPEG encoder [2], sharing hardware accelerators among these tasks could reduce area consumption and may not degrade the system throughput if execution of these tasks is insignificant compared to other tasks.

The rest of the paper is organized as follow. Section II presents the related works in HW/SW partitioning. Section III describes the modeling of streaming application where as Section IV formulates the HW/SW partitioning problem. Software-oriented and hardware-oriented greedy heuristic algorithms are proposed and presented in Section V. Section VI shows the experimental results of the proposed greedy algorithms against ground truth brute force solution. Section VII concludes this paper.

II. RELATED WORKS

HW/SW partitioning improves embedded SoCs performances in different aspects as the complexity of SoCs increases. Traditionally, HW/SW partitioning is carried out manually based on system designer's experience [3], [4]. Many approaches have been proposed recently to fulfil and optimize multiple objectives and costs. Although there exist a wide variety of problem formulation and cost definition, these are highly dependent on targeted system architectures.

The most common problem described in literature [5]–[11] is the optimization of execution time, hardware area and communication cost, targeted for simple single-software single-hardware system. There are fairly little works conducted for throughput optimization compared to overall execution time due to the difficulty of its estimation and formulation. Hence, several previous works [1], [12]–[15] have integrated pipeline scheduling of tasks in order to construct throughput formulation together with other co-design constraints. Apart from that, other works have also incorporated different cost metrics and objectives in HW/SW partitioning, including power consumption [2], [16]–[18], and software memory usage [2], [16], [17], as summarized in Table I.

Other features have also been explored to further enhance the efficiency and optimize HW/SW partitioning for different modern SoC system. References [13], [14], [18] have proposed the HW/SW partitioning for multi-processor system by considering software parallelism. Reference [20] has proposed hardware accelerator sharing (among hardware tasks) to reduce area cost. Multiple hardware architecture choices is discussed in [21], allowing tasks to be mapped to different hardware accelerator alternatives with different performance and cost.

This paper addresses HW/SW partitioning for optimization of throughput subject to a predefined hardware area constraint considering software parallelism (multi-processor system) and hardware sharing capability. Two greedy algorithms (software-oriented and hardware-oriented) are proposed to give a fast near-optimal solutions for the targeted objectives and features.

III. MODELLING OF STREAM APPLICATIONS AS DIRECTED ACYCLIC GRAPH (DAG)

Task graph of an application can be characterized by a Directed Acyclic Graph (DAG), $G = (T, E)$, $T = \{t_1, t_2, ..., t_n\}$ and $E = \{e_1, e_2, ..., e_n\}$ where $s, h, a : T \rightarrow \mathbb{R}^+$ and

978-1-4673-9141-2/15 $31.00 © 2015 IEEE

TABLE I: Related works in HW/SW partitioning

Previous works	Objectives/Costs						Features			
	Execution time	Throughput	HW area	Communication	SW memory	Power consumption	Multi-processors	HW Accelerator sharing	Multi-choices HW	NoC
[1]		✓	✓	✓			✓			✓
[5]–[11],	✓		✓	✓						
[2], [16], [17]	✓		✓		✓	✓				
[12]		✓	✓							
[13], [14]		✓	✓				✓			
[15]		✓	✓			✓				
[18]	✓		✓			✓	✓			
[19]	✓		✓							
[20]	✓		✓	✓					✓	
[21]	✓		✓	✓					✓	
Proposed		✓	✓				✓	✓		

$c : E \to \mathbb{R}^+$ as shown in Fig. 1. Each task, t_i contains a tupple $< s_i, h_i, a_i >$ denoting software execution time, hardware execution time and hardware area cost, respectively. On the other hand, each edge e_i includes c_{ij} which represents the communication cost between tasks t_i and t_j. All tasks are to be bi-partitioned into hardware executable, H or software executable, S, satisfying $H \cup S = T$ and $H \cap S = \emptyset$. Software processors are assumed to be available as in any modern embedded SoCs. It is common for modern embedded SoCs to have multiple processors. The area for software processors are not considered as additional area overhead in the HW/SW partitioning model. All software processors are also assumed to be identical, thus a software task's execution time is identical in any processor.

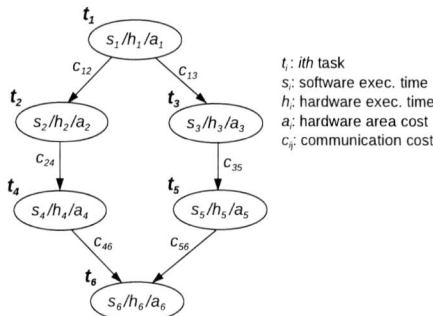

Fig. 1: Different type of costs in task graph specifying an application.

In most stream applications such as multimedia programs, some tasks are split into several tasks with similar functions to increase the system throughput by executing these tasks concurrently using different cores (such as MP3 Decoder task in [1] and JPEG Encoder task in [2]). The existence of these similar tasks motivates the necessity of hardware cores sharing as an option to utilize area costs, giving rise to a bigger search space in HW/SW partitioning problem.

As this paper focuses on throughput optimization in multiprocessor system with hardware sharing capability, HW/SW partitioning is aiming to distribute tasks among all available software cores and hardware cores. Any task can be assigned to any software core. However, only tasks with similar func-

tionality can share the same hardware cores. For instance, task graph in Fig. 2 is mapped to a system consisting two software processors (PE_7 and PE_8). Core PE_5 is assigned to execute both similar tasks t_4 and t_5. Tasks that are mapped to software or shared hardware will render their respective hardware cores unused and unimplemented, thus at the same time, reducing hardware area cost.

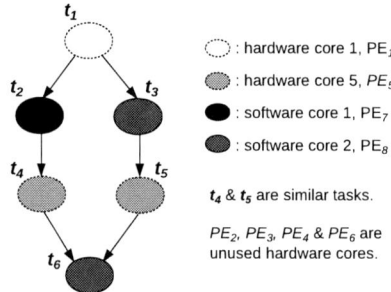

Fig. 2: HW/SW partitioning of tasks for system consisting two software processors with hardware sharing capability.

As different communication interfaces have different latency and throughput, exact HW/SW interconnection delay cannot be directly estimated. With the possibility of having high throughput communications in all connections, this paper assumes that communication overheads are insignificant (similar as [2], [12]–[19]) compared to computation time and thus, does not incorporate any communication costs in HW/SW partitioning problem.

Assuming all tasks are scheduled and pipelined-executed in the same time step, the overall system throughput is defined by the reciprocal of the time step, which is determined by the minimum allowable task processing time (critical time). Fig. 3 illustrates the task executions timeline for a stream application based on task partitioning in Fig. 2. Data are streamed through hardware (PE_1 to PE_6) and software cores (PE_7 and PE_8) according to a certain partitioning order. These data are required to be processed in six tasks by different cores at different time steps. PE_8 emerges as the slowest core with execution of two tasks, t_3 and t_6 every time step, thus resulting in the slowest task completion time (critical time) in each time step to be $s_3 + s_6$.

978-1-4673-9141-2/15 $31.00 © 2015 IEEE

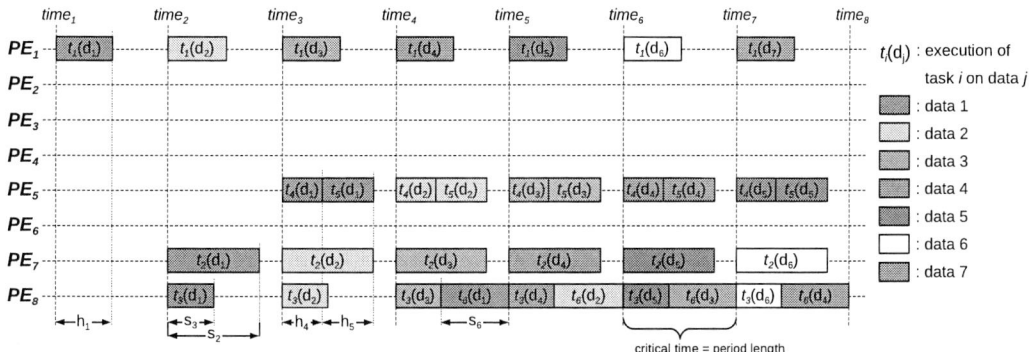

Fig. 3: HW/SW pipelining of tasks with core sharing capability.

Hardware sharing of the tasks can be purely beneficial in area reduction with no trade-off in throughput. For instance, PE_5 is shared by t_4 and t_5 as they are similar in function and both tasks are executed in serial. Although serial execution is slower compared to parallelism in their specialized hardware cores, the reduction in execution speed of these tasks does not affect the overall throughput as they are not the slowest (critical time) in the system.

IV. HW/SW PARTITIONING

Given a DAG representation of a stream application with N_t tasks and a multi-core system with N_s software processor, HW/SW partitioning partitions these tasks to N_s software cores and N_h hardware cores which maximizes the system throughput, subject to an area constraint, A. Assuming all tasks have their own specialized hardware cores, the number of hardware cores is equal to number of tasks ($N_h = N_t$). Thus, the number of total cores, N_c for a given problem is $N_c = N_s + N_t$.

As hardware sharing is possible in this partitioning problem, the allowable task-to-core mapping is defined by all available software cores and allowable hardware cores. The allowable task-to-core mapping can be represented in a mask M, constructed of a $N_c \times N_t$ matrix with column index i and row index j indicate task identifier and core identifier respectively, and $m_{ij} \subset M$. If $m_{ij} = 1$, then t_i are permitted to be assigned to core PE_j. For instance, Equation (1) illustrates an example of mask M for the tasks graph described in Fig. 2, assuming the targeted system contains two software processors.

$$
M = \begin{array}{c} \\ PE_1 \\ PE_2 \\ PE_3 \\ PE_4 \\ PE_5 \\ PE_6 \\ PE_7 \\ PE_8 \end{array}
\begin{array}{c} \begin{array}{cccccc} t_1 & t_2 & t_3 & t_4 & t_5 & t_6 \end{array} \\
\begin{bmatrix}
1 & 0 & 0 & 0 & 0 & 0 \\
0 & 1 & 0 & 0 & 0 & 0 \\
0 & 0 & 1 & 0 & 0 & 0 \\
0 & 0 & 0 & 1 & 0 & 0 \\
0 & 0 & 0 & 1 & 1 & 0 \\
0 & 0 & 0 & 0 & 0 & 1 \\
1 & 1 & 1 & 1 & 1 & 1 \\
1 & 1 & 1 & 1 & 1 & 1
\end{bmatrix} \end{array} \quad (1)
$$

The mask M is sorted in increasing row index j from hardware cores to software cores. Since all tasks are assumed to have their own specialized hardware cores, for simplicity, these cores are allocated with the core identifiers similar to

their respective task identifiers ($i = j$). All software cores are assumed to be able to execute any task. As a result, an identity matrix is obtained for $1 \leq j < N_t$ while an all-ones matrix for $N_t \leq j < N_c$.

On the other hand, similar tasks may have two or more admissible spaces (HW cores) due to the possibility of hardware sharing. The feasible way for hardware sharing can be formulated as a combination selection of $\sum_{r=1}^{n-1} {}^nC_r$, where n is the number of similar tasks and r is the number of shared hardware cores. To reduce the search space by eliminating similar hardware sharing selections, each task t_i is only assigned to a core PE_j whose identifier is greater or equal to the task identifier, $j \geq i$. The total possible permutation search space can be easily determined from M and is evaluated as $\prod_{i=1}^{N_t} \sum_{j=1}^{N_c} m_{ij}$.

In addition, mask M is capable of distinguishing differences in costs for each task t_i assigned in different cores PE_j. Along the increasing j, each task can be mapped to a different core, starting from its own specialized hardware core, followed by sharing hardware core, and finally on software cores. Thus, it is undeniable that assigning tasks to greater j reduces more hardware area cost while increasing its execution time.

Similarly, HW/SW assignment action can be represented mathematically by a $N_c \times N_t$ assignment matrix X. $x_{ij} = 1$, denotes the mapping of t_i to core PE_j. Since each task can only be assigned to one available core and all tasks must be mapped at the end of the partition, the summation of each column in X must equal to one, formulated as $\sum_{j=1}^{N_c} x_{ij} = 1$, $\forall i = 1, 2, ..., N_t$. Equation (2) illustrates an example of assignment matrix X for the mapping in Fig.2. Only one task assignment appears in each column while multiple assignments appear in row with software cores (PE_8) or shared hardware core (PE_5). Rows without any assignment indicate unused (unimplemented) cores.

$$
X = \begin{array}{c} \\ PE_1 \\ PE_2 \\ PE_3 \\ PE_4 \\ PE_5 \\ PE_6 \\ PE_7 \\ PE_8 \end{array}
\begin{array}{c} \begin{array}{cccccc} t_1 & t_2 & t_3 & t_4 & t_5 & t_6 \end{array} \\
\begin{bmatrix}
1 & 0 & 0 & 0 & 0 & 0 \\
0 & 0 & 0 & 0 & 0 & 0 \\
0 & 0 & 0 & 0 & 0 & 0 \\
0 & 0 & 0 & 0 & 0 & 0 \\
0 & 0 & 0 & 1 & 1 & 0 \\
0 & 0 & 0 & 0 & 0 & 0 \\
0 & 1 & 0 & 0 & 0 & 0 \\
0 & 0 & 1 & 0 & 0 & 1
\end{bmatrix} \end{array} \quad (2)
$$

With the aid of assignment matrix X, the partitioning problem discussed in this paper is formulated as a minimization problem in Equation (3), aiming to find maximum throughput (minimum critical time) subject to area constraint A.

$$\text{minimize} \quad \max\left(H_t, S_t\right)$$
$$\text{subject to} \quad \sum_{j=1}^{N_t}\left(\max_{1\leq i\leq N_t} x_{ij}.a_i\right) \leq A, \, x_{ij} \in \{0,1\}$$
$$(3)$$

where H_t and S_t denote the execution time of the slowest hardware and software respectively, which are formulated in Equation (4) and Equation (5) respectively.

$$H_t = \max_{1\leq j\leq N_t}\left(\sum_{i=1}^{N_t} h_i.x_{ij}\right) \qquad (4)$$

$$S_t = \max_{N_t < j\leq N_c}\left(\sum_{i=1}^{N_t} s_i.x_{ij}\right) \qquad (5)$$

Based on Equation (3), the system throughput and hardware area usage for any given mapping can be easily enumerated. Critical time is determined by identifying the longest execution time of each core while execution time of each core is total execution time of the assigned tasks. Hardware area is computed by summation of all hardware areas except for unused cores.

V. GREEDY ALGORITHMS

Two greedy heuristic algorithms, *Alg-greedy1* and *Alg-greedy2* are proposed in this paper to optimize HW/SW partitioning problem described in Section IV. The former is software-oriented while the latter is hardware-oriented, both aiming to maximize throughput (i,e: minimize critical time) for a given area constraint A, without exhaustively exploring all possible partitionings.

Similar to conventional greedy algorithm [6], both algorithms exploit the profit-to-cost ratio to make a decision. As the proposed greedy algorithms are intended to maximize throughput, different profit-to-cost ratio is proposed. As each task is allowed to be assigned to more than two cores with different profit or cost, the profit-to-cost ratio of each task has to be calculated specifically based on the execution speed of the tasks on each allowable cores. Assuming that a task t_i is pre-assigned to core PE_j, the profit-to-cost ratio, PCR of the task to move to another core PE_k, is formulated in Equation (6). If the task is moved among hardware cores, the PCR is calculated by dividing its hardware speed to area. However, if a software core is involved either as a source or a destination core, PCR is calculated as software speed-to-area ratio.

$$PCR_{ijk} = \begin{cases} \frac{h_i}{a_i} & \text{if } j \leq N \text{ and } k \leq N \\ \frac{s_i}{a_i} & \text{otherwise} \end{cases} \qquad (6)$$

$\forall i = \{1, 2, ..., N_t\}$, $\forall j = \{1, 2, ..., N_c\}$, $\forall k = \{1, 2, ..., N_c\}$, where PCR_{ijk} is profit-to-cost ratio for moving task t_i from PE_j to PE_k.

A. Software-Oriented Greedy (Alg-greedy1)

This algorithm first considers all-software system. Then, tasks are converted to hardware one-by-one to improve throughput until no more tasks could be repartitioned or mapped to hardware cores as long as the partition preserves the area constraint A, as shown in Algorithm 1.

Algorithm 1 *Alg-greedy1*
---
generate *PCR* matrix
distribute all tasks to SW cores
while 1 **do**
   best mapping= current mapping
   find slowest core
   for t_i= tasks in slowest core sorted in descending *PCR* **do**
      if upward(-j) move is available **then**
         move t_i to next upward(-j) core
         if area after move $\leq$ area constraint **then**
            break for loop
         else
            undo move
         end if
      end if
   end for
   if no more available move **then**
      break while loop
   end if
end while
return best mapping

It is obvious that tasks can only be moved in upward (-j) direction in mask M to gain higher throughput by investing area cost as a trade-off. Each task will have a predefined move one step at a time from software to shareable hardware, and finally as dedicated hardware core. Thus, by identifying sources and destination cores for each step, *PCR* can be pre-generated in a matrix as summarized in Equation (7) based on Equation (6).

$$PCR = \begin{matrix} & \begin{matrix} i=1 & \cdots & i=N_t \end{matrix} \\ \begin{matrix} j=1 \\ \vdots \\ j=N_t \\ j=N_{t+1} \\ \vdots \\ j=N_c \end{matrix} & \begin{bmatrix} \frac{h_i}{a_i} & \cdots & \frac{h_i}{a_i} \\ \vdots & \ddots & \vdots \\ \frac{h_i}{a_i} & \cdots & \frac{h_i}{a_i} \\ \frac{s_i}{a_i} & \cdots & \frac{s_i}{a_i} \\ \vdots & \ddots & \vdots \\ \frac{s_i}{a_i} & \cdots & \frac{s_i}{a_i} \end{bmatrix} \end{matrix} \cdot M \qquad (7)$$

All tasks are initially distributed among software cores as equal as possible. This is achieved by sorting all tasks based on execution time in descending order, followed by assigning them one-by-one to the least utilized core in each step.

As system throughput is always determined by the slowest core, only tasks assigned in this core will be examined and remapped at each step. Based on the obtained *PCR* matrix, the algorithm iterates and reassigns the highest *PCR* task upward (-j) with the condition that area cost after reassignment does not exceed the predefined area constraint A. Termination of the loop occurs when there is no further move which could satisfy the area constraint, returning the best mapping as the greedy result.

978-1-4673-9141-2/15 $31.00 © 2015 IEEE

B. Hardware-Oriented Greedy (Alg-greedy2)

Alg-greedy2 initializes all tasks as dedicated hardware cores, then reassigns tasks one-by-one to software to reduce area cost by sacrificing throughput until area constraint is satisfied, as illustrated in Algorithm 2.

Similarly, a *PCR* matrix is pre-generated based on Equation (6) before tasks reassignments. Starting by assigning all tasks in uppermost location of matrix M, the moveable task with lowest *PCR* is mapped downward $(+j)$ in each iteration until area constraint is attained. Lastly, all software tasks are distributed as equal as possible for all software cores.

Algorithm 2 *Alg-greedy2*

---

generate *PCR* matrix
assign all tasks to dedicated HW cores
while area$\geq$ area constraint **do**
    best mapping= current mapping
    for t_i= all tasks sorted in ascending *PCR* **do**
        if downward $(+j)$ move for t_i is available **then**
            if downward $(+j)$ move is SW core **then**
                move t_i to least utilized SW core
            else
                move t_i to next downward $(+j)$ core
            end if
            break for loop
        end if
    end for
end while
redistribute all SW tasks equally
return best mapping

VI. RESULTS AND DISCUSSIONS

The proposed greedy algorithms are coded in Octave in Linux environment. A common data intensive streaming multimedia applications, MP3 decoder (see [1]) is selected as test case for HW/SW partitioning of two software cores and three software cores systems. The task graph describing the specifications of MP3 decoder is illustrated in Fig. 4. There are several similar tasks in MP3 decoder applications which can be determined by the similar costs (SW time, HW time, HW area) in the tasks, providing hardware sharing possibility. The experiment has been conducted using both proposed greedy algorithms to maximize system throughput (minimize critical time) for different area constraints.

The results of critical time-to-area constraints of greedy algorithms are shown in Fig. 5. When area constraint is too low to be occupied by any of the hardware cores, the system throughput is determined by equal distribution of total software execution time to the number of software cores. As area constraint increases, the proposed algorithms attempt to utilize the available resources effectively. This improves throughput by choosing the correct tasks for hardware execution as well as hardware sharing. Hence, even with infinite area constraints, all tasks do not necessary consume hardware area. Several tasks can still be performed in software or in shared hardware, resulting in the maximum achievable throughput (minimum possible critical time) attained without the need for full hardware acceleration.

Apart from MP3 Decoder application, random task graphs are also used to benchmark and determine the performance for different sizes of task graphs. Random graphs are generated

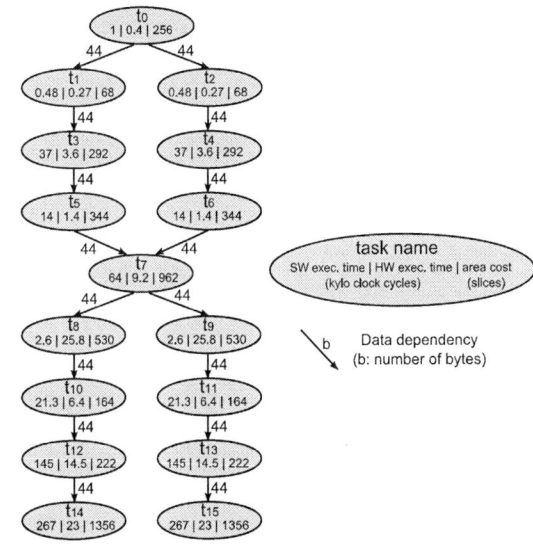

Fig. 4: Task graph of MP3 decoder application (see [1]).

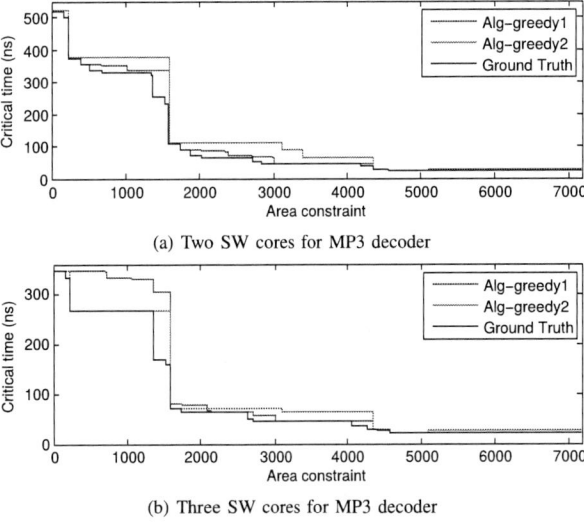

(a) Two SW cores for MP3 decoder

(b) Three SW cores for MP3 decoder

Fig. 5: Critical time-to-area constraint result of MP3 Decoder streaming applications.

using TGFF [22], with 100 ± 90 hardware speed, 1000 ± 900 software speed and 100 ± 50 hardware area with 2 and 3 software processors. Both greedy algorithms are used to perform HW/SW partitioning for all graphs, optimizing throughput under area constraint.

Solutions from the greedy algorithms are compared with ground truth result from exhaustive brute force to obtain their optimality. *Optimality* is defined as the average ratio of ground truth result to the obtained greedy result for all DAGs, and is illustrated in Equation (8). An obtained result is 100% optimal if it is identical with ground truth result. Optimality of combined greedy is also determined, where the combined greedy chooses the best result from both *Alg-greedy1* and *Alg-*

greedy2 for each HW/SW partitioning. The optimality of the proposed greedy heuristics are shown in Table II.

$$\text{Optimality} = 100 \times \underset{all\ DAG}{\text{Average}} \left(\frac{R_t}{R_g} \right) \quad (8)$$

where R_g and R_t are critical time results from greedy and brute force ground truth respectively.

TABLE II: Optimality of proposed greedy algorithms

Test Case	No. of tasks	SW cores	Optimality (%)		
			Alg-greedy1	*Alg-greedy2*	Combined
MP3	16	2	94.2	77.1	94.2
MP3	16	3	92.1	84.1	95.2
Random1	14	2	93.2	90.3	96.6
Random1	14	3	87.3	93.0	97.6
Random2	7	2	81.7	84.7	88.0
Random2	7	3	85.2	91.9	95.8
Random3	10	2	91.7	76.9	95.2
Random3	10	3	89.0	82.1	94.6
Random4	7	2	88.2	77.7	89.7
Random4	7	3	92.0	87.4	98.6
Random5	13	2	76.7	79.4	85.6
Random5	13	3	83.6	81.3	89.0
Random6	9	2	89.6	76.5	91.6
Random6	9	3	76.5	89.7	99.4

Both *Alg-greedy1* and *Alg-greedy2* results are near-optimal with 87.2% and 83.7% from ground truth solution respectively. The combined greedy performs better than individual greedy algorithms as it is 93.6% optimal.

VII. CONCLUSION

Efficient HW/SW partitioning produces different cost-effective solutions in throughput-to-area trade-off. This paper formulated the optimization problem of streaming throughput in embedded SoCs with hardware sharing capability given a predefined constraint of hardware area cost. A software-oriented and a hardware-oriented greedy heuristic algorithms were proposed to optimize throughput without performing full search. Both algorithms were empirically tested on MP3 decoder test case, that is a well-known streaming multimedia applications. Results show that *Alg-greedy1* and *Alg-greedy2* give near-optimal solution with 87.2% and 83.7% respectively in comparison with brute force ground truth. If both greedy are combined by choosing best result for each area constraint A, the accuracy is 93.6%. The proposed approach may be potentiality extended to more specific on-chip interconnect architecture with formulated communication costs.

REFERENCES

[1] S. Le Beux, G. Bois, G. Nicolescu, Y. Bouchebaba, M. Langevin, and P. Paulin, "Combining mapping and partitioning exploration for noc-based embedded systems," *Journal of Systems Architecture*, vol. 56, no. 7, pp. 223–232, 2010.

[2] T.-Y. Lee, Y.-H. Fan, Y.-M. Cheng, C.-C. Tsai, and R.-S. Hsiao, "Enhancement of hardware-software partition for embedded multiprocessor fpga systems," in *Third International Conference on Intelligent Information Hiding and Multimedia Signal Processing, 2007. IIHMSP 2007.*, vol. 1. IEEE, 2007, pp. 19–22.

[3] B. Mei, P. Schaumont, and S. Vernalde, "A hardware-software partitioning and scheduling algorithm for dynamically reconfigurable embedded systems," in *Proceedings of ProRISC*. Citeseer, 2000, pp. 405–411.

[4] F. Clouté, J.-N. Contensou, D. Esteve, P. Pampagnin, P. Pons, and Y. Favard, "Hardware/software co-design of an avionics communication protocol interface system: an industrial case study," in *Proceedings of the Seventh International Workshop on Hardware/Software Codesign, 1999.(CODES'99)*. IEEE, 1999, pp. 48–52.

[5] P. Arató, Z. Á. Mann, and A. Orbán, "Algorithmic aspects of hardware/software partitioning," *ACM Transactions on Design Automation of Electronic Systems (TODAES)*, vol. 10, no. 1, pp. 136–156, 2005.

[6] W. Jigang, T. Srikanthan, and G. Chen, "Algorithmic aspects of hardware/software partitioning: 1d search algorithms," *IEEE Transactions on Computers*, vol. 59, no. 4, pp. 532–544, 2010.

[7] G. Li, J. Feng, C. Wang, and J. Wang, "Hardware/software partitioning algorithm based on the combination of genetic algorithm and tabu search," *Engineering Review*, vol. 34, no. 2, pp. 151–160, 2014.

[8] G. Lin, W. Zhu, and M. M. Ali, "A tabu search-based memetic algorithm for hardware/software partitioning," *Mathematical Problems in Engineering*, vol. 2014, 2014.

[9] J. Wu, P. Wang, S.-K. Lam, and T. Srikanthan, "Efficient heuristic and tabu search for hardware/software partitioning," *The Journal of Supercomputing*, vol. 66, no. 1, pp. 118–134, 2013.

[10] X. Zhao, H. Zhang, Y. Jiang, S. Song, X. Jiao, and M. Gu, "An effective heuristic-based approach for partitioning," *Journal of Applied Mathematics*, vol. 2013, 2013.

[11] J. Henkel and R. Ernst, "An approach to automated hardware/software partitioning using a flexible granularity that is driven by high-level estimation techniques," *IEEE Transactions on Very Large Scale Integration (VLSI) Systems*, vol. 9, no. 2, pp. 273–289, 2001.

[12] K. S. Chatha and R. Vemuri, "Hardware-software partitioning and pipelined scheduling of transformative applications," *IEEE Transactions on Very Large Scale Integration (VLSI) Systems*, vol. 10, no. 3, pp. 193–208, 2002.

[13] S. Bakshi and D. D. Gajski, "Hardware/software partitioning and pipelining," in *Proceedings of the 34th annual Design Automation Conference*. ACM, 1997, pp. 713–716.

[14] ——, "Partitioning and pipelining for performance-constrained hardware/software systems," *IEEE Transactions on Very Large Scale Integration (VLSI) Systems*, vol. 7, no. 4, pp. 419–432, 1999.

[15] N. N. Bình, M. Imai, A. Shiomi, and N. Hikichi, "A hardware/software partitioning algorithm for designing pipelined asips with least gate counts," in *Design Automation Conference Proceedings 1996, 33rd*. IEEE, 1996, pp. 527–532.

[16] M. B. Abdelhalim and S.-D. Habib, "An integrated high-level hardware/software partitioning methodology," *Design Automation for Embedded Systems*, vol. 15, no. 1, pp. 19–50, 2011.

[17] P. K. Nath and D. Datta, "Multi-objective hardware–software partitioning of embedded systems: A case study of jpeg encoder," *Applied Soft Computing*, vol. 15, pp. 30–41, 2014.

[18] E. Sha, L. Wang, Q. Zhuge, J. Zhang, and J. Liu, "Power efficiency for hardware/software partitioning with time and area constraints on mpsoc," *International Journal of Parallel Programming*, pp. 1–22, 2013.

[19] A. Bhattacharya, A. Konar, S. Das, C. Grosan, and A. Abraham, "Hardware software partitioning problem in embedded system design using particle swarm optimization algorithm," in *International Conference on Complex, Intelligent and Software Intensive Systems, 2008. CISIS 2008*. IEEE, 2008, pp. 171–176.

[20] J. Jeon and K. Choi, "Loop pipelining in hardware-software partitioning," in *Design Automation Conference 1998. Proceedings of the ASP-DAC'98. Asia and South Pacific*. IEEE, 1998, pp. 361–366.

[21] J. Wu, Q. Sun, and T. Srikanthan, "Algorithmic aspects for multiple-choice hardware/software partitioning," *Computers & Operations Research*, vol. 39, no. 12, pp. 3281–3292, 2012.

[22] R. P. Dick, D. L. Rhodes, and W. Wolf, "Tgff: task graphs for free," in *Proceedings of the 6th international workshop on Hardware/software codesign*. IEEE Computer Society, 1998, pp. 97–101.

Exploiting Scalable CGRA Mapping of LU for Energy Efficiency using the *LAYERS* Architecture

Zoltán Endre Rákossy[†], Dominik Stengele[†],
Gerd Ascheid[†], Rainer Leupers[†] and Anupam Chattopadhyay[*]
[†] Institute for Communication Technologies and Embedded Systems (ICE)
RWTH University Aachen, Germany
[*] School of Computer Engineering, Nanyang Technological University, Singapore
Email: rakossy@ice.rwth-aachen.de; anupam@ntu.edu.sg

Abstract—A scalable and highly efficient numerical linear algebra kernel mapping for coarse-grained reconfigurable architectures is proposed and applied to a 3D reconfigurable architecture, *Layers*, which exploits functional parallelism and a functional reconfiguration-based programming model to achieve flexibility, scalability and low energy. Instead of solving the complex problem of mapping an application to fit architectural constraints, in our approach we tailor the mapping scheme for efficiency and scalability and exploit architectural flexibility and reconfigurability to adapt the architecture to match the derived mapping. Thus, kernel execution reaches asymptotically optimal efficiency for various architectural parameters and input matrix sizes, without modification of the derived mapping. Detailed performance and power evaluations were done with input data sets with matrix sizes ranging from 64×64 to 16384×16384. Twelve architectural variants with up to 10×10 processing elements were used to explore scalability of the mapping and the architecture, achieving <10% energy increase for architectures up to 8×8 PEs, coupled with performance speed-ups of more than an order of magnitude.

Keywords—*Coarse-Grained Reconfigurable Architecture (CGRA); 3D Architecture; Numerical Linear Algebra; Energy-Efficient Architecture;*

I. INTRODUCTION

Rising computational demand of applications and stringent power consumption constraints force latest System-on-Chip designs to adopt solutions which rely on domain-specific accelerators for complex tasks such as baseband signal processing/wireless MIMO, cryptography, media and data processing. Accelerating applications using coarse-grained reconfigurable architectures (CGRA) has been shown to be successful by several designs [1]–[8] making use of both computational density [9] and flexibility [10] advantages. However, *scalability* is also important to consider when exploring the design space. Optimal trade-off points need to be explored for maximizing performance while minimizing energy. Changing architectural or application parameters can ruin a carefully crafted manual mapping. In this paper, we propose scalable and highly efficient mapping solutions and apply them on the *Layers* CGRA originally proposed in [11] and recently improved for scalability [12] in order to achieve low energy consumption.

To analyze the effect of mapping efficiency and scalability on energy consumption, we focus on Numerical Linear Algebra (NLA), which due to its highly parallel and regular nature is a well-suited application domain for CGRAs. However, each kernel from this domain requires special care in order to find a mapping that runs efficiently on a CGRA, more so if

the underlying architecture is scalable. This is due to kernel-specific parallelism requiring specific storage access patterns and varying bandwidths, a problem which is amplified by the inherent lack of memory access bandwidth of architectures with a large number of processing elements (PEs). An inefficiently mapped kernel does not fully tap the potential of all execution units, increasing energy usage and decreasing performance. Especially architectural features such as number of PEs and memory ports are variable, further complexity arises in the mapping process. Finding the best mapping for a given application under architectural constraints is one of most complex problems in CGRA development.

The key idea of our proposal is to approach the problem from a different angle: first, instead of searching for a mapping to fit given architectural constraints, the algorithm of the kernel is thoroughly analyzed, to extract opportunities for parallelism and scalability. Next, a manually optimized scheduling focused on execution efficiency and scalability is derived, using the analysis results. This derivation considers only a high-level view of architectural constraints, such as number of processing elements and their location, available memory bandwidth (loads/stores per cycle), entered as parameters, with efficiency and scalability as the primary optimization rule. Finally, for the detailed mapping to the architecture, the manual scheduling and coarse mapping is considered as fixed and immutable in time/space and architectural flexibility is exploited to *adapt* the architecture to these mapping constraints and *construct* the missing mapping elements using a reconfigurable data path. In essence, instead of adapting the mapping to the architecture, we adapt the architecture via reconfigurability to a mapping that we construct for efficiency and scalability.

This complex interplay between application mapping, scaling and architectural features and how it influences energy efficiency is discussed in the next sections for the LU decomposition kernel. Extensive data is provided to aid comparison with other CGRA architectures coupled with results analysis, since a literature survey has produced no exact comparison points. LAC [6] [13] [14] is another CGRA targeting the linear algebra domain, although performance/power results are based on high-level estimates. Experiments with NLA kernels have been conducted also on the REDEFINE CGRA [5], and we are aware of FPGA-based solutions such as LAPACKrc [15], which promise >100× speed-up versus CPU-based solutions, but no directly comparable results for LU are published. GPGPU solutions such as CULA [16] and CUBLAS [17] show impressive aggregated results, but there is no sufficient detailed data to make a detailed performance comparison possible,

978-1-4673-9141-2/15 $31.00 © 2015 IEEE

especially in terms of energy efficiency.

II. THE *Layers* ARCHITECTURE

The *Layers* architecture is a new multi-layered scalable and parameterizable CGRA [12], developed initially as a 4×4 fixed architecture [11], using a high-level design methodology proposed in [18]. The core philosophy aligns with the functional separation into layers of control (Q), computation (L0), communication (L1) and memory access (L2) with dedicated hardware structures for each of task class, maximizing parallelism and computational efficiency to achieve low energy. Conceptually, for each of the task classes the architecture contains a pool of elementary hardware structures, which can be rearranged by means of reconfiguration. The application mapping is reconstructed in the architecture by rearranging the existing elementary structures to adapt the data-path. The application data/control path is thus tailored to the application, allowing to be arranged differently when the application changes.

The architecture is organized in a 3D pipeline structure, where data flows from the memory banks via the memory and communication layers to the computation layer in a vertical pipeline, while control flow is processed in the horizontal pipeline, shown in Fig. 1. The *computation layer (L0)* contains an array of N^2 processing elements in a mesh interconnect, similar to generic CGRA architectures. All data processing happens in this layer, while L1 and L2 provide necessary inputs. The key consideration for energy efficiency here is to keep all units in this layer always busy, to exploit maximum parallelism. Targeting the NLA domain, DesignWare floating point (fp) modules provided by Synopsys are employed, supporting $op(PE_n) = \{+, -, *\}$ in all elements. One pipelined fp divider is added to PE0, providing one 32-bit fp division result in 4 cycles. Each PE reads input data from 6 sources for each input port $src(PE_n^{a,b}) = \{N, S, E, W, U, X\}$, where U represents upstream L1 connection and X takes its own output, captured in a register. Each data layer can work at a $\frac{1}{2^n}$ speed ratio in relation to the control layer to balance out bottle-necks. In our experiments, due to relatively slow L0 fp units, L0 works at $\frac{1}{8} \times$ the speed of other layers.

The *communication layer (L1)* is a complex network of buses with register clusters per L0 element for data sharing, broadcasting and partial storage. The main purpose of this layer is to arrange and prepare data that will be processed by the next L0 cycle in advance, such that L0 has as few idle cycles as possible and plays a major role in the adaptability of the architecture to a given mapping. A set of elementary data movement patterns provide the building blocks of complex application-specific data movement patterns, such as broadcasts, delays, copying, etc. Such patterns can be created by means of reconfiguration, i.e. composing a larger pattern from small patterns in different combinations. The pattern handle is then made available at assembly level as a simple instruction call.

To alleviate the memory bandwidth problem of most CGRA-like structures, the *memory access layer (L2)* handles distribution of P memory ports to the N^2 interface lines to connect with L1. To avoid the necessity of a full crossbar from N^2 elements to P ports and ensure scalability, P *hubs* are introduced in-between. Thus each hub has access to each port, needing only a $P \times P$ crossbar, which is reasonable since usually memory ports are scarce ($N^2 \geq P$). From the hubs, a static modulo $N^2 \% P$ distribution is employed, uniformly distributing hubs across downstream elements, e.g. if $PE(n)\% P = 0$, the n-th PE is connected to hub 0, each hub having $\lceil \frac{N^2}{P} \rceil$ connections. Another role of the hub is to use the memory's protocol to forward access requests and select the correct port based on the desired data address, create pipelined load/store operations and overall streamline memory access. This layer provides all building blocks for memory-access.

The *control layer (Q)* uses Algorithmic State Machine (ASM) concepts to implement control flow components such as for loops, if-else branches and keep the current execution state in q_registers, shown in the Q-decode pipeline stage of Fig.1. ASM-based control flow has been proposed for use in high-level synthesis earlier [19] [20]. Here we use only the core ideas of this concept, combining and reconfiguring a set of small hardware components (counters, comparators, registers) to assemble the *qualifiers*, by which the complete control flow of each kernel can be captured in a state machine, controlling data path components. Qualifiers update the current execution state, calculate addresses and forward valid configurations to each of the data layers every cycle. The Q layer also does (partial) predication of the other structures of the architecture.

Programming is done via calling the respective higher-order constructs in assembly code, which is fast and efficient. Automation of this manual process is also under development, a force-directed scheduler and mapper for the L0 layer being proposed in [21], which will be adapted to our mapping approach in future work.

III. MAPPING LU DECOMPOSITION

A. General considerations

The key idea, as stated in Section I, is to derive an *efficient and scalable* scheduling with coarse mapping elements, based on available algorithmic features (parallelism, dependencies) and then *reconstruct* this mapping in the architecture by reconfiguring the data path. Generally, if the architecture is scalable with varying architectural parameters, efficiency can be gained only if the application mapping perfectly fits every variant – a scalable mapping. For CGRAs with banked memories, all mappings must additionally respect the constraint of not loading data on the column (keeping one access per memory port per cycle). Especially for matrix-matrix operations, accessing column and rows of values from either matrix makes scalable mapping difficult. Rectangular or square windows with a height of more than 1 are not possible either since this would require to work on columns in either matrix. Block-based approaches which work best for CPUs, are not scalable when modifying CGRA size N or memory port amount P and produce complex addressing problems. An efficient block-based scheduling and mapping solution has been discussed in [11], however only for a fixed 4×4 PEs and 8 ports configuration, yielding a fixed mapping. This work shows superior results due to improved design, programmability, scalability and mapping performance.

The mapping derivation considers several architectural and algorithmic parameters coarsely at first, such as available bandwidth per cycle, architectural computation:memory speed ratio, array size, dependency, common data and progress through the data, avoiding column-based memory loads due to access conflicts on modulo-P based memory bank distributions, etc. These coarse constraints help to derive the optimal execution window from the algorithmic point of view, which should match the optimal execution window relative to the capabilities

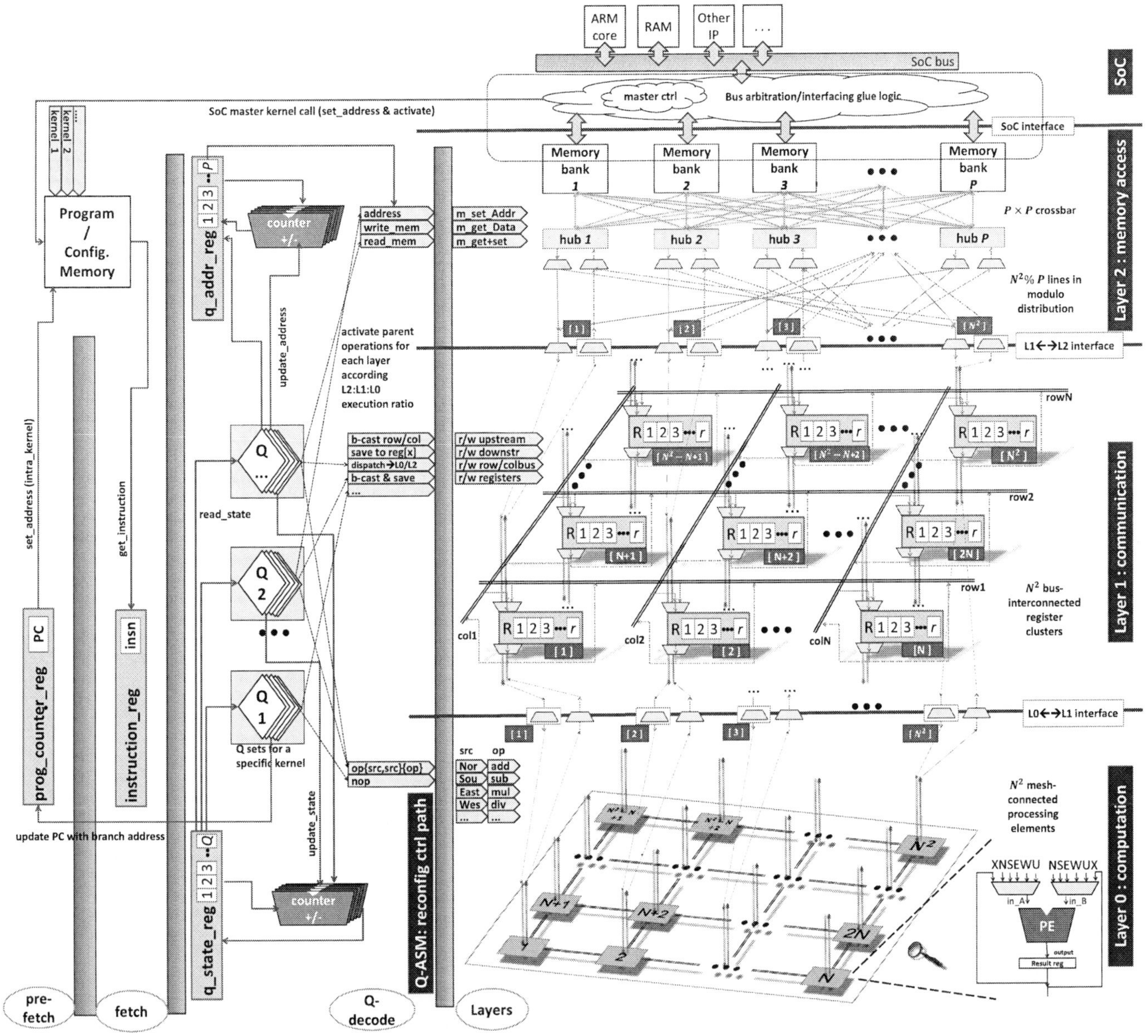

Fig. 1. The *Layers* architecture recreates the desired application function using elementary hardware functions in each of the functional classes of computation, communication memory access and control, in a 3D pipeline. Data flow (vertical) and control flow (horizontal) are happening in parallel.

of the architecture (number parallel PEs, memory bandwidth, etc). These can then be parameterized to ensure scalability by extracting relevant variables from both algorithm and architecture. Additional considerations of processing element timing and capabilities are considered, e.g. floating-point divider needs 4 cycles for completion of one division, while other operations finish in one cycle.

B. Scheduling and mapping

The input to the *LU*-factorization is a square matrix $A \in \Re^{n \times n}$ with $|A| \neq 0$. The output for our implementation[1]

---

[1]It should be mentioned here, that there are other implementations of the *LU*-factorization, which do not require the diagonal elements of L to be 1. Furthermore, our implementation does not deal with the fact that not every input matrix A can be processed using this algorithm, exceptions for which pre-processing can be done. An advanced implementation of LU, possibly using techniques like *Pivoting*, is subject to future work.

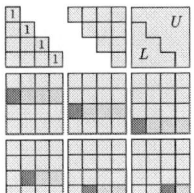

Fig. 2. LU execution window progress over the input matrix. One access per memory port per cycle is guaranteed.

consists of two matrices, $L \in \Re^{n \times n}$ and $U \in \Re^{n \times n}$, such that values above and below the diagonal are zero, respectively.

Keeping the above general considerations in mind, we illustrate, for clarity, the proposed mapping for the LU-algorithm using an example with $n = 3$, hence $A_3 = L_3 \cdot U_3$, however the mapping holds for any $n > 3$.

$$\begin{bmatrix} a_{11} & a_{12} & a_{13} \\ a_{21} & a_{22} & a_{23} \\ a_{31} & a_{32} & a_{33} \end{bmatrix} = \begin{bmatrix} 1 & 0 & 0 \\ l_{21} & 1 & 0 \\ l_{31} & l_{32} & 1 \end{bmatrix} \cdot \begin{bmatrix} u_{11} & u_{12} & u_{13} \\ 0 & u_{22} & u_{23} \\ 0 & 0 & u_{33} \end{bmatrix}$$

where L_3 and U_3 can be computed as

$$\begin{bmatrix} 1 & 0 & 0 \\ l_{21} & 1 & 0 \\ l_{31} & l_{32} & 1 \end{bmatrix} = \begin{bmatrix} 1 & 0 & 0 \\ a_{21}u_{11}{}^{-1} & 1 & 0 \\ a_{31}u_{11}{}^{-1} & (a_{32} - l_{31}u_{12})u_{22}{}^{-1} & 1 \end{bmatrix}$$

$$\begin{bmatrix} u_{11} & u_{12} & u_{13} \\ 0 & u_{22} & u_{23} \\ 0 & 0 & u_{33} \end{bmatrix} = \begin{bmatrix} a_{11} & a_{12} & a_{13} \\ 0 & a_{22} - l_{21}u_{12} & a_{23} - l_{21}u_{13} \\ 0 & 0 & a_{33} - l_{31}u_{13} - l_{32}u_{23} \end{bmatrix}$$

For more efficient memory usage, we combine the matrices L_3 and U_3 to matrix Q_3, superposing the matrices such that

$$Q_3 := \begin{bmatrix} u_{11} & u_{12} & u_{13} \\ l_{21} & u_{22} & u_{23} \\ l_{31} & l_{32} & u_{33} \end{bmatrix} =$$

$$\begin{bmatrix} a_{11} & a_{12} & a_{13} \\ a_{21}u_{11}{}^{-1} & a_{22} - l_{21}u_{12} & a_{23} - l_{21}u_{13} \\ a_{31}u_{11}{}^{-1} & (a_{32} - l_{31}u_{12})u_{22}{}^{-1} & a_{33} - l_{31}u_{13} - l_{32}u_{23} \end{bmatrix}$$

which we will compute in the following. First, we realize that $l_{11} := 1$ implies $u_{1n} = a_{1n}$, which results in

$$= \begin{bmatrix} u_{11} & u_{12} & u_{13} \\ a_{21}u_{11}{}^{-1} & a_{22} - l_{21}u_{12} & a_{23} - l_{21}u_{13} \\ a_{31}u_{11}{}^{-1} & (a_{32} - l_{31}u_{12})u_{22}{}^{-1} & a_{33} - l_{31}u_{13} - l_{32}u_{23} \end{bmatrix}$$

Then, we compute $\boxed{l_{21}}$, which is used to compute $\boxed{u_{22} \dots u_{2n},}$ where the multiplications and subtractions are parallel. This gives a flexible execution window of size $N^2 \times 1$, sliding to the right on the row until the end, where possible size mismatches are truncated by predication in the Q-layer.

$$= \begin{bmatrix} u_{11} & u_{12} & u_{13} \\ \boxed{a_{21}u_{11}{}^{-1}} & \boxed{a_{22} - \boxed{l_{21}u_{12}}} & \boxed{a_{23} - \boxed{l_{21}u_{13}}} \\ a_{31}u_{11}{}^{-1} & (a_{32} - l_{31}u_{12})u_{22}{}^{-1} & a_{33} - l_{31}u_{13} - l_{32}u_{23} \end{bmatrix}$$

For the next row, l_{31} is computed yielding the partial results $\boxed{l_{32}' \dots u_{3n}',}$ using again all execution units.

The procedure is repeated until l_{n1} is reached.

$$= \begin{bmatrix} u_{11} & u_{12} & u_{13} \\ l_{21} & u_{22} & u_{23} \\ \boxed{a_{31}u_{11}{}^{-1}} & \boxed{(a_{32} - \boxed{l_{31}u_{12}})u_{22}{}^{-1}} & \boxed{a_{33} - \boxed{l_{31}u_{13}}} - l_{32}u_{23} \end{bmatrix}$$

Please note the dependencies on the row on l_{mn} and on the column for u_{mn}. The higher m and n are, the longer the string of dependencies, which are rank-nested within the matrix. In the proposed mapping, we repeatedly iterate over the sub-matrices of decreasing rank, as shown in Fig. 2, until the complete Q_n matrix is computed. In the above example, only the last row remains, after which the final form of Q_3 is reached.

In general, while the the first row of Q, i.e. $u_{11}, \dots, u_{1n}$, needs no computation, there are $n-1$ values which can be computed in parallel in any other row within the first iteration, $n-2$ in the second iteration and so on until the computation of u_{nn} is a single value. Because only one single l_{ij}-value has to be computed to perform computations in all the following values in row i in parallel, which might result in final $u_{ik}{}'{}^{\cdots}$ or partial results $l_{ik}{}'{}^{\cdots}$ respectively $u_{ik}{}'{}^{\cdots}$, the scaling performs well on large matrix sizes. Having a window of height 1 makes the window fit efficiently regardless of the input matrix size, because there can only be a maximum of one window in any row which needs truncation.

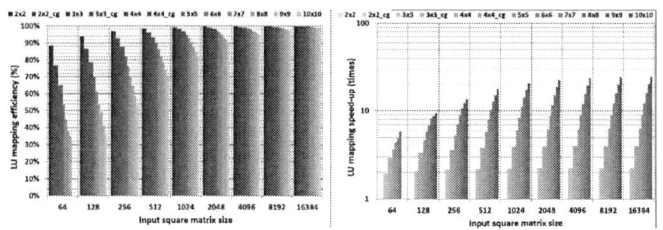

Fig. 3. LU mapping efficiency and expected mapping speedup for various architectures (N=2..10) and data sizes (64..16384).

C. Complexity and Efficiency

Efficiency evaluation is performed by taking the ratio between the theoretically required number of execution cycles $c_{k_{min}}$ and the number of actual cycles executed by the architecture c_k for each array size N, yielding $\eta_k(\{\cdot\}, N) := \frac{c_{k_{min}}(\{\cdot\}, N)}{c_k(\{\cdot\}, N)}$. Complexity can be divided in 3 parts: division complexity $op_{\mathtt{lu}_{div}}$, complexity for lower $op_{\mathtt{lu}_l}$ and upper factorization $op_{\mathtt{lu}_u}$, then summed up to $op_{\mathtt{lu}}$. In order to compute L, one multiplication per element in the lower diagonal part, excluding the diagonal itself, is needed, i.e. $\frac{n}{2}(n-1)$. Additionally, we need 0 pairs of subtractions/multiplications in the first column, 1 pair in the second column and so on, hence

$$op_{\mathtt{lu}_l}(\{n\}) = \frac{n}{2}(3n - 7) + 2. \tag{1}$$

Similarly, in order to compute U, we need 0 pairs of subtractions/multiplications in row 1, 1 pair in row 2 and so on, resulting in

$$op_{\mathtt{lu}_u}(\{n\}) = 2\sum_{i=1}^{n-1}(n-i)i = \frac{n}{3}(n^2 - 1). \tag{2}$$

Finally, adding these to $op_{\mathtt{lu}_{div}}(\{n\}) = n-1$ the minimum cycle complexity on Layers with $r_{L0:L1:L2} = 1 : 8 : 8$ and 4 cycles for a division:

$$c_{\mathtt{lu}_{min}}(\{n\}, N) = 8\left(\frac{1}{N^2}\left(\frac{n^3}{3} + \frac{3n^2}{2} - \frac{23n}{6} + 2\right) + 4(n-1)\right)$$

Mapping efficiency is shown in Fig. 3 (left). With enough data to fill up the CGRA of size N, the efficiency is very close to optimal. Expected speedup from mapping when scaling N, is shown on the right side of Fig. 3, very close to the expected theoretical value i.e. scaling by x amount of elements a speedup of close to x is achieved.

IV. EVALUATION AND RESULTS

A. General considerations

Layers has been coded completely in the LISA ADL of Synopsys Processor Designer, completely parametrized for easy scalability. Simulations have been conducted for random square input matrices of size 64..16384, for different combinations of $P = 2..32$ and $N = 2..10$. The assembly program for LU has been coded in a scalable way, needing 12 L0-cycles for LU, highlighting the programmability of *Layers*. Memory banks are dual port SRAMs, with 1-cycle read/write latency. One port of each bank is reserved for SoC integration and is not used in the mapping/processing of the kernels, providing identical bandwidth towards the system side as well (Fig.1). Values are for single-precision floating point (32-bit). For these configurations RTL code has been generated and synthesized with DesignCompiler I-2013-SP5 for Faraday $65nm$ technology library. PowerCompiler and backward switching activity files were used for power estimation. For lower size designs, clock-gating has been enabled at synthesis, marked with _cg in the results. For larger designs the additional clock-gating circuitry used more power than it actually saved, hence those results are removed for clarity. Also, using more memory ports P than the minimal amount required for mapping yielded greater power and area usage with no advantages, the power gained from more relaxed L1/L2 data handling did not compensate for the power and area used for additional structures and this data is omitted for clarity.

B. Time and energy

Detailed results are provided in Fig. 4 and 5, where the overall time and energy values are depicted for each configuration and input matrix size, highlighting the scalability of our approach. Execution time spreads over several orders of magnitude with varying input data size, while an order of magnitude speed-up can be maintained between the smallest and largest array for large input data sizes. The architecture and mapping scale with almost constant energy ($<10\%$ variance), translating into a clean trade-off between area and speed, without affecting energy. Thus, designs that need to respect certain requirements in the amount of memory ports or a certain amount of performance or area, can be easily picked from the scalable set, without needing to consider the energy impact of the choice, as energy stays constant for a given workload. The loss of frequency with increasing size, power scaling and speed-up scaling are balanced out to a constant energy requirement.

For the largest designs ($N = \{9, 10\}$), the critical path of the L1 structures severely affects frequency and thus energy., the sudden increase in area and critical path due to the upgrade from 3-bit multiplexers sufficient for $N \leq 8$ interconnects to 4-bit ones. If required, small architectural optimizations could compensate this, by splitting L1 buses into shorter ones, without affecting mapping. Clock-gated designs for the smaller sizes provide superior results especially for power (Fig. 7).

C. Comparisons with related work

Table I and Fig. 6 provide some area, frequency and performance density data. The frequency of each architecture is limited by the control flow complexity in the q-decode stage for small N, and by L1 critical path for larger N at $r = 1 : 8 : 8$. Choosing lower inter-layer speed ratio r, the fp PEs limit overall frequency, while sacrificing memory bandwidth, especially when requiring a divider. When comparing to the clean version of *Layers*, it is interesting to note that the reconfigurable control path slows the architecture down, although by not operating near maximum frequency has great advantages in power consumption.

Most recent results on LAC [14], a CGRA targeting linear algebra, show comparable numbers for LU with partial pivoting. A fixed block-based mapping is used and architectural enhancements for pivoting are employed to reach excellent performance. Although LAC is considering a more complex algorithm, it lacks the seamless scalability provided by our architecture and mapping solution. While performance in terms of aggregate energy efficiency is similar (Table I), the estimated area of LAC is 6.2× larger than a similar-sized *Layers* core. An exact comparison is not possible due to the different process libraries and the estimated nature of the results reported in [14]. It would be interesting to compare actual post-synthesis and post-physical design results in the same technology node.

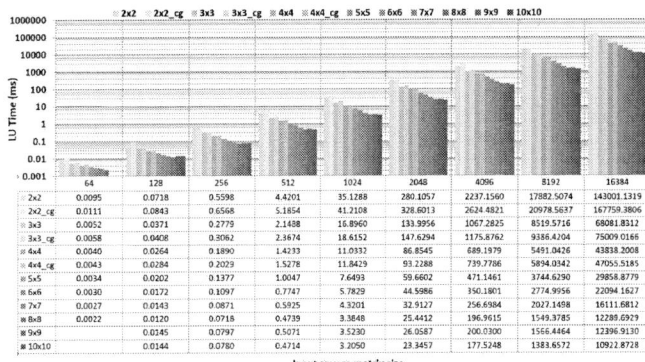

Fig. 4. Timing results for LU. More than an order of magnitude speedup can be observed.

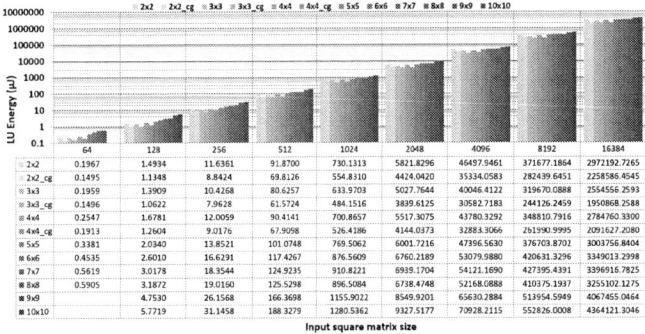

Fig. 5. Energy results for LU. Except the largest arrays energy stays constant when scaling N. Clock-gated designs *_cg perform better. Constant energy is required for the same problem size across variants, giving a clean area:performance trade-off.

A recently enhanced REDEFINE CGRA for NLA [5] shows comparable values for 65nm, and shows better performance density if scaled to 45nm with custom DOT product units, however no execution times for large matrices could be found. Based on the reported latencies for 60×60 and 120×120 data size running matrix multiplication in [5], the slowest variant of *Layers* performs 4.2× and 2.8× faster on a 64×64 and 128×128 data set. Unfortunately we found no other works in the literature which provide detailed time/energy results, but general comparisons with other platforms can be summarized in Table I.

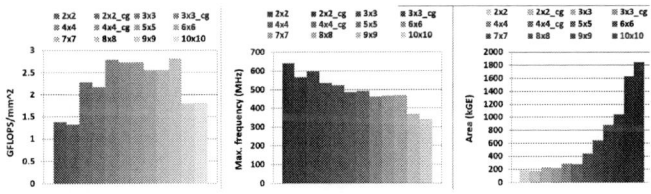

Fig. 6. Area, frequency and performance density for the new *Layers* architecture.

V. CONCLUSIONS

Exploiting high-level design methodologies, in this paper we meticulously explored the design space and highlighted the advantages of scalable algorithm-hardware co-design over 12 architectural variants for the LU kernel. Constant energy when trading off area and performance is enabled by a scalable architecture with a adaptive, scalable mapping. Extensive data is provided to allow similar designs implement these mappings and compare energy/performance results. In the future, we

TABLE I.   COMPARISON WITH SIMILAR ARCHITECTURES IN THE LINEAR ALGEBRA DOMAIN

Architecture	Area (kGE or mm²)	Frequency (MHz)	Power (mW)	GFlops/W	GFlops/mm²	TechLib [refs]
2×2 Layers r=1:8:8 (clean)	84kGE (0.12mm²)	990	17.74	27.95	5.84	65nm
2×2 Layers r=1:2:2 (clean)	84kGE (0.12mm²)	990	17.74	111.83	23.37	65nm
3×3 Layers r=1:8:8 (@SGEMM)	219kGE (0.21mm²)	543	25.42	23.48	2.17	65nm
4×4 Layers_cg r=1:8:8 (@LU)	278kGE (0.35mm²)	488	44.45	21.94	2.78	65nm
30 LAC cores (@SGEMM)(sim. estim.)	115mm²	1400	N/A	30-55	6-11	45nm [6]
4×4 LAC (@LU)(sim. estim.)	2.2mm²	1000	N/A	25-30	1-2.6	45nm [14]
REDEFINE with CFU (@SGEMM)	0.16mm²	416	N/A	N/A	2.54	65nm [5]
REDEFINE with CFU+DOT (@SGEMM)	0.23mm²	416	N/A	N/A	12.24	45nm [5]
Nvidia GTX280(@SGEMM)	576mm²	1300	236000 (max)	0.001	0.63	65nm [22]
Nvidia GTX280(@QR)	576mm²	1300	236000 (max)	0.001	0.63	65nm [23]
Nvidia 9800GTX (@GEMM)	324mm²	1670	140000 (max)	0.77	5.2	65nm [22]
Nvidia 8800 Ultra (@CUBLAS)	480mm²	575	175000 (max)	1.45	4	90nm [17]
Tesla S1070 (@SGEMM)	610mm²	470	800000 (max)	0.73	0.42	55nm [24]
Core2 Quad QX6850 (@SGEMM)	286mm²	3000	130000 (max)	0.0006	0.28	65nm [22]
TMS320C6678 (@L3BLAS)	NA	1000	10000 (max)	8	NA	40nm [25]

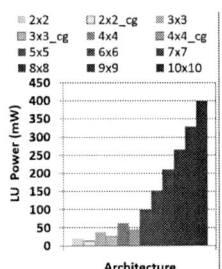 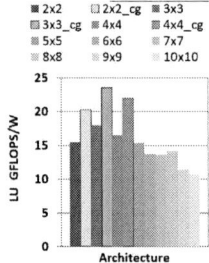

Fig. 7.   Power results for LU. Clock gated designs perform much better.

seek to create a library of highly optimized, scalable linear algebra kernels and provide SoC integration and (3D silicon) place&route details.

REFERENCES

[1] F. Bouwens, M. Berekovic, A. Kanstein, and G. Gaydadjiev, "Architectural Exploration of the ADRES Coarse-Grained Reconfigurable Array," in *Reconfigurable Computing: Architectures, Tools and Applications.* Washington, DC, USA: Springer Berlin Heidelberg, 2007, vol. 4419, pp. 1–13.

[2] V. Govindaraju, C.-H. Ho, T. Nowatzki, J. Chhugani, N. Satish, K. Sankaralingam, and C. Kim, "Dyser: Unifying functionality and parallelism specialization for energy-efficient computing," *IEEE Micro*, vol. 32, no. 5, pp. 0038–51, 2012.

[3] Z. A. Ye, A. Moshovos, S. Hauck, and P. Banerjee, *CHIMAERA: a high-performance architecture with a tightly-coupled reconfigurable functional unit.* ACM, 2000, vol. 28, no. 2.

[4] D. Burger, S. W. Keckler, K. e. McKinley, M. Dahlin, L. K. John, C. Lin, C. R. Moore, J. Burrill, R. G. McDonald, and W. Yoder, "Scaling to the end of silicon with edge architectures," *Computer*, vol. 37, no. 7, pp. 44–55, 2004.

[5] F. Merchant, A. Maity, M. Mahadurkar, K. Vatwani, I. Munje, M. Krishna, S. Nalesh, N. Gopalan, S. Raha, S. Nandy *et al.*, "Microarchitectural Enhancements in Distributed Memory CGRAs for LU and QR Factorizations," in *VLSI Design (VLSID), 2015 28th International Conference on.* IEEE, 2015, pp. 153–158.

[6] A. Pedram, R. A. van de Geijn, and A. Gerstlauer, "Codesign tradeoffs for high-performance, low-power linear algebra architectures," *IEEE Trans. Comput.*, vol. 61, no. 12, pp. 1724–1736, 2012.

[7] A. Parashar, M. Pellauer, M. Adler, B. Ahsan, N. Crago, D. Lustig, V. Pavlov, A. Zhai, M. Gambhir, A. Jaleel *et al.*, "Triggered instructions: A control paradigm for spatially-programmed architectures," in *Proceedings of the 40th Annual International Symposium on Computer Architecture.* ACM, 2013, pp. 142–153.

[8] Y. Huang, P. Ienne, O. Temam, Y. Chen, and C. Wu, "Elastic CGRAs," in *Proceedings of the ACM/SIGDA international symposium on Field programmable gate arrays.* ACM, 2013, pp. 171–180.

[9] A. DeHon, "The density advantage of configurable computing," *Computer*, vol. 33, no. 4, pp. 41–49, 2000.

[10] A. Chattopadhyay, "Ingredients of adaptability: a survey of reconfigurable processors," *VLSI Design*, vol. 2013, p. 10, 2013.

[11] Z. E. Rákossy, T. Naphade, and A. Chattopadhyay, "Design and analysis of layered coarse-grained reconfigurable architecture." in *Reconfigurable Computing and FPGAs (ReConFig), 2012 International Conference on*, 2012.

[12] Z. E. Rákossy, F. Merchant, A. Acosta Aponte, S. Nandy, and A. Chattopadhyay, "Scalable and Energy-Efficient Reconfigurable Accelerator for Column-wise Givens Rotation," in *22nd International Conference on Very Large Scale Integration (VLSI-SoC).* IEEE, 2014.

[13] A. Pedram, A. Gerstlauer, and R. A. Van De Geijn, "Floating point architecture extensions for optimized matrix factorization," in *Computer Arithmetic (ARITH), 2013 21st IEEE Symposium on.* IEEE, 2013, pp. 49–58.

[14] A. Pedram, A. Gerstlauer, and R. van de Geijn, "Algorithm, architecture, and floating-point unit codesign of a matrix factorization accelerator," *IEEE Trans. Comput.*, vol. 63, no. 8, 2014.

[15] J. Gonzalez and R. C. Núñez, "LAPACKrc: Fast linear algebra kernels/solvers for FPGA accelerators," in *Journal of Physics: Conference Series*, vol. 180, no. 1. IOP Publishing, 2009, p. 012042.

[16] J. R. Humphrey, D. K. Price, K. E. Spagnoli, A. L. Paolini, and E. J. Kelmelis, "CULA: hybrid GPU accelerated linear algebra routines," in *SPIE Defense, Security, and Sensing.* International Society for Optics and Photonics, 2010, pp. 770 502–770 502.

[17] S. Barrachina, M. Castillo, F. D. Igual, R. Mayo, and E. S. Quintana-Orti, "Evaluation and tuning of the level 3 CUBLAS for graphics processors," in *Parallel and Distributed Processing, 2008. IPDPS 2008. IEEE International Symposium on.* IEEE, 2008, pp. 1–8.

[18] Z. E. Rákossy, A. Acosta Aponte, and A. Chattopadhyay, "Exploiting architecture description language for diverse IP synthesis in heterogeneous MPSoC," in *Reconfigurable Computing and FPGAs (ReConFig), 2013 International Conference on.* IEEE, 2013, pp. 1–6.

[19] A. Kuehlmann and R. A. Bergamaschi , "High-Level State Machine Specification and Synthesis," in *ICCD '92*, pp. 536–539.

[20] J. David and E. Bergeron, "An Intermediate Level HDL for System Level Design," in *Proceedings of FDL'2004*, pp. 526–536.

[21] A. Fell, Z. E. Rákossy, and A. Chattopadhyay, "Force-Directed Scheduling for Data-Flow Graph Mapping on Coarse-Grained Reconfigurable Architectures ," in *Reconfigurable Computing and FPGAs (ReConFig), 2014 International Conference on.* IEEE, 2014.

[22] V. Volkov and J. W. Demmel, "Benchmarking GPUs to tune dense linear algebra," in *Proceedings of the 2008 ACM/IEEE conference on Supercomputing.* IEEE Press, 2008, p. 31.

[23] A. Kerr, D. Campbell, and M. Richards, "QR decomposition on GPUs," in *Proceedings of 2nd Workshop on General Purpose Processing on Graphics Processing Units.* ACM, 2009, pp. 71–78.

[24] V. Allada, T. Benjegerdes, and B. Bode, "Performance analysis of memory transfers and GEMM subroutines on NVIDIA Tesla GPU cluster," in *Cluster Computing and Workshops, 2009. CLUSTER'09. IEEE International Conference on.* IEEE, 2009, pp. 1–9.

[25] M. Ali, E. Stotzer, F. D. Igual, and R. A. van de Geijn, "Level-3 BLAS on the TI C6678 multi-core DSP," in *Computer Architecture and High Performance Computing (SBAC-PAD), 2012 IEEE 24th International Symposium on.* IEEE, 2012, pp. 179–186.

Dynamic Migratory Selection Strategy for Adaptive Routing In Mesh NoCs

John Jose
Indian Institute of Technology
Guwahati, India
johnjose@iitg.ernet.in

Joe Augustine
Rajagiri School of Engineering
and Technology, Kochi, India
joe00thayil@gmail.com

Sijin Sebastian
Rajagiri School of Engineering
and Technology, Kochi, India
sijinsebastian@gmail.com

Abstract—NoC architectures are the most commonly used communication framework for multicore processors. Few factors that affect the performance of an on-chip network include the efficiency of the routing algorithm and the effectiveness of the output selection strategy used. All popular selection strategies use a static technique that behaves uniformly across various traffic patterns. This paper proposes a cost effective adaptive model of Regional Congestion Awareness (RCA) selection strategy. The traffic analyser incorporated in the proposed model learns the flit-flow pattern at each router and makes RCA behaves like the local best selection strategy under local traffic. Under non-local traffic, the normal RCA selection strategy works as it is. This switching (migration) between two selection strategies is done by proper controlling of aggregation and propagation mechanisms of RCA. As only in-router information is used for this switching, the design has no additional communication overhead. This dynamic switching decreases average packet latency and effectively optimises the network resources depending on traffic pattern, thereby, reducing power consumption. Our experiments on 8×8 mesh NoC with various synthetic and real traffic patterns show promising improvements compared to the existing baseline adaptive selection strategies.

Keywords—*flit-flow pattern, switching, traffic analysis.*

I. Introduction

Traditionally, System on Chip (SoC) has been designed with dedicated point-to-point bus based interconnections. For large designs, this has several limitations from a physical design view point. Buses are not only non-scalable, but also consume significant area and power on the chip. Network on Chip (NoC) replaces traditional bus based communication system with an on-chip packet-switched interconnection network. NoC is the most popular communication framework used in modern multicore designs [11].

Figure 1 shows a 9-core multiprocessor SoC that uses a two dimensional 3x3 mesh NoC topology for on-chip communication. From the figure, we can see that each core is connected to a router. Routers are interconnected using bi-directional links. NoC based system uses packet based switching for inter-core communication. Packets are divided into flow control units called flits. Buffers in the routers and handshaking signals between routers enable flow control and smooth movement of packets from the source router to the destination router.

Each core consists of an out-of-order superscalar processor, a private L1 cache and a shared distributed L2 cache. Whenever there is a cache miss, a miss request is generated from

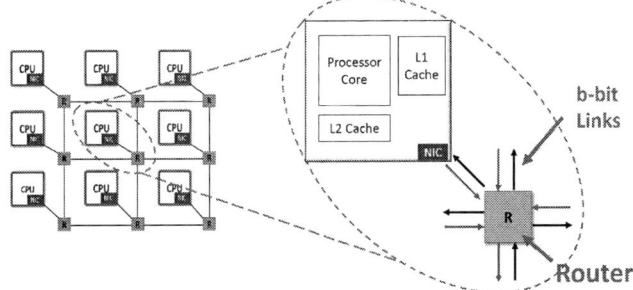

Fig. 1. Core-router interaction in a 3x3 mesh topology.

the source core in the form of a packet. Packet is forwarded through various intermediate routers to reach the destination core. Routing algorithm employed in the router determines the next outgoing link for every incoming packet. Similarly the reply packet also traverses through the on-chip network. Typically the cache miss request is a single flit packet whereas the reply packet can have multiple flits based on cache block size and inter-router link width.

The miss penalty of a cache miss depends on the round-trip latency of the cache miss request and the cache miss reply packet. Since the latency of cache miss request packet is a very critical performance parameter, the underlying NoC which carries the cache miss request and reply packet has to deliver them in the minimum time. In this context, the performance of an NoC framework is highly critical in determining the throughput of applications running on the cores.

The performance of an NoC depends mainly on the routing algorithm and selection strategy used in the routers [2]. The routing algorithm implemented on the router computes the output link for every incoming packet. If the routing algorithm returns more than one output link, a selection strategy is used to choose the most suitable link. By developing cost effective output channel selection techniques, the average packet latency and power consumption can be reduced.

The basic and the most simple selection strategy employed is the local best selection strategy [11]. Here, each router receives the number of free buffers in each of its downstream neighbor through a dedicated 4-bit wide control channel. It then chooses the router with higher number of free buffers. This can take greedy decisions at each hop,

978-1-4673-9141-2/15 $31.00 © 2015 IEEE   343

without considering the congestion scenario beyond neighbor nodes [6]. A better and more efficient selection strategy is implemented using the Regional Congestion Awareness (RCA) algorithm [3]. In this strategy, congestion information from all the downstream neighbors (local as well as non-local) of a router is aggregated and propagated to its upstream neighbors. RCA uses 9-bit wide control channel for transferring this aggregated congestion information. This causes extra overhead in the network, even though it delivers good performance.

We identify few limitations of RCA selection strategy and propose a cost effective dynamic model of RCA that can improve the performance of an NoC under varying traffic patterns. Conventional RCA performs best under non-local traffic patterns. We modify RCA such that it delivers improved performance under local traffic also. We incorporate a traffic analyser on each router to analyse the run time traffic patterns. Based on the traffic, a switching technique chooses the normal RCA under non-local traffic load and cut down the resource over-head of RCA when the traffic is local. This migration (switching) makes RCA behave as local best selection strategy in routers where traffic is mostly towards local (nearby) destinations.

The rest of the paper is organised as follows. We describe the related works in Section II. In Section III, the motivation for the proposed work is explained. The architectural details of the proposed model is given in Section IV. Experimental methodology and result analysis are covered in Section V followed by the conclusion of our work in Section VI.

II. RELATED WORK

The packets generated at the source core need to be directed towards the destination core through the network. Routing algorithms are used to determine the sequence of channels (inter-router links) a packet traverses from the source to the destination. Minimal Odd-Even (MOE) Routing [4] is a commonly used adaptive deadlock free routing algorithm. It restricts the location where certain turns that a packet can take while moving to the downstream routers. Once a packet reaches a router, the MOE routing may return more than one admissible output ports. To enhance the performance of a routing function, output selection functions are employed on top of MOE routing. Selection function captures the congestion metric of the reachable downstream neighbors and chooses that neighbor which is less congested, thus reducing the delay in movement of packets towards their respective destinations. Several parameters of the network as well as routers are used as congestion metric. Neighbors-on-Path [5] checks the Free Virtual Channels (FVCs) of reachable neighbors of adjacent downstream routers. TRACKER [6] uses the history of flow of flits through all the output ports of reachable downstream routers. In BOFAR [7] , the cycles spend by a flit in a buffer is taken as the congestion metric. Global Congestion Awareness (GCA) [8] is yet another technique that uses local as well as non-local status information for computing congestion metric.

RCA is an effective path selection technique that improves the load balance in the network. It aggregates and propagates congestion information about a region of the network beyond the adjacent routers to the upstream routers. This helps the upstream routers in estimating the best path with minimal

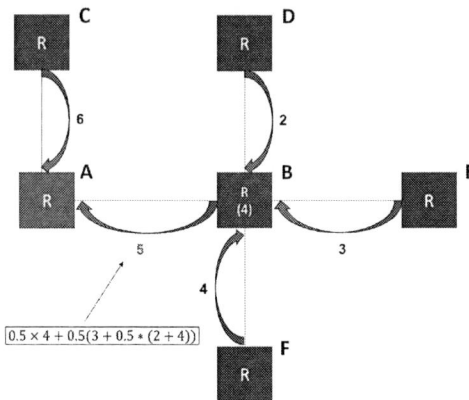

Fig. 2. Illustration of aggregation and propagation in RCA

congestion. The RCA unit of a router basically consists of two modules. The aggregation module combines the local as well as the non-local congestion metric by assigning appropriate weights to each of them. The propagation module sends the aggregated congestion metric to respective neighbors through dedicated control network.

The working of RCA is illustrated in Figure 2. The figure shows the aggregation and propagation of the congestion metric values in RCA-FanIn [3] selection strategy through a link. Here A, B, C, D, E and F are routers in an 8x8 mesh network. The curved edges connecting the routers are control channels. Assume that a flit in router A has two permissible output neighbors (B or C) after MOE routing. In RCA, the propagated congestion information from the downstream neighbors B and C is used for choosing one of the neighbor. The congestion metric from B is computed as follows:

B adds the local congestion metric (i.e., 4) and the average congestion metric from its neighbors (i.e., 2, 3, and 4 from D, E and F, respectively) in the ratio 0.5:0.25:0.5:0.25. This calculation is illustrated in the bottom left side of the figure. This gives a higher weightage for local information over non-local information. Also, the congestion metric given by the east neighbor of B (i.e., value 3 passed by E to B) is given more weightage than north and south directions. This is because both north-east and south-east destinations include east channel. The congestion metric given to B by routers D, E and F are also an aggregated value of a similar computation done in those routers. Thus A is given the value 5 from B as shown in the figure. The congestion value is a combination of the number of free buffers and crossbar demand of routers. Similarly, an aggregated value is obtained from C also. Based on these values, A chooses the next router to send out the flit.

Local best selection strategy uses the number of FVCs [11] in the adjacent downstream routers. If local best selection strategy was used in the above case, the router B and C transmit their local free virtual channel count to A. This information alone will be used for the output port selection. Even though it has a simpler circuitry and smaller control channel, it takes greedy decision without considering the congestion status beyond downstream routers.

978-1-4673-9141-2/15 $31.00 © 2015 IEEE 344

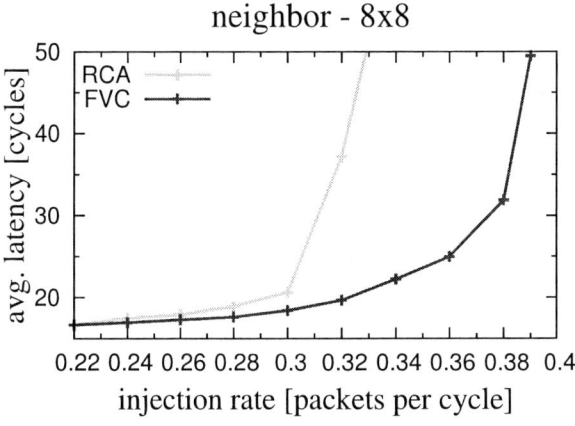
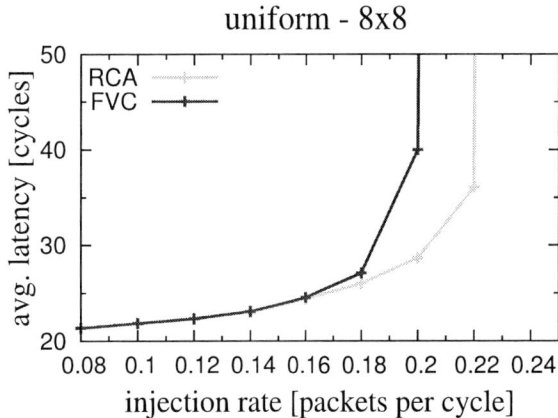

Fig. 3. Comparative analysis of average packet latency versus injection rate for neighbor and uniform traffic patterns in 8×8 mesh network.

III. MOTIVATION

Booksim [9] is a cycle accurate simulator specialized for NoCs with a highly precise underlying network model. Using *Booksim*, we model an 8x8 mesh network that employs MOE routing. We studied the impact of the FVC selection strategy under uniform and neighbor traffic patterns. Synthetic traffic patterns are abstract models of message passing in NoCs. Simple synthetic traffic patterns like uniform, tornado, bit-complement and neighbor traffics allow a network to be stressed with a regular, predictable pattern which aid NoC designers in acquiring new insights [1], [12]. In uniform traffic, each node sends messages to other nodes with an equal probability (i.e., destination nodes are chosen randomly using a uniform probability distribution function). In neighbor traffic, the source and destination are nodes at 2 hop distance that differ in one cordinate both row-wise and column-wise. Similarly, *Booksim* is modified to model the RCA-FanIn architecture as mentioned in [3]. We obtained the average packet latency for varying injection rates from zero to saturation using uniform and neighbor traffic. The load vs. latency graph is shown in Figure 3.

From the figure, we can observe that under neighbor traffic, FVC technique outperforms RCA by a significant margin. In neighbor traffic, every packet's source core and destination core are at a two hop distance (except for packets originating from edge and corner routers). i.e., a packet generated into the network will travel only through two intermediate routers before reaching its destination. So, a selection strategy like RCA that aggregates non-local congestion metric (congestion metric of routers beyond two hops) is meaningless. In our experiments using neighbor traffic, in many cases we observed that RCA selection strategy selects output channels that are not in favour for a packet whose destination is within two hops. Local best selection strategy is a simple technique which requires only 4 bit-lines for communicating the free buffer count of neighbors. It performs well in local traffic loads compared to other techniques.

But RCA was significantly outperforming FVC technique under uniform traffic. In uniform traffic the average hop-length of a packet (in an 8x8 mesh network) is more than six, which indicates that uniform traffic is an example of a non-local traffic. So as expected RCA outperformed FVC technique.

These two contradicting observations emphasizes the fact that selection strategy as such cannot improve performance. Certain selection strategy is meaningful and productive under particular traffic patterns only. For a fabricated chip containing a collection of NoC routers, we cannot change the selection strategy on individual routers based on run time traffic pattern.

RCA requires a 9-bit lines in the case of an 8×8 mesh for propagating the congestion feedbacks [3]. These lines are always used irrespective of the traffic load and pattern. When the traffic is mostly towards neighboring nodes (local traffic), congestion information about an entire region is irrelevant. Thus RCA uses equal amount of power and area in the feedback network even when the signals through them are not productively used.

IV. THE PROPOSED WORK

We propose that the performance of an NoC can be improved irrespective of the traffic pattern, if selection strategy is decided based on run-time traffic. We put forward a technique that can take dynamic decision based on the traffic pattern in the network. We implement a runtime traffic pattern analyser for each router that checks whether on-chip traffic is local or non-local. Based on the nature of the traffic, the appropriate selection function is used. If the traffic is non-local, the RCA technique is used and it is made to behave as FVC selection strategy when the traffic is local by proper weight adjustment in the RCA aggregation module.

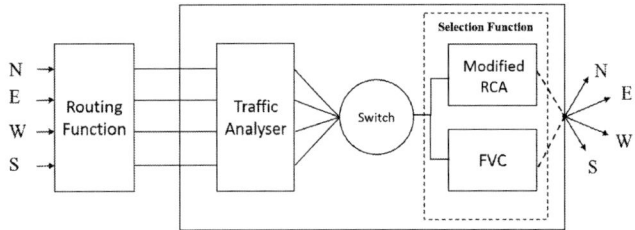

Fig. 4. Proposed Router Architecture

Figure 4 shows the router architecture of our proposed system. We use MOE routing algorithm to obtain the admissible output ports for the incoming packets. The traffic analyser decides the nature of the traffic at each router. The

978-1-4673-9141-2/15 $31.00 © 2015 IEEE   345

switching circuit enables a suitable selection strategy that can yield favourable results.

A. Traffic Analyser

The traffic analyser, extracts the destination address of each packet passing through a router and analyses this information for every T clock cycles. Two 5-bit saturating counters L (local) and N (Non-local) are used to represent the traffic pattern through each router.

If a packet's destination is more than 2-hops away from the current router, it is considered as a non-local packet, otherwise a local packet. The analyser calculates the number of hops to the destination and updates the L and N counters. This information is delivered to the switch for deciding the selection strategy. The counters are cleared at the end of the T cycles.

B. Switch

We implement a switching technique that enables either the RCA or the local best selection strategy at a given time in a router. The switching condition is checked at every T clock cycles. If T value is high, the network's response to change in traffic pattern will be slow. If it is low, frequent switching can affect throughput. We study the performance of the system for various switching intervals at T = 8, 16, 32, 64, 128 and 256 clock cycles. Based on these experiments, we fix T at 32 clock cycles, since it give better results compared others.

At the end of 32 clock cycles, the traffic pattern is decided using the analyser output. If the traffic trend is prominent towards non-local destinations, the customised RCA-FanIn selection strategy is enabled. Otherwise, aggregation portion of RCA circuitry is shutdown to make RCA behave as local best selection strategy. A router is evaluated to be local or non-local based on the ratio of the traffic through it.

For a non-edge router, ratio of number of local nodes to non-local nodes in an 8×8 mesh network is 12:51. i.e., the influence of the non-local packets will be more prominent in deciding the overall performance of the network. So, in routers, having a fixed minimum ratio of non-local traffic through it, the RCA selection strategy (which has higher visibility) was to be used. Thus a decision parameter value (i.e., a ratio of non-local packets to local packet through that particular router) must be slightly in-favour of choosing the RCA over FVC. We carried out a study by varying this value (x) in the range 0.25 to 0.45 under various traffic patterns. After detailed analysis of the results, we decided to fix this value as 0.40 to obtain a consistent performance. That is, if at least 40% of the router traffic is towards non-local destinations, then the router should work in RCA. Algorithm 1 explains the switching logic.

In an extended study, we apply different switching points (decision value) for different regions of the mesh network. Better results were obtained when switching is done at x=0.35 for the two outermost layer routers and x=0.45 for the inner layer routers of an 8×8 mesh NoC (48 out of 64 routers are positioned in the outer 2 layers).

The switching logic is implemented on every router, so that, independent switching of each routers rather than a collective switching of the entire network takes place. Thus different routers in the NoC operate in different selection strategies at

Data: L=Local value; N=Non-Local value
Result: Chooses the appropriate selection strategy
for *every T clock cycles* **do**
  Compute x= N/(L+N)
  if *x > 0.4* **then**
    Switch to modified RCA;
  else
    Switch to FVC;
  end
end

Algorithm 1: Algorithm for switching

a given time. The analyser and switch takes only data within each router, so no additional network communication overhead is added for implementation of this migration.

C. Cost effective RCA module

The original RCA transfers the congestion metric periodically to the upstream routers. In our design, all the routers need not be operating in RCA. This can affect the network performance if the current version of RCA is used, as some routers will not respond to a status request. So we customize RCA, so that, the latest status information is transferred to the neighbors before the router shutdown few RCA feedback lines. This helps in propagating a fair congestion information across the network than fixing a static value for the RCA-dormant routers.

The local best selection strategy is implemented using the resource subset of RCA itself. The design is such that, turning off a certain portion of RCA circuit is in fact the local best selection strategy itself. This is done by weight adjustment in the aggregation module of RCA. RCA gives 50-50 weightage for local and non-local information. When FVC is to be used in our design, rather than using a different circuitry, 100% weightage is given to local congestion metric and zero weightage to non-local congestion metric. This ensures that the use of two selection strategies does not increase the router hardware cost.

V. EXPERIMENTAL ANALYSIS

We customize the simulator as per our proposed architecture and analysed for 8x8 mesh network using various standard synthetic traffic patterns. We implement our system in view of combining the advantages of both FVC and RCA techniques. So we perform a comparative study with respect to these two selection strategies.

A. Analysis of Average Packet Latency

Latency of a packet is defined as the number of cycles needed for the packet to travel from its source to destination. It is a crucial factor for evaluating the performance of an NoC based system. Lower the average latency of the packets, faster the cache miss will be serviced for the application running on the source core. Hence for better performance, the average latency should be as low as possible.

The average latency of RCA, FVC and our selection strategy at different injection rates for an 8x8 mesh network is obtained. Figure 5 shows the injection rate vs average

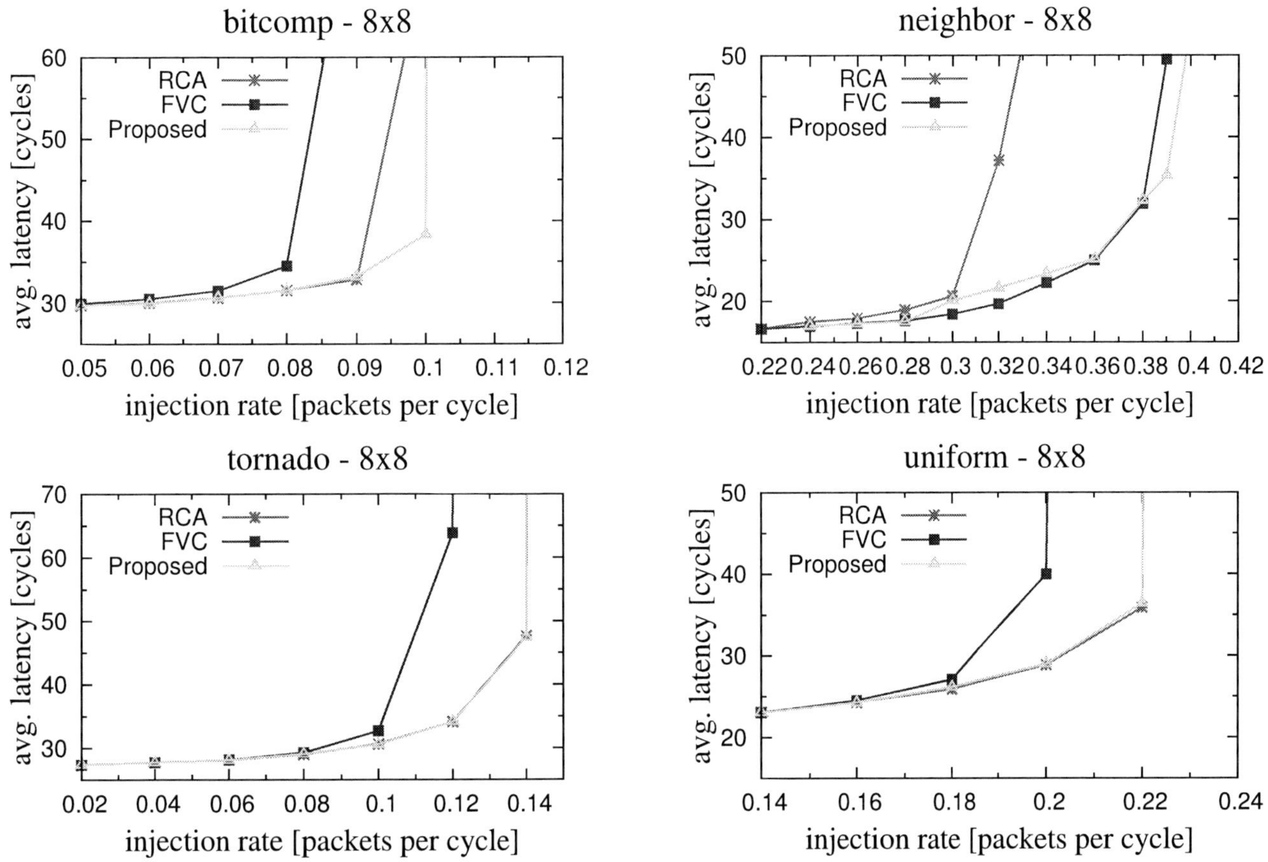

bitcomp - 8x8

neighbor - 8x8

tornado - 8x8

uniform - 8x8

Fig. 5.   Comparison of average packet latency in 8×8 mesh network using various synthetic traffic patterns.

packet latency graphs for bit-complement, neighbor, tornado and uniform traffics. From the graphs it is clear that across all synthetic traffic patterns our proposed technique shows lower latency values than FVC and conventional RCA.

In neighbor traffic, our technique shows values closer to FVC. In tornado, bit-complement and uniform traffics the RCA performs better than FVC, as the traffic pattern in most of the routers are non-local as evident from Figure 6. So, the latency curve of our system is close to that of RCA. This validates the claim that, if for a traffic RCA outperforms FVC, then our technique has latency values close to that of RCA. Otherwise, if FVC is better than RCA, our system gives latency near to that of FVC. i.e., based on the traffic pattern it adapts to the best performing strategy to deliver the least average latency. The migration across selection strategies helps our system to achieve a robust performance across traffics.

B. Control Network Design

The additional network resource needed for any selection strategy is the communication channels required for transferring the status information across the network. RCA-FanIn requires 9 bit control channel, whereas, FVC needs only 4 bit-lines for an 8×8 mesh with 16 virtual channels per router port.

Our design proposes to hardwire the resource requirements of RCA, but use them only when required. That is when the router has to operate in conventional RCA, it uses the entire

9-bit control channel, but as the traffic is either very low or local, only the 4 bit-lines are needed to operate as FVC. So, in the worst case (all the routers handling non-local heavy load) the network resource requirement of our system is equal to that of RCA.

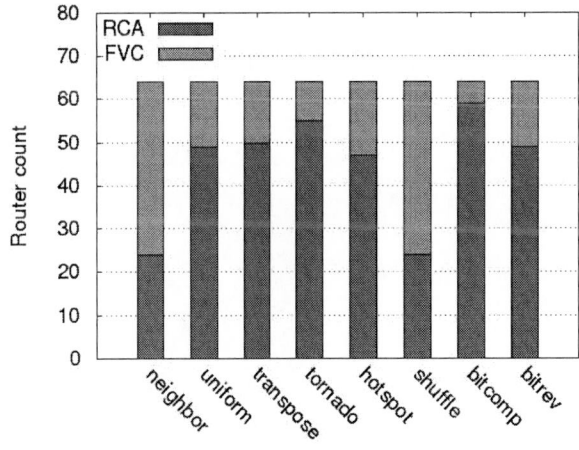

Fig. 6.   Count of routers working in RCA and FVC

Figure 6 shows the number of routers that are operating in RCA and FVC for various traffics in an 8x8 mesh network at saturation load. This is also a representation of the average communication channel utilisation. We see that many of the

978-1-4673-9141-2/15 $31.00 © 2015 IEEE   347

routers are working in FVC only, hence, the additional RCA bit lines associated with it will be shutdown. Thus the average network resource requirement is considerably less compared to that of RCA. This in-turn reduces the dynamic power consumption of the system.

For example, consider the uniform traffic. When RCA alone is used the average bit-lines per port in each router is 9. From the graph it is clear that 49 routers are in RCA (9 bit-lines) and the rest is in FVC (4 bit-lines). So the average bit-line utilisation can be calculated as follows: (49/64×9+ 15/64×4)/9 =0.8697. That is 13% of the entire communication channels are turned off saving a fraction of dynamic power dissipation.

C. Analysis of Router Complexity

It may seem that use of 2 selection strategy along with a traffic analyser and switching technique makes the router architecture complex. In our system, the implementation is done so that, the FVC is obtained by turning off a part of RCA, rather than as a separate selection logic. Hence, the architecture complexity of selection strategy is comparable to that of RCA. The analysis and switching techniques requires only 2 counters and a comparator in addition to an existing RCA router design. This adds a negligible overhead. Thus the routers do not significantly increase the overall system requirements. Verilog synthesis of the proposed router using Synopsys Design Compiler at 65nm shows an area overhead of 3.2% and static power overhead of 2.1% w.r.t RCA design. This overhead is due to the additional hardware units.

D. Analysis using Real Workloads

Apart from synthetic traffic, we evaluate the performance of our proposed system using traces of multi-programmed workloads also. We use Multi2sim [10] simulator to model a 64-core CMP set up with CPU cores, cache hierarchy, and coherence protocols in detail and accuracy. Each core consists of an out-of-order x86 processing unit with a 64KB, 4-way set-associative, 32 byte block, private L1 cache and a 512KB, 16-way set associative, 64 byte block, shared distributed L2 cache. Each core is assigned with a SPEC 2006 CPU benchmark application for running on it.

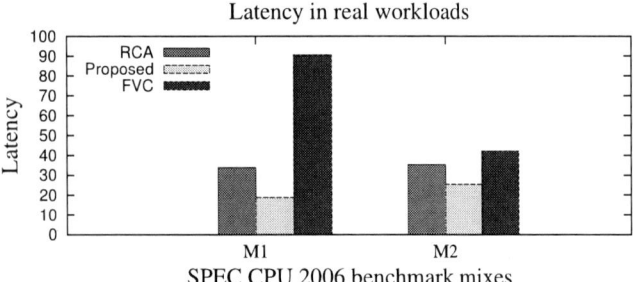

Fig. 7.   Comparison of Average packet latency using real workloads traffic traces.

We prepared 2 mixes of 64 core multiprogram workloads, M1 consists of 16 instances of medium *misses per kilo instructions* (MPKI) applications like *bwaves*, *bzip2*, *gamess*, and *gcc* and M2 consists of 16 instances of low MPKI applications like *calculix*, *gobmk* and 16 instances of high MPKI applications

like *mcf*, and *leslie3d*. After sufficient fast forwarding, we capture the L1 cache misses that generate network traffic and feed it to the modified Booksim model to simulate the network operations.

Figure 7 shows the performance comparison graph of the proposed selection strategy with RCA and FVC. In both mixes the proposed technique has slightly lower latency than RCA, and much lower than FVC. This establishes the fact that a performance equivalent to that of the better performing strategy will be delivered by our technique in real workloads also.

VI.   CONCLUSION

A refined NoC router architecture with a dynamic traffic analyser and a run-time switching is implemented to make the best use of RCA and FVC techniques. Routers uses RCA as such or a cost reduced version of it (effectively behaving as FVC), based on the real-time traffic patterns. Experiments on 8×8 mesh NoCs showed that the proposed design has less latency values consistently across various traffics compared to RCA and FVC. The overall energy utilisation is also reduced compared to RCA as many routers will in FVC. Hence the on-chip network which uses our selection logic can minimize network resource utilisation without affecting performance. As the FVC was designed to be obtained by a weight adjustment on RCA, it has only a very small hardware overhead. Thus using RCA and local best selection strategy in an equilibrium can bring down the power consumption and deliver stable performance. Hence, we conclude that our proposed design will be a good design alternative to future NoCs.

REFERENCES

[1] W. Dally and B. Towles *Principles and Practices of Interconnection Networks* USA: Morgan Kaufmann Publishers Inc., 2003.

[2] W. Dally and B. Towles, *Route packets, not wires: on-chip interconnection networks* in DAC, pp. 684-689, 2001.

[3] P.Gratz, B.Grot, and S.W. Keckler, *Regional Congestion Awareness for Load Balance in Network-on-Chip* in HPCA, pp. 203-214,February 2008.

[4] G. M. Chiu., *The odd-even turn model for adaptive routing* in IEEE TPDS,11 (7):729-738, July 2000.

[5] Giuseppe Ascia, Vincenzo Catania, Maurizio Palesi and Davide Patti, *Implementation and analysis of a new selection strategy for adaptive routing in NoC* in IEEE TOC, 57 (6), pp. 809-820, 2008.

[6] John Jose, K.V. Mahathi, J. Shiva Shankar and Madhu Mutyam, *TRACKER: A low overhead adaptive NoC router with load balancing selection strategy* in ICCAD, pp. 564-568, 2012.

[7] John Jose, J. Shiva Shankar, K.V. Mahathi, Damarla Kranthi Kumar and Madhu Mutyam, *BOFAR: Buffer occupancy factor based adaptive router for mesh NoC* in NoCArc, pp. 23-28, 2011.

[8] Mukund Ramakrishna, Paul V. Gratz and Alexander Sprintson, *GCA: Global congestion awareness for load balance in Networks-on-Chip* in NOCS, pp. 21-24, 2013.

[9] Nan Jiang, Daniel U. Becker, George Michelogiannakis, James Balfour, Brian Towles, John Kim and William J. Dally, *A detailed and flexible cycle-accurate Network-on-Chip simulator* in ISPASS, pp. 86 - 96, 2013.

[10] R Ubal et al., *Multi2sim: A simulation framework to evaluate multicore-multithreaded processors* in SBAC-PAD, 2007, pp. 62-68.

[11] J. Kim, et al., *A low latency router supporting adaptivity for on-chip interconnects* in DAC, pp. 559-564, 2005.

[12] K. Lahiri et al., *Evaluation of the traffic-performance characteristics of system-on-chip communication architectures*, Proceedings of the International Conference on VLSI Design (2000), pp. 2935.

A Cluster-Based Reliability- and Thermal- Aware 3D Floorplanning using redundant STSVs

Ying-Jung Chen
National Taiwan University of Science and Technology
43,Sec.4,Keelung Rd.,Taipei,106,Taiwan,R.O.C
Email: m10102122@mail.ntust.edu.tw

Shanq-Jang Ruan
National Taiwan University of Science and Technology
43,Sec.4,Keelung Rd.,Taipei,106,Taiwan,R.O.C
Email:sjruan@mail.ntust.edu.tw

Abstract—**In 3D-IC architecture, vertically stacked multiple layer impedes heat dissipation and exacerbates thermal problem especially for reliability degradation. In this paper, we propose a cluster-based reliability- and thermal- aware 3D floorplanning to place modules. We derive a cost function considering both reliability and thermal factors with the balance among the chip of area, wire length, power density and density of STSV in the floorplan. Then we insert rectangle-STSVs and double-STSVs for improving reliability by modified Ford-Fulkerson method. Furthermore, we enhance redundant STSVs insertion rate by RSI algorithm. The experimental results show that more than 80% of single-STSVs can be replaced by rectangle-STSVs or double-STSVs, improving reliability accordingly. After STSV insertion, we construct a precise thermal conduction model to compute temperature distribution and insert TTSVs.Our framework is able to reduce the peak temperature effectively and maintain around 80 °C with minimal TTSVs based on a precise temperature computation model.**

Keywords—*3D floorplanning, reliability, double-STSV, rectangle-STSV.*

I. INTRODUCTION

3D IC has recently been a popular research topic because it has several advantages over 2D ICs such as smaller chip area, higher performance, and lower power consumption. To connect different layers of dies vertically, a special device called through silicon via (TSV) is thus introduced. An example of 3D-IC structure and TSV is shown in Fig. 1.

However, yield of TSV is a challenge under current manufacturing process. Thus, to improve TSV yield and reliability, several researchers attempt to solve yield of TSV issue using spare redundant TSVs scheme [1] [2].

Thermal issue is another problem when using 3D-IC since the power density increased dramatically especially in 3D-IC system [3]. Several researchers have shown high-density vertical interconnection and high performance operation adversely affect the heat dissipation of 3D-ICs [4] [5]. Vertical integration of layers exacerbates thermal problem especially for reliability degradation [6]. Several thermal-aware floorplanning techniques [7] [8] [9] have been presented to improve wire length, interconnect delays, and thermal problem. Thermal through silicon via (TTSV) planning schemes have been demonstrated for thermal management in [7] [10].

Unfortunately, previous works do not consider the problems of both reliability and thermal simultaneously in the floorplanning. In order to mitigate the problems of both

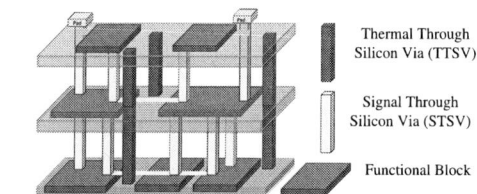

Fig. 1. STSVs and TTSVs in 3D IC

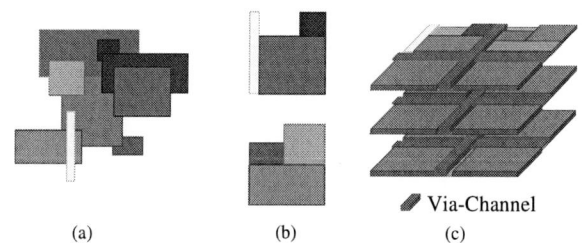

Fig. 2. (a) 2D modules (b)The clustering mechanism is used to group 2D modules. (c) The each cluster of modules is placed the region partitioned by via-channel.

reliability and thermal in the meantime, we present a set of strategies to address the issues of thermal and reliability for a 3D-IC design.

In this paper, we propose a cluster-based reliability- and thermal- aware 3D floorplanning, using redundant STSVs to improve reliability and TTSVs insertion for thermal dissipation. The purpose is to insert rectangle-STSVs, double-STSVs, and arranged TTSVs so that the yield of STSVs can be improved significantly and the thermal distribution can be suppressed after modules are placed.

Experimental results show the reliability of chip can be improved by double-STSVs and rectangle-STSVs. The insertion rate of redundant STSVs can be obtained over 80%. The temperature of 3D-IC systems can be maintained around 80 °C with minimal number of TTSVs.

The remainder of this paper is organized as follows. Section II presents problem formulation.Section III presents the whole design flow. We propose our reliability- and thermal-aware floorplanning with cost function and algorithm in Section IV. Experimental results are reported in Section V. Section VI is the conclusion.

978-1-4673-9141-2/15 $31.00 © 2015 IEEE

II. Problem Formulation

Given a set of n blocks, $\mathbf{B} = \{b_1, b_2, b_3, ..., b_n\}$, whose repective width, heigh, and area are denoted by w_i, h_i and a_i. The corresponding power density is denoted as Pd_i for each block b_i in $\mathbf{B}$, $1 \leq i \leq n$. Note that a set of each block b_i whose $STSVd_i$ is STSV density of i-th block. The objective is to improve thermal dissipation and yield of STSV in 3D-IC design. Thus, a novel repair framework considering both thermal issue and reliability STSV is proposed in this paper. In this scheme, we cluster these modules and place clusters layer by layer based on power density of clusters. Note that this scheme gives the priority to dissipate the heat produced since the high temperature would affect reliability of 3D circuit. Then, we place modules with a cost function. The cost function is proposed for considering thermal, reliability factors and cost in the 3D floorplanning. After floorplanning, redundant STSVs and TTSVs are inserted for improving reliability and thermal dissipation. To raise the insertion rate of redundant STSV in limited chip area , we propose a novel algorithm RSI in this scheme. Additionally, to insert least TTSVs to reduce the peak temperature T_p so that $T_p \leq T_{th}$ in the 3D floorplanning, we apply the concept of center of mass with weight method. Note that T_{th} is the threshold temperature and the maximum temperature should be less than T_{th} in the chip.

The cost function considers area, wire length, power density and STSV density based on [11] and [12]. Let T_{fi} be a polynomial that has three variables : area, wire length, and power density, respectively. Note that $Area$ is total area of floorplan, Wl is the wire length of floorplan and the power density is Pd. Let R_{fi} be a reliability factor that has a variable $STSVd$. $STSVd$ is STSV density of each module. The equation can be written below.

$$cost = \alpha \times T_{fi}(Area, Wl, Pd) + \beta \times (1-(R_{fi}(STSVd)), \quad (1)$$

where the two parameters α and β can be defined by users. We can use this formula to evaluate a design alternative if we know its impact on these two parameters. We separate the two key factors heat and reliability that affect cost.

$$T_{fi}(Area, Wl, Pd) = a \times Area + b \times Wl + c \times Pd, \quad (2)$$

where the three weighted parameters a, b and c can be defined by users.

In thermal part, this formula makes it clear that designers can improve cost by keeping the area of each module almost the same size, reducing the power density of each module and reducing the wire length between the blocks. In the floorplanning, considering the size of area, wire length and power density can dissipate thermal efficiently since large area can reduce power density. However, the long wire length between the blocks will generate a lot of heat. Although using TTSV can reduce temperature, a large number of TTSVs cost high in manufacturing. Thus, it is important to cope with thermal dissipation problem during the floorplanning with the proposed function.

In reliable part, we consider $STSVd$, since we improve reliable STSV by inserting more redundant STSVs. As aforementioned, the modules with high density STSV can increase

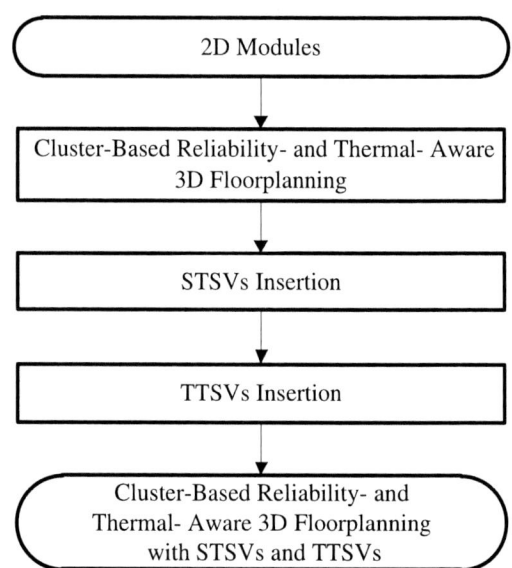

Fig. 3. Overall flow chart

the reliability of STSV. Thus, we need more space to insert more redundant STSVs in a module with high density STSV if some STSVs are failed. However, the high density of STSV is proportional to the cost. The detailed treatment will be described in the following subsections.

III. Design Flow

The whole design flow is shown in Fig. 3. In the beginning, 2D modules are the input into clustered-based reliability- and thermal- aware 3D floorplanning stage. Then, we apply a clustering technique [13] to group modules and area of each cluster would be approximate as shown in Fig. 2(b). In the modules clustering step, the heat reduction is our major consideration. Thus, the cluster with higher power density would be placed at the layer near to the heat sink. Second, to determine positions of these modules, these modules of each cluster would balance the factors of area, wire length, power density and STSV density based on Equ. (1). Then, using simulated annealing (SA) with the purpose of minimum area, wire length and STSV density of each cluster, the result determines the position of each module. These white space from the adjacent layers would be divided into not only several locations of STSV candidate for STSVs insertion but also the via-channel [13] for TTSVs insertion. Finally, we place cluster layer by layer based on the specification of via-channel as shown in Fig. 2(c).

In the end of the floorplanning stage, as the positions of these modules are fixed, the locations of the STSVs are determined. After assigning the STSVs, we construct flow network graph. Then, we insert rectangle-STSVs, and double-STSVs to a suitable location for improving STSVs reliability. Based on [12], the STSV with lager area can make better reliability, so we insert double-STSVs first and then rectangle-STSVs. We use modified Ford-Fulkerson algorithm to insert redundant STSVs and propose an algorithm to improving reliability at edge of the each layer. Finally, we compute the

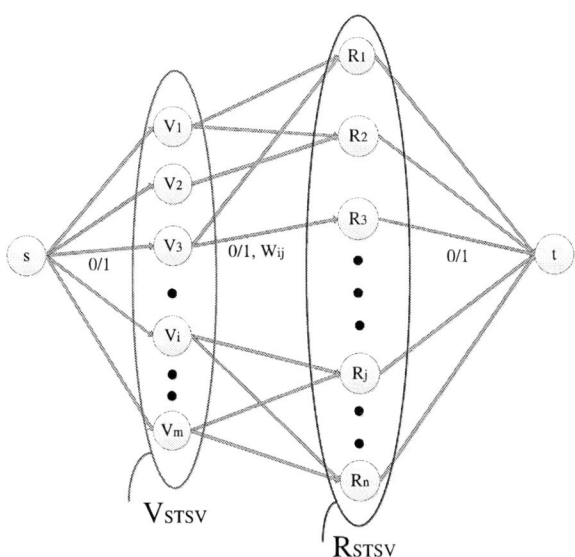

Fig. 4. Flow network graph

$V = V_{STSV} \cup R_{STSV}$, where V_{STSV} and R_{STSV} are disjoint and all edges in set **E** go between set **V** and set **R**. We further assume that every vertex in **V** has at east one incident edge [15]. To solve the STSVs assignment problem, the network graph is constructed as shown in Fig. 4. The STSVs assignment problem can be formulated as follows. Let $G = (V, E)$ be a flow network with source s and the sink t and let f be a flow in G.
In the graph, $G(V, E)$,

$$V = \{s, t\} \cup V_{STSV} \cup R_{STSV},$$
$$E = \{s, V_i | V_i \in V_{STSV}\}$$
$$\cup \{(V_i, R_i) | V_i \in V_{STSV}, R_i \in R_{STSV}\}$$
$$\cup \{(R_i, t) | R_i \in R_{STSV}\}.$$

Edge capacity:

$$C(s, V_i) = 1,$$
$$C(V_i, R_i) = 1,$$
$$C(R_i, t) = 1.$$

Given a set of m STSVs, $V_{STSV} = \{V_1, V_2, V_3, ..., V_m\}$ and the set $R_{STSV} = \{R_1, R_2, R_3, ..., R_n\}$, corresponding to n candidate locations. According to [12], the better the reliability of STSV is and the lager the area of STSV is. Therefore, we give the different weights for edges $E(V_i, R_i)$ in the graph to solve the maximum flow as shown in Fig. 4.

$$w(V_i, R_i) = \begin{cases} 3, if \text{ R } is \text{ double-STSV candidate,} \\ 2, if \text{ R } is \text{ rectangle-STSV candidate,} \\ 1, \text{otherwise.} \end{cases}$$

All vertices V_i are single STSV, all R_i are locations of candidates, and the $w(V_i, R_i)$ indicates the edge between STSV and their candidates with weight. To perform the maximum flow, if the weight from V to R is double-STSV candidate, the weight is assumed to be 3, the weight of rectangle-STSV candidate is assumed to be 2 and otherwise is set to be 1. Using flow network on **G**, redundant STSVs are assigned to preserve feasible locations so that the capacity is maximum. The Ford-Fulkerson algorithm can be modified to deal with the flow network as shown in Fig. 4. The proposed the redundant STSV assignment algorithm will be introduced in the subsequent section.

cost and the reliability $R_{current}$ for present assignments and input the 3D floorplan with STSVs to next stage.

According to the result of floorplanning, we will insert and delete some TTSVs in this stage. To calculate temperature, we construct precise thermal model [13] and the corresponding thermal distribution. After computing temperature, if the peak temperature is higher than T_{th}, we will insert more TTSVs at via-channel; if not, we will delete some TTSVs for cost reduction. Note that T_{th} is set at 80 °C on the basis of [11]. In the end of our framework, we can guarantee the peak temperature around T_{th} and insert TTSVs at least as possible.

IV. Proposed Method

In this section, a cluster-based reliability- and thermal-aware 3D floorplanning is developed that considers the thermal and reliability issues based on the implicit heat flow analysis and density of STSV. In the floorplanning, the SA with the cost function is used to calculate positions of modules iteratively. After floorplanning, a modified Ford-Fulkerson algorithm is proposed for rectangle-STSV and double-STSV insertion. With new kinds of STSV and TTSV, we present a framework to improve reliability and thermal dissipation.

A. Redundant STSV Assignment Algorithm

When the 3D placement of STSV is obtained, the positions of STSVs are determined. However, the positions of the redundant STSVs still need to be determined. The redundant STSVs are assigned to appropriate locations in the limited area after 3D placement. Previous researches have demonstrated that the STSV assignment can be efficiently solved based on the flow network problem [14] as shown in Fig. 4. Therefore, the STSV assignment can be modeled as the maximum flow problem which can be solved by Ford-Fulkerson method.

1) **Flow Network Construction:** Unlike the conventional approaches, our flow network graph conform to the restriction that the vertex set can be partitioned into

---

Algorithm 1 Modified Ford-Fulkerson Algorithm

1: **for** each edge $(u, v) \in E$ **do**
2:     $f(u, v) \leftarrow 0$;
3: **end for**
4: **while** exist path p from s to t in the residual network G_f **do**
5:     $c_f(p) \leftarrow min\{c_f : (u, v) \in p, u, v \in PRS, c_f(u, v) \neq 1\}$;
6:     **for** each edge $(u, v) \in p$ **do**
7:         $f(u, v) \leftarrow f(u, v) + c_f(p)$;
8:     **end for**
9: **end while**

978-1-4673-9141-2/15 $31.00 © 2015 IEEE

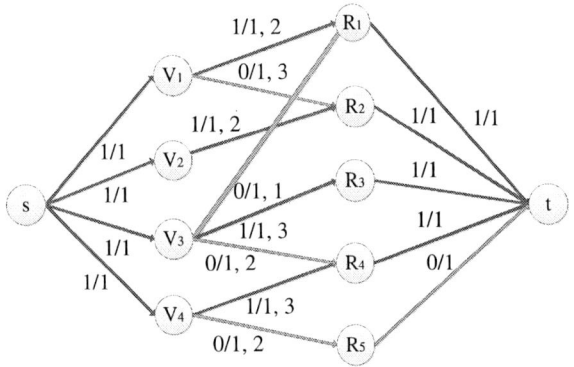

Fig. 5. Flow network graph operation

Algorithm 2 Redundant STSV Insertion

1: construct the adjacency-list graph of $RSTSV[i]$
2: // $RSTSV\_Adj[i]$ is $RSTSV[i]$ neighbourhood and adjacency-list graph
3: **for** each u $\in$ $RSTSV$ **do**
4:    **if** $RSTSV[u].state ==$ "rSTSV " **then**
5:       **for** each $v \in RSTSV\_Adj[u]$ **do**
6:          **if** $RSTSV[v].state ==$ "dSTSV"&& $RSTSV[v].dist$ is less than two unit **then**
7:             // the layer is set three units, far, middle and near of the center.
8:             // the less than two unit mean the RSTSV locate middle or near area.
9:             $RSTSV[u].state$ exchange $RSTSV[v].state$
10:          **end if**
11:       **end for**
12:    **end if**
13: **end for**

---

2) Operation of Flow Network:

2) Operation of Flow Network: We modify the Ford-Fulkerson algorithm and apply it to flow network as illustrated in Algorithm 1. As aforementioned our flow network strict some conditions, so the graph G is simpler than a normal flow network as shown in Fig. 5.

We start with $f(u, v) = 0$ for all $u, v \in V$, giving an initial flow of value 0. In lines 3-8, we apply depth first search (DFS) to find each path p in G and flow f along p by the residual capacity $c_f(p)$. In lines 6-8, the flow can be updated in each case appropriately. In the modified Ford-Fulkerson algorithm, the back edges are used to replace candidates with the highest weight for the lower weight, if it can not get the result with maximum capacity. Otherwise, those candidates with high weight are still chosen first. When no augmenting paths exist in the graph, the flow f is a maximum flow. Consequently, we can get the maximum candidates locations of STSV.

The running time of modified Ford-Fulkerson method depends on how we find the augmenting path p. In our construction, we find the augmenting path by depth first search (DFS) and the DFS runs in $O(E)$. We define the corresponding flow network $G' = (V', E')$ for the flow network graph G. The vertex partition of G is $V = V_{STSV} \cup R_{STSV}$, and let $G' = (V', E')$ be its corresponding flow network. Moreover, we assign unit capacity to each edge in E'. Thus we can find a maximum-flow by creating the flow network G', running the modified Ford-Fulkerson method so that any value of maximum-flow in a flow network graph has cardinality at most $\min(V_{STSV}, R_{STSV}) = O(V)$ and the value of maximum-flow in G' is $O(V)$. The **while** loop takes $O(E)$ time and we can find a maximum flow in our flow network graph in time $O(VE') = O(VE)$ since $|E'| = \Theta(E)$.

B. Redundant STSV Insertion

The redundant STSVs at the edge of the matrix have lower yield than those at center [16]. Thus, we propose a RSI algorithm to enhance the yield of redundant STSVs at the edge of layer. After the modified Ford-Fulkerson method is used to insert redundant STSVs, we consider the positions of redundant STSVs and surrounding area of redundant STSVs. In each layer, the distance from center to edge of layer is divided three units, center, middle and far. It is possible to change the state of redundant STSV at edge layer from rectangle to double, if the rectangle-STSV is next to

another double-STSV which is in the middle of layer. The data structure V_s of Algorithm 1 and 2 is utilized in each STSV.

structure Vs {

   state;   //sSTSV, rSTSV or dSTSV

   coord;   //Xo, Yo, X, Y

   area;   //single-STSV area

   dist;   //distance between STSV and the center of layer

   $RSTSV_{Adj}$;   //RSTSV neighbourhood
};

For each rectangle-STSV or double-STSV originates from a single-STSV $\in V_{STSV}\_SET$. The rectangle-STSVs or double-STSVs are in R_{STSV}_SET after executing the Algorithm 1. R_{STSV}_SET is constructed to get each $RSTSV_i$. Then the adjacency-list graph of $RSTSV_i$ is constructed for the Algorithm 2. The work done within the **for** loop of the Algorithm 2 can be fulfilled efficiently to enhance reliability and the maximum number of double-STSV can be determined at edge of the layers.

The contribution of this paper is that we use a double- and rectangle-STSV mechanism for redundant STSV insertion in cluster-based reliability- and thermal- aware 3D floorplanning. The objective is to prevent the interconnect failures by using double- and rectangle-STSV given limited resources in the workflow and concurrently optimizing the redundant STSV insertion rate by RSI algorithm. The novel RSI flow generally overcomes the limitation of redundant STSV insertion with both double-STSV and rectangle-STSV in our scheme. Compared to the conventional approach [11], the proposed scheme considerably improves the redundant STSV insertion rate and failed STSVs by using double- and rectangle-STSV. Previous researches do not use double-STSVs and rectangle-STSVs in thermal- and reliability-aware 3D floorplanning. The proposed scheme not only addresses the problem of failed STSVs but also increases redundant STSV insertion rate.

V. Experimental Results

Our framework is implemented in C++ and compiled with g++ 4.1.2. All experiments are conducted on an Intel(R) Core(TM) 3.40GHz workstation with 32GB memory. In our experiments, we apply four MCNC benchmarks (hp, xerox, ami33 and ami49) [17]. The block number of four MCNC benchmarks ranges between 11 and 49. The size of single TSV is set between 10 μm and 30 μm, depending on the ratio of area of each benchmark. The net size is from 83 to 408. The input floorplans are all generated by the 3D floorplanner proposed in [18]. The power density of each block is randomly generated between 30 and 1000 (W/cm^2) in the floorplans. With these previously defined parameters, we rearrange blocks with clusters by clustering technique and inserting STSVs and TTSVs.

The experimental results are listed in Table I. In Table I, column 1 and 2 show the name of benchmarks and the corresponding module numbers. Column 3 lists the net size of each benchmark. In column 4, the insertion rate which is the rate of how high percentage of STSVs can be enhanced as rectangle-STSVs and double-STSVs. Column 5, 6 and 7 list the number of total number of STSVs, rectangle-STSVs and double-STSVs we inserted in each case. The temperature cool down after the insertion of redundant STSVs is shown in Table III. Under the precise thermal model, we calculate the temperature before TTSVs insertion and the number of inserted TTSV at via-channel is illustrated in column 2 and 3. Column 4 lists the temperature after TTSVs insertion.

The results are shown our method can improve reliability of STSV and increase insertion rate of redundant STSV effectively. Note that the insertion rate of redundant STSV is more than 80% in each case. Additionally, the synergy of our clustering-based techniques and TTSVs insertion successfully reduce the thermal effects and temperature can be maintained around 80 °C with the minimal TTSVs.

Table II presents the MCNC benchmark circuits considering reliability and thermal situation compared with the result which Hsu et al. presented [11]. It is observed our floorplanning has better reliability of STSV for all MCNC benchmarks in term of rectangle-STSV and double-STSV. Furthermore, the proposed method has better insertion rate of redundant STSV by modified Ford-Fulkerson and RSI algorithm. The result shows our method is better than the [11] method.

In comparison to the 3D floorplanner without consideration of reliability and thermal, the area is optimized the peak temperature is apparently over-high and the circuits yield rate is uncertainly. After our reliability- and thermal- aware floorplanning, we can ensure that the circuits are more reliable by inserting rectangle-STSVs and double-STSVs with maximum number of redundant STSVs in limited area. Moreover, after inserting STSVs, we proceed to the analytic TTSV insertion positions according to the distribution of temperature. Note that these rectangle-STSVs, double-STSVs and TTSVs only occupied slight portion of the whole chip.

To know the detail of temperature distribution before and after TTSV insertion, we dissect each layer of the benchmark ami33. Figure 6 illustrates the temperature distribution in each layer. Figure 6 (a)-(c) represent the temperature distribution without STSVs and TTSVs, and Fig. 6 (d)-(f) show the distribution after inserting STSVs and TTSVs.

The temperature distribution of Layer 1 illustrates that the heat is easily stuck at the layer which is the farthest from the heat sink. The chip is cooled down to an acceptable temperature for all layers and it is more reliable for the whole chip, after inserting STSVs for reliability and TTSVs for heat dissipation at via channels.

VI. Conclusions

In this paper, we proposed a cluster-based reliability- and thermal- aware 3D floorplanning which does include heat dissipation and consideration of reliability with insertion of rectangle-STSVs and double-STSVs. Based on the cluster technique and specification of via-channel, we insert rectangle-STSVs and double-STSVs into 3D-IC to improve reliability. Since we insert more rectangle-STSVs and double-STSVs to get better reliability and decrease failed chip cost under the area constraint, we proposed a cost function to balance reliability and area. The experiment result showed that insertion rate are higher than 80%. It means we can get the suitable positions of redundant STSVs for improving reliability in limited area. After our work of STSVs insertion, we insert TTSVs at via-channel and the resultant peak temperature is acceptable under the precise thermal model check iteratively. The experiment result showed that after TTSVs insertion, the temperature can be kept around 80 °C with minimal TTSVs. We believe the floorplanner can make chip more reliability based on considerations of both reliability and thermal.

References

[1] I. Loi, S. Mitra, T. H. Lee, and S. Fujita, "A low-overhead fault tolerance scheme for tsv-based 3d network on chip links," in *Computer-Aided Design, 2008. ICCAD 2008. IEEE/ACM International Conference on*, Nov. 2008, pp. 598–602.

[2] A. C. Hsieh, T. Hwang, M. T. Chang, M. H. Tsai, C. M. Tseng, and H. C. Li, "Tsv redundancy: Architecture and design issues in 3dic," in *Design, Automation & Test in Europe Conference & Exhibition (DATE)*, Mar. 2010, pp. 166–171.

[3] J. Park, J. Jung, K. Yi, and C. M. Kyung, "Static energy minimization of 3d stacked l2 cache with selective cache compression," in *Very Large Scale Integration (VLSI-SoC), 2013 IFIP/IEEE 21st International Conference on*, 2013, pp. 228–233.

[4] G. Katti, M. Stucchi, D. Velenis, B. S., K. D. Meyer, and W. Dehaene, "Temperature-dependent modeling and characterization of through-silicon via capacitance," in *IEEE Electron Device Lett*, vol. 32, no. 4, Apr. 2011, pp. 563–565.

[5] M. Lee, J. Cho, J. Kim, and J. S. Pak, "Thermal effects on through-silicon via (tsv) signal integrity," in *Electronic Components and Technology Conference (ECTC), 2012. IEEE 62nd*, Jun. 2012, pp. 816–821.

[6] M. M. Sabry, A. K. Coskun, and D. Atienza, "Fuzzy control for enforcing energy efficiency in high-performance 3d systems," in *2010 IEEE/ACM International Conference on Computer-Aided Design (IC-CAD)*, 2010, pp. 642–648.

[7] B. Goplen and S. S. Sapatnekar, "Placement of thermal vias in 3-d ics using various thermal objectives," in *Computer-Aided Design of Integrated Circuits and Systems, IEEE Transactions on*, vol. 25, no. 4, Apr. 2006, pp. 692–709.

[8] S. Aroonsantidecha, S. S. Y. Liu, C. Y. Chin, and H. M. Chen, "A fast thermal aware placement with accurate thermal analysis based on green function," in *Design Automation Conference (ASP-DAC), 2012 17th Asia and South Pacific*, Jan. 2012, pp. 425 – 430.

TABLE I.   COMPARISONS FOR RELIABILITY, BEFORE AND AFTER REDUNDANT STSV INSERTION

MCNC Benchmark	#blocks	#Net	#STSV	#Rectangle-STSV	#Double-STSV	Insertion Rate
ami33	33	123	88	50	32	93%
ami49	49	408	232	173	18	82%
xerox	10	203	99	52	32	85%
hp	11	83	42	28	8	86%

TABLE II.   COMPARISON OF RESULT ON MCNC BENCHMARK CIRCUITS

MCNC benchmark	#STSV	#Rectangle STSV		#Double STSV		Insertion rate		Run time	
		[11]	Our method	[11]	Our method	[11]	Our method	[11]	Our method
ami33	88	50	50	31	32	92%	93%	$O(E|f^*|)$	O(VE)
ami49	232	171	173	15	18	80%	82%		
xeror	99	50	52	31	32	82%	85%		
hp	42	28	28	7	8	83%	86%		

TABLE III.   COMPARISONS FOR TEMPERATURE, BEFORE AND AFTER TTSV INSERTION

MCNC benchmark	Temperature before TTSV insertion	#TTSV	Temperature after TTSV insertion
ami33	115.675	25	79.839
ami49	129.953	73	79.733
xerox	107.234	58	79.426
hp	88.913	182	79.782

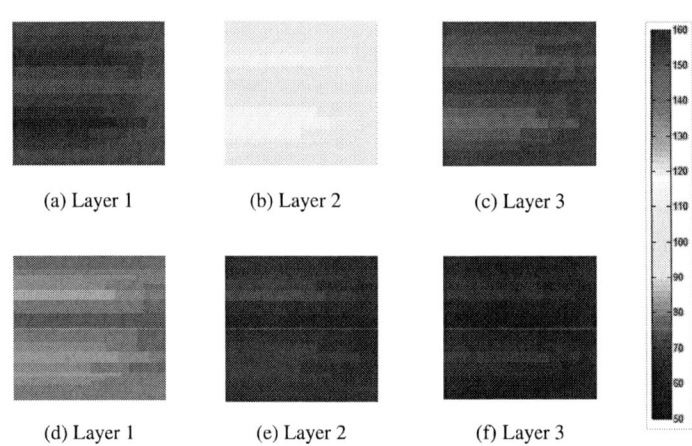

(a) Layer 1   (b) Layer 2   (c) Layer 3

(d) Layer 1   (e) Layer 2   (f) Layer 3

Fig. 6.   The distribution of temperature of the test case $ami33$ before thermal-via insertion (a-c) and after thermal-via insertion (d-f). The via-channels separate the whole space into 3×1 subspaces and the size of thermal-via is $20 \times 20 \mu m^2$

[9]   Y. Chen, E. Kursun, D. Motschman, C. Johnson, and Y. Xie, "Through silicon via aware design planning for thermally efficient 3-d integrated circuits," in *Computer-Aided Design of Integrated Circuits and Systems, IEEE Transactions on*, vol. 32, no. 9, Sep. 2013, pp. 1335–1346.

[10]   J. Cong and Y. Zhang, "Thermal via planning for 3-d ics," in *Computer-Aided Design, 2005. ICCAD-2005. IEEE/ACM International Conference on*, Nov 2005, pp. 745 – 752.

[11]   C. H. Hsu, S. J. Ruan, Y. J. Chen, and T. C. Kan, "Reliability consideration with rectangle- and double-signal through silicon vias insertion in 3d thermal-aware floorplanning," in *Quality Electronic Design (ISQED), 2013 14th International Symposium on*, Mar. 2013, pp. 316 – 321.

[12]   A. Shayan, X. Hu, H. Peng, and C. K. Cheng, "Reliability aware through silicon via planning for 3d stacked ics." in *Design, Automation & Test in Europe Conference & Exhibition, 2009. DATE '09.*, vol. 1, Apr. 2009, pp. 288 – 291.

[13]   S. Hu, Y. Y. G. Hoe, H. Li, D. Zhao, J. Shi, Y. Han, K. H. Teo, Y. Z. Xiong, J. He, and M. Je, "A thermal isolation technique using through-silicon vias for three-dimensional ics," in *Electron Devices, IEEE Transactions on*, vol. 60, no. 3, Mar. 2013, pp. 1282–1287.

[14]   L. Zhou, C. Wakayama, and C. R. Shi, "Cascade: A standard super-cell design methodology with congestion-driven placement for three-dimensional interconnect-heavy very large-scale integrated circuits," in *Computer-Aided Design of Integrated Circuits and Systems, IEEE Transactions on*, vol. 26, no. 7, Jul. 2007, pp. 1270 – 1282.

[15]   T. H. Cormem, C. E. Leiserson, R. L. Rivest, and C. Stein, *Introduction To Algorithms*.   The MIT Press, 2009.

[16]   G. V. der Plas, P. Limaye, I. Loi, and A. Mercha, "Design issues and considerations for low-cost 3-d tsv ic technology," in *IEEE Journal of Solid-State Circuits*, vol. 46, no. 1, Jan 2011, pp. 293 – 307.

[17]   "http://vlsicad.cs.binghamton.edu/benchmarks.html."

[18]   W. Hung, G. Link, Y. Xie, N. Vijaykrishnan, and M. J. Irwin, "Interconnect and thermal-aware floorplanning for 3d microprocessors," in *2006. ISQED '06. 7th International Symposium on Quality Electronic Design*, Mar. 2006, pp. 98–104.

Trace Buffer Attack: Security versus Observability Study in Post-Silicon Debug

Yuanwen Huang*, Anupam Chattopadhyay[†], Prabhat Mishra*

*University of Florida, Gainesville, Florida, USA

[†]Nanyang Technological University, Singapore

{yuanwen, prabhat}@cise.ufl.edu, anupam@ntu.edu.sg

Abstract—Since the standardization of AES/Rijndael symmetric-key cipher by NIST in 2001, it gained widespread acceptance in various protocols and withstood intense scrutiny from the theoretical cryptanalysts. From the physical implementation point of view, however, AES remained vulnerable. Practical attacks on AES via fault injection, differential power analysis, scan-chain and cache-access timing have been demonstrated so far. Along this line, in this paper, we propose a novel and effective attack, termed *Trace Buffer Attack*. Trace buffers are extensively used for post-silicon debug of digital designs. We identify this as a source of information leakage and show that, unless proper countermeasure is taken, *Trace Buffer Attack* is capable of partially recovering the secret keys of different AES implementations. We report the detailed process of trace-buffer attack with experimental results. We also propose a countermeasure in order to avoid such attack.

Keywords: Cryptography, Cryptanalysis, Trace Buffer, Post-silicon Debug, AES.

I. INTRODUCTION

As the human civilization is collectively progressing towards an ubiquitous information age, the corresponding stakes on ensuring confidentiality, integrity and authenticity are also rising higher. Advanced Encryption Standard (AES) algorithm with various key lengths (128, 192 and 256) is widely used. The fact that AES stood the intense scrutiny from attackers over the last 15 years itself makes it an important benchmark for cryptography and cryptanalysis. So far, the best-known attempt against full AES-128, by *algebraic cryptanalysis*, has a computational complexity of $2^{126.1}$, which is slightly better than the brute-force attack and practically infeasible [1]. However, the perspective of *physical cryptanalysis* changes this scenario completely.

In practice, one routinely faces a situation where the cryptographic schemes are deployed in different adversarial setting, where keys are compromised, and the internal memory is not fully opaque. This situation leads to a set of physical cryptanalysis techniques, commonly known as *side channel attacks*. Side channel attacks exploit the physical implementation of cryptographic algorithms. The physical implementation might enable *leakage*, i.e., observations and measurements on the implementation details, as well as tampering with them. Such attacks have broken systems with mathematical security proof. In this scenario, secure implementation is rapidly becoming as

This work was partially supported by the NSF grants (CCF-1218629 and CNS-1441667) and SRC grant (2014-TS-2554).

important as the mathematical security proofs. For example, an AES implementation with protection against a first-order side-channel attack is presented here [2]. The protected design is still vulnerable to more sophisticated attacks and even then, incurs $4.6\times$ area- and $3.6\times$ power-overhead, respectively, compared to the unprotected implementation.

In light of these developments, it is of utmost importance to remain fully aware of the design vulnerabilities, in the form of precise information leakage. In this paper, we introduce **Trace Buffer Attack** (TBA), a novel attack that can be mounted with the help of post-silicon debug facilities present in a chip. System-on-Chip (SoC) designs have in-built trace buffer (described in Section 2) that traces a small set of internal signals during execution, and the traced signal values are used during post-silicon (off-line) debug. There is an inherent conflict between security and observability. While debug engineers would like to have better observability, the security experts would like to enforce limited or no visibility with respect to the security modules in a SoC design. A trade-off is typically made where trace signals are carefully selected to maintain security while providing reasonable debug capability. To the best of our knowledge, the vulnerability of trace buffers in cryptographic implementation has not been studied in the literature. We conclusively show that to achieve a certain quantifiable level of debugging ability, security is compromised. We consider AES as the benchmark algorithm for demonstrating the efficacy of this attack though, the attack can be mounted on other ciphers following the same principles outlined in this work. Our experimental results demonstrate that we can fully recover the secret key for AES-128 (iterative) implementation whereas we can partially recover the secret key for various pipelined AES implementations.

The rest of this paper is organized as follows. A background on AES and trace buffer is provided in Section II. Section III surveys related work on AES attack and trace buffer. Section IV describes the details of trace buffer attack. Section V presents the experimental studies. To prevent TBA, a countermeasure is proposed in Section VI. The paper is concluded with outline of future work in Section VII.

II. BACKGROUND

A. AES Specification

AES works on a block size of 128 bits and a key size of 128, 192 or 256 bits, which are referred to as AES-128, AES-

192 and AES-256, respectively[1]. We briefly review AES-128 here, for further details readers can refer to [3].

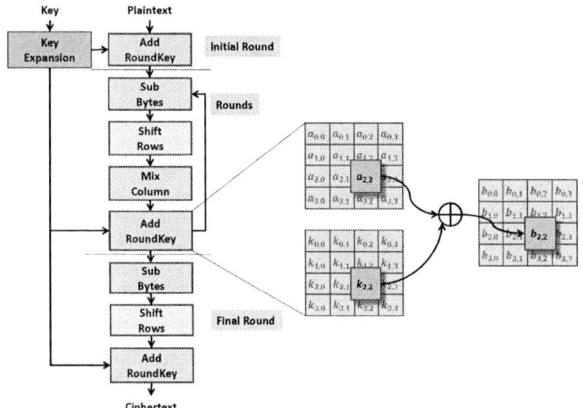

Fig. 1: AES Encryption Flow

The encryption flow of AES is shown in the Figure 1. AES accepts a 128-bit plaintext, 128-bit user key and generates 128-bit ciphertext. The encryption proceeds through an initial round and subsequent 10 round repetition of 4 steps. These steps are *SubBytes*, *ShiftRows*, *MixColumns* and *AddRoundKey*. In the final round, MixColumns step is skipped. For each of these rounds, separate 128-bit round subkeys are needed. The round subkeys are generated from the initial user key via a key expansion step. The key expansion uses Rijndael's key schedule.

The plaintext is organized as a 4×4 column-major order matrix, which is operated through the AES rounds. The SubBytes step uses a non-linear transformation on every element of the matrix. The non-linear transformation is defined by an 8-bit substitution box, also known as Rijndael S-box. The ShiftRows step cyclically shifts the bytes in each row by a certain offset. In the MixColumns step, each column is multiplied by a fixed matrix. In the AddRoundKey step, each byte of the matrix is exclusive-OR-ed with each byte of the current round subkey. This is shown graphically in the Figure 1.

B. Trace Buffer

One of the major challenges in post-silicon validation and debug is the limited controllability and observability of the fabricated integrated circuit. Trace buffer is widely used to improve the observability of circuit and thus assist post-silicon debug and analysis. It is a buffer that traces (records) some of the internal signals in a silicon chip during runtime. If an error is encountered, the content of trace buffer would be dumped out through JTAG interface for off-line debug and error analysis. Due to design overhead constraints, the number of trace signals is only a small fraction of all internal signals in the design. The size of the trace buffer directly affects the observability that we can get from the trace buffer.

Figure 2 illustrates how the trace buffer is used during post-silicon validation and debug. Signal selection is done during

[1]For the rest of the paper, unless explicitly specified, we will use AES-128 and AES interchangeably.

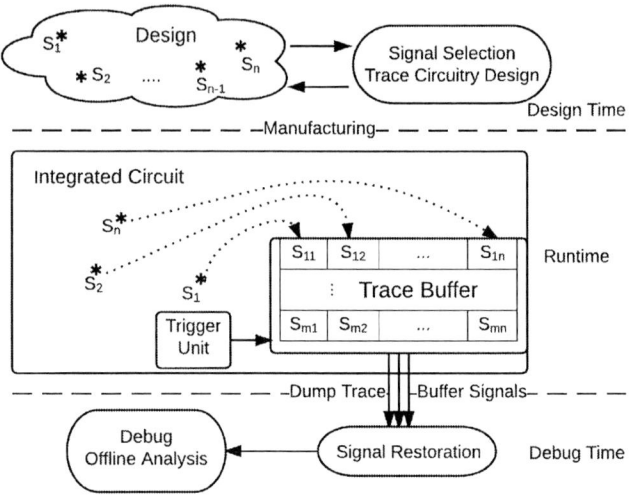

Fig. 2: Overview of trace buffer in system validation and debug

the design time (pre-silicon phase). Let us assume that S_1, S_2, ..., S_n are the selected trace signals. Figure 2 shows a trace buffer with a total size of $n \times m$ bits, which traces n signals (buffer width) for m cycles (buffer depth). For example, the ARM ETB [4] trace buffer provides buffer sizes ranging from 16Kb to 4Mb. In this case, a 16Kb buffer can trace 32 signals for 512 cycles (i.e., $n=32$ and $m=512$). Once the trace signals are selected, they need to be routed to the trace buffer. A trigger unit is also needed that decides when to start and stop recording the trace signals based on specific (error) events. The trace buffer records the states of the traced signals during runtime. During debug time, the states of traced signals will be dumped out through the standard JTAG interface. Signal restoration is performed to restore as many states as possible, which is to maximize the observability of the internal signals in the chip. The off-line debug and analysis would be based on the traced signals and the restored signals.

III. RELATED WORK AND MOTIVATION

A. AES Attack

Since the pioneering works on differential power analysis [5], numerous side-channel attacks have been developed. Side-channel attacks are classified into *passive*, *semi-invasive* and *invasive* attacks depending on the level of intrusion necessary for the attacker. The side-channels are of varied forms ranging from the software execution pattern such as cache timing [6] to more detailed hardware-oriented information leakages such as electromagnetic waves [7], acoustic waves [8] and optical fault injections [9]. Recent surveys on timing channels and invasive fault attacks are available in [10] and [11], respectively. Another approach of constructing an invasive attack originates from a malicious hardware, secretly inserted into a chip. These are commonly known as hardware Trojans [12].

Considering the impact that AES has on our everyday communications, many of the attack techniques report their efficacy by demonstrating an attack on AES, which is also the target cipher for the current work. Among the hardware

978-1-4673-9141-2/15 $31.00 © 2015 IEEE

side-channel attacks reported against AES, attacks based on scan-chain [14] and external fault injections [15] are most prominent. For all these attacks, effective countermeasures are proposed and the inherent resilience of various design points [16] is studied. It is also shown that there exists an interplay between the countermeasures of one attack and the consequently increased vulnerability against another attack [17].

B. Trace Buffer Observability versus Security

Trace buffer is widely used to improve the observability of circuit and thus assist post-silicon debug and analysis. The quality of selected trace signals will directly affect the observability that we can get from the trace buffer. The goal of trace signal selection is to obtain a set of signals, which can restore the maximum number of internal states in the chip. Basu et al. [18] proposed a metric based algorithm that employs total restorability for selecting the most profitable signals. Chatterjee et al. [19] proposed a simulation based algorithm which is shown to be more promising than metric based approaches. Li and Davoodi [20] proposed a hybrid approach which combines the advantages of metric and simulation based approaches.

While it is accepted in the research community that there is a strong link between observability/testability and security, it is surprising that the vulnerability of trace buffers in cryptographic implementation is not studied so far. This forms the core motivation of our work. We show that an effective security attack is possible by analyzing the trace buffer content during post-silicon debug.

IV. TRACE BUFFER ATTACK

The proposed trace buffer attack proceeds in two phases. In the first phase, we attempt to establish the correspondence between the signal values in trace buffer and variables in the AES design. In the second phase, depending on the trace buffer size and the number of cycles for which each signal is dumped, the signal values are fed to the restoration algorithm. The restoration algorithm attempts to recover the user-specified key. Details of each step are elaborated in the following sections.

A. Attack Step 1: Determine Trace Buffer Signals

If an attacker wants to steal the primary key, signal values in the trace buffer are the starting point of hacking. Unless the traced data is encrypted or debugging is authentication based, the attacker can easily dump traced data through JTAG interface. The challenge for *Trace Buffer Attack* is that the attacker does not know what signals recorded in the trace buffer. We assume that the attacker has access to a few test chips and the RTL description of the AES design. The one-to-one mapping between the traced signals and the registers in RTL description can be established by running some test chips and matching with RTL simulation.

1) Simulate the RTL implementation of the AES design with a random key k and a random input plaintext t for c cycles. During simulation, all the internal register values are stored.
2) Run the test chip with the same key k and the same input plaintext t for c cycles. Each traced signal will have a vector of c values stored in the trace buffer.
3) Dump out the values in trace buffer through JTAG. For each traced signal, we compare its value vector with all the register value vectors from RTL simulation. If a unique match is found in the RTL simulation, this traced signal is identified in the RTL description. Repeat the process until all the traced signals are uniquely identified.

For the two case studies in Section V, the above mapping process can be finished in no more than 512 cycles. For the iterative AES-128 in Section V-A, it takes 24 cycles (2 runs, each run takes 12 cycles) to uniquely identify the 32 traced signals. For the pipelined AES ciphers in Section V-B, 512 cycles suffice to uniquely identify all the 64 traced signals in each design.

B. Attack Step 2: Signal Restoration

Let us assume that the attacker has finished the preparation in Attack Step 1 and successfully identified the signals in the trace buffer. The next step is to run the chip in the working mode with the secret primary key and take advantage of the trace buffer to initialize the attack. The attacker dumps out the signal states recorded in the buffer during online encryption, and tries to analyze the design so as to recover as many other signals as possible, and eventually obtain the primary key. In post-silicon debug, restoration of unknown signals based on trace buffer data is a crucial step in debugging. This section will detail the approach for signal restoration based on trace buffer.

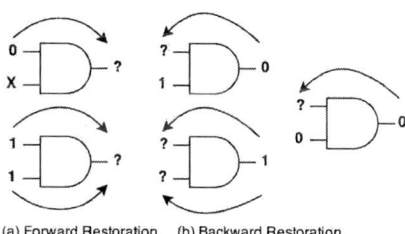

(a) Forward Restoration   (b) Backward Restoration

Fig. 3: Illustration of signal restoration for an AND gate

The signals can be reconstructed from the traced signals in two directions: forward and backward restoration. Forward restoration pushes the restoration of signals from input to output, which is the process of inferring output values if some inputs are known. Backward restoration infers input values if some outputs are known. Figure 3 illustrates forward and backward restoration with a simple example of AND gate. Figure 3(a) shows forward restoration: if one of the inputs is 0, the output can be inferred to be 0; if both of the inputs are 1, the output can be inferred to be 1. Figure 3(b) and (c) shows backward restoration: if the output is 1, both of the inputs can be inferred to be 1. However, if the output is 0, backward restoration might not be successful as shown in (c). The restoration process for other logic components is similar to AND gate. The restoration for registers (flip-flops) is that the state at current cycle is related to the state at previous cycle as specified by their truth tables. Algorithm 1 outlines the major steps in a typical restoration algorithm. It assigns the signal values (based on trace buffer content) of the design and performs forward and backward restorations to construct value assignments for un-traced signals. This process continues until no new assignments are created. Although this algorithm has exponential complexity, in reality, it completes the process very fast (as demonstrated in Section V) since the number of new values created decreases significantly after each iteration.

Algorithm 1 SIGNAL RESTORATION ALGORITHM

Input: Traced signals (nodes) with values at each cycle

Output: Restored signal (node) values

Put all traced nodes into $UnderProcess$ queue

Update the traced nodes with their known values (0/1)

Update all other nodes with unknown values (X)

while $UnderProcess$ *is not empty* **do**

    Take a node N from $UnderProcess$ to restore its neighbors

    for *each node in N's BackwardNeighbors* **do**

        Backward Restoration for this neighbor node

        if *value at any cycle is restored* **then**

            Push this neighbor into $UnderProcess$

    for *each node in N's ForwardNeighbors* **do**

        Forward Restoration for this neighor node

        if *value at any cycle is restored* **then**

            Push this neighbor into $UnderProcess$

return

V. EXPERIMENTAL RESULTS

We use the AES Verilog implementations (the iterative AES-128 [22], and the pipelined AES-128, AES-192 and AES-256 [23]) from the OpenCores website. The Synopsys Design Compiler is used to synthesize the RTL implementation into a gate-level netlist. We develop C++ code to simulate the gate-level circuits and run the signal restoration algorithm. We use [21] to select trace signals for the trace buffers since it produces signals that can maximize observability compared to the other signal selection techniques. The signal selection algorithm picks the best signals for debugging purpose without consideration of security, which means it is ignorant of the fact that the AES design contains security secrets (keys). The experiments were conducted on a computer with AMD Opteron 2.4GHz core and 32GB memory.

A. Case Study 1: iterative AES-128

The iterative AES-128 design has 530 flip-flops and about 25,000 basic logic gates. The 530 flip-flops (registers) include:

- *ld_r*, which is a one-bit control signal.
- *dcnt[0..3]*, which is a 4-bit register keeping track of the encryption rounds.
- *text_in_r[0..127]*, which is a 128-bit register holding the plaintext.
- *w0[0..31]*, *w1[0..31]*, *w2[0..31]*, and *w3[0..31]*, which are 32-bit each, holding the round keys.
- *u0.rcon[24..31]* and *u0.r0.rcnt[0..3]*, which are 8 temporary registers in the key expansion unit.
- *text_out[0..127]*, which is a 128-bit register holding the ciphertext.

The signals recorded in the trace buffer are identified by using methods detailed in Section IV-A. The selected signals for each buffer width is as follows:

- BufferWidth=8: {dcnt[2], ld_r, w3[2], w3[1], w3[30], w3[27], w3[17], w3[13]}
- BufferWidth=16: {dcnt[2], ld_r, w3[4], w3[29], w3[27], w3[23], w3[22], w3[18], w3[16], w3[15], w3[14], w3[13], w3[12], w3[10], w1[9], w3[8]}
- BufferWidth=32: {dcnt[2], ld_r, sa03[7], sa13[7], w3[7], w3[6], w3[3], w3[2], w3[1], w3[31], w3[30], w2[29], w3[27], w3[26], w3[25], w3[24], w3[23], w3[22], w3[21], w3[20], w3[18], w2[17], w3[16], w3[15], w0[14], w3[13], w3[12], w3[11], w3[10], w3[9], w3[8], w3[0]}

TABLE I: Iterative AES-128: Number of bits in the key recovered and memory/time requirements for signal restoration.

BufferWidth	BufferDepth	64	128	256	512
8	**leaked key** (bits)	6	6	6	6
	memory (MB)	116.4	161.4	252.0	432.0
	time (mm:ss)	0:27.75	0:56.07	1:50.35	3:43.26
16	**leaked key** (bits)	18	25	28	28
	memory (MB)	116.4	161.4	252.0	432.0
	time (mm:ss)	0:27.82	0:55.94	1:51.00	3:44.10
32	**leaked key** (bits)	98	128	128	128
	memory (MB)	116.4	161.4	252.0	432.0
	time (mm:ss)	0:28.01	0:55.98	1:52.81	3:51.38

We explore different trace buffer sizes with buffer widths of 8, 16, and 32, buffer depth (traced cycles) of 64, 128, 256 and 512 in our experiments, which should be suitable for the AES-128 design. Table I shows our results of restoring the primary key from the trace buffer content on the iterative AES-128 cipher. The trace buffers with a buffer width of 32 and a buffer depth no less than 128 are able to recover the full primary key in a few minutes.

Figure 4(a) shows the number of bits in the user key leaked with different buffer sizes. Figure 4(b) shows the total number of internal states restored (debug observability) during restoration. The number of restored primary key bits increases with bigger buffer width. For the same buffer width, the number of restored key bits increases slightly as the trace cycles increase, and it will be saturated after buffer depth is big enough (256 cycles or more). The 8 × 512, 16 × 512 and 32 × 512 trace buffer can respectively restore 6, 28 and 128 bits of the primary key.

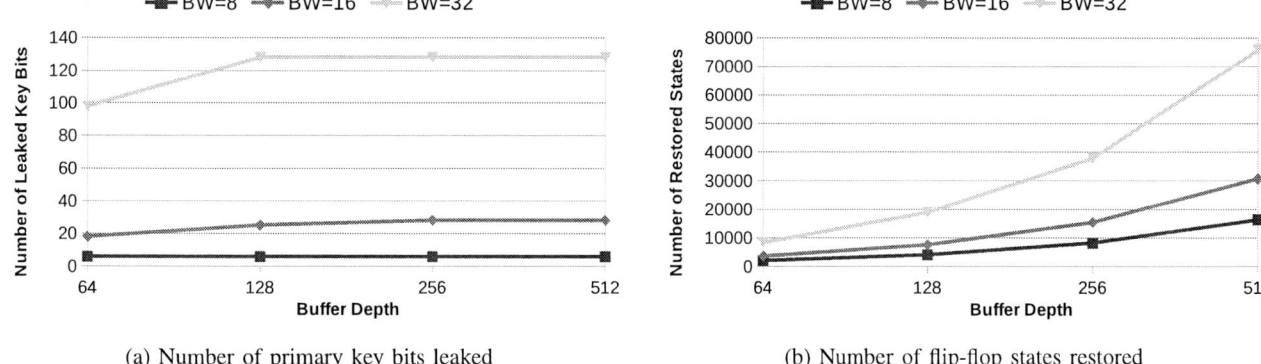

(a) Number of primary key bits leaked   (b) Number of flip-flop states restored

Fig. 4: Iterative AES-128: security and observability trade-off using Buffer Widths (BW) of 8, 16 and 32, and Buffer Depths of 64, 128, 256 and 512. The 32×128, 32×256, and 32×512 trace buffers are able to recover the full primary key.

The fact that the 32 × 512 trace buffer can restore all 128-bit primary key is not surprising. The iterative AES-128 design[2] has relatively short pathways with only 530 flip-flops in total. The 32 signals selected out of the 530 flip-flops is the set of signals which could offer best observability to the debugger. As shown above, 2 signals are from the control unit; 2 signals are from the intermediate result register; the other 28 signals are from the round key register in the key expansion unit. The selected signals from the key expansion unit are most responsible for giving away information to restore the primary key. The success of recovering the full primary key is due to the observability provided by the trace buffer.

B. Case Study 2: pipelined AES ciphers

The main difference from the iterative version is that the pipelined implementation unrolls all the encryption rounds to be independent hardware units, which makes the pipelined version about 10-15 times as large as the iterative. For example, the pipelined AES-128 cipher has 6720 flip-flops and about 290,000 logic gates, which is roughly 10 times (10 encryption rounds) as large as the iterative AES-128. This poses a greater challenge for the restoration process, because many signal values are not inferable due to the long pathways between the known signals. Only signals that are very close to the input can be propagated backward and possibly restore the primary key bits.

We explore different trace buffer sizes with buffer widths of 8, 16, 32 and 64, buffer depth of 512 in our experiments. We set the buffer depth to be 512 cycles, which should be suitable for the pipelined AES ciphers. Table II shows the experimental results on the pipelined implementation of AES-128, AES-192, and AES-256 ciphers. For a buffer width of 64, we are able to respectively restore 20, 19 and 44 bits of the primary key for AES-128, AES-192 and AES-256 in a few hours.

Figure 5 shows our experimental results of pipelined AES ciphers as we increase the trace buffer width. As the trace buffer width increases, both observability and the leaked

[2]For iterative implementation, the restoration is clearly able to recover the key and we expect the same trend to follow for AES-192 and AES-256.

number of key bits increase. The restoration algorithm is not able to restore the full primary key for any of the pipelined AES ciphers. Nevertheless, considerable knowledge about the key is gained, which does not suffice to recover the secret though, can aid other modes of cryptanalysis.

TABLE II: Pipelined AES-128, AES-192 and AES-256: Number of bits in the key recovered and memory/time requirements for signal restoration.

BufferWidth	AESciphers	AES-128	AES-192	AES-256
8	leaked key (bits)	4	1	8
	memory (GB)	4.66	5.37	6.56
	time (h:mm:ss)	3:51:45	4:29:05	6:38:06
16	leaked key (bits)	6	4	16
	memory (GB)	4.66	5.37	6.56
	time (h:mm:ss)	3:44:14	4:12:22	6:22:59
32	leaked key (bits)	11	8	32
	memory (GB)	4.66	5.37	6.56
	time (h:mm:ss)	3:19:12	4:10:25	6:31:08
64	leaked key (bits)	20	19	44
	memory (GB)	4.66	5.37	6.56
	time (h:mm:ss)	3:42:02	4:08:43	6:03:15

VI. PROPOSED COUNTERMEASURE

To prevent the trace-buffer attack, we propose a countermeasure based on built-in randomness assumption. Physically Unclonable Functions (PUFs) [24] are used as the chosen source of randomness. PUF provides a challenge-response mechanism, where the mapping from a challenge to a response is controlled by the manufacturing process as well as the nature of the Integrated Circuit (IC). This complex control makes PUF structures hard to clone and at the same time a unique device identification can be obtained. Compared to the look-up table-based storage of key, PUF provides a large set of challenge-response keys with a storage requirement that increases linearly with the number of challenge bits. Only a valid user is aware of the challenge-response sets.

The countermeasure is graphically shown in the Figure 6. During the recording of the trace signals, the signals from consecutive clock cycles are XOR-ed according to a PUF response. Since the PUF response is only known to the valid user, he/she can recover the trace signal easily. For a

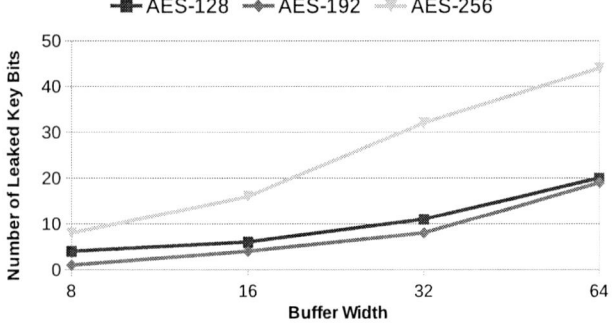
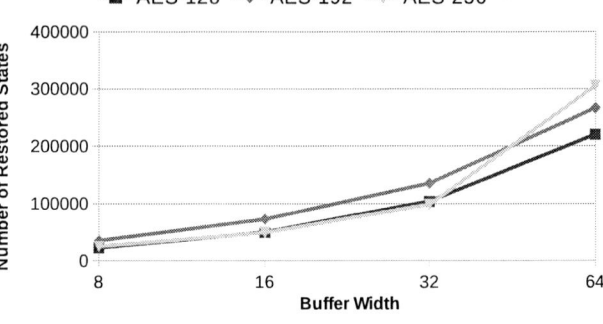

(a) Number of primary key bits leaked (b) Number of flip-flop states restored

Fig. 5: Pipelined AES ciphers: security and observability trade-off.

malicious user, recovering the original trace signals is hard. The idea of this countermeasure closely follows a similar countermeasure proposed for scan-chain attacks [25]. Note that the countermeasure is described in generic fashion as it can be scaled to larger bit-widths as needed.

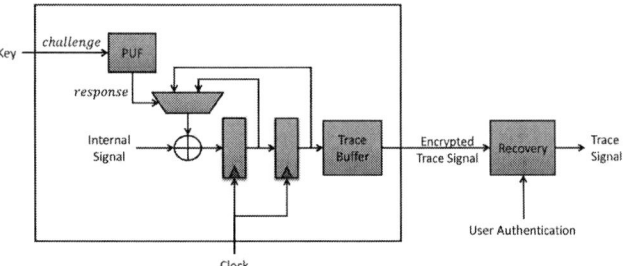

Fig. 6: PUF-based Countermeasure

VII. SUMMARY AND OUTLOOK

In this paper *Trace Buffer Attack* is introduced. We study the current practice in trace buffer signal selection and restoration algorithms. Based on that, it is experimentally demonstrated that AES, the currently dominant block cipher, is vulnerable. With a trace buffer size of 32×128, the full key of the iterative AES-128 can be restored with a computation time of one minute. For pipelined AES, partial key can be restored in a few hours. This leads to a trade-off between security and debug observability. An efficient PUF-based countermeasure is proposed to prevent the trace buffer attack. This work can be extended in multiple directions. One may take up further experiments to determine the actual PUF overhead, to account for the PUF modeling attacks, to construct hybrid attacks involving trace buffers as well as study the vulnerability of other ciphers apart from AES.

REFERENCES

[1] A. Bogdanov et al., "Biclique cryptanalysis of the full AES," in *Advances in Cryptology: ASIACRYPT 2011*.

[2] A. Moradi et al., "Pushing the limits: A very compact and a threshold implementation of AES," in *Advances in Cryptology: EUROCRYPT 2011*.

[3] *FIPS 197, Advanced Encryption Standard*, 2001. [Online]. Available: csrc.nist.gov/publications/fips/fips197/fips-197.pdf

[4] *ARM Embedded Trace Buffer*, [Online]. Available: http://infocenter.arm. com/help/index.jsp?topic=/com.arm.doc.dai0168b/ar01s03s03.html

[5] P. C. Kocher et al., "Differential power analysis," in CRYPTO , 1999. D. J. Bernstein, *Cache-timing attacks on AES*, 2005. [Online]. Available: http://cr.yp.to/papers.html#cachetiming

[6] D. Osvik et al., "Cache attacks and countermeasures: The case of AES," CT-RSA, 2006.

[7] D. Genkin et al., "Stealing keys from PCs using a radio: Cheap electromagnetic attacks on windowed exponentiation," Cryptology ePrint Archive, Report 2015/170, 2015.

[8] D. Genkin et al., "RSA key extraction via low-bandwidth acoustic cryptanalysis," CRYPTO, 2014.

[9] S. Skorobogatov and R. Anderson, "Optical fault induction attacks,"Springer, 2003.

[10] C. Rebeiro et al., *Timing Channels in Cryptography: A Micro-Architectural Perspective*, Springer, 2015 edition.

[11] A. Barenghi et al., "Fault injection attacks on cryptographic devices: Theory, practice, and countermeasures," *Proceedings of the IEEE*, 2012.

[12] S. Bhunia et al., "Hardware trojan attacks: Threat analysis and counter-measures," *Proceedings of the IEEE*, 2014.

[13] B. Ege et al., "Differential scan attack on AES with x-tolerant and x-masked test response compactor," in *Digital System Design (DSD), 2012*.

[14] S. Ali et al., "Test-mode-only scan attack using the boundary scan chain," in ETS, 2014.

[15] D. Mukhopadhyay, "An improved fault based attack of the advanced encryption standard," in *Progress in Cryptology: AFRICACRYPT 2009*.

[16] S. Ali et al., "AES design space exploration new line for scan attack resiliency," in VLSI-SoC, 2014.

[17] F. Regazzoni et al., "Interaction between fault attack countermeasures and the resistance against power analysis attacks," in *Fault Analysis in Cryptography*, 2012.

[18] K. Basu and P. Mishra, "RATS: restoration-aware trace signal selection for post-silicon validation," *IEEE Trans. VLSI Syst.*, 2013.

[19] D. Chatterjee et al., "Simulation-based signal selection for state restoration in silicon debug," in ICCAD, 2011.

[20] M. Li and A. Davoodi, "A hybrid approach for fast and accurate trace signal selection for post-silicon debug," *IEEE Trans. on CAD of Integrated Circuits and Systems*, 2014.

[21] K. Rahmani et al., "Efficient trace signal selection using augmentation and ILP techniques," in ISQED 2014.

[22] *OpenCores AES-128 cipher*, [Online]. Available: http://opencores.org/ project,aes_core

[23] *OpenCores AES ciphers (all key sizes)*, [Online]. Available: http:// opencores.org/project,tiny_aes

[24] I. Verbauwhede and R. Maes, "Physically unclonable functions: Man-ufacturing variability as an unclonable device identifier," in GLSVLSI 2011.

[25] S. Banik et al., "Cryptanalysis of the double-feedback xor-chain scheme proposed in Indocrypt 2013" in *INDOCRYPT 2014*.

Author Index

Abbasizadeh, Hamed   241

Abdessaied, Nabila   286

Agosta, Giovanni   25

Ahn, Junwhan   183, 195

Aktouf, Chouki   307

Al Khatib, Chadi   307

Ali, Sk Subidh   264

Amagasaki, Motoki   110

Amaravati, Anvesha   231

Ascheid, Gerd   337

Augustine, Joe   343

Aupetit, Claire   307

Balasubramanian, Anusuya   37

Banerjee, Pritha   7

Barke, Erich   313

Bashizade, Ramin   201

Bayon, Pierre   19

Beckett, Paul   31

Benini, Luca   25

Beretta, Michele   25

Berisford, Daniel   134

Bilal, Muhammad   142

Billoint, Olivier   116

Blanc, Sebastien   219

Bocca, Alberto   57

Boguslawski, Bartosz   116

Bossuet, Lilian   19

Butzen, Paulo F.   1

Byun, Kyungjin   292, 303

Capoccia, Raffaele   207

Carlson, Robert   134

Champac, Victor   165, 297

Chattopadhyay, Anupam   337, 355

Chen, Harry H.   177

Chen, Ying-Jung   349

Cherng, Jun-Fei   177

Chevalier, Cyril   307

Choi, Kiyoung   74, 183

Choudhary, Shridhar   258

Chugh, Manan   231

Chung, Myung-Ae   237

Clermidy, Fabien   116

Cronin, Patrick   92

Danese, Alessandro   246

de Gyvez, Jose Pineda   159

Di Federico, Alessandro   25

Drechsler, Rolf   286

Elfadel, Ibrahim M.   225

Eum, Nak-Woong   292

Eum, Nakwoong   303

Fant, Karl   31

Fesquet, Laurent   307

Filini, Francesca   246

Fischer, Viktor   19

Flach, Guilherme   1

Fogaça, Mateus   1

Fujita, Masahiro   13, 258, 280

Gautschi, Michael   25

Gharehbaghi, Amir Masoud   258, 280

Ghasempouri, Tara   325

Giri, Chandan   122

Gomez Chacon, Andres   297

Gomez, Andres   165

Guo, Jun   252

Hand, Kevin   134

Hau, Yuan Wen   331

Heitzmann, Frédéric   116

Hsieh, Meng-Ta   177

Hu, Jingtong   92

Huang, Yuanwen   355

Iida, Masahiro   110

Iturbe, Xabier   134

Jaiswal, Manish Kumar   213

Jayakrishnan, Mini   159

Jelemenská, Katarína   63

Jerome, Dr. Jovitha   154

Johann, Marcelo   1

Jose, John   343

Jürimägi, Lembit   171

Kang, Seokhyeong   69

Kang, Sungweon   149, 237

Kang, Taewook   237

Keymeulen, Didier   134

Khan, Asim   142

Khan, Muhammad Umar Karim   142

Kim, Chan   303

Kim, Chanha   319

Kim, Gain   207

Kim, Matthew   31

Kim, Namhyung   183

Kim, Seungwon   69

Kim, Sung-Eun   237

Kim, Whan-Woo   149

Kim, Youngmin   69

Kuga, Morihiro   110

Kyung, Chong-Min   142

Leblebici, Yusuf   207, 219

Lee, Hyung Gyu   98, 104

Lee, Jae-Jin   303

Lee, Jaemin   69

Lee, Kang-Yoon   241

Leupers, Rainer   337

Li, Hai   52

Li, Zheng   52

Lim, Ingi   149

Liou, Jing-Jia   177

Liu, Chenchen   52

Liu, Peng   252

Low, Qiong Wei   270

Macii, Alberto   57

Macii, Enrico   57

Macko, Dominik   63

Mandal, Supriyo   122

Mani, Dr. Geetha   154

Mani, Geetha   43

Marsono, Muhammad Nadzir   331

Matsumoto, Takeshi   258

Mazumdar, Bodhisatwa   264

Mishra, Prabhat   355

Monteiro, Jucemar   1

Murthy, C Siva Ram   37

Muzaffar, Shahzad   225

Nicolas Nicolaz, Pierre   74

Olbrich, Markus   313

Orasson, Elmet   171

Ozer, Emre   134

Pandiyan, Manikandan   43, 154

Park, Chanik   319

Park, Eunhyeok   195

Park, Hyung-Il   149

Park, Hyunsun   195, 319

Park, Jaehyun   98, 104

Park, Kyunghwan   237

Park, Seongmo   292

Parthasarathy, Ananthanarayanan   274

Paul, Sudipta   7

Poncino, Massimo   57

Pravadelli, Graziano   246, 325

Puget, Julia   1

Pullini, Antonio   25

Quiring, Artur   313

Rahman, Hafizur   122

Raik, Jaan   171

Raychowdhury, Arijit   231

Reis, Ricardo   1

Reyes, Alejandra Nicte-Ha   165

Roy, Surajit   122

Ruan, Shanq-Jang   349

Ruiz Varela, Maria   86

Rákossy, Zoltán Endre   337

S, Natarajan   154

Saeed, Samah   264

Samadpoor Rikan, Behnam   241

Sarbazi-Azad, Hamid   201

Sarhan, Hossam   116

Sassone, Alessandro   57

Scandale, Michele   25

Sebastian, Sijin   343

Seguin, Fabrice   116

Seo, Jae-Sun   49

Seo, Woong   183

Seok, Mingoo   49

Seyid, Kerem   219

Shim, Seongbo   80

Shin, Donghwa   57, 98, 104

Shin, Youngsoo   80

Sicard, Gilles   307

Siek, Liter   270

Sinanoglu, Ozgur   264

So, Hayden K.-H   213

Soeken, Mathias   286

Srinivasan, Manikantan   37

Stengele, Dominik   337

Su, Hung-Cheng   189

Sueyoshi, Toshinori   110

Sur-Kolay, Susmita   7

Tae-Hyoung, Kim   159

Takeuchi, Yuto   110

Tang, Jia Wei   331

Thuries, Sebastien   116

Traber, Andreas   25

Tsai, Chun-Jen   189

Ubar, Raimund   171

W. Dueck, Gerhard   286

Wang, Weidong   252

Wang, Yandan   52

Wei, Lin   128

Wu, Tsung-Han   189

Xue, Yuan   92

Yan, Bonan   52

Yang, Chaofei   52

Yang, Chengmo   86, 92

Yang, Jianlei   52

Yiu, Patrick   134

Yoo, Sungjoo   195, 319

Zhao, Qian   110

Zhou, Lei   128

Zhou, Mi   270

Čičák, Pavel   63

IEEE
445 Hoes Lane
Piscataway, NJ 08854-4141

ISBN 978-1-4673-9141-2